Third Edition

Statistical Methods for the Social Sciences

D1002324

Alan Agresti
University of Florida

Barbara Finlay
Texas A & M University

Pearson Education

PRENTICE HALL INTERNATIONAL, INC.

If you purchased this book within the United States
or Canada you should be aware that it has been wrongfully
imported without the approval of the Publisher or the Author.

Acquisitions Editor: Ann Heath
Editorial Assistant: Mindy Ince
Editorial Director: Tim Bozik
Editor-in-Chief: Jerome Grant
Assistant Vice President of Production and Manufacturing: David W. Riccardi
Editorial/Production Supervision: Robert C. Walters
Managing Editor. Linda Mihatov Behrens
Executive Managing Editor: Kathleen Schiaparelli
Manufacturing Buyer: Alan Fischer
Manufacturing Manager: Trudy Pisciotti
Marketing Manager: John Tweeddale
Marketing Assistant: Diana Penha
Creative Director: Paula Maylahn
Cover Director: Jayne Conte
Cover Designer: Bruce Kenselaar
Cover Photo: Sharon McGinley-Nally
Quilt, festooned with a globe and statistical symbol patches, surrounded by students, people from different ethnic backgrounds, animals, and other whimsical shapes.

© 1997 by Prentice-Hall, Inc.
Pearson Education
Upper Saddle River, New Jersey 07458

Previous edition copyright @1986 by
Dellen Publishing Company, a division of Macmillan, Inc.

All rights reserved. No part of this book may be
reproduced, in any form or by any means,
without permission in writing from the publisher.

Printed in the United States of America
10 9 8 7 6 5 4

ISBN 0-13-622515-2

PRENTICE-HALL OF AUSTRALIA PTY. LIMITED, *Sydney*
PRENTICE-HALL CANADA, INC., *Toronto*
PRENTICE-HALL HISPANOAMERICANA, S.A., *Mexico*
PRENTICE-HALL OF INDIA PRIVATE LIMITED, *New Delhi*
PRENTICE-HALL OF JAPAN, INC., *Tokyo*
SIMON & SCHUSTER ASIA PTE. LTD., *Singapore*
EDITORA PRENTICE-HALL DO BRASIL, LTDA., *Rio de Janeiro*
PRENTICE-HALL, *Upper Saddle River, New Jersey*

Contents

Preface

When my co-author and I undertook the first edition of this book nearly two decades ago, our goal was to introduce both basic and advanced statistical methods in a style that emphasized their application to the social sciences rather than the mathematics behind them. We did this by focusing on how the methods are used and interpreted rather than their theoretical derivations.

In this third edition of the book, there is an even stronger emphasis on applications, with greater attention to "real data" both in the examples and exercises. We continue to downplay mathematics, in particular probability, which is all too often a stumbling block for students. We present only the basic probabilistic concepts needed for statistical inference, concentrating on the fundamental concept of sampling distributions. On the other hand, the text is not a cookbook. Reliance on an overly simplistic recipe-based approach to statistics is not the route to good statistical practice.

Changes in the Third Edition

In the nearly twenty years since the first edition, the tremendous increase in computer power coupled with the continued improvement and accessibility of statistical software has had a major impact on the way social scientists analyze data. This is reflected in many of the major changes in this edition. This edition de-emphasizes even further the traditional shortcut hand-computational formulas and approximations. The presentation of computationally complex methods, such as regression, devotes more attention to interpreting computer output than to using formulas to perform the analysis. The second half of the text contains numerous sample SAS and SPSS printouts, both in chapter text and homework problems. This edition also adds a detailed Appendix explaining how to apply SAS and SPSS to conduct the methods of each chapter.

Exposure to realistic but simple examples and to numerous homework exercises is vital to student learning. This edition has increased greatly the number of exercises, with major emphasis on real data. The data come from all the social and behavioral sciences, but especially sociology, political science, education, and psychology. Each chapter's homework set is divided into two parts, straightforward exercises on the text material in *Practicing the Basics*, and exercises dealing with open-ended data analyses, understanding of concepts, and advanced material in *Concepts and Applications*. The large data sets in the examples and exercises, including a data set introduced in Exercise 1.7 that appears again in exercises at the end of each chapter, are available on the World Wide Web at the address

`http://www.stat.ufl.edu/users/aa/social/data.html`

This edition contains several changes and additions in content, directed toward a more modern approach. The main changes are as follows:

- As already mentioned, there is much stronger focus on ***real examples*** and on the integration of ***computer software***.
- The book has a somewhat ***lower technical level*** in the first nine chapters, to make the book more easily accessible to undergraduate students.
- Chapter 8, on contingency tables, is reorganized to introduce modern methods, such as measuring association using ***odds ratios*** and describing patterns of association using cell ***residuals***.
- The material on regression presents new types of ***plots*** in Chapter 11, such as scatterplot matrices, conditioning plots, and partial regression plots, and a new section on ***model diagnostics*** in Chapter 14.
- Chapter 12, on analysis of variance methods, now includes two sections on ***repeated measures*** analyses.
- Chapter 14 presents the ***generalized linear model*** unification of regression models for continuous and discrete response variables. Ordinary regression models and alternative models for categorical responses or count data, or responses with nonconstant variance, are special cases of a single model that can be fitted with a single software procedure.
- Chapter 15 provides stronger emphasis on ***logistic regression*** and introduces its extension for ordinal response variables.
- Chapter 16 includes a section on ***event history analysis***.

Use of Text in Introductory Statistics Courses

Like the first two editions, this edition is appropriate for introductory statistics courses at either the undergraduate or beginning graduate level, and for either a single term or a two-term sequence. Chapters 1–9 are the basis for a single term course. If the instructor wishes to go further than Chapter 9 or wishes to cover some material in greater depth, sections that can easily be omitted without disturbing continuity include 2.4, 5.5, 6.6–6.7, 7.3–7.5, and 8.5–8.7. Also, Chapters 7–9 and Sections 12.1–12.2 are self-contained, and the instructor could move directly into any of these after covering the fundamentals in Chapters 1–6. Four possible paths for a one-term course are as follows:

- Chapters 1–9 (possibly omitting sections noted above): Standard cross-section of methods, including basic descriptive and inferential statistics, two-sample procedures, contingency tables, and linear regression
- Chapters 1–6, 9, 11: Emphasis on regression
- Chapters 1–7, 12: Emphasis on group comparisons and analysis of variance
- Chapters 1–8, 10: Emphasis on categorical data, basic issues of multivariate relationships

My view is that, regardless of the type of data, a modeling paradigm emphasizing parameter estimation is more useful than the artificial hypothesis-testing approach of many statistics texts. Thus, the second half of this text, starting in Chapter 9, is primarily concerned with model building. This material forms the basis of a second course. These chapters naturally link with an introduction to computer software, using printouts that the text provides. Appendix A explains how to use SAS and SPSS for methods of each chapter.

Some material appears in sections, subsections, or exercises marked by asterisks. This material is optional, having lesser importance for introductory courses. The text does not attempt to present every available method, since it is meant to be a teaching tool, not an encyclopedic cookbook. It does cover the most important methods for social science research, however, and it includes topics not usually discussed in introductory statistics texts, such as:

- Methods for contingency tables that are more informative than chi-squared, such as cell residuals and analyses that utilize category orderings
- Controlling for variables, and testing for causal relationships
- The generalized linear modeling approach, encompassing ordinary regression, analysis of variance and covariance, gamma regression for nonnegative responses with standard deviation proportional to the mean, logistic regression for binary and ordinal data, and loglinear association models for contingency tables

I believe that the student who works through this book successfully will acquire a solid foundation in applied statistical methodology.

Acknowledgments

I am grateful to Barbara Finlay for her contributions to the first two editions of this text. I hope that the combination of our respective fields of expertise provides a book that is statistically sound as well as relevant to the social sciences.

I owe special thanks to Burke Grandjean for providing detailed comments about the second edition and the revised material for this edition. Many improvements are directly due to his suggestions. Also, many thanks to Alfred DeMaris, Thomas Severini, Robert Franz, Monika Ardelt, Ronda Priest, David Mitchell, and Janet Wilmoth for their advice in preparing this edition. Other individuals who provided advice or data sets include Abe Goldman, John Henretta, Harry Khamis, Ming-Long Lam, Jacki Levine, Jane Myers, Anupama Narayanan, Michael Radelet, Bonnie Seegmiller, and Larry Winner. Thanks to Susan Reiland, Maureen Hallinan, Shirley Scritchfield, Sonja Wright, Douglas Zahn, Jeff Witmer, E. Jacquelin Dietz, Dorothy K. Davidson, and Mary Sue Younger for their comments that helped in the preparation of the first two editions.

My special gratitude extends to seven individuals who invested a considerable amount of time in helping this book to reach fruition. Cathy Knudsen took on the enormous job of preparing a LaTeX file from the second edition, which I used in making my revisions. Brent Coull used his S-Plus expertise to compose the figures. Sharon McGinley-Nally, a wonderfully creative and cooperative artist, painted the cover. Ann Heath, my editor, was a source of much wisdom and support and is as fine an editor as one could hope to have. Robert Walters of Prepress Management, Inc. served as project manager in the production of this text. Jeremy Hayhurst at Chrysalis Productions created this polished final copy from my text files.

Finally, extra special thanks to my wife, Jacki Levine, for assistance with editing and style and with overall encouragement throughout this revision project.

Alan Agresti
Gainesville, Florida

Chapter 1

Introduction

1.1 Introduction to Statistical Methodology

In the last several years, all social science disciplines have seen an increase in the use of statistical methods. There are many reasons for this. Research in the social sciences has taken on a more quantitative orientation. Like research in other sciences, it is becoming more strongly oriented toward analyzing empirical data. The computer revolution has made a greater variety of information easily available to researchers as well as to students and the general public. Computers have also made statistical methods themselves easier to use.

The increase in the use of statistics is evident in the changes in the content of articles published in major journals and reports prepared for use in government and private industry, in the styles of textbooks, and in the increasingly common requirement of academic departments that their majors take courses in statistics. A quick glance through recent issues of *American Political Science Review*, *American Sociological Review*, or other leading social science journals will reveal the fundamental role of statistics in social science research.

Job advertisements for social scientists commonly list a knowledge of statistics as an important work tool. A student preparing for a career as a social scientist needs to become familiar with basic statistical methodology. Or, as the joke goes that we recently heard, "What did the sociologist who passed statistics say to the sociologist who failed it? 'I'll have a Big Mac, fries, and a Coke.' "

In today's world, an understanding of statistics is essential in many professions, across the spectrum from medicine to business. Physicians and other health-related professionals evaluate results of studies investigating new drugs and therapies for treating disease. Managers analyze quality of products, determine factors that help predict sales of various products, and measure employee performance.

But statistics is an important tool today even for those who do not use statistical methods as part of their job. Every day we are exposed to an explosion of information, from advertising, news reporting, political campaigning, surveys about opinions on controversial issues, and other communications containing statistical arguments. Statistics helps us make sense of this information and better understand our world. Even if you never use statistical methods in your career, we think that you will find many of the ideas in this text helpful in understanding the information that you will encounter.

We realize that you are probably not reading this book in hopes of becoming a statistician, and you may not even plan on working in social science research. In fact, you may suffer from math phobia and feel fear at what lies ahead. Please be assured that you can read this book and learn the major concepts and methods of statistics with very little knowledge of mathematics. To understand this book, logical thinking and perseverance are much more important than mathematics. And don't be frustrated if learning comes slowly and you need to read a chapter a few times before it starts to make sense. Just as you would not expect to take a single course in a foreign language and be able to speak that language fluently, the same is true with the language of statistics. On the other hand, once you have completed even a portion of this text, you will have a much better understanding of how to make sense of quantitative information.

Data

Information-gathering is at the heart of all sciences. The social sciences use a wide variety of information-gathering techniques that provide the ***observations*** used in statistical analyses. These techniques include questionnaire surveys, telephone surveys, content analysis of newspapers and magazines, planned experiments, and direct observation of behavior in natural settings. In addition, social scientists often analyze information already observed and recorded for other purposes, such as police records, census materials, and hospital files.

The observations gathered through such processes are collectively called ***data***. Data consist of measurements on the characteristics of interest. We might measure, for instance, characteristics such as political party affiliation, annual income, marital status, race, and opinion about the legalization of abortion.

Statistics

Statistics consists of a body of methods for collecting and analyzing data.

These methods for collecting and analyzing data help us to evaluate the world in an objective manner.

Purposes of Using Statistical Methods

Let's now be more specific about the objectives of using statistical methods. Statistics provides methods for

1. ***Design***: Planning and carrying out research studies.
2. ***Description***: Summarizing and exploring data.
3. ***Inference***: Making predictions or generalizing about phenomena represented by the data.

Design refers to ways of determining how best to obtain the required data. The design aspects of a study might consider, for instance, how to conduct a survey, including the construction of a questionnaire and selection of a sample of people to participate in it. ***Description*** and ***inference*** are the two elements of ***statistical analysis***—ways of analyzing the data obtained as a result of the design.

This book deals primarily with statistical analysis. This is not to suggest that statistical design is unimportant. If a study is poorly designed or if the data are improperly collected or recorded, then the conclusions may be worthless or misleading, no matter how good the statistical analysis. Methods for statistical design are covered in detail in textbooks on research methods (e.g., Babbie, 1995).

Description—describing and exploring data—includes ways of summarizing and exploring patterns in the data using measures that are more easily understood by an observer. The main purpose is to take what, to the untrained observer, are meaningless reams of data and present them in an understandable and useful form.

The raw data are a complete listing of measurements for each characteristic under study. For example, an analysis of family size in New York City might start with a list of the sizes of all families in the city. Such bulks of data, however, are not easy to assess—we simply get bogged down in numbers.

For presentation of statistical information to readers, instead of listing *all* observations we use numbers that summarize the *typical* family size in the collection of data. Or we present a graphical picture of the data. These summary descriptions, called ***descriptive statistics***, are much more meaningful for most purposes than the complete data listing. In addition, these descriptions and related explorations of the data may reveal patterns to be investigated more fully in future studies.

Inference consists of ways of making predictions based on the data. For instance, in a recent survey of 750 Americans conducted by the Gallup organization, 24% indicated a belief in reincarnation. Can we use this information to predict the percentage of the entire population of 260 million Americans that believe in reincarnation? A method presented later in this book allows us to predict that for this much larger group, the percentage believing in reincarnation falls between 21% and 27%. Predictions made using data are called ***statistical inferences***.

Social scientists use descriptive and inferential statistics to answer questions about social phenomena. For instance, "Are women politically more liberal than men?" "Is the imposition of the death penalty associated with a reduction in violent crime?" "Does student performance in secondary schools depend on the amount of money spent per student, the size of the classes, or the teachers' salaries?" Statistical methods help us study such issues.

1.2 Description and Inference

We have seen that statistics consists of methods for *designing* studies and methods for *analyzing* data collected for the studies. Statistical methods for analyzing data include descriptive methods for summarizing the data and inferential methods for making predictions. A statistical analysis is classified as ***descriptive*** or ***inferential***, according to whether its main purpose is to describe the data or make predictions. To explain this distinction in more detail, we next define the *population* and *sample*.

Populations and Samples

The objects on which one makes measurements are called the ***subjects*** for the study. Usually the subjects are people, but might instead be families, schools, cities, or companies, for instance.

Population and Sample

The ***population*** is the total set of subjects of interest in a study. A ***sample*** is the subset of the population on which the study collects data.

The ultimate goal of any study is to learn about populations. But it is usually necessary, and more practical, to study only samples from those populations. For example, the Gallup and Harris polling organizations usually select samples of 750–1500 Americans to collect information about political and social beliefs of the population of *all* Americans.

Descriptive Statistics

Descriptive statistical methods summarize the information in a collection of data.

We use descriptive statistics to summarize basic characteristics of a sample. We might, for example, describe the typical family size in New York City by computing the *average* family size. The main purpose of descriptive statistics is to explore the data and to reduce them to simpler and more understandable terms without distorting or losing much of the available information. Summary graphs, tables, and numbers such as averages and percentages are easier to comprehend and interpret than are long listings of data.

Inferential Statistics

Inferential statistical methods provide predictions about characteristics of a population, based on information in a sample from that population.

Example 1.1 illustrates the use of inferential statistics.

Example 1.1 Opinion About Handgun Control

The first author of this text is a resident of Florida, a state with a relatively high crime rate. He would like to know the percentage of Florida residents who favor controls over the sales of handguns. The population of interest is the collection of more than 10 million adult residents in Florida. Since it is impossible for him to discuss the issue with everyone, he can study results from a poll of 834 residents of Florida conducted in 1995 by the Institute for Public Opinion Research at Florida International University. In that poll, 54% of the sampled subjects said that they favored controls over the sales of handguns.

This poll collected data for 834 residents. He is interested, however, not just in those 834 people but in the *entire population* of all adult Florida residents. Inferential statistics can provide a prediction about this larger population using the sample data. An inferential method presented in Chapter 5 predicts that the population percentage favoring control over sales of handguns falls between 50% and 58%. Even though the sample is very small compared to the population size, he can conclude, for instance, that probably a slim majority of Florida residents favor handgun control. □

Using inferential statistical methods with properly chosen samples, we can determine characteristics of entire populations quite well by selecting samples that are small relative to the size of the population.

Parameters and Statistics

Parameters and Statistics

A ***parameter*** is a numerical summary of the population. A ***statistic*** is a numerical summary of the sample data.

Example 1.1 dealt with estimating the percentage of Florida residents who support gun control. The parameter of interest was the true, but unknown, population percentage favoring gun control. The inference about this parameter was based on a statistic—the percentage of the 834 Florida residents in the sample who favor gun control, namely, 54%. Since this number *describes* a characteristic of the sample, it is an example of a descriptive statistic. The value of the parameter to which the inference refers, namely, the population percentage in favor of gun control, is unknown. In summary, we use known sample statistics in making inferences about unknown population parameters.

(Students should note that, in statistical usage, the term *parameter* does not have its usual meaning of "limit" or "boundary.")

The primary focus of most research studies is the parameters of the population, not the statistics calculated for the particular sample selected. The sample and statistics describing it are important only insofar as they provide information about the unknown parameters. We would want a prediction about *all* Floridians, not only the 834 subjects in the sample.

An important aspect of statistical inference involves reporting the likely *accuracy* of the sample statistic that predicts the value of a population parameter. An inferential statistical method predicts how close the *sample* value of 54% is likely to be to the true (unknown) percentage of the *population* favoring gun control. A method from Chapter 5 determines that a sample of size 834 yields accuracy within about 4%; that is, the true population percentage favoring gun control falls within 4% of the sample value of 54%, or between 50% and 58%.

When data exist for an entire population, there is no need to use inferential statistical methods, since one can then calculate exactly the parameters of interest. For example, place of residence and home ownership are observed for virtually all Americans during census years. When the population of interest is small, we would normally study the records of the entire population instead of only a sample. In studying the voting records of members of the U.S. Senate on bills concerning defense appropriations, for example, we could obtain data on votes for all senators on all such bills.

In most social science research, it is impractical to collect data for the entire population, due to monetary and time limitations. It is usually unnecessary to do so, in any case, since good precision for inferences about population parameters results from relatively small samples, such as the 750–1500 subjects that most polls take. This book explains why this is so.

Defining Populations

Inferential statistical methods require specifying clearly the population to which the inferences apply. Sometimes the population is a clearly defined set of subjects. In Example 1.1, it was the collection of adult Florida residents. Often, however, the generalizations refer to a *conceptual* population—a population that does not actually exist but that one can hypothetically conceptualize.

For example, suppose a team of researchers tests a new drug designed to relieve severe depression. They plan to analyze results for a sample of patients suffering from depression to make inferences about the conceptual population of all individuals who might suffer depressive symptoms now or sometime in the future. Or a consumer organization may evaluate gas mileage for a new model of an automobile by observing the average number of miles per gallon for five sample autos driven on a standardized 100-mile course. Inferences then refer to the performance on this course for the conceptual population of all autos of this model that will be or could hypothetically be manufactured.

A caution is due here. Investigators often try to generalize to a broader population than the one to which the sample results can be statistically extended. A psychologist may conduct an experiment using a sample of students from an introductory psychology course. With statistical inference, the sample results generalize to the population of all students in the class. For the results to be of wider interest, however, the psychologist might claim that the conclusions generalize to *all* college students, to *all* young adults, or even to a more heterogeneous group. These generalizations may well be wrong, since the sample may differ from those populations in fundamental ways, such as in racial composition or average socioeconomic status.

For instance, in her 1987 book *Women in Love*, Shere Hite presented results of a survey she conducted of adult women in the United States. One of her conclusions was that 70% of women who had been married at least five years have extramarital affairs. She based this conclusion on responses to questionnaires returned from a sample of 4500 women, which sounds impressively large. However, the questionnaire was mailed to about 100,000 women. We cannot know whether this sample of 4.5% of the women who responded is representative of the 100,000 who received the questionnaire, much less the entire population of adult American women. Thus, it is dangerous to try to make an inference to the larger population.

You should carefully assess the scope of conclusions in research articles, political and government reports, advertisements, and the mass media. Evaluate critically the basis for the conclusions by noting the makeup of the sample upon which the inferences are built. Chapter 2 discusses some desirable and undesirable types of samples.

In the past quarter century, social scientists have increasingly recognized the power of inferential statistical methods. Presentation of these methods occupies a large portion of this textbook, beginning in Chapter 5.

1.3 The Role of Computers in Statistics

Even as you read this book, the computer industry continues its ceaseless growth. New and more powerful personal computers and workstations are reaching the market, and these computers are becoming more accessible to people who are not technically trained. An important aspect of this expansion is the development of highly specialized software.

Versatile and user-friendly software is now readily available for analyzing data using descriptive and inferential statistical methods. The development of this software has provided an enormous boon to the use of sophisticated statistical methods.

Statistical Software

Statistical Package for the Social Sciences (SPSS), SAS (SAS Institute, Inc.), and Minitab are among popular statistical software found on college campuses. It is much easier to apply statistical methods using these software than using old-fashioned hand calculation. The accuracy of computations is greatly improved, since hand calculations often result in mistakes or crude answers, especially when the data set is large. Moreover, some modern statistical methods presented in this text are too complex to be done by hand.

The first seven chapters of this text present some fundamental ideas and methods of statistics. The remaining chapters deal with more advanced methods that provide more complete analyses. The presentation of these more advanced methods shows examples of the output of statistical software. The calculations for the methods of Chapters 8–16 are so complex that they would almost always be done by computer. One purpose of this textbook is to teach you what to look for in a computer printout and how to interpret the information provided. Knowledge of computer programming is not necessary for using statistical software or for reading this book.

The appendix contains many examples of the use of SAS and SPSS statistical software for conducting the analyses, organized by chapter. If you use SAS or SPSS, refer to this appendix as you read each chapter to see how they perform the analyses of that chapter.

A Data File

Table 1.1 is an example of part of a file of data as organized for analysis by computer software. Each row contains the measurements for a different subject in the sample. Each column contains the measurement for each characteristic we observe. Table 1.1 shows data for the first ten subjects in the sample, for the characteristics: subjects' gender (F = female, M = male), racial group (B = black, H = Hispanic, W = white), marital status (1 = married, 0 = unmarried), age (in years), and annual income (in thousands of dollars). Some of the data are numerical, and some consist simply of labels. Chapter 2 introduces the various types of data that occur in files of this type. Chapter 3 presents descriptive statistics for summarizing the information about each characteristic.

TABLE 1.1 Example of Part of a Data File

Subject	Gender	Racial Group	Marital Status	Age	Annual Income
1	F	W	1	23	18.3
2	F	B	0	37	21.9
3	M	W	1	47	64.0
4	F	W	1	61	46.2
5	M	H	1	30	16.5
6	M	W	0	21	14.0
7	M	W	1	55	26.1
8	F	W	0	27	59.8
9	F	H	1	61	21.5
10	M	B	0	47	20.0

Uses and Misuses

A note of caution: The easy accessibility of complex statistical methods by means of computer software has dangers as well as benefits. It is simple, for example, to apply inappropriate methods to the data. A computer performs the analysis requested whether or not the assumptions required for its proper use are satisfied.

Many incorrect analyses result when researchers take insufficient time to understand the nature of a statistical method, the assumptions for its use, or its application to the specific problem. It is vital to understand the method before using it. Even if you read about similar studies that used the same type of analysis you are considering, do not assume that the analysis was correct without understanding the reasons. It may well be that a different approach is more appropriate.

Just knowing how to use statistical software does not guarantee a proper analysis. You'll need a good background in statistics to understand which method to select, which options to choose in that method, and how to make valid conclusions based on the computer output. The main purpose of this text is to give you this background.

1.4 Chapter Summary

The discipline of statistics includes methods for

- designing and carrying out research studies,
- exploring and describing data,
- making predictions (inferences) using the data.

We normally apply statistical methods to measurements in a ***sample*** selected from a ***population*** of interest. ***Statistics*** describe samples, while ***parameters*** describe populations. The two types of statistical analyses are ***descriptive methods*** for summarizing the sample and population and ***inferential methods*** for making predictions about population parameters using sample statistics. Statistical methods are easy to apply using computer software, relieving us of computational drudgery and helping us focus on the proper application and interpretation of the methods.

PROBLEMS

Practicing the Basics

1. Distinguish between *description* and *inference* as two purposes for using statistical methods. Illustrate the distinction using an example.
2. Give an example of a situation in which descriptive statistics would be helpful but inferential statistical methods would not be needed.
3. The Environmental Protection Agency (EPA) uses a few new automobiles of each brand every year to collect data on pollution emission and gasoline mileage performance. For a particular brand, identify the (a) subject, (b) sample, (c) population.
4. **a)** Distinguish between a *statistic* and a *parameter*.
   b) An article in a Florida newspaper (*The Gainesville Sun*) in 1995 reported that 66.5% of Floridians believe that state government should not restrict access to abortion. Is 66.5% most likely the value of a statistic, or of a parameter? Why?

5. The student government at the University of Wisconsin conducts a study about alcohol abuse among students. One hundred of the 40,000 members of the student body are sampled and asked to complete a questionnaire. One question asked is "On how many days in the past week did you consume at least one alcoholic drink?"
a) Describe the population of interest.
b) For the 40,000 students, suppose that one characteristic of interest is the percentage who would respond "zero" to this question. This value is computed for the students sampled. Is it a parameter or a statistic? Why?

6. The Current Population Survey of about 60,000 households in the United States in 1992 indicated that 10.3% of whites, 31.0% of blacks, and 26.7% of Hispanics in the United States have annual income below the poverty level (*Statistical Abstract of the United States, 1994*).
a) Are these numbers statistics or parameters? Explain.
b) Using a method from this text, we would conclude that the percentage of *all* black households in the United States having income below the poverty level is at least 30% but no greater than 32%. What type of statistical method does this illustrate—descriptive, or inferential?

Concepts and Applications

7. Your instructor will help the class create a data file consisting of the values for class members of characteristics such as *GE* = gender, *AG* = age in years, *HI* = high school GPA (on a four-point scale), *CO* = college GPA, *DH* = distance (in miles) of the campus from your home town, *DR* = distance (in miles) of the classroom from your current residence, *NE* = number of times a week you read a newspaper, *TV* = average number of hours per week that you watch TV, *SP* = average number of hours per week that you participate in sports or have other physical exercise, *VE* = whether you are a vegetarian (yes, no), *AB* = opinion about whether abortion should be legal in the first three months of pregnancy (yes, no), *PI* = political ideology (1 = very liberal, 2 = liberal, 3 = slightly liberal, 4 = moderate, 5 = slightly conservative, 6 = conservative, 7 = very conservative), *PA* = political affiliation (D = Democrat, R = Republican, I = independent), *RE* = how often you attend religious services (never, occasionally, most weeks, every week), *LD* = belief in life after death (yes, no), *AA* = support affirmative action (yes, no), *AH* = number of people you know who have died from AIDS or who are HIV+. Alternatively, your instructor may ask you to use a data file of this type already prepared in fall 1996 with a class of social science graduate students at the University of Florida, available on the World Wide Web at

`http://www.stat.ufl.edu/users/aa/social/data.html`

Using a spreadsheet program or the statistical software the instructor has chosen for your course, create a computer data file containing this information. Each row of the file should contain data for a particular student, and each column should contain values of a particular characteristic. Print the data. What are some questions one might ask about these data? Homework exercises in each chapter will refer to these data.

8. A sociologist is interested in estimating the average age at marriage for women in New England in the early eighteenth century. She finds within her state archives reasonably complete marriage records for a large Puritan village for the years 1700–1730. She then

takes a sample of those records, noting the age of the bride for each. The average age of the brides in the sample is 24.1 years. Using a statistical method from Chapter 5, the sociologist then estimates the average age of brides at marriage for the population to be between 23.5 and 24.7 years.

a) What part of this example is descriptive?

b) What part of this example is inferential?

c) To what population does the inference refer?

9. The 1994 General Social Survey of adult Americans asked subjects whether astrology—the study of star signs—has some scientific truth. Of 1245 sampled subjects who had an opinion, 651 responded definitely or probably true, and 594 responded definitely or probably not true. The proportion responding definitely or probably true was 651/1245 = .523.

a) Describe the population of interest.

b) For what population parameter might we want to make an inference?

c) What sample statistic could be used in making this inference?

d) Does the value of the statistic in (c) necessarily equal the parameter in (b)? Explain.

10. Look at a few recent issues of a major social science journal, such as *American Sociological Review* or *American Political Science Review*. About what proportion of the articles seem to use statistics? Find some examples of descriptive statistics.

11. Find out what statistical software is available to you while taking this course, either on PCs or workstations in a computer lab or on an institution-wide mainframe computer. Find out how to access the software, enter data, and print any files you create. As an exercise, create a data file using the data in Table 1.1 in Section 1.3, and print it.

Chapter 2

Sampling and Measurement

The ultimate goals of social science research are to understand, explain, and make inferences about social phenomena. To do this, we need data. *Descriptive* statistical methods provide ways of summarizing the data. *Inferential* statistical methods use sample data to make predictions about populations. To make inferences, we must decide which subjects of the population to sample. Selecting a sample that is likely to be representative of the population is a primary topic of this chapter.

We must convert our ideas about social phenomena into actual data through measurement. The development of ways to measure abstract concepts such as prejudice, love, intelligence, and status is one of the most difficult problems of social research. Moreover, the problems related to finding valid and reliable measures of concepts have consequences for statistical analysis of the data. In particular, invalid or unreliable data-gathering instruments render the statistical manipulations of the data meaningless.

The first section of this chapter discusses some statistical aspects of measurement, such as the different types of data. The second and third sections discuss the principal methods for selecting the sample that provides the measurements.

2.1 Variables and Their Measurement

Statistical methods provide a way to deal with *variability*. Variation occurs among people, schools, towns, and the various subjects of interest to us in our everyday lives. For instance, variation occurs from person to person in characteristics such as income, IQ, political party preference, religious beliefs, marital status, and musical talent. We shall see that the nature and the extent of the variability has important implications both on descriptive and inferential statistical methods.

Variables

A characteristic measured for each subject in a sample is called a *variable*. The name refers to the fact that values of the characteristic vary among subjects in a sample or population.

> **Variable**
>
> A ***variable*** is a characteristic that can vary in value among subjects in a sample or population.

Each subject has a particular value for a variable, but different subjects may possess different values. Examples of variables are gender (with values female and male), age at last birthday (with values 0, 1, 2, 3, and so on), religious affiliation (Protestant, Roman Catholic, Jewish, Other, None), number of children in a family (0, 1, 2, . . .), and political party preference (Democrat, Republican, Independent). The possible values the variable can assume form the ***scale*** for measuring the variable. For gender, for instance, that scale consists of the two labels, female and male.

The valid statistical methods for analyzing a variable depend on the scale for its measurement. We treat a numerical-valued variable such as annual income (in thousands of dollars) differently than a variable with a scale consisting of labels, such as political preference (with scale Democrat, Republican, Independent). We next introduce two ways to classify variables that determine the valid statistical methods. The first refers to whether the measurement scale consists of labels or numbers. The second refers to the number of levels in that scale.

Qualitative and Quantitative Data

Data are called ***qualitative*** when the scale for measurement is a set of unordered categories. For example, marital status, with categories (single, married, divorced, widowed), is qualitative. For Canadians, the province of one's residence is qualitative, with the categories Alberta, British Columbia, and so on. Other qualitative variables are religious affiliation (with categories such as Catholic, Jewish, Muslim, Protestant, Other, None), gender (female, male), political party preference (Democrat, Republican, Independent), and marriage form of a society (monogamy, polygyny, polyandry). For each variable, the categories are unordered; the scale does not have a "high" or "low" end.

For qualitative variables, distinct categories differ in quality, not in quantity or magnitude. Although the different categories are often called the *levels* of the scale, no level is greater than or smaller than any other level. Names or labels such as "Alberta" and "British Columbia" identify the categories, but those names do not represent different magnitudes of the variable.

When the possible values of a variable do differ in magnitude, the variable is called ***quantitative***. Each possible value of a quantitative variable is *greater than* or *less than* any other possible value. Such comparisons result from variables having a numerical scale. Examples of quantitative variables are a subject's annual income, number of years of education completed, number of siblings, and number of times arrested.

The set of categories for a qualitative variable is called a ***nominal scale***. For instance, a variable pertaining to one's mode of transportation to work might use the nominal scale consisting of the categories (car, bus, subway, bicycle, walk). A set of numerical values for a quantitative variable is called an ***interval*** scale. Interval scales have a specific numerical distance or "interval" between each pair of levels. Annual income is usually measured on an interval scale; the interval between $40,000 and $30,000, for instance, equals $10,000. We can compare outcomes in terms of how much larger or how much smaller one is than the other, a comparison that is not relevant for a nominal scale.

A third type of scale falls, in a sense, between nominal and interval. It consists of categorical scales having a natural *ordering* of values, but undefined interval distances between the values. Examples are social class (classified into upper, middle, lower), political philosophy (measured as very liberal, slightly liberal, moderate, slightly conservative, very conservative), and government spending on the environment (classified as too little, about right, too much). These scales are not nominal, because the categories are naturally ordered. The levels are said to form an ***ordinal*** scale.

Ordinal scales consist of a collection of ordered categories. Although the categories have a clear ordering, the distances between them are unknown. For example, a person categorized as very liberal is *more* liberal than a person categorized as slightly liberal, but there is no numerical value for *how much more* liberal that person is.

Both nominal and ordinal scales consist of a set of categories. Each observation falls into one and only one category. Variables having categorical scales are called ***categorical variables***. While the categories have a natural ordering for an ordinal scale, they are unordered for a nominal scale. For the categories (Catholic, Jewish, Muslim, Protestant, Other, None) for religious affiliation, it does not make sense to think of one category as being higher or lower than another.

The various scales refer to the actual measurement of social phenomena and not to the phenomena themselves. *Place of residence* may indicate the geographic place name of one's residence (nominal), the distance of that residence from a point on the globe (interval), the size of one's community (interval or ordinal), or other kinds of sociological variables.

Quantitative Nature of Ordinal Data

As we've discussed, data from nominal scales are qualitative—distinct levels differ in quality, not in quantity. Data from interval scales are quantitative: distinct levels have differing magnitudes of the characteristic of interest. The position of ordinal scales on the quantitative–qualitative classification is fuzzy. Because their scale consists of a set of categories, they are often treated as qualitative, being analyzed using methods for nominal scales. But in many respects, ordinal scales more closely resemble interval scales. They possess an important quantitative feature: each level has a *greater* or *smaller* magnitude of the characteristic than another level.

Some statistical methods apply specifically to ordinal variables. Often, though, statisticians take advantage of the quantitative nature of ordinal scales by assigning numerical scores to categories. That is, they often treat ordinal data as interval in order to use the more sophisticated methods available for quantitative data. For instance, course grades (such as A, B, C, D, E) are ordinal, but we treat them as interval when we assign numbers to the grades (such as 4, 3, 2, 1, 0) to compute a grade point average. Treating ordinal data as interval requires good judgment in assigning scores, and it is often accompanied by a "sensitivity analysis" of checking whether substantive results differ for differing choices of the scores. The quantitative treatment of ordinal data has benefits in the variety of methods available for data analysis, particularly for data sets with many variables.

Statistical Methods and Type of Measurement

The main reason for distinguishing between qualitative and quantitative data is that different different statistical methods apply to each type of data. Some methods are designed for qualitative variables and others are designed for quantitative variables.

It is not possible to analyze qualitative data using methods for quantitative variables. If a variable has only a nominal scale, for instance, one cannot use methods for interval data, since the levels of the scale do not have numerical values. One cannot apply quantitative statistical methods based on interval scales to qualitative variables such as religious affiliation or county of residence. For instance, the *average* is a statistical summary for quantitative data, since it uses numerical values; one can compute the average for a variable having an interval scale, such as income, but not for a variable having a nominal scale, such as religious affiliation.

On the other hand, it is always possible to treat a variable in a less quantitative manner. For example, suppose age is measured using the ordered categories under 18, 18–40, 41–65, over 65. This variable is quantitative, but one could treat it as qualitative either by ignoring the ordering of these four categories or by using unordered levels such as working age, nonworking age. Normally, though, we apply statistical methods specifically appropriate for the actual scale of measurement, since they use the characteristics of the data to the fullest. You should measure variables at as high a level as possible, because a greater variety of methods apply with higher-level variables.

Discrete and Continuous Variables

We now present one other way of classifying variables that helps determine which statistical method is most appropriate for a data set. This classification refers to the number of values in the measurement scale.

Discrete and Continuous Variables

A variable is ***discrete*** if it can take on a finite number of values and ***continuous*** if it can take an infinite continuum of possible real number values.

Examples of discrete variables are number of children (measured for each family), number of murders in the past year (measured for each census tract), and number of visits to a physician in past year (measured for each subject). Any variable phrased as "the number of ..." is discrete, since one can list all the possible values $\{0, 1, 2, 3, 4, \ldots\}$ for the variable. (Strictly speaking, there could be an infinite number of values for such a variable, namely, all the nonnegative integers. As long as the possible values do not form a continuum, the variable is still said to be discrete.)

Examples of continuous variables are height, weight, age, and the amount of time it takes to read a passage of a book. It is impossible to write down all the distinct potential values of a continuous variable, since they form a continuum. The amount of time needed to read a book, for example, could take on the value 8.6294473... hours.

With discrete variables, one cannot subdivide the basic unit of measurement. For example, 2 and 3 are possible values for the number of children in a family, but 2.571 is not. On the other hand, a collection of values for a continuous variable can always be refined; that is, between any two possible values, there is always another possible value. For example, an individual does not age in discrete jumps. Between 20 and 21 years of age, there is 20.5 years (among other values); between 20.5 and 21, there is 20.7. At some well-defined point during the year in which a person ages from 20 to 21, that person is 20.3275 years old, and similarly for every other real number between 20 and 21. A continuous, infinite collection of age values occurs between 20 and 21 alone.

Qualitative variables are discrete, having a finite set of unordered categories. In fact, all categorical variables, nominal or ordinal, are discrete. Quantitative variables can be discrete or continuous; age is continuous, and number of times arrested is discrete.

The distinction between discrete and continuous variables is often blurry in practice, because of the way variables are actually measured. Continuous variables must be rounded when measured, so we measure them as though they are discrete. We usually say that an individual is 20 years old whenever that person's age is somewhere between 20 and 21. Other variables of this type are prejudice, intelligence, motivation, and other internalized attitudes or orientations. Such variables are assumed to vary continuously, but measurements of them describe, at best, rough sections of the underlying continuous distributions. A scale of prejudice may have discrete units from 0 to 10, but each discrete value is assumed to include all values within a certain continuous range of the degree of prejudice.

On the other hand, some variables, though discrete, may take on a very large number of different values. In measuring annual family income in thousands of dollars, the potential values are 0, 1, 2, 3, ..., up to some very large highest value. Statistical methods for continuous variables are often simpler than methods for discrete variables. Thus, statisticians treat discrete variables that can assume many different values as if they were continuous. For example, they treat variables such as income and college entrance examination score as continuous variables. The discrete–continuous distinction is, in practice, a distinction between variables that can take lots of values, such as income, and variables that take relatively few values, such as number of times married.

You need to understand the discrete–continuous classification, qualitative–quantitative classification, and nominal–ordinal–interval scale classification, because each statistical method refers to a particular type of data. For instance, some methods (such as summarizing data using an average) require quantitative data, and some of these methods also require the variable to be continuous.

Figure 2.1 summarizes the types of data and their connections. Variables having a nominal scale are qualitative. Variables having an interval scale are quantitative. Variables having an ordinal scale are sometimes treated as quantitative and sometimes as qualitative. Variables having a nominal or ordinal scale take values in a set of categories, and are categorical. Categorical variables are discrete. Variables having an interval scale can be either discrete or continuous.

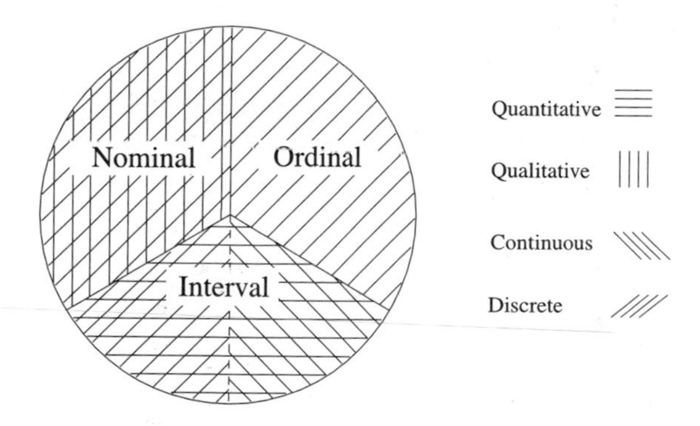

Note: Ordinal data are treated sometimes as qualitative and sometimes as quantitative

Figure 2.1 Summary of Quantitative–Qualitative, Nominal–Ordinal–Interval, Continuous–D Classifications

2.2 Randomization

Inferential statistical methods use sample statistics to make predictions tion parameters. The quality of the inferences depends crucially on how represents the population. This section introduces an important sampl incorporates ***randomization***, the mechanism for ensuring that the sa tion is adequate for inferential methods.

Simple Random Sampling

Subjects of a population to be sampled could be individuals, families, schools, houses, cities, hospitals, records of reported crimes, and so on. *Simple random sampling* is a method of sampling for which every possible sample has equal chance of occurring. This provides fairness and also permits inference about the population sampled. In fact, most inferential statistical methods assume randomization of the sort provided by simple random sampling.

Let n denote the number of subjects in the sample, called the ***sample size***.

> **Simple Random Sample**
>
> A ***simple random sample*** of n subjects from a population is one in which each possible sample of that size has the same probability of being selected.

For instance, suppose that a survey interviewer administers a questionnaire to one randomly selected adult subject of each of several separate households. A particular household contains four adults—mother, father, aunt, and uncle—identified as M, F, A, and U. A simple random sample of $n = 1$ of the adults is one in which each of the four adults is equally likely to be interviewed. The selection might be made, for example, by placing the four names on four identical ballots and selecting one blindly from a hat. For a simple random sample of $n = 2$ adults, each possible sample of size two is equally likely. The six potential samples are (M, F), (M, A), (M, U), (F, A), (F, U), and (A, U). To select the sample, we blindly select two ballots from the hat.

A simple random sample is often just called a ***random sample***. The "simple" adjective distinguishes this type of sampling from more complex sampling schemes presented in Section 2.4 that also have elements of randomization.

How to Select a Simple Random Sample

Before we can select a random sample, we need a list of all subjects in the population. This list is called the ***sampling frame***. The most common method for selecting a random sample from the sampling frame uses a *random number table* to ensure that each subject has an equal chance of selection.

> **Random Number Table**
>
> A ***random number table*** is a table containing a sequence of numbers that is computer generated according to a scheme whereby each digit is equally likely to be any of the integers $0, 1, 2, \ldots, 9$.

Table 2.1 shows a section of a random number table. The numbers fluctuate according to no set pattern. Any particular number has the same chance of being a 0, 1, 2, . . ., or 9. The numbers are chosen independently, so any one digit has no influence on any other. If the first digit in a row of the table is a 9, for instance, the next digit is still just as likely to be a 9 as a 0 or 1 or any other number.

TABLE 2.1 Part of a Table of Random Numbers

Line/Col.	(1)	(2)	(3)	(4)	(5)	(6)	(7)	(8)
1	10480	15011	01536	02011	81647	91646	69179	14194
2	22368	46573	25595	85393	30995	89198	27982	53402
3	24130	48360	22527	97265	76393	64809	15179	24830
4	42167	93093	06243	61680	07856	16376	39440	53537
5	37570	39975	81837	16656	06121	91782	60468	81305
6	77921	06907	11008	42751	27756	53498	18602	70659
7	99562	72905	56420	69994	98872	31016	71194	18738
8	96301	91977	05463	07972	18876	20922	94595	56869
9	89579	14342	63661	10281	17453	18103	57740	84378
10	85475	36857	53342	53988	53060	59533	38867	62300
11	28918	69578	88231	33276	70997	79936	56865	05859
12	63553	40961	48235	03427	49626	69445	18663	72695
13	09429	93969	52636	92737	88974	33488	36320	17617
14	10365	61129	87529	85689	48237	52267	67689	93394
15	07119	97336	71048	08178	77233	13916	47564	81056
16	51085	12765	51821	51259	77452	16308	60756	92144
17	02368	21382	52404	60268	89368	19885	55322	44819
18	01011	54092	33362	94904	31273	04146	18594	29852
19	52162	53916	46369	58586	23216	14513	83149	98736
20	07056	97628	33787	09998	42698	06691	76988	13602

Source: Abridged from William H. Beyer, ed., *Handbook of Tables for Probability and Statistics*, 2nd ed., © The Chemical Rubber Co., 1968, Used by permission of the Chemical Rubber Co.

We illustrate the use of this table for selecting a simple random sample of $n = 100$ students from a university student body of size 30,000. The sampling frame is a list of these students, such as a student directory. We select the students by using five-digit sequences to identify them, as follows:

1. Assign the numbers 00001 to 30000 to the students in the sampling frame, using 00001 for the first student in the list, 00002 for the second student in the list, and so on.
2. Starting at any point in the random number table, choose successive five-digit numbers until you obtain 100 distinct numbers between 00001 and 30000.
3. Include in the sample the students with assigned numbers equal to the random numbers selected.

For example, using the first column of five-digit numbers in Table 2.1, the first three random numbers are 10480, 22368, and 24130; thus, the first three students selected are those numbered 10480, 22368, and 24130 in the listing.

In selecting the 100 five-digit numbers, we skip numbers greater than 30000, such as the next seven five-digit numbers in Table 2.1, since no student in the sampling frame has an assigned number that large. After using the first column of five-digit numbers, we move to the next column of numbers and continue. If the population size were between 1000 and 9999, we would use only four digits at a time. The column (or row) from which we begin selecting the numbers does not matter, since the numbers have no set pattern.

The reason for using random sampling is that it reduces the chance of selecting a sample that is seriously biased in some way, thus leading to inaccurate inferences about the population. Everyone has the same chance of inclusion in the sample.

Probability and Nonprobability Sampling

Simple random sampling is a type of ***probability sampling*** method. Such methods can specify the probability that any particular sample will be selected. For simple random sampling, each distinct possible sample of n subjects has the same probability of selection. With probability samples one can apply inferential statistical methods, since the derivation of those methods requires knowing the probabilities of the possible samples. ***Nonprobability sampling*** methods are ones for which it is not possible to specify the probabilities of the possible samples. Inferences using such samples are of unknown reliability.

One of the most common nonprobability sampling methods is ***volunteer sampling***. As the name implies, in this method subjects volunteer themselves for the sample. Inherent in this method is the danger that the sample will poorly represent the population and will yield misleading conclusions.

For instance, a mail-in questionnaire published in *TV Guide* posed the question, "Should the President have the Line Item Veto to eliminate waste?" Of those who responded, 97% said yes. For the same question posed to a random sample, 71% said yes (D. M. Wilbur, *The Public Perspective*, 1993).

A good example of volunteer sampling is visible almost any day on U.S. television. It's become a trend on many TV news and entertainment programs to ask viewers to offer their opinions on an issue of the moment by calling a 900 number. The problem is that the viewers who respond are unlikely to be a representative cross section, but will be those people who watch that program and who happen to feel strongly enough to call. Individuals possessing one specific opinion on that issue might be much more likely to respond than individuals holding a different opinion.

For instance, one night the ABC program *Nightline* asked viewers whether the United Nations should continue to be located in the United States. Of more than 186,000 callers, 67% wanted the United Nations out of the United States. At the same time, a scientific poll using a random sample of about 500 respondents estimated the true percentage wanting the United Nations out of the United States to be about 28%. Even though the random sample is much smaller, it is far more trustworthy since it greatly reduces the chance of bias. From statistical inferential methods (from Chapter 5) with this random sample, the true percentage of the population of all Americans who want the United Nations out of the United States is between about 24% and 32%.

A large sample does not help with volunteer sampling—the bias remains. In 1936, the newsweekly *Literary Digest* sent over 10 million questionnaires in the mail to predict the outcome of the presidential election. The questionnaires went to a relatively wealthy segment of society (those having autos or telephones), and fewer than 25% were returned. The journal used these to predict an overwhelming victory by Alfred Landon over Franklin Roosevelt. The opposite result was predicted by George Gallup with a much smaller sample in the first scientific public poll taken for this purpose. (In fact, Roosevelt won with 62% of the vote.)

Another example of nonprobability sampling is what we might call the *streetcorner interview*: An interviewer stands at a specific location and conducts interviews by stopping whoever passes by. Severe biases may arise as a function of the time and location of the interview and the judgment of the interviewer in deciding whom to interview. For example, working people might be underrepresented if the interviews are conducted on weekdays between 9:00 A.M. and 5:00 P.M., and the racial or socioeconomic composition of the sample may be biased if the interviewer conducts the interviews in an upscale shopping mall.

Inferential statistical methods utilize assumptions about the probability that any particular sample is selected. They are not valid for data obtained with nonprobability samples, for which such probabilities are unknown.

Samples Based on Experimental Designs

In many sciences, data result from a planned *experiment*. The scientist has experimental control over the subjects' values on factors that can influence the variable of interest in the study, such as temperature, pressure, humidity, and so forth. Data obtained in such studies are called ***experimental data***. By contrast, data obtained from surveys are called ***observational data***. The researcher measures subjects' responses on the variables of interest, but has no experimental control over the subjects.

A major purpose of many experiments is to compare responses of subjects on some outcome measure, under different conditions. For instance, those conditions might be different drugs for treating some illness or different combinations of chemicals for manufacturing some product. The conditions are called ***treatments***. To obtain experimental data, the researcher needs a plan for assigning subjects to the different conditions being compared. These plans are called ***experimental designs***.

For instance, in the late 1980s, the Physicians' Health Study Research Group at Harvard Medical School designed a five-year randomized study to analyze whether regular intake of aspirin reduces mortality from cardiovascular disease. Of about 22,000 physicians, half were randomly chosen to take an aspirin every other day, and the remaining half took a placebo, which had no active agent. After five years, rates of heart attack were compared for the two groups.

By using randomization to determine who received the placebo and who received the aspirin, the researchers knew that the groups would roughly balance on factors that could affect heart attack rates, such as age and overall quality of health. If the physicians could decide on their own which treatment to take, the groups might have been out of balance on some important factor. Perhaps, for instance, younger physicians would have been more likely to select the aspirin regimen; then, a lower heart attack rate among the aspirin group could simply represent younger subjects being less likely to suffer heart attacks.

Imbalance between groups is always a danger with observational studies, making it difficult to compare groups. For instance, if white students have a higher average score than black students on some standardized test, a variety of other unmeasured variables might account for that difference, such as parents' education or income or other environmental factors.

In social research, unfortunately, it is rarely possible to conduct controlled experiments. One cannot randomly assign subjects to treatments such as race or gender. This is also sometimes true in other sciences, particularly medical sciences. Consider, for instance, a study of whether passive smoking (being exposed to secondhand cigarette smoke on a regular basis) leads to higher rates of lung cancer. An experimental study might take a sample of children, randomly select half of them for placement in an environment where they are passive smokers, and place the other half in an environment where they are not exposed to smoke. Then, perhaps 60 years later the observation is whether each has developed lung cancer. Clearly, for many reasons, including time and ethics, it is not possible to conduct such an experimental study.

Regardless of whether a study is observational or experimental, randomization is an important feature in any study that involves making inferences. This randomization could take the form of randomly selecting a sample for an observational study, or randomly allocating subjects to different conditions for an experimental study.

2.3 Sampling and Nonsampling Variability

Even if a study wisely uses randomization in selecting a sample, the results of the study still depend on which sample of subjects is actually selected. Two researchers who select separate random samples from some population may have very little overlap, if any, between the two sample memberships. Therefore, the values of sample statistics will differ for the two samples, and the respective inferences based on these samples may differ.

Sampling Error

For instance, the Gallup and Harris organizations might each take a random sample of 1000 Americans, in order to estimate the percentage of Americans who give the president's performance in office a favorable rating. Based on the samples they select, Gallup might report an approval rating of 53%, whereas Harris might report one of 58%. This difference could reflect slightly different question wording, but even if the questions are worded exactly the same, the percentages would probably differ somewhat because the samples are different. For conclusions based on statistical methods to be worthwhile, one must determine the potential *sampling error*—the extent to which the value of a statistic may differ from the parameter it predicts because of the way results vary from sample to sample.

Sampling Error

The ***sampling error*** of a statistic is the error that occurs when a statistic based on a sample estimates or predicts the value of a population parameter.

Suppose that the true population percentage giving the president a favorable rating is 56%. Then the Gallup organization, which predicted 53%, had a sampling error of $53\% - 56\% = -3\%$; the Harris organization, which predicted 58%, had a sampling error of $58\% - 56\% = 2\%$. In practice, of course, the sampling error is unknown, since the true values of population parameters are unknown. However, the methods of this text allow us to predict the size of the sampling error. For samples of size 1000, for instance, the sampling error for estimating percentages is usually no greater than 3% or 4%.

Random sampling guards against a systematic bias in the sampling error, such as a tendency to underestimate consistently or overestimate consistently the true parameter values. It also allows us to gauge the likely size of the sampling error. Variability also occurs in the values of sample statistics with nonrandom sampling, but the extent of that variability is not predictable as it is with random samples.

Other Sources of Variability

Other factors besides sampling error can introduce variability into results from samples and possibly cause bias. For instance, the sampling frame may suffer from ***undercoverage***. It may lack representation from some groups in the population of interest to us. A telephone survey will not reach homeless people or prison inmates or people not having a telephone. If its sampling frame consists of the names in a telephone directory, it will not reach those having an unlisted number. For many variables, responses by the homeless or by those with unlisted numbers might well tend to be considerably different from those actually sampled, leading to biased results.

Some subjects who are supposed to be in the sample may refuse to participate, or it may not be possible to reach them. This results in the problem of ***nonresponse***, which is a serious one for many surveys. If only half the intended sample was actually observed, we should worry about whether the half not observed differ from those observed in a way that causes bias in the overall results. Even if we select the sample randomly, the results are questionnable if there is substantial nonresponse, say, over 20%. Even in censuses, which are supposed to record data for all people in the country, some people are not observed or simply fail to cooperate.

In an interview, characteristics of the interviewer or other factors may affect the response in a way that introduces ***response bias***. Respondents might lie if they think their response to a question is socially unacceptable. They may be more likely to give the response that they think the interviewer would prefer. An example is provided by a study by Lynn Sanders, a political scientist at the University of Chicago, on the effect of the race of the interviewer. Following a phone interview, respondents were asked whether they thought the interviewer was black or white (all were actually black). Perceiving a white interviewer resulted in more conservative opinions. For example, 14% agreed that "American society is fair to everyone" when they thought the interviewer was black, but 31% agreed to the same statement when posed by an interviewer the respondent thought was white (*Washington Post*, National Weekly Edition, June 26, 1995).

The way a variable is measured can have a large impact on the types of results observed. For instance, the wording of a question in a survey can greatly affect the responses. A Roper Poll was designed to determine the percentage of Americans who express some doubt that the Holocaust occurred. In response to the question, "Does it seem possible or does it seem impossible to you that the Nazi extermination of the Jews never happened?" 22% said it was possible the Holocaust never happened. The Roper organization later admitted that the question was worded in a confusing manner. When they asked, "Does it seem possible to you that the Nazi extermination of the Jews never happened, or do you feel certain that it happened?" only 1% said it was possible it never happened (*Newsweek*, July 25, 1994).

Finally, even the order in which questions are asked can influence the results dramatically. Crosson (1994) described a study that, during the Cold War, asked "Do you think the U.S. should let Russian newspaper reporters come here and send back whatever they want?" and "Do you think Russia should let American newspaper reporters come in and send back whatever they want?" The percentage of yes responses to the first question was 36% when it was asked first and 73% when it was asked second.

In a fine book summarizing the potential difficulties with conducting and interpreting survey research, Crosson (1994) makes several recommendations. In particular, she notes that any newspaper report or TV story should say who sponsored and conducted the research, indicate how the questions were worded, and tell how the sample was selected and how large it was. She notes, "As a general rule, the less information that is available about the way a poll was conducted, the less it can be trusted."

Missing Data

A problem encountered in almost all large studies is ***missing data***. Some subjects do not provide responses for some of the variables measured. Standard software ignores cases for which observations are missing for at least one of the variables used in an analysis. This can result in much wasted information, however, and statisticians have recently developed methods that replace missing observations by predicted values based on patterns in the data. Most of this work is beyond the scope of this text, but we refer readers to Little and Rubin (1989) for a good introduction to this important topic.

2.4 Other Probability Sampling Methods*,[1]

Section 2.2 explained the importance to statistical inference of randomization, such as occurs in probability sampling. It introduced an example of probability sampling, called ***simple random sampling***. In practice, other probability sampling methods that also have elements of randomness are sometimes preferable to simple random sampling, or are easier to obtain. This section introduces some of these methods.

Systematic Sampling

A probability sampling method called ***systematic random sampling*** is often easier to implement than simple random sampling but just as good for inference-making. With this method, one selects a subject near the beginning of the sampling frame list, skips several names and selects another subject, skips several more names and selects the next subject, and so forth. The number of names skipped at each stage depends on the desired sample size. Specifically, here's how it's done:

Systematic Random Sample

Denote the sample size by n and the population size by N. Let $k = N/n$, the population size divided by the desired sample size. A ***systematic random sample*** (1) selects a subject at random from the first k names in the sampling frame, and (2) selects every kth subject listed after that one. The number k is called the *skip number.*

For a systematic random sample of 100 students from a population of 30,000 students listed in a campus directory, $N = 30{,}000$, $n = 100$, and $k = 30{,}000 / 100 = 300$. This means that the population size is 300 times the desired sample size, so the method selects one out of every 300 students. It selects one student at random, using a random number table, from the first 300 students listed in the directory. Then it selects every 300th student after the one selected randomly. This results in a sample of size 100. The first three digits in Table 2.1 are 104, which falls between 001 and 300, so the first student selected is the one numbered 104. The numbers of the other students selected are $104+300 = 404$, $404+300 = 704$, $704+300 = 1004$, $1004+300 = 1304$, and so on. The 100th student selected is listed in the last 300 names in the directory.

A systematic random sample is simpler to select than a simple random sample, since it uses only one random number. Yet, it typically provides just as good a representation of the population. With this method, one can use statistical formulas based on simple random sampling. A systematic random sample is not a simple random sample, because all samples of size n are not equally likely. For instance, unlike in a simple random sample, two subjects listed next to each other on the list cannot both appear in the sample.

---

[1]Sections or subsections marked with an asterisk have less importance for a first introduction to statistics and are optional.

A systematic random sample has one potential disadvantage: Bias may occur in studying a variable if regular cyclical fluctuation occurs in its values throughout the sampling frame, with the period of the cycle equal to the skip number k. Suppose, for example, that a content analysis of a daily newspaper studies the percentages of newspaper space devoted to foreign news, national news, state news, local news, sports, and advertising. The sampling frame might consist of the daily editions of the newspaper for the previous year. For a systematic sample of these editions with k equal to 7 or a multiple of 7, only one day of the week is sampled. This could lead to bias in interpreting the amount of space devoted to each subject, since the format of the paper would vary regularly with period 7; that is, all Sunday papers would be similar in format, all Monday papers would be similar, and so forth. If only Thursdays were sampled and if the newspaper carried special sections of advertisements for grocery stores on that day, then this would affect the results.

For alphabetic listings, such as directories of names of individuals, cities, or counties, values of most variables fluctuate randomly through the list. So, systematic random samples from a sampling frame with alphabetically ordered sampling units are usually acceptable.

Stratified Sampling

Another probability sampling method used in social science research is the *stratified random sample*.

Stratified Random Sample

A ***stratified random sample*** divides the population into separate groups, called ***strata***, and then selects a simple random sample from each stratum.

Stratified random samples are useful when the strata are groups that the study will compare. A study in Onondaga County, New York, plans to compare white and black subjects in terms of their access to quality health care facilities. Stratifying according to race of subject, the study selects a random sample of white subjects and another random sample of black subjects. The sampling is called ***proportional*** if the proportions of the sample chosen in the various strata are the same as those existing in the entire population. For example, if 80% of the population of interest is white and 20% is black, then the sampling is proportional if the sample size for the white sample is four times the sample size for the black sample. The sampling is called ***disproportional*** if the sampled proportions differ from the population proportions.

Disproportional stratified sampling is useful when the population size of at least one stratum is relatively small. A group that comprises a small part of the population may not gain enough representation in a simple random sample to allow precise inferences, especially if the total sample size is not very large. It is not possible to compare accurately Native Americans to other Americans on the percentage favoring legalized gambling, for example, if only 3 of the 200 people in the sample are Native Americans. Disproportional stratified sampling, on the other hand, controls the number of observations from each stratum. The study might, for instance, randomly sample 100 Native Americans and 100 other Americans.

This consideration becomes especially important when the strata are formed by combining levels of several variables. A simple random sample might yield too few observations to analyze characteristics of the population subset who are black and registered Republicans.

One problem with stratification is that, to implement it, one must know the stratum into which each subject of the sampling frame belongs. This usually restricts the variables that can be used for forming the strata. The variables must have strata that are easily identifiable. For example, it is fairly easy to select a stratified sample of a school population using grade level as the stratification variable. On the other hand, it is difficult to prepare an adequate sampling frame of city households stratified by income of household head.

Cluster Sampling

Simple, systematic, and stratified random sampling are very expensive or even impossible to implement in many situations. Each method requires access to a complete sampling frame. Such lists are easy to obtain for sampling units such as cities from the collection of all cities in a nation; however, for sampling individuals or families, it is more difficult. A sample of families that is based on listings in a telephone directory, for example, has no representation of newcomers to the community, families with unlisted numbers, or families too poor to afford telephones. If these nonrepresented people differ from the rest of the population on the variables being studied, statistical inferences may be faulty. The *cluster sampling* technique is useful when a complete listing of the population is not available.

Cluster Sample

Divide the population into a large number of groups, called ***clusters***. A ***cluster sample*** is one for which the sampling units are the subjects in a random sample of the clusters.

For instance, a study might plan to sample about 1% of the families in a city, using city blocks as clusters. Using a map to identify city blocks, one could select a simple random sample of 1% of the blocks and then sample every family on each block. Or, to study patient care in mental hospitals in Ontario, one could first sample mental hospitals (the clusters) and then collect data for patients within those hospitals.

The clusters in a cluster sample are often defined geographically, so the sampling units within a cluster are geographically close. If the measurement process requires interviewing the subjects in person, and if the geographic area covered by the population is large, cluster sampling has lower travel time between interviews than other sampling methods. For example, by interviewing every family living in a particular city block, one obtains many observations quickly and with little travel.

Stratified sampling draws a sample from *within every* stratum, and the strata are the groups compared. By contrast, in cluster sampling, clusters are simply ways of easily identifying groups of subjects. The goal is not to compare the clusters but simply to use them to obtain a sample. Sampling occurs from *among* the clusters, with most clusters not represented at all in the eventual sample. This causes the main disadvantage of cluster sampling: One may need a larger sample to achieve as much accuracy in making inferences about populations as other types of random samples provide. Observations within clusters tend to be homogeneous, because of the tendency of subjects living near one another to be similar on economic and demographic variables such as age, income, race, occupation, and opinion issues. Thus, n observations from a cluster do not provide n times as much information as a single observation from that cluster.

Multistage Sampling

Multistage sampling methods use combinations of various sampling techniques. These are common in social science research, often because they are simpler to implement than simple random sampling but provide a broader sampling of the population than a single method, such as cluster sampling, provides. To study various characteristics of adult residents in the United States, one could treat counties (or census tracts) as clusters and select a random sample of a certain number of them. Within each county selected, one could take a cluster sample of square-block regions. Within each region selected, one could systematically sample every tenth house. Within each house selected, one could select one adult at random for the sample.

For statistical inference procedures, stratified samples, cluster samples, and multistage samples require different formulas from the ones in this book. The formulas are complex even for basic statistical methods. In short, for a fixed sample size, the results for stratified samples may be more precise than those stated in this textbook for simple random samples, whereas the results for cluster samples are less precise. Books specializing in sampling methodology provide further details (Kalton, 1983; Kish, 1965; Scheaffer et al., 1996; Sudman, 1976; Thompson, 1992).

2.5 Chapter Summary

Statistics is concerned with the study of ***variables***, which are characteristics that vary among subjects. The statistical methods used for a particular variable depend on whether it is ***qualitative*** or ***quantitative***.

- Numerically measured variables, such as income and number of arrests, are quantitative. They are said to be measured on an *interval* scale.
- Variables having a scale of unordered categories, such as religious affiliation and province of residence, are qualitative. They are said to be measured on a *nominal* scale.

- Variables having a scale of ordered categories, such as social class and political philosophy, have an *ordinal* scale of measurement. They are often treated in a quantitative manner by assigning scores to the categories.
- Variables are also classified as ***discrete*** or ***continuous***. Categorical variables, whether nominal or ordinal, are discrete. Quantitative variables can be of either type; they are continuous if there is a continuous, infinite set of possible values for the variable, such as age.
- We select samples from populations to collect quantitative or qualitative data. For a ***simple random sample***, every possible sample has the same chance of selection. Random sampling allows control over the amount of ***sampling error***. Random samples are more likely to be representative of the population than are nonrandom samples, and one can determine their sampling error by methods presented later in this text.
- Inferential statistical methods are appropriate only for ***probability samples***, which incorporate randomization in some way. Examples of probability sampling include ***simple random*** sampling, ***systematic*** sampling, ***stratified*** sampling, ***cluster*** sampling, and multistage combinations of these. The formulas in this text refer to simple random sampling.

Chapter 3 introduces basic statistics for describing samples, as well as the corresponding parameters for describing populations. Hence, our focus changes next to the *descriptive* aspects of statistical methodology.

PROBLEMS

Practicing the Basics

1. Explain the difference between
   a) Discrete and continuous variables
   b) Qualitative and quantitative variables
   c) Nominal and ordinal variables
   Why do these distinctions matter for statistical analysis?
2. Identify each of the following variables as qualitative or quantitative.
   a) Number of pets in family
   b) County of residence
   c) Choice of auto (domestic or import)
   d) Distance (in miles) commute to work
   e) Choice of diet (vegetarian, nonvegetarian)
   f) Time spent in previous month browsing the World Wide Web
   g) Ownership of personal computer (yes, no)
   h) Number of people you have known with AIDS (0, 1, 2, 3, 4 or more)
3. Which scale of measurement (nominal, ordinal, or interval) is most appropriate for
   a) Attitude toward legalization of marijuana (favor, neutral, oppose)
   b) Gender (male, female)
   c) Number of siblings (0, 1, 2, ...)
   d) Political party affiliation (Democrat, Republican, Independent)

e) Religious affiliation (Roman Catholic, Baptist, Methodist, ...)
f) Political philosophy (very liberal, somewhat liberal, moderate, somewhat conservative, very conservative)
g) Years of school completed (0, 1, 2, 3, ...)
h) Highest degree obtained (none, high school, bachelor's, master's, doctorate)
i) College major (education, anthropology, physics, sociology, ...)
j) Scholastic Aptitude Test score (200–800 range for scores)
k) Employment status (employed, unemployed)

4. Which scale of measurement is most appropriate for each of the following variables?
a) Occupation (plumber, teacher, secretary, etc.)
b) Occupational status (blue collar, white collar)
c) Social status (lower, middle, upper class)
d) Statewide murder rate (number murders per 1000 population)
e) County population size (number of people)
f) Population growth rate (in percentages)
g) Community size (rural, small town, large town, small city, large city)
h) Annual income (thousands of dollars per year)
i) Attitude toward affirmative action (favorable, neutral, unfavorable)

5. Which scale of measurement is most appropriate for "attained education" when it is measured as
a) Number of years (0, 1, 2, 3, ...)
b) Grade level (elementary school, middle school, high school, college, graduate school)
c) School type (public school, private school)

6. Give an example of a variable that is
a) Qualitative
b) Quantitative
c) Ordinal scale
d) Categorical
e) Discrete
f) Continuous
g) Quantitative and discrete

7. A survey asks subjects to rate five issues according to their importance in determining voting intention for U.S. senator, using the scale very important, somewhat important, unimportant. The issues are foreign policy, unemployment, inflation, the arms race, and civil rights. The evaluations can be treated as five variables: "foreign policy evaluation," "unemployment evaluation," and so on. These variables represent what scale of measurement?

8. Which of the following variables are continuous when the measurements are as fine as possible?
a) Age of mother
b) Number of children in family
c) Income of spouse
d) Population of cities
e) Latitude and longitude of cities
f) Distance of home from place of employment

9. Which of the following variables could theoretically be measured on a continuous scale?
a) Method of contraception used

b) Length of time of residence in a state
c) Task completion time
d) Intelligence
e) Authoritarianism
f) Alienation
g) State of residence

10. A class has 50 students. Use the column of the first two digits in the random number table (Table 2.1) to select a simple random sample of three students. If the students are numbered 1 to 50, what are the numbers of the three students selected?

11. In the 1995 Miss America beauty pageant, television viewers could cast their vote on whether to cancel the swimwear parade by phoning a number the network provided. About 1 million viewers called and registered their opinion, of whom 79% said they wanted to see the contestants dressed as bathing beauties. Is this sample a probability sample of all viewers of this program in 1995? Why, or why not?

12. In fall 1995, the BBC in Britain requested viewers to call the network and indicate their favorite poem. Of more than 7500 callers, more than twice as many voted for Rudyard Kipling's *If* than for any other poem. The BBC then reported that this was the clear favorite. Was their sample a probability sample? Explain. How might they have obtained a more reliable sample, even if much smaller in size?

13. A local telephone directory has 300 pages with 120 names per page.
a) Explain how you would choose a simple random sample of 5 names. Using the second column of Table 2.1, select 5 numbers to identify subjects for the sample.
b) Select five numbers to identify subjects for a systematic sample of five names from the same directory.

14. Refer to the preceding problem. Is cluster sampling applicable? How could it be carried out, and what would be the advantages and disadvantages?

Concepts and Applications

15. Refer to the data file created in Problem 1.7. For each variable in the data set, indicate whether it is
a) Qualitative or quantitative
b) Nominal, ordinal, or interval

16. Repeat the previous exercise for the World Wide Web (WWW) data set (Problem 1.7).

17. Table 2.2 shows the result of the 1992 Presidential election along with the vote predicted by several organizations in the days before the election. The sample sizes were typically about 2000. The percentages for each poll do not sum to 100 because of voters reporting as undecided. What factors cause the results to vary somewhat among organizations? How accurate would you characterize these results, considering the relative sizes of the sample and the population?

18. A simple random sample of size n is one in which (select the best response):
a) Every nth member is selected from the population.
b) Each possible sample of size n has the same chance of being selected.
c) There is exactly the same proportion of women in the sample as is in the population.
d) One keeps sampling until obtaining a fixed number of people having various characteristics (e.g., males, females, or whites, blacks).

TABLE 2.2

	Predicted Vote		
Poll	Clinton	Bush	Perot
Gallup	44	36	14
Harris	44	38	17
ABC	44	37	16
CBS	44	35	15
NBC	44	36	15
Washington Post	43	35	16
Actual vote	43	38	19

Source: *The New York Times*, November 3, 1992.

e) A particular minority group member of the population is less likely to be chosen than a particular majority group member.
f) All of the above
g) None of the above

19. You are assigned to direct a study on your campus to determine the factors that relate to strong academic performance at the school.
a) Describe how you might select a sample of 100 students for the study.
b) List some variables that you would measure in your study. For each, provide the scale you would use to measure it, and indicate whether statistical analysis could treat it as (i) qualitative or quantitative, (ii) nominal, ordinal, or interval, (iii) continuous or discrete.

20. ***Quota sampling*** is a method used when, for example, an interviewer stands at a street corner and conducts interviews until obtaining a quota in various categories representing the relative sizes of various groups in the population of interest. For instance, the quota might be 50 factory workers, 100 housewives, 60 elderly people, 30 blacks, and so forth.
a) Is this a probability or nonprobability sampling method? Explain, and discuss potential advantages or disadvantages of this method. (Even professional pollsters such as Gallup used this method until 1948, when they incorrectly predicted that Dewey would defeat Truman in a landslide in the presidential election.)
b) Explain how quota sampling differs from (i) cluster sampling, (ii) stratified sampling.

21. In 1995 in the United Kingdom, the Equality Code used by the legal profession added a section to make members more aware of the dangers of sexual harassment. It states that "research for the Bar found that over 40 percent of female junior tenants said they had encountered sexual harassment during their time at the Bar." This was based on a study conducted at the University of Sheffield. The academics sent a questionnaire to 334 junior tenants at the Bar, of whom 159 responded. Of the 159, 67 were female. Of the females, 3 said they had experienced sexual harassment as a major problem, and 24 had experienced it as a slight problem. Comment on ways in which the quoted statement might be misleading.

22. When the Yankelovich polling organization asked, "Should laws be passed to eliminate all possibilities of special interests giving huge sums of money to candidates?" 80% of the sample answered *yes*. When they posed the question, "Should laws be passed to prohibit interest groups from contributing to campaigns, or do groups have a right to contribute to the candidate they support?" only 40% said *yes* (*Source*: *A Mathematician Reads the Newspaper*, by J. A. Paulos, New York: Basic Books, 1995, p. 15). Explain what this example illustrates, and use your answer to differentiate between sampling error and other types of variability in survey results.

23. Crosson (1994, p. 168) describes an analysis of published medical studies involving treatments for heart attacks. In the studies having randomization and strong controls for bias, the new therapy provided improved treatment 9% of the time. In studies without randomization or other controls for bias, the new therapy provided improved treatment 58% of the time. Select the correct response(s).
a) This result suggests it is better not to use randomization in medical studies, because it is harder to show that new ideas are beneficial.
b) Many newspaper articles that suggest that a particular food, drug, or environmental agent is harmful or beneficial should be viewed skeptically, unless we learn more about the statistical design and analysis for the study.
c) This result suggests that you should be skeptical about published results of medical studies that are not randomized, controlled studies.
d) Controlling for biases, both suspected and unsuspected, is necessary in medical research but not in social research, because the social sciences deal in subjective rather than objective truth.

24. In each of the following situations, evaluate whether the method of sample selection is appropriate for obtaining information about the population of interest. How would you improve the sample design?
a) A newspaper wants to determine whether its readers believe that government expenditures should be reduced by cutting welfare programs. They provide a telephone number for readers to call to give their opinions. Based on 1434 calls that they receive, they report that 93% of the city's residents believe that welfare payments should be reduced.
b) A congresswoman reports that letters to her office are running 3 to 1 in opposition to the passage of stricter gun control laws. She concludes that approximately 75% of her constituents oppose stricter gun control laws.
c) An anthropology professor wanted to compare attitudes toward premarital sex of physical science majors and social science majors. She administered a questionnaire to her large class of Anthropology 437, Comparative Human Sexuality. She found no appreciable difference between her physical science and social science majors in their attitudes, so she concluded that the two student groups were about the same in their relative acceptance of premarital sex.
d) A questionnaire was mailed to a simple random sample of 500 household addresses in a city. Ten were returned as bad addresses, 63 were returned completed, and the rest were not returned. The researcher analyzed the 63 cases and reported that they represent a "random sample of city households."

25. Refer to the previous exercise. Repeat for the following scenarios:
a) A principal in a large high school is interested in student attitudes toward a proposed general achievement test to determine whether a student should graduate. She lists all of the first-period classes, assigning a number to each. Then, using a random number table, she chooses a class at random and interviews every student in that class about the proposed test.
b) A new restaurant opened in January. In June, after six months of operation, the owner applied for a loan to improve the building. The loan application requested an estimate of the yearly gross income of the business. The owner's record book contains receipts for each day of operation since opening. She decided to calculate the average daily receipt based on a sample of the daily records and multiply that by the number of days of opera-

tion in a year. She took a systematic sample, with $k = 7$, sampling every Friday's record. The average daily receipt for this sample was then used to estimate the yearly receipts.

26. You plan to sample from the 5000 students at your college in order to compare the proportions of men and women who believe that women should have the right to an abortion.
a) Explain how you would proceed, if you want a simple random sample of 100 students.
b) How would you proceed if you want a systematic random sample?
c) You use a random number table to select students, but you stop selecting females as soon as you have 50, and you stop selecting males as soon as you have 50. Is the resulting sample a simple random sample? Why or why not?

27. In a systematic random sample, every subject has the same chance of selection, but the sample is not a simple random sample. Explain why, by showing that every possible sample of size n is not equally likely.

28. I need to collect data for a sample of residents of registered nursing homes in my state. I obtain from the state a list of all nursing homes, which I number from 1 to 317. Beginning randomly, I choose every tenth home on the list, ending up with 31 homes. I then obtain lists of residents from those 31 homes, and I select a simple random sample from each list. What kinds of sampling have I used?

Bibliography

Babbie, E. (1995). *The Practice of Social Research*, 7th ed. Belmont, CA: Wadsworth.

Bailey, K. (1994). *Methods of Social Research*, 4th ed. New York: Free Press.

Crosson, C. (1994). *Tainted Truth: The Manipulation of Fact in America*. New York: Simon & Schuster.

Kalton, G. (1983). *Introduction to Survey Sampling*. Newbury Park, CA: Sage.

Kish, L. (1965). *Survey Sampling*. New York: Wiley.

Little, R. J. A., and Rubin, D. B. (1989). The analysis of social science data with missing values. *Sociological Methods and Research*, 18, 292-326.

Scheaffer, R. L., Mendenhall, W., and Ott, L. (1996). *Elementary Survey Sampling*, 5th ed. Belmont, CA: Wadsworth.

Sudman, S. (1976). *Applied Sampling*. New York: Academic Press.

Thompson, S. K. (1992). *Sampling*. New York: Wiley.

Chapter 3

Descriptive Statistics

One of the two primary reasons for using statistical methods is to summarize and describe data, to make the information easier to assimilate. This chapter presents common methods of data description. The first section discusses statistical description through the use of tables and graphs. These tools provide a summary picture of the data.

We then present ways of describing the data with numerical measures. Section 3.2 defines statistics that describe the *center* of a collection of data—in other words, a "typical" measurement in the sample. Section 3.3 introduces statistics that describe the *variation* of the data about that center. The final section distinguishes between statistics describing samples and related parameters describing populations.

3.1 Tabular and Graphical Description

Example 3.1 State Murder Rates

We use the data in Table 3.1 to illustrate descriptive methods. This table lists all 50 states in the United States and their 1993 murder rates. The murder rate measures the number of murders in that state in 1993 per 100,000 population. For instance, if a state had 120 murders and a population size of 2,300,000, its murder rate was (120/2,300,000) $\times$ 100,000 = 5.2. It is difficult to learn much by simply reading through the murder rates. We will use tables, graphs, and numerical measures to understand these data more fully. □

TABLE 3.1 List of States with 1993 Murder Rates Measured as Number of Murders per 100,000 Population

Alabama	11.6	Louisiana	20.3	Ohio	6.0
Alaska	9.0	Maine	1.6	Oklahoma	8.4
Arizona	8.6	Maryland	12.7	Oregon	4.6
Arkansas	10.2	Massachusetts	3.9	Pennsylvania	6.8
California	13.1	Michigan	9.8	Rhode Island	3.9
Colorado	5.8	Minnesota	3.4	South Carolina	10.3
Connecticut	6.3	Mississippi	13.5	South Dakota	3.4
Delaware	5.0	Missouri	11.3	Tennessee	10.2
Florida	8.9	Montana	3.0	Texas	11.9
Georgia	11.4	Nebraska	3.9	Utah	3.1
Hawaii	3.8	Nevada	10.4	Vermont	3.6
Idaho	3.5	New Hampshire	2.0	Virginia	8.3
Illinois	11.4	New Jersey	5.3	Washington	5.2
Indiana	7.5	New Mexico	8.0	West Virginia	6.9
Iowa	2.3	New York	13.3	Wisconsin	4.4
Kansas	6.4	North Carolina	11.3	Wyoming	3.4
Kentucky	6.6	North Dakota	1.7		

Frequency Distributions

Rather than simply listing all the separate observations, as Table 3.1 does, we can summarize the data. A common summary method divides the measurement scale into a set of intervalsand totals the number of observations in each interval. A *frequency distribution*, defined next, does this.

Frequency Distribution

A ***frequency distribution*** is a listing of intervals of possible values for a variable, together with a tabulation of the number of observations in each interval.

To construct a frequency distribution for murder rate, for example, we divide the possible murder rate values into separate intervals. We then count the number (frequency) of states in each interval.

We must first select a set of intervals for murder rate. Or computer statistical software such as SAS or SPSS chooses them for us. SAS, for instance, uses the intervals {0–2.9, 3.0–5.9, 6.0–8.9, 9.0–11.9, 12.0–14.9, 15.0–17.9, 18.0–20.9} for the number of murders per 100,000 population. Counting the number of states with murder rates in each interval, we get the frequency distribution shown in Table 3.2. It is clear from looking at this frequency distribution that considerable variability exists in statewide murder rates, with one state being considerably higher than the rest. As with any summary method, some information is lost as the cost of achieving some clarity. The fre-

TABLE 3.2 Frequency Distribution of Murder Rates for the 50 States

Murder Rate (No. Murders per 100,000)	Frequency (No. States)
0.0–2.9	5
3.0–5.9	16
6.0–8.9	12
9.0–11.9	12
12.0–14.9	4
15.0–17.9	0
18.0–20.9	1
Total	50

quency distribution does not identify which states have low or high murder rates, nor are the exact murder rates known.

The intervals of values for the categories in frequency distributions are usually of equal width; the width equals 3 in Table 3.2. The intervals should include all possible values of the variable. In addition, any possible value must fit into one and only one interval; that is, they should be ***mutually exclusive***.

The number of intervals in a frequency distribution depends both on the judgment of the researcher and on the number of observations to be classified. Usually, the larger the number of observations, the greater the number of intervals used. If too many intervals are used (say, more than 15), they are so narrow that the information presented is difficult to digest, and an overall pattern in the results may be obscured. If very few intervals are used, however, too much information may be lost through pooling together observations that are not very similar. Follow this general guideline: The interval should not be so wide that two measurements included in it have a difference between them that is considered major. To summarize annual income, for example, if a difference of $5000 in income is not considered especially important, but a difference of $10,000 is somewhat notable, we might choose intervals of width less than $10,000, such as 0–$7999, $8000–$15,999, $16,000–$23,999, and so forth.

Relative Frequencies

Frequency distributions are informative, but it is easier to make comparisons between different intervals using ***relative frequencies***.

Relative Frequency

The ***relative frequency*** for an interval is the proportion of the sample observations that fall in that interval.

TABLE 3.3 Relative Frequency Distribution and Percentages for Murder Rates

Murder Rate	Frequency	Relative Frequency	Percentage
0.0–2.9	5	.10	10.0
3.0–5.9	16	.32	32.0
6.0–8.9	12	.24	24.0
9.0–11.9	12	.24	24.0
12.0–14.9	4	.08	8.0
15.0–17.9	0	.00	0.0
18.0–20.9	1	.02	2.0
Total	50	1.00	100.0

The relative frequency equals the number of observations in an interval divided by the total number of observations. For instance, for the murder rates, the relative frequency for the first interval in Table 3.2 is $5/50 = .10$; that is, 5 states out of 50, for a relative frequency of .10, had murder rates between 0 and 2.9. The relative frequency is a proportion—a number between 0 and 1 that expresses the share of the observations falling in that interval. A listing of these, by interval, provides a ***relative frequency distribution***. We construct the relative frequency distribution for the data on murder rates in Table 3.2 by dividing each frequency by 50, the total number of states. Table 3.3 shows it.

More often, relative frequencies are recorded as percentages rather than proportions. A percentage is simply a relative frequency multiplied by 100; that is, the decimal place is moved two positions to the right. For example, $5/50 = .10$ is the relative frequency for the interval 0–2.9, and $100(.10) = 10$ is the percentage. Table 3.3 also shows the relative frequency distribution as a percentage distribution.

The total sum of the proportions equals 1.00, and the sum of the percentages equals 100. The process of rounding may lead to slightly different totals, such as 100.1 or 99.9. When presenting relative frequencies in a table, always include the total number of cases upon which they are based. Obviously, the statement that 60% of a sample of 1000 individuals favor a decrease in the national defense budget is much more striking than the same statement derived from a sample of 5 individuals.

Histograms and Bar Graphs

A graph of a frequency distribution for a quantitative variable is called a ***histogram***. A bar is drawn over each interval of numbers, with height of the bar representing the relative number of observations in that interval. Figure 3.1 is a histogram for the murder rates, using the intervals in Table 3.2.

Although guidelines exist for drawing histograms (see Tufte, 1983), it is primarily a matter of common sense. As with frequency distributions, if too few intervals are used, too much information is lost or obscured. For example, Figure 3.2 is a histogram

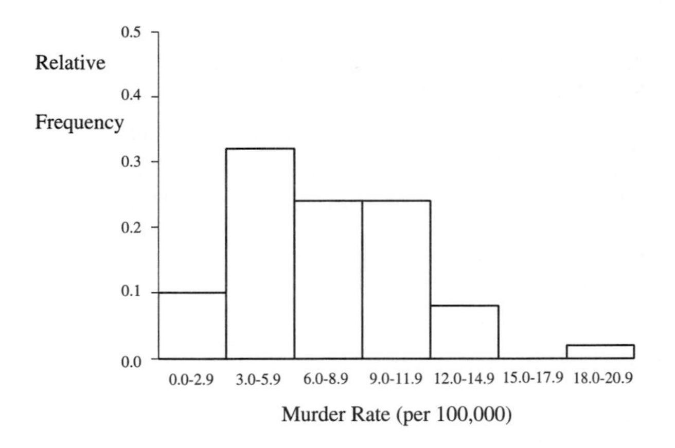

Figure 3.1 Relative Frequency Histogram for Murder Rates

of murder rates using the intervals 0.0–6.9, 7.0–13.9, 14.0–20.9. This is too crude to be very informative. On the other hand, the histogram is very irregular if too many intervals are used relative to the size of the data set. Most statistical software makes it simple to request histograms of data, and the software automatically chooses intervals that are sensible.

Relative frequencies are useful for data of any type. For categorical (nominal or ordinal) variables, instead of intervals of numbers we use the categorical scale for the variable. In that case, the graph of the relative frequencies for those categories is called a ***bar graph***.

Example 3.2 Bar Graph of Family Household Structure

Table 3.4 lists percentages of different types of family households in the United States in 1994. It is sufficient in such a table to report just the percentages and the total sample size, since each frequency equals the corresponding proportion multiplied by the

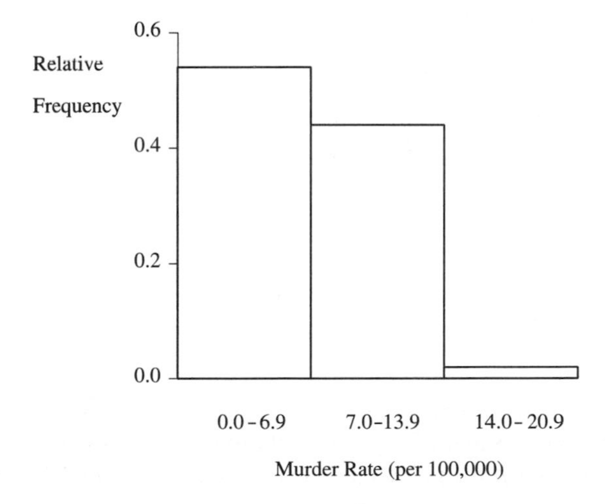

Figure 3.2 Relative Frequency Histogram for Murder Rates, Using Crude Intervals

TABLE 3.4 Family Structure, U.S. Families, 1994

Type of Family	Number (millions)	Percentage
Married couple with children	25.1	36.6
Married couple, no children	28.1	41.0
Single mother with children	7.6	11.1
Single father with children	1.3	1.9
Other families	6.4	9.3
Total	68.5	99.9

Source: U.S. Bureau of the Census, *Current Population Reports.*

total sample size. For instance, the frequency of single-mother families with children equals $.111(68.5) = 7.6$ million. Figure 3.3 presents the same data in a bar graph. Since family structure is a nominal variable, the order of the bars is not determined. By convention, they are usually ordered by frequency, except possibly for an "other" category, which is listed last. The order of presentation for an ordinal classification is the natural ordering of the levels of the variable. The bars in a bar graph, unlike in a histogram, are separated to emphasize that the variable is categorical rather than interval (quantitative). □

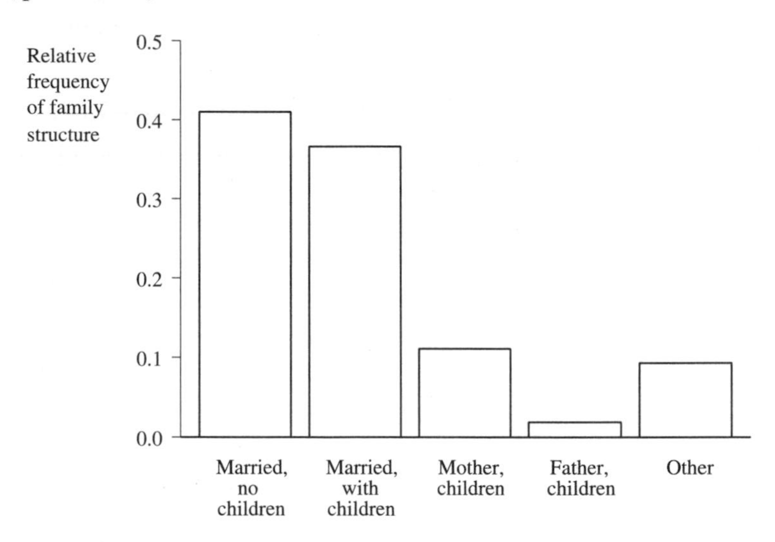

Figure 3.3 Relative Frequency of Family Structure Types, U.S. Families, 1994

Stem and Leaf Plots

Figure 3.4 shows an alternative graphical representation of the murder rate data. This figure, called a ***stem and leaf plot***, represents each observation by its leading digit(s) (the *stem*) and by its final digit (the *leaf*). In Figure 3.4, each stem is a number to the left of the vertical bar and a leaf is a number to the right of it. For the murder rates, the stem is the whole part of a number, and the leaf is the fractional part. For instance, on the first line, the stem of 1 and the leaves of 6 and 7 represent the murder rates 1.6 and 1.7. On the second line, the stem of 2 has leaves of 0, 3, 9, representing the murder rates 2.0, 2.3, and 2.9.

Stem and leaf plots arrange the leaves in order on each line, from smallest to largest. Two-digit stems refer to double-digit numbers; for instance, the last line has a stem of 20 and a leaf of 3, representing the murder rate 20.3.

A stem and leaf plot conveys much of the same information as a histogram. Turned on its side, it has the same shape as the histogram. In fact, since one can recover the sample measurements from the stem and leaf plot, it displays information that is lost with a histogram. For instance, from Figure 3.4, the largest murder rate for a state was 20.3 and the smallest was 1.6. It is not possible to determine these exact values from the histograms in Figures 3.1 and 3.2.

Stem	Leaf									
1	6	7								
2	0	3	9							
3	0	1	4	4	4	6	8	9	9	9
4	4	6								
5	0	2	3	8						
6	0	3	4	6	8	9				
7	5									
8	0	3	4	6	9					
9	0	8								
10	2	2	3	4						
11	3	3	4	4	6	9				
12	7									
13	1	3	5							
14										
15										
16										
17										
18										
19										
20	3									

Figure 3.4 Stem and Leaf Plot for Murder Rate Data in Table 3.1

Stem and leaf plots are useful for quick portrayals of small data sets. As the sample size increases, you can accommodate the increase in leaves by splitting the stems. For instance, you might list each stem twice, putting leaves of 0 to 4 on one line and leaves of 5 to 9 on another. When a number has several digits, it is simplest for graphical portrayal to drop the last digit or two. For instance, for a stem and leaf plot of annual income in thousands of dollars, a value of $27.1 thousand has a stem of 2 and a leaf of 7 and a value of $106.4 thousand has a stem of 10 and leaf of 6.

Comparing Groups

Many studies compare different groups with respect to their distribution on some variable. Relative frequency distributions, histograms, and stem and leaf plots are useful for describing differences between the groups.

Example 3.3 Comparing Canadian and U.S. Murder Rates

Table 3.5 shows recent annual murder rates for the provinces of Canada. The rates are all less than 3.0, so they would all fall in the first category of Table 3.2 or the first bar of the histogram in Figure 3.1.

TABLE 3.5 Canadian Provinces and Their Murder Rates (Number of Murders per 100,000 Population)

Alberta	2.7	British Columbia	2.6
Manitoba	2.9	New Brunswick	1.1
Newfoundland	1.2	Nova Scotia	1.3
Ontario	2.0	Prince Edward Island	0.7
Quebec	2.3	Saskatchewan	2.2

Source: *Canada Year Book*, 1992.

Stem and leaf plots can provide simple visual comparisons of two relatively small samples on a quantitative variable. For ease of comparison, the results are plotted "back to back"; each plot uses the same stem, with leaves for one sample to its left and leaves for the other sample to its right. To illustrate, Figure 3.5 shows back-to-back stem and leaf plots of the murder rate data for the United States and Canada. From this figure, it is clear that the murder rates tend to be much lower in Canada. □

Sample and Population Distributions

Frequency distributions and histograms for a variable apply both to a population and to samples from that population. The first type is called the ***population distribution*** of the variable, and the second type is called a ***sample distribution***. In a sense, the sample distribution is a blurry photograph of the population distribution. As the sample size increases, the sample proportion in any interval gets closer to the true population proportion. Thus, the photograph gets clearer, and the sample distribution looks more like the population distribution.

When a variable is continuous, one can choose the intervals for a histogram as narrow as desired. Now, as the sample size increases indefinitely and the number of intervals simultaneously increases, with their width narrowing, the shape of the sample histogram gradually approaches a smooth curve. This text uses such curves to represent population distributions. Figure 3.6 shows two sample histograms, one based on

Canada	Stem	United States
7	0	
3 2 1	1	6 7
9 7 6 3 2 0	2	0 3 9
	3	0 1 4 4 4 6 8 9 9 9
	4	4 6
	5	0 2 3 8
	6	0 3 4 6 8 9
	7	5
	8	0 3 4 6 9
	9	0 8
	10	2 2 3 4
	11	3 3 4 4 6 9
	12	7
	13	1 3 5
	14	
	15	
	16	
	17	
	18	
	19	
	20	3

Figure 3.5 Back-to-Back Stem and Leaf Plots for Murder Rate Data from U.S. and Canada

a sample of size 100 and the second based on a sample of size 500, and also a smooth curve representing the population distribution. Even if a variable is discrete, a smooth curve often approximates well the population distribution, especially when the number of possible values of the variable is large.

One way to summarize a sample or population distribution is to describe its shape. A group for which the distribution is bell-shaped is fundamentally different from a group for which the distribution is U-shaped, for example. See Figure 3.7. In the U-shaped distribution, the highest points (representing the largest frequencies) are at

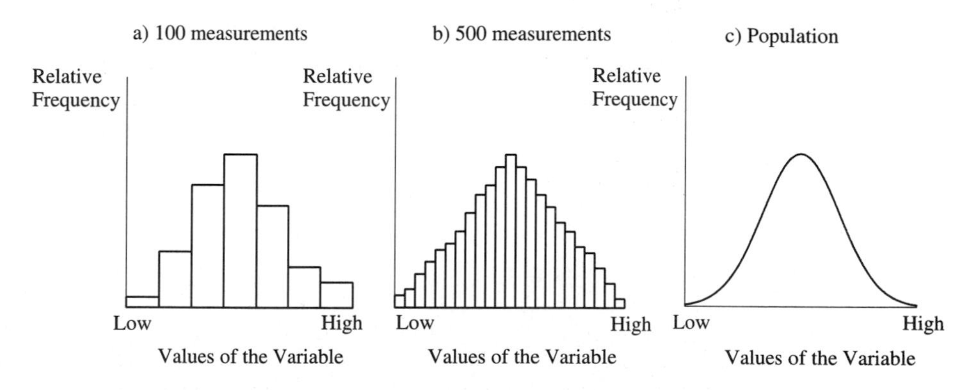

Figure 3.6 Histograms for a Continuous Variable

the lowest and highest scores, whereas in the bell-shaped distribution, the highest point is near the middle value of the variable. A U-shaped distribution indicates a polarization on the variable between two segments of the group, whereas a bell-shaped distribution indicates that most subjects tend to fall close to a central value.

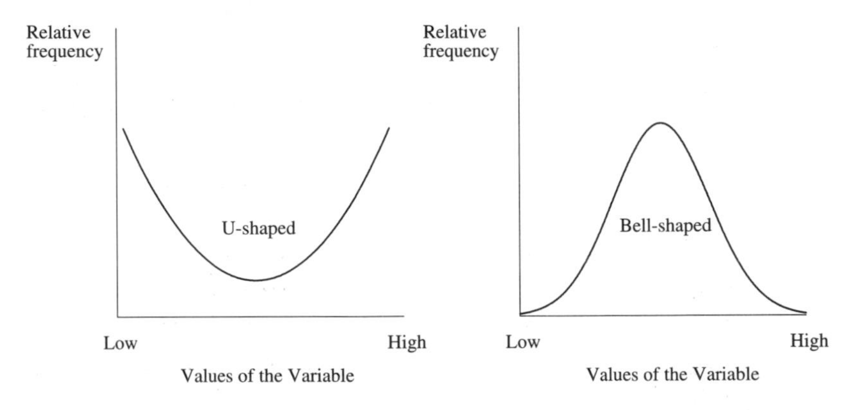

Figure 3.7 U-Shaped and Bell-Shaped Frequency Distributions

The bell-shaped and U-shaped distributions in Figure 3.7 are ***symmetric***. Most distributions of variables studied in the social sciences are not exactly symmetric. Figure 3.8 illustrates. The parts of the curve for the lowest values and the highest values are called the ***tails*** of the distribution. A nonsymmetric distribution is said to be ***skewed to the right*** or ***skewed to the left***, according to which tail is longer.

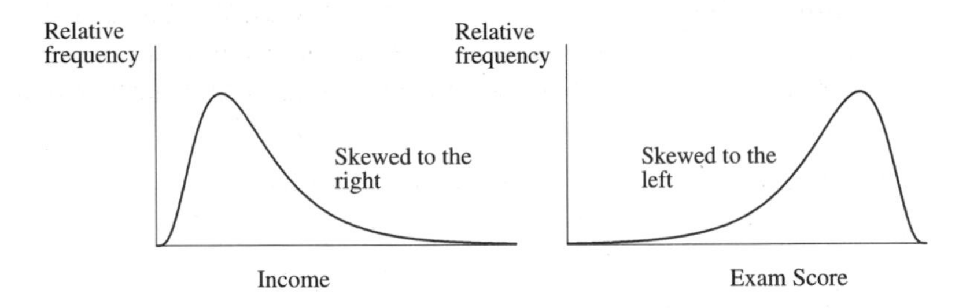

Figure 3.8 Skewed Frequency Distributions

A histogram for a sample approximates the corresponding population histogram. It is simpler to describe the difference between the two histograms, or the difference between sample distributions for two groups, using numerical descriptive methods. With these methods, one can make comparisons such as "On the average, the murder rate for U.S. states is 5.4 higher than the murder rate for Canadian provinces." We now turn our attention to ways of numerically describing data.

3.2 Measuring Central Tendency—The Mean

The next two sections present statistics that describe the center of a frequency distribution. The statistics show what a *typical* measurement in the sample is like. They are called ***measures of central tendency***.

The Mean

The best known and most frequently used measure of central tendency is the ***mean***, a description of the *average* response.

Mean

The ***mean*** is the sum of the measurements divided by the number of subjects.

The mean is often called the ***average***. We illustrate the mean and its calculation with the following example.

Example 3.4 Female Economic Activity in Europe

Table 3.6 shows an index of female economic activity for the countries of Western and Eastern Europe in 1994 (data were not available for Germany). The number reported refers to female employment, as a percentage of male employment. In Austria, for instance, the number of females in the work force was 60% of the number of males in the work force.

The table lists six observations for Eastern Europe. For these data, the sum of the measurements equals 88 + 84 + 70 + 77 + 77 + 81 = 477. The mean economic activity for these countries equals 477/6 = 79.5. By comparison, you can check that the mean for the Western European countries equals 722/13 = 55.5, considerably lower. (The values in the United States and Canada were 65 and 63.) □

We now introduce notation for the mean. We use this notation in a formula for the mean and in formulas for other statistics that use the mean.

Notation for Observations and Sample Mean

The sample size is symbolized by n. For a variable denoted by Y, its observations are denoted by $Y_1, Y_2, \ldots, Y_n$. The sample mean is denoted by $\bar{Y}$.

Throughout the text, n denotes the sample size. The n sample observations on a variable Y are denoted by Y_1 for the first observation, Y_2 the second, and so forth up to Y_n, the last observation made. For example, for female economic activity in Eastern Europe, $n = 6$, and the observations are $Y_1 = 88$, $Y_2 = 84, \ldots, Y_n = Y_6 = 81$.

TABLE 3.6 Female Economic Activity in Europe; Female Employment as a Percentage of Male Employment

Western Europe		Eastern Europe	
Country	Activity	Country	Activity
Austria	60	Bulgaria	88
Belgium	47	Czech Republic	84
Denmark	77	Hungary	70
France	64	Poland	77
Ireland	41	Romania	77
Italy	44	Slovakia	81
Netherlands	42		
Norway	68		
Portugal	51		
Spain	31		
Sweden	77		
Switzerland	60		
United Kingdom	60		

Source: Human Development Report 1995, United Nations Development Programme.

The symbol $\bar{Y}$ for the sample mean is read as "Y-bar." Other symbols are also sometimes used for variables, such as X or Z. A bar over the symbol represents the sample mean of data for that variable. For instance, $\bar{X}$ represents the sample mean for a variable denoted by X.

The definition of the sample mean implies that it equals

$$\bar{Y} = \frac{Y_1 + Y_2 + \cdots + Y_n}{n}$$

The symbol Σ (uppercase Greek letter sigma) represents the process of summing. For instance, ΣY_i represents the sum $Y_1 + Y_2 + \cdots + Y_n$. This symbol stands for the sum of the Y-values, where the index i represents a typical value in the range 1 to n. To illustrate, for the Eastern European data,

$$\sum Y_i = Y_1 + Y_2 + Y_3 + Y_4 + Y_5 + Y_6 = 477$$

The symbol is sometimes even further abbreviated as ΣY. Using this summation symbol, we have the shortened expression for the sample mean of n measurements,

$$\bar{Y} = \frac{\Sigma Y_i}{n}$$

Properties of the Mean

Before presenting additional examples, we consider some basic properties of the mean.

- The formula for the mean assumes numerical values $Y_1, Y_2, \ldots, Y_n$ for the observations. Because of this, the mean is appropriate only for quantitative data. It is not sensible to compute the mean for observations on a nominal scale. For instance, for religion measured with categories such as (Protestant, Catholic, Jewish, Other), the mean religion does not make sense, even though these levels may sometimes be coded by numbers for convenience. Similarly, we cannot find the mean of observations on an ordinal rating such as excellent, good, fair, and poor, unless we assign numbers such as 4, 3, 2, 1 to the ordered levels, treating it as quantitative.

- The mean can be highly influenced by an observation that falls far from the rest of the data, called an ***outlier***.

Example 3.5 Effect of Outlier on Mean Income

The owner of a small store reports that the mean annual income of employees in the business is $37,900. Upon closer inspection, we find that the annual incomes of the seven employees are $10,200, $10,400, $10,700, $11,200, $11,300, $11,500, and $200,000. The $200,000 income is the salary of the owner's son, who happens to be an employee. The value $200,000 is an outlier. The mean computed for the other six observations alone equals $10,883, quite different from the mean of $37,900 including the outlier. □

This example shows that the mean is not always representative of the measurements in the sample. This is fairly common with small samples when one or more measurements is much larger or much smaller than the others, such as in highly skewed distributions.

- The mean is pulled in the direction of the longer tail of a skewed distribution, relative to most of the data.
  In Example 3.5, the large observation $200,000 results in an extreme skewness to the right of the income distribution. This skewness pulls the mean above six of the seven measurements. In general, the more highly skewed the frequency distribution, the less representative the mean is of a typical observation.
- The mean is the point of balance on the number line when an equal weight occurs at each measurement point. For example, Figure 3.9 shows that if an equal weight is placed at each observation from Example 3.4, then the line balances by placing a fulcrum at the point 79.5. The mean is the *center of gravity* of the observations. This property implies that the sum of the distances to the mean from the observations above the mean equals the sum of the distances to the mean from the observations below the mean.
- Denote the sample means for two sets of data with sample sizes n_1 and n_2 by $\bar{Y}_1$ and $\bar{Y}_2$. The overall sample mean for the combined set of (n_1+n_2) measurements is the ***weighted average***

$$\bar{Y} = \frac{n_1\bar{Y}_1 + n_2\bar{Y}_2}{n_1 + n_2}$$

The numerator $n_1\bar{Y}_1 + n_2\bar{Y}_2$ is the total sum of all the measurements, since $n\bar{Y} = \sum Y$ for each set of measurements. The denominator is the total sample size.

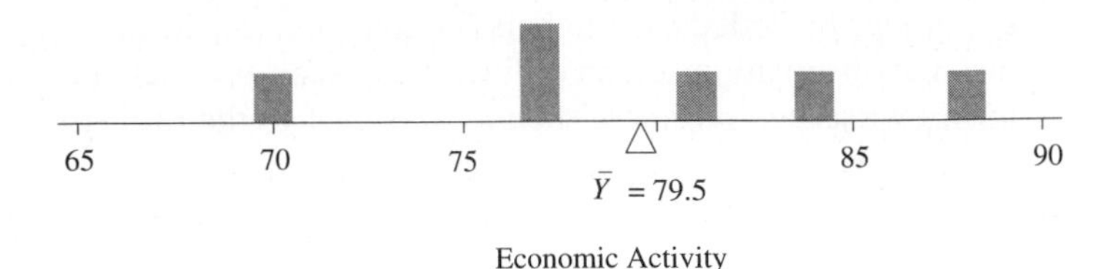

Figure 3.9 The Mean as the Center of Gravity

To illustrate, for the female economic activity data in Table 3.6, the Western European measurements have $n_1 = 13$ and $\bar{Y}_1 = 55.5$, and the Eastern European measurements have $n_2 = 6$ and $\bar{Y}_2 = 79.5$. The overall mean economic activity for the 19 nations equals

$$\bar{Y} = \frac{n_1\bar{Y}_1 + n_2\bar{Y}_2}{n_1 + n_2} = \frac{13(55.5) + 6(79.5)}{13 + 6} = \frac{(722 + 477)}{19} = \frac{1199}{19} = 63.1$$

The weighted average of 63.1 is closer to 55.5, the value for Western Europe, than to 79.5, the value for Eastern Europe, because most of the 19 observations for the overall sample come from Western Europe.

3.3 The Median and Other Measures of Central Tendency

Although the mean is a simple measure of central tendency, other measures are also informative and occasionally more appropriate than the mean.

The Median

The *median* splits the sample into two parts with equal numbers of subjects, when the subjects' observations are ordered from lowest to highest. It is a measure of central tendency that better describes a typical value when the sample distribution of measurements is highly skewed.

Median

The ***median*** is the measurement that falls in the middle of the ordered sample. When the sample size n is odd, a single measurement occurs in the middle. When the sample size is even, two middle measurements occur, and the median is the midpoint between the two.

To illustrate, the ordered income measurements for the seven employees in Example 3.5 are \$10,200, \$10,400, \$10,700, \$11,200, \$11,300, \$11,500, and \$200,000. The median is the middle measurement, \$11,200. This is a much more typical value for this sample than the sample mean of \$37,900. In this case, the median better describes central tendency than does the mean. In Table 3.6, the ordered economic activity values for the Eastern European nations are 70, 77, 77, 81, 84, and 88. Since $n = 6$ is even, the median is the midpoint between the two middle values, 77 and 81, which is $(77+81)/2 = 79.0$. This is close to the sample mean of 79.5, since this small data set has no outliers.

Since a stem and leaf plot arranges the observations in order, it is easy to determine the median using such a plot. For the data in Table 3.1 on murder rates, Figure 3.4 shows the stem and leaf plot. Since the sample size $n = 50$ is even, the median is the midpoint between the middle measurements, the 25th and 26th smallest. Counting down 25 leaves from the top of the plot, we find that 25th and 26th smallest values are 6.6 and 6.8. So, the median is $(6.6 + 6.8)/2 = 6.7$. The mean is $\bar{Y} = 7.3$, somewhat larger than the median. This is partly due to the outlier observation of 20.3 for Louisiana, which is considerably higher than the other observations. Turning Figure 3.4 on its side, we see that the murder rate values are skewed to the right.

The middle observation is the one having index $(n+1)/2$. That is, the median is the value of the $(n+1)/2$nd measurement in the ordered sample. For instance, when $n = 7$, $(n+1)/2 = (7+1)/2 = 4$, so the median is the fourth smallest, or equivalently fourth largest, observation. When n is even, $(n + 1)/2$ falls halfway between two numbers, and the median is the midpoint of the measurements with those indices. For instance, when $n = 50$, $(n + 1)/2 = 25.5$, so the median is the midpoint between the 25th and 26th smallest observations.

Example 3.6 Median for Grouped or Ordinal Data

Table 3.7 summarizes data on the highest degree completed for a sample of subjects taken recently by the U.S. Bureau of the Census. The measurement scale grouped the possible responses into an ordered set of categories. The sample size is $n = 177{,}618$. The median score is the $(n + 1)/2 = (177,618 + 1)/2 = 88{,}809.5$th lowest. Now, 38,012 responses fall in the first category, $(38{,}012 + 65{,}291) = 103{,}303$ in the first two,

TABLE 3.7 Highest Degree Completed, for a Sample of Americans

Highest Degree	Frequency	Percentage
Not a high school graduate	38,012	21.4%
High school only	65,291	36.8%
Some college, no degree	33,191	18.7%
Associate's degree	7,570	4.3%
Bachelor's degree	22,845	12.9%
Master's degree	7,599	4.3%
Doctorate or professional	3,110	1.7%

and so forth. The 38,013rd to 103,303rd lowest scores fall in category 2, which therefore contains the 88,809.5th lowest, which is the median. The median response is "High school only." Equivalently, from the percentages in the last column of the table, 21.4% fall in the first category and (21.4% + 36.8%) = 58.2% fall in the first two, so the 50% point falls in the second category. □

Properties of the Median

- The median, like the mean, is appropriate for interval data. Since it requires only ordered observations to compute it, it is also valid for ordinal data, as illustrated in the previous example. It is not appropriate for nominal data, since the observations cannot be ordered.
- For symmetric distributions, such as in Figure 3.7, the median and the mean are identical. To illustrate, the sample of measurements 4, 5, 7, 9, 10 is symmetric about 7; 5 and 9 fall equally distant from it in opposite directions, as do 4 and 10. Thus, 7 is both the median and the mean.
- For skewed distributions, the mean lies toward the direction of skew (the longer tail) relative to the median, as Figure 3.10 shows. Income distributions tend to be skewed to the right, though usually not as severely as in Example 3.5. The mean household income in the United States in 1993, for example, was about $8000 higher than the median household income of $31,000 (U.S. Bureau of the Census, *Current Population Reports*).

  Length of prison sentences tend to be highly skewed to the right. For instance, in 1994, for 67 sentences for murder imposed using U.S. Sentencing Commission guidelines, the mean length was 251 months and the median was 160 months. The distribution of grades on an exam tends to be skewed to the left when some students perform considerably poorer than the others. In this case, the mean is less than the median. For instance, suppose that an exam scored on a scale of 0 to 100 has a median of 88 and a mean of 76. Then most students performed quite

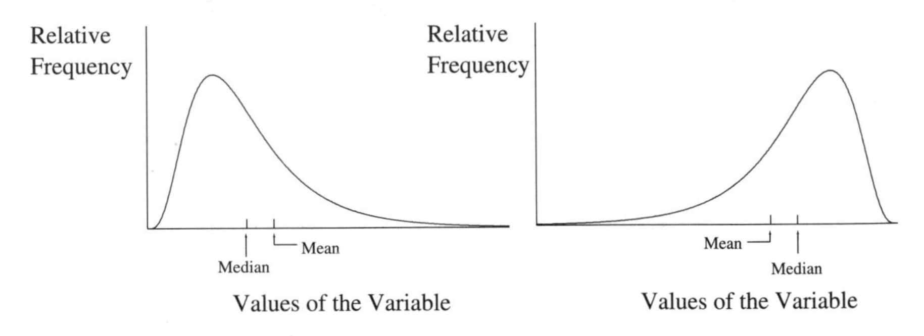

Figure 3.10 The Mean and the Median for Skewed Distributions

well (half being over 88), but apparently some scores were very much lower than the majority of students in order to bring the mean down to 76.

- The median is insensitive to the distances of the measurements from the middle, since it uses only the ordinal characteristics of the data. For example, the following four sets of measurements all have medians of 10:

Set 1:	8,	9,	10,	11,	12
Set 2:	8,	9,	10,	11,	100
Set 3:	0,	9,	10,	10,	10
Set 4:	8,	9,	10,	100,	100

- The median is unaffected by outliers. For instance, the incomes of the seven employees in Example 3.5 have a median of $11,200 whether the largest observation is $20,000, $200,000, or $2,000,000.

Example 3.7 Effect of Extreme Outlier for Murder Rate Data

Table 3.1 contains murder rates for the 50 states and has a mean of 7.3 and a median of 6.7. The data set does not include the District of Columbia (D.C.), which had a murder rate in 1993 of 78.5, nearly four times that of Louisiana. This is certainly an extreme outlier. If we include this observation in the data set, then $n = 51$. The median, the 26th largest observation, has 25 smaller and 25 larger observations. This is 6.8, so the median is barely affected by including this outlier. On the other hand, the mean changes from 7.3 to 8.7, being considerably affected by the outlier. The effect of an outlier tends to be even greater when the sample size is small, as Example 3.5 showed. □

Median Compared to Mean

The median has certain advantages, compared to the mean. For instance, the median is usually more appropriate when the distribution is highly skewed, as we have seen in Examples 3.5 and 3.7. The mean can be greatly affected by outliers, whereas the median is not.

The mean requires quantitative data, whereas the median also applies for ordinal scales (see Example 3.6). By contrast, using the mean for ordinal data requires assigning scores to the categories. In Table 3.7, if we assign scores 10, 12, 13, 14, 16, 18, 20 to the categories of highest degree, representing approximate number of years of education, we get a sample mean of 12.8.

The median also has disadvantages, compared to the mean. For discrete data that take on relatively few values, quite different patterns of data can give the same result. For instance, consider Table 3.8, from the General Social Survey of 1991. This survey, conducted annually by the National Opinion Research Center (NORC) at the University of Chicago, asks a sample of adult American subjects about a wide variety of issues. Table 3.8 summarizes the 1514 responses in 1991 to the question, "Within the past 12

TABLE 3.8 Number of People You Know Who Have Committed Suicide

Response	Frequency	Percentage
0	1344	88.8
1	133	8.8
2	25	1.7
3	11	.7
4	1	.1

months, how many people have you known personally that have committed suicide?" Only five distinct responses occur, and 88.8% of those are 0. Since $(n + 1)/2 = 757.5$, the median is the midpoint between the 757th and 758th smallest measurements. But those are both 0 responses, so the median response is 0.

To calculate the sample mean for Table 3.8, it is unnecessary to add the 1514 separate measurements to obtain ΣY_i for the numerator of $\bar{Y}$, since most values occurred several times. To sum the 1514 observations, we multiply each possible value by the frequency of its occurrence, and then add; that is,

$$\sum Y_i = 1344(0) + 133(1) + 25(2) + 11(3) + 1(4) = 220$$

The sample size is $n = 1344+133+25+11+1 = 1514$, so the sample mean response is

$$\bar{Y} = \frac{\Sigma Y_i}{n} = \frac{220}{1514} = .15$$

If the distribution of the 1514 observations among these categories were (758, 133, 25, 11, 587) (i.e., we shift 586 responses from 0 to 4), then the median would still be 0, but the mean would shift to 1.69. The mean uses the numerical values of all the observations, not just their ordering.

A more extreme form of this problem occurs for *binary data*. Such data can take only two values, such as (0, 1) or (low, high). The median equals the most common outcome, but gives no information about the relative number of observations at the two levels.

Quartiles and Other Percentiles

The median is a special case of a more general set of measures of location called *percentiles*.

Percentile

The ***pth percentile*** is a number such that $p\%$ of the scores fall below it and $(100 - p)\%$ fall above it.

Substituting $p = 50$ in this definition gives the 50th percentile. This is simply the median. That is, the median is larger than 50% of the measurements and smaller than the other $(100 - 50) = 50\%$. Two other commonly used percentiles are the *lower quartile* and *upper quartile*.

Lower and Upper Quartiles

The 25th percentile is called the ***lower quartile***. The 75th percentile is called the ***upper quartile***.

These refer to $p = 25$ and $p = 75$ in the percentile definition. One quarter of the data fall below the lower quartile, and one quarter fall above the upper quartile. The lower quartile is the median for the observations that fall below the median, that is, for the bottom half of the data. The upper quartile is the median for the observations that fall above the median, that is, for the upper half of the data. The quartiles together with the median split the distribution into four parts, each containing one-fourth of the measurements, as Figure 3.11 shows.

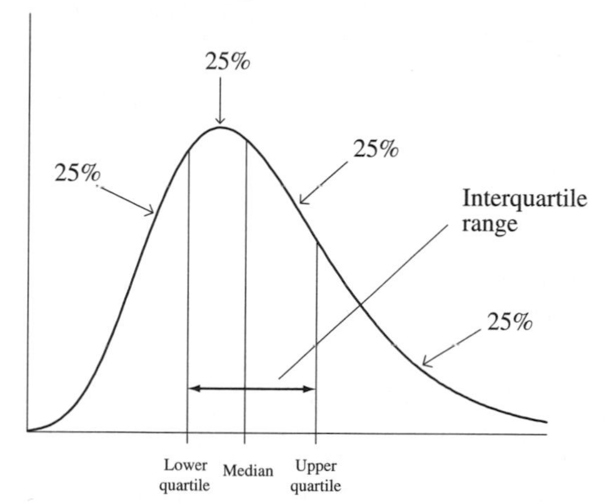

Figure 3.11 The Quartiles and Interquartile Range

We illustrate with the murder rates from Table 3.2. The sample size is $n = 50$, and the median equals 6.7. As with the median, the quartiles can easily be found from a stem and leaf plot, such as Figure 3.4. The lower quartile is the median for the 25 observations below the median, which is the 13th smallest observation, or 3.9. The upper quartile is the median for the 25 observations above the median, which is the 13th largest observation, or 10.3. This means that a quarter of the states had murder rates above 10.3. Similarly, a quarter of the states had murder rates below 3.9, between 3.9 and the median of 6.7, and between 6.7 and 10.3. The distance between the upper quartile and the median is $10.3 - 6.7 = 3.6$, which exceeds the distance $6.7 - 3.9 = 2.8$

between the lower quartile and the median. This commonly happens when the distribution is skewed to the right.

We can summarize this information by reporting a *five-number summary*, consisting of the three quartiles and the minimum and maximum values. For instance, a popular software package reports these as follows:

```
100% Max     20.3
 75% Q3      10.3
 50% Med      6.7
 25% Q1       3.9
  0% Min      1.6
```

The five-number summary provides a simple-to-understand description of a data set.

The difference between the upper and lower quartiles is called the ***interquartile range***. The middle half of the observations fall within that range. This measure describes variability of the data and is described further in Section 3.4. For the U.S. murder rates, the interquartile range equals $10.3 - 3.9 = 6.4$. The middle half of the murder rates fall within a range of 6.4.

Percentiles other than the quartiles and the median are usually reported only for fairly large data sets, and we omit rules for their calculation in this text.

The Mode

Another measure, the *mode*, describes a typical sample measurement in terms of the most common outcome.

Mode

The ***mode*** is the value that occurs most frequently.

In Table 3.8 on the suicide data, the mode is 0. The mode is more commonly used with categorical data or grouped frequency distributions than with ungrouped observations. The mode is then the category or interval with the highest frequency. In the data of Table 3.7 on the highest degree completed, for instance, the mode is "High school only," since the frequency for that category is higher than the frequency for any other rating.

The mode need not be near the center of the distribution. In fact, it may be the largest or the smallest value, if that is most common. Thus, it is somewhat inaccurate to call the mode a measure of central tendency. Many quantitative variables studied in the social sciences, though, have distributions in which the mode is near the center, such as in bell-shaped distributions and in slightly skewed distributions such as those in Figures 3.10 and 3.11.

Properties of the Mode

- The mode is appropriate for all types of data. For example, we might measure the modal religion (nominal level) in the United Kingdom, the modal rating (ordinal level) given a teacher, or the modal number of years of education (interval level) completed by Hispanic Americans.
- A frequency distribution is called ***bimodal*** if two distinct mounds occur in the distribution. Bimodal distributions often occur with attitudinal variables, when responses tend to be strongly in one direction or another, leading to polarization of the population. For instance, Figure 3.12 shows the relative frequency distribution of responses in the 1991 General Social Survey to the question, "Do you personally think it is wrong or not wrong for a woman to have an abortion if the family has a very low income and cannot afford any more children?" The relative frequencies in the two extreme categories are higher than those in the middle categories.
- The mean, median, and mode are identical for a unimodal, symmetric distribution, such as a bell-shaped distribution.

The mode is not as popular as the mean or median for describing central tendency of quantitative variables. It is useful when the most frequently occurring level of a variable is relevant, which is often true for categorical variables. The mean, median, quartiles, and mode are complementary measures. They describe different aspects of the data. In any particular example, some or all of their values may be useful.

Finally, these statistics are sometimes misused, as in Example 3.5. People who present statistical conclusions often choose the statistic giving the impression they wish to convey. Other statistics that might provide somewhat different interpretations are ignored. You should be on the lookout for misleading statistical analyses. For instance, be wary of the mean when you think that the distribution may be highly skewed.

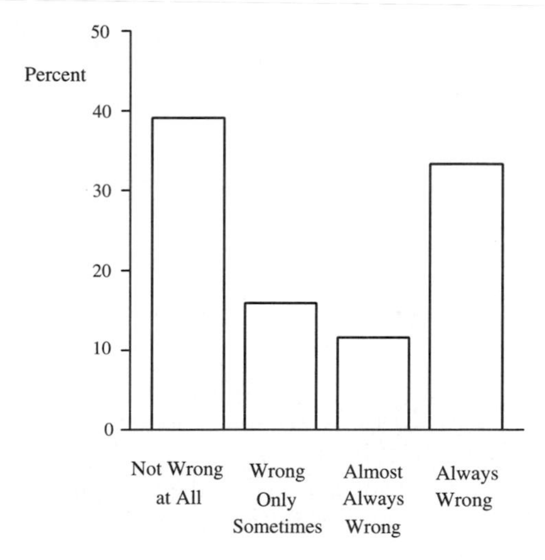

Figure 3.12 Bimodal Distribution for Opinion about Abortion

3.4 Measures of Variation

A measure of central location alone is not adequate for numerically describing a frequency distribution. It describes a typical value, but not the spread of the data about that value. The two distributions in Figure 3.13 illustrate. The citizens of nation A and the citizens of nation B have the same mean annual income ($25,000). The distributions of those incomes differ fundamentally, however, nation B being much more homogeneous. An income of $30,000 is extremely large for a resident of nation B, though not especially large for a resident of nation A. This section introduces statistics that describe the variability of a data set. These statistics are called ***measures of variation***.

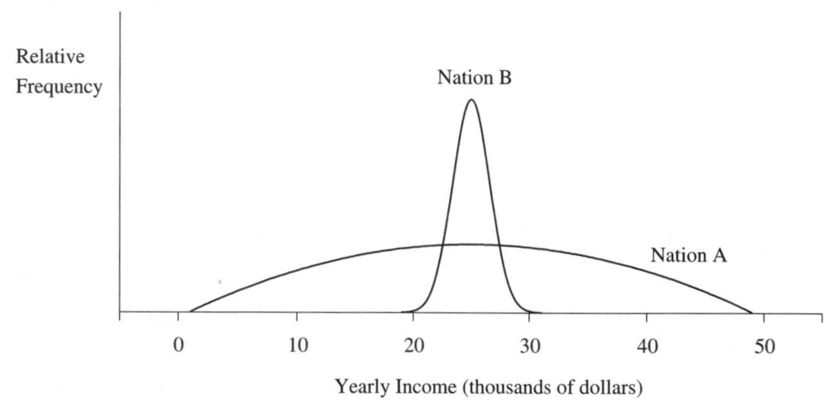

Figure 3.13 Distributions with the Same Mean but Different Variability

The Range

The difference between the largest and smallest observations in a sample is a simple measure of variation.

> **Range**
>
> The ***range*** is the difference between the largest and smallest observations.

For nation A, Figure 3.13 indicates that the range of income values is about $50,000 − 0 = $50,000; for nation B, the range is about $30,000−$20,000 = $10,000. Nation A has greater variation of incomes than nation B.

The range is not, however, sensitive to other characteristics of data variability. The three distributions shown in Figure 3.14 all have the same mean ($25,000) and range ($50,000), yet they differ in variation about the center of the distribution. In terms of distances of measurements from the mean, nation A is the most disperse, and nation B is the least. The incomes in nation A tend to be farthest from the mean, and the incomes in nation B tend to be closest.

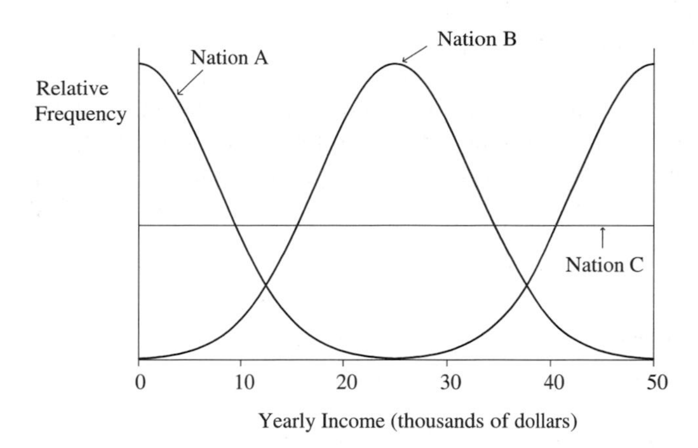

Figure 3.14 Distributions with the Same Mean and Range, but Different Variations About the Mean

Variance and Standard Deviation

Other measures of variation are based on the deviations of the data from a measure of central tendency, usually their mean.

> **Deviation**
>
> The ***deviation*** of the ith observation Y_i from the sample mean $\bar{Y}$ is $(Y_i - \bar{Y})$, the difference between them.

Each observation has a deviation. The deviation is positive when the observation falls above the sample mean and negative when it falls below it. The interpretation of $\bar{Y}$ as the center of gravity of the data implies that the sum of the positive deviations equals the negative of the sum of negative deviations; that is, the sum of all the deviations about the mean, $\Sigma(Y_i - \bar{Y})$, equals 0 for any sample. Because of this, summary measures of variation use either the absolute values or the squares of the deviations. The two measures we present incorporate the squares. The first measure is the ***variance***.

> **Variance**
>
> The ***variance*** of n observations $Y_1, \ldots, Y_n$ is
>
> $$s^2 = \frac{\Sigma\,(Y_i - \bar{Y})^2}{n-1} = \frac{(Y_1 - \bar{Y})^2 + (Y_2 - \bar{Y})^2 + \cdots + (Y_n - \bar{Y})^2}{n-1}$$

The variance is approximately an average of the squared deviations. That is, it approximates the average of the squared distances from the mean. The units of measurement are the squares of those for the original data, since it uses squared deviations. This makes the variance difficult to interpret. The square root of the variance, called the ***standard deviation***, is better for this purpose.

Standard Deviation

The ***standard deviation s*** is the positive square root of the variance:

$$s = \sqrt{\frac{\Sigma \left(Y_i - \bar{Y}\right)^2}{n-1}}$$

The expression $\Sigma \left(Y_i - \bar{Y}\right)^2$ in the formulas for the variance and standard deviation is called a ***sum of squares***. It represents squaring the deviations and then adding those squares. It is incorrect to first add the deviations and then square that sum; this gives a value of 0. The larger the deviations about the mean, the larger the sum of squares and the larger s and s^2 tend to be.

Example 3.8 Comparing Variability of Quiz Scores

Each of the following sets of quiz scores for two small samples of students has a mean of 5 and a range of 10:

Sample 1: 0, 4, 4, 5, 7, 10
Sample 2: 0, 0, 1, 9, 10, 10

By inspection, the scores in sample 1 show less variability about the mean than those in sample 2. Most scores in sample 1 are close to the mean of 5, whereas all the scores in sample 2 are quite far from 5.

For sample 1,

$$\Sigma \left(Y_i - \bar{Y}\right)^2 = (0-5)^2 + (4-5)^2 + (4-5)^2 + (5-5)^2 + (7-5)^2 + (10-5)^2 = 56$$

so that the variance equals

$$s^2 = \frac{\Sigma \left(Y_i - \bar{Y}\right)^2}{n-1} = \frac{56}{6-1} = \frac{56}{5} = 11.2$$

Likewise, you can verify that for sample 2, $s^2 = 26.4$. The average squared distance from the mean is about 11 in sample 1 and 26 in sample 2. The standard deviation for sample 1 equals $s = \sqrt{11.2} = 3.3$, whereas for sample 2 it equals $s = \sqrt{26.4} = 5.1$.

Since 5.1 > 3.3, the performances in sample 2 were more variable than those in sample 1, as expected. □

Similarly, if s_A, s_B, and s_C denote the standard deviations of the three distributions in Figure 3.14, then $s_B < s_C < s_A$; that is, s_B is less than s_C, which is less than s_A.

Statistical software and many hand calculators can calculate the standard deviation for you. You should do the calculation yourself for a few small data sets to help you understand what this measure represents. The answer you get may differ slightly from the value reported by computer software, depending on how much you round off the mean before inserting it into the sum of squares part of the calculation.

Properties of the Standard Deviation

- $s \geq 0$.
- $s = 0$ only when all observations have the same value. For instance, if the ages in a sample of five students are 19, 19, 19, 19, 19, then the sample mean equals 19, each of the five deviations equals 0, and $s^2 = s = 0$. This is the minimum possible variation for a sample.
- The greater the variation about the mean, the larger is the value of s. Example 3.8 illustrated this property. For another example, we refer back to the U.S. and Canadian murder rates shown in Figure 3.5. The plot suggests that murder rates are much more variable in the U.S. In fact, the standard deviations are $s = 4.0$ for the United States and $s = .8$ for Canada.
- The reason for using $(n - 1)$, rather than n, in the denominator of s and s^2 is a technical one (discussed later in the text) concerning the use of these statistics to estimate population parameters. In the (rare) instances when we have data for the entire population, we replace $(n - 1)$ in these definitions by the actual population size. In this case, the standard deviation can be no larger than half the size of the range.
- Problem 3.64 at the end of this chapter presents two properties of standard deviations that refer to the effect of rescaling the data. Basically, if the data are rescaled, the standard deviation is also rescaled. For instance, if we double the scores, thus doubling the variation, then s doubles. If we change data on annual incomes from dollars (such as 34,000) to thousands of dollars (such as 34.0), the standard deviation also changes by a factor of 1000 (such as from 11,800 to 11.8).

Interpreting the Magnitude of *s*

Thus far, we have not discussed the magnitude of the standard deviation s other than in a comparative sense. A distribution with $s = 5.1$ has greater variation than one with $s = 3.3$, but how do we interpret *how large* $s = 5.1$ is? A very rough answer to this question is that s is a type of *average distance* of an observation from the mean. To illustrate, suppose the first exam in this course is graded on a scale of 0 to 100, and the sample mean for the students is 77. A value of $s = 0$ in very unlikely, since every student must then score 77; a value of $s = 50$ seems implausibly large for a typical distance from the mean; values of s such as 8 or 11 or 14 seem much more realistic.

More precise ways to interpret s require further knowledge of the mathematical form of a frequency distribution. The following rule provides an approximate interpretation for many data sets.

Empirical Rule

If the histogram of the data is approximately bell-shaped, then

1. About 68% of the data fall between $\bar{Y} - s$ and $\bar{Y} + s$.
2. About 95% of the data fall between $\bar{Y} - 2s$ and $\bar{Y} + 2s$.
3. All or nearly all the data fall between $\bar{Y} - 3s$ and $\bar{Y} + 3s$.

The rule is called the Empirical Rule because many distributions encountered in practice (that is, *empirically*) are approximately bell-shaped. Figure 3.15 is a graphical portrayal of the rule.

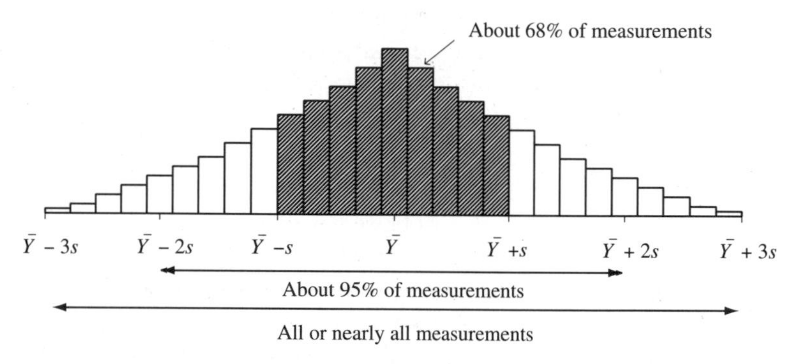

Figure 3.15 Empirical Rule: Interpretation of the Standard Deviation for a Bell-Shaped Distribution

Example 3.9 Describing the Distribution of SAT Scores

The distribution of scores on the verbal or math portion of the Scholastic Aptitude Test (SAT) is now scaled so it is approximately bell-shaped with mean of 500 and standard deviation of 100, as portrayed in Figure 3.16. By the Empirical Rule, about 68% of the scores fall between 400 and 600 on each test, since 400 and 600 are the numbers that are one standard deviation below and above the mean of 500. Similarly, about 95% of the scores fall between 300 and 700, the numbers that are two standard deviations from the mean. The remaining 5% fall either below 300 or above 700. The distribution is roughly symmetric about 500, so about 2.5% of the scores fall above 700 and about 2.5% fall below 300. □

The percentages stated in the Empirical Rule are approximate and refer only to distributions that are approximately bell-shaped. In the bell-shaped case, for instance,

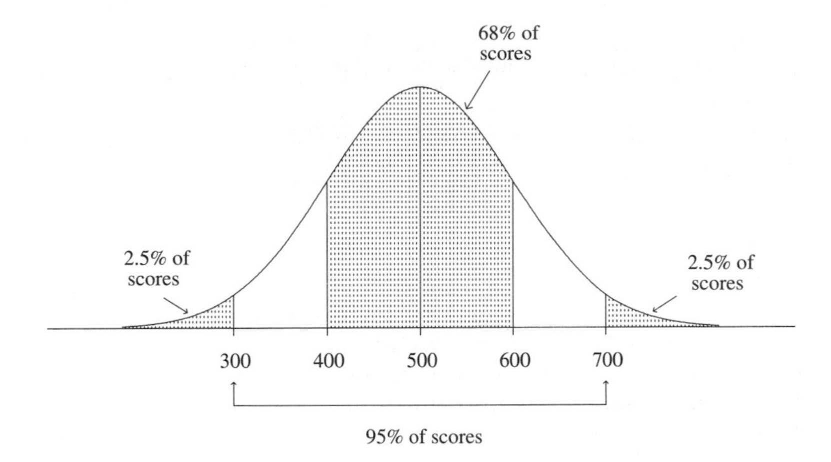

Figure 3.16 A Bell-Shaped Distribution of Test Scores with Mean 500 and Standard Deviation 100

the percentage of the distribution falling within two standard deviations of the mean is 95%, but this could change to as low as 75% or as high as 100% for other distributions. The Empirical Rule may not work well if the distribution is highly skewed or if it is highly discrete, with the variable taking relatively few values. The exact percentages depend on the form of the distribution, as Example 3.10 demonstrates.

Example 3.10 Familiarity with AIDS Victims

The 1993 General Social Survey asked "How many people have you known personally, either living or dead, who came down with AIDS?" Table 3.9 shows part of a computer printout for summarizing the 1598 responses on this variable. It indicates that 76% of the responses are 0, so that the lower quartile (Q1), median, and upper quartile (Q3) all equal 0.

The mean and standard deviation are $\bar{Y} = 0.47$ and $s = 1.09$. The values 0 and 1 both fall within one standard deviation of the mean. Now, 88.8% of the distribution falls at these two points, or within $\bar{Y} \pm s$. This is considerably larger than the 68% that the Empirical Rule predicts for bell-shaped distributions. The Empirical Rule does not apply to this frequency distribution, since it is not even approximately bell-shaped. Instead, it is highly skewed to the right, as you can check by sketching a histogram for Table 3.9. The smallest value in the distribution (0) is less than one standard deviation below the mean; the largest value in the distribution (8) is nearly seven standard deviations above the mean. □

Whenever the smallest or largest observation is less than a standard deviation from the mean, this is evidence of severe skew. For instance, a recent exam one of us gave having scale from 0 to 100 had $\bar{Y} = 86$ and $s = 15$. Since the upper bound of 100 was less than one standard deviation above the mean, we surmised that the distribution of scores was highly skewed to the left.

TABLE 3.9 Frequency Distribution of the Number of People Known Personally With AIDS

AIDS	Frequency	Percent
0	1214	76.0
1	204	12.8
2	85	5.3
3	49	3.1
4	19	1.2
5	13	0.8
6	5	0.3
7	8	0.5
8	1	0.1

```
Analysis Variable : AIDS
 N              1598         Quartiles               Range    8
 Mean          0.473         100% Max      8         Q3-Q1    0
 Std Dev       1.089          75% Q3       0         Mode     0
                              50% Med      0
                              25% Q1       0
                               0% Min      0
```

The standard deviation, like the mean, can be greatly affected by an outlier, particularly for small data sets. For instance, the murder rate data in Table 3.1 for the 50 states have $\bar{Y} = 7.33$ and $s = 3.98$. The distribution is somewhat irregular, but you can check that 68% of the states have murder rates within one standard deviation of the mean and 98% within two standard deviations. Now, suppose we include the murder rate for the District of Columbia in the data set, which equals 78.5. Then $\bar{Y} = 8.73$ and $s = 10.72$. The standard deviation more than doubles, and now 96.1% of the murder rates (all except D.C. and Louisiana) fall within one standard deviation of the mean.

Interquartile Range

The ***interquartile range***, denoted by IQR, is another range–type statistic for describing variation. It is defined as the difference between the upper and lower quartiles. An advantage of the IQR over the ordinary range or the standard deviation is that it is not sensitive to extreme outlying observations.

To illustrate, we use the U.S. murder rate data shown in the stem and leaf plot in Figure 3.4. The rates range from 1.6 to 20.3, with a lower quartile of 3.9, a median of 6.7, and an upper quartile of 10.3. For these data, IQR $= 10.3 - 3.9 = 6.4$. When we add the observation of 78.5 for D.C. to the data set, the IQR changes only from 6.4 to 6.5. By contrast, the range changes from 18.7 to 76.9 and the standard deviation changes from 4.0 to 10.7.

Like the range and standard deviation, the IQR increases as the variability increases, and it is useful for comparing variation of different groups. To illustrate, we compare variability in U.S. and Canadian murder rates using the data shown in the back-to-back stem and leaf plots of Figure 3.5. The Canadian data has IQR = 2.6 − 1.2 = 1.4, showing much less variability than the IQR value of 6.4 for the U.S. data.

For bell-shaped distributions, the distance from the mean to either quartile is roughly 2/3rd of a standard deviation, and IQR is very roughly about $(4/3)s$. The insensitivity of the IQR to outliers has recently increased its popularity, though in practice the standard deviation is still much more common.

Box Plots

We conclude this section by presenting a graphical summary of both the central tendency and variation of a data set. This graphic, called a ***box plot***, portrays the range and the quartiles of the data, and possibly some outliers.

The *box* contains the central 50% of the distribution, from the lower quartile to the upper quartile. The median is marked by a line drawn within the box. The lines extending from the box are called *whiskers*. These extend to the maximum and minimum values, unless there are outliers.

Figure 3.17 shows the box plot for the U.S. murder rates, in the format of box plots provided with SAS software (with the PLOT option in PROC UNIVARIATE). The upper whisker and upper half of the central box are longer than the lower ones. This indicates that the right tail of the distribution, which corresponds to the relatively large values, is longer than the left tail. The plot reflects the skewness to the right of the distribution of U.S. murder rates.

Box plots are particularly useful for comparing two distributions side by side. Figure 3.17 also shows the box plot for the Canadian murder rate data. These side-by-side

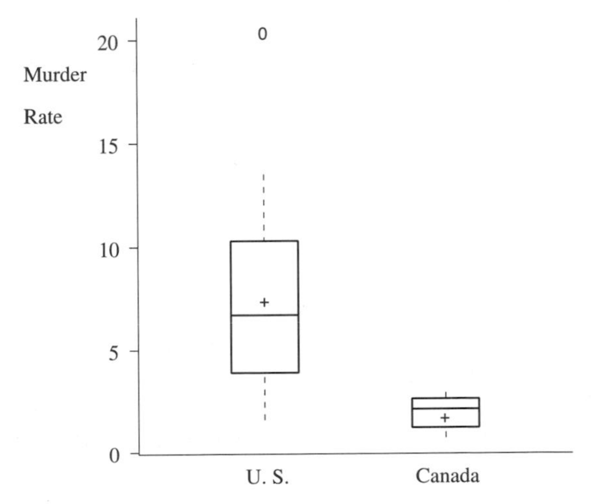

Figure 3.17 Box Plots for U.S. and Canadian Murder Rates

plots reveal that the murder rates in the U.S. tend to be much higher and have much greater variability.

Box plots identify outliers separately. To explain this, we now present a formal definition of an outlier.

Outlier

An observation is an ***outlier*** if it falls more than 1.5 IQR above the upper quartile or more than 1.5 IQR below the lower quartile.

In box plots, the whiskers extend to the smallest and largest observations only if those values are not outliers; that is, if they are no more than 1.5 IQR beyond the quartiles. Otherwise, the whiskers extend to the most extreme observations within 1.5 IQR, and the outliers are specially marked. For instance, SAS marks by an O (O for outlier) a value between 1.5 and 3.0 IQR from the box and by an asterisk (*) a value even farther away. Figure 3.17 shows one outlier with a very high murder rate, which is the murder rate of 20.3 for Louisiana. The distance of this observation from the upper quartile is $20.3 - 10.3 = 10.0$, which is greater than $1.5\ \text{IQR} = 1.5(6.4) = 9.6$.

The outliers are shown separately because they do not provide much information about the shape of the distribution, particularly for large data sets. SAS also plots the mean on the box plot, representing it by a + sign; these equal 7.3 for the United States and 1.9 for Canada. Comparing the mean to the median, which is the line within the box, helps show any skewness.

3.5 Sample Statistics and Population Parameters

Of the measures introduced in this chapter, the mean $\bar{Y}$ and the standard deviation s are the most commonly reported. We shall refer to them frequently in the rest of the text. The formulas that define $\bar{Y}$ and s refer to sample measurements. Since their values depend on the sample selected, they vary in value from sample to sample. In this sense, they are variables, sometimes called ***random variables*** to emphasize that their values vary according to the (random) sample selected. Their values are unknown before the sample is chosen. Once the sample is selected and they are computed, they become known sample statistics.

We shall regularly distinguish between sample statistics and the corresponding measures for the population. Section 1.2 introduced the term *parameter* for a summary measure of the population. A statistic describes a sample, while a parameter describes the population from which the sample was taken. In this text, lowercase Greek letters usually denote population parameters and Roman letters denote the sample statistics.

Notation for Parameters

Let μ (Greek mu) and σ (Greek sigma) denote the mean and standard deviation of a variable for the population.

We call μ and σ the ***population mean*** and ***population standard deviation***. The population mean is the average of the population measurements. The population standard deviation describes the variation of the population measurements about the population mean.

Whereas the statistics $\bar{Y}$ and s are variables, with values depending on the sample chosen, the parameters μ and σ are constants. This is because μ and σ refer to just one particular group of measurements, namely, the measurements for the entire population. Of course, the parameter values are usually unknown, which is the reason for sampling and calculating sample statistics as estimates of their values. Much of the rest of this text deals with ways of making inferences about unknown parameters (such as μ) using sample statistics (such as $\bar{Y}$). Before studying these inferential methods, though, we must introduce some basic ideas of *probability*, which serves as the foundation for the methods. Probability is the subject of Chapter 4.

3.6 Chapter Summary

This chapter introduced ***descriptive statistics***—ways of *describing* a sample. Data sets in social science research are often large, and it is imperative to summarize the important characteristics of the information.

Overview of Tabular and Graphical Methods

- A ***frequency distribution*** of the sample measurements summarizes the counts of responses for a set of intervals of possible values. A ***relative frequency*** distribution reports this information in the form of percentages or proportions.
- A ***histogram*** provides a picture of this distribution. It is a bar graph of the relative frequencies. The histogram shows whether the distribution is approximately bell-shaped, U-shaped, skewed to the right (longer tail pointing to the right), or whatever.
- The ***stem and leaf plot*** is an alternative way of portraying the data, grouping together all observations having the same leading digits (stem), and showing also their final digit (leaf). Turned on its side, it shows the shape of the distribution, like a histogram, but it also presents the individual scores.
- The ***box plot*** portrays the quartiles, the extreme values, and any outliers. This plot and the stem and leaf plot are useful for back-to-back comparisons of two groups.

Stem and leaf plots and box plots, simple as they are, are relatively recent innovations in statistics (Tukey, 1977). See Cleveland (1985, 1993) and Tufte (1983, 1990) for even more recent and innovative ways to present data graphically.

Overview of Measures of Central Tendency

- ***Measures of central tendency*** describe the center of the collection of measurements, in terms of the "typical" score.
- The ***mean*** is the sum of the measurements divided by the sample size. It is the center of gravity of the data.
- The ***median*** divides the ordered data set into two parts of equal numbers of subjects, half scoring below and half above that point. It is less affected than the mean by outliers or extreme skew.
- The lower quarter of the observations fall below the ***lower quartile***, and the upper quarter fall above the ***upper quartile***. These are the 25th and 75th percentiles, and the median is the 50th percentile. The quartiles and median split the data into four equal parts.
- The ***mode*** is the most commonly occurring value. It is valid for any type of data, though usually used with categorical data.

Overview of Measures of Variation

- ***Measures of variation*** describe the variability of the measurements.
- The ***range*** is the difference between the largest and smallest measurements. The ***interquartile range*** is the difference between the upper and lower quartiles; it is less affected by extreme outliers.

TABLE 3.10 Measures of Central Tendency and Variation

	Measure	Definition	Interpretation
Central Tendency	Mean	$\bar{Y} = \Sigma Y_i/n$	Center of gravity
	Median	Middle measurement of ordered sample	50th percentile
	Mode	Most frequently occurring value	Most likely outcome, valid for all types of data
Variability	Variance	$s^2 = \Sigma(Y_i - \bar{Y})^2/(n-1)$	Greater with more variability, average squared distance from mean
	Standard deviation	$s = \sqrt{\Sigma(Y_i - \bar{Y})^2/(n-1)}$	Empirical Rule: If bell-shaped, 68%, 95% within s, $2s$ of $\bar{Y}$
	Range	Difference between largest and smallest measurement	Greater with more variability
	Interquartile range	Difference between upper and lower quartiles	Encompasses middle half of data

- The ***variance*** averages the squared deviations about the mean. Its square root, the ***standard deviation***, is easier to interpret. The Empirical Rule states that for a sample with a bell-shaped distribution, about 68% of the measurements fall within one standard deviation of the mean and about 95% of the measurements fall within two standard deviations. Nearly all, if not all, the measurements fall within three standard deviations of the mean.

Table 3.10 summarizes the measures of central tendency and variation. A ***statistic*** summarizes a sample. A ***parameter*** summarizes a population. It is usually more relevant than the particular value of the statistic, which depends on the sample chosen. ***Statistical inference*** uses statistics to make predictions about parameters.

PROBLEMS

Practicing the Basics

1. According to the Bureau of the Census (*Current Population Reports*), in 1994 in the United States there were 23.6 million households with one person, 31.2 million with two persons, 16.9 million with three persons, 15.1 million with four persons, 6.7 million with five persons, 2.2 million with six persons, and 1.4 million with seven or more persons.
   a) Construct a relative frequency distribution.
   b) Construct a histogram. What is its shape?
   c) Using a score of 8 for the final category, find the mean number of persons per household.
   d) Report and interpret the median and mode of household size.
2. According to News America Syndicate, in 1986 the number of followers of the world's major religions were 835 million for Christianity, 420 million for Islam, 322 million for Hinduism, 300 million for Confucianism, 210 million for Buddhism, 79 million for Shinto, 50 million for Taoism, and 12 million for Judaism.
   a) Construct a relative frequency distribution for these data.
   b) Construct a bar graph for these data.
   c) Can you calculate a mean, median, or mode for these data? If so, do so and interpret.
3. Refer to Table 3.1. Use software to construct a histogram for these data, using its default method of forming intervals. Describe the shape of the distribution, and construct the corresponding relative frequency distribution.
4. Table 3.11 shows the number (in millions) of the foreign-born population of the United States in 1990, by place of birth.
   a) Construct a relative frequency distribution.
   b) Plot the data in a bar graph.
   c) Is "Place of birth" quantitative or qualitative? How, if at all, can you describe these data using numerical measures?
5. A researcher in an alcoholism treatment center, interested in summarizing the length of stay in the center for first-time patients, randomly selects ten records of individuals institutionalized within the previous two years. The lengths of stay in the center, in days, are as follows: 11, 6, 20, 9, 13, 4, 39, 13, 44, and 7.
   a) Construct a stem and leaf plot.

TABLE 3.11

Place of Birth	Number
Europe	4.0
Former Soviet Union	.3
Asia	5.0
Canada	.7
Mexico	4.3
Caribbean	1.9
Central America	1.1
South America	1.0
Africa	.4
Oceania	.1
Total	18.8

Source: Statistical Abstract of the United States, 1994

b) Find the mean, and interpret.
c) Find the median, and interpret.
d) Find the standard deviation, and interpret.
e) For a similar study 25 years ago at the same institution, lengths of stay for ten sampled individuals were 32, 18, 55, 17, 24, 31, 20, 40, 24, and 15 days. Compare results to those in the new study using (i) a back-to-back stem and leaf plot, (ii) the mean, (iii) the median, (iv) the standard deviation. Interpret any differences you find.
f) Actually, the new study also selected one other record. That patient is still institutionalized after 40 days. Thus, that patient's length of stay is at least 40 days, but the actual value is unknown. Can you calculate the mean or median for the complete sample of size 11 including this partial observation? Explain. (An observation such as this is said to be *censored*, meaning that the measured value is "cut short" of its true, unknown value.)

6. The 1994 General Social Survey asked respondents "How often do you read the newspaper?" The possible responses were (every day, a few times a week, once a week, less than once a week, never), and the counts in those categories were (969, 452, 261, 196, 76).
a) Identify the median response.
b) Identify the mode.
c) Consider the variable, Y = number of times reading the newspaper in a week, measured as described above. Can you calculate $\bar{Y}$? Why? What would you need to do to approximate its value?

7. Table 3.12 summarizes responses of 1250 subjects in the 1991 General Social Survey to the question, "About how often did you have sex during the last 12 months?"
a) Construct a bar graph, and interpret.
b) Report the median and the mode. Interpret.
c) Treat this scale in a quantitative manner by assigning the scores 0, .1, 1.0, 2.5, 4.3, 10.8, and 17 to the categories, representing approximate monthly frequency. Calculate the sample mean, and interpret.

8. The 1991 General Social Survey asked respondents, "How many sex partners have you had in the last 12 months?" Table 3.13 shows results for 637 respondents.
a) Calculate and interpret the median and the mode.

TABLE 3.12

How Often Had Sex	Frequency
Not at all	292
Once or twice	99
About once a month	108
2 or 3 times a month	181
About once a week	233
2 or 3 times a week	265
More than 3 times a week	72

b) For the highest 11 values, we know only an interval within which the observation fell. To approximate these values, we could use midpoint scores. For instance, for interval 5–10, use (5+10)/2 = 7.5. We must choose an arbitrary score over 100 for the final interval. Using 120 for that observation, calculate the mean response. Compare to the median, and interpret.
c) Suppose the highest two observations were misrecorded, and the actual values were 4 for each. Recompute the mean and median, and use this example to describe potential effects of outliers on these measures.

TABLE 3.13

Number of Sex Partners	Frequency
0	146
1	418
2	39
3	15
4	8
5–10	7
11–20	2
21–100	1
More than 100	1

9. For 1992, the statewide number of abortions per 1000 women 15 to 44 years of age, for states in the Pacific region of the United States, were: Washington, 33; Oregon, 16; California, 304; Alaska, 2; and Hawaii, 11 (*Statistical Abstract of the United States, 1994*).
a) Calculate the mean.
b) Calculate the median. Why is it so different from the mean?

10. For 1993, Table 9.1 in Chapter 9 shows data on the statewide violent crime rate per 100,000 population. In this exercise, do not use the observation for D.C.
a) Using the intervals 0-100, 100-200, 200-300, and so forth, tally the frequencies and construct a frequency distribution.
b) Find the relative frequencies.
c) Sketch a histogram. How would you describe the shape of the distribution?
d) Drop the final digit of each crime rate. Then, construct a stem and leaf plot on this set of modified values. How does this plot compare to the histogram in (c)?

11. Refer to the preceding problem. Table 3.14 shows part of a computer printout for analyzing the data; the first column refers to the entire data set, and the second column deletes the observation for D.C.
a) Report and interpret the mean and median of the first set of crime rates. Explain what their relative values suggest about the shape of the distribution.
b) For each statistic reported, evaluate the effect of including the outlying observation for D.C.

TABLE 3.14

```
Variable=VIOLENT
   N                  51            N                   50
   Mean           612.84            Mean            566.66
   Std Dev        441.10            Std Dev         295.88

        Quartiles                        Quartiles
    100% Max    2922                100% Max    1206
     75% Q3      780                 75% Q3      766
     50% Med     515                 50% Med   509.5
     25% Q1      326                 25% Q1      326
      0% Min      82                  0% Min      82

     Range      2840                 Range      1124
     Q3-Q1       454                 Q3-Q1       440
     Mode        208                 Mode        208
```

12. In 1992 in the United States, the median family income was $38,909 for white families, $21,161 for black families, and $23,901 for Hispanic families (U.S. Bureau of the Census, *Current Population Reports*, P-60-184). In constant 1992 dollars, the median family incomes in 1975 were $35,619 for white families, $21,916 for black families, and $23,844 for Hispanic families. Interpret the medians in 1992 and the changes in their values between 1975 and 1992.

13. Table 3.15 shows 1994 female economic activity for countries in South America.
a) Construct a back-to-back stem and leaf plot of these values contrasted with those from Eastern Europe in Table 3.6. What is your interpretation?
b) Compare the means for the two sets of nations, and interpret.
c) Compare the medians, and interpret.

14. According to the U.S. Bureau of the Census, *Current Population Reports*, in 1994 the median household income was $32,368 for whites and $18,660 for blacks, whereas the mean household income was $40,708 for whites and $25,409 for blacks. Does this sug-

TABLE 3.15

Country	Activity	Country	Activity	Country	Activity
Argentina	38	Ecuador	24	Colombia	28
Uruguay	44	Paraguay	26	Peru	32
Chile	39	Bolivia	31	Brazil	38
Venezuela	39	Guyana	34		

Source: *Human Development Report, 1995,* United Nations Development Programme.

gest that the distribution of income is symmetric, or skewed to the right, or skewed to the left? Explain.

15. Refer to the previous exercise. The results refer to 57.9 million white households and 8.0 million black households.
a) Find the overall mean income.
b) If the mean income equals $30,291 for 5.9 million Hispanic families, find the overall mean for the three groups combined.

16. For towns with population size 2500 to 4599 in the U.S. Northeast in 1994, the mean salary of chiefs of police was $37,527, and the median was $30,500 (*The Municipal Year Book 1995*. Washington, D.C.: International City/County Management Association, 1995). Does this suggest that the distribution of salary was skewed to the left, symmetric, or skewed to the right? Explain.

17. According to the National Association of Home Builders, the U.S. nationwide median selling price of homes sold in 1995 was $118,000.
a) Would you expect the mean to be larger, smaller, or equal to $118,000? Explain.
b) Which of the following is the most plausible value for the standard deviation: (i) −15,000, (ii) 1,000, (iii) 45,000, (iv) 1,000,000? Why?

18. The 1990 General Social Survey asked respondents, "During the past 12 months, how many people have you known personally that were victims of homicide." Table 3.16 shows a computer printout from analyzing responses for 1370 subjects.

TABLE 3.16

VICTIMS	Frequency	Percent
0	1244	90.8
1	81	5.9
2	27	2.0
3	11	0.8
4	4	0.3
5	2	0.1
6	1	0.1

N	Mean	Std Dev	100% Max	75% Q3	50% Med	25% Q1	0% Min
1370	0.146	0.546	6	0	0	0	0

a) Report the relative frequency distribution.
b) Sketch a histogram. Is the distribution bell-shaped, skewed to the right, or skewed to the left?
c) Calculate the mean, median, and mode, and interpret their values.
d) Report and interpret the standard deviation. Does the Empirical Rule apply to this distribution. Why or why not?

19. The Human Development Index (HDI) has three components: life expectancy at birth, educational attainment, and income. It ranges from 0 to 1, with higher values representing greater development. In 1992, the HDI ratings for eight Central American countries were .884 for Belize, .884 for Costa Rica, .579 for El Salvador, .591 for Guatemala, .578 for Honduras, .842 for Mexico, .611 for Nicaragua, and .856 for Panama.
a) Construct a stem and leaf plot. Drop the final digit, and split the values into two parts;

that is, have two lines for responses with first digit 8, putting entries with second digit 0 to 4 on one line and 5 to 9 on the second, have two lines for 7, two lines for 6, and two lines for 5. What is the shape of the distribution?
b) Calculate and interpret the mean, median, and range.

20. According to *Statistical Abstract of the United States, 1995*, average salary (in dollars) of secondary school classroom teachers in 1994 in the United States varied among states with a five-number summary of:

```
100% Max      51,700 (Connecticut)
 75% Q3       38,500
 50% Med      33,900
 25% Q1       29,800
  0% Min      25,300 (South Dakota)
```

a) Find and interpret the range and the interquartile range.
b) Construct a box plot.
c) Based on (b), predict the direction of skew for this distribution. Explain.
d) If the distribution, though skewed, is approximately bell-shaped, which of the following values would you expect for the standard deviation:
(i) 100, (ii) 1000, (iii) 6000, (iv) 15,000? Explain.

21. Consider the data in Table 3.8 on the number of people you know who have committed suicide. The mean equals .145, and the standard deviation equals .457. From the results reported in the table, what percentage of measurements fall within one standard deviation of the mean? Is the Empirical Rule appropriate for this distribution? Why or why not?

22. Why is the median sometimes preferred over the mean as a measure of central tendency? Give an example to illustrate your answer.

23. Why is the mean sometimes preferred over the median? Give an example to illustrate your answer.

24. Give an example of a variable for which the mode applies, but not the mean or median.

25. A group of high school students takes an exam. The mean score for the boys is 65, and the median is 75. Both the mean and the median score for the girls is 70. How can you explain the large difference between the two summary measures for the boys?

26. During the spring semester of 1995 at the University of Florida, computer usage of students having accounts on a mainframe computer at the university was summarized by a mean of 1921 and a standard deviation of 11,495 kilobytes of drive usage.
a) Does the Empirical Rule apply to this distribution? Why?
b) Would you expect this distribution to be symmetric, skewed to the right, or skewed to the left? Explain.
c) What could cause the standard deviation to be so large compared to the mean? (Data supplied by Dr. Michael Conlon, University of Florida.)

27. Refer to Problem 3.26. The five-number summary of these data was minimum = 4, $Q1$ = 256, median = 530, $Q3$ = 1105, and maximum = 320,000. What does this suggest about the shape of the distribution? Why?

28. Residential electrical consumption in March 1994 in Gainesville, Florida, had a mean of 780 and a standard deviation of 506 kilowatt-hours (Kwh). The minimum usage was 3 Kwh and the maximum was 9390 Kwh. (Data supplied by N. Todd Kamhoot, Gainesville Regional Utilities.)
a) What shape do you expect this distribution to have? Why?
b) Do you expect this distribution to have any outliers? Explain.

29. Residential water consumption in March 1994 in Gainesville, Florida, had a mean of 7.1 and a standard deviation of 6.2 (thousand gallons). What shape do you expect this distribution to have? Why? (Data supplied by N. Todd Kamhoot, Gainesville Regional Utilities.)

30. For each of the following, sketch roughly what you expect a histogram to look like, and explain whether the mean or the median would be greater. Also sketch box plots for cases (a) and (c) that are consistent with the histograms.
a) The selling price of new homes in 1997
b) The number of children ever born per woman age 40 or over
c) The score on an easy exam (mean = 88, standard deviation = 10, maximum possible = 100)
d) The number of cars owned per family
e) Number of months in which subject drove a car last year

31. For each of the following variables, indicate whether you would expect its relative frequency histogram to be bell-shaped, U-shaped, skewed to the right, or skewed to the left. For parts (a), (b), and (g), sketch a box plot that would be plausible for that variable.
a) Exam score (scores fall between 0 and 100, with a mean of 90 and a standard deviation of 10)
b) IQ
c) Number of times arrested in past year
d) Time needed to complete difficult exam (maximum time is 1 hour)
e) Assessed value of home
f) Age at death
g) Weekly church contribution (median is $10 and mean is $17)
h) Number of years lived in present home (mode = 0 to 1 year)
i) Attitude toward legalization of abortion

32. Give examples of social science variables having a distribution that you would expect to be
a) Approximately symmetric
b) Skewed to the right
c) Skewed to the left
d) Bimodal
e) Skewed to the right, with a mode and median of 0 but a positive mean

33. A recent Roper organization survey asked: "How far have environmental protection laws and regulations gone?" For the possible responses not far enough, about right, and too far, the percentages of responses were 51%, 33%, and 16%.
a) Which response is the mode?
b) Can you compute a mean or a median for these data? If so, do so; if not, explain why not.

34. A company conducts a study of the number of miles traveled using public transportation by its employees during a typical day. A random sample of ten employees yields the following values (in miles):

0, 0, 4, 0, 0, 0, 10, 0, 6, 0

a) Calculate and interpret the mean, median, mode, range, variance, and standard deviation of these measurements.
b) The next person sampled lives in a different city and travels 90 miles a day on public transport. Recompute the mean, median, and standard deviation, and note the effect of this outlying observation.

35. To measure variation:
a) Why is the standard deviation s usually preferred over the range?
b) The IQR is sometimes preferred to s when there are some extreme outliers. Why?

36. In the mid-1980s, the General Social Survey asked respondents how many close friends they had. For a sample of size 1467, the mean was 7.4 and the standard deviation was 11.0. The distribution had a median of 5 and a mode of 4. Based on these statistics, what would you surmise about the shape of the distribution? Why?

37. In 1994 the General Social Survey asked, "On the average day, about how many hours do you personally watch television?" Of 1964 responses, the mode was 2, the median was 2, the mean was 2.8, and the standard deviation was 2.4. Based on these statistics, what would you surmise about the shape of the distribution?

38. For an exam given to a class, the students' scores ranged from 35 to 98, with a mean of 74. Which of the following is the most realistic value for the standard deviation? 1,12, 60, −10? Why?

39. The sample mean for a data set equals 80. Which of the following is an impossible value for the standard deviation? 200, 0, −20.

40. According to a recent report from the U.S. National Center for Health Statistics, females with age between 25 and 34 years have a bell-shaped distribution on height, with mean of 65 inches and standard deviation of 3.5 inches.
a) Give an interval within which about 95% of the heights fall.
b) What is the height for a female who is three standard deviations below the mean in height. Would this be a rather unusual height? Why?

41. In a large northern city, monthly payments to people on welfare last year were observed to have approximately a bell shape with mean $700 and standard deviation $100. Give a range of values within which all or nearly all the payments fell.

42. For the WWW data on number of times a week reading a newspaper, referred to in Problem 1.7, Figure 3.18 shows a computer printout of the stem and leaf plot and the box plot.
a) From the box plot, identify the minimum, lower quartile, median, upper quartile, and maximum.
b) Identify these five numbers using the stem and leaf plot.
c) Do the data appear to contain any outliers? If so, identify.
d) Based on the box plot, indicate the approximate value of the mean. The standard deviation is one of the following values—.5, 3, 10, 20. Which do you think it is, and why?

43. Suppose the distribution of the prices of new homes built in the United States in 1996 was approximately bell-shaped, with a mean of $120,000 and a standard deviation of $40,000.
a) Describe the distribution using properties of the standard deviation.
b) If your new house was priced half a standard deviation above the mean in 1996, how much did it cost?
c) If the distribution is not actually bell-shaped, what shape would you expect it to have? Why?

44. In 1993, the five-number summary for the statewide percentage of people without health insurance had a minimum of 8.7% (Wisconsin), $Q1 = 11.9$, Med = 13.4, $Q3 = 17.8$, and maximum of 23.9% (Louisiana) (*Statistical Abstract of the United States, 1995*).
a) Construct a box plot for these data.
b) Do you think that the distribution is symmetric, skewed to the right, or skewed to the left? Why?

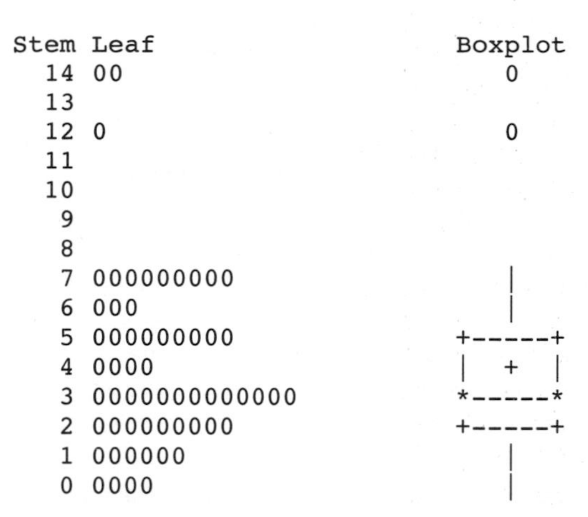

Figure 3.18

c) Which of the following is the most plausible value for the standard deviation of this distribution: 0, 4, 13, 22? Why?

45. Refer to Problem 3.20. Construct a box plot for these data. Are there any apparent outliers?

46. The distribution of high school graduation rates in the United States in 1993 had a minimum value of 64.3 (Mississippi), lower quartile of 73.9, median of 76.75, upper quartile of 80.1, and maximum value of 86.6 (Alaska) (*Statistical Abstract of the United States, 1995*).

a) Report and interpret the 50th percentile.
b) Report the range and the interquartile range.
c) Sketch a box plot. Are there any outliers?
d) Provide a guess for the standard deviation. Justify.

47. In your library, find the percentage of the vote that Bill Clinton received in each state in the 1996 presidential election.

a) Prepare a stem and leaf plot. Are there any apparent outliers?
b) Construct a box plot. Are there any outliers?
c) Construct back-to-back stem and leaf plots or side-by-side box plots for Northeastern and West Coast states versus other states. Interpret.

48. Refer to Problem 3.10.

a) Using the data set without D.C., find the quartiles and the interquartile range.
b) According to the definition of an outlier in terms of the IQR, are any of the observations outliers?
c) Construct a box plot for the distribution.
d) Repeat the analyses, including the D.C. observation, and compare results.

49. What is the difference between the descriptive measures symbolized by
a) $\bar{Y}$ and μ? **b)** s and σ?

Concepts and Applications

50. For the WWW data file (Problem 1.7), use computer software to conduct graphical and numerical summaries for **a)** distance from home town, **b)** weekly hours of TV watching, **c)** weekly number of times reading a newspaper, and **d)** number of HIV–AIDS victims known. Describe the shapes of the distributions, and summarize your findings.

51. Refer to the data file you created in Problem 1.7. For variables chosen by your instructor, conduct descriptive statistical analyses. Prepare a report, interpreting and summarizing your findings.

52. Refer to the data in Table 9.1 on poverty rates. Using methods of this chapter, summarize these data. Prepare a report, graphically displaying the data and summarizing the central tendency and variation. In your report, discuss whether there any outliers, and if there are, analyze their influence on the results.

53. The number of therapeutic abortions in 1988 in Canada, per 100 live births, is shown in Table 3.17. Using methods of this chapter, present a descriptive statistical analysis of these data, interpreting your results.

TABLE 3.17

Alberta	15.0	British Columbia	25.5
Manitoba	16.6	New Brunswick	4.9
Newfoundland	6.3	Nova Scotia	14.2
Ontario	20.9	Prince Edward Island	3.5
Quebec	14.7	Saskatchewan	7.7
Yukon	22.6	Northwest Territories	17.9

Source: Canada Year Book, 1991.

54. Refer to Problem 3.19. Table 3.18 shows the HDI ratings for African countries. Using graphical and numerical methods of this chapter, summarize HDI for these countries, and compare to the distribution of HDI for Central American countries.

55. Obtain data on statewide murder rates from the latest edition of *Statistical Abstract of the United States*.

a) Analyze the data using the graphical and numerical methods of this chapter.

b) Use graphical and numerical methods to compare the murder rate distribution to the one for the data in Table 3.1.

TABLE 3.18

Libya	.768	Tunisia	.763	Algeria	.732	S. Africa	.705
Egypt	.613	Namibia	.611	Gabon	.579	Morocco	.554
Zimbabwe	.539	Congo	.538	Swaziland	.522	Cameroon	.503
Ghana	.482	Kenya	.481	Zambia	.425	Nigeria	.406
Zaire	.384	Sudan	.379	Ivory Coast	.369	Tanzania	.364
Cen. Af. Rep.	.361	Senegal	.340	Malawi	.330	Uganda	.329
Liberia	.325	Gambia	.299	Chad	.296	Angola	.291
Burundi	.286	Somalia	.246	Mozambique	.246	Ethiopia	.227
Mali	.222						

Source: U.N. Human Development Report, 1995.

56. During the strike of professional baseball players in 1994, two quite different numbers were reported for the central tendency of players' annual salaries. One was $1.2 million and the other was $500,000. One of these was the median and one was the mean. Which value do you think was the mean? Why?

57. In 1986, the U.S. Federal Reserve sampled about 4000 households to estimate overall net worth of a family. Excluding some outliers of extremely wealthy individuals, they reported the summaries $44,000 and $145,000. One of these was the mean, and one was the median. Which do you think was the median? Why?

58. According to a recent report from the U.S. National Center for Health Statistics, for males with age 25–34 years, 2% of their heights are 64 inches or less, 8% are 66 inches or less, 27% are 68 inches or less, 39% are 69 inches or less, 54% are 70 inches or less, 68% are 71 inches or less, 80% are 72 inches or less, 93% are 74 inches or less, and 98% are 76 inches or less. These are called *cumulative percentages*.
a) Find the median height.
b) Nearly all the heights fall between 60 and 80 inches, with less than 1% falling outside that range. If the heights are approximately bell-shaped, give a rough approximation for the standard deviation of the heights. Explain your reasoning.

59. Grade point averages of graduating seniors at the University of Rochester are approximately bell-shaped in distribution, ranging from 2.0 to 4.0 with a mean of about 3.0. Using the fact that all or nearly all measurements for this form of distribution occur within three standard deviations of their mean, give an approximation for the value of the standard deviation.

For the following two multiple-choice items, select the correct response(s).

60. In Canada in 1981, for the categories Catholic, Protestant, Eastern Orthodox, Jewish, None, Other for religious affiliation, the relative frequencies were 47.3%, 41.2%, 1.5%, 1.2%, 7.3%, 1.5% (*Canada Year Book, 1992*).
a) The median religion is Protestant.
b) The distribution is bimodal.
c) Only 2.7% of the subjects fall within one standard deviation of the mean.
d) The mode is Catholic.
e) The "Other" response is an outlier.

61. The 1991 General Social Survey asked whether having sex before marriage is always wrong, almost always wrong, wrong only sometimes, not wrong at all. The response counts in these four categories were 274, 98, 186, 435. This distribution is
a) Skewed to the right.
b) Approximately bell-shaped.
c) Bimodal.
d) Shape does not make sense, since the variable is nominal.

62. Ten families are randomly selected in Florida and another ten families are randomly selected in Alabama. Table 3.19 provides summary information on mean family income. The mean is higher in Alabama both in rural areas and in urban areas. Which state has the larger overall mean income? (The reason for this apparent paradox is that mean urban incomes are larger than mean rural incomes for both states and the Florida sample has a higher proportion of urban residents than the Alabama sample.)

TABLE 3.19

State	Rural		Urban	
Florida	\$26,000	$(n = 3)$	\$39,000	$(n = 7)$
Alabama	\$27,000	$(n = 8)$	\$40,000	$(n = 2)$

63. Refer to Problem 3.10. Explain why the mean of these 50 measurements is not necessarily the same as the violent crime rate for the entire U.S. population.

64. The mean and standard deviation of a sample may change if data are rescaled. For a sample with mean $\bar{Y}$, adding a constant c to each observation changes the mean to $\bar{Y} + c$, and the standard deviation s is unchanged. Multiplying each observation by c changes the mean to $c\bar{Y}$ and the standard deviation to $|c|s$.

a) Scores on a difficult exam have a mean of 57 and a standard deviation of 20. The teacher boosts all the scores by 20 points before awarding grades. Report the mean and standard deviation of the boosted scores.

b) Suppose that annual income has a mean of \$39,000 and a standard deviation of \$15,000. Values are converted to British pounds for presentation to a British audience. If one British pound equals \$1.50, report the mean and standard deviation in British currency.

65. The results of the study described in Problem 3.34 are to be reported in a French newspaper. The ten measurements are converted to kilometer units (1 mile = 1.6 kilometers). Report the mean and standard deviation of the converted measurements. (*Hint*: You do not need to convert the original scores; you just need to convert the original mean and standard deviation.)

66. *,[1] The crude death rate is the number of deaths in a year, per size of the population, multiplied by 1000. According to the U.S. Bureau of the Census, in 1995 Mexico had a crude death rate of 4.6 (i.e., 4.6 deaths per 1000 population) while the United States had a crude death rate of 8.4. Explain how this overall death rate could be higher in the United States even if the United States had a lower death rate than Mexico for people of each specific age.

67. * The sample means for k sets of data with sample sizes $n_1, n_2, \ldots, n_k$ are $\bar{Y}_1, \bar{Y}_2, \ldots, \bar{Y}_k$. Show that the overall sample mean for the combined data set is

$$\bar{Y} = \frac{n_1\bar{Y}_1 + n_2\bar{Y}_2 + \cdots + n_k\bar{Y}_k}{n_1 + n_2 + \cdots + n_k}$$

Interpret $\bar{Y}$ as a weighted average of $\bar{Y}_1, \bar{Y}_2, \ldots, \bar{Y}_k$.

68. * Show that $\Sigma(Y_i - \bar{Y})$ must equal 0 for any collection of measurements $Y_1, Y_2, \ldots, Y_n$.

69. The Russian mathematician Tchebysheff proved that for any real number $k > 1$, the proportion of the measurements that fall more than k standard deviations from the mean can be no greater than $1/k^2$. Moreover, this holds for *any* distribution, not just bell-shaped ones.

a) Find the upper bound for the proportion of measurements falling (i) more than two standard deviations from the mean, (ii) more than three standard deviations from the mean, (iii) more than ten standard deviations from the mean.

b) Compare the upper bound for $k = 2$ to the approximate proportion falling more than

[1] Problems marked with an asterisk are of slightly greater difficulty or else introduce new and optional material

two standard deviations from the mean in a bell-shaped distribution. Why is there a difference?

70. * The ***least squares property*** of the mean states that the data fall closer to $\bar{Y}$ than to any other real number c, in the sense that the sum of squares of deviations of the data about their mean is smaller than the sum of squares of their deviations about c. That is,

$$\sum(Y_i - \bar{Y})^2 < \sum(Y_i - c)^2$$

If you have studied calculus, prove this property by treating $f(c) = \Sigma(Y_i - c)^2$ as a function of c and deriving the value of c that provides a minimum. (*Hint*: Take the derivative of $f(c)$ with respect to c and set it equal to zero.)

Bibliography

Cleveland, W. S. (1985). *The Elements of Graphing Data*. Belmont, CA: Wadsworth.

Cleveland, W. S. (1993). *Visualizing Data*. Summit, NJ: Hobart Press.

Huff, D. (1954). *How to Lie with Statistics*. New York: W. W. Norton

Tufte, E. R. (1983). *The Visual Display of Quantitative Information*. Cheshire, CT: Graphics Press.

Tufte, E. R. (1990). *Envisioning Information*. Cheshire, CT: Graphics Press.

Tukey, J. W. (1977). *Exploratory Data Analysis*. Reading, MA.: Addison-Wesley.

Chapter 4

Probability Distributions

Compared to most mathematical sciences, statistics is quite young. Most methods discussed in this book were developed within the past century. By contrast, probability, the subject of this chapter, has a long history. For instance, mathematicians used probability in seventeenth-century France to help evaluate various gambling strategies. Probability by itself is a large and highly developed subject, and one could devote several courses to a discussion of its philosophical foundations, theory, and application. This chapter limits attention to the basics of probability needed to use and understand methods of statistical inference.

Section 4.1 introduces *probability distributions*, which provide likelihoods for the possible outcomes of a variable. Section 4.2 introduces the *normal distribution*, a bell-shaped curve that is the most important probability distribution for statistical analysis. Section 4.3 introduces *sampling distributions*, which are distributions for possible values of a sample statistic such as the sample mean. Section 4.4 explains the main reason for the importance of the normal distribution, namely, the remarkable result that a bell-shaped curve approximates most distributions used in statistical inference. For instance, the normal distribution helps us predict how close a sample mean $\bar{Y}$ falls to the population mean, μ,that it estimates. The final section reviews the types of distributions introduced in the first four chapters of this text.

Connection Between Probability and Statistical Inference

Inferential statistical methods use sample data to make predictions about the values of useful summary descriptions, called *parameters*, of the population of interest. This chapter treats parameters as *known* numbers. Admittedly, this is artificial, since parameter values are normally unknown or we would not need inferential methods. However, many inferential statistical methods involve comparing observed sample statistics to the values expected if the parameter values equaled particular numbers. If the data are inconsistent with particular parameter values, then we infer that the actual parameter values are somewhat different. Let's take a look at an example that illustrates the connection between statistical inference and probability calculations with known parameter values.

Television networks sample voters on election day to help them predict the winners early. After polling 1000 voters in an election race between gubernatorial candidates Sam Stone and Betty McGee, the Statistics Channel finds that 40% of the sample voted for Stone and 60% voted for McGee. The unknown parameters are the percentages of the entire population that voted for each candidate. In projecting the winner, the network makes an inference about which parameter is larger.

To do this, the network can consider the question, "If actually more of the population of voters prefers Stone than McGee, would it be unusual that only 40% of the sampled individuals preferred Stone?" In other words, if Stone actually has enough support to win the election, is it plausible that only 40% of the voters in a sample of size 1000 would prefer him, as occurred in this sample? If such data would be unlikely, the network infers that Stone will lose the election. The inference about the election outcome is based on finding the probability of the sample result under the supposition that the population parameter, the percentage of voters preferring Stone, truly exceeds 50%.

4.1 Probability Distributions for Discrete and Continuous Variables

We first define the term *probability*, using a *relative frequency* approach. Imagine a hypothetical experiment consisting of a very long sequence of repeated observations on some random phenomenon. Each observation may or may not result in some particular outcome. The *probability* of that outcome is defined to be the relative frequency of its occurrence, in the long run.

Probability

The ***probability*** of a particular outcome is the proportion of times that outcome would occur in a long run of repeated observations.

A simplified representation of such an experiment is a very long sequence of flips of a coin, the outcome of interest being that a head faces upwards. Any one flip may or may not result in a head. If the coin is balanced, a basic result in probability, called the ***law of large numbers***, implies that the proportion of flips resulting in a head tends toward 1/2 as the number of flips increases. Thus, the probability of a head in any single flip of the coin equals 1/2.

Similarly, if the weather forecaster says that the probability of rain today is 70%, this means that in a long series of days with atmospheric conditions like those today, rain occurs on 70% of the days. Being a relative frequency, a probability is a number between 0 and 100 when expressed as a percentage and between 0 and 1 when expressed as a proportion. This book uses the proportion scale, expressing, for instance, the probability of rain as .70.

Probability Distributions

A variable can take at least two different values. Some outcomes may be more likely than others. The ***probability distribution*** of the variable lists the possible outcomes together with their probabilities.

The probability distribution of a *discrete* variable Y assigns a probability to each possible value of the variable. Each probability is a number between 0 and 1, and the sum of the probabilities of all possible values equals 1. Let y denote a possible outcome for the variable Y, and let $P(y)$ denote the probability of that outcome. Then

$$0 \leq P(y) \leq 1 \text{ and } \Sigma_{\text{all } y} P(y) = 1$$

since each probability falls between 0 and 1 and since the total probability equals 1. Example 4.1 illustrates a probability distribution for a discrete variable.

Example 4.1 Number of People Who Were Victims of Homicide

Let Y denote the number of people you have known personally who were victims of homicide within the past 12 months. This is a discrete variable, taking the possible values 0, 1, 2, 3, and so forth. According to results from recent General Social Surveys, the probability distribution of Y is approximately the one shown in Table 4.1. The table displays the four recorded values for Y and their probabilities: $P(0) = .91$, $P(1) = .06$, $P(2) = .02$, $P(3) = .01$. For instance, $P(0)$, the probability that you knew $Y = 0$ people personally that were victims of homicide, equals .91. Each probability in Table 4.1 is between 0 and 1, and the sum of the probabilities equals 1. (All potential y values higher than 3 had a total probability of only .005 and are ignored in Table 4.1.) □

Continuous variables have a continuum of possible values. Probability distributions of continuous variables assign probabilities to *intervals* of numbers. The probability that a variable falls in any particular interval is between 0 and 1, and the probability of the interval containing all the possible values equals 1. For instance, a recent study about proximity between the elderly and their children measured, for each elderly person in a sample, Y = travel time (in minutes) to the nearest child. The probability distribution of this variable provides probabilities such as $P(Y < 10)$, the probability that the travel time is less than 10 minutes, or $P(60 < Y < 180)$, the probability that the

TABLE 4.1 Probability Distribution of Y = Number of People You've Known That Were Victims of Homicide Within the Past 12 Months

y	$P(y)$
0	.91
1	.06
2	.02
3	.01
Total	1.0

travel time is between 60 and 180 minutes. Section 4.2 shows how to calculate such probabilities for the most important probability distribution.

The *population distribution* of a variable introduced in Section 3.1 is, equivalently, the probability distribution for the value of that variable for a subject selected randomly from the population. For example, if .06 is the proportion of individuals in a certain population who knew one victim of homicide in the past 12 months, then the probability that an individual selected randomly from that population knew one homicide victim is also .06.

Graphs for Probability Distributions

Most probability distributions have formulas for calculating probabilities. For others, tables or graphs provide the probabilities.

One can construct a *histogram* to plot the probability distribution of a discrete variable. The rectangular bar over a possible value of the variable has height equal to the probability of that value. For example, Figure 4.1 is a histogram for the probability distribution of the number of people you knew who were victims of homicide in the past 12 months, from Example 4.1. The bar over the value 0 has height .91, the probability of the outcome 0.

A graph of the probability distribution of a continuous variable is a smooth, continuous curve. The *area* under the curve for an interval of values represents the probability that the variable takes a value in that interval. For example, Figure 4.2 shows a possible graph for the probability distribution of Y = travel time to nearest child. The shaded area in the figure refers to the region of values above 60. If this area equals one-tenth of the total area under the curve, then the probability is .10 that the nearest child lives more than 60 minutes from home. Those regions in which the curve has relatively high height are the ones having values most likely to be observed.

Parameters Describing Probability Distributions

Like a population distribution, a probability distribution has parameters describing central tendency and variability. The *mean* describes central tendency and the *standard deviation* describes variability. The parameter values are the values these measures would

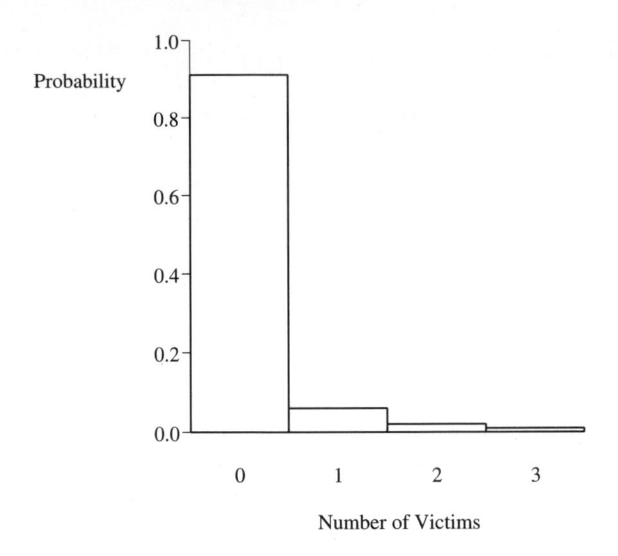

Figure 4.1 Histogram for the Probability Distribution of the Number of People You've Known That Were Victims of Homicide Within the Past 12 Months

assume, in the long run, if we repeatedly took random observations on the variable Y having that probability distribution.

For instance, suppose we take observations from the distribution in Table 4.1, for which $P(0) = .91$, $P(1) = .06$, $P(2) = .02$, $P(3) = .01$. Over the long run, we expect $y = 0$ to occur 91% of the time, $y = 1$ to occur 6% of the time, $y = 2$ to occur 2% of the time, and $y = 3$ to occur 1% of the time. In 100 observations, for instance, we expect about ninety-one 0's, six 1's, two 2's, and one 3; in that case, since the mean

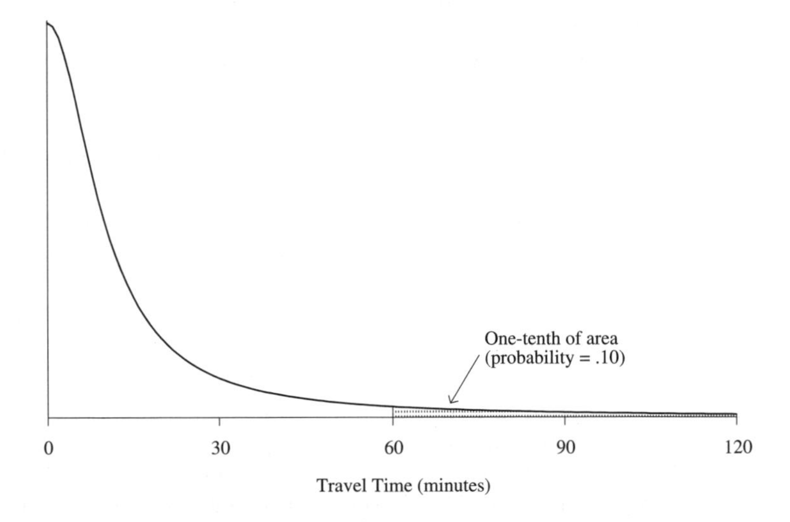

Figure 4.2 Probability Distribution of Travel Time of Elderly Parent from Nearest Child

equals the total of the observations divided by the sample size, the mean equals

$$\frac{(91)0 + (6)1 + (2)2 + (1)3}{100} = \frac{13}{100} = .13$$

This calculation has the form $0(.91)+1(.06)+2(.02)+3(.01)$, the sum of the possible outcomes times their probabilities. In fact, for any discrete variable Y, the mean of its probability distribution has this form.

Mean of a Probability Distribution

The ***mean of a probability distribution*** of a discrete variable ***Y*** is

$$\mu = \sum yP(y)$$

where the sum is taken over all possible values ***y*** of the variable. This parameter is also called the ***expected value*** of ***Y*** and is denoted by ***E(Y)***.

For Table 4.1, for instance,

$$\begin{aligned} \mu &= \sum yP(y) = 0P(0) + 1P(1) + 2P(2) + 3P(3) \\ &= 0(.91) + 1(.06) + 2(.02) + 3(.01) \\ &= .13 \end{aligned}$$

This is also the *expected value* of Y, $E(Y) = \mu = .13$, the terminology reflecting that $E(Y)$ represents what we expect for the average value of Y in a long run of observations.

The ***standard deviation*** of a probability distribution, denoted by σ, measures its variability. The larger the value of σ, the more spread out the distribution. In a very rough sense, σ describes how far the variable Y tends to fall from the mean of its distribution. The Empirical Rule (Section 3.4) helps us to interpret σ. If a probability distribution is approximately bell-shaped, about 68% of the probability falls between $\mu - \sigma$ and $\mu + \sigma$, about 95% falls between $\mu - 2\sigma$ and $\mu + 2\sigma$, and all or nearly all falls between $\mu - 3\sigma$ and $\mu + 3\sigma$.

We refer again to the example at the beginning of the chapter on polling a random sample of 1000 voters. Let Y denote the sample percentage who voted for Sam Stone. Then, Y is a variable, since the outcome for Y would vary from sample to sample of size 1000. Suppose that the *population* percentage voting for Stone is 50%. Then, the *sample* percentage Y has approximately a bell-shaped probability distribution with a mean of 50.0 and a standard deviation of 1.6. (These numbers follow from results in the next chapter.) Since 3σ is roughly 5, almost surely the sample percentage voting for Stone will fall within about 5 of 50, or between 45% and 55%. Hence, if only 40% of the sample voted for Stone, and that sample is a random sample of all voters, it seems unlikely that as many as 50% of the population voted for Stone. Our inference is that Stone will lose the election.

The standard deviation is the square root of the ***variance*** of the probability distribution. The variance, denoted by σ^2, measures the average squared deviation of an observation from the mean of the distribution. That is, it is the average value of $(Y - \mu)^2$. The variance is also denoted by $E(Y - \mu)^2$, the expected value of the squared distance between Y and its mean. We shall not need to compute this measure, so we do not study its formula here (Problem 4.55 shows the formula for the discrete case).

Some probability distributions are important because they approximate well the distributions of variables in the real world. Some are important because of their uses in statistical inference. The next section introduces the *normal* probability distribution, which is important for both reasons.

4.2 The Normal Probability Distribution

A continuous variable graphically described by a certain bell-shaped curve is said to have the ***normal probability distribution***. This distribution is the most important one in statistics. It is important partly because it approximates well the distributions of many variables. Histograms of sample data often tend to be approximately bell-shaped. In such cases, we say that the variable is *approximately normally distributed.* The main reason for its prominence, however, is that most inferential statistical methods make use of properties of the normal distribution even when the sample data are not bell-shaped.

Each normal distribution has two parameters—the mean μ and the standard deviation σ. We now define this distribution and show how to calculate probabilities for it.

Normal Distributions

The ***normal distribution*** is a symmetric, bell-shaped curve, characterized by its mean μ and standard deviation σ. For each fixed number z, the probability concentrated within z standard deviations of μ is the same for all normal distributions; in particular, the probability equals .68 within $z = 1$ standard deviation, .95 within $z = 2$ standard deviations, and .997 within $z = 3$ standard deviations.

The exact form of a normal distribution is determined by its mean μ and standard deviation σ. For example, heights of adult females in North America have a mean of $\mu = 65.0$ inches with standard deviation $\sigma = 3.5$. Only one normal distribution has those values. For any real number for μ and any nonnegative number for σ, there is a normal distribution having that mean and standard deviation.

For a normal distribution with particular values of μ and σ, the probability falling within z standard deviations of the mean depends only on the value of z. This is the area under the bell-shaped normal curve between $\mu - z\sigma$ and $\mu + z\sigma$. For every normal distribution, this probability is .68 for $z = 1$, .95 for $z = 2$, and nearly 1.0 for $z = 3$. That is, essentially the entire distribution falls between $\mu - 3\sigma$ and $\mu + 3\sigma$. Figure 4.3 illustrates. These probabilities hold precisely for the normal distribution and approximately for other bell-shaped distributions. They are the basis of the Empirical Rule (Section 3.4).

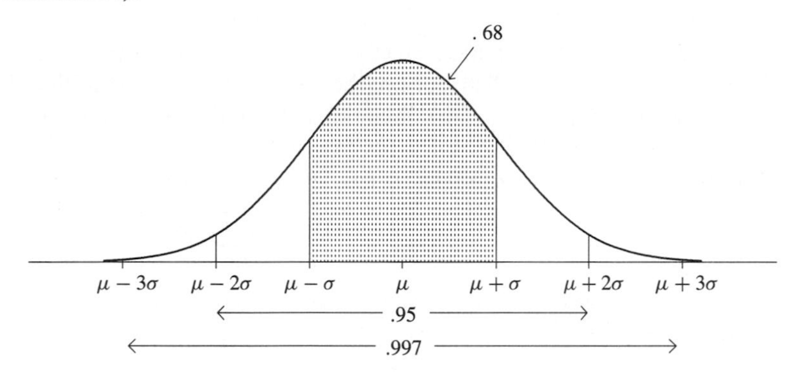

Figure 4.3 For Every Normal Distribution, the Probability Equals .68 Within σ of μ, .95 Within 2σ of μ, and .997 Within 3σ of μ

To illustrate, adult female height has the normal distribution with mean $\mu = 65.0$ and standard deviation $\sigma = 3.5$ inches. The probability equals .95 that a randomly selected female has height between $\mu - 2\sigma = 65.0 - 2(3.5) = 58$ inches and $\mu + 2\sigma = 65.0 + 2(3.5) = 72.0$ inches (6 feet). Adult male height has a normal distribution with mean $\mu = 70.0$ and standard deviation $\sigma = 4.0$ inches, so the probability equals .95 that a randomly selected male has height between $\mu - 2\sigma = 70.0 - 2(4.0) = 62$ inches and $\mu + 2\sigma = 70.0 + 2(4.0) = 78$ inches (6.5 feet). See Figure 4.4.

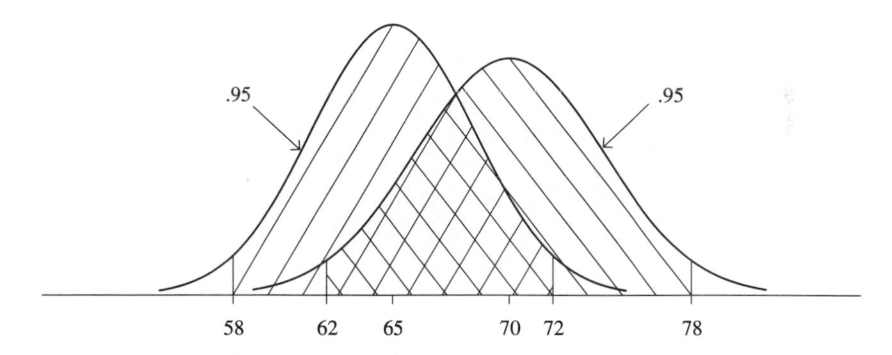

Figure 4.4 Normal Distributions for Women's Height ($\mu = 65$, $\sigma = 3.5$) and for Men's Height ($\mu = 70$, $\sigma = 4.0$)

The number z need not be a whole number (1, 2, or 3). The probability concentrated within $z\sigma$ of μ is the same for all normal curves even if z is, say, 1.43 or 0.67. Table A in Appendix A, also shown on the inside back cover, determines probabilities for any region of values for any normal distribution. It tabulates the normal curve probability for the values falling in the right-hand tail, at least z standard deviations above the mean. The left-hand column of the table lists the values for z to one decimal point, with the second decimal place listed above the columns.

To illustrate, Table 4.2 displays a small excerpt from Table A. The tabulated probability for $z = 1.43$ falls in the row labeled 1.4 and in the column labeled .03; it equals .0764. For every normal distribution, the right-hand tail probability above $\mu + 1.43\sigma$ (that is, more than 1.43 standard deviations above the mean) equals .0764.

TABLE 4.2 Part of Table A Displaying Normal Right-Hand Tail Probabilities

	Second Decimal Place of z									
z	.00	.01	.02	.03	.04	.05	.06	.07	.08	.09
0.0	.5000	.4960	.4920	.4880	.4840	.4801	.4761	.4721	.4681	.4641
						....				
						....				
1.3	.0968	.0951	.0934	.0918	.0901	.0885	.0869	.0853	.0838	.0823
1.4	.0808	.0793	.0778	.0764	.0749	.0735	.0722	.0708	.0694	.0681
1.5	.0668	.0655	.0643	.0630	.0618	.0606	.0594	.0582	.0571	.0559
						....				
						....				

Since the entries in the main body of Table A are probabilities for the right half of the normal distribution above $\mu + z\sigma$, they fall between 0 and .5. By the symmetry of the normal curve, the probabilities in Table A also refer to the left tail of the normal distribution below $\mu - z\sigma$. For instance, the probability below $\mu - 1.43\sigma$ also equals .0764.

Normal Probabilities and the Empirical Rule

Using Table A, we can determine the probabilities in the Empirical Rule (Section 3.4) and the definition of the normal distribution. For instance, the distance of two standard deviations from the mean refers to a z-value of 2.00. The normal curve probability listed in Table A opposite $z = 2.00$ is .0228. The right-hand tail probability above $\mu + 2\sigma$ equals .0228 for every normal distribution. The left-hand tail probability below $\mu - 2\sigma$ also equals .0228, by symmetry (see Figure 4.5). The total probability more than two standard deviations from the mean is $2(.0228) = .0456$. Since the total probability equals 1 and since the probability more than two standard deviations from the mean equals .0456, the probability between $\mu - 2\sigma$ and $\mu + 2\sigma$ (i.e., within two standard deviations of the mean) equals $1 - .0456 = .9544$. When a variable is normally distributed, 95% of the observations fall within two standard deviations of the mean.

We could also have calculated this as follows. The probability equals .5 above the mean, since the normal distribution is symmetric about μ. The probability concentrated between μ and $\mu + 2\sigma$ equals $.5 - .0228 = .4772$, shown in Figure 4.5. The probability between $\mu - 2\sigma$ and μ also equals .4772, by symmetry. Thus, the total probability within two standard deviations of the mean equals $2(.4772) = .9544$, or about 95%.

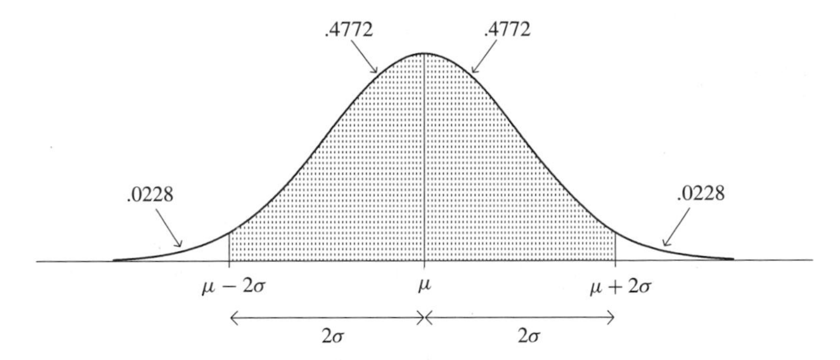

Figure 4.5 The Probability Within Two Standard Deviations of the Mean for a Normal Distribution

Regardless of the approach used for the calculation, it is helpful to sketch a normal curve and shade in the region of interest to clarify the steps and logic.

The approximate percentages that the Empirical Rule lists are the actual percentages for the normal distribution, rounded to two decimal places. For instance, using Table A, you should verify that the probability within one standard deviation of the mean of a normal distribution equals .68 (*Hint:* Let $z = 1.00$.). The probability within three standard deviations of the mean ($z = 3.00$) equals .997, or essentially 1.0.

Not all bell-shaped curves are normal distributions, only those described by the mathematical formula shown in Problem 4.57 at the end of the chapter. This formula is unimportant for our purposes, but we will use the probabilities tabulated for it in Table A throughout the rest of the text.

Finding z-Values for Certain Tail Probabilities

Some inferential statistical methods use z-values corresponding to certain normal curve probabilities. This entails the reverse use of Table A. We start with the probability falling beyond some number, which is listed in the body of Table A, and we find the z-value that provides the number of standard deviations that that number falls from the mean.

To illustrate, let us find the z-value corresponding to a right-hand tail probability of .025. We look up the probability of .025 in the body of Table A. It corresponds to $z = 1.96$. This means that a probability of .025 lies above $\mu + 1.96\sigma$. Similarly, a probability of .025 lies below $\mu - 1.96\sigma$, so a total probability of .050 lies more than 1.96σ from μ. 95% of a normal distribution falls within two standard deviations of the mean; more precisely, .9544 falls within 2.00 standard deviations, and .950 falls within 1.96 standard deviations.

To check that you understand this reasoning, verify that the z-value for a right-hand tail probability of (1) .05 is $z = 1.64$, (2) .10 is $z = 1.28$, (3) .01 is $z = 2.33$, (4) .005 is $z = 2.58$. Show that 90% of a normal distribution falls between $\mu - 1.64\sigma$ and $\mu + 1.64\sigma$.

Example 4.2 Finding the z-Value for the 99th Percentile

For a normal distribution, how many standard deviations from the mean is the 99th percentile? In other words, what is the value of z such that $\mu + z\sigma$ falls above 99% of the distribution?

To find the z-value for the 99th percentile, we use the normal curve probability in the right-hand tail beyond the 99th percentile. Then, we can use Table A to find the z-value corresponding to that probability. Now, for $\mu + z\sigma$ to represent the 99th percentile, the probability below $\mu + z\sigma$ must equal .99, by the definition of a percentile. In other words, 1% of the distribution is above the 99th percentile, so the right-hand tail probability equals .01, as shown in Figure 4.6.

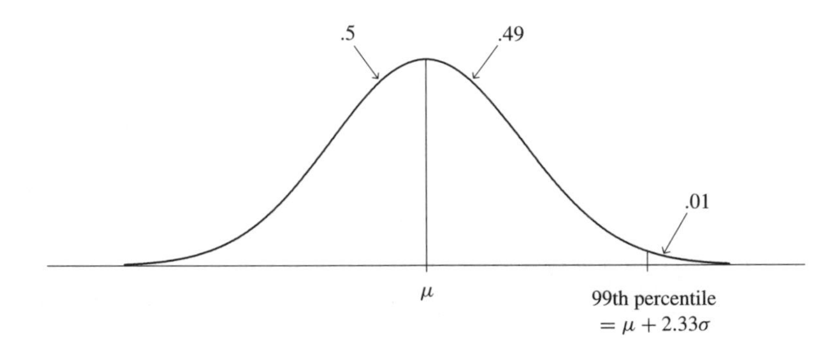

Figure 4.6 The 99th Percentile for a Normal Distribution

The body of Table A does not contain the normal curve probability of exactly .0100. Notice, though, that the probability .0102 corresponds to $z = 2.32$, and the probability .0099 corresponds to $z = 2.33$. We could interpolate, but it is sufficient to use the z-value rounded to two decimal places, so we select the one having probability closer to the desired probability. Thus, the 99th percentile is 2.33 standard deviations above the mean. In summary, 1% of any normal distribution is located above $\mu + 2.33\sigma$, or more than 2.33 standard deviations above the mean.

For instance, IQ scores are approximately normally distributed with a mean of 100 and a standard deviation of 16. The 99th percentile equals

$$\mu + 2.33\sigma = 100 + 2.33(16) = 137$$

That is, about 1% of IQ scores are higher than 137. By symmetry, about 1% of IQ scores are less than $\mu - 2.33\sigma = 100 - 2.33(16) = 63$, which is the 1st percentile. The remaining 98% of the IQ scores fall between 63 and 137, which is the region within 2.33 standard deviations of the mean. □

To check that you understand the reasoning above, show that the 95th percentile of a normal distribution is $\mu + 1.64\sigma$, and show that the 95th percentile for the IQ distribution equals 126.

Example 4.3 z-Value for Two-Tail Probability Totaling .01

What is the z-value such that the total probability is only .01 that a normally distributed variable falls more than z standard deviations above or below the mean? This is the z-value so that .01 is the total probability below $\mu - z\sigma$ or above $\mu + z\sigma$. Since the probability below $\mu - z\sigma$ equals the probability above $\mu + z\sigma$ by symmetry, the probability in each tail is $.01/2 = .005$, as shown in Figure 4.7. Now, the single-tail probability of .005 is the type listed in the body of Table A. A value of about .005 in that table corresponds to $z = 2.58$. Thus, the probability is .01 that a normally distributed variable falls more than 2.58 standard deviations from its mean.

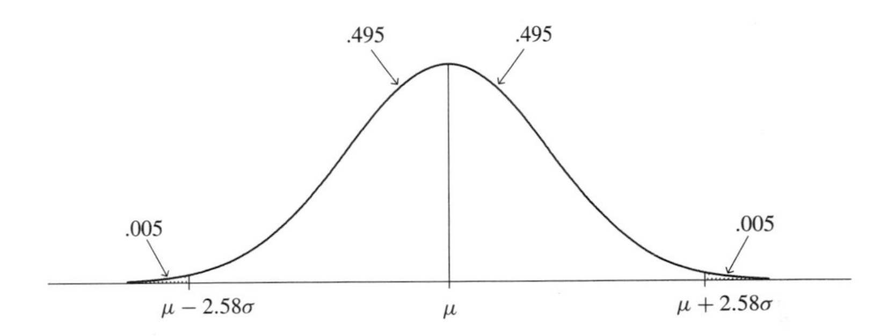

Figure 4.7 Two-Tail Normal Curve Probability of .01; The Probability Equals .005 in Each Tail

In other words, 99% of the observations for a normally distributed variable fall within 2.58 standard deviations of the mean. For the IQ distribution with mean $\mu = 100$ and standard deviation $\sigma = 16$, about 99% of the IQ scores fall between

$$\mu - 2.58\sigma = 100 - 2.58(16) = 59 \text{ and } \mu + 2.58\sigma = 100 + 2.58(16) = 141$$

□

To check that you understand this reasoning, show that the total probability of .050 falls more than $z = 1.96$ standard deviations above or below the mean.

z-Scores

Looking up a value for z in Table A provides the probability falling at least z standard deviations above the mean of a normal distribution. More generally, the z symbol refers to the distance between a possible value of a variable Y and the mean μ of its probability distribution, in terms of the number of standard deviations that Y falls from μ.

z-Score

The z-score for a value Y on a variable is the number of standard deviations that Y falls from μ.

Given a value for Y, we now show how to calculate the z-score. We also show that z-scores provide a useful way to compare distances for different normal distributions.

Example 4.4 z-scores for the SAT

Scores on the verbal or math portion of the Scholastic Aptitude Test (SAT), a college entrance examination, are approximately normally distributed with mean $\mu = 500$ and standard deviation $\sigma = 100$. The test score of $Y = 650$ has a z-score of $z = 1.50$, since 650 is 1.50 standard deviations above the mean. In other words, $Y = 650 = \mu + z\sigma = 500 + z(100)$, where $z = 1.50$. □

The distance between Y and μ equals $Y - \mu$. The z-score, which expresses this difference in units of standard deviations, has the formula

$$z = \frac{Y - \mu}{\sigma}$$

To illustrate, when $\mu = 500$ and $\sigma = 100$, an observation of $Y = 650$ has the z-score of

$$z = \frac{Y - \mu}{\sigma} = \frac{650 - 500}{100} = 1.50$$

Positive z-scores occur when the number Y falls above the mean μ. Negative z-scores occur when the number Y falls below the mean. For instance, $Y = 400$ has a z-score of $(400 - 500)/100 = -1.00$. Referring the z-scores to Table A yields tail probabilities—right-hand tail probabilities for positive z-scores and left-hand tail probabilities for negative z-scores. For instance the z-score of 1.50 has a right-hand tail probability of .0668, so 6.68% of the SAT test scores fall above 650.

Calculating a z-Score

A number Y is $z = (Y - \mu)/\sigma$ standard deviations from μ. The probability in the tail of a normal distribution beyond a number Y is the tail probability for that z-score in Table A.

For SAT scores with $\mu = 500$ and $\sigma = 100$, a value of $Y = 350$ has a z-score of

$$z = \frac{Y - \mu}{\sigma} = \frac{350 - 500}{100} = -1.50$$

The test score of 350 is 1.50 standard deviations below the mean. The value $Y = 350$ falls below the mean, so the z-score is negative. Now, Table A contains only positive z-values. Since the normal distribution is symmetric about the mean, the left-hand tail probability below $-z$ equals the right-hand tail probability above $+z$. Looking up $z = 1.50$ in Table A, we see that the probability that a score falls below 350 is .0668, as illustrated in Figure 4.8.

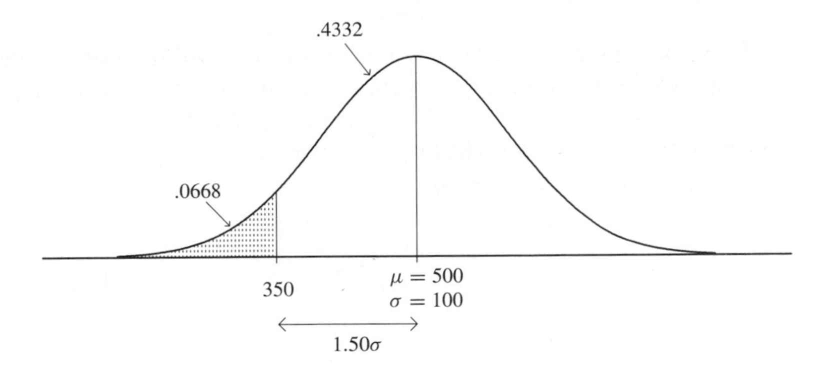

Figure 4.8 Normal Distribution for SAT Scores

Example 4.5 Comparing Standardized Test Scores

Suppose that when you applied to college, you took a SAT exam, scoring 550. Your friend took the ACT, scoring 30. Which score is better?

We cannot compare the test scores of 550 and 30 directly, since they have different scales. We must first convert them to z-scores, analyzing how many standard deviations each falls from the mean. Since the SAT has $\mu = 500$ and $\sigma = 100$, a SAT test score of $Y = 550$ converts to a z-score of

$$z = \frac{(Y - \mu)}{\sigma} = \frac{(550 - 500)}{100} = .5$$

The ACT has approximately $\mu = 18$ and $\sigma = 6$, so an ACT score of 30 converts to a z-score of $(30 - 18)/6 = 2.0$.

The ACT score of 30 is relatively higher than the SAT score of 650, since 30 is 2.0 standard deviations above its mean whereas 550 is .5 standard deviations above its mean. The SAT and ACT scores both have approximate normal distributions, so from Table A, $z = 2.0$ has a tail probability of .0228 and $z = .5$ has a tail probability of .3085. Of all students taking the ACT, only about 2% scored higher than 30, whereas of all students taking the SAT, about 31% scored higher than 550. The ACT score falls at the 98th percentile of its scale, while the SAT score falls at the 69th percentile of its scale. In this relative sense, the ACT score is higher. □

The Standard Normal Distribution

Inferential statistical methods frequently make reference to a particular normal distribution, called the ***standard normal distribution***.

Standard Normal Distribution

The ***standard normal distribution*** is the normal distribution with mean $\mu = 0$ and standard deviation $\sigma = 1$.

For the standard normal distribution, the number falling z standard deviations above the mean is $\mu + z\sigma = 0 + z(1) = z$; it is simply the z-score itself. For instance, the value of 2 is two standard deviations above the mean, and the value of -1.3 is 1.3 standard deviations below the mean. The original values are the same as the z-scores. See Figure 4.9.

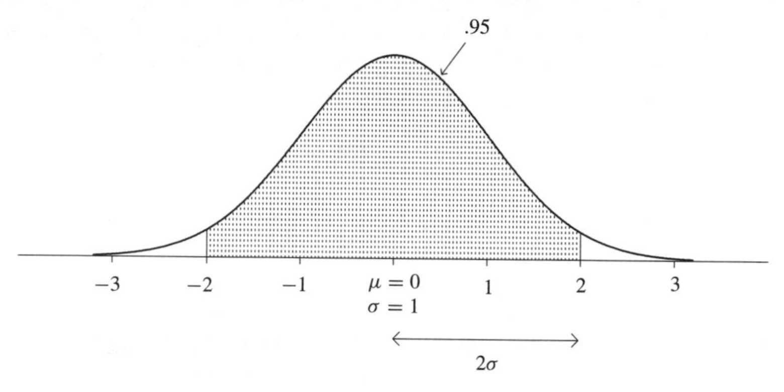

Figure 4.9 The Standard Normal Distribution

When the values for an arbitrary normal distribution are converted to z-scores, those z-scores are centered around 0 and have a standard deviation of 1; that is, the z-scores have the standard normal distribution.

> **z-Scores and the Standard Normal Distribution**
>
> If a variable has a normal distribution, and if its values are converted to z-scores by subtracting the mean and dividing by the standard deviation, then the distribution of the z-scores is the standard normal distribution.

Suppose we convert each SAT score Y to a z-score by using $z = (Y - 500)/100$. For instance, $Y = 650$ converts to $z = 1.50$, and $Y = 350$ converts to $z = -1.50$. Then, the entire set of z-scores has a normal distribution with a mean of 0 and a standard deviation of 1, which is the standard normal distribution.

Many inferential statistical methods convert values of statistics to z-scores and normal curve probabilities. We use z-scores and normal probabilities throughout the rest of the book. It is crucial that you understand the use of Table A in the examples in this section and that you can do the related exercises at the end of the chapter.

4.3 Sampling Distributions

The first two sections of this chapter used probability distributions to summarize probabilities of possible outcomes for a variable. In practice, the distributions of most variables are unknown. Instead, using sample data we estimate characteristics of the distributions such as their parameters.

For an upcoming senatorial election in Massachusetts, the probability that a randomly selected adult resident of that state prefers the Republican (or the Democratic) candidate would be unknown. A polling organization may survey 1500 residents, however, and predict that the proportion favoring the Democratic candidate is .54 and the proportion favoring the Republican candidate is .46. How can they possibly know their estimate is a good one? After all, the state has several million residents, and they sampled a minuscule percentage of that population.

Simulating the Estimation Process

The following device can show how close sample estimates are likely to be to unknown population parameters. Simulate the process of selecting a person at random from the population by picking a two-digit number from the random number table, Table 2.1. Suppose that, in fact, exactly 50% of the residents favor each candidate. Identify all 50 two-digit numbers between 00 and 49 as Republicans and all 50 two-digit numbers between 50 and 99 as Democrats, so that each party has a 50% chance of selection on each choice. For instance, the first two digits of the first column of Table 2.1 provide the random numbers 10, 22, 24, 42, 37, 77, and so forth. Of the first 6 people selected, 5 are Republicans (i.e., have numbers between 00 and 49). Selecting 1500 two-digit numbers simulates the process of randomly sampling 1500 residents of the much larger population, which is actually treated as infinite in size.

Using a computer, we selected 1500 random two-digit numbers and got 726 Democrats and 774 Republicans. The sample proportion of Republicans is $774/1500 = .516$,

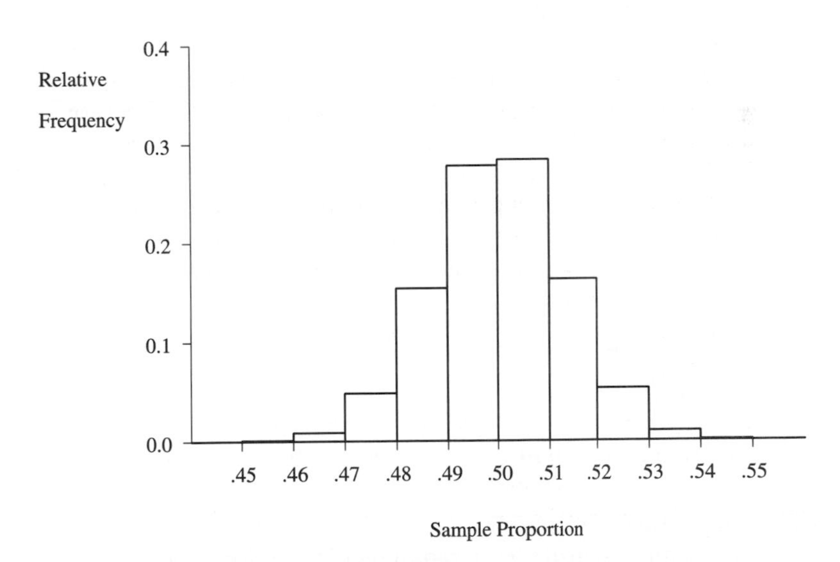

Figure 4.10 Results of Simulating the Sample Proportion Favoring the Republican Candidate, for Random Samples of 1500 Subjects from a Population in Which Half Favor Each Candidate

quite close to the "true" proportion of .50. This particular estimate was good. Were we simply lucky? We repeated the process and used the computer to select 1500 more two-digit random numbers. This time the sample proportion of Republicans was .491, also quite good. Computer time is cheap these days, so we performed this process of picking 1500 people a million times so we could search for a pattern in the results. Figure 4.10 shows a histogram of the million values of the sample proportion. Nearly all the simulated proportions fell between .46 and .54, that is, within .04 of the true population value of .50. Apparently a sample of size 1500 provides quite a good estimate of a population proportion.

Now, in practice, we would not know that the population proportion preferring each candidate is .50. However, we could perform this simulation for other proportion values, to see if similar results occur. For instance, we could simulate sampling when the population proportion favoring the Republican candidate is .45 by letting the 45 random numbers between 00 and 44 represent Republicans and the 55 between 45 and 99 represent Democrats. In fact, from results of the next section, the sample proportion for a random sample of size 1500 is very likely to fall within .04 of the true population proportion, regardless of its value. Thus, when we obtain the sample data and calculate the sample proportion, we know the likely accuracy of that estimate.

Representing Sampling Variability by a Sampling Distribution

When a sample statistic estimates a population parameter, it is essential to predict how accurate that estimate is. When a polling organization samples a certain number of people, for instance, the quality of the sample proportion as an estimate of the population proportion depends on the answers to such questions as, "How likely is it that the sample proportion falls within .03 of the population proportion? Within .05? Within .10?"

Now, just as candidate preference is a variable, varying among subjects, so is the sample proportion preferring a given candidate, varying among samples. In other words, before the sample is obtained, the value of the sample proportion is unknown and is a variable. If several random samples of size $n = 1500$ each were selected, a certain predictable amount of variation would occur in the sample proportion values. A probability distribution with appearance similar to Figure 4.10 describes the variation that occurs from repeatedly selecting samples of a certain size n and forming a particular statistic. This distribution is called a *sampling distribution*. It provides probabilities of the possible values of the statistic for a *single* sample of size n.

Sampling Distribution

A ***sampling distribution*** is a probability distribution that determines probabilities of the possible values of a sample statistic.

Each sample statistic has a sampling distribution. There is a sampling distribution of a sample mean, a sampling distribution of a sample proportion, a sampling distribution of a sample median, and so forth. A sampling distribution is simply a type of probability distribution. Unlike the distributions studied so far, a sampling distribution refers not to individual observations but to the values of a statistic computed from those observations, in sample after sample. Knowledge of the relevant sampling distribution allows us to calculate, for example, probabilities about

- The sample proportion of individuals who prefer the Republican candidate, for a random sample of size 1500
- The sample percentage of women in a sample of size 12 randomly selected for a jury in a celebrity murder trial
- The sample mean number of previous convictions for 100 inmates selected at random from the California prison system

Before the samples are obtained, all of these are variables. They have sampling distributions that describe the probabilities of their possible values.

The sampling distribution is important in inferential statistics because it enables us to predict how close a statistic falls to the parameter it estimates. In particular, it determines the probability that the statistic falls within any given distance of that parameter. From Figure 4.10, for instance, the probability is apparently high that a sample proportion falls within .04 of the population proportion, for a sample of size 1500. The sampling distribution for a sample of size 4 is much more spread out, and such accuracy would not occur.

Example 4.6 Constructing a Sampling Distribution

It is sometimes possible to construct the sampling distribution without resorting to simulation or complex mathematical derivations. To illustrate, we construct the sampling distribution of the sample proportion in sampling $n = 4$ people from a population in which half favor each candidate. For each subject, define the Y variable representing preference of candidate as follows:

$$Y = 1, \text{ subject's preference is the Republican}$$
$$Y = 0, \text{ subject's preference is the Democrat}$$

We use a symbol with four entries to represent the Y values for a potential sample. For instance, (1, 0, 0, 1) represents a sample in which the first and fourth subjects sampled prefer the Republican candidate, and the second and third subjects prefer the Democrat. The sixteen possible samples are

(1, 1, 1, 1) (1, 1, 1, 0) (1, 1, 0, 1) (1, 0, 1, 1)
(0, 1, 1, 1) (1, 1, 0, 0) (1, 0, 1, 0) (1, 0, 0, 1)
(0, 1, 1, 0) (0, 1, 0, 1) (0, 0, 1, 1) (1, 0, 0, 0)
(0, 1, 0, 0) (0, 0, 1, 0) (0, 0, 0, 1) (0, 0, 0, 0)

The samples are equally likely if half the population favors each candidate.

Now we construct the sampling distribution of the proportion of the sample that favors the Republican candidate. For a sample of size 4, that proportion can be 0, .25, .50, .75, or 1.0. The proportion 0 occurs with only one of the 16 possible samples, (0, 0, 0, 0), so its probability equals $1/16 = .0625$. The proportion .25 occurs for four samples, (1, 0, 0, 0), (0, 1, 0, 0), (0, 0, 1, 0), and (0, 0, 0, 1), so its probability equals $4/16 = .25$. Similarly, one can construct the probability for each possible sample proportion value, as shown in Table 4.3.

TABLE 4.3 Sampling Distribution of Sample Proportion, for Sample of Size $n = 4$ from Population with True Proportion of .50

Sample Proportion	Probability
0.0	.0625
.25	.2500
.50	.3750
.75	.2500
1.0	.0625

Figure 4.11 portrays the sampling distribution of the sample proportion for this case of $n = 4$. This sampling distribution, with a range of 0 to 1.0, is much more spread out than the one in Figure 4.10 for samples of size $n = 1500$, which concentrates between .45 and .55. With such a small sample ($n = 4$), the sample proportion need not be near the true population proportion. This is not surprising, and in practice, samples are usually much larger than $n = 4$; we used a small value in this example so it was simpler to construct the potential samples and calculate probabilities for the sampling distribution. □

When we denote two possible outcomes by 0 and 1, the proportion of times that 1 occurs is simply the sample mean of the data. For instance, for the sample (0, 1, 0, 0) in which only the second subject preferred the Republican, the sample mean equals $(0+1+0+0)/4 = 1/4 = .25$, the sample proportion favoring the Republican. Thus, *the sample proportion is a special case of a sample mean.* This is a useful result for future sections. It also implies that Table 4.2 and Figure 4.11 portray an example of a sampling distribution of a sample mean, as well as a sample proportion.

Repeated Sampling Interpretation of Sampling Distributions

Sampling distributions reflect the sampling variability that occurs in collecting data and using sample statistics to estimate parameters. If four different polling organizations each take their own sample and estimate the population proportion preferring the Republican candidate, they will get four different estimates, since the samples consist of different sets of people. More generally, Figure 4.10 describes the variability in sample proportion values that occurs in selecting a huge number of samples of size $n = 1500$

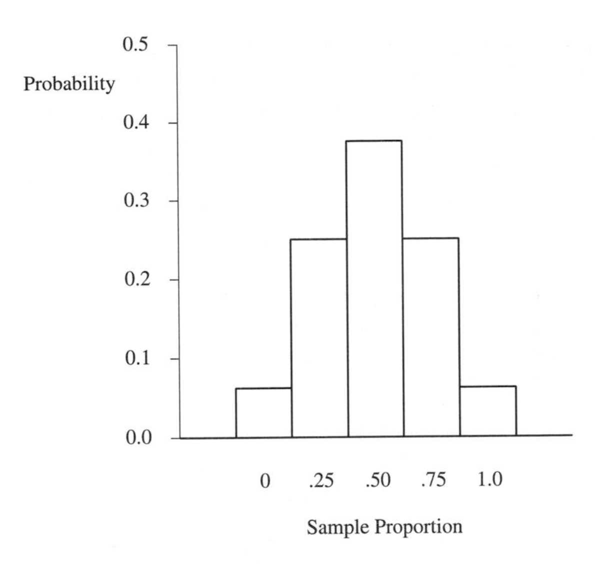

Figure 4.11 Sampling Distribution of Sample Proportion, for Sample of Size $n = 4$ from Population with True Proportion of .50

and constructing a histogram of the sample proportions. By contrast, Figure 4.11 describes the variability for a huge number of samples of size $n = 4$.

A sampling distribution of a statistic based on n observations is the relative frequency distribution for that statistic resulting from repeatedly taking samples of size n, each time calculating the statistic value. One can form such a distribution empirically, as in Figure 4.10, by repeated sampling. In practice, it is not necessary to take repeated samples and empirically generate sampling distributions. The form of sampling distributions is often known theoretically, as shown in the previous example and in the next section. We can then make probabilistic statements about the value of the sample statistic for one sample of some fixed size n.

4.4 Sampling Distributions of Sample Means

Because the sample mean $\bar{Y}$ is used so much, its sampling distribution merits special attention. When we collect the sample and observe $\bar{Y}$, we do not know how close $\bar{Y}$ falls to μ, since we do not know the value of μ. With the sampling distribution, though, we can predict the error of estimation. For example, the sampling distribution might indicate that with high probability, $\bar{Y}$ falls within 10 units of μ.

This section discusses the sampling distribution of the sample mean. There are two main results. One provides formulas for its mean and standard deviation, and the other describes its shape.

Mean and Standard Error of Sampling Distribution of $\bar{Y}$

The sample mean $\bar{Y}$ is a variable, varying from sample to sample. For random samples, it fluctuates around the population mean μ, sometimes being smaller and sometimes being larger. In fact, the mean of the sampling distribution of $\bar{Y}$ equals μ. If we repeatedly took samples of size n, then in the very long run, the mean of the sample means would simply equal the population mean μ.

Next, we describe the spread of the sampling distribution of $\bar{Y}$ by providing a formula for its standard deviation.

Standard Error

Let $\sigma_{\bar{Y}}$ denote the standard deviation of the sampling distribution of $\bar{Y}$. This is called the ***standard error*** of the sampling distribution of $\bar{Y}$, or simply the ***standard error of*** $\bar{Y}$.

The standard error describes the spread of the sampling distribution. It refers to the variability in the value of $\bar{Y}$ from sample to sample. The value $\sigma_{\bar{Y}}$ is the number that results from repeatedly selecting samples of size n from the population, finding $\bar{Y}$ for each set of n observations, and calculating the standard deviation of the $\bar{Y}$-values. The symbol $\sigma_{\bar{Y}}$ (instead of σ) and the terminology *standard error* (instead of *standard deviation*) distinguish this measure from the standard deviation σ of the population distribution.

For a random sample of size n, the standard error of $\bar{Y}$ is related to the population standard deviation σ by

$$\sigma_{\bar{Y}} = \frac{\sigma}{\sqrt{n}}$$

That is, the spread of the sampling distribution depends on the sample size n and the spread of the population distribution.

Figure 4.12 displays a population distribution and shows the sampling distribution of $\bar{Y}$ for $n = 100$. As the sample size n increases, the standard error $\sigma_{\bar{Y}} = \sigma/\sqrt{n}$ decreases. The reason for this is that the denominator ($\sqrt{n}$) of the ratio increases as n increases, whereas the numerator is the population standard deviation (σ), which is a constant and is not dependent on the value of n.

For instance, suppose the population standard deviation equals $\sigma = 10$. Then the standard error when $n = 25$ equals

$$\sigma_{\bar{Y}} = \frac{\sigma}{\sqrt{n}} = \frac{10}{\sqrt{25}} = 2.0$$

and the standard error when $n = 100$ equals

$$\sigma_{\bar{Y}} = \frac{\sigma}{\sqrt{n}} = \frac{10}{\sqrt{100}} = 1.0$$

Both sampling distributions are much less spread out than the population distribution.

This result shows that the sample mean $\bar{Y}$ tends to be closer to the population mean μ for larger values of n, since the sampling distribution becomes less spread out about

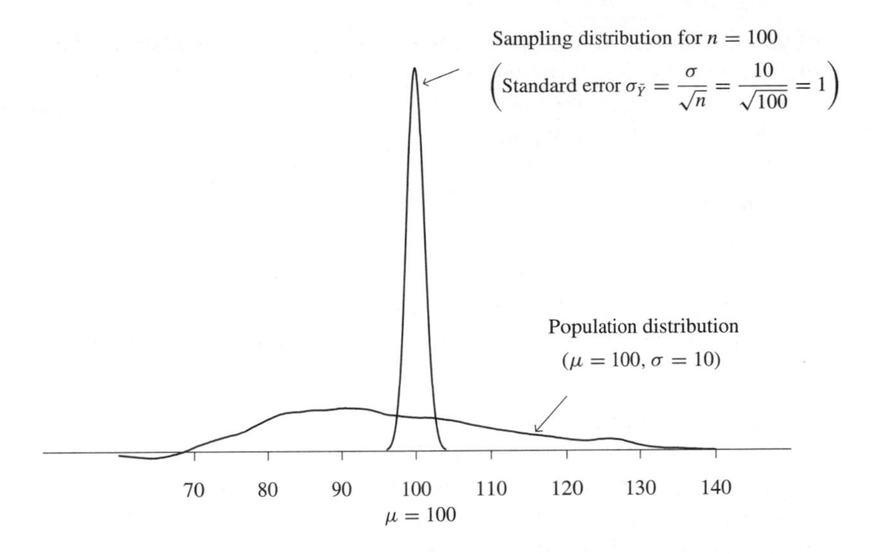

Figure 4.12 A Population Distribution and the Sampling Distributions of $\bar{Y}$ for $n = 100$

μ. This agrees with our intuition that larger samples provide more precise estimates of population characteristics. The error that results from estimating μ by $\bar{Y}$, which occurs because we sampled only part of the population, tends to decrease as the sample size n increases. This error is called the ***sampling error***. The standard error is fundamental to inferential procedures that predict how much sampling error occurs in using $\bar{Y}$ to estimate μ.

In summary, the following result describes the central location and variability of the sampling distribution of $\bar{Y}$.

Mean and Standard Error of $\bar{Y}$

For a random sample of size n from a population having mean μ and standard deviation σ, the sampling distribution of $\bar{Y}$ has mean μ and standard error $\sigma_{\bar{Y}} = \sigma/\sqrt{n}$.

Example 4.7 Standard Error of Sample Proportion

We return to the example of the previous section on sampling residents to determine their preference for the Democrat or Republican candidate in a Massachusetts senatorial election. The variable Y representing candidate choice equals 1 for Republican and 0 for Democrat. The probability distribution for Y is highly discrete, consisting of only two values for Y and their probabilities.

If half the population prefer each candidate, the probability that $Y = 0$ and the probability that $Y = 1$ both equal .5. Figure 4.13 shows this distribution. The mean of the distribution then equals .5, which is simply the population proportion preferring the Republican. (Or, more formally, $\mu = \Sigma y P(y) = 0(.5) + 1(.5) = .5$.) The squared deviation of Y from the mean, $(Y - \mu)^2$, equals $(0 - .5)^2 = .25$ when $Y = 0$ and it equals $(1 - .5)^2 = .25$ when $Y = 1$. Thus, the variance, which is the average squared deviation, equals $\sigma^2 = .25$, and the standard deviation of the population distribution of Y equals $\sigma = \sqrt{.25} = .50$.

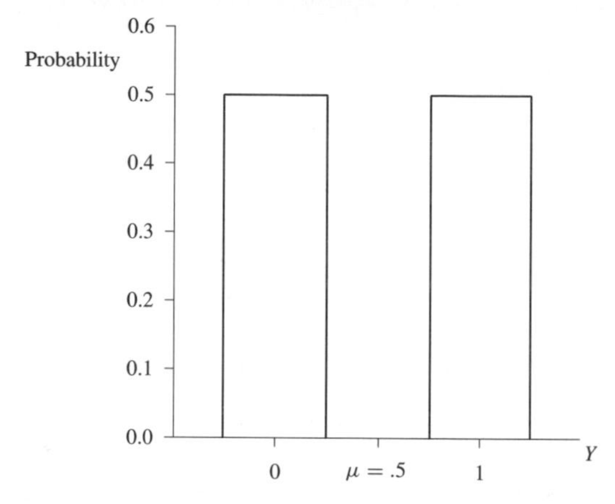

Figure 4.13 The Population Distribution When $Y = 0$ or 1, with Probability .5 Each

For a sample from this distribution, the sample mean of the 0 and 1 values is the sample proportion of subjects preferring the Republican. This sample proportion has a sampling distribution. The mean of that distribution is the mean of the population distribution of Y, namely, $\mu = .50$. For repeated samples of a fixed size n, the sample proportions fluctuate around .50, being larger about half the time and smaller half the time. The standard deviation of this distribution is the standard error, $\sigma_{\bar{Y}} = \sigma/\sqrt{n} = .50/\sqrt{n}$.

For instance, for a sample of size 100, the standard error equals $.50/\sqrt{100} = .05$. If samples of size 100 were repeatedly selected and the sample proportion computed for each one, the proportion values would fluctuate around .50 with a standard deviation of .05. Moreover, a result from the next subsection says that this sampling distribution is bell-shaped. Thus, with very high probability the sample proportion falls within three standard errors of μ, that is, within $3(.05) = .15$ of .50, or between .35 and .65.

By contrast, when the sample has size $n = 1500$, the standard error equals

$$\sigma_{\bar{Y}} = \frac{\sigma}{\sqrt{n}} = \frac{.50}{\sqrt{n}} = \frac{.50}{\sqrt{1500}} = .013.$$

Since three standard errors equals $3(.013) = .04$, the probability is very high that the sample proportion falls within .04 of .50, or between .46 and .54. Figure 4.14 shows the theoretical sampling distributions of the sample proportion when $n = 100$ and when $n = 1500$. As n increases, the standard error decreases and the curve gets narrower, the sample proportion tending to fall closer to the population proportion. □

Normal Sampling Distribution of Sample Mean

The population distribution shown in Figure 4.13 is highly discrete, having probability concentrated at only two points. Nevertheless, the two sampling distributions shown in Figure 4.14 have bell shapes. This is a consequence of the second main result of this section, which refers to the shape of the sampling distribution of $\bar{Y}$. This result can be proven mathematically, and it is often called the *Central Limit Theorem.*

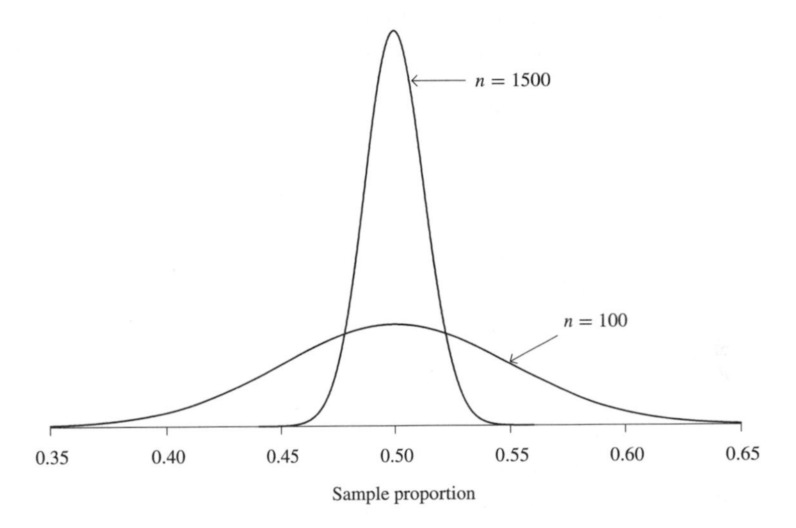

Figure 4.14 The Sampling Distributions of the Sample Proportion, When $n = 100$ and When $n = 1500$, for Sampling from the Distribution in Figure 4.13

Central Limit Theorem

For random sampling, as the sample size n grows, the sampling distribution of $\bar{Y}$ approaches a normal distribution.

Some implications and interpretations of this result and the formula for the standard error of $\bar{Y}$ follow:

- The approximate normality of the sampling distribution applies *no matter what the shape* of the population distribution. This is quite remarkable. For large sample sizes, the sampling distribution of $\bar{Y}$ is approximately normal even if the population distribution is highly discrete, highly skewed, or U-shaped. Figure 4.15 displays sampling distributions of $\bar{Y}$ for four different shapes for the population distribution. The population shapes are shown at the top of the figure. Below them are portrayed the sampling distributions for samples of sizes $n = 2$, 5, and

30. Even if the population distribution itself is not normal, the sampling distribution has approximately a normal shape, sometimes for n as small as 5. In addition, the spread of the sampling distribution noticeably decreases as n increases, as discussed previously.

- The sampling distribution takes a more normal shape as n increases. How large n must be before the shape is close to normal largely depends on the skewness of the population distribution. More skewed distributions require larger sample sizes. For most cases, a sample size of about 25 or 30 is adequate for a good approximation. If the *population* distribution is approximately bell-shaped, then the sampling distribution is approximately normal for all sample sizes. Refer to the rightmost panel of Figure 4.15 for an example.

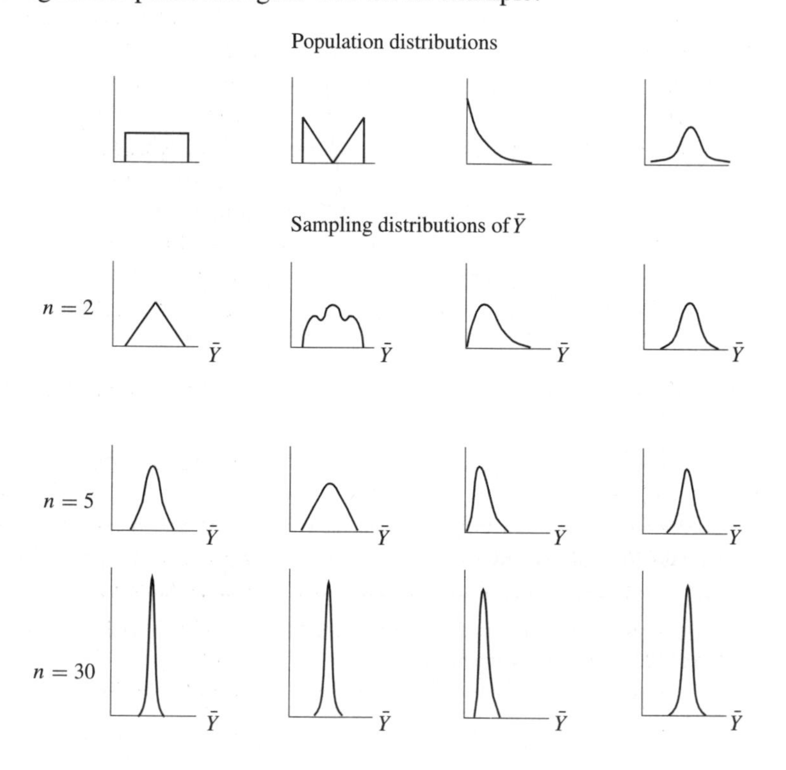

Figure 4.15 Four Different Population Distributions and the Corresponding Sampling Distributions of $\bar{Y}$

- We could verify the Central Limit Theorem empirically. We could repeatedly select random samples of fixed size n from a distribution with mean μ and standard deviation σ, calculating $\bar{Y}$ for each sample of n observations. Then, the histogram of the $\bar{Y}$-values would be approximately a normal curve about μ with standard error equal to $\sigma/\sqrt{n}$, the population standard deviation divided by the square root of the sample size of each sample.

- Since the sampling distribution of $\bar{Y}$ is approximately normal about μ with standard deviation $\sigma_{\bar{Y}}$, Table A provides probabilities for $\bar{Y}$. For instance, $\bar{Y}$ falls within $2\sigma_{\bar{Y}} = 2\sigma/\sqrt{n}$ units of μ with probability close to .95, since 95% of a normal distribution falls within two standard deviations of the mean; similarly, $\bar{Y}$ almost certainly falls within $3\sigma_{\bar{Y}}$ of μ. Reasoning of this nature is vital to inferential statistical methods.
- Though a sample size of about 30 is usually sufficiently large for the sampling distribution to be close to normal, this does not mean that only 30 observations are adequate for precise statistical inference. How precisely $\bar{Y}$ estimates μ is determined by the size of the standard error, $\sigma_{\bar{Y}}$. Depending on the precision desired, the required sample size may be less than 30, or it may be several thousand. What this does mean is that if n is at least about 30, then the sampling distribution of $\bar{Y}$ is approximately normal, no matter what the form of the population distribution.
- For random sampling, the sampling distribution of $\bar{Y}$ has mean μ and standard error $\sigma_{\bar{Y}} = \sigma/\sqrt{n}$ for *any* value of n. However, the shape of the sampling distribution of $\bar{Y}$ is normal for small n only if the population distribution is also normal, as the rightmost panel of Figure 4.15 shows. The power of the Central Limit Theorem is that, for *large* n, the sampling distribution of $\bar{Y}$ is approximately normal even if the population distribution is not. This enables us to make inferences when the population distribution is highly irregular. This is helpful, since many social science variables are either very skewed or highly discrete.

Example 4.8 How Close Is Sample Mean Income of Migrant Workers to True Mean ?

For the population of migrant workers in California, suppose that weekly income has a distribution that is skewed to the right with a mean of $\mu = \$250$ and a standard deviation of $\sigma = \$80$. A researcher, unaware of these values, plans to select a random sample of 100 migrant workers and use the sample mean annual income $\bar{Y}$ to estimate μ. Where is this sample mean likely to fall, relative to the population mean of $250?

By the Central Limit Theorem, the sampling distribution of the sample mean $\bar{Y}$ is approximately normal, even though the population distribution is skewed. The sampling distribution has the same mean as the population distribution, namely, $\mu = \$250$. Its standard error is

$$\sigma_{\bar{Y}} = \frac{\sigma}{\sqrt{n}} = \frac{\$80}{\sqrt{100}} = \$8.0$$

Thus, the probability is about .68 that the sample mean falls within $8 of $\mu = \$250$ (i.e., within one standard error), the probability is about .95 that $\bar{Y}$ falls within $16 (two standard errors) of μ, and it is almost certain that $\bar{Y}$ falls within $24 (three standard errors) of μ. The researcher's sample mean should not fall far from the population mean. □

This last example is unrealistic, since it used the value of the population mean μ, which is typically unknown. However, the sampling distribution of $\bar{Y}$ provides the probability that the sample mean falls within a certain distance of the population mean

μ, even when μ is unknown. We illustrate for the planned study of income of California migrant workers, using a random sample of size 100. We shall calculate the probability that the sample mean weekly income $\bar{Y}$ falls within a certain distance, say, \$10, of the true mean income μ for all such workers.

Now, the sampling distribution of $\bar{Y}$ is approximately normal in shape and is centered about μ, with standard error (when $n = 100$) of $\sigma_{\bar{Y}} = \$8.0$. Hence, the probability that $\bar{Y}$ falls within \$10 of μ is the probability that a normally distributed variable falls within $10/8 = 1.25$ standard deviations of its mean. That is, the number of standard errors that $\mu + 10$ (or $\mu - 10$) falls from μ is

$$z = \frac{(\mu + 10) - \mu}{8} = \frac{10}{8} = 1.25$$

as shown in Figure 4.16.

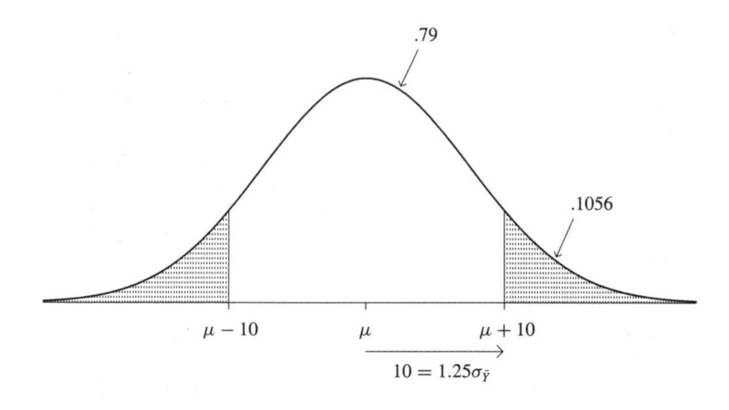

Figure 4.16 Sampling Distribution of $\bar{Y}$ for Unknown μ and Standard Error $\sigma_{\bar{Y}} = 8$

From the normal curve table (Table A), the probability that $\bar{Y}$ falls *more than* 1.25 standard errors from μ (in either direction) is $2(.1056) = .21$. Thus, the probability that $\bar{Y}$ falls no more than \$10 from μ equals $1 - .21 = .79$.

This example is still unrealistic, because it assumes knowledge of the population standard deviation σ. In practice, one might substitute a guess for this value based on the standard deviation from a previous study. The next chapter shows that, to conduct inference, one can substitute the sample standard deviation s for σ.

The sampling distribution is fundamental to statistical inference and the methodology presented in the rest of this text. Because of this, the next section discusses in greater detail the distinction between it and the other types of distributions presented in the past two chapters.

4.5 Review: Population, Sample, and Sampling Distributions

At this point, we review three quite distinct types of distributions discussed in the past two chapters —the ***population*** distribution, the distribution of the ***sample***, and the ***sampling*** distribution. You need to understand the distinctions among them before proceeding to the next chapter. In particular, the logic of the sampling distribution is fundamental to statistical inference, so you need to understand the difference between the distribution of the sample data and the sampling distribution of a statistic such as the sample mean. The terms *sample distribution* and *sampling distribution* seem similar, but the distributions look completely different and serve different purposes.

Here is a capsule description of the three types of distribution:

- The ***population distribution***: This is the distribution from which we select the sample. The distribution is usually unknown and we make inferences about its characteristics, such as the parameters μ and σ that describe its central tendency and variation. Denote the population size by N.
- The ***sample distribution***: This is the distribution of data that we actually observe, that is, the sample observations $Y_1, Y_2, \ldots, Y_n$. The sample distribution may be graphically displayed as a histogram of the data, or numerically described by statistics such as the sample mean $\bar{Y}$ and the sample standard deviation s. The larger the sample size n, the closer the sample distribution resembles the population distribution, and the closer the sample statistics such as $\bar{Y}$ fall to the population parameters, such as μ.
- The ***sampling distribution*** of a statistic: This is the probability distribution of a sample statistic, such as $\bar{Y}$. A sampling distribution describes the variability that occurs in the value of the statistic among samples of a certain size. This distribution determines the probability that the statistic falls within a certain distance of the population parameter.

Example 4.9 Three Distributions for Polling Example

We consider, once again, the variable Y = preference for senatorial candidate in Massachusetts, with $Y = 1$ for Republican and $Y = 0$ for Democrat. Of the 4 million adult residents of Massachusetts, suppose half favor each candidate. Then, as we saw in Example 4.7, the probability distribution for Y has probability 1/2 at $Y = 0$ and probability 1/2 at $Y = 1$. The mean of this distribution is $\mu = .50$, and the standard deviation also equals $\sigma = .50$.

The population distribution of candidate preference consists of $N = 4$ million values of Y, half of which are 0 and half of which are 1. This distribution is described by the parameters $\mu = .50$ and $\sigma = .50$. Figure 4.13 portrayed this distribution, which is highly discrete and not at all bell-shaped.

In practice, not knowing the population distribution, we select a sample. Consider a random sample of $n = 100$ adult residents of Massachusetts. A histogram of the 100 preference values in the sample provides a sample distribution, described by sample statistics such as $\bar{Y}$ and s. For example, of the 100 subjects sampled, perhaps 54 say they prefer the Democrat (i.e., have $Y = 0$) and 46 prefer the Republican ($Y = 1$).

Figure 4.17 displays the histogram of these sample values as well as the histogram for the population distribution. Like the population distribution, the sample distribution concentrates at the two values $Y = 0$ and $Y = 1$. The sample values are summarized by $\bar{Y} = .460$ and $s = .498$. The larger the sample size, the more this sample distribution tends to resemble the population distribution, since the sample observations are a subset of the population values. If the entire population is sampled, then the two distributions are identical.

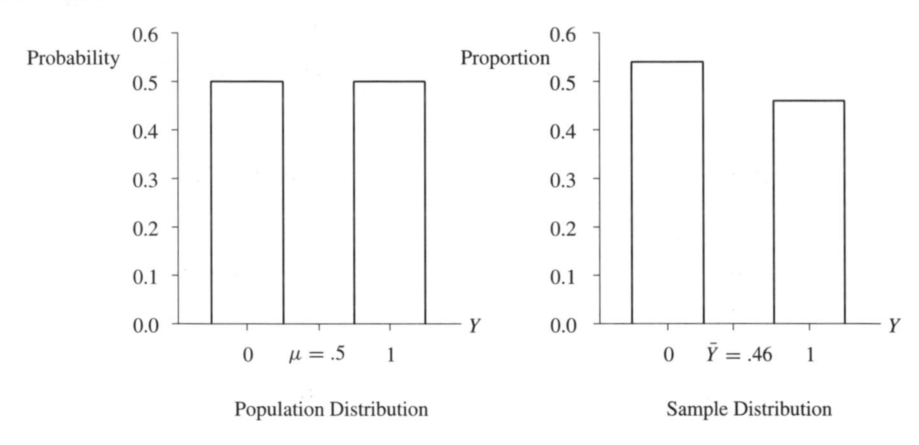

Figure 4.17 The Population ($N = 4$ million) and Sample ($n = 100$) Distributions of Candidate Preference

Before sampling, we know that the sampling distribution of $\bar{Y}$ is approximately normal about $\mu = .50$ with a standard error of

$$\sigma_{\bar{Y}} = \frac{\sigma}{\sqrt{n}} = \frac{.50}{\sqrt{n}}$$

For a random sample of size $n = 100$, the sampling distribution is approximately the normal distribution with mean .50 and standard error $\sigma_{\bar{Y}} = \sigma/\sqrt{n} = .50/\sqrt{100} = .050$. Figure 4.18 portrays this sampling distribution, relative to the population distribution of preference values.

By contrast, the population distribution of preference values is concentrated at the values 0 and 1, and it has a standard deviation of .50. The sampling distribution looks completely different from the population and sample distributions, being much less spread out and, for practical purposes, continuous. The population and sample distributions of candidate preference are not bell-shaped. They are highly discrete, concentrated at the two values 0 and 1. Nevertheless, when $n \geq 30$ the sample proportion can

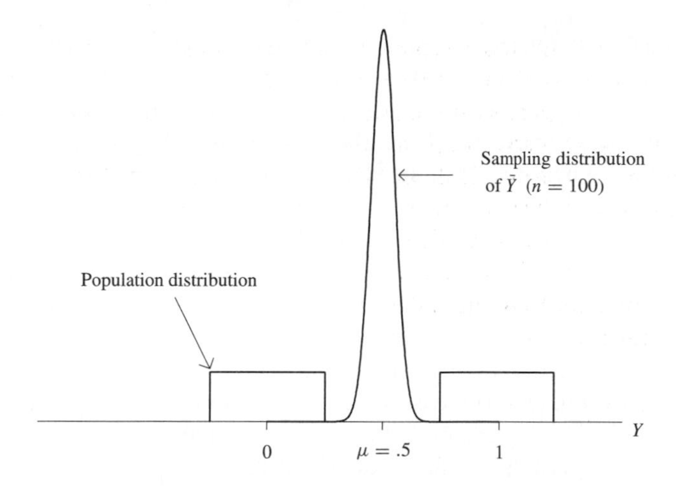

Figure 4.18 The Population Distribution and the Sampling Distribution of $\bar{Y}$ for $n = 100$

take a large number of values between 0 and 1, and its sampling distribution is approximately normal, by the Central Limit Theorem. □

Effect of Sample Size on Sampling Distribution

The sampling distribution is more nearly normal in shape for larger values of n. For sampling only one observation ($n = 1$), $\bar{Y} = Y_1$, and the sampling distribution of $\bar{Y}$ is the same as the probability distribution for one observation on Y. This is simply the population distribution of Y, which need not be the least bit normal. As n increases from 1 to 30, the sampling distribution of $\bar{Y}$ assumes more of a bell shape. For $n \geq 30$, it is usually well approximated by the normal distribution. As the sample size n approaches the population size N, the normal sampling distribution of $\bar{Y}$ gets narrower and eventually converges to a spike at the single number μ. When the entire population is sampled, $\bar{Y} = \mu$ with probability 1 (i.e., the two measures are the same), and the sampling distribution concentrates at the point μ.

Figure 4.18 shows a large difference in dispersion between the population distribution and the sampling distribution of $\bar{Y}$ (for $n = 100$). You should also note the large difference between the sample distribution (Figure 4.17) of the measurements and the sampling distribution (Figure 4.18). The sample distribution is a rough approximation for the population distribution, and it looks more like that distribution as the sample size increases. The sampling distribution, on the other hand, has a bell-shaped appearance and gets narrower as the sample size increases.

In particular, the sampling distribution can be theoretically stated without observing the data and is much less disperse. As shown in Figure 4.17, the sample values of Y can only be 0 or 1. On the other hand, sample mean values (which are sample proportions) can fall anywhere between 0 and 1. According to the sampling distribution of $\bar{Y}$ for $n = 100$, it is practically impossible that a random sample of that size has a sample mean anywhere near 0 or 1; nearly all the probability falls between .35 and .65.

In fact, when $n = 100$, the standard error is $\sigma_{\bar{Y}} = .050$ and a value of $\bar{Y} = 1$ has a z-score of

$$z = \frac{1 - .50}{.05} = 10.0$$

So, the potential $\bar{Y}$-value of 1.0 is 10 standard errors above the mean of the sampling distribution and is therefore an extremely unlikely event! Although any particular individual sampled has preference value of 0 or 1, the sample mean $\bar{Y}$ calculated from a random sample of 100 measurements will almost certainly fall between about .35 and .65. That is, $\bar{Y}$ will very probably occur within about three standard errors ($3 \times .05$) of the mean (.50) of the sampling distribution.

Sampling Distributions and Statistical Inference

Probability statements about $\bar{Y}$ are based on the Central Limit Theorem. If σ is known, they can be stated *before* observing the actual sample measurements. In practice, of course, we rarely know the values of parameters such as μ and σ. Nevertheless, the next two chapters will show that inferences about μ rely on this theorem.

The result about sample means having approximately normal sampling distributions is important not only for the reasons discussed in this section but also because similar results hold for most other statistics. For instance, most sample statistics used to estimate population parameters have approximately normal sampling distributions, for large random samples. The remainder of the text states analogous results for the other statistics we study. The primary reason for the key role of the normal distribution is that so many statistics have approximately normal sampling distributions, at least for large random samples.

4.6 Chapter Summary

- The ***probability*** of an outcome is the proportion of times that the outcome would occur, in the long run, in repeated observations.
- A ***probability distribution*** for a variable lists the possible values of the variable together with their probabilities. Probability distributions are described by parameters, such as the mean μ and standard deviation σ. Letting $P(y)$ denote the probability of a possible outcome y for a variable, the mean of the distribution is $\mu = \sum yP(y)$, also called the ***expected value*** of Y.
- The ***normal distribution*** has a graph that is a symmetric bell-shaped curve with form specified by the parameters μ and σ. For any z, the probability falling within z standard deviations of the mean is the same for every normal distribution. The normal is the most important probability distribution for statistical methodology.

- The *z-score* for an observation measures the number of standard deviations that it falls from the mean of its distribution. The z-score is positive when the observation falls above the mean and negative when it falls below the mean. If the variable has a normal distribution, then the z-scores have a *standard normal distribution*—the normal distribution with a mean of 0 and a standard deviation of 1.
- A ***sampling distribution*** is a probability distribution of a sample statistic. It determines probabilities about the possible values of that statistic.
- The sampling distribution of the sample mean $\bar{Y}$ is centered around the true mean μ, with spread described by the ***standard error***, $\sigma_{\bar{Y}} = \sigma/\sqrt{n}$. The spread of the sampling distribution decreases as the sample size increases, so the sample mean tends to be closer to the population mean for larger samples.
- The ***Central Limit Theorem*** states that for large random samples, the sampling distribution of the sample mean is approximately normal. This is true even if the population distribution is far from normal, such as highly skewed or highly discrete. This result is the main reason for the importance of the normal distribution, because similar results hold for other statistics; most statistics that estimate population parameters have large-sample normal distributions. The result applies even to qualitative data, since the sample proportion is a special case of the sample mean. The next two chapters show how this theorem is the basis of methods of large-sample statistical inference.

PROBLEMS

Practicing the Basics

1. Let Y = the number of people you have known personally that have committed suicide within the past 12 months. According to recent General Social Surveys, the probabilities for the potential values of Y are approximately $P(0) = .90$, $P(1) = .08$, $P(2) = .02$.
   a) Construct a probability distribution and its histogram for Y.
   b) Find the mean of the distribution of Y.
2. Let Y = number of people you have known personally who came down with AIDS. Suppose Table 3.9 refers to the population of interest.
   a) Construct the probability distribution for Y.
   b) Find the mean of the distribution of Y.
3. Five students, Ann, Betty, Clint, Douglas, and Edward, are rated equally qualified for admission to law school, ahead of other applicants. However, all but two positions have been filled for the entering class. Since the admissions committee can admit only two more students, it decides to randomly select two of these five candidates. For this strategy, let Y = number of females admitted. Using the first letter of the name to denote a student, the different combinations that could be admitted are (A, B), (A, C), (A, D), (A, E), (B, C), (B, D), (B, E), (C, D), (C, E), and (D, E).
   a) Construct the probability distribution for Y.
   b) Find the expected value of Y.

4. Refer to the previous exercise. Construct the probability distribution of X = number of females admitted, when three of the five students are randomly selected for admission. Find the mean of the distribution.

5. Let Y be the outcome of selecting a single digit from a random number table.
a) List the possible values for Y.
b) Construct the probability distribution for Y. (This type of distribution is called a *uniform* distribution, because of the uniform spread of probabilities across the possible outcomes.)
c) Find the mean of this distribution.
d) The standard deviation σ of this distribution is one of the following: 0.7, 2.9, 5.0, 9.0. Pick the correct value, and justify.

6. In a statewide lottery, one can buy a ticket for $1. With probability .0000001, one wins a million dollars ($1,000,000), and with probability .9999999 one wins nothing ($0).
a) Let Y denote the winnings from buying one ticket. Construct the probability distribution for Y. Show that the mean of the distribution equals .10, corresponding to an expected return of 10 cents for the dollar paid.
b) The profit X from buying a single $1 ticket equals the winnings Y from playing the lottery minus the dollar paid for the ticket; that is, $X = Y - 1$. Construct the probability distribution of X, and find its mean. Interpret.

7. For a normal distribution, find the probability that an observation is
a) At least one standard deviation above the mean.
b) At least one standard deviation below the mean.
c) At least .67 standard deviations above the mean.
d) At least 2.33 standard deviations above the mean.

8. For a normally distributed variable, verify that the probability between
a) $\mu - \sigma$ and $\mu + \sigma$ equals .68.
b) $\mu - 1.96\sigma$ and $\mu + 1.96\sigma$ equals .95.
c) $\mu - 3\sigma$ and $\mu + 3\sigma$ equals .997.
d) $\mu - .67\sigma$ and $\mu + .67\sigma$ equals .50.

9. Find the z-value for which the probability that a normal variable exceeds $\mu + z\sigma$ equals
a) .01 **b)** .025 **c)** .05 **d)** .10 **e)** .25 **f)** .50

10. Find the z-value such that the interval from $\mu - z\sigma$ to $\mu + z\sigma$ contains **a)** 50%, **b)** 90%, **c)** 95%, **d)** 98%, **e)** 99% of the probability for a normal distribution.

11. Find the z-values corresponding to the **a)** 90th, **b)** 95th, **c)** 98th, and **d)** 99th percentiles of a normal distribution.

12. Show that if z is the number such that the interval from $\mu - z\sigma$ to $\mu + z\sigma$ contains 90% of a normal distribution, then $\mu + z\sigma$ equals the 95th percentile.

13. If z is the positive number such that the interval from $\mu - z\sigma$ to $\mu + z\sigma$ contains 50% of a normal distribution, then which percentile is **a)** $\mu + z\sigma$, **b)** $\mu - z\sigma$? Find this value of z. Using this result, explain how to find the upper quartile, lower quartile, and interquartile range of a normal distribution.

14. An observation is .50 standard deviations below the mean on a normally distributed variable. What proportion of the data fall below that observation? Above it?

15. What proportion of a normal distribution falls in the following ranges?
a) Above a z-score of 2.10.
b) Below a z-score of -2.10.

c) Above a z-score of -2.10.
d) Between z-scores of -2.10 and 2.10.

16. Find the z-score for the number that is less than only 1% of the values of a normal distribution.

17. Find the z-scores for the lower and upper quartiles of a normal distribution.

18. According to *Current Population Reports*, self-employed individuals in the United States in 1990 worked an average of 44.6 hours per week, with a standard deviation of 14.5. Assuming this variable is approximately normally distributed, find the proportion who averaged more than 40 hours per week.

19. The Mental Development Index (MDI) of the Bayley Scales of Infant Development is a standardized measure used in longitudinal follow-up of high-risk infants. It has approximately a normal distribution with a mean of 100 and a standard deviation of 16.
a) What proportion of children have a MDI of at least 120?
b) What proportion of children have a MDI of at least 80?
c) Find the MDI score that is the 90th percentile.
d) Find the MDI score such that 10% of the population have MDI scores below that value.
e) Find and interpret the lower quartile, median, and upper quartile of MDI.
f) Find the interquartile range of MDI scores, and, using the definition of outlier in terms of the IQR, find the ranges of MDI scores that would be considered outliers.

20. For 5459 pregnant women using Aarhus University Hospital in Denmark in a two-year period who reported information on length of gestation until birth, the mean was 281.9 days, with standard deviation 11.4 days. A baby is classified as premature if the gestation time is 258 days or less (*British Medical Journal*, Vol. 307, July 24, 1993, p. 234).
a) If gestation times are normally distributed, what proportion of babies would be born prematurely?
b) The actual proportion born prematurely during this period was .036. Based on this information, how would you expect the distribution of gestation time to differ from normal?

21. Suppose that the weekly use of gasoline for motor travel by adults in North America is approximately normally distributed, with a mean of 16 gallons and a standard deviation of 5 gallons.
a) What proportion of adults use more than 20 gallons per week?
b) Find and interpret the lower and upper quartiles and interquartile range of gasoline use.
c) Assuming that the standard deviation and the normal form would remain constant, to what level must the mean reduce so that only 5% use more than 20 gallons per week?
d) If the distribution of gasoline use is not actually normal, how would you expect it to deviate from normal?

22. In 1992, the murder rates (per 100,000 residents) for 74 major U.S. cities had a mean of 20.8 and a standard deviation of 14.6 (*Statistical Abstract of the United States, 1994*).
a) Could the distribution of murder rates be well approximated by a normal distribution? Why or why not?
b) One of these cities, Washington, D.C., had a murder rate of 75.2. Find its z-score. If the distribution were roughly normally distributed, would this be an unusually high value? Why?
c) What is the murder rate of a city that has a z-score of 0.0?

23. Suppose that monthly rental rates for one-bedroom unfurnished apartments in Ann Arbor, Michigan, are approximately normal in distribution, with a mean of \$750 and a standard deviation of \$150.

a) What proportion of the rental rates are at least $1000 per month?
b) What proportion of rates are less than $600 per month?
c) What proportion are between $500 and $1000 per month?

24. Suppose that property taxes on homes in Gainesville, Florida, are approximately normal in distribution, with a mean of $1400 and a standard deviation of $600. The property tax for one particular home is $1700.
a) Find the z-score corresponding to that value.
b) What proportion of the property taxes exceed $1700?
c) In fact, the true distribution is moundshaped but not normal. How would you expect it to deviate from normal?

25. An energy study in Gainesville, Florida, found that in a typical month, household use of electricity has a mean of 800 and a standard deviation of 500 (kilowatt-hours). Each observation is converted to a z-score. Each household with a z-score greater than 1.0 (i.e., electricity use more than one standard deviation above the mean) is sent a notice suggesting a reduction in electrical use, for conservation purposes.
a) If the distribution were normal, what percentage of the households would receive this notice?
b) Do you think the distribution is truly normal? Why or why not?

26. Refer to Problem 4.3. For randomly selecting two of the five students, construct the sampling distribution of the *proportion* of the students selected who are female. What is the probability that the proportion equals 0?

27. Construct the sampling distribution of the proportion of heads, for flipping a balanced coin
a) Once.
b) Twice. (*Hint*: The possible samples are (H, H), (H, T), (T, H), (T, T).)
c) Three times. (*Hint*: There are 8 possible samples.)
d) Four times. (*Hint*: There are 16 possible samples.)
e) Describe how the shape of the sampling distribution seems to be changing as the number of flips increases.

28. The probability distribution associated with the outcome of rolling a balanced die has probability $1/6$ attached to each integer, $\{1, 2, 3, 4, 5, 6\}$. Consider the experiment of rolling a balanced die twice.
a) Enumerate the 36 possible pairs of numbers, taking into account the order of rolling (e.g., (1, 2), representing a 1 followed by a 2, is different from (2, 1)).
b) Treating these 36 pairs as equally likely, construct the sampling distribution for the sample mean of the two numbers rolled.
c) Construct a histogram of the sampling distribution in (b). What is its shape?
d) Calculate the mean of the sampling distribution in (b), and show that it is identical to the mean of the population distribution.

29. According to *Current Population Reports*, the population distribution of number of years of education for self-employed individuals in the United States in 1990 had a mean of 13.6 and a standard deviation of 3.0. Find the mean and standard deviation of the sampling distribution of $\bar{Y}$ for a random sample of **a)** 9 residents, **b)** 36 residents, **c)** 100 residents. Describe the pattern as n increases.

30. Refer to Problem 4.6. The mean and standard deviation of the probability distribution for the lottery winnings Y are $\mu = .10$ and $\sigma = 316.23$. Suppose you play the lottery 1 million times. Let $\bar{Y}$ denote your average winnings.
a) Find the mean and standard deviation of the distribution of $\bar{Y}$.

b) How likely is it that you would "come out ahead," with your average winnings exceeding \$1, the amount you paid to play each time?

31. Suppose that weekly incomes of migrant workers in California have a mean of \$250 and a standard deviation of \$75.
a) If the distribution is normal, find the probability that a migrant worker has income over \$300.
b) For a random sample of nine migrant workers, report the sampling distribution of the sample mean income. Find the probability that the sample mean exceeds \$300.

32. According to recent General Social Surveys, in the United States the distribution of Y = number of sex partners you have had in the past 12 months has a mean of about 1.0 and a standard deviation of about 1.0.
a) Does Y have a normal distribution? Explain.
b) For a random sample of 100 adults, find the sampling distribution of $\bar{Y}$.
c) Refer to **b)**. Report an interval within which the sample mean would almost surely fall.

33. According to *Statistical Abstract of the United States, 1995*, household size has a mean of 2.6 and a standard deviation of 1.5. For a random sample of 225 homes, find the probability that the sample mean household size falls within .1 of the population mean.

34. The distribution of family size in a particular tribal society is skewed to the right, with $\mu = 5.2$ and $\sigma = 3.0$. These values are unknown to an anthropologist, who takes a sample of families in this society in order to estimate mean family size. Let $\bar{Y}$ denote the sample mean family size she obtains, for a random sample of 36 families.
a) Specify the sampling distribution of $\bar{Y}$.
b) Find the probability that her sample mean falls within .5 of the true mean.
c) Suppose she takes a random sample of size 100. Find the probability that the sample mean falls within .5 of the true mean, and compare the answer to that obtained in part **b)**.
d) Refer to part **c)**. If the sample were truly random, would you be surprised if the anthropologist obtained $\bar{Y} = 4.0$? Why? (This could well happen if the sample were not random.)

35. The scores on the Psychomotor Development Index (PDI), a scale of infant development, are approximately normally distributed with mean 100 and standard deviation 15. An infant is selected at random.
a) Find the probability that the infant's PDI score is at least 100.
b) Find the probability that PDI exceeds 103.
c) Find the probability that PDI is between 97 and 103.
d) Find the z-score for a PDI value of 90. Would you be surprised to observe a value of 90?
e) Suppose we convert all the PDI observations to z-scores; that is, for each infant, subtract 100 from the value of PDI and divide by 15. Then, what is the distribution of the z-scores called? What are the mean and standard deviation of these z-scores?

36. Refer to the previous exercise. A study uses a random sample of 225 infants.
a) Describe the sampling distribution of the sample mean PDI.
b) Find the probability that the sample mean exceeds 103.
c) Find the probability that the sample mean falls between 97 and 103.
d) Find the z-score from the sampling distribution corresponding to a sample mean of $\bar{Y} = 90$, when the sample size is 225. Would you be surprised to observe a sample mean PDI of 90? Why?
e) Compare the results of parts (a)–(d) with those in the previous exercise, and interpret the differences.

f) Repeat parts (a)–(d) for a random sample size of $n = 25$ and compare the results to those for $n = 225$.

37. Refer to the previous two exercises.
a) Sketch the population distribution for PDI.
b) Superimpose, on the sketch of part (a), a sketch of the sampling distribution of the sample mean, for a random sample of 225 infants.
c) Superimpose, on the above sketches, a sketch of the sampling distribution of the sample mean for a random sample of 25 infants.

38. We need an estimate of the mean acreage of Canadian farms. We plan to measure acreage for a random sample of 100 farms. Results from an earlier study suggest that 200 acres is a reasonable guess for the standard deviation of farm size. Find the probability that the sample mean acreage falls within 10 acres of the population mean acreage.

39. An executive in a savings and loan association decided to estimate the mean amount of money lent to individuals for financing higher education in the past year. From past experience, she believes that $240 is a reasonable guess for the standard deviation of the distribution of loan amounts. She would like her estimate of the mean to be within $50 of the actual mean. Find the probability that this happens under the following conditions.
a) She takes a random sample of 36 loan records.
b) She takes a random sample of 64 loan records.
c) She takes a random sample of 144 loan records.

40. A study investigating the relationships among voting patterns, political opinions, and age takes a random sample of 100 individuals in a typical precinct.
a) If the standard deviation of the ages of all individuals in the precinct is $\sigma = 15$, find the probability that the mean age of the individuals sampled is within 2 years of the mean age for all individuals in the precinct.
b) Would the probability be larger, or smaller, if $\sigma = 10$?

Concepts and Applications

41. (*Class Exercise*) Refer to Problem 1.7. Using the population defined by your class or using the WWW data set, the instructor will select a variable, such as weekly time watching television.
a) Construct a histogram or stem and leaf plot of the population distribution of the variable for the class.
b) Using a random number table, each student should select five students at random and compute the sample mean response for those students. (Each student should use different random numbers.) Plot a histogram of the sample means obtained by all the students. How do the spread and shape compare to the histogram in (a)? What does this illustrate?

42. (*Class Exercise*) Table 4.4 provides the ages of all 50 heads of households in a small Nova Scotian fishing village. The distribution of these ages is characterized by $\mu = 47.18$ and $\sigma = 14.74$.
a) Construct a stem and leaf plot of the population distribution of the ages of all heads of households.
b) Using a random number table, each student should select nine random numbers between 01 and 50. Using these numbers, each student should sample nine heads of households and compute their sample mean age. Using the intervals 31.01–34.00, 34.01–37.00,

37.01–40.00, and so forth to 61.01–64.00, construct the empirical sampling distribution of the $\bar{Y}$-values. Compare it to the distribution in part (a).
c) Find the mean of the $\bar{Y}$-values generated in part (b). What value do you expect for this mean in a long run of repeated samples of size 9?
d) Find the standard deviation of the $\bar{Y}$-values generated in part (b). What value do you expect for this standard deviation in a long run of repeated samples of size 9?

TABLE 4.4

Name	Age	Name	Age	Name	Age	Name	Age
Alexander	50	Griffith	66	McTell	49	Staines	33
Bell	45	Grosvenor	51	MacLeod	30	Stewart	36
Bell	23	Ian	57	McNeil	28	Stewart	25
Bok	28	Jansch	40	McNeil	31	Thames	29
Clancy	67	Kagan	36	McNeil	45	Thomas	57
Cochran	62	Lavin	38	McNeil	43	Todd	39
Fairchild	41	Lunny	81	Mitchell	43	Trickett	50
Finney	68	MacColl	27	Muir	54	Trickett	64
Fisher	37	McCusker	37	Oban	62	Tyson	76
Fraser	60	McCusker	56	Reid	67	Watson	63
Fricker	41	McDonald	71	Renbourn	48	Young	29
Gaughan	70	McDonald	39	Rogers	32		
Graham	47	McDonald	46	Rush	42		

43. (*Class Exercise*) For a single toss of a coin, let $Y = 1$ for a head and $Y = 0$ for a tail.
a) Assuming the coin is balanced, construct the probability distribution for Y, and calculate its mean.
b) The coin is flipped ten times, yielding six heads and four tails. Construct the sample relative frequency distribution.
c) Each student in the class should flip a coin ten times and calculate the proportion of heads in the sample. Summarize the empirical sampling distribution by constructing a histogram or stem and leaf plot of the proportions obtained. Describe the shape and spread of the sampling distribution compared to the distributions in (a) and (b). Compute the mean and standard deviation of the sample proportion values.
d) If we performed the experiment in (c) an indefinitely large number of times, what values would we get for the (i) mean and (ii) standard deviation of the sample proportion values?

44. (*Class Exercise*) The previous exercise assumes the coin is balanced. Suppose instead that the probability of a head equals .75 and the probability of a tail equals .25.
a) Construct the probability distribution of Y, the outcome of a single flip, and calculate its mean.
b) Each student in the class should simulate the results of flipping this coin ten times by selecting ten two-digit numbers from the random number table; a number between 00 and 74 represents a head and a number between 75 and 99 represents a tail. Again, construct a plot of the sample proportions, which is an empirical approximation for the true sampling distribution. Compute the mean and standard deviation of the sample proportion values. Report the theoretical value for the mean of the sampling distribution. (From Problem 4.55(c), the theoretical standard deviation is .137.)

45. (*Class Exercise*) Each student should bring ten coins to class. The observation for each coin is its age, the difference between the current year and the year on the coin.
a) Using all the students' observations, the class should construct a stem and leaf plot of the ages. What is its shape?
b) Now each student should compute the mean for that student's ten coins, and the class should construct a plot of the means. What type of distribution is this, and how does it compare to the one in (a)?
c) What concepts does this exercise illustrate?
(*Note*: For a wide variety of activities that illustrate basic concepts of statistics, see the highly innovative text by Scheaffer et al., 1996.)

46. **a)** Which distribution does the sample distribution tend to resemble more closely—the sampling distribution or the population distribution? Explain.
b) Explain carefully the difference between a *distribution of sample measurements* and the *sampling distribution of* $\bar{Y}$. Illustrate your answer for a variable Y that can only take values of 0 and 1.

Select the correct response(s) in the following multiple-choice questions.

47. The standard error of a statistic describes
a) The standard deviation of the sampling distribution of that statistic.
b) The standard deviation of the sample measurements.
c) How close that statistic is likely to fall to the parameter that it estimates.
d) The variability in the values of the statistic for repeated random samples of size n.
e) The error that occurs due to nonresponse and measurement errors.

48. The Central Limit Theorem implies that
a) All variables have approximately bell-shaped sample distributions if a random sample contains at least 30 observations.
b) Population distributions are normal whenever the population size is large.
c) For large random samples, the sampling distribution of $\bar{Y}$ is approximately normal, regardless of the shape of the population distribution.
d) The sampling distribution looks more like the population distribution as the sample size increases.
e) All of the above.

49. The numbers in the body of Table A provide tail probabilities
a) For the normal distribution.
b) For any symmetric distribution.
c) For any distribution.
d) Only for the standard normal distribution.

50. True or False: As the sample size increases, the standard error of the sampling distribution of $\bar{Y}$ increases.

51. * Lake Wobegon Junior College admits students only if they score above 400 on a standardized achievement test. Applicants from group A have a mean of 500 and a standard deviation of 100 on this test, and applicants from group B have a mean of 450 and a standard deviation of 100. Both distributions are approximately normal, and both groups have the same size.
a) Find the proportion not admitted for each group.
b) Of the students who are not admitted, what proportion are from group B?
c) A state legislator proposes that the college lower the cutoff point for admission to 300,

thinking that the proportion of the students who are not admitted who are from group B would decrease. If this policy is implemented, determine the effect on the answer to (b), and comment.

52. For a normal distribution, show that
a) The upper quartile equals $\mu + .67\sigma$.
b) The interquartile range, IQR, is related to σ by IQR $= 1.35\sigma$.
c) For sample data that are approximately normal, an outlier (according to its definition for box plots) is an observation falling more than 2.7 standard deviations below or above the mean, and this happens for only 0.7% of the data.

53. Sunshine City was designed to attract retired people, and its current population of 50,000 residents has a mean age of 60 years and a standard deviation of 16 years. The distribution of ages is skewed to the left, reflecting the predominance of older individuals. A random sample of 100 residents of Sunshine City has $\bar{Y} = 58.3$ and $s = 15.0$.
a) Describe the population distribution.
b) Describe the sample distribution. What shape does it probably have?
c) Describe the sampling distribution of $\bar{Y}$ for $n = 100$. What shape does it have?
d) Describe the sampling distribution of $\bar{Y}$ for a sample of size $n = 1$.
e) Describe the sampling distribution of $\bar{Y}$ for a sample of size 50,000.
f) Suppose random samples of size $n = 100$ were repeatedly selected, so that each resident has an equal chance of being chosen for any one observation in each sample of size 100. Describe the likely appearance of the histogram of $\bar{Y}$ values.
g) Would it be highly unusual to observe a subject of age under 40 in Sunshine City? Would it be highly unusual to observe a sample mean under 40, for a sample size of 100? Explain.

54. Refer to Example 4.7. Compute the standard error when $n = 10$, 100, 1000, and 10,000. In each case, use the Empirical Rule to predict an interval within which the sample proportion is almost certain to fall. Notice that the interval shrinks in width as the sample size increases. This is a consequence of the ***law of large numbers***, which states that the sample proportion tends to eventually converge to the population proportion as the sample size increases indefinitely.

55. * The variance of a discrete probability distribution is

$$\sigma^2 = \sum (y - \mu)^2 \, P(y)$$

the average squared deviation. This also equals

$$\sigma^2 = \left[\sum y^2 P(y)\right] - \mu^2$$

a) Suppose $Y = 1$ with probability .5 and $Y = 0$ with probability .5, such as in Example 4.7 on preference for Democratic and Republican candidates. Using each formula, show that $\sigma^2 = .25$, and hence the standard deviation $\sigma = .50$.
b) Suppose $Y = 1$ with probability π and $Y = 0$ with probability $1 - \pi$, where π denotes a fixed number between 0 and 1. Show that $\mu = \pi$ and that $\sigma^2 = \pi(1 - \pi)$.
c) Refer to (b). Show that the standard error of a sample proportion for a random sample of size n equals $\sqrt{\pi(1 - \pi)/n}$. (*Hint*: Recall that a proportion equals a mean for a variable taking values 0 and 1.)

56. * Refer to the formula for the variance of a probability distribution in the previous exercise. Find the standard deviation for the distribution in Problem 4.1. Can one use the Empirical Rule to interpret this standard deviation? Explain.

57. * The curve for a normal distribution with mean μ and standard deviation σ has mathematical formula

$$f(y) = \frac{1}{\sqrt{2\pi}\sigma} e^{-(y-\mu)^2/(2\sigma^2)}$$

(The integral of this function with respect to y between $\mu + z\sigma$ and ∞ equals the tail probability tabulated in Table A). Show that this curve is symmetric; that is, for any constant c, the curve has the same value for $y = \mu + c$ as for $y = \mu - c$.

58. * The standard error formula $\sigma_{\bar{Y}} = \sigma/\sqrt{n}$ actually treats the population size N as *infinitely* large relative to the sample size n. The formula for $\sigma_{\bar{Y}}$ for a *finite* population size N is

$$\sigma_{\bar{Y}} = \sqrt{\frac{N-n}{N-1}}\left(\frac{\sigma}{\sqrt{n}}\right)$$

The term $\sqrt{(N-n)/(N-1)}$ is called the ***finite population correction***.

a) When $n = 300$ students are selected from a college student body of size $N = 30,000$, show that $\sigma_{\bar{Y}} = .995\sigma/\sqrt{n}$. (In practice, n is usually small relative to N, so the correction has little influence.)

b) If $n = N$ (i.e., we sample the entire population), show that $\sigma_{\bar{Y}} = 0$. In other words, no sampling error occurs, since $\bar{Y} = \mu$.

Bibliography

Goldberg, S. (1982). *Probability in Social Science*. Boston: Birkhauser.

Moore, D., and McCabe, G. (1993). *Introduction to the Practice of Statistics*. New York: Freeman.

Olkin, I., Gleser, L., and Derman, C. (1994). *Probability Models and Applications*, 2nd ed. New York: Macmillan.

Scheaffer, R. L. (1995). *Introduction to Probability and its Applications*, 2nd ed. Belmont, CA: Wadsworth.

Scheaffer, R. L., Gnanadesikan, M., Watkins, A., and Witmer, J. (1996). *Activity–Based Statistics*. New York: Springer.

Chapter 5

Statistical Inference: Estimation

This chapter shows how to use sample data to estimate population parameters. With quantitative variables, studies usually estimate the population mean. A study dealing with health care issues in Texas, for example, might select a sample of residents to estimate such population parameters as the mean amount of money spent on health care during the past year, the mean number of visits to a physician, and the mean number of days of work missed due to illness. With qualitative variables, studies usually estimate the population proportions of measurements in the various categories. For example, the health care study might estimate the proportions of people who (have, do not have) medical insurance, the proportions who are (satisfied, not satisfied) with their access to health care, and the proportions who have (experienced, not experienced) an illness requiring hospitalization in the past year.

Statistical inference uses sample data to form two types of estimators of parameters. A ***point estimate*** consists of a single number, calculated from the data, that is the best single guess for the parameter. For example, in a recent General Social Survey, 1359 subjects were asked "Do you believe in Hell?" The point estimate for the proportion of all Americans who would respond "yes" equals .63. An ***interval estimate*** consists of a range of numbers around the point estimate, within which the parameter is believed to fall. For example, for the data just mentioned, an interval estimate predicts that the population proportion responding "yes" falls between .59 and .67; that is, it predicts that the point estimate of .63 falls within .04 of the true value. Thus, an interval estimate helps us gauge the probable accuracy of a sample point estimate.

Section 5.1 introduces point estimation. Sections 5.2 and 5.3 present interval estimates for population means and proportions. Section 5.4 shows how to determine the sample size needed to achieve the desired accuracy, and Section 5.5 discusses interval estimation of medians.

5.1 Point Estimation

The process of predicting a parameter value reduces the sample data to a single number that is the best guess about that value. The statistic that provides the prediction is called a *point estimator* of the parameter.

Point Estimator

A ***point estimator*** of a parameter is a sample statistic that predicts the value of that parameter.

For instance, to estimate a population mean, which we have denoted by μ, an obvious point estimator is the sample mean $\bar{Y}$. A good point estimator of a parameter is one with a sampling distribution that (1) is centered around the parameter and (2) has as small a standard error as possible. An estimator with property (1) is said to be *unbiased*, and an estimator with property (2) is said to be *efficient*. For simplicity, it is common to use the term "estimate" in place of point estimator.

Unbiased and Efficient Point Estimators

A point estimator is ***unbiased*** if its sampling distribution centers around the parameter, in the sense that the parameter is the mean of the distribution. For any particular sample, the point estimator may underestimate the parameter or overestimate it. If that point estimator were used repeatedly in different situations with different samples, however, it would not tend to overestimate or underestimate the parameter systematically; the overestimates would tend to counterbalance the underestimates.

From Section 4.4, for random sampling the mean of the sampling distribution of $\bar{Y}$ equals μ. Thus, $\bar{Y}$ is an unbiased estimator of the population mean μ, as Figure 5.1 illustrates. Sometimes $\bar{Y}$ falls below μ, sometimes it falls above, but it does well in this average sense.

A ***biased*** estimator, on the other hand, tends to either underestimate or overestimate the parameter, on the average. Figure 5.1 also portrays the sampling distribution of an estimator that is biased, tending on the average to underestimate μ. For instance, the sample median is a biased estimate of the population mean when the population distribution is skewed to the right. The population median is less than the population mean in that case, and the sample median also tends to be less than the population mean, on the average.

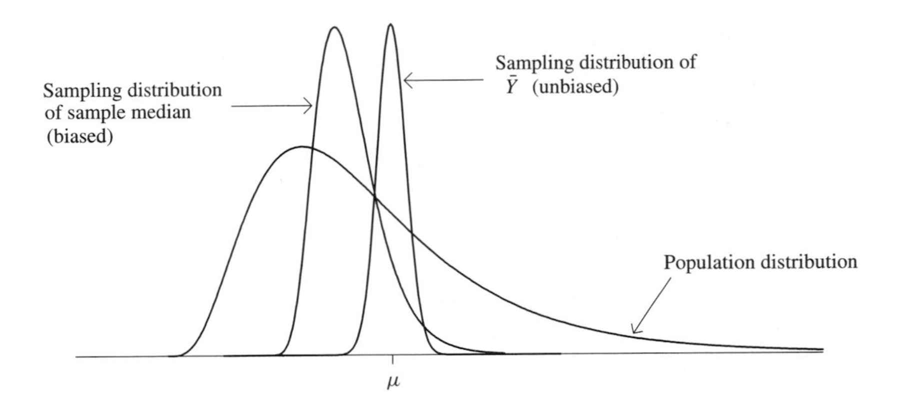

Figure 5.1 Sampling Distributions of Two Point Estimators of the Population Mean, for a Skewed Population Distribution

The concept of bias refers to the estimator's behavior in repeated sampling, not in one particular sample. Estimators are evaluated in terms of their theoretical performance in a long run of repeated samples. In practice, however, we select a single sample of fixed size to estimate a particular parameter. Statistical methods use estimators that are unbiased or for which the bias is negligible and disappears as the sample size increases.

A second preferable property for an estimator is a small sampling error compared with other estimators. An estimator whose standard error is smaller than those of other potential estimators is said to be ***efficient***. An efficient estimator is desirable because, on the average, it falls closer than other estimators to the parameter.

For instance, suppose a population distribution is normal, and we want to estimate its center, which is its mean, median, and mode. We could use the sample mean as the estimate. Or, we could use the sample median. Section 5.5 shows, though, that in sampling from a normal distribution, the sample median has a standard error that is 25% larger than the standard error of the sample mean. Thus, the sample mean tends to be closer than the sample median to the population center. In this case, the sample mean is an efficient estimator, but the sample median is inefficient.

In summary, a good estimator of a parameter is *unbiased*, or nearly so, and *efficient*. The point estimates presented in this text possess these properties.

Point Estimators of the Mean and Standard Deviation

The sample mean $\bar{Y} = \sum Y_i/n$ is the obvious point estimator of a population mean μ. In fact, $\bar{Y}$ is unbiased, and it is relatively efficient for most population distributions. It is the point estimator used in this text.

The symbol " ^ " over a parameter symbol represents an estimate of that parameter. The symbol " ^ " is called a *caret*, and is usually read as "hat." For example, $\hat{\mu}$ is read as "mu-hat." Thus, $\hat{\mu}$ denotes the estimate $\bar{Y}$ of the population mean μ, and $\hat{\sigma}$ denotes an estimate of the population standard deviation σ. The sample standard deviation s is the most popular point estimate of σ. That is,

$$\hat{\sigma} = s = \sqrt{\frac{\Sigma(Y_i - \bar{Y})^2}{n - 1}}$$

For qualitative data, the population proportion of observations falling in some category is relevant. The usual point estimator is the sample proportion. Similarly, the point estimator of a population percentage is the sample percentage. It is common, though not necessary, to use the sample analog of a population parameter as its point estimator.

Point estimates are the most common form of inference reported by the mass media. For example, a survey in May 1996 reported that 55% of the American public approved of President Clinton's performance in office. This is a point estimate rather than parameter, since it is based on sample data rather than the entire population.

Maximum Likelihood Estimation *

As mentioned earlier, compared to other mathematical sciences, statistical science is young. Most methods described in this book were developed in the twentieth century. For instance, interval estimation methods were introduced in a series of articles beginning in 1928 by Jerzy Neyman (1894–1981) and Egon Pearson (1895-1980).

The most important contributions to modern statistical science were made by the British statistician and geneticist R. A. Fisher (1890–1962). While working at an agricultural research station north of London, he developed much of the theory of point estimation as well as methodology for the design of experiments and data analysis.

For point estimation, Fisher advocated using the ***maximum likelihood estimate***. This estimate is the value of the parameter that is most consistent with the observed data, in the following sense: if the parameter equaled that number (i.e., the value of the estimate), the observed data would have had greater chance of occurring than if the parameter equaled any other number. For instance, a recent survey of about 1000 adult Americans reported that the maximum likelihood estimate of the population proportion who believe in astrology is .37. Then, the observed sample would have been more likely to occur if the population proportion equals .37 than if it equaled any other possible value.

For many population distributions, such as the normal, the maximum likelihood estimate of a population mean is the sample mean. Fisher showed that, for large samples, maximum likelihood estimators have three desirable properties:

- They are efficient. One cannot find other estimators that have smaller standard errors and tend to fall closer to the parameter.
- They have little, if any, bias, with the bias diminishing as the sample size increases.
- They have approximately normal sampling distributions.

The point estimates presented in this book are, under certain population assumptions, maximum likelihood estimates or essentially identical to such estimates for moderate to large samples. For small samples, however, not all statisticians agree that maximum likelihood estimates are the best, particularly for problems with several parameters. Some interesting research in the past quarter century has shown conditions under which biased estimators may be better than the usual estimators such as sample means and proportions, when there are several means or proportions to estimate. See, for instance, Efron and Morris (1977).

5.2 Confidence Interval for a Mean

To be truly informative, an inference about a parameter should provide not only a point estimate but should also indicate the probable accuracy of the estimate. That is, it should describe how close that estimate is likely to fall to the true parameter value. If a study with 100 college seniors reports that the estimated mean number of sex partners that college seniors have had equals 5, we'd like to know whether that estimate of 5 is likely to be within 1 of the actual population mean, within 2, within 4, or whatever.

The accuracy of a point estimator depends on characteristics of the sampling distribution of that estimator. For example, the sampling distribution determines the probability that the estimator falls within a certain distance of the parameter. If the sampling distribution is approximately normal, then with high probability (about .95), the estimator falls within two standard errors of the parameter, and almost certainly it falls within three standard errors. The estimated standard error helps us determine the likely accuracy of the estimator. The smaller the standard error, the more accurate the estimator tends to be.

Confidence Intervals

The information about the likely accuracy of a point estimator determines the width of an ***interval estimate*** of the parameter. This consists of a range of numbers that contains the parameter with some fixed probability close to 1. Interval estimates are called ***confidence intervals***.

Confidence Interval

A ***confidence interval*** for a parameter is a range of numbers within which the parameter is believed to fall. The probability that the confidence interval contains the parameter is called the ***confidence coefficient***. This is a chosen number close to 1, such as .95 or .99.

A confidence interval is based on a point estimator and the spread of the sampling distribution of that estimator. When the sampling distribution is approximately normal, we construct a confidence interval by adding to and subtracting from the point estimate some multiple (a z-score) of its standard error. This section shows how to do this for a mean. The confidence interval has the Central Limit Theorem as its foundation, so it is appropriate whenever the sample size is large enough to apply that result, say, $n \geq 30$. (The reason for 30 as the cutoff point will be more apparent in Section 6.5, which presents an analogous method for smaller samples.)

Large-Sample Confidence Interval for a Mean

The Central Limit Theorem states that, for large random samples, the sampling distribution of $\bar{Y}$ is approximately normal. The mean of the sampling distribution equals the population mean, μ, and the standard error equals

$$\sigma_{\bar{Y}} = \frac{\sigma}{\sqrt{n}}$$

Now, 95% of a normal distribution falls within two standard deviations of the mean, or, more precisely, 1.96 standard deviations. So, with probability .95, $\bar{Y}$ falls within $1.96\sigma_{\bar{Y}}$ units of the parameter μ, that is, between $\mu - 1.96\sigma_{\bar{Y}}$ and $\mu + 1.96\sigma_{\bar{Y}}$, as Figure 5.2 shows.

Now, once the sample is selected, if $\bar{Y}$ does fall within $1.96\sigma_{\bar{Y}}$ units of μ, then the interval from $\bar{Y} - 1.96\sigma_{\bar{Y}}$ to $\bar{Y} + 1.96\sigma_{\bar{Y}}$ contains μ. See line 1 of Figure 5.2. In other words, with probability .95 a $\bar{Y}$ value occurs such that the interval $\bar{Y} \pm 1.96\sigma_{\bar{Y}}$ contains the population mean μ.

On the other hand, the probability is .05 that $\bar{Y}$ does not fall within $1.96\sigma_{\bar{Y}}$ of μ. If that happens, then the interval from $\bar{Y} - 1.96\sigma_{\bar{Y}}$ to $\bar{Y} + 1.96\sigma_{\bar{Y}}$ does *not* contain μ (see

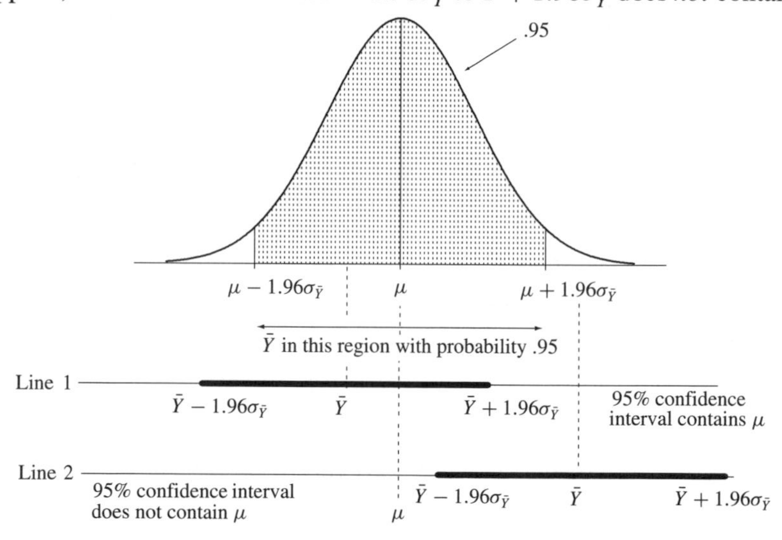

Figure 5.2 Sampling Distribution of $\bar{Y}$ and Possible 95% Confidence Intervals for μ

Figure 5.2, line 2). Thus, the probability is .05 that $\bar{Y}$ is such that $\bar{Y} \pm 1.96\sigma_{\bar{Y}}$ does *not* contain μ.

The interval $\bar{Y} \pm 1.96\sigma_{\bar{Y}}$ is an interval estimate for μ with confidence coefficient .95, called a ***95% confidence interval***. Unfortunately, the value of the standard error $\sigma_{\bar{Y}} = \sigma/\sqrt{n}$ in this formula is unknown, since the population standard deviation σ is an unknown parameter. For $n \geq 30$, a good approximation for $\sigma_{\bar{Y}}$ results from substituting the sample standard deviation s for σ in this formula. Then,

$$\hat{\sigma}_{\bar{Y}} = \frac{s}{\sqrt{n}}$$

estimates the true standard error. One can insert this estimated standard error in the formula for a confidence interval. The error in substituting the point estimate s for σ is small when $n \geq 30$. The resulting 95% confidence interval equals

$$\bar{Y} \pm 1.96\hat{\sigma}_{\bar{Y}}, \quad \text{which is } \bar{Y} \pm 1.96\frac{s}{\sqrt{n}}$$

Example 5.1 Estimating Mean Number of Sex Partners

Recent General Social Surveys have asked respondents how many female partners they have had sex with since their 18th birthday. Over half the respondents answered 0, presumably because the question was asked of both the male and female respondents. In 1994, of those 1055 respondents who responded with a number higher than 0, the distribution was highly skewed to the right with a sample mean of 10.2 and standard deviation of 10.1. Let μ denote the mean for the population represented by this sample.

When $s = 10.1$ and $n = 1055$, the estimated standard error of the sampling distribution of $\bar{Y}$ is

$$\hat{\sigma}_{\bar{Y}} = \frac{s}{\sqrt{n}} = \frac{10.1}{\sqrt{1055}} = .31$$

A 95% confidence interval for μ is

$$\bar{Y} \pm 1.96\hat{\sigma}_{\bar{Y}} = 10.2 \pm 1.96(.31) = 10.2 \pm .6, \quad \text{or } (9.6, 10.8)$$

We can be 95% confident that this interval contains μ, the population mean number of female sex partners. The point estimate of μ is 10.2, and the interval estimate predicts that μ is no smaller than 9.6 and no greater than 10.8.

The survey also asked for the number of male sex partners since the 18th birthday. Of the 1431 subjects responding with a positive number, the mean was 4.8 and the standard deviation was 6.2. You can check that the 95% confidence interval for that population mean equals (4.5, 5.1).

Keep in mind that the error allowed in these intervals refers only to sampling error. Other errors relevant for these parameters include those due to nonresponse (e.g., for the number of female partners, 270 subjects provided no response, 34 responded "don't know," and 23 refused to answer) or measurement error (lying or giving an inaccurate response). Certainly we are suspicious here, since the results are so different for males and females. The inferences may apply to a population that differs somewhat from the one of actual interest. □

Controlling the Confidence Coefficient and Error Probability

The inference just presented had a confidence coefficient of .95. In some applications, a 5% chance of an incorrect prediction is unacceptable. Increasing the chance that the confidence interval contains μ requires a larger confidence coefficient. For instance, one might construct a 99% confidence interval for μ. Now, 99% of a normal distribution occurs within 2.58 standard deviations of the mean, so the probability is .99 that $\bar{Y}$ falls within $2.58\sigma_{\bar{Y}}$ of μ. A 99% confidence interval for μ is $\bar{Y} \pm 2.58\hat{\sigma}_{\bar{Y}}$.

For the data in Example 5.1, the 99% confidence interval for μ is

$$\bar{Y} \pm 2.58\hat{\sigma}_{\bar{Y}} = 10.2 \pm 2.58(.31) = 10.2 \pm .8, \quad \text{or } (9.4, 11.0)$$

Compared to the 95% confidence interval of (9.6, 10.8), this interval estimate is less precise, being wider. This is the sacrifice for greater assurance of a correct inference.

Why do we settle for anything less than 100% confidence? To be absolutely 100% certain of a correct inference, the interval must contain all possible values for μ. A 100% confidence interval for the mean number of female sex partners goes from 0 to infinity. This is not informative, and in practice we settle for a little less than perfection in order to focus more tightly on the true parameter value.

The general form for the large-sample confidence interval for the mean is

$$\bar{Y} \pm z\hat{\sigma}_{\bar{Y}}$$

where z depends on the confidence coefficient. The higher the confidence coefficient, the greater the chance that the confidence interval contains the parameter. High confidence coefficients are used in practice, so that the chance of error is small. The most common confidence level is .95, with .99 used when it is more crucial not to make an error. In summary, we have the following result:

Large-Sample Confidence Interval for μ

A large-sample confidence interval for μ is

$$\bar{Y} \pm z\hat{\sigma}_{\bar{Y}} = \bar{Y} \pm z\left(\frac{s}{\sqrt{n}}\right)$$

The z-value is such that the probability under a normal curve within z standard errors of the mean equals the confidence coefficient. For 95% and 99% confidence intervals, z equals 1.96 and 2.58.

Let's study in greater detail this formula. One multiplies the estimated standard error $\hat{\sigma}_{\bar{Y}}$ by a z-value and then adds and subtract it from $\bar{Y}$. The z-value is such that the probability within z standard errors of the mean of the normal sampling distribution equals the confidence coefficient. For example, let's find z for a 98% confidence interval. When the probability .98 falls within z standard errors of the mean, .02 falls in the two tails and .01 in the right-hand tail. Looking up .01 in the body of Table A, we find $z = 2.33$. A 98% confidence interval equals $\bar{Y} \pm 2.33\hat{\sigma}_{\bar{Y}}$, since the probability equals .98 that $\bar{Y}$ falls within 2.33 standard errors of μ.

The probability that a confidence interval does *not* contain the parameter is called the ***error probability***. This equals 1 minus the confidence coefficient. For confidence coefficient .95, the error probability equals .05. In general, the z-score for a confidence interval is the one for which the error probability falls in the two tails of a normal curve. Half the error probability falls in each tail. For instance, for a 95% confidence interval, the error probability equals .05; the z-score is the one with probability $.05/2 = .025$ in each tail, which is $z = 1.96$.

Let α denote the error probability. Then, $1 - \alpha$ is the confidence coefficient. For instance, for an error probability of $\alpha = .05$, the confidence coefficient equals $1 - \alpha = .95$. The z-value for the confidence interval is such that the probability is $1 - \alpha$ that $\bar{Y}$ falls within z standard errors of μ. Equivalently, the probability is α that $\bar{Y}$ falls more than z standard errors from μ. The z-value refers to a total probability of α in the two tails of a normal distribution, or $\alpha/2$ in each tail.

In reality, the probability that the confidence interval contains μ is *approximately* equal to the chosen confidence coefficient. The approximation improves for larger samples, as the sampling distribution of $\bar{Y}$ is more closely normal in form and the estimated standard error $\hat{\sigma}_{\bar{Y}}$ gets closer to the true standard error $\sigma_{\bar{Y}}$.

Properties of the Confidence Interval for a Mean

We next study the properties of confidence intervals for means. These properties also apply to confidence intervals for other parameters.

The confidence level associated with confidence intervals has a long-run relative frequency interpretation. The unknown mean μ is a fixed number. A confidence interval constructed from any particular sample either does or does not contain μ. However, if we repeatedly selected random samples of that size and each time constructed a 95% confidence interval, then in the long run about 95% of the intervals would contain μ. This happens because about 95% of the sample means would fall within $1.96\sigma_{\bar{Y}}$ of μ, as does the $\bar{Y}$ in line 1 of Figure 5.2. Saying that a particular interval contains μ with "95% confidence" signifies that *in the long run* 95% of such intervals would contain μ; that is, 95% of the time the inference is correct.

Figure 5.3 shows the results of selecting ten separate samples and calculating the sample mean for each and a 95% confidence interval for the population mean. The confidence intervals jump around because $\bar{Y}$ varies from sample to sample, but nine of

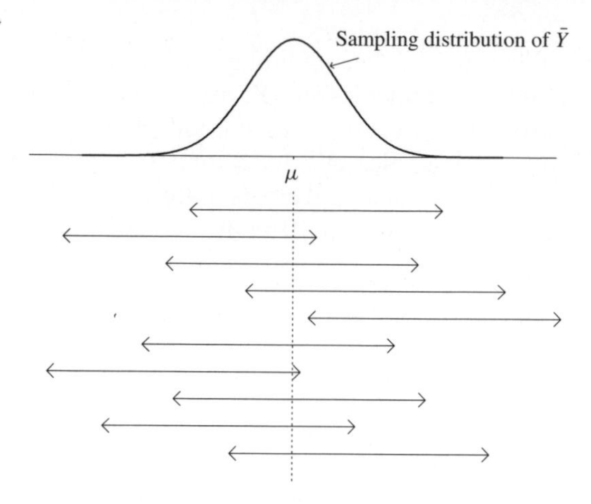

Figure 5.3 Ten 95% Confidence Intervals for μ; In the Long Run, Only 5% of the Intervals Fail to Contain μ

the ten intervals contain the population mean μ. On the average, only about 1 out of 20 times does a 95% confidence interval fail to enclose the population mean.

In practice, of course, we select just *one* sample of some fixed size n and construct one confidence interval using the observations in that sample. We do not know whether any particular 95% confidence interval truly contains μ. Our 95% confidence in that interval is based on long-term properties of the procedure. We can, though, control by our choice of the confidence coefficient the chance that the interval contains μ. If an error probability of .05 makes us nervous, we can instead form a 99% confidence interval.

Unfortunately, the greater the confidence level, the wider the confidence interval. This happens because the z-value in the formula is larger—for instance, $z = 1.96$ for 95% confidence and $z = 2.58$ for 99% confidence. To be more sure of enclosing μ, we must sacrifice precision of estimation by permitting a wider interval. In forming a confidence interval, we must often compromise between the desired precision of estimation and the desired confidence that the inference is correct; as one gets better, the other gets worse. This is why you would not typically see a 99.9999% confidence interval. Although it sounds very safe and nearly error free, it would usually be too wide to tell us much about where the population mean falls (its z-value is 4.9).

Intuitively, one should be able to estimate μ better with a larger sample size. The plus and minus part of a confidence interval is $zs/\sqrt{n}$, which is inversely proportional to the square root of the sample size. The larger the value of n, the narrower is the interval. Thus, one can improve the precision by increasing the sample size.

To illustrate, suppose that $\bar{Y} = 10.2$ and $s = 10.1$ in Example 5.1 were based on a sample of size $n = 4220$, four times the actual sample size of $n = 1055$. Then, the estimated standard error $\hat{\sigma}_{\bar{Y}}$ of the sampling distribution of $\bar{Y}$ is

$$\hat{\sigma}_{\bar{Y}} = \frac{s}{\sqrt{n}} = \frac{10.1}{\sqrt{4220}} = .155$$

half as large as in that example. The resulting 95% confidence interval is

$$\bar{Y} \pm 1.96\hat{\sigma}_{\bar{Y}} = 10.2 \pm 1.96(.155) = 10.2 \pm .3, \quad \text{or } (9.9, 10.5)$$

This is half as wide as the confidence interval formed from the sample of size $n = 1055$ in Example 5.1. A confidence interval based on $n = 4220$ is half as wide as one based on $n = 1055$.

Since the width of a confidence interval for μ is inversely proportional to the square root of n, and since $\sqrt{4n} = 2\sqrt{n}$, one must *quadruple* the sample size in order to *double* the precision (i.e., halve the width). Section 5.4 shows how to calculate the sample size needed to achieve a certain precision.

The width of a confidence interval
1. Increases as the confidence coefficient increases.
2. Decreases as the sample size increases.

Some statistical software can calculate confidence intervals for you. All such software reports the basic ingredients you need to construct an interval. For instance, one package reports the sample size, sample mean, sample standard deviation, and estimated standard error for Example 5.1 as:

N	Mean	Std Dev	Std Err
1055	10.233	10.069	0.310

5.3 Confidence Interval for a Proportion

The last section dealt with estimating the population mean, a summary parameter for quantitative data. We now present interval estimation for qualitative data, in which each observation occurs in one of a set of categories. This type of measurement occurs when the variable is nominal, such as preferred candidate (Democrat, Republican, Independent), or ordinal, such as opinion about government spending (increase, keep the same, decrease). It also occurs when inherently continuous variables are measured with categorical scales, such as when annual income has categories \$0–20,000, \$20,000–40,000, \$40,001–75,000, over \$75,000.

One can summarize categorical data by counting the number of observations in each category. Equivalently, one can record the *proportions* of observations in the categories, which are the counts divided by the total sample size. For example, a study might provide a point or interval estimate of

- The proportion of registered voters who voted in the previous presidential election
- The proportion of Canadians who favor independent status for Quebec
- The proportion of Hispanic adults who have attended college
- The proportion of American families with income below the poverty level

Large-Sample Estimation for a Proportion

Let π denote the parameter representing a population proportion. Then, π falls between 0 and 1 (here, π is *not* the mathematical constant, 3.1415...). The point estimate of the population proportion π is the *sample proportion.* We denote the sample proportion by $\hat{\pi}$, since it estimates π. The sample proportion $\hat{\pi}$ is, in fact, an unbiased and efficient point estimator of π.

Section 4.3 noted that the proportion is a type of mean. Denote an observation by 1 if it falls in the category of interest and by 0 otherwise. Then, the sample mean is the sample proportion $\hat{\pi}$ for that category. Since the sample proportion $\hat{\pi}$ is a sample mean, its large-sample sampling distribution is approximately normal about the parameter π it estimates. Figure 5.4 illustrates.

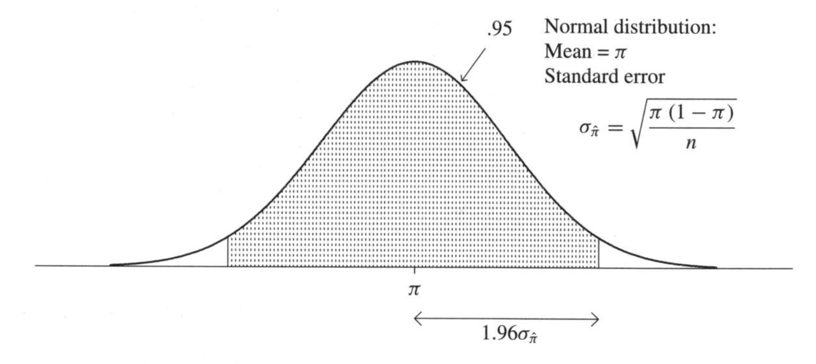

Figure 5.4 Sampling Distribution of $\hat{\pi}$

Similarly, the population proportion π is the mean μ of the probability distribution having probability π for 1 and $(1-\pi)$ for 0. The standard deviation of this probability distribution is $\sigma = \sqrt{\pi(1-\pi)}$. (Problem 4.55 derives this formula.) Since the standard error of a sample mean equals $\sigma_{\bar{Y}} = \sigma/\sqrt{n}$, the formula for the standard error $\sigma_{\hat{\pi}}$ of the sample proportion $\hat{\pi}$ is

$$\sigma_{\hat{\pi}} = \sigma/\sqrt{n} = \sqrt{\frac{\pi(1-\pi)}{n}}$$

That is, the standard error $\sigma_{\hat{\pi}}$ of the sampling distribution of $\hat{\pi}$ is a special case of $\sigma_{\bar{Y}}$, the standard error of the sampling distribution of the sample mean $\bar{Y}$. Again, the standard error is inversely proportional to the square root of the sample size. As the sample size increases, the standard error gets smaller, and the sample proportion tends to fall closer to the population proportion.

Like the formula $\sigma_{\bar{Y}} = \sigma/\sqrt{n}$ for the standard error of $\bar{Y}$, the formula for the standard error of $\hat{\pi}$ depends on an unknown parameter, in this case, π. In practice, we estimate this standard error using

$$\hat{\sigma}_{\hat{\pi}} = \sqrt{\frac{\hat{\pi}(1-\hat{\pi})}{n}}$$

This estimated standard error appears in confidence intervals. From the same reasoning shown in the previous section for the mean, a 95% confidence interval for π is

$$\hat{\pi} \pm 1.96\hat{\sigma}_{\hat{\pi}} = \hat{\pi} \pm 1.96\sqrt{\frac{\hat{\pi}(1-\hat{\pi})}{n}}$$

Example 5.2 Estimating Proportion Favoring Legalized Abortion

The 1994 General Social Survey asked respondents, "Please tell me whether or not you think it should be possible for a pregnant woman to obtain a legal abortion if the woman wants it for any reason." Of 1934 respondents, 895 said yes and 1039 said no. We shall estimate the population proportion that would respond yes to this question.

Let π represent the population proportion that would respond yes. Of the $n = 1934$ respondents, 895 said yes, so $\hat{\pi} = 895/1934 = .46$, and $1 - \hat{\pi} = .54$. That is, 46% of those sampled said yes and 54% of those sampled said no.

The estimated standard error of the estimate $\hat{\pi}$ of π equals

$$\hat{\sigma}_{\hat{\pi}} = \sqrt{\frac{\hat{\pi}(1-\hat{\pi})}{n}} = \sqrt{\frac{(.46)(.54)}{1934}} = \sqrt{.00013} = .011$$

A 95% confidence interval for π is

$$\hat{\pi} \pm 1.96\hat{\sigma}_{\hat{\pi}} = .46 \pm 1.96(.011) = .46 \pm .02, \quad \text{or } (.44, .48)$$

The population percentage that supports unrestricted access to abortion appears to be at least 44% but no more than 48%.

All numbers in the confidence interval (.44, .48) fall below .50. Thus, apparently fewer than half the population supports unrestricted access to abortion. Results in this survey varied greatly depending on the question wording. For instance, when asked whether abortion should be available if the woman becomes pregnant as a result of rape, 1616 said yes and 318 said no; you can check that the 95% confidence interval for the population proportion saying yes equals (.82, .85).

When $n = 1934$ and $\hat{\pi} = .46$, the estimated standard error of $\hat{\pi}$ is

$$\hat{\sigma}_{\hat{\pi}} = \sqrt{\hat{\pi}(1-\hat{\pi})/n} = .011$$

Similarly, the estimated standard error for $1 - \hat{\pi}$, the proportion of voters who say no to legalized abortion, is

$$\hat{\sigma}_{1-\hat{\pi}} = \sqrt{(1-\hat{\pi})\hat{\pi}/n} = \sqrt{(.54)(.46)/1934} = .011$$

A 95% confidence interval for the population proportion of negative responses is

$$.54 \pm 1.96(.011) = .54 \pm .02, \quad \text{or } (.52, .56)$$

Now $.52 = 1 - .48$ and $.56 = 1 - .44$, where (.44, .48) is the 95% confidence interval for π. Thus, inferences for the proportion $1 - \pi$ follow directly from those for the proportion π by subtracting each endpoint of the confidence interval from 1.0. □

Effect of Confidence Coefficient and Sample Size

The formula for the large-sample confidence interval for a proportion is $\hat{\pi} \pm z\hat{\sigma}_{\hat{\pi}}$. The z-value depends on the confidence coefficient in the same way as a confidence interval for the mean μ.

To illustrate, to be more cautious about possibly incorrectly predicting the population proportion favoring unrestricted abortion, we might instead use a 99% confidence interval. This equals

$$\hat{\pi} \pm 2.58\hat{\sigma}_{\hat{\pi}} = .46 \pm 2.58(.011) = .46 \pm .03, \quad \text{or } (.43, .49)$$

The confidence interval is slightly wider, (.43, .49) instead of (.44, .48), as the cost of achieving greater confidence.

Like the width of the confidence interval for a mean, the width of a confidence interval for a proportion depends on the sample size n as well as the confidence coefficient. To illustrate, suppose that 46% of a random sample of size $n = 30$ supported unrestricted abortion. Then $\hat{\sigma}_{\hat{\pi}} = \sqrt{(.46)(.54)/30} = .091$, and a 99% confidence interval for π is

$$\hat{\pi} \pm 2.58\hat{\sigma}_{\hat{\pi}} = .46 \pm 2.58(.091) = .46 \pm .23, \quad \text{or } (.23, .69)$$

In other words, if the sample proportion referred to a sample of size 30 instead of 1934, it would provide a very imprecise prediction of the population proportion. Since the interval contains values both well below and well above .50, it is plausible that a strong majority or that a weak minority of the population would support unrestricted abortion. Our conclusion from such a small sample would be ambiguous, while the conclusion from a sample as large as the General Social Survey provides is much more clear cut.

Summary of Formula and Sample Size Validity

The confidence interval for a proportion, like the one for a mean, applies for large samples. When the proportion is between about .30 and .70, the usual sample size criterion for a mean works fine. That is, one can use the method if n is at least about 30. When the proportion is less than .30 or higher than .70, the sampling distribution is skewed and requires a larger sample size to achieve normality. In this case, there should be at least ten observations both in the category of interest and not in it. When neither of these are satisfied, estimating the proportion is complex, though Problem 5.57 shows a method that usually works quite well. In Example 5.2, the sample proportion is .46 and the sample size is $895 + 1039 = 1934$. The sample size requirement is easily satisfied.

We complete this section by summarizing the large-sample confidence interval for a population proportion.

Let α denote the error probability that the interval does not contain the parameter. As in the confidence interval for a mean, the z-value refers to a total probability of α in the two tails, with $\alpha/2$ in each tail.

Large-Sample Confidence Interval for Proportion π

A large-sample confidence interval for a population proportion π, based on a sample proportion $\hat{\pi}$, is

$$\hat{\pi} \pm z\hat{\sigma}_{\hat{\pi}} = \hat{\pi} \pm z\sqrt{\frac{\hat{\pi}(1-\hat{\pi})}{n}}$$

The z-value is such that the probability under a normal curve within z standard errors of the mean equals the confidence coefficient. For 95% and 99% confidence intervals, z equals 1.96 and 2.58. The sample size n should exceed 30, with a somewhat larger sample needed if the proportion is relatively small or large—at least ten observations in the category and at least ten not in it.

5.4 Choice of Sample Size

Samples taken by professional polling organizations, such as the Gallup poll, typically contain 1000–2000 subjects. This is large enough to obtain a sample proportion that theoretically falls within about .03 of the population value. These organizations use sampling methods that are usually more complex than simple random samples; however, the formulas for standard errors of estimates under their sampling plans are approximated reasonably well by the ones for random samples.

At first glance, it seems astonishing that a sample on the order of 1000 from a population of perhaps many millions is adequate for predicting outcomes of elections, summarizing opinions on controversial issues, showing relative sizes of television audiences, and so forth. The basis for this inferential power lies in the formulas for the standard errors of the sample point estimates (which actually treat the population size as infinite; see Problem 4.58). As long as the sampling is properly executed, excellent estimates result from relatively small samples, no matter how large the population size.

Before data collection begins, most studies attempt to determine the size of the sample needed to achieve a certain degree of accuracy in estimation. A relevant measure is the value of n for which a confidence interval for the parameter is no wider than some specified width. This section studies sample size determination for estimating a mean or proportion with random samples. We use the facts that (1) the width of the confidence interval depends directly on the standard error of the sampling distribution of the estimator and (2) the standard error itself depends on the sample size.

Sample Size for Estimating Proportions

Before computing the sample size, we must first decide on the degree of *precision* desired, that is, how close the estimate should fall to the parameter. In some studies, highly precise estimation is not as important as in others. A study conducted to estimate the proportions of voters who intend to vote for each candidate in a close election requires an accurate estimate to predict the winner. If, on the other hand, the goal is to estimate the proportion of residents of Syracuse, New York, who have rural origins, a larger margin of error might be acceptable. So, we must first decide whether the error should be no more than .04 (four percentage points), .05, .10, or whatever.

Second, we must set the *probability* with which the specified precision is achieved. For instance, we might decide that the error in estimating a population proportion should not exceed .04, with .95 probability. This probability must be stated, since with any sample size one can have an error of no more than .04 with *some* probability, though perhaps a very small one.

The next example illustrates sample size determination for estimating a population proportion.

Example 5.3 Sample Size for a Survey on Single-Parent Children

A group of social scientists wanted to estimate the proportion of school children in Boston who were living with only one parent. Since their report was to be published, they wanted a reasonably accurate estimate. However, since their funding was limited, they did not want to collect a larger sample than necessary. They decided to use a sample size such that, with probability .95, the error would not exceed .04. In other words, they wanted the sample proportion to fall within .04 of the true value, with probability .95. Thus, they wanted to determine n such that a 95% confidence interval for π equals $\hat{\pi} \pm .04$.

Since the sampling distribution of the sample proportion $\hat{\pi}$ is approximately normal, the sample proportion $\hat{\pi}$ falls within $1.96\sigma_{\hat{\pi}}$ of π with probability .95. Thus, if the sample size is such that $1.96\sigma_{\hat{\pi}} = .04$, then with probability .95, $\hat{\pi}$ falls within .04 units of π and the error of estimation does not exceed .04. See Figure 5.5.

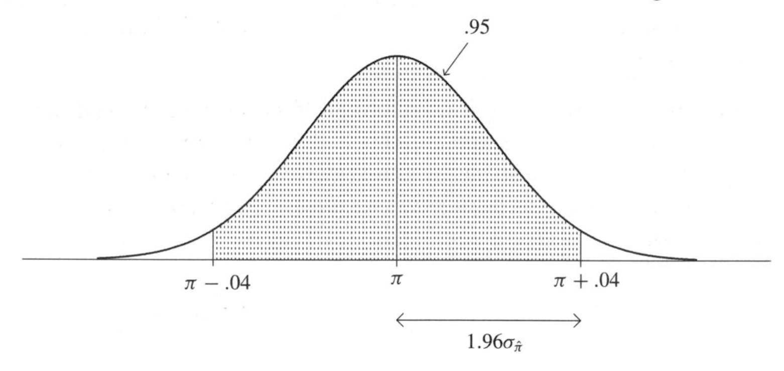

Figure 5.5 Sampling Distribution of $\hat{\pi}$ with the Error of Estimation No Greater than .04, with Probability .95

We must solve algebraically for the value of n that provides a value of $\sigma_{\hat{\pi}}$ for which $.04 = 1.96\sigma_{\hat{\pi}}$; that is, we must solve for n in the expression

$$.04 = 1.96\sqrt{\frac{\pi(1-\pi)}{n}}$$

Multiplying both sides of the expression by $\sqrt{n}$ and dividing both sides by .04, we get

$$\sqrt{n} = 1.96 \frac{\sqrt{\pi(1-\pi)}}{.04}$$

Squaring both sides, we obtain the formula

$$n = \frac{(1.96)^2 \pi(1-\pi)}{(.04)^2}$$

Now, we face a problem. We want to select n for the purpose of estimating the parameter π, but this formula requires the value of π. This is because the spread of the sampling distribution depends on the value of π. The distribution is less spread out, and it is easier to estimate π, if π is close to 0 or 1 than if it is near .5. Since π is unknown, we must substitute an educated guess for it in this equation to obtain a numerical solution for n.

Alternatively, the largest possible value for $\pi(1-\pi)$ is .25, which occurs when $\pi = .5$. In fact, $\pi(1-\pi)$ is fairly close to .25 unless π is quite far from .5. For example, $\pi(1-\pi) = .24$ when $\pi = .4$ or $\pi = .6$, and $\pi(1-\pi) = .21$ when $\pi = .7$ or $\pi = .3$. Thus, a sample of size

$$n = \frac{(1.96)^2(.25)}{(.04)^2} = 600$$

ensures that the error will not exceed .04, with a probability of *at least* .95, no matter what the value of π. □

Obtaining n by setting $\pi(1-\pi) = .25$ is the safe and cautious approach. This n value is excessively large if π is not close to .5. Suppose, for example, that based on other studies, the social scientists believed that the proportion π of school children in Boston who were living with only one parent was no more than .25. Then an adequate sample size is

$$n = \frac{(1.96)^2 \pi(1-\pi)}{(.04)^2} = \frac{(1.96)^2(.25)(.75)}{(.04)^2} = 450$$

A sample size of 600 would be larger than needed. With it, the probability would actually exceed .95 that the sample proportion falls within .04 of the true proportion.

We next provide a general formula for sample size. Let B denote the desired bound on the error (B = bound). This is the maximum distance preferred between the sample proportion and the true value, which is $B = .04$ in the example. The formula also uses a general z-value (in place of 1.96) determined by the probability with which the error is no greater than B.

Sample Size Required for Estimating a Proportion π

Let B denote the chosen bound on error. The sample size n ensuring that, with fixed probability, the error of estimation of π by the sample proportion $\hat{\pi}$ is no greater than B, is

$$n = \pi(1 - \pi)\left(\frac{z}{B}\right)^2$$

The z-score is the one for a confidence interval with confidence coefficient equal to the fixed probability; for instance, $z = 1.96$ for probability .95 and $z = 2.58$ for probability .99. Using this formula requires guessing π or taking the safe but conservative approach of setting $\pi(1 - \pi) = .25$.

To illustrate, suppose the study about single-parent children wanted to estimate the proportion to within .08 with a probability of at least .95. Then the bound on error equals $B = .08$, and $z = 1.96$, the z-value for a 95% confidence interval. The required sample size using the safe approach is

$$n = .25\left(\frac{z}{B}\right)^2 = .25\left(\frac{1.96}{.08}\right)^2 = 150$$

This sample size of 150 is one-fourth the sample size of 600 necessary to guarantee a 95% confidence bound of $B = .04$. Reducing the bound on error by a factor of one-half requires quadrupling the sample size.

Sample Size for Estimating Means

We next present an analogous result for quantitative data and estimating a population mean. Let μ and σ denote the population mean and standard deviation for the variable of interest. Figure 5.6 illustrates the basic problem. We want to determine how large n needs to be so that the sampling distribution of $\bar{Y}$ is sufficiently narrow that $\bar{Y}$ is very likely to fall within B units of μ. A derivation using this sampling distribution yields the following result:

Sample Size Required for Estimating a Mean μ

Let B denote the desired bound on error. The sample size n ensuring that, with fixed probability, the error of estimation of μ by $\bar{Y}$ is no greater than B, is

$$n = \sigma^2\left(\frac{z}{B}\right)^2$$

The z-score is the one for a confidence interval with confidence coefficient equal to the fixed probability.

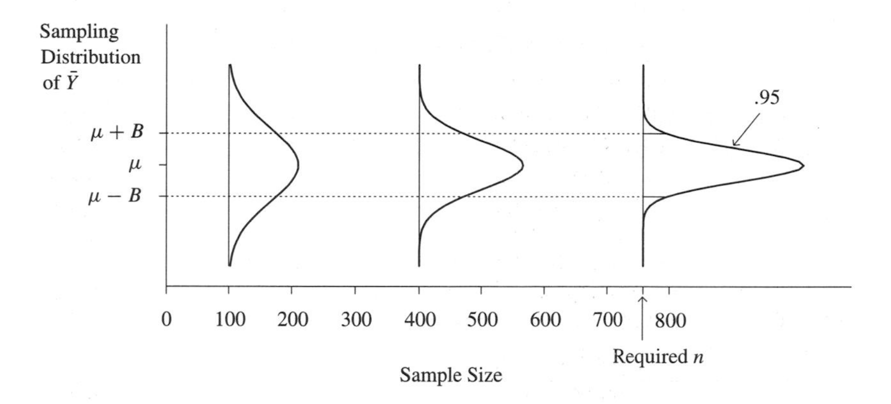

Figure 5.6 Determining n So That $\bar{Y}$ Has Probability .95 of Falling Within B Units of μ

The greater the spread of the population distribution, as measured by the standard deviation σ, the larger the sample size needed to achieve a certain accuracy. If subjects show little variation (i.e., σ is small), we need less data than if they are highly heterogenous. In practice, σ is unknown. One substitutes an educated guess for it, perhaps based on results of a previous study.

Example 5.4 Estimating Mean Educational Level of Native Americans

A study is planned of elderly Native Americans. Variables to be studied include educational level. How large a sample size is needed to estimate the mean number of years of attained education correct to within 1 year with probability .99?

Suppose the study has no prior information about the standard deviation of educational attainment for Native Americans. As a crude approximation, they might guess that nearly all values of this variable fall within a range of about 15 years, such as between 5 and 20 years. If this distribution is approximately normal, then since the range from $\mu - 3\sigma$ to $\mu + 3\sigma$ contains nearly all of a normal distribution, the range of 15 would equal about 6σ. Then, 15/6 = 2.5 is a crude guess for σ. This seems plausible, since it means that about 68% of the education values would fall within 2.5 years of the mean, or within a span of 5 years.

Now, for 99% confidence, the z-score is the one with probability .01/2 = .005 in each tail, or $z = 2.58$. Since the desired bound on error equals $B = 1$ year, the required sample size is

$$n = \sigma^2 \left(\frac{z}{B}\right)^2 = (2.5)^2 \left(\frac{2.58}{1}\right)^2 = 42 \text{ subjects}$$

A more cautious approach would select for σ a number quite sure to be an upper bound for its value. For example, it is reasonable to predict that σ is no greater than 3.5, since a range of six standard deviations then extends from 0 to 21. This yields

$n = (3.5)^2(2.58/1)^2 = 81$ families. Then, if σ is actually less than 3.5, the estimate $\bar{Y}$ will fall within 1 of μ with probability even greater than .99. □

These sample size formulas apply to simple and systematic random sampling. Cluster samples and complex multistage samples must usually be larger to achieve the same precision, whereas stratified samples can usually be smaller. In such cases, determination of sample size is complex, and you should seek guidance from a statistical consultant.

Other Considerations in Determining Sample Size

From a practical point of view, determining sample size is not a simple matter of plugging numbers into a formula. Several other considerations affect the number of observations needed in a study. We have just discussed two, *precision* and *confidence*. Precision refers to the width of a confidence interval, while confidence refers to the probability that the interval actually contains the estimated parameter.

A third characteristic affecting the sample size decision is the *variability* in the population for the variables measured. We have already seen this for estimating means, where the required sample size increases as σ increases. The more heterogeneous the population, the larger the sample needs to be. In the extreme case in which all population elements are alike (zero variability), a sample size of 1 can accurately represent the population. On the other hand, if there are 15 ethnic groups, age variation from 18 to 85, and wide variation in income, we would need a large sample to reflect accurately the variation in these variables. In most social surveys, large samples (1000 or more) are necessary, while for more homogeneous populations (e.g., residents of nursing homes) smaller samples are often adequate, due to reduced population variability.

A fourth consideration is the *complexity of analysis* planned. The more complex the analysis, such as the more variables one analyzes simultaneously, the larger the sample needed to make an adequate analysis. If one is to analyze a single variable using a simple measure such as a mean, a relatively small sample might be adequate. On the other hand, planned comparisons of several groups using complex multivariate methods require a larger sample. For instance, Example 5.4 showed that one can estimate mean educational attainment quite well using a sample of only 42 people. On the other hand, if one also wanted to compare the mean for several ethnic and racial groups and study how the mean depends on other variables such as gender, parents' income and education, IQ, and size of the community, a much larger sample would be needed, probably a thousand or more. One reason for the increase in the typical sample size of studies in recent years is the greater complexity of statistical analyses used in social science research.

Finally, a fifth consideration concerns time, money, and other *resources*. Larger samples are more expensive and more time consuming, and may require more resources than the study has available. Time, cost, and resource limitations are often the major constraints on sample size. For example, sample size formulas might suggest that 1000 cases provide the desired accuracy. Perhaps, however, we can afford to gather only 500.

Should we go ahead with the smaller sample and sacrifice precision and/or confidence, or should we give up unless we find additional resources? We often must face such questions as "Is it better to have some knowledge that is not very precise, or no knowledge at all?" or "Is it really crucial to study all population groups, or can I reduce the sample by focusing on some subsets?" The costs and benefits of large samples must be weighed against the importance of the study, the need for accuracy, and the complexity of the problem and statistical analysis.

In summary, no simple formula can always determine the proper sample size. While sample size is an important matter, its choice depends on an assessment of needs and resources and requires careful judgment.

A final caveat: When we say that a sample of size 600 is adequate for estimating a proportion to within .04 with .95 confidence, this precision of .04 is a theoretical target that takes into account only sampling error. Practical problems often imply that the actual accuracy is somewhat less. If the study is carried out poorly, or if data are never obtained for a substantial percentage of the target sample, or if some subjects in the study lie, or if some observations are incorrectly recorded by the data collector or by the statistical analyst, then the actual probability of accuracy to within .04 may be substantially less than .95. When someone claims to achieve a certain accuracy in estimating a parameter, always be skeptical unless you know that the study was substantially free of such problems.

5.5 Confidence Intervals for a Median*

The past two chapters have emphasized sampling properties of the sample mean. Chapter 3 showed, though, that other statistics are also useful for describing data. These other statistics also have sampling distributions. Moreover, for large random samples, their sampling distributions are usually approximately normal. One can use sample data to form confidence intervals for population values of the measures. We illustrate in this section for the median.

Inefficiency of Median for Normal Data

Let M denote the sample median. When the population distribution is normal and the sample is random, the standard error of M has formula similar to the one for the sample mean. Namely, the standard error equals $1.25\sigma/\sqrt{n}$. A large-sample confidence interval for the population median then has form

$$M \pm z\frac{(1.25s)}{\sqrt{n}}$$

where the z-score depends on the confidence coefficient in the usual way.

Since the population median for a normal distribution equals the population mean μ, the sample median and sample mean are both point estimates of the same number.

The confidence interval using the sample median is 25% wider than the one using the sample mean. The sample median is not as efficient an estimator. When the population distribution is approximately normal, the sample mean is a better estimator of the center of that distribution than the sample median. This is one reason the mean is more commonly used than the median in statistical inference.

If the population distribution is highly skewed, however, the population median may be a more useful summary measure than the population mean. Then, the sample median is still a relevant estimate of the population median, but the sample mean is not appropriate. In that case, a different confidence interval applies, since the one just described is valid only when the population distribution is approximately normal.

Large-Sample Confidence Interval for Median

The confidence interval for the median discussed next is valid for large samples (at least about 20–30), but requires no assumption about the form of the population distribution other than it is essentially continuous. Its logic utilizes ideas of this chapter.

By definition, the probability π that a randomly selected observation falls below the median is .50. So, for a random sample of size n, the sample proportion $\hat{\pi}$ falling below the median has mean .50 and standard error $\sigma_{\hat{\pi}} = \sqrt{\pi(1-\pi)/n} = \sqrt{.5(.5)/n} = .5/\sqrt{n}$. In particular, the probability is about .95 that the sample proportion of observations falling below the median is within two standard errors, or $1/\sqrt{n}$, of .50. Equivalently, since the sample *number* of observations falling below the median is n times the sample proportion, the probability is about .95 that the number of observations falling below the median is within $n(1/\sqrt{n}) = \sqrt{n}$ of half the sample, and that the number of observations falling *above* the median is within $n(1/\sqrt{n}) = \sqrt{n}$ of half the sample.

Now, for a sample of size n that is ordered from smallest to largest, the median is the middle measurement, which has index $(n+1)/2$. One can use the above calculation to show that the observation with index $(n+1)/2 - \sqrt{n}$ is the lower endpoint of a 95% confidence interval for the median, and the observation with index $(n+1)/2 + \sqrt{n}$ is the upper endpoint. More generally, for arbitrary confidence level, one uses the usual z-score and adds and subtracts $z(.5)\sqrt{n}$ from $(n+1)/2$ to obtain the appropriate indices.

Example 5.5 Estimating Median Shelf Time in a Library

Recently a librarian at the University of Florida asked for help in estimating various characteristics of the books in one of the university's special collections. Among the questions of interest were "How old is a typical book in the collection?" and "How long has it been since a typical book has been checked out?" We suspected that the distributions of variables of this type may be heavily skewed to the right, so we used the median to describe central tendency.

Table 5.1 shows data on the variables P = number of years since publication of book and C = number of years since book checked out, for a systematic random sample of 54 books from the collection. We shall construct a 95% confidence interval for the median of the distribution of P.

TABLE 5.1 Number of Years Since Publication (P) and Number of Years Since Checked Out (C) for 54 Books

C	P	C	P	C	P	C	P	C	P
1	3	9	9	4	4	1	18	1	5
30	30	0	17	2	7	0	12	1	13
7	19	5	5	47	47	3	15	9	17
11	140	2	19	5	8	2	10	11	18
1	5	1	22	1	11	5	19	2	3
2	97	0	10	1	21	7	7	4	19
4	4	11	11	5	20	14	14	5	43
2	19	10	10	10	10	0	18	10	17
4	13	17	71	8	19	0	17	48	48
2	19	11	11	6	6	7	20	4	4
92	92	4	44	1	5	1	54		

For $n = 54$, the endpoints of a 95% confidence interval have indices

$$\frac{n+1}{2} \pm \sqrt{n} = \frac{54+1}{2} \pm \sqrt{54} = 27.5 \pm 7.3, \text{ or } (20.2, 34.8)$$

That is, the confidence interval consists of the 20th smallest and 35th smallest (20th largest) values of the variable.

For a small sample such as this, it is simple to identify ordered values from a stem and leaf plot of the data. Table 5.2 shows the part of this plot for the smallest 44 of the 54 observations, splitting stems into two. Thc 20th smallest observation equals 11 and the 35th smallest observation equals 19. The 95% confidence interval equals (11, 19). We can be 95% confident that the median time since publication is at least 11 years and no greater than 19 years. To get a narrower interval than this, we need a larger data set. □

TABLE 5.2 Lower Part of Stem and Leaf Plot for Number of Years Since Publication

Stem	Leaf
0	3 3 4 4 4
0	5 5 5 5 6 7 7 8 9
1	0 0 0 0 1 1 1 2 3 3 4
1	5 7 7 7 7 8 8 8 9 9 9 9 9 9 9
2	0 0 1 2

The Bootstrap

Confidence intervals are relatively simple to construct for means and medians. For some other parameters, it is not possible to write down a confidence interval formula that works well regardless of the population distribution or sample size. For such cases, a recent computational invention called the ***bootstrap*** is often useful. With this approach, one treats the sample distribution as if it were the true population distribution. One samples n observations from this distribution, where each of the original n data points has probability $1/n$ of selection for each "new" observation. For this "new" sample of size n, one then constructs the point estimate of the parameter. One repeats this sampling process a large number of times, for instance selecting 1000 separate samples of size n.

The generated sampling distribution of the point estimate values provides information about the true parameter. For instance, with the *percentile* method of bootstrapping, the 95% confidence interval for the parameter is the 95% central set of estimate values, which are those falling between the 2.5th percentile and 97.5th percentile of the generated sampling distribution. This is a computationally intensive process, but one that is quite feasible with modern computing power.

5.6 Chapter Summary

This chapter presented methods of estimation, focusing on the population mean μ for quantitative variables and the population proportion π for qualitative variables.

- A ***point estimate*** is a number, based on sample data, that is the best single guess for the parameter value. The point estimates of the population mean μ, standard deviation σ, and proportion π are the sample values, $\bar{Y}$, s, and $\hat{\pi}$.
- An ***interval estimate***, called a ***confidence interval***, is a range of numbers within which the parameter is believed to fall. Confidence intervals presented for a mean μ and for a proportion π apply whenever the sample size n is relatively large. They have the form

  $$\text{Estimate} \pm z\text{-score (standard error)}$$

- The probability that the confidence interval contains the parameter is called the ***confidence coefficient***. This is controlled by the choice of the z-score in the formula. To be more certain that the confidence interval contains the parameter, select a higher confidence coefficient. Increasing the confidence entails the use of a larger z-score and, hence, the sacrifice of a wider interval. This is why the most commonly used coefficients are large, such as .95 or .99, but not extremely large, such as .99999.
- The width of a confidence interval also depends on the standard error of the sampling distribution of the point estimate. Larger sample sizes produce smaller standard errors and narrower confidence intervals and, hence, more precise estimates.

The reason these confidence intervals require a large sample size is that their formulas use z-scores based on a normal sampling distribution of the point estimate, and the Central Limit Theorem guarantees normality for large n. Large-sample inference requires no assumption about the population distribution, since the sampling distribution is roughly normal even if the population is highly nonnormal. For instance, the confidence interval for the mean is valid when $n > 30$, even if the population distribution is highly skewed. Table 5.3 summarizes large-sample estimation methods. Section 6.5 in the next chapter presents a small-sample ($n \leq 30$) confidence interval for μ.

TABLE 5.3 Summary of Estimation Methods for Means and Proportions

Parameter	Point Estimate	Estimated Std. Error	Confidence Interval	Sample Size to Estimate to Within B
Mean μ	$\bar{Y}$	$\hat{\sigma}_{\bar{Y}} = \frac{\hat{\sigma}}{\sqrt{n}}$	$\bar{Y} \pm z\hat{\sigma}_{\bar{Y}}$	$n = \sigma^2 \left(\frac{z}{B}\right)^2$
Proportion π	$\hat{\pi}$	$\hat{\sigma}_{\hat{\pi}} = \sqrt{\frac{\hat{\pi}(1-\hat{\pi})}{n}}$	$\hat{\pi} \pm z\hat{\sigma}_{\hat{\pi}}$	$n = \pi(1-\pi)\left(\frac{z}{B}\right)^2$

Note: $z = 1.96$ for 95% confidence; for error probability α and confidence level $(1-\alpha)$, z-score is the one having single-tail probability $\alpha/2$ in Table A (e.g., $\alpha/2 = .025$ for 95% confidence).

By working backward, one can determine the sample size n giving a standard error small enough to provide a sufficiently narrow confidence interval. Table 5.1 also shows the sample size formulas. To use them, one must select (1) a bound B on the error of estimation, and (2) a probability with which the error is less than that bound, which determines the z-score in the formula. In addition, one must substitute a guess for the population standard deviation σ to determine the sample size for estimating a population mean μ, and one must substitute a guess for the true proportion π to determine the sample size for estimating that proportion. In the latter case, substituting $\pi = .5$ guarantees that the sample size is large enough to give the desired precision and confidence.

PROBLEMS

Practicing the Basics

1. The numbers of children in six families selected at random from 1870 census records in a southern town are 1, 4, 0, 0, 6, 7. Calculate and interpret point estimates of the population mean and standard deviation.
2. Find and interpret the 95% confidence interval for μ, if $\bar{Y} = 70$ and $s = 10$, based on a sample size of (a) $n = 25$, (b) $n = 100$.
3. In the General Social Survey, respondents are asked to rate their political views on a seven-point scale, where 1 = extremely liberal, 4 = moderate, and 7 = extremely conservative. For the 1994 survey, a computer printout reports:

```
-------------------------------------
  N          Mean     Std Dev    Std Err
2879        4.171     1.390      0.0259
-------------------------------------
```

a) Construct and interpret a 95% confidence interval for the mean political ideology.
b) Show the effect of increasing the confidence level, by constructing a 99% confidence interval.

c) What assumption are you making about the scale of measurement for political ideology, when you use the sample mean and standard deviation?

4. Refer to Example 5.1 and the responses about the number of male sex partners. Of the 1431 subjects responding with a positive number, the mean was 4.8 and the standard deviation was 6.2.
a) Verify that the 95% confidence interval for that population mean equals (4.5, 5.1).
b) What statistical factors cause the interval in (a) to be narrower than the 95% confidence interval reported in Example 5.1 for the mean number of female partners?

5. Refer to the data in Table 3.13 regarding the subjects' number of sex partners in the last 12 months. A computer printout shows the following information:

```
---------------------------------------------
No. Cases       Mean        SD        SE of Mean
   637         1.314      5.418         0.215
---------------------------------------------
```

a) Let μ denote the population mean of this distribution. Report a point estimate of μ and a point estimate of the population standard deviation.
b) Based on the reported sample size and standard deviation, verify the reported value for the standard error.
c) Construct a 95% confidence interval for μ. Interpret.
d) Construct a 99% confidence interval for μ. Compare to part (c).

6. A survey is taken to estimate the mean annual family income for families living in public housing in Chicago. For a random sample of 30 families, the annual incomes (in hundreds of dollars) are as follows:

83	90	77	100	83	64	78	92	73	122
96	60	85	86	108	70	139	56	94	84
111	93	120	70	92	100	124	59	112	79

a) Construct a stem and leaf plot of the incomes. What do you predict about the shape of the population distribution?
b) Construct and interpret point estimates of μ and σ, the mean and standard deviation of the family incomes of all families living in public housing in Chicago.
c) Construct and interpret a 95% confidence interval for μ.
d) Construct a 99% confidence interval for μ. Interpret the interval and compare it to the one in part (c).

7. A hospital administrator wants to estimate the mean length of stay for all inpatients who use that hospital. Based on a systematic random sample of 100 records of patients for the previous year, she reports that "The sample mean was 5.3. In repeated random samples of this size, the sample mean could be expected to fall within 1.0 of the true mean about 95% of the time."
a) Construct and interpret a 95% confidence interval for the mean.
b) The administrator decides that this interval is too wide, and she prefers one of only half this width. How large a sample size is needed?

8. Find the z-score used in the formula for a confidence interval with confidence coefficient **a)** .98, **b)** .90, **c)** .50, **d)** .9973.

9. For the WWW data on number of times a week reading a newspaper, referred to in Problem 1.7, the observations for the 60 University of Florida social science students have a mean of 4.1 and standard deviation of 3.0.
a) For the population that this sample represents, construct and interpret a 95% confidence interval for the mean number of times a week reading a newspaper. Would a 99%

confidence interval for μ be wider, narrower, or the same width? Explain.

b) Suppose that the sample standard deviation had been 6.0. Find a 95% confidence interval for the mean, and compare it to the one obtained in part (a).

c) Suppose that the sample size had been 240, with $\bar{Y} = 4.1$ and $s = 3.0$. Again, find a 95% confidence interval for μ, and compare it to the one obtained in part (a).

10. Recent General Social Surveys asked subjects, "How long have you lived in the city, town or community where you live now?" The possible responses were (less than one year, 1, 2, 3, 4, ...). The responses of 1415 subjects had a mode of "less than one year," a median of 16 years, a mean of 20.3 and a standard deviation of 18.2.

a) Do you think that the population distribution is normal? Why or why not?

b) Based on your answer in (a), is it valid to construct a 99% confidence interval for the true mean for the population represented by this sample? If not, explain why not. If it is valid, do so and interpret.

11. In the 1994 General Social Survey, responses of 1964 subjects to the question, "On the average day, about how many hours do you personally watch television?" had a mean of 2.8. The standard error of the mean was .05.

a) Calculate a 98% confidence interval for the mean daily time spent watching television. Interpret.

b) Find the standard deviation of the time spent watching television. Do you think that the distribution was bell-shaped? Why or why not?

12. A study is conducted of the geographic distribution of the residences of the employees at a large factory, in order to determine the suitability of initiating busing to that factory. One variable considered is the distance the employee lives from the factory. For a random sample of 100 employees, the mean distance is 6.3 miles and the standard deviation is 4.0.

a) Find and interpret a 90% confidence interval for the mean residential distance from the factory of all employees.

b) About how large a sample would have been adequate if we merely needed to estimate the mean to within 1.0, with 90% confidence?

13. The 1991 General Social Survey asked, "During the last year, did anyone take something from you by using force—such as a stickup, mugging, or threat?" Of 987 subjects, 17 answered yes and 970 answered no.

a) Find the point estimate of the proportion of the population who were victims of this type.

b) Construct and interpret a 95% confidence interval for the population proportion. Can you conclude that fewer than 5% were victims of this type?

14. In the 1991 General Social Survey, respondents were asked whether people convicted of murder should receive the death penalty. 1078 responded yes and 336 responded no. Construct a 99% confidence interval for the proportion of American adults who would answer yes. Interpret. Can you conclude that more than half of all American adults would answer yes? Why?

15. The 1994 General Social Survey asked subjects "Do you believe there is a life after death?" Of 1958 respondents, 1425 answered yes. A report based on these data stated that "73% of Americans believe in life after death. The margin of error for this result is plus or minus 2%." Explain how the researchers obtained this result.

16. An anthropologist wants to estimate the proportion of children in a tribe in the Philippines who die before reaching adulthood. For families she knew who had children born between 1970 and 1975, 3 of 30 children died before reaching adulthood.

a) Construct a 90% confidence interval for the corresponding population proportion.

b) Explain why this interval is highly approximate and perhaps inappropriate. (*Hint*: Check the sample size and sampling assumptions.)

17. A study in 1991 by the U.S. National Center for Health Statistics provided a point estimate of 25.5% for the percentage of adult Americans who were currently smokers. The sample size was 42,000. Assuming that this sample has the characteristics of a random sample, construct and interpret a 99% confidence interval for the proportion of the population who were smokers. (Note: When the sample size is very large, even confidence intervals with large confidence coefficients are narrow.)

18. Refer to the previous problem. The same study provided a point estimate of 5.5% for the percentage of Americans who averaged two or more alcoholic drinks per day.
a) Construct and interpret a 99% confidence interval for the proportion of the population who drink that much.
b) Compare the width of this interval to the one in the previous problem.
c) By computing values of the standard error of $\hat{\pi}$ for $n = 42{,}000$ when $\pi = .1, .3, .5, .7, .9$, explain why it is more difficult to estimate π when it is near .5 than when it is near 0 or 1.

19. A national television network samples 1400 voters after each has cast a vote in a state gubernatorial election. Of these 1400 voters, 742 claim to have voted for the Democratic candidate and 658 for the Republican candidate. There are only two candidates in the election.
a) Assuming that each sampled voter actually voted as claimed and that the sample is a random sample from the population of all voters, is there enough evidence to predict the winner of the election? Base your decision on a 95% confidence interval.
b) Base your decision on a 99% confidence interval. Explain why it requires greater evidence to make a prediction when we require greater confidence of being correct.

20. *Newsweek* magazine (March 27, 1989) reported results of a poll about religious beliefs, conducted by the Gallup organization. Of 750 American adults, 24% believed in reincarnation. Treating this as a random sample, construct and interpret a 95% confidence interval for the true proportion of American adults believing in reincarnation.

21. Of 577,006 people involved in motor vehicle accidents in Florida in a recent year, 412,878 were wearing seat belts (Florida Department of Highway Safety and Motor Vehicles).
a) Find a point estimate of the population proportion of Florida motorists wearing seat belts.
b) Construct a 99% confidence interval for the population proportion wearing seat belts at that time. Can we conclude that a majority wore seat belts? Explain.
c) For the inferences in (a) and (b) to be sensible, explain what you must assume about the sample data.

22. A random sample is taken of students at a large university to determine the proportion who own automobiles. If 66% of the students in the sample own automobiles, form a 99% confidence interval for the proportion of all students owning automobiles under the following conditions:
a) Sample size is 50.
b) Sample size is 100.
c) Sample size is 400.

23. Out of an election day sample of 400 individuals who voted in a gubernatorial election, 160 voted for Jones and 240 for Smith.
a) Assuming this is a random sample of all voters, construct a 99% confidence interval for the proportion of votes that Jones will receive. Do you think that Jones will lose the election? Why?

b) Suppose, instead, that the sample size had been 40, of whom 16 voted for Jones. Again, find the 99% confidence interval and, if possible, predict the winner. How does the result compare to (a)?

24. A city council votes to appropriate funds for a new civic auditorium. The mayor of the city threatens to veto this decision unless it can be shown that a majority of citizens would use it at least twice a year. The council commissions a poll of city residents. For a random sample of 400 residents, 230 say they would use the facility at least twice a year. Find a 98% confidence interval for the proportion of all residents of the town who would say they would use the proposed auditorium at least twice a year. Interpret the interval and advise the mayor.

25. Refer to Problem 5.16. Before obtaining the sample, the anthropologist wanted to determine the size of sample needed to estimate π to within .07 with 95% confidence. Assuming nothing about the value of π, calculate the sample size that ensures at least this degree of accuracy.

26. To estimate the proportion of traffic deaths in Florida last year that were alcohol related, determine the necessary sample size for the estimate to be accurate to within .06 with probability .90. Based on results of a previous study, we expect the proportion to be about .30.

27. A study is being planned for estimating the proportion of married women with living parents in the United States who live in the same state as their parents. Approximately how large a sample size is needed to guarantee estimating this proportion correct to within
a) .10 with probability .95
b) .05 with probability .95
c) .05 with probability .99
d) .01 with probability .99
Compare sample sizes for parts (a) and (b), (b) and (c), and (c) and (d).

28. A public health unit wants to sample death records for the past year in New York City to estimate the proportion of the deaths that were due to accidents. They want the estimate to be accurate to within .02 with probability .95.
a) Find the necessary sample size if, based on previous studies, they believe that this proportion does not exceed .10.
b) Suppose that in determining the necessary sample size, they use the safe approach that sets $\pi = .5$ in the appropriate formula. Then how many records need to be sampled? Compare the result to the answer in part (a), and note the reduction in sample size that occurs by making an educated guess for π.

29. A tax assessor wants to estimate the mean property tax bill for all homeowners living in Madison, Wisconsin. A survey completed three years ago indicated that the mean and standard deviation were \$1400 and \$1000. Assuming that the standard deviation is still about \$1000, how large a sample of tax records should the tax assessor take in order to estimate the mean correct to within \$100 with probability .90?

30. An estimate is needed of the mean acreage of farms in Canada. The estimate must be correct to within 25 acres with probability .95. A preliminary study suggests that 200 acres is a reasonable guess for the standard deviation of farm size. How large a sample of farms is required?

31. Refer to the previous exercise. A sample is selected of the size believed to be needed to estimate the mean correct to within 25 acres with probability .95. Suppose, however, that the sample has a standard deviation of 400 acres, rather than 200. Then, how close can we expect the sample mean to be to the true mean? In other words, what amount is added to and subtracted from the sample mean to get a 95% confidence interval for the mean acreage of farms?

32. How large a sample size is needed to estimate the mean annual income of Native Americans correct to within \$1000 with probability .99? Suppose there is no prior information about the standard deviation of annual income of Native Americans, but we guess that about 95% of their incomes are between \$6000 and \$50,000 and that this distribution of incomes is approximately mound shaped.

33. A random sample of $n = 20$ private companies having at least 100 employees each is chosen to form the basis of a study of political contributions by industry. In particular, a 95% confidence interval is wanted for the median yearly total contribution to political campaigns by all companies with at least 100 employees. The interest in the median instead of the mean is due to the belief that this distribution may be very skewed to the right. The reported total contributions for the past year for the 20 companies sampled are, in thousands of dollars

10	17	1	15	14	0	5	2	8	10
0	3	4	100	3	0	25	12	9	36

Show that a 95% confidence interval for the median contribution is (\$3000, \$14,000).

34. Refer to Problem 5.6. Construct a 95% confidence interval for the median annual income of the public housing residents. Interpret.

35. Refer to Example 5.5. Construct a (a) 90%, (b) 99% confidence interval for the median time since a book was last checked out. Interpret.

Concepts and Applications

36. Refer to the WWW data set (Problem 1.7). Using computer software, construct and interpret a 95% confidence interval for the mean weekly number of hours spent watching TV.

37. For the WWW data set, find a 95% confidence interval for the proportion believing in life after death. Interpret.

38. Refer to the data file created in Problem 1.7. For variables chosen by your instructor, conduct inferential statistical analyses using basic methods of point and interval estimation. Summarize and interpret your findings.

39. Based on data from the National Survey of Families and Households, a recent study (by S. South and G. Spitze, *American Sociological Review*, Vol. 59, 1994, pp. 327-347) reported the descriptive statistics in Table 5.4 for the hours spent in housework and in employment per week. Analyze these data, and summarize your conclusions.

TABLE 5.4

		Housework Hours		Employment Hours	
Gender	Sample Size	Mean	Standard Dev.	Mean	Standard Dev.
Men	4252	18.1	12.9	31.8	22.6
Women	6764	32.6	18.2	18.4	20.0

40. The 1994 General Social Survey asked subjects about government spending on the environment. Table 5.5 shows the responses. Analyze and interpret these data.

TABLE 5.5

Response	Frequency
Spending too much	130
About right	370
Spending too little	880

41. The 1988 General Social Survey asked respondents, "Did you live with your husband/wife before you got married?" The responses were 176 yes, 566 no, 725 not applicable, and 14 no answer. Analyze these data.

42. Explain what it means for an estimator to be (a) unbiased, (b) efficient.

43. Explain the distinction between *point* estimation and *interval* estimation. Why is an interval estimate more informative than a point estimate?

44. What is the purpose of forming a confidence interval for a population mean? What does one learn that one could not otherwise learn by calculating a point estimate, such as the sample mean?

45. Explain why confidence intervals are wider when we use (a) larger confidence coefficients, (b) smaller sample sizes.

46. Why would it be unusual to see a 99.9999% confidence interval?

47. Give an example of a study in which it would be important to have

a) A high degree of confidence

b) A high degree of precision

c) Both a high degree of confidence and a high degree of precision

48. How does population heterogeneity affect the sample size required to estimate a population mean? Illustrate with an example.

Select the best response(s) in Problems 49–51.

49. Increasing the confidence coefficient causes the width of a confidence interval to (a) increase, (b) decrease, (c) stay the same.

50. Other things being equal, quadrupling the sample size causes the width of a confidence interval to (a) double, (b) halve, (c) be one quarter as wide, (d) stay the same.

51. Based on responses of 1467 subjects in General Social Surveys in the mid-1980s, a 95% confidence interval for the mean number of close friends equals (6.8, 8.0). Which of the following interpretations is (are) correct?

a) We can be 95% confident that $\bar{Y}$ is between 6.8 and 8.0.

b) We can be 95% confident that μ is between 6.8 and 8.0.

c) Ninety-five percent of the values of $Y =$ number of close friends (for this sample) are between 6.8 and 8.0.

d) If random samples of size 1467 were repeatedly selected, then 95% of the time $\bar{Y}$ would be between 6.8 and 8.0.

e) If random samples of size 1467 were repeatedly selected, then in the long run 95% of the confidence intervals formed would contain the true value of μ.

52. A random sample of 50 records yields a 95% confidence interval for the mean age at first marriage of women in a certain county of 21.5 to 23.0 years. Explain what is wrong with each of the following interpretations of this interval.

a) If random samples of 50 records were repeatedly selected, then 95% of the time the sample mean age at first marriage for women would be between 21.5 and 23.0 years.

b) Ninety-five percent of the ages at first marriage for women in the county are between 21.5 and 23.0 years.
c) We can be 95% confident that $\bar{Y}$ is between 21.5 and 23.0 years.
d) If we repeatedly sampled the entire population, then 95% of the time the population mean would be between 21.5 and 23.5 years.

53. Refer to the previous exercise. Provide the proper interpretation.

54. For a random sample of n subjects, explain why it is about 95% likely that the sample proportion has an error of no more than $1/\sqrt{n}$ as an estimate of the population proportion. (*Hint*: To show this "$1/\sqrt{n}$ Rule," find two standard errors when $\pi = .5$, and explain how this compares to two standard errors at other values of π.) Using this result, show that $n = 1/B^2$ is a safe sample size for estimating a proportion π to within B units with 95% confidence.

55. * An article reports that (4.0, 5.6) is a 95% confidence interval for the mean length of stay, in days, of patients in hospital for a particular operation. The article reports the sample size of 50, but not the sample mean or standard deviation. Find them.

56. * To encourage subjects to make responses on sensitive questions, the method of ***randomized response*** is often used. The subject is asked to flip a coin, in secret. If it is a head, the subject tosses the coin once more and reports the outcome, head or tails. If, instead, the first flip is a tail, the subject reports instead the response to the sensitive question, for instance reporting the response *head* if the true response is *yes* and reporting the response *tail* if the true response is *no*. Let π denote the true probability of the *yes* response on the sensitive question.
a) Explain why the numbers in Table 5.6 are the probabilities of the four possible outcomes.
b) Let p denote the sample proportion of subjects who report *head* for the second response. Explain why $\hat{\pi} = 2p - .5$ estimates π.
c) Using this approach, 200 subjects are asked whether they have ever knowingly cheated on their income tax. Report the estimate of π if the number of reported heads equals (i) 50, (ii) 70, (iii) 100, (iv) 150.

TABLE 5.6

	Second Response	
First Coin	Head	Tail
Head	.25	.25
Tail	$\pi/2$	$(1-\pi)/2$

57. * To construct a large-sample confidence interval for a proportion π, it is not necessary to substitute $\hat{\pi}$ for the unknown value of π in the formula for the standard error of $\hat{\pi}$. A less approximate method for constructing a 95% confidence interval finds the endpoints by determining the π values that are 1.96 standard errors from the sample proportion. That is, one solves for π in the equation

$$|\hat{\pi} - \pi| = 1.96\sqrt{\frac{\pi(1-\pi)}{n}}$$

One can solve this by trial and error, using the endpoints of the usual confidence interval as initial guesses. Or one can square both sides of the equation and solve the resulting quadratic equation.

a) Use this method for the data in Problem 5.16, and compare the result to that obtained with the usual method.
b) Explain what happens with the usual method when the sample proportion equals 0 or 1. (The two methods give similar results for very large samples, but otherwise can be quite different if the sample proportion is near 0 or 1.)

58. * Derive the formula for determining how large a sample size is needed to estimate a population mean to within B units with probability .95.

Bibliography

Box, J. F. (1978). *R. A. Fisher, The Life of a Scientist.* New York: Wiley.

Efron, B., and Morris, C. (1977). Stein's paradox in statistics. *Scientific American*, 236, 119–127.

Eliason, S. (1993). *Maximum Likelihood Estimation: Logic and Practice.* Beverly Hills, CA: Sage.

Fisher, R. A. (1970). *Statistical Methods for Research Workers*, 14th ed. New York: Hafner.

King, G. (1989). *Unifying Political Methodology: The Likelihood Theory of Statistical Inference.* Cambridge: Cambridge University Press.

Reid, C. (1982). *Neyman–From Life.* New York: Springer-Verlag.

Chapter 6

Statistical Inference: Significance Tests

A common aim in many studies is to check whether the data agree with certain predictions. These predictions are *hypotheses* about variables measured in the study.

Hypothesis

A ***hypothesis*** is a statement about some characteristic of a variable or a collection of variables.

Hypotheses arise from the theory that drives the research. When a hypothesis relates to characteristics of a population, such as population parameters, one can use statistical methods with sample data to test its validity. Examples of hypotheses that might be tested statistically are the following: "A majority of Canadians are satisfied with their national health service," "The mean age at marriage for men in colonial America was the same in rural and urban areas," "For workers in service jobs, the mean income is lower for women than for men," and "There is a difference between Democrats and Republicans in the probabilities that they vote with their party leadership."

A ***significance test*** is a way of statistically testing a hypothesis by comparing the data to values predicted by the hypothesis. Data that fall far from the predicted values provide evidence against the hypothesis. The following example illustrates ideas behind significance tests.

Example 6.1 Testing for Gender Bias in Selecting Managers

A large supermarket chain in Florida occasionally selects some of its employees to receive management training. A group of women employees recently claimed that males are picked at a disproportionally high rate for such training. The company denied this claim (*Tampa Tribune*, April 6, 1996).

Let's consider how the women employees could statistically back up their assertion. Suppose the employee pool for potential selection for management training is half male and half female. Then, the company's claim of a lack of gender bias is a hypothesis. It states that, other things being equal, at each choice the probability of selecting a female equals 1/2 and the probability of selecting a male equals 1/2. If the employees truly are selected for management training randomly in terms of gender, about half the employees picked should be females and about half should be male. The women's claim is an alternative hypothesis that the probability of selecting a male exceeds 1/2.

Since this program began, suppose that nine of the ten employees chosen for management training have been male. Based on this evidence, we might be inclined to support the women's claim. However, we should check first to see if these results would be unlikely, if there were no gender bias. Would it be highly unlikely that at least nine of the ten employees chosen would have the same gender, if they were truly selected at random from the employee pool? Due to sampling variation, it need not happen that exactly 50% of the ten people in the sample are male. We need guidelines about how large the percentage of males must be before we can support the women's hypothesis. □

This chapter introduces statistical methods for obtaining evidence and making decisions about hypotheses. The process is statistical in the sense that it uses sample data to make inferences. In doing so, it can control the probability of an incorrect decision.

The first section of the chapter describes the elements of a significance test. The remainder of the chapter deals with significance tests about a population mean μ or a population proportion π. Sections 6.2 and 6.3 discuss the large-sample case, and Sections 6.5 and 6.6 present small-sample significance tests. Sections 6.4 and 6.7 show how to control the probability of an incorrect decision.

6.1 Elements of a Significance Test

Now let's take a closer look at what we mean by a significance test. All significance tests have five elements: assumptions, hypotheses, test statistic, P-value, and conclusion.

Assumptions

All significance tests require certain assumptions for the tests to be valid. These assumptions refer to

- The ***type of data***: Like other statistical methods, each test applies for either quantitative data or qualitative data.
- The form of the ***population distribution***: For some tests, the variable must have a particular form of distribution, such as the normal. This is primarily true for small-sample tests.
- The ***method of sampling***: The tests presented in this book require simple random sampling.
- The ***sample size***: The validity of many tests improves as the sample size increases. These tests require a certain minimum sample size for the analyses to work well.

Hypotheses

A significance test considers two hypotheses about the value of a parameter.

Null Hypothesis, Alternative Hypothesis

The ***null hypothesis*** is the hypothesis that is directly tested. This is usually a statement that the parameter has value corresponding to, in some sense, *no effect*. The ***alternative hypothesis*** is a hypothesis that contradicts the null hypothesis. This hypothesis states that the parameter falls in some alternative set of values to what the null hypothesis specifies.

Notation for Hypotheses

The symbol H_0 represents the null hypothesis, and the symbol H_a represents the alternative hypothesis.

A significance test analyzes the strength of sample evidence against the null hypothesis. The test is conducted to investigate whether the data contradict the null hypothesis, hence suggesting that the alternative hypothesis is true. The approach taken is the indirect one of *proof by contradiction*. The alternative hypothesis is judged acceptable if the sample data are inconsistent with the null hypothesis. In other words, the alternative hypothesis is supported if the null hypothesis appears to be incorrect.

The researcher usually conducts the test to gauge the amount of support for the alternative hypothesis. Thus, the alternative hypothesis is often called the ***research hypothesis***. The hypotheses are formulated *before* collecting or analyzing the data.

To illustrate, we refer to Example 6.1 about possible gender discrimination in the selection of employees of a supermarket chain for management training. The company claims that the probability that any given employee selected is male equals 1/2. This is an example of a null hypothesis, *no effect* referring to a lack of gender bias. The alternative hypothesis reflects the skeptical women employees' belief that this probability actually exceeds 1/2. We conduct the test by checking whether the sample data are inconsistent with the null hypothesis probability value of 1/2.

Test Statistic

The ***test statistic*** is a statistic calculated from the sample data to test the null hypothesis. This statistic typically involves a point estimate of the parameter to which the hypotheses refer.

For instance, to test a hypothesis about an unknown probability, one could use as test statistic the sample estimator of that probability. If nine out of ten selected trainees are male, the estimator is the sample proportion, 9/10 = .90.

P-Value

Using the sampling distribution of the test statistic, we calculate the probability that values of the statistic like the one observed would occur if the null hypothesis were true. This provides a measure of how unusual the observed test statistic value is compared to what H_0 predicts.

Specifically, we consider the set of possible test statistic values that provide *at least as much evidence* against the null hypothesis as the observed test statistic. This set is formed with reference to the alternative hypothesis; the values providing stronger evidence *against* the null hypothesis are those providing stronger evidence *in favor of* the alternative hypothesis. The ***P*-value** is the probability, if H_0 were true, that the test statistic would fall in this collection of values.

_P_-Value

The ***P*-value** is the probability, when H_0 is true, of a test statistic value at least as contradictory to H_0 as the value actually observed. The smaller the P-value, the more strongly the data contradict H_0. The P-value is denoted by P.

The P-value summarizes the evidence in the data about the null hypothesis. A moderate to large P-value means that the data are consistent with H_0. For instance, a P-value such as .26 or .83 indicates that the observed data would not be unusual if H_0 were true. On the other hand, a P-value such as .001 means that such data would be very unlikely, if H_0 were true. This provides strong evidence against H_0.

For the gender bias example, the alternative hypothesis states that the probability of selecting a male for the managerial track exceeds 1/2. The test statistic is the sample proportion of males in the ten trainees selected; the observed value equals 9/10 = .90. The values of this test statistic providing this much or even stronger evidence against the null hypothesis and in favor of the alternative hypothesis are sample proportion values of .90 and higher. A formula from Section 6.6 calculates this probability as .011. Thus, the P-value equals $P = .011$, as shown in Figure 6.1. If the selections truly are random with respect to gender, the chance is only .011 of such an extreme sample result, namely, that nine or all ten selections for management training would be males. Other things being equal, this small P-value provides considerable, though not overwhelming, evidence of gender bias.

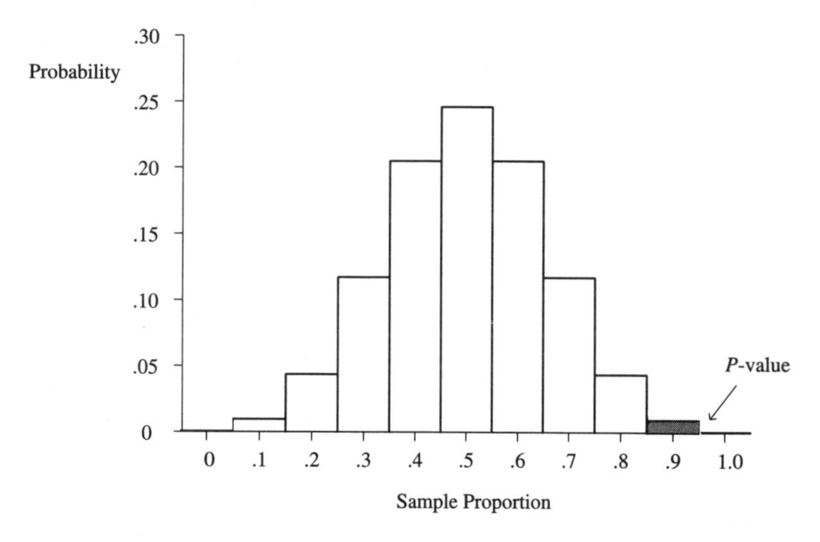

Figure 6.1 The P-value Refers to the Probability of the Observed Data or Even More Extreme Results

Generally, when a null hypothesis is true, the P-value is equally likely to fall anywhere between 0 and 1; for instance, it has a .01 chance of falling between 0.0 and .01, a .01 chance of falling between .01 and .02, a .01 chance of falling between .02 and .03, and so forth up to a .01 chance of falling between .99 and 1.00. In that case, the P-value tends to vary around an expected value of .50. By contrast, when H_0 is false, the P-value is more likely to be close to 0 than close to 1.

Conclusion

The P-value is the primary reported result of a significance test. An observer of the test results can then judge the extent of the evidence against H_0. Sometimes it is necessary to make, in addition, a formal decision about the validity of H_0. If the P-value is sufficiently small, one rejects H_0 and accepts H_a. In either case, the conclusion should include an *interpretation* of what the P-value or decision about H_0 tells us about the original question motivating the test.

Most studies require very small P-values, such as $P \leq .05$, before concluding that the data sufficiently contradict H_0 to reject it. In such cases, results are said to be *sig-*

TABLE 6.1 The Five Elements of a Statistical Significance Test

1.	Assumptions Type of data, form of population, method of sampling, sample size
2.	Hypotheses Null hypothesis, H_0 (parameter value for "no effect") Alternative hypothesis, H_a (alternative parameter values)
3.	Test statistic Compares point estimate to null hypothesized parameter value
4.	P-value Weight of evidence about H_0; smaller P is more contradictory
5.	Conclusion Report P-value Formal decision (optional; see Section 6.4)

nificant at the .05 level. This means that if the null hypothesis were true, the chance of getting such extreme results as in the sample data would be no greater than 5%.

The process of making a formal decision by rejecting or not rejecting a null hypothesis is an optional part of the significance test. We defer further discussion of it until Section 6.4. Table 6.1 summarizes the elements of a significance test.

6.2 Significance Test for a Mean

We now present a significance test about the population mean μ for quantitative variables. This test assumes that the sample size n is at least 30. It uses the fact that, for large random samples, the sampling distribution of the sample mean $\bar{Y}$ is approximately normal, no matter what distribution the variable has. The five elements of the significance test follow:

Elements of a Large-Sample Test for a Mean

1. *Assumptions*

The test requires a random sample of size $n \geq 30$. The variable measured is quantitative, and the test refers to the population mean of the variable, μ.

2. *Hypotheses*

The null hypothesis has form

$$H_0: \mu = \mu_0$$

where μ_0 is some particular number. In other words, the hypothesized value of μ in H_0 is a single value. This usually refers to *no effect* or *no change* compared to some standard.

The alternative hypothesis refers to alternative parameter values from the one in the null hypothesis. The most common form of alternative hypothesis is

$$H_a : \mu \neq \mu_0$$

This alternative hypothesis is called ***two-sided***, since it includes values falling both below and above the value μ_0 listed in H_0.

The hypotheses $H_0 : \mu = 0$ and $H_a : \mu \neq 0$ illustrate these forms. The null hypothesis states that the population mean equals 0, and the alternative hypothesis states that the population mean equals some value other than 0.

3. *Test Statistic*

The sample mean $\bar{Y}$ estimates the population mean. When $n \geq 30$, the sampling distribution of $\bar{Y}$ is approximately normal about μ, with standard error $\sigma_{\bar{Y}} = \sigma/\sqrt{n}$. If H_0: $\mu = \mu_0$ is true, then the center of the sampling distribution is the number μ_0, as shown in Figure 6.2. The evidence about H_0 is the distance of the sample value $\bar{Y}$ from the null hypothesis value μ_0, relative to the standard error. A value of $\bar{Y}$ falling far out in the tail of this sampling distribution casts doubt on the validity of H_0, because it would be unlikely to observe $\bar{Y}$ very far from μ_0 if truly $\mu = \mu_0$.

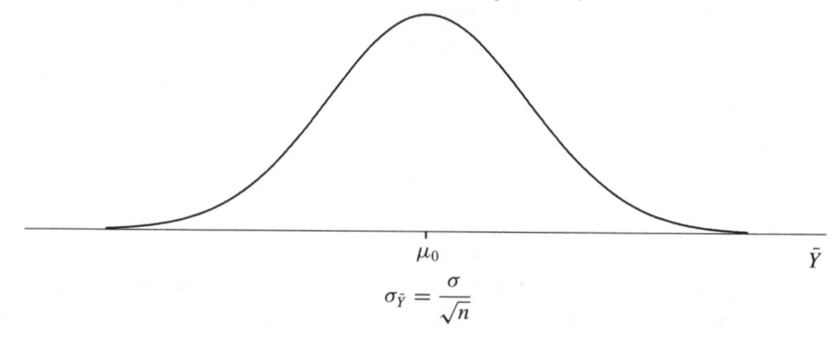

Figure 6.2 Sampling Distribution of $\bar{Y}$ if $H_0 : \mu = \mu_0$ Is True. For large random samples, it is approximately normal, centered at the null hypothesis value, μ_0.

The test statistic is the z-score

$$z = \frac{\bar{Y} - \mu_0}{\hat{\sigma}_{\bar{Y}}} = \frac{\bar{Y} - \mu_0}{s/\sqrt{n}}$$

As in Chapter 5, we substitute the sample standard deviation s for σ to get an estimated standard error, $\hat{\sigma}_{\bar{Y}} = s/\sqrt{n}$. The test statistic counts the number of estimated standard errors that $\bar{Y}$ falls from the hypothesized value μ_0. When H_0 is true, the sampling distribution of this test statistic is approximately the *standard normal* distribution; that is, normal with mean equal to 0 and standard deviation equal to 1, as presented in Section 4.2. The farther $\bar{Y}$ falls from μ_0, the larger the absolute value of the z test statistic. Hence, the larger the value of $|z|$, the stronger the evidence against H_0.

One reason for placing a single number μ_0 in the null hypothesis H_0 should now be apparent. The calculation of the test statistic, and hence the result of the test, refers to that one value.

4. *P-Value*

The test statistic summarizes the sample evidence. Different tests use different test statistics, though, and it is easier to interpret the test statistic by transforming it to the probability scale of 0 to 1. The P-value does this. It describes whether the observed test statistic value is consistent with the null hypothesis, small values of P indicating inconsistency.

We calculate the P-value under the assumption that H_0 is true. That is, we give the benefit of the doubt to the null hypothesis, analyzing how likely the observed data would be if that hypothesis were true. For the alternative hypothesis H_a: $\mu \neq \mu_0$, the P-value is the probability that the z test statistic is at least as large in absolute value as the observed test statistic. This means that P is the probability of a $\bar{Y}$ value at least as far from μ_0 *in either direction* as the observed value of $\bar{Y}$. The P-value refers to the probability of the observed result or any other result that provides even stronger evidence against the null hypothesis.

A z test statistic value of 0 results when $\bar{Y} = \mu_0$. This is the z-value most consistent with H_0. The P-value is the probability of a z test statistic value at least as far from this consistent value as the one observed. In other words, P is the probability of those $\bar{Y}$ values that are at least as contradictory to $H_0 : \mu = \mu_0$ and at least as favorable to H_a: $\mu \neq \mu_0$ as the observed $\bar{Y}$, that is, at least as many standard errors distant from μ_0.

Figure 6.3 shows the sampling distribution of the z test statistic when H_0 is true. To illustrate the calculation of P, suppose $z = -1.5$. This is the z-score resulting from a sample mean $\bar{Y}$ that is 1.5 standard errors below μ_0. The P-value is the probability that $z \geq 1.5$ or $z \leq -1.5$ (i.e., $|z| \geq 1.5$). From Table A, the probability in one tail above $z = +1.5$ is .0668, so the probability in both tails, beyond $|z| = 1.5$, equals $2(.0668) = .1336$. This is the probability that the sample mean falls at least 1.5 standard errors from the true mean.

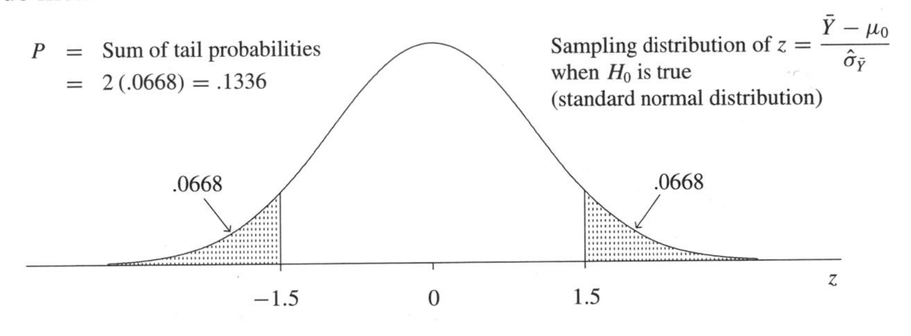

Figure 6.3 Calculation of P When $z = -1.5$, for Testing H_0: $\mu = \mu_0$ Against H_a: $\mu \neq \mu_0$. The P-Value is the two-tail probability of a more extreme result than the observed one.

One should round the calculated P-value such as .1336 to .134 or .13 before reporting it. Reporting the P-value as .1336 makes it seem as if more accuracy exists than actually does, since the sampling distribution is only *approximately* normal.

5. *Conclusion*

Finally, the study should report the P-value, so others can view the strength of evidence. The smaller P is, the stronger the evidence against H_0 and in favor of H_a.

Example 6.2 Political Conservatism and Liberalism

Many political commentators have remarked that since the Reagan presidential years, there has been an upsurge of political conservatism. One way to summarize political ideology in the United States is to analyze results of various items on the General Social Survey. For instance, every year that survey asks, "I'm going to show you a seven-point scale on which the political views that people might hold are arranged from extremely liberal, point 1, to extremely conservative, point 7. Where would you place yourself on this scale?" Table 6.2 shows the seven-point scale and the distribution of 627 responses among the levels for a recent survey.

TABLE 6.2 Responses of 627 Subjects to a Seven-Point Scale of Political Ideology

Response	Count
1. Extremely liberal	12
2. Liberal	66
3. Slightly liberal	109
4. Moderate, middle of road	239
5. Slightly conservative	116
6. Conservative	74
7. Extremely conservative	11

Political ideology is an ordinal scale. In some cases, the main interest in such a scale may refer to category proportions; for instance, is the population proportion who are extremely liberal different from the population proportion who are extremely conservative? More commonly, such scales are treated in a quantitative manner by assigning scores to the categories. One can then summarize responses by quantitative measures such as means, allowing us to detect the extent to which observations tend to gravitate toward the conservative or the liberal end of the scale.

If we assign the category scores shown in Table 6.2, then a mean below 4 shows a propensity toward liberalism, and a mean above 4 shows a propensity toward conservatism. We can test whether these data show much evidence of either of these by conducting a significance test about how the population mean compares to the middle value of 4.

1. *Assumptions*: The sample is randomly selected and the sample size exceeds 30, so these assumptions for a large-sample test about a mean are satisfied. We are treating political ideology as quantitative with equally-spaced scores.
2. *Hypotheses*: Let μ denote the population mean ideology, for this seven-point scale. The null hypothesis contains one specified value for μ. Since we conduct the analysis to check how, if at all, the population mean departs from the moderate response of 4, the null hypothesis is

$$H_0 : \mu = 4.0$$

The alternative hypothesis is then

$$H_a : \mu \neq 4.0$$

The null hypothesis states that, on the average, the population response is politically "moderate, middle of road." The alternative states that the mean falls in the liberal direction ($\mu < 4$) or in the conservative direction ($\mu > 4$).
3. *Test statistic*: The 627 responses in Table 6.2 are summarized by $\bar{Y} = 4.032$ and $s = 1.257$. The estimated standard error of the sampling distribution of $\bar{Y}$ is

$$\hat{\sigma}_{\bar{Y}} = \frac{s}{\sqrt{n}} = \frac{1.257}{\sqrt{627}} = .050$$

The value of the test statistic is, therefore,

$$z = \frac{\bar{Y} - \mu_0}{\hat{\sigma}_{\bar{Y}}} = \frac{4.032 - 4.0}{.050} = .64$$

The sample mean falls .64 estimated standard errors above the null hypothesis value of the mean.
4. *P-value*: The P-value is the two-tail probability that z would exceed .64 in absolute value, if H_0 were true. Figure 6.4 portrays the P-value. From the normal probability table (Table A), this two-tail probability equals $P = 2(.2611) = .52$. If the population mean ideology were 4.0, then the probability equals .52 that a sample mean for $n = 627$ subjects would fall at least as far from 4.0 as the observed $\bar{Y}$ of 4.032. That is, P is the probability that $\bar{Y}$ is at least as contradictory to H_0 as the observed $\bar{Y}$.
5. *Conclusion*: To summarize the evidence about the null hypothesis, we report the P-value of $P = .52$. This value is not small, so it does not contradict the null hypothesis. If H_0 were true, the data we observed would not be unusual. It is plausible that the population mean response is 4.0, showing no tendency in the conservative or liberal direction. Generally, researchers do not regard the evidence against H_0 as strong unless P is very small, say, $P < .05$ or $P < .01$.

□

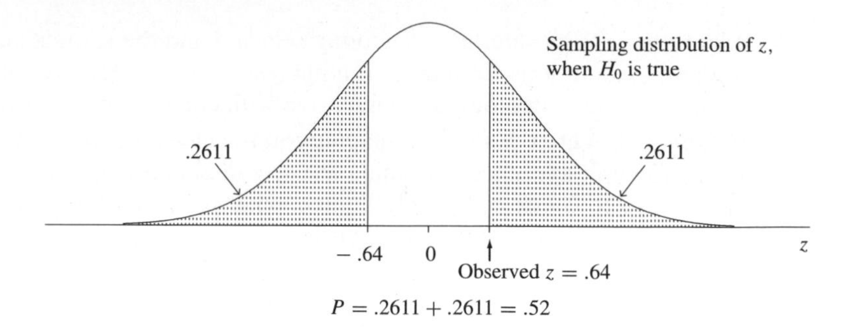

Figure 6.4 Calculation of P-value for Example 6.2. For two-sided alternatives, P is a two-tail probability.

Effect of Sample Size on *P*-values

In Example 6.2, suppose $\bar{Y} = 4.032$ and $s = 1.257$ were based on a sample of size $n = 6270$ instead of $n = 627$. The standard error then decreases, with

$$\hat{\sigma}_{\bar{Y}} = \frac{s}{\sqrt{n}} = \frac{1.257}{\sqrt{6270}} = .0159$$

The test statistic increases to

$$z = \frac{\bar{Y} - \mu_0}{\hat{\sigma}_{\bar{Y}}} = \frac{4.032 - 4.0}{.0159} = 2.01$$

This has P-value equal to $P = 2(.0222) = .044$. The same difference between $\bar{Y}$ and μ_0 based on a larger sample size results in a smaller P-value. Naturally, the larger the sample size, the more certain we can be that sample deviations from H_0 are indicative of true population deviations. For a given size of effect, smaller P-values result from larger sample sizes. In particular, notice that even a small departure of the sample mean from the value in the null hypothesis can yield a small P-value if the sample size is large.

A related phenomenon holds with estimation methods. As n increases, confidence intervals for means decrease in width, leading to improved precision in estimating μ.

Correspondence Between Results of Tests and Confidence Intervals

Conclusions using significance tests are consistent with conclusions using confidence intervals. If a test says that a particular value is plausible for the parameter, then so does a confidence interval.

Example 6.3 Confidence Interval for Mean Political Ideology

An alternative inferential approach to the significance test in Example 6.2 constructs a confidence interval for the population mean political ideology. Since $\bar{Y} = 4.032$ and

$\hat{\sigma}_{\bar{Y}} = .050$, a 95% confidence interval for μ is

$$\bar{Y} \pm 1.96\hat{\sigma}_{\bar{Y}} = 4.032 \pm 1.96(.050) = 4.03 \pm .10, \quad \text{or } (3.93, 4.13)$$

At the 95% confidence level, these are the plausible values for μ.

This confidence interval indicates that 4.0 is a plausible value for μ, since it falls inside the confidence interval. Thus, it is not surprising that the P-value ($P = .52$) in testing H_0: $\mu = 4.0$ against H_a: $\mu \neq 4.0$ in Example 6.2 was not small. In fact, whenever $P > .05$ in a test of H_0: $\mu = \mu_0$ against H_a: $\mu \neq \mu_0$, a 95% confidence interval for μ necessarily contains the null hypothesis value μ_0 of μ. Similarly, suppose that a confidence interval suggests that a particular number is implausible for μ, that number falling outside the confidence interval. Then, a small P-value results from testing the null hypothesis that μ equals that number. In this sense, results of confidence intervals and of two-sided significance tests are consistent. Section 6.4 discusses further the connection between the two methods. □

One-Sided Significance Tests

Two other forms of alternative hypotheses are sometimes used. They have the directional form

$$H_a : \mu > \mu_0 \qquad \text{and} \qquad H_a : \mu < \mu_0$$

The alternative hypothesis H_a: $\mu > \mu_0$ applies when the purpose of the test is to detect whether μ is *larger* than the particular number μ_0, whereas H_a: $\mu < \mu_0$ refers to detecting whether μ is *smaller* than that value.

The alternative hypotheses H_a: $\mu > \mu_0$ and H_a: $\mu < \mu_0$ are called ***one-sided***. They apply when the researcher predicts a deviation from H_0 in a particular direction. By contrast, the two-sided alternative H_a: $\mu \neq \mu_0$ applies when the researcher wishes to detect *any* type of deviation of μ from μ_0. This choice is made before analyzing the data.

For the one-sided alternative $H_a : \mu > \mu_0$, P is the probability of a z-score above the observed z-score (i.e., to the right of it on the real number line) when H_0 is true. Equivalently, P is the probability of a sample mean above the observed value of $\bar{Y}$. These $\bar{Y}$ values are the ones that provide at least as much evidence in favor of $H_a : \mu > \mu_0$ as the observed value. So, P equals the tail probability to the right of the observed z-score under the standard normal curve, as Figure 6.5 portrays. A z-score of .64 results in $P = .26$ for this alternative.

For H_a: $\mu < \mu_0$, P is the tail probability to the left of the observed z-score under the standard normal curve. A z-score of $-.64$ results in $P = .26$ for this alternative, and a z-score of .64 results in $P = 1 - .26 = .74$.

Example 6.4 Mean Weight Change in Anorexic Girls

This example refers to a study that compared various treatments for young girls suffering from anorexia. (The data, courtesy of Prof. Brian Everitt, Institute of Psychiatry, London, are shown in Table 12.19 in Chapter 12, where they are analyzed more fully.)

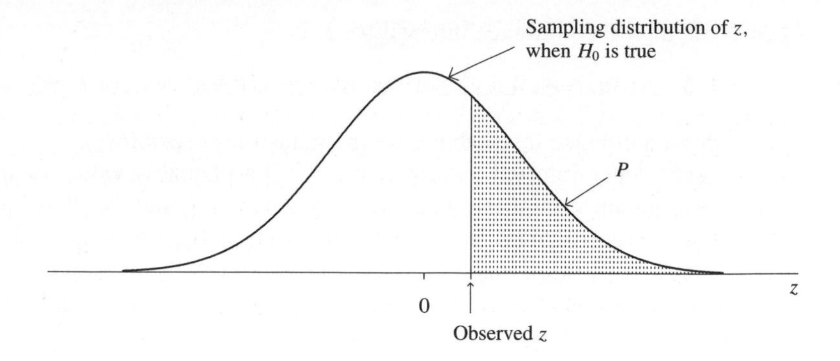

Figure 6.5 Calculation of P-value in Testing H_0: $\mu = \mu_0$ Against H_a: $\mu > \mu_0$. The P-value is the probability of values to the right of the observed test statistic.

For each girl, weight was measured before and after a fixed period of treatment. The variable of interest was the change in weight, that is, weight at the end of the study minus weight at the beginning of the study. The change in weight was positive if the girl gained weight and negative if she lost weight. The treatments were designed to aid weight gain.

Let μ denote the population mean change in weight for the cognitive behavioral treatment, for the population represented by this sample of girls. If this treatment has beneficial effect, as expected, then μ is positive. To test for no effect of treatment versus a positive mean weight change, we test H_0: $\mu = 0$ against H_a: $\mu > 0$.

Software (SPSS) used to analyze the data reports the summary results:

Variable	Number of Cases	Mean	SD	SE of Mean
CHANGE	29	3.007	7.309	1.357

Thus, $n = 29$ girls received this treatment, and the mean weight change was $\bar{Y} = 3.01$ pounds and the sample standard deviation (SD) was $s = 7.31$. The sample mean had an estimated standard error (SE) of $\hat{\sigma}_{\bar{Y}} = s/\sqrt{n} = 7.31/\sqrt{29} = 1.357$. The test statistic equals

$$z = \frac{\bar{Y} - \mu_0}{\hat{\sigma}_{\bar{Y}}} = \frac{3.01 - 0}{1.36} = 2.22$$

From Table A, the one-tail P-value above 2.22 equals .0132. A P-value of .013 provides relatively strong evidence against the null hypothesis and in favor of the alternative that the mean change in weight is positive.

The sample size here ($n = 29$) is borderline for using large-sample methods. Section 6.5 shows that the same conclusion results from small-sample methods. □

Implicit One-Sided Null Hypotheses

Example 6.4 showed that if $\mu = 0$, then the probability equals .013 of observing a sample mean weight gain of 3.01 or greater for a sample of size 29. Now, suppose $\mu < 0$; that is, the true mean weight change is negative. Then the probability of observing $\bar{Y} \geq 3.01$ would be even smaller than .013. For example, a sample value of $\bar{Y} = 3.01$ is even less likely when $\mu = -5$ than when $\mu = 0$, since the sample value of 3.01 is farther out in the tail of the sampling distribution of $\bar{Y}$ when $\mu = -5$ than when $\mu = 0$. Thus, rejection of H_0: $\mu = 0$ in favor of H_a: $\mu > 0$ also inherently rejects the broader null hypothesis of H_0: $\mu \leq 0$. In other words, one concludes that $\mu = 0$ is false *and* that $\mu < 0$ is false.

The Choice of One-Sided Versus Two-Sided Tests

In practice, two-sided tests are much more common than one-sided tests. Even if a researcher predicts the direction of an effect, two-sided tests permit the detection of an effect that falls in the opposite direction. This practice coincides with the ordinary approach in estimation. Confidence intervals are two-sided, obtained by adding and subtracting some quantity from the point estimate. One can form one-sided confidence intervals, for instance concluding that a population mean is *at least* equal to 7 (i.e., between 7 and ∞). In practice, though, two-sided intervals are much more common.

In deciding whether to use a one-sided or a two-sided alternative hypothesis in a particular exercise, consider the purpose of the test. A statement such as "Test whether the mean has *changed*" suggests a two-sided alternative, to allow for increase or decrease. "Test whether the mean has *increased*" suggests the one-sided alternative, H_a: $\mu > \mu_0$.

In either the one-sided or two-sided case, both hypotheses refer to the population mean μ, not the sample mean $\bar{Y}$. Hypotheses always refer to population parameters, not sample statistics. There is no uncertainty or need to conduct statistical inference about sample statistics, since we can calculate their values exactly once we have the data.

Table 6.3 summarizes the elements of large-sample significance tests for population means.

6.3 Significance Test for a Proportion

For a qualitative variable, each measurement falls in one of a set of categories. Statistical inference refers to the proportions in the categories. For instance, one might test a hypothesis about the population proportion π planning to vote for the Democratic candidate for President. This section presents a large-sample significance test for population proportions. The test is similar to the test for a mean. It utilizes the approximate normal sampling distribution of the sample proportion $\hat{\pi}$.

Elements of the Test

1. ***Assumptions***

As usual, the method assumes random sampling. The size of the sample must be sufficiently large that the sampling distribution of $\hat{\pi}$ is approximately normal. (The discussion following the examples presents sample size guidelines.)

TABLE 6.3 The Five Elements of Large-Sample Significance Tests for Population Means

1. Assumptions
   $n \geq 30$
   Random sample
   Quantitative variable
2. Hypotheses
   H_0: $\mu = \mu_0$
   H_a: $\mu \neq \mu_0$ (or H_a: $\mu > \mu_0$ or H_a: $\mu < \mu_0$)
3. Test statistic
   $z = \dfrac{\bar{Y} - \mu_0}{\hat{\sigma}_{\bar{Y}}}$ where $\hat{\sigma}_{\bar{Y}} = \dfrac{s}{\sqrt{n}}$
4. P-value
   In standard normal curve, use
   P = Two-tail probability for H_a: $\mu \neq \mu_0$
   P = Probability to right of observed z-value for H_a: $\mu > \mu_0$
   P = Probability to left of observed z-value for H_a: $\mu < \mu_0$
5. Conclusion
   Report P-value. Smaller P provides stronger evidence against H_0 and supporting H_a

2. ***Hypotheses***

The null hypothesis has form

$$H_0 : \pi = \pi_0$$

where π_0 denotes a particular proportion value between 0 and 1. The most common alternative hypothesis is

$$H_a : \pi \neq \pi_0$$

This two-sided alternative states that the true proportion differs from the value in the null hypothesis. Other forms, less common, are the one-sided alternatives

$$H_a : \pi > \pi_0 \text{ and } H_a : \pi < \pi_0$$

These apply when the researcher predicts a deviation of π from π_0 in a certain direction.

3. *Test Statistic*

From Section 5.3, the sampling distribution of the sample proportion $\hat{\pi}$ has mean π and standard error $\sigma_{\hat{\pi}} = \sqrt{\pi(1-\pi)/n}$. When H_0 is true, $\pi = \pi_0$, so the sampling distribution has mean π_0 and standard error $\sigma_{\hat{\pi}} = \sqrt{\pi_0(1-\pi_0)/n}$.

The test statistic is

$$z = \frac{\hat{\pi} - \pi_0}{\sigma_{\hat{\pi}}} = \frac{\hat{\pi} - \pi_0}{\sqrt{\pi_0(1-\pi_0)/n}}$$

This measures the distance of the sample proportion from the null hypothesis value, in standard error units.

The z test statistic has the same form as in large-sample tests for a mean, namely,

Form of *z* Test Statistic

$$z = \frac{\text{Estimate of parameter} - \text{null hypothesis value of parameter}}{\text{Standard error of estimator}}$$

Here, the estimate $\hat{\pi}$ of the proportion replaces the estimate $\bar{Y}$ of the mean, the hypothesized proportion π_0 replaces the hypothesized mean μ_0, and the standard error $\sigma_{\hat{\pi}}$ of the sample proportion replaces the standard error $\sigma_{\bar{Y}}$ of the sample mean. For large samples, the sampling distribution of the z test statistic is the standard normal distribution, when H_0 is true.

4. *P-Value*

The calculation of the P-value is the same as in tests for a mean. For the alternative H_a: $\pi \neq \pi_0$, P is the two-tail standard normal probability that z has absolute value larger than the absolute value of the observed z-value. See Figure 6.6. This probability is double the single-tail probability beyond the observed z-value. For a one-sided alternative, the P-value is a one-tail probability. For instance, H_a: $\pi > \pi_0$ predicts that the true proportion is larger than the null hypothesis value; its P-value is the probability to the right of the observed value of z under the standard normal curve. For $H_a : \pi < \pi_0$, the P-value is the probability to the left of the observed z-value.

5. *Conclusion*

One summarizes the test by reporting the P-value. As usual, the smaller the P-value, the more strongly the data contradict H_0 and support H_a.

As you read the examples in this section, notice the parallel between each element of the test and the corresponding element for a test about a mean.

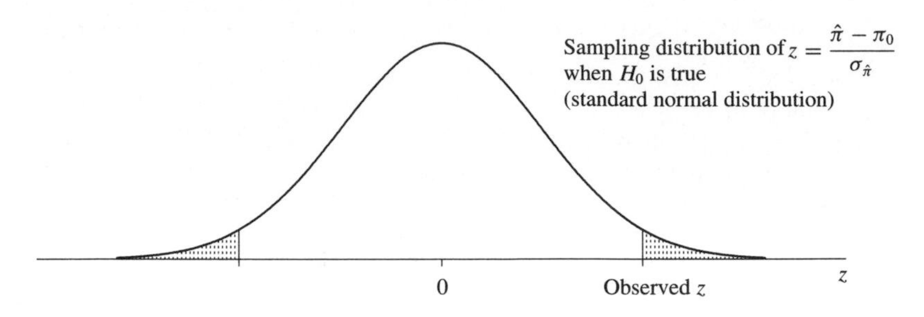

Figure 6.6 Calculation of P-value in Testing H_0: $\pi = \pi_0$ Against H_a: $\pi \neq \pi_0$. The two-sided alternative has a two-tail probability.

Example 6.5 Government Responsibility for Income Inequality

Do you think it should or should not be the government's responsibility to reduce income differences between the rich and poor? Let π denote the population proportion of American adults who believe it should be. If $\pi < .5$, this is a minority of the population, whereas if $\pi > .5$, it is a majority. One can analyze whether the sample data indicate that π is in either of these ranges by testing $H_0 : \pi = .5$ against $H_a : \pi \neq .5$.

In the 1991 General Social Survey of 1227 adults, 591 people responded that it should be the government's responsibility to reduce income differences. The estimate of π equals $591/1227 = .482$. The standard error of $\hat{\pi}$ when H_0: $\pi = .5$ is true is

$$\sigma_{\hat{\pi}} = \sqrt{\frac{\pi_0(1 - \pi_0)}{n}} = \sqrt{\frac{(.5)(.5)}{1227}} = .0143$$

The value of the test statistic is, therefore,

$$z = \frac{\hat{\pi} - \pi_0}{\sigma_{\hat{\pi}}} = \frac{.482 - .50}{.0143} = -1.28$$

From Table A, the two-tail P-value for testing H_0: $\pi = .5$ against H_a: $\pi \neq .50$ is $P = 2(.1003) = .20$. If H_0 is true (i.e., if $\pi = .50$), the probability equals .20 that sample results would be as extreme in one direction or the other as in this sample. This P-value is not small, so it does not provide much evidence against H_0. It seems plausible that $\pi = .50$. With this sample, one cannot determine whether the population proportion is less than, equal to, or greater than .50. □

In calculating the standard error, we substituted the null hypothesis value π_0 for the population proportion π in the formula for the true standard error. This differs from confidence intervals, in which the sample proportion $\hat{\pi}$ substitutes for π. The parameter values of sampling distributions in tests are based on the assumption that H_0 is true, since the P-value is calculated under that assumption. This is why one uses π_0 in standard errors for tests. By contrast, the confidence interval method does not have a hypothesized value for π, so that method substitutes the point estimate $\hat{\pi}$ for π in the standard error.

Theoretically, it is not incorrect to substitute the sample proportion in the standard error for the test. One simply obtains a slightly different answer for the test statistic and P-value, but both approaches work well for large n. If one does the test that way, an advantage is that the result necessarily agrees with conclusions from confidence intervals. A disadvantage is that the normal approximation for the sampling distribution is somewhat poorer, especially for proportions close to 0 or 1.

Never "Accept H_0"

A small P-value provides evidence against H_0, since the observed sample result would be unlikely if H_0 were true. On the other hand, if the P-value is not small, the null hypothesis is plausible. In this case, the conclusion is sometimes reported as "Do not reject H_0," since the data do not contradict H_0.

When the P-value is not small, failure to reject H_0 does not mean one can "accept H_0." The population proportion has other plausible values besides the number in the null hypothesis. In addition, the failure to obtain a small P-value may be due to the sample size being too small to estimate the true proportion precisely.

For instance, Example 6.5 showed that $\hat{\pi} = .482$ for $n = 1227$ provides a P-value of .20 in testing H_0: $\pi = .5$ against H_a: $\pi \neq .50$. Thus, it is plausible that $\pi = .50$, but other values are also plausible. For instance, a 95% confidence interval for π is

$$\hat{\pi} \pm 1.96\sqrt{\frac{\hat{\pi}(1-\hat{\pi})}{n}} = .482 \pm 1.96\sqrt{\frac{(.482)(.518)}{1227}} = .482 \pm .028, \text{ or } (.45, .51)$$

This interval shows a range of plausible values for π. Even though insufficient evidence exists to conclude that $\pi \neq .5$, it is improper to conclude that necessarily $\pi = .5$. The data do not contradict H_0, but we need a larger sample size to determine whether a majority or minority of the population believe that government has the responsibility to reduce income differences between the rich and poor. For instance, if $\hat{\pi} = .482$ had been based on $n = 5000$ instead of $n = 1227$, you can verify that the test statistic $z = -2.55$ and the P-value = .01. That P-value provides strong evidence against $H_0 : \pi = .50$ and suggests that fewer than half believe it is government's responsibility to reduce income differences. In that case, though, the 95% confidence interval for π equals (.468, .496), indicating that π is quite close to .50 in practical terms.

Of course, we could have used the confidence interval approach from the start, rather than a significance test, to gather information about the value of π. The confidence interval is more informative, since it displays the entire set of plausible values for π rather than merely indicating whether $\pi = .50$ is plausible.

Sample Size Requirement for Test

We next present a guideline about how large the sample size should be to use the large-sample test for a proportion. When π_0 is between .3 and .7, the familiar rule for means of $n \geq 30$ ensures an adequate sample size. A more general rule that applies for all π_0 is based on the normal approximation for the sampling distribution of $\hat{\pi}$, under H_0.

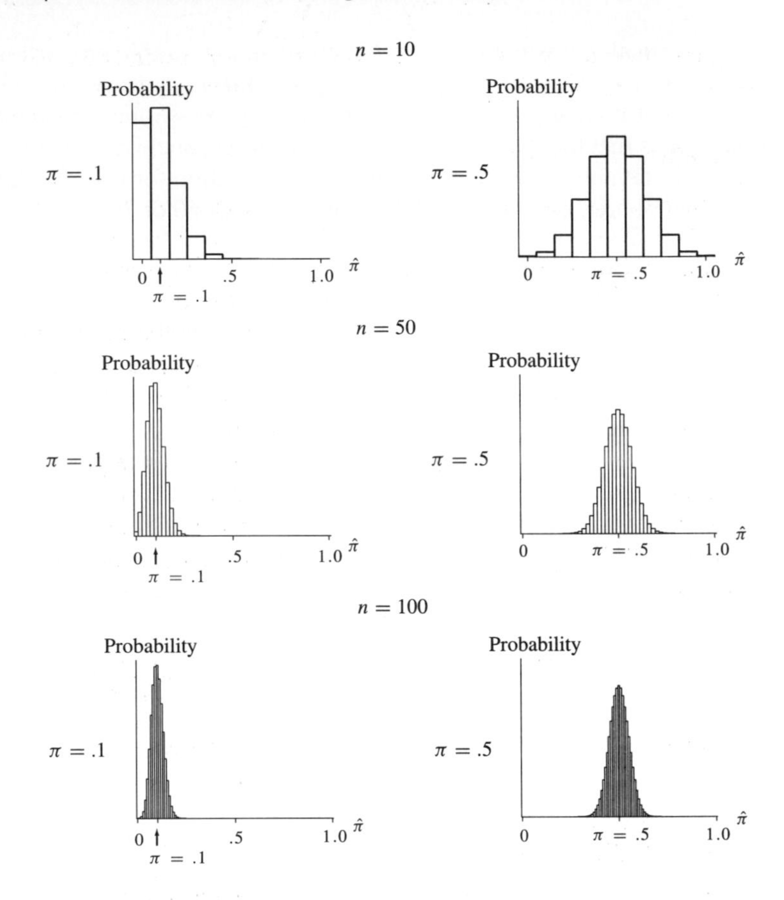

Figure 6.7 Sampling Distribution of $\hat{\pi}$ When $\pi = .1$ or $.5$, for $n = 10, 50, 100$

This approximation is good when

$$n > \frac{10}{\min(\pi_0, 1 - \pi_0)}$$

where the notation $\min(\pi_0, 1-\pi_0)$ denotes the minimum of the numbers π_0 and $1-\pi_0$. For example, for testing H_0: $\pi = .5$, we need $n > 10/.5 = 20$. For testing H_0: $\pi = .9$ or $H_0 : \pi = .1$, we need $n > 10/.1 = 100$.

The sampling distribution of $\hat{\pi}$ is more skewed when π is near 0 or near 1 than when π is near the middle of the range. Figure 6.7 illustrates. For instance, when $\pi = .1$, the sample proportion $\hat{\pi}$ can't fall much below .1 since it must be positive, but it could fall considerably above .1. The sample size requirement reflects the fact that a symmetric bell shape for the sampling distribution of $\hat{\pi}$ requires larger sample sizes when π is near 0 or 1 than when π is near .5. In Example 6.5, the sample size of $n = 1227$ was more than adequate to test H_0: $\pi = .5$. One can use a small-sample test introduced in Section 6.6 when the sample size requirement is not satisfied.

Interpreting the *P*-Value

In summary, tests describe whether the data are consistent with H_0 by reporting the P-value. This is the one- or two-tail probability beyond the observed result, calculated under the assumption that H_0 is true. When the P-value is small, the data contradict H_0; the observed data would be unusual if H_0 were true.

A common error is to misinterpret the P-value as the probability that H_0 is true. Classical statistical methods apply probability statements to variables and to statistics, not to parameters. In reality, the null hypothesis H_0 is not a matter of probability; it is either true or not true, and we simply do not know which is the case. In Example 6.5, either π equals .50, or π does not equal .50. A proper interpretation for $P = .20$ is as follows: *If* H_0 *were true*, the probability would be .20 that the sample proportion $\hat{\pi}$ would fall at least as many standard errors from the null hypothesis value of .50 as the observed $\hat{\pi}$ does. That is, P is the probability that $\hat{\pi}$ is at least as contradictory to H_0 as the observed value, *under the assumption that* H_0 *is true.*

6.4 Decisions and Types of Errors in Tests of Hypotheses

In significance tests, the P-value summarizes the evidence about H_0. The smaller the P-value, the more strongly the data contradict H_0.

α-Level

It is sometimes necessary to decide whether the evidence against H_0 is strong enough to reject it. The usual approach bases the decision on whether the P-value falls below a prespecified cutoff point.To illustrate, one might reject H_0 if $P \leq .05$, but conclude that the evidence is not strong enough to reject H_0 if $P > .05$. The boundary value .05 is called the α-*level* of the test.

α-Level

The **α-*level*** is a number such that one rejects H_0 if the P-value is less than or equal to it. The α-level is also called the *significance level* of the test. The most common α-levels are .05 and .01.

Like the choice of a confidence coefficient for a confidence interval, the choice of the α-level for a test reflects how cautious the researcher wants to be. The smaller the α-level, the stronger the evidence must be to reject H_0. To avoid bias in the decision-making process, one selects the α-level *before* analyzing the data.

Table 6.4 summarizes the two possible conclusions for a test with α-level .05. The null hypothesis is either "rejected" or "not rejected." If H_0 is rejected, then H_a is accepted; in this case, H_a seems more valid than H_0. If H_0 is not rejected, then H_0 is plausible, but other parameter values are also plausible. Thus, H_0 is never "accepted." In this case, results are inconclusive, and the test does not identify either hypothesis as more valid.

TABLE 6.4 Possible Conclusions in a Test of Hypothesis with α-Level .05

	Conclusion	
P-Value	H_0	H_a
$P \leq .05$	Reject	Accept
$P > .05$	Do not reject	Do not accept

Example 6.6 Adding Decisions to Previous Examples

Example 6.2 tested the hypothesis $H_0 : \mu = 4.0$ about mean political ideology. We now use an α-level of $\alpha = .05$ to guide us in making a decision about H_0. Since the P-value equaled $P = .52$, we have $P > .05$ and insufficient evidence to reject H_0. In other words, we cannot conclude that the mean ideology in the population differs from the moderate value of 4.0.

Now consider Example 6.4 on the hypothesis $H_0 : \mu = 0$ about the mean weight gain for a sample of women suffering from anorexia. The P-value was .013. This is less than .05, so this result provides sufficient evidence to reject H_0 in favor of $H_a : \mu > 0$; we conclude that the treatment does produce an increase in mean weight. This type of conclusion is sometimes phrased as "The increase in the mean is *statistically significant* at the .05 level." Since $P = .013$ is not less than .010, the result is not significant at the .01 level. In fact, *the P-value is the smallest level for α at which the results are significant.* That is, we would reject H_0 if α were any level above .013. □

In our opinion, it is preferable to report the P-value rather than to indicate simply whether the result is significant at a particular α-level. Reporting the P-value has the advantage that the reader can tell whether the result is significant at any level. The P-values of .049 and .001 are both "significant at the .05 level," but the second case provides much stronger evidence than the first case. Likewise, P-values of .049 and .051 provide, in practical terms, the same amount of evidence about H_0. It is artificial to call one result "significant" and the other "nonsignificant."

Rejection Regions

The null hypothesis contains a single possible value for the parameter. Using the terminology "Do not reject H_0" instead of "Accept H_0" emphasizes that that value is merely one of many plausible values. Because of sampling error, there is a range of plausible values rather than just the H_0 value, so one can never accept a null hypothesis. The reason "accept H_a" terminology is permissible for the alternative hypothesis is that when the P-value is sufficiently small, the entire range of plausible values for the parameter fall within the broad range of numbers contained in H_a.

The collection of test statistic values for which the test rejects H_0 at a particular α-level is called the ***rejection region***. For example, the rejection region for a test of level $\alpha = .05$ is the set of test statistic values for which $P \leq .05$.

For large-sample two-sided tests, for instance, the two-tail probability that forms the P-value is $\leq .05$ whenever the test statistic satisfies $|z| \geq 1.96$. In other words, the rejection region for an $\alpha = .05$ level test consists of values of z for which $|z| \geq 1.96$, that is, values of z resulting from the estimate of the parameter falling at least 1.96 standard errors from the null hypothesized value.

Type I and Type II Errors

Because of sampling error, decisions in tests of hypotheses always have some uncertainty. The decision could be erroneous, just as a confidence interval can falsely predict where the parameter falls. There are two types of potential errors, conventionally called *Type I* and *Type II* errors.

Type I and Type II Errors

A ***Type I error*** occurs when H_0 is rejected, even though it is true.
A ***Type II error*** occurs when H_0 is not rejected, even though it is false.

A decision in a test has four possible results. These refer to the two possible decisions combined with the two possible conditions for H_0. Table 6.5 shows these four results.

TABLE 6.5 The Four Possible Results of Making a Decision in a Test; Two of These Refer to Incorrect Decisions

		Decision	
		Reject H_0	*Do not reject H_0*
Condition of H_0	H_0 true	Type I error	Correct decision
	H_0 false	Correct decision	Type II error

Suppose we test H_0 at the $\alpha = .05$ level, rejecting H_0 if $P \leq .05$. For example, for the large-sample test about a mean or proportion with two-sided alternative, we reject H_0 when $|z| \geq 1.96$. These z values form the rejection region. For a continuous sampling distribution such as the normal distribution, the probability of rejecting H_0 when it is true is exactly .05, since the probability of the values in the rejection region is .05. But this is precisely the α-level.

The probability of a Type I error is the α-level for the test.

With $\alpha = .05$, if the null hypothesis is true, the probability equals .05 of making a Type I error and rejecting that (true) null hypothesis. One controls the probability of a Type I error by the choice of the α-level.

The more serious the consequences of a Type I error, the smaller α should be. For exploratory research conducted for data snooping—scanning several hypotheses to see which might warrant further investigation—one would not be too stringent (say, $\alpha = .10$). On the other hand, suppose that the decision has serious policy implications. For instance, suppose that the alternative hypothesis states that a newly developed drug is better than the one currently used to treat a particular illness; if we reject H_0, then the new drug will be prescribed instead of the current one to treat the illness. Then, we might prefer tougher standards, such as $\alpha = .01$. In that case, the data must contradict H_0 very strongly in order to reject it, to lessen the chance of Type I error.

Relation Between *P*(Type I Error) and *P*(Type II Error)

A Type II error occurs in not rejecting H_0 even though it is false. We shall see in Section 6.7 that the probability this happens depends on just how far the actual value of the parameter falls from H_0. If the true parameter value is nearly equal to the value hypothesized in H_0, the probability of a Type II error might be quite high, whereas it would be smaller for more distant values of the parameter. The farther the true value of the parameter falls from the value specified in H_0, the less likely the sample is to fail to detect the difference and result in a Type II error.

The probability of Type I error and the probability of Type II error are inversely related. The smaller the α-level and hence the probability of Type I error, the larger the probability of Type II error. In other words, the stronger the evidence required to reject H_0 (i.e., the smaller the α-value), the more likely it becomes that we will fail to detect a real difference. If we tolerate only an extremely small chance of a Type I error, then the test may be unlikely to reject the null hypothesis even if it is false.

For a fixed probability of Type I error, we can decrease the probability of Type II error by selecting a larger sample. That is, the larger the sample size, the more likely we are to reject a false null hypothesis at a particular α-level. To keep both the probabilities of Type I and Type II errors at low levels, it may be necessary to use a very large sample size.

In making a decision in a test, we do not know whether we have made a Type I or Type II error, just as we do not know whether a particular confidence interval truly contains the unknown parameter value. However, we can control the probability of an incorrect decision for either type of inference. Although we do not know whether the conclusion in a particular test is correct, we justify the procedure in terms of the long-run proportions of Type I and Type II errors.

Except in Section 6.7, we shall not study calculation of the probability of a Type II error, since these calculations are quite complex. In practice, making a decision in a test only requires setting α, the probability of Type I error. One should realize, though, that the probability of a Type II error may be quite large when the sample size is small. In other words, the reason for not rejecting H_0 may be that for that sample size, the test simply does not have a very high chance of detecting the actual deviation from H_0.

Equivalence Between Confidence Intervals and Tests of Hypotheses

We now elaborate on the equivalence between decisions from two-sided tests and conclusions from confidence intervals, first alluded to in Example 6.3. Consider the large-sample test of

$$H_0 : \mu = \mu_0 \qquad \text{versus} \qquad H_a : \mu \neq \mu_0$$

When $P \leq .05$, H_0 is rejected at the $\alpha = .05$ level. This implies that the test statistic $z = (\bar{Y} - \mu_0)/\hat{\sigma}_{\bar{Y}}$ is at least 1.96 in absolute value. That is, $\bar{Y}$ falls more than $1.96\hat{\sigma}_{\bar{Y}}$ from μ_0. But if this happens, then the 95% confidence interval for μ, namely, $\bar{Y} \pm 1.96\hat{\sigma}_{\bar{Y}}$, does not contain the null hypothesis value μ_0. See Figure 6.8.

In other words, rejecting H_0: $\mu = \mu_0$ at the $\alpha = .05$ level is equivalent to the 95% confidence interval for μ not containing μ_0. These two inference procedures are consistent. If a confidence interval indicates that a particular number μ_0 is not plausible for the value of μ, then we would reject H_0: $\mu = \mu_0$ in favor of H_a: $\mu \neq \mu_0$ with the test. The null hypothesis H_0 is rejected at the α-level equal to one minus the confidence coefficient; for instance, $\alpha = .05$ for a 95% confidence interval. This level is both the probability of Type I error for the test and the probability that the confidence interval does not contain the parameter.

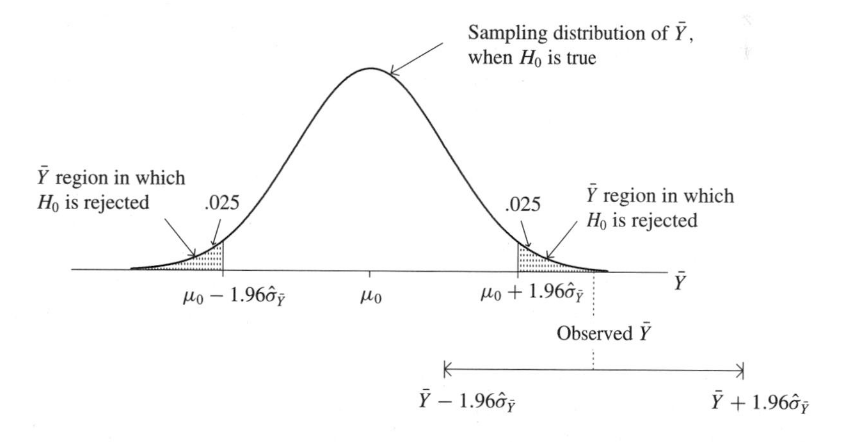

Figure 6.8 Relationship Between Confidence Interval and Hypothesis Test

In testing H_0: $\mu = \mu_0$ against H_a: $\mu \neq \mu_0$, suppose one rejects H_0 at the .05 α-level. Then, the 95% confidence interval for μ does not contain μ_0. The 95% confidence interval for μ consists of those μ_0 values for which one does not reject H_0: $\mu = \mu_0$ at the .05 α-level.

Example 6.2 tested a hypothesis about mean political ideology. The P-value for testing $H_0 : \mu = 4.0$ against $H_a : \mu \neq 4.0$ was $P = .52$. At the $\alpha = .05$ level, H_0: $\mu = 4.0$ is not rejected; it is plausible that $\mu = 4.0$. Example 6.3 showed that a 95% confidence interval for μ is (3.93, 4.13), which contains $\mu_0 = 4.0$. On the other hand, we saw that if the results had been based on $n = 6270$ instead of $n = 627$, then $P = .044$. In that case, since $P = .044$ is less than .05, we can reject H_0 at the $\alpha = .05$ level; that is, 4.0 is not a plausible value for μ. In fact, you can verify that a 95% confidence interval in this case equals (4.001, 4.063), not containing 4.0.

Statistical and Practical Significance

Anyone who uses significance tests should understand the distinction between *statistical* and *practical* significance. A very small P-value, such as $P = .001$, does not imply an "important" finding in any practical sense. This merely means that if H_0 were true, the observed results would be very unusual. Even if H_0 is false, though, the true value of the parameter may be close to the null hypothesized value. In this case, the difference may not be significant in practical terms. If the sample size is very large, small P-values can occur even though the difference is small.

Example 6.7 Mean Political Ideology in 1994

The mean of 4.03 for political ideology in Example 6.2 refers to a sample taken in 1978. For the General Social Survey of 1994, the counts in the seven categories were (71, 328, 378, 1049, 472, 478, 103). For a scoring of 1 through 7 for this seven-point scale, these 2879 observations have a mean of 4.17 and a standard deviation of 1.39. It appears that the mean level of conservatism increased only slightly between 1978 and 1994.

As in Example 6.2, we test whether the population mean differs from the moderate ideology score of 4.0. Then, $\hat{\sigma}_{\bar{Y}} = s/\sqrt{n} = 1.39/\sqrt{2879} = .026$, and

$$z = \frac{\bar{Y} - \mu_0}{\hat{\sigma}_{\bar{Y}}} = \frac{4.17 - 4.0}{.026} = 6.6$$

The P-value is $P = .000000000005$. There is *extremely* strong evidence that the true mean exceeds 4.0, that is, that the true mean falls on the conservative side of moderate. On the other hand, on a scale of 1 to 7, 4.17 is quite close to the moderate score of 4.0. Although the difference of .17 between the sample mean of 4.17 and the null hypothesis mean of 4.0 is highly significant statistically, the magnitude of this difference is small in practical terms. We can conclude that, on the average, the mean response on political ideology is essentially still a moderate one. □

In Example 6.2, the sample mean of 4.03 based on $n = 627$ had $P = .52$, not much evidence against H_0. On the other hand, we noted earlier that the same evidence based on $n = 6270$ yields $P = .044$. This is statistically significant at the .05 level, but not practically significant. For practical purposes, a mean political ideology of 4.03 does not differ from 4.00.

One point of this example is that larger sample sizes can provide more powerful inferences; thus, test statistics can detect deviations of smaller magnitude from H_0 than they can for smaller samples. The size of P merely measures the extent of evidence about the truth of H_0, not how far from the truth H_0 happens to be. One should always inspect the difference between the sample estimate and the hypothesized value of the parameter (e.g., between $\bar{Y}$ and μ_0, between $\hat{\pi}$ and π_0) to gauge the practical implications of a test result.

Limitations of Significance Tests Compared to Estimation

Null hypotheses such as $H_0 : \mu = \mu_0$ and $H_0 : \pi = \pi_0$ are rarely true in the social sciences. That is, rarely is the true value of the parameter *exactly* equal to the value listed in H_0. With sufficiently large samples, so that a Type II error is unlikely, these hypotheses will normally be rejected. What is more relevant is whether the true parameter value is sufficiently different from the null hypothesis value to be of importance.

Although tests of hypotheses can be useful, many social scientists and nearly all statisticians believe that significance testing is greatly overemphasized in social science research. By contrast, confidence intervals are underutilized. It is preferable to construct confidence intervals for parameters instead of performing only significance tests. A test merely indicates whether a particular parameter value is plausible; a confidence interval displays additional information, showing the entire set of plausible values. When a P-value is small, the test indicates that the parameter value in the null hypothesis is not plausible, but it tells us nothing about which potential parameter values *are* plausible. The confidence interval, on the other hand, displays those plausible values. It shows the extent to which H_0 may be false by showing whether the values in the interval are very far from the null hypothesis value. Thus, it helps us to determine whether rejection of the null hypothesis has practical importance.

To illustrate, for Example 6.7, a 95% confidence interval for μ is $\bar{Y} \pm 1.96\hat{\sigma}_{\bar{Y}} = 4.17 \pm 1.96(.026)$, or (4.12, 4.22). This indicates that the difference from the moderate score of 4.0 is of small magnitude. Although the P-value of $P = .000000000005$ provides extremely strong evidence against H_0, in practical terms the confidence interval shows that the departure from the null hypothesis is minor. On the other hand, if $\bar{Y}$ had been 5.17 (instead of 4.17), the 95% confidence interval would equal (5.12, 5.22). This indicates a more substantial difference from 4.0, the average response being close to the slightly conservative category rather than the moderate category.

A confidence interval displays the set of values that are plausible parameter values. When the P-value is not small, the confidence interval indicates whether the lack of evidence against H_0 may be due to a lack of power. A wide confidence interval containing the null hypothesis value of the parameter indicates a strong possibility of a Type II error in the test. In that case, the lack of precision of the interval estimate also indicates why it does not make sense to accept H_0, as we discussed previously. For small to moderate sample sizes, it is not unusual for a confidence interval to be wide, and this forces us to recognize the lack of precision that any inference involves.

The remainder of the text presents significance tests for a variety of situations. It is important to become familiar with the elements of these tests, if for no other reason than their frequent use in the social science literature. However, we shall also introduce confidence intervals for parameters that describe how far reality is from the hypothesized condition.

6.5 Small-Sample Inference for a Mean—The *t* Distribution

The confidence interval for means presented in Section 5.2 and the significance test for means presented in Section 6.2 both apply for large sample sizes. The large-sample assumption ensures that the sampling distribution of $\bar{Y}$ is approximately normal. It also ensures that the sample standard deviation estimate s is close to the unknown population standard deviation σ; this ensures that the estimated standard error $\hat{\sigma}_{\bar{Y}} = s/\sqrt{n}$ is sufficiently close to the true standard error $\sigma_{\bar{Y}} = \sigma/\sqrt{n}$ that one can substitute the estimate in confidence intervals and test statistics. As a rough guideline, these approximations are adequate if the sample size is at least 30.

Some studies are limited, however, to smaller sample sizes. For example, each observation may result from a long or expensive experimental procedure. A consumer group that decides to evaluate the mean repair cost resulting when a new-model automobile crashes into a brick wall at 30 miles per hour would probably not want to restrict itself to large-sample statistical methods!

The *t* Distribution

This section introduces inferential methods for small samples. The derivation of the methods assumes that the *population distribution* of the variable Y is normal. In that case, the sampling distribution of $\bar{Y}$ and of the statistic $(\bar{Y} - \mu)/\sigma_{\bar{Y}}$ is normal even for small sample sizes. (Figure 4.15 illustrated this.) However, substitution of s for σ in $\sigma_{\bar{Y}}$, as is done in practice, introduces additional variability in the sampling distribution. It is then no longer normal, but has the ***t distribution***.

_t_ Statistic; _t_ Distribution

Suppose the population distribution of a variable is normal, with parameters μ and σ. Then, for a random sample of size *n*, the sampling distribution of the ***t statistic***

$$t = \frac{\bar{Y} - \mu}{\hat{\sigma}_{\bar{Y}}} = \frac{\bar{Y} - \mu}{s/\sqrt{n}}$$

is called the ***t* distribution with $(n - 1)$ degrees of freedom.**

The t statistic resembles the z statistic of Section 6.2. The different symbol for the statistic here emphasizes that the sampling distribution is different and that it is valid in a different situation—when the population distribution is normal. The *degrees of freedom* for the t distribution, denoted by df, determine the precise shape of the distribution, as discussed shortly. This quantity refers to the divisor in the point estimate $s^2 = \left[\sum (Y - \bar{Y})^2\right]/(n-1)$ of σ^2, namely, $df = n - 1$.

The t distribution was discovered in 1908 by the statistician and chemist W. S. Gosset. At the time, Gosset was employed in the experimental unit of Guinness Breweries in Dublin, Ireland. He had only small samples available for several of his analyses for determining the best varieties of barley and hops for the brewing process. Due to company policy forbidding the publishing of trade secrets, Gosset used the pseudonym Student in articles he wrote about this result. The t statistic is often called *Student's t*.

Properties of the *t* Distribution

Before presenting methods of statistical inference for small samples, we list the major properties of the t distribution.

- The t distribution is bell-shaped and symmetric about 0. This property is shared by the sampling distribution of the z statistic, the standard normal distribution.
- The spread of the t distribution depends on the degrees of freedom. The standard deviation of the t distribution always exceeds 1, but decreases toward 1 as df (and hence n) increases. (The standard deviation equals $\sqrt{df/(df-2)}$. This value exceeds 1.0, but it decreases toward 1.0 as df increases.)
- Though the t distribution is bell-shaped about 0, the probability falling in the tails is higher than for the standard normal distribution. The larger the df value, however, the more closely it resembles the standard normal distribution, as Figure 6.9

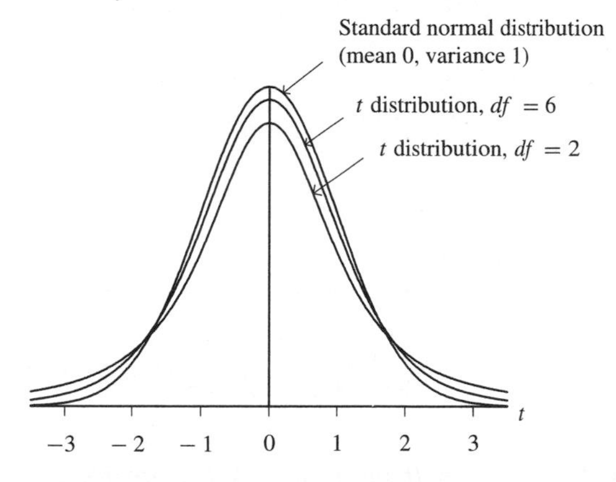

Figure 6.9 *t* Distribution Relative to Standard Normal Distribution. The t gets closer to the normal as the degrees of freedom (df) increase, and the two distributions are practically identical when $df > 30$.

illustrates. In the limit as df increases indefinitely, the two distributions are identical.

The change in the form of the t distribution as df increases is due to the increasing precision of s as a point estimate of σ in the standard error formula $\hat{\sigma}_{\bar{Y}} = s/\sqrt{n}$. Because s is a less accurate estimator of σ when $df < 30$, its presence in the denominator of the t statistic produces additional sampling error. This additional sampling error for small samples results in the t sampling distribution being more spread out than the standard normal sampling distribution of the large-sample z statistic, in which s is nearly identical to σ. As the sample size increases, s becomes a more accurate estimator of σ, and the t distribution becomes less disperse. When $df \geq 30$, the t distribution is so similar to the standard normal distribution that inference procedures for the mean using the t distribution are practically equivalent to those using the standard normal distribution.

- Table B at the end of the text lists values from the t distribution with various tail probabilities. Since the t distribution has a slightly different shape for each distinct value of df, different t-values apply for each df value. Table B lists the t-values only for the one-tail probabilities of .100, .050, .025, .010, and .005. The table denotes these by $t_{.100}$, $t_{.050}$, $t_{.025}$, $t_{.010}$, and $t_{.005}$. These same values refer to two-tail probabilities of .20, .10, .05, .02, and .01.

To illustrate Table B, suppose $df = 6$. Then, since $t_{.025} = 2.447$, 2.5% of the t distribution falls in the right-hand tail above 2.447. Figure 6.10 illustrates. By symmetry, 2.5% also lies in the left-hand tail below $-t_{.025} = -2.447$. When $df = 6$, the probability equals .05 that the absolute value of the t statistic exceeds 2.447.

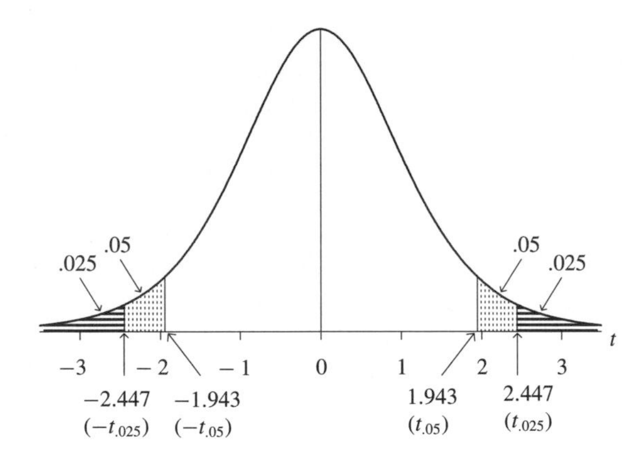

Figure 6.10 t Distribution with $df = 6$

Table B shows that as df increases, the t-score decreases to the z-score for a standard normal distribution. This reflects the t distribution becoming less disperse and more similar in appearance to the standard normal distribution as df increases. For instance, when df increases from 1 to 29 in Table B, the t-score with right-tail probability equal to .025 decreases from 12.706 to 2.045. The z-score with this right-tail probability for the standard normal distribution is $z = 1.96$. For df of about 30 or higher, the t-score is similar to the z-score of 1.96.

The last row of Table B lists the z-values for one-tail probabilities, opposite $df = \infty$ (infinity). The t-values are not printed for $df \geq 30$, since they are close to the z-values. Whenever $df \geq 30$ for a method using the t distribution, one can refer to the normal probability table (Table A) and proceed as if it uses the standard normal distribution. Computer software has the t distribution in memory for all df values, so such software does not need to use the normal approximation when $df \geq 30$.

Small-Sample Confidence Interval for a Mean

Small-sample confidence intervals and significance tests for a mean resemble those for large samples, except that they use the t distribution instead of the standard normal. We first present the confidence interval.

Small-Sample Confidence Interval for μ

For a random sample from a normal population distribution, a 95% confidence interval for μ is

$$\bar{Y} \pm t_{.025}\hat{\sigma}_{\bar{Y}} = \bar{Y} \pm t_{.025}\left(\frac{s}{\sqrt{n}}\right)$$

where $df = n - 1$ for the t-value.

The intervals use the $t_{.025}$-value, which is the t value for a right-tail probability of .025. This is because 95% of the probability for a t distribution falls between $-t_{.025}$ and $t_{.025}$. Let α denote the error probability that the confidence interval does not contain μ. For instance, for a 95% confidence interval, $\alpha = .05$. A confidence interval uses the t-score with tail probability $\alpha/2$ in each tail. For a 99% confidence interval, for instance, $\alpha = .01$, and the appropriate t-score is $t_{.005}$ for the specified df value.

Like the confidence interval in Section 5.2 for large samples, this confidence interval equals the point estimate of μ plus and minus a table value multiplied by the estimated standard error. The only difference in the formula is the substitution of the t-table value for the normal-table value, to reflect the small sample size. The t method also makes the additional assumption of a normal population distribution, which is needed for small samples. In practice, the distribution is typically not normal, and we discuss the importance of this assumption later in the section.

Example 6.8 Estimating Mean Weight Change for Anorexic Girls

Example 6.4 discussed a study that compared various treatments for young girls suffering from anorexia. The variable of interest was the change in weight from the beginning to the end of the study. For the sample of 29 girls receiving the cognitive behavioral treatment, the changes in weight were summarized by $\bar{Y} = 3.01$ and $s = 7.31$ pounds.

Example 6.4 used large-sample methods, which are of borderline acceptability with $n = 29$. Here, we use small-sample methods.

Let μ denote the population mean change in weight for this treatment. Since $n = 29$, $df = n - 1 = 28$. For a 95% confidence interval, we use $t_{.025} = 2.048$. The estimated standard error equals $\hat{\sigma} = s/\sqrt{n} = 7.31/\sqrt{29} = 1.357$. The 95% confidence interval is

$$\bar{Y} \pm t_{.025}\hat{\sigma}_{\bar{Y}} = 3.01 \pm 2.048(1.357) = 3.0 \pm 2.8, \text{ or } (0.2, 5.8)$$

We infer with 95% confidence that this interval contains the true mean weight change for this treatment. It appears that the true mean change in weight is positive, but rather small. □

Elements of a *t* Test for a Mean

We next list the five elements of a small-sample significance test for a mean.

1. *Assumptions*

A random sample is selected. The variable is quantitative and has a normal population distribution. (The method is designed for $n \leq 30$, but can be used with any size n.)

2. *Hypotheses*

The hypotheses are the same as in the large-sample test for a mean. The null hypothesis has form H_0: $\mu = \mu_0$, and the two-sided alternative hypothesis has form H_a: $\mu \neq \mu_0$. The one-sided alternative hypotheses are H_a: $\mu > \mu_0$ and H_a: $\mu < \mu_0$.

3. *Test Statistic*

The test statistic is the t statistic with $\mu = \mu_0$, namely,

$$t = \frac{\bar{Y} - \mu_0}{\hat{\sigma}_{\bar{Y}}} = \frac{\bar{Y} - \mu_0}{s/\sqrt{n}}$$

Like the z statistic, this statistic measures the distance between the sample mean and the null hypothesis value, divided by the estimated standard error of $\bar{Y}$. If H_0 is true, the sampling distribution of the t test statistic is the t distribution with $df = n - 1$.

4. *P-Value*

The calculation of the P-value uses one or two tails in the same way as the large-sample calculation, but it uses the t distribution (Table B) instead of the standard normal distribution. For the two-sided alternative hypothesis H_a: $\mu \neq \mu_0$, P is the two-tail probability of a t-value at least as large in absolute value as the observed one, if H_0 were true. Figure 6.11 depicts the two-sided P-value. As usual, the smaller the P-value, the stronger the evidence against H_0 and in favor of H_a.

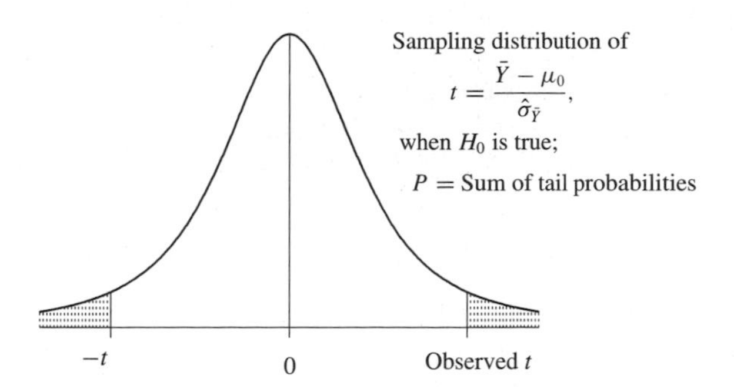

Figure 6.11 Calculation of P in Testing H_0: $\mu = \mu_0$ Against H_a: $\mu \neq \mu_0$, for Small Samples. The P-value is the total two-tail probability beyond the observed test statistic.

5. *Conclusion*

Normally, we report the P-value. For a formal decision, as in the large-sample case, reject H_0 if the P-value is no greater than some fixed α-level, such as .05 or .01.

Example 6.9 Small-Sample Test for Anorexia Data

We illustrate with a t test for the anorexia data. For the 29 observations, $\bar{Y} = 3.01$ and $\hat{\sigma}_{\bar{Y}} = 1.357$. As in Example 6.4, one might test for no effect of treatment versus a positive average weight change, by testing H_0: $\mu = 0$ against H_a: $\mu > 0$. More commonly, in practice, one would use the two-sided alternative $H_a : \mu \neq 0$. The test statistic equals

$$t = \frac{\bar{Y} - \mu_0}{\hat{\sigma}_{\bar{Y}}} = \frac{3.01 - 0}{1.357} = 2.22$$

precisely the same as the large-sample z statistic.

Now, for $n = 29$ and $df = 28$, $t = 2.048$ yields $P = .025$ for the one-sided alternative hypothesis. Since the observed $t = 2.22 > 2.048$, the one-tail P-value is $P < .025$, since a value farther out in the tail has a smaller tail probability. Moreover, $P > .01$, since Table B indicates that $t = 2.467$ has a tail probability of .01. Figure 6.12 illustrates. Table B is not detailed enough to provide the exact value of P. We could summarize the P-value for the one-sided test by reporting that $.01 < P < .025$. Table B provides enough information to determine whether the one-tailed P-value is greater than or less than .10, .05, .025, .01, and .005. If two of the tabled t-scores bracket the observed t statistic, above and below, their tail probabilities bracket the actual tail probability. For a two-sided alternative, we double the results. For instance, for these data we double the bounds of .01 and .025 to report $.02 < P < .05$.

When computer software performs the analysis, the output reports the actual P-value rather than bounds for it. Most software reports the P-value for a two-sided alternative. Table 6.6 shows how SPSS reports results for tests and confidence intervals.

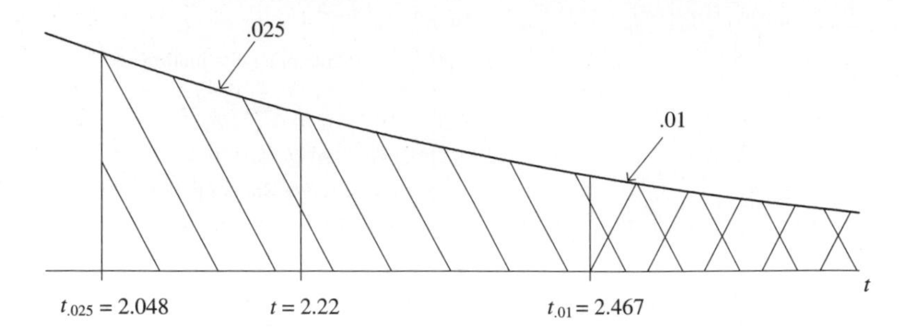

Figure 6.12 For $df = 28$, $t = 2.22$ Has a Tail Probability Between .01 and .025

For $t = 2.22$ with $df = 28$, it reports $P = .035$. The one-sided P-value is half this, or about $P = .017$. There is strong evidence against the hypothesis that the treatment has no effect. The small-sample one-sided P-value of .017 does not differ much from the value of .013 reported in Example 6.4 using large-sample methods.

TABLE 6.6

Variable	Number of Cases	Mean	SD	SE of Mean
CHANGE	29	3.0069	7.309	1.357

Mean	95% CI Lower	Upper	t-value	df	2-Tail Sig
3.01	.227	5.787	2.22	28	.035

The significance test concludes that the mean weight gain was not equal to 0. The 95% confidence interval of (0.2, 5.8) is more informative, showing just how different from 0 the true mean change is likely to be. □

Robustness for Violations of Normality Assumption

A basic assumption of the confidence interval and test using the t distribution is that the population distribution is normal. For instance, to be valid, the confidence interval in Example 6.8 and the significance test in Example 6.9 require the assumption that the true distribution of weight change for that treatment is normal. It would be impossible to verify this assumption. A histogram or stem and leaf plot can provide some information about the shape of the population distribution, but it is not precise when n is smaller than 30. In most practical problems, one has only a rough notion of the form of the population distribution when the sample size is small.

A statistical method is said to be ***robust*** if it performs adequately even when an assumption is violated. The study of the robustness of methods is important, because rarely in practice are all assumptions perfectly satisfied. Statisticians have shown that small-sample two-sided inferences for a mean using the t distribution are quite robust against violations of the assumption that the population distribution is normal. Even if the population is not normally distributed, two-sided tests and confidence intervals based on the t distribution still work quite well. The P-values and confidence coefficients are fairly accurate, the accuracy being quite good when n exceeds about 15. The test does not work so well for a one-sided test with small n when the population distribution is highly skewed. There is evidence of such skewness if you see outliers in one direction.

The results of t tests and confidence intervals are *not* robust to violations of the random sample assumption. The results may be completely invalid if the sample is not random.

Computer Software and Inference for Means

We have used the t distribution for small-sample inference about a mean and the normal distribution for large-sample inference, with $n = 30$ being a rather arbitrary dividing line. This is partly because the t table in this book (Table B) has df values only below 30, and for larger values the t-scores are practically identical to z-scores.

Computer software does not distinguish between the two cases. It uses the t distribution for all cases. It has the t distribution stored in memory for all possible df values, so it is not limited to $n < 30$. The advantage of using the t methods is that they account for the extra variability due to estimating σ by s. Though they make the extra assumption of a normal population, this is unneeded for $n > 30$; the sampling distribution of $\bar{Y}$ is then approximately normal regardless of the shape of the population, by the Central Limit Theorem (Sec. 4.4). Of course, when $n > 30$, you will get nearly identical results if you use z-scores instead of t.

Significance tests that a parameter equals a particular value μ_0 are often artificial. It is rare that we learn much by testing a hypothesis about a single population mean. The next chapter presents more realistic tests, involving comparisons of means for two populations. In most applications, we learn more by constructing a confidence interval than by performing a test. In particular, with small samples, confidence intervals are usually wide, forcing us to recognize that estimates of parameters are imprecise.

6.6 Small-Sample Inference for a Proportion—The Binomial Distribution*

The confidence interval for a population proportion π presented in Section 5.3 and the significance test presented in Section 6.3 are valid for large samples. The sampling distribution of the sample proportion $\hat{\pi}$ is then approximately normal, and one can use z-scores in tests and confidence intervals. The closer the true parameter π is to 0 or 1 for a given sample size, the more skewed the actual sampling distribution becomes, and the normal approximation may be poor (refer back to Figure 6.7). For instance, when π equals .05 or .95, a sample size of about $n = 200$ is needed before the normal distribution provides a good approximation.

In addition, for small n, the sampling distribution of $\hat{\pi}$ is highly discrete; it is concentrated at very few points, and a continuous approximation such as the normal distribution is inappropriate. If $n = 5$, for example, the only possible values for the sample proportion $\hat{\pi}$ are 0, 1/5, 2/5, 3/5, 4/5, and 1. This section introduces a small-sample test for proportions that is valid when the normal approximation is not.

The Binomial Distribution

For a categorical variable, let $\hat{\pi}$ be the sample proportion in the category of interest. This proportion equals $\hat{\pi} = X/n$, where X denotes the number of observations in that category. For example, when $n = 5$, the possible values for X are 0, 1, 2, 3, 4, and 5, which provide the sample proportion values $\hat{\pi} = X/n = 0, 1/5, 2/5, 3/5, 4/5, 1$. Small-sample tests about proportions use the sampling distribution of X, called the ***binomial distribution***.

For observations on a categorical variable with two categories, the binomial distribution applies when the following three conditions hold:

1. For a fixed number of observations n, each falls into one of two categories.
2. The probability of falling in each category, π for the first category and $(1 - \pi)$ for the second category, is the same for every observation.
3. The outcomes of successive observations are independent; that is, the category that occurs for one observation does not depend on the outcomes of other observations.

An idealized representation of a binomial variable refers to a sequence of flips of a coin. For each of n flips, we observe whether the outcome is head (category 1) or tail (category 2). The probability of each outcome is the same for each flip (e.g., the probability of a head equals .5 if the coin is balanced). The outcome of a particular flip does not depend on the outcome of other flips.

Now, let X denote the number of the n observations that occur in the first category, such as X = the number of heads. Each of the n observations either is or is not in that category, and the number X falling in that category is a variable. For instance, for $n = 5$ coin flips, X = number of heads could equal 0, 1, 2, 3, 4, or 5. For n observations satisfying the three binomial conditions, the sampling distribution of X is the binomial distribution.

The binomial variable X is discrete, taking one of the integer values $0, 1, 2, \ldots, n$. The formula for the binomial probabilities follows:

Probabilities for a Binomial Distribution

Denote the probability of the first category, for each observation, by π. For a sample of n independent observations, the probability that X of them fall in that category equals

$$P(X) = \frac{n!}{X!(n-X)!}\pi^X(1-\pi)^{n-X}, \quad X = 0, 1, 2, \ldots, n$$

The symbol $n!$ is called ***n factorial*** and represents $n! = 1 \cdot 2 \cdot 3 \cdots n$.

That is, $n!$ denotes the product of all integers from 1 to n; $1! = 1,\ 2! = 1 \cdot 2 = 2,\ 3! = 1 \cdot 2 \cdot 3 = 6,\ 4! = 1 \cdot 2 \cdot 3 \cdot 4 = 24$, and so forth. Also, $0!$ is defined to be 1.

For particular values for π and n, substituting the possible values for X into the formula for $P(X)$ provides the probabilities of the various possible outcomes. The sum of the probabilities, $P(0) + P(1) + P(2) + \cdots + P(n)$, equals 1.0.

Example 6.10 Gender and Selection of Managerial Trainees

Example 6.1 discussed a case involving potential bias against females in selection of management trainees for a large supermarket chain. The pool of employees for potential selection is half female and half male. Ten trainees are supposedly selected at random from this pool. If they are truly selected at random, how many females would we expect to be selected?

The probability that any one person selected is a female is $\pi = .50$, the proportion of available trainees who are female. Similarly, the probability that any one person selected is male is $(1 - \pi) = .50$. The variable X, the number of females selected for management training, has the binomial distribution with $n = 10$ and $\pi = .50$. For each X between 0 and 10, the probability that X of the ten people selected are female equals

$$P(X) = \frac{10!}{X!(10-X)!}(.50)^X(.50)^{10-X}, \quad X = 0, 1, 2, \ldots, 10$$

For example, the probability that no females are chosen ($X = 0$) equals

$$P(0) = \frac{10!}{0!10!}(.50)^0(.50)^{10} = (.50)^{10} = .001.$$

Recall that any number raised to the power of 0 equals 1. Also, $0! = 1$, and the $10!$ terms in the numerator and denominator cancel, leaving $P(0) = (.50)^{10}$. Similarly, the probability that exactly one female is chosen equals

$$P(1) = \frac{10!}{1!9!}(.50)^1(.50)^{11} = 10(.50)(.50)^{11} = .010$$

This computation simplifies considerably by using $10!/9! = 10$, since $10!$ is just $9!$ multiplied by 10. Table 6.7 lists the entire binomial distribution for $n = 10$, $\pi = .50$.

TABLE 6.7 The Binomial Distribution for $n = 10$, $\pi = .50$; the Binomial Variable X Can Take Any Value Between 0 and 10

x	$P(x)$	x	$P(x)$
0	.001	6	.205
1	.010	7	.117
2	.044	8	.044
3	.117	9	.010
4	.205	10	.001
5	.246		

In Table 6.7, the probability is about .98 that X falls between 2 and 8, inclusive. The least likely values for X are 0, 1, 9, and 10, which have a combined probability of only .022. Thus, if the sample were randomly selected, somewhere between about two and eight females would probably be selected. It is especially unlikely that none or ten would be selected.

The probabilities for females determine those for males. For instance, the probability that nine of the ten people selected are male equals the probability that one of the ten selected are female. □

Properties of the Binomial Distribution

In the preceding example, the three binomial conditions are satisfied. Each observation has only two possible outcomes, female or male. The probability of each outcome is the same for each selection, .50 for selecting a woman and .50 for selecting a man. For random sampling, the outcome of any one selection does not depend on any other one.

The population size must be large for the binomial distribution to apply, even approximately, in situations such as this. To illustrate, suppose a population contains only five persons, of whom one is a woman. If we randomly sample two separate individuals from this group, the second observation has different probabilities than the first. For example, if the first person selected was female, then the probability that the second person selected is female equals 0, since the population has only one female. Thus, the probabilities are not the same for each selection, which the binomial requires. For the successive observations to have the same probabilities and be independent, the population size must be much larger than the sample size. A rough guideline is that the sample size should be at most 10% as large as the minimum of the population numbers of subjects in the two categories. This is usually easily satisfied in practice. The population size is actually treated as infinite in the binomial formula.

When the population size is not much larger than the sample size, the probability distribution of X is the *hypergeometric*, which is not discussed in this text. One way in which the binomial distribution still applies in this case is when the sampling is done *with replacement*: When a person is sampled on a given selection from the population, they are put back into the sampling frame so they can be chosen again for successive selections for the sample. In that case, the population to be sampled is the same for each selection, and the probability of a particular outcome is the same for each selection.

The binomial distribution of X for sample size n and parameter π has mean and standard deviation

$$\mu = n\pi, \ \sigma = \sqrt{n\pi(1-\pi)}$$

For instance, the distribution with $n = 10$ and $\pi = .50$ has

$$\mu = 10(.50) = 5.0, \ \sigma = \sqrt{10(.50)(.50)} = 1.58$$

The binomial distribution is perfectly symmetric only when $\pi = .50$. In this example, for instance, since the proportion of women equals .50, $X = 10$ has the same probability as $X = 0$. When $\pi \neq .50$, the binomial distribution is skewed, the degree of skew increasing as π gets closer to 0 or 1. For any value of π, the binomial distribution is approximately bell-shaped for large n, specifically when $n > 10/\min(\pi, 1-\pi)$.

The Binomial Test

Refer to Example 6.10. Suppose the ten trainees selected contained only one woman, even though 50% of the available employees were women. We would be highly suspicious that the method of selection was not random. One can construct a formal significance test to check this, using the binomial distribution for calculating the P-value. This is called a ***binomial test***.

Let π denote the probability that any one person selected for management training by the company is female. For random sampling, $\pi = .50$. Thus, we can test random sampling by testing

$$H_0 : \pi = .50 \qquad \text{versus} \qquad H_a : \pi \neq .50$$

In H_0, the parameter π is hypothesized to equal the probability corresponding to random sampling. A small P-value will provide strong evidence against it.

Of the ten employees chosen for management training, let X denote the number of women. Under the assumption of random sampling, the sampling distribution of X is the binomial distribution with $n = 10$ and $\pi = .50$, already tabulated in Table 6.7. In this sample, $X = 1$. The P-value is then the two-tail probability of an outcome at least this extreme; that is, $X = 1$ or 0 or $X = 9$ or 10. This is the outcome that one or fewer of either sex is selected.

Figure 6.13 shows the formation of this P-value. From Table 6.7, the P-value equals

$$P = P(0) + P(1) + P(9) + P(10) = .001 + .010 + .010 + .001 = .022$$

If the company selected trainees randomly, the probability of choosing one or fewer people of either sex is only .022. This result provides evidence against the null hypothesis of a random selection process. We can reject H_0 for $\alpha = .05$, though not for α

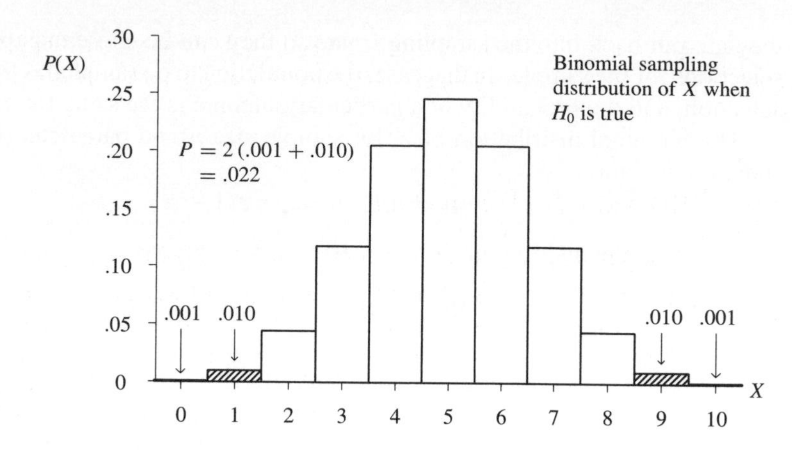

Figure 6.13 Calculation of P-value in Testing H_0: $\pi = .50$ Against H_a: $\pi \neq .50$, When $n = 10$ and $X = 1$

= .01. The outcome $X = 0$ yields a P-value of .002, which would be much stronger evidence.

One could test H_0 specifically against the alternative of bias against women by using the alternative hypothesis, H_a: $\pi < .50$. The P-value is then the left-hand tail probability. For example, $X = 1$ woman selected has P-value equal to $P(0) + P(1) = .011$. Even if we suspect bias in a particular direction, though, the most even-handed way to perform the test uses a two-sided alternative. For the two-sided alternative H_a: $\pi \neq \pi_0$, the P-value is a two-tail sum of the probabilities that are smaller than or equal to the observed probability.

In summary, the assumptions for the binomial test are the three conditions for the binomial distribution. The hypotheses take the same form as in the large-sample test for a proportion. The test statistic is the binomial variable X, and the P-value is based on its sampling distribution.

Small-Sample Confidence Intervals for Proportions

Confidence intervals for proportions are complex when n is small, and no single approach always works well. Problem 5.57 showed a rough approximation that does not require estimating π in the standard error. Though specialized small-sample methods exist, the discreteness of the binomial distribution causes them to be "conservative." This means that the actual confidence level can be much greater than the intended one. For instance, for a particular parameter value, the probability that a "95%" confidence interval contains it may actually be .98 or .99. Because of this, the small-sample confidence intervals tend to be overly wide.

When $n > 10/\min(\pi, 1 - \pi)$, the binomial sampling distribution of X is approximately normal. The bell-shaped approximation has mean $\mu = n\pi$ and standard deviation $\sqrt{n\pi(1 - \pi)}$. Equivalently, the sampling distribution of the sample proportion $\hat{\pi} = X/n$ is then approximately normal with mean π and standard error $\sigma_{\hat{\pi}} = \sqrt{\pi(1-\pi)/n}$. This approximation is the basis of the large-sample methods of Sections 5.3 and 6.3.

6.7 Calculating *P*(Type II Error)*

A decision about a null hypothesis H_0 in a statistical test can result in two types of error. A Type I error results from rejecting H_0 when it is actually true. The probability of Type I error is the α-level of the test. When $\alpha = .05$, for instance, one rejects H_0 when $P \leq .05$. Then, if H_0 is true, the probability of improper rejection equals .05.

A Type II error results from failing to reject H_0 even though it is false. The probability of Type II error depends on the actual value of the parameter. The farther the true parameter value falls from the value in H_0, the less likely a Type II error. This section shows how to calculate the probability of Type II error.

Example 6.11 *P*(Type II Error) for Detecting Stress Changes

Recent findings have suggested that neonatal sex differences exist in behavioral and physiological reactions to stress. For instance, a recent study (M. Davis and E. Emory, *Child Development*, Vol. 66, 1995, pp. 14–27) evaluated changes in the Neonatal Behavior Assessment Scale (NBAS) and changes in heart rate for a sample of 37 to 42 week-old infants placed in a stressful situation. The sample mean change in heart rate was small for males compared to females: -1.2 compared to 10.7, each with standard deviations of about 18. However, the sample size was only 15 for each group.

Suppose one is skeptical of the result for males and plans a larger-scale experiment to reanalyze males' reactions. The plan is to test whether the mean heart rate increases when male infants undergo the stressful experience. Let μ denote the population mean of the difference in heart rate, after versus before the stress. The null hypothesis
H_0: $\mu = 0$ will be tested against H_a: $\mu > 0$, at the $\alpha = .05$ level. The experiment will use 36 infant males. If the standard deviation of the sample measurements is about 18, then the standard error of $\bar{Y}$ will be about $\hat{\sigma}_{\bar{Y}} = 18/\sqrt{36} = 3.0$.

To calculate the probability of Type II error, we select a parameter value from the range in the alternative hypothesis. Suppose it is important to reject H_0 if μ is actually 10 or larger, 10 being the value for females in the previous study. Then, we calculate the probability of Type II error for the alternative value of $\mu = 10$.

Now, for testing H_0: $\mu = 0$ against the one-sided alternative H_a: $\mu > 0$, a test statistic of $z = 1.645$ has a P-value (right-hand tail probability) of .05. Thus, if $z \geq 1.645$, then $P \leq .05$, and we reject H_0 at the .05 α-level. In other words, rejection of H_0 results from $\bar{Y}$ falling at least 1.645 standard errors above $\mu_0 = 0$, or when

$$\bar{Y} \geq 0 + 1.645\hat{\sigma}_{\bar{Y}} = 0 + 1.645(3.0) = 4.94$$

Similarly, H_0 is not rejected if $\bar{Y} < 4.94$. See Figure 6.14. When H_0 is false, a Type II error occurs if $\bar{Y} < 4.94$.

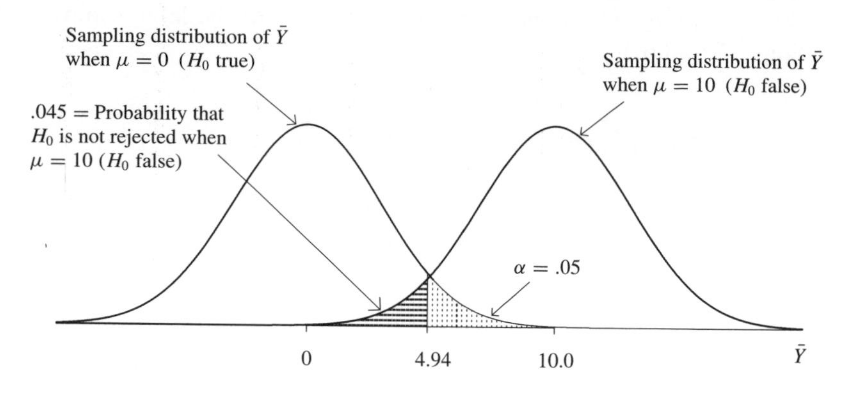

Figure 6.14 Calculation of Probability of Type II Error for Testing H_0: $\mu = 0$ Against H_a: $\mu > 0$ at $\alpha = .05$ Level, When True Mean is $\mu = 10$. A Type II error occurs if $\bar{Y} < 4.94$, since then $P > .05$ even though H_0 is false.

If the true value of μ is 10, then the probability of a Type II error is the probability that $\bar{Y} < 4.94$ when $\mu = 10$. For a normal sampling distribution with mean 10 and standard error 3.0, the $\bar{Y}$ value of 4.94 has a z-score of

$$z = \frac{4.94 - 10}{3.0} = -1.69$$

The left-hand tail probability below 4.94 for the distribution of $\bar{Y}$ equals the left-hand tail probability below -1.69 for the standard normal distribution. From Table A, this equals .045. So, for a sample of size 36, the probability of not rejecting H_0: $\mu = 0$ is .045, if in fact $\mu = 10$. The probability of a Type II error, when $\mu = 10$, is .045. □

Using a similar argument, you can verify that the probability of Type II error equals .49 at $\mu = 5$. That is, if the true mean is not very far from the null hypothesis value of 0, there may be a substantial chance of failing to reject the null hypothesis. You can also verify that the probability of Type II error is less than .001 at $\mu = 15$. The probabilities decrease for alternative values farther above 0. That is, the farther the true parameter falls from the null hypothesis, the less likely a Type II error. Figure 6.15 shows a plot of P(Type II error) for the various μ values in H_a.

Tests with Smaller α Have Greater *P*(Type II error)

The smaller the probability of Type I error, α, in a test, the larger the probability of Type II error. To illustrate, suppose Example 6.11 used $\alpha = .01$ instead of .05. Then, you can verify that a Type II error occurs if the sample mean falls less than 2.33 standard errors above the null hypothesis value, or $\bar{Y} < 7.0$. When $\mu = 10$, this happens with probability .16, compared to probability .045 when $\alpha = .05$. Hence, the smaller the α-level, the more likely we are to fail to detect a real difference.

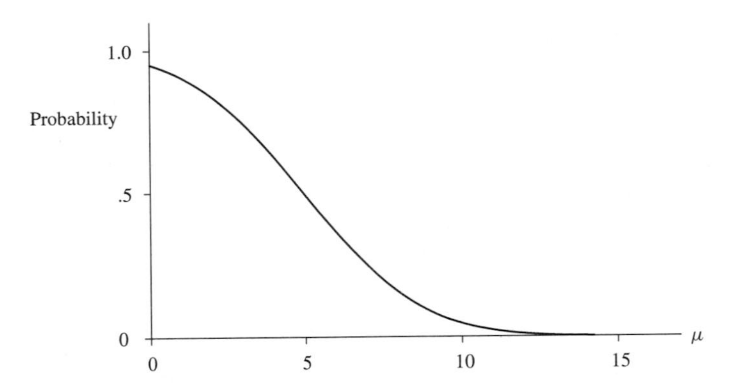

Figure 6.15 Probability of Type II Error for Testing H_0: $\mu = 0$ Against H_a: $\mu > 0$ at $\alpha = .05$ Level, Plotted for the Potential μ Values in H_a

For a fixed α-level, the probability of Type II error decreases when the sample size increases. For instance, quadrupling the sample size in this example from 36 to 144 has the impact of halving the standard error, from 3.0 to 1.5. Then you can check that when $\mu = 10$, an $\alpha = .01$ level test has probability of Type II error equal to .02 rather than .16.

The Power of a Test

The probability of rejecting the null hypothesis when it is false is called the ***power*** of the test. It is desirable for a test to have high power.

For a particular value of the parameter from the alternative hypothesis,

$$\text{Power} = 1 - P(\text{Type II error})$$

In Example 6.11, for instance, the $\alpha = .05$ level test for the alternative $\mu = 10$ has P(Type II error) = .045. Therefore, the power of the test at $\mu = 10$ equals $1 - .045 = .955$. The power increases for values of the parameter falling farther from the value in the null hypothesis. Just as the curve for P(Type II error) in Figure 6.15 decreases as μ gets farther from $\mu_0 = 0$, the curve for the power would increase.

For a fixed α-level, power increases as sample size increases. For a fixed parameter value and sample size, power decreases as α decreases. The reason that extremely small values are not normally used for α, such as $\alpha = .001$, is that the power of the test is too low; we may be unlikely to reject the null hypothesis even if the true parameter is far from the null hypothesis.

Later chapters omit calculation of the power or the probability of Type II error, since these calculations are complex for most significance tests. Keep in mind, however, that P(Type II error) may be fairly large for important alternatives, especially if the sample size is small. In other words, the reason for not obtaining a small P-value and rejecting H_0 may be that the test simply does not have enough power for the size of sample used. For further details about calculating P(Type II error) or power, see Cohen (1988).

6.8 Chapter Summary

Statistical inference involves predictions about population parameters using sample data. Chapters 5 and 6 have introduced the two inference methods—***estimation*** and ***significance tests***. The estimation method of *confidence intervals* provides a range of the most plausible values for a parameter. A significance test provides a way of judging whether a particular value for the parameter is plausible. Both methods are based on the foundations of probability, through the sampling distribution of the estimator of the parameter. Large-sample tests and confidence intervals, for instance, treat sample means and sample proportions as normally distributed, which the Central Limit Theorem justifies.

Significance tests have five elements:

1. ***Assumptions*** about the sampling method, the type of variable, the sample size, and the form of the population distribution:
   - Nearly all tests require *random sampling*.
   - Tests for quantitative variables refer to *means*, whereas tests for qualitative variables refer to *proportions*.
   - Large-sample tests for proportions ($n > 10/[\min(\pi_0, 1 - \pi_0)]$) and for means ($n \geq 30$) require no assumption about the population distribution; the Central Limit Theorem implies approximate normality of the sampling distribution, regardless of the population distribution.
   - Small-sample tests for means use the ***t distribution***, and small-sample tests for proportions use the ***binomial distribution***. Methods using the t distribution assume that, for small samples, the population distribution is normal. In practice, two-sided tests and confidence intervals are *robust* to violations of the normality assumption.
2. ***Null and alternative hypotheses*** about the parameter: Null hypotheses have the form $H_0 : \mu = \mu_0$ for a mean and $H_0 : \pi = \pi_0$ for a proportion, where μ_0 and π_0 denote values hypothesized for the parameters. The most common alternative is the two-sided one, such as $H_a : \mu \neq \mu_0$. One-sided hypotheses such as $H_a : \mu > \mu_0$ and $H_a : \mu < \mu_0$ are also sometimes used.
3. A ***test statistic*** describes how far the data fall from the null hypothesis value. The large-sample z statistic for means and proportions and the small-sample t statistic measure the number of standard errors that the sample estimate ($\bar{Y}$ or $\hat{\pi}$) falls from the null hypothesis value.
4. The ***P-value*** describes the weight of evidence that the data provide about the null hypothesis. This equals the probability of a test statistic value at least as contradictory to H_0 as the value actually observed, calculated under the assumption that H_0 is true.
   - When the P-value is small, it would be unusual to observe such data if H_0 were true. The smaller the P-value, the stronger the evidence against H_0.

- The results that are "more contradictory to H_0" are determined by the alternative hypothesis. For two-sided alternative hypotheses, the P-value is a two-tail probability.
- For large samples, we calculate the P-value using the standard normal distribution. For small samples, we use the t distribution for tests about means and the binomial distribution for tests about proportions.

5. A ***conclusion*** based on the sample evidence about the null hypothesis: Usually we simply report the P-value. Sometimes it is necessary to make a decision. If the P-value is less than or equal to a fixed α-level (such as $\alpha = .05$), we reject H_0; if the P-value is larger than that level, we cannot reject it.

When one makes a decision, two types of errors can occur.

- A Type I error results from rejecting the null hypothesis when it is true.
- A Type II error results from failure to reject H_0 when it is false.

The choice of α, the cutoff point for the P-value in making a decision, determines the probability of Type I error. Normally, we choose small values such as $\alpha = .05$ or .01 to keep the probability of Type I error small. For fixed α, the probability of Type II error decreases as the distance increases between the true parameter and the null hypothesis value, and it decreases as the sample size increases.

Table 6.8 summarizes the five elements of the tests presented in this chapter. In each test, small P-values result when the estimate of the parameter falls far from the null hypothesis value. When that happens, the test statistic is large, being far from the mean of the sampling distribution.

TABLE 6.8 Summary of Significance Tests for Means and Proportions

Parameter	Mean		Proportion	
	$n > 30$	$n \leq 30$	$n > \frac{10}{\min(\pi_0, 1-\pi_0)}$	$n \leq \frac{10}{\min(\pi_0, 1-\pi_0)}$
1. Assumptions	Random sample, quantitative variable	Random sample, normal population distribution, quantitative variable	Random sample, qualitative variable	Same
2. Hypotheses	H_0: $\mu = \mu_0$ H_a: $\mu \neq \mu_0$ H_a: $\mu > \mu_0$ H_a: $\mu < \mu_0$	Same	H_0: $\pi = \pi_0$ H_a: $\pi \neq \pi_0$ H_a: $\pi < \pi_0$ H_a: $\pi > \pi_0$	Same
3. Test statistic	$z = \frac{\bar{Y} - \mu_0}{\hat{\sigma}_{\bar{Y}}}$	$t = \frac{\bar{Y} - \mu_0}{\hat{\sigma}_{\bar{Y}}}$	$z = \frac{\hat{\pi} - \pi_0}{\sigma_{\hat{\pi}}}$	Binomial variable $X = n\hat{\pi}$
4. P-value	Two-tail probability in sampling distribution for two-sided test (H_0: $\mu \neq \mu_0$ or H_a: $\pi \neq \pi_0$); One-tail probability in sampling distribution for one-sided test			
5. Conclusion	Reject H_0 if $P \leq \alpha$-level			

Testing a hypothesis about a parameter value is not as informative as estimating that parameter using a confidence interval. A test merely indicates whether a particular parameter value is plausible, whereas a confidence interval displays the range of plausible values. Even if you use a significance test, you should also construct a confidence interval to determine which parameter values are plausible.

Sample size is a critical factor in both estimation and significance tests. With small sample sizes, confidence intervals are wide, making estimation imprecise. Small sample sizes also make it difficult to reject false null hypotheses unless the true parameter is far from the null hypothesis value. With small sample sizes, the probability of Type II error may be depressingly high for many parameter values of interest.

Where We're Going

To introduce estimation and hypothesis testing, Chapters 5 and 6 have presented methods for making inference about a single parameter for a single variable. In practice, it is usually artificial to have a particular fixed number for the null hypothesis value of a parameter such as the mean. One of the few times this happens is when the response score results from taking a difference of two values, such as the change in weight in Example 6.4 and the change in heart rate in Example 6.11; in that case, $\mu_0 = 0$ is a natural baseline. In fact, significance tests much more commonly refer to comparisons of means for two samples than to a fixed value of a parameter for a single sample. The next chapter shows how to compare means or proportions for two different groups.

In addition, studying single variables in isolation rarely is adequate. More commonly, analyses refer to two or more variables. Rather than simply estimating an overall mean μ, we might want to analyze how the mean changes according to the level of another variable, such as gender or race or educational level. The next three chapters present ***bivariate*** methods, dealing with relationships between two variables. Then, Chapters 10–16 present ***multivariate*** statistical methods, which deal with relationships among *several* variables. The greater the number of variables, the more complex the analysis. Increasing the number of variables has the advantage, however, of adding to our understanding of the world by representing more realistically its complexity.

PROBLEMS

Practicing the Basics

1. For a large-sample test of H_0: $\mu = 0$ against H_a: $\mu \neq 0$, the z test statistic equals 1.04.
a) Find the P-value, and interpret.
b) Suppose $z = -2.50$ rather than 1.04. Find the P-value. Does this provide stronger, or weaker, evidence against the null hypothesis? Explain.

2. Complete the previous exercise for the alternatives (i) H_a: $\mu > 0$, (ii) H_a: $\mu < 0$.

3. The P-value for a large-sample test about a mean is $P = .10$.
a) If the alternative hypothesis is H_a: $\mu \neq \mu_0$, report the value of the test statistic.
b) Does this P-value provide stronger, or weaker, evidence against the null hypothesis than $P = .010$? Explain.

4. Repeat (a) of the previous exercise for (i) H_a: $\mu > \mu_0$, (ii) H_a: $\mu < \mu_0$.
5. Find and interpret the P-value for testing $H_0 : \mu = 100$ against $H_0 : \mu \neq 100$, if a sample of 400 observations has $s = 40$ and **a)** $\bar{Y} = 103$, **b)** $\bar{Y} = 97$.
6. Refer to the previous exercise. Repeat part a) if **a)** $n = 1600$ instead of 400, **b)** $s = 20$ instead of 40. Comment on the effects of n and s on the results of a significance test.
7. In response to the statement "A preschool child is likely to suffer if his or her mother works," the response categories (Strongly agree, Agree, Disagree, Strongly disagree) had counts (91, 385, 421, 99) for the 996 responses in the 1991 General Social Survey. With scores (2, 1, −1, −2) for the four categories, computer software reported the following results:

```
--------------------------------------
N          Mean       Std Dev     Std Err
996        -.052      1.253       0.0397
--------------------------------------
```

a) Set up null and alternative hypotheses to test whether the population mean response differs from the neutral value, 0.
b) Find the test statistic and P-value. Interpret.
c) Explain what this choice of scoring assumes about relative distances between levels of the scale.
8. The mean score for all high school seniors taking a college entrance exam equals 500. A study is conducted to see whether a different mean applies to those students born in a foreign country. For a random sample of 100 of such students, the mean and standard deviation on this exam equal 508 and 100.
a) Set up hypotheses for a significance test.
b) Compute the test statistic.
c) Report the P-value, and interpret.
d) Based on (c), can you conclude that the population mean for students born in a foreign country equals 500? Why or why not?
e) Make a decision about H_0, using $\alpha = .05$.
f) Construct a 95% confidence interval for μ. Show the correspondence between whether 500 falls in the interval and whether $H_0 : \mu = 500$ is rejected in favor of $H_a : \mu \neq 500$.
9. A social psychologist plans to conduct an experiment with a random sample of 49 children from some school district. Before conducting the experiment, the psychologist checks how this sample compares to national norms on several variables that could affect the results of the experiment. The IQ scores for the 49 children have $\bar{Y} = 103$ and $s = 14$. The national population mean IQ equals 100. Is it plausible that the mean μ of the population of children in the school district from which these students were sampled equals 100?
a) Conduct a test of H_0: $\mu = 100$ against H_a: $\mu \neq 100$. Report the P-value, and interpret.
b) Make a decision about H_0 using $\alpha = .05$.
c) Construct a 95% confidence interval for μ, and contrast the types of information obtained with the two inference methods, tests and confidence intervals.
d) Does it make sense to "accept H_0" in part (b)? Why or why not?
e) What conclusion applies for each of the following α-levels:
(i) $\alpha = .20$, (ii) $\alpha = .10$, (iii) $\alpha = .01$.
Why is $\alpha = .20$ not typically used in practice?
10. Refer to Example 6.2. Suppose we instead use the scores (−3, −2, −1, 0, 1, 2, 3), subtracting 4 from each of the original scores. We then test $H_0 : \mu = 0$. Explain the effect of

the change in scores on **a)** the sample mean and standard deviation, **b)** the test statistic, **c)** the P-value and interpretation.

11. The mean age at first marriage for married men in a New England community was 28.0 in 1790. For a random sample of 40 married men in that community in 1990, the sample mean and standard deviation of age at first marriage were $\bar{Y} = 26.0$ and $s = 9.0$.
a) State the hypotheses, and find the test statistic and P-value for testing whether the mean has changed. Interpret.
b) Make a decision, using $\alpha = .05$.
c) If the decision in (b) is an error, what type of error is it, Type I or Type II?
d) Report the P-value for the alternative $H_a : \mu < 28$. Interpret.
e) Report the P-value for the alternative $H_a : \mu > 28$. Interpret.

12. A joint *USA Today*/CNN/Gallup Poll in July 1995 indicated that of 832 white adults, 53% thought affirmative action has been good for the country and 37% thought it had not been good; the remaining 10% were undecided. Let π denote the population proportion of white adults who believe that affirmative action has been good for the country.
a) Test $H_0 : \pi = .50$ against $H_a : \pi \neq .50$. Report the P-value, and interpret.
b) Can you reject H_0 using $\alpha = .05$? Indicate whether you can conclude that a majority or minority of the population think affirmative action has been good.
c) Construct a 95% confidence interval for π. Explain how this provides greater information than the conclusion of the test in (b).
d) The same poll also interviewed 299 black adults. Of them, 66% thought that affirmative action had been good and 23% thought it had not been good. Conduct statistical inference for black adults.

13. An experiment consists of giving everyone in some group a lottery ticket, and then later asking if they would be willing to exchange their ticket for another one, plus a small monetary incentive. Let π denote the population proportion who would agree to the exchange. When this experiment was conducted with 26 students in an Israeli classroom recently, only 7 students agreed to the exchange (M. Bar-Hillel and E. Neter, *J. Personality and Social Psychology*, Vol. 70, 1996, pp. 17–27).
a) Using these data, test $H_0 : \pi = .50$ against $H_a : \pi \neq .50$, and interpret the P-value.
b) In a related experiment, 31 students in an Israeli statistics class were given a new pen and then later asked to exchange it for another pen and a small monetary incentive. All 31 agreed. Repeat the test for these data, and interpret.

14. Let π denote the proportion of Floridians who think that government environmental regulations are too strict. Test H_0: $\pi = .5$ against H_a: $\pi \neq .5$ using data from a telephone poll of 834 people conducted in June 1995 by the Institute for Public Opinion Research at Florida International University, in which 26.6% said regulations were too strict.
a) Calculate the test statistic.
b) Find the P-value, and interpret.
c) Using $\alpha = .01$, can you determine whether a majority or minority think that environmental regulations are too strict, or is it plausible that $\pi = .50$?
d) Construct a 99% confidence interval. Explain the advantage of the confidence interval over the test.

15. The 1994 General Social Survey asked, "Do you think it should be possible for a pregnant woman to obtain a legal abortion if the family has a very low income and cannot afford any more children?" Let π denote the population proportion who would answer yes to this question. An anti-abortion activist claims that a minority of Americans support legalized abortion in such cases.

a) Set up hypotheses for a significance test that can analyze the claim.
b) In this survey, 971 answered yes and 954 answered no. Calculate the test statistic and P-value for the hypotheses presented in part (a). Interpret.
c) Using $\alpha = .05$, what is your decision? Interpret.

16. A mayoral election in Madison, Wisconsin, has two candidates. Exactly half the residents of the city currently prefer each candidate.
a) Find the probability that, for a random sample of 400 residents, at least 230 indicate a preference for one of the candidates or the other? This is also the P-value for testing H_0: $\pi = .5$ against H_a: $\pi \neq .5$, where π denotes the probability that a randomly selected voter prefers a particular candidate.
b) For a random sample of 400 voters, 230 voted for a particular candidate. Are you willing to predict the outcome of the election? Why?
c) For a random sample of 40 voters, 23 voted for a particular candidate. Would you be willing to predict the outcome of the election? Why?

17. The authorship of an old document is in doubt. A historian hypothesizes that the author was a journalist named Jacalyn Levine. Upon a thorough investigation of Levine's known works, it is observed that one unusual feature of her writing was that she consistently began 6% of her sentences with the word *whereas*. To test the historian's hypothesis, it is decided to count the number of sentences in the disputed document that begin with the word *whereas*. Out of the 300 sentences in the document, none begin with that word. Let π denote the probability that any one sentence written by the unknown author of the document begins with the word *whereas*. Conduct a test of the hypothesis H_0: $\pi = .06$ against H_a: $\pi \neq .06$. What conclusion can you make? What assumptions are needed for that conclusion to be valid? (Mosteller and Wallace (1964) conducted an investigation similar to this to determine whether Alexander Hamilton or James Madison was the author of 12 of the *Federalist Papers*.)

18. Refer to the ordinal scale in Table 6.2. Treat this as a qualitative variable, as follows: Ignore the subjects responding in the *moderate* category. Of the remainder not responding in that category, let π denote the population proportion who make one of the three liberal responses and let $1 - \pi$ denote the population proportion making one of the three conservative responses. Test the hypothesis that these proportions are identical. Find the P-value and interpret.

19. The owner of a department store in Rochester, New York, initiates a week-long newspaper advertising campaign to increase awareness of the store. Before investing any additional money in advertising, the owner takes a phone survey to check whether potential customers are more inclined or less inclined to shop in this store after seeing the advertising. Consider the population of residents who have seen the ad and are either more inclined or less inclined to shop there than before seeing the ad. Of these people, let π denote the proportion who are more inclined. For a random sample of 100 names selected from the telephone book, 25 express an opinion. Of these 25 subjects, 18 say they are more inclined to shop at the store, and 7 say they are less inclined.
a) Report the P-value for testing H_0: $\pi = .5$ against H_a: $\pi \neq .5$. Interpret.
b) Make a decision using $\alpha = .05$. Can the owner conclude that, of those with an opinion, a majority are more likely to shop at the store?
c) If you made an error with the decision in (b), would it be a Type I error or a Type II error?
d) Consider the proportion of residents who have not seen the ad or who are neither more nor less likely to shop at the store as a result of the ad. Can you determine whether this is

a majority or minority of the population? (Answer by performing a test or constructing a confidence interval that uses all 100 observations.)

20. A multiple-choice test question has four possible responses. The question is designed to be very difficult, with none of the four responses being obviously wrong, yet with only one correct answer. It first occurs on an exam taken by 400 students. The designers test whether more people answer the question correctly than would be expected just due to chance (i.e., if everyone randomly guessed the correct answer).
a) Set up the hypotheses for the test.
b) Of the 400 students, 125 correctly answer the question. Find the P-value, and interpret.
c) Make a decision about H_0, using $\alpha = .05$. Based on this decision, what can you conclude about the parameter?
d) As an alternative inference, construct a 95% confidence interval for the parameter. Contrast what you learn using these two methods.

21. Two researchers conduct separate studies to test $H_0 : \mu = 500$ against $H_0 : \mu \neq 500$. The first researcher gets $\bar{Y} = 519.5$, with a standard error of 10.0. The second researcher gets $\bar{Y} = 519.7$, with a standard error of 10.0.
a) Using $\alpha = .05$, test in each case whether the result is "statistically significant."
b) Explain the dangers of reporting the result of a test as "$P \leq .05$" versus "$P > .05$," or as "reject H_0" versus "Do not reject H_0."

22. A study considers whether the mean score on a college entrance exam for students in 1996 is any different from the mean score of 500 for students who took the same exam in 1966. Let μ represent the mean score for all students who took the exam in 1996. Test H_0: $\mu = 500$ against H_a: $\mu \neq 500$, if for a nationwide random sample of 10,000 students who took the exam in 1996, $\bar{Y} = 497$ and $s = 100$. Show that the result is highly significant statistically, but not practically significant.

23. Report the t-score that multiplies by the standard error to form a
a) 95% confidence interval with 5 observations.
b) 95% confidence interval with 15 observations.
c) 95% confidence interval with 25 observations.
d) 95% confidence interval with $df = 25$.
e) 99% confidence interval with $df = 25$.

24. A t test for a mean uses a sample of 15 observations. Find the t test statistic value that has a P-value of $P = .05$ when the alternative hypothesis is **a)** H_a: $\mu > 0$, **b)** H_a: $\mu \neq 0$, **c)** H_a: $\mu < 0$?

25. A school board commissions a study of the absentee rate of students at a particular high school in Vancouver. For a random sample of 20 student records from the previous year, the number of days absent in the previous year is recorded, yielding $\bar{Y} = 4.0$ and $s = 4$.
a) Stating the necessary assumptions, calculate a 95% confidence interval for the mean number of days absent for all students at that high school during the previous year.
b) Does the normality assumption seem plausible for these data? Explain, and discuss the implications of the term "robustness" for your analysis.

26. Refer to Example 6.11. For the 15 female infants, Table 6.9 shows an SPSS printout for the data on the change in heart rate.
a) Explain how to interpret all the results in this table.
b) Explain how to test whether the true mean is 0. Report the P-value for (i) a one-sided alternative of a positive mean change, (ii) two-sided alternative, interpreting in each case.

TABLE 6.9

Variable	Number of Cases	Mean	SD	SE of Mean
CHANGE	15	10.70	17.70	4.570

Mean	95% CI Lower	95% CI Upper	t-value	df	2-Tail Sig
10.70	.898	20.501	2.341	14	.0346

27. Explain how the sample size affects the width of a small-sample confidence interval through its effect on the **a)** standard error, **b)** t-score that multiplies the standard error.

28. According to a union agreement, the mean income for all senior-level assembly-line workers in a large company equals \$500 per week. A representative of a women's group decides to analyze whether the mean income for female employees matches this norm. For a random sample of nine female employees, $\bar{Y} = \$410$ and $s = 90$. Conduct a significance test of whether the mean income of female employees differs from \$500 per week. Include all assumptions, the hypotheses, test statistic, and P-value, and interpret the result.

29. Refer to the previous problem.
a) For which α-levels in Table B (the t distribution) can you reject H_0?
b) For which confidence coefficients would the confidence interval contain 500?
c) Use (a) and (b) to illustrate the correspondence between results of tests and results of confidence intervals.

30. By law, an industrial plant can discharge no more than 500 gallons of waste water per hour, on the average, into a neighboring lake. Based on other infractions they have noticed, an environmental action group believes this limit is being exceeded. Monitoring the plant is expensive, and only a small sample is possible. A random sample of four hours are selected over a period of a week. A computer printout shows the results:

Variable	Number of Cases	Mean	SD	SE of Mean
WASTE	4	1000.0	400.0	200.0

a) Test the null hypothesis that the mean discharge equals 500 gallons per hour against the alternative that the limit is being exceeded. Find the P-value, and interpret.
b) Explain why the result of this test may be invalid, or at least very highly approximate, if the population distribution of discharge is far from normal.
c) Explain how your one-sided analysis implicitly tests the broader null hypothesis that $\mu \leq 500$.
d) Make a decision, using $\alpha = .05$. Interpret.
e) If the decision in (d) was incorrect, what type of error was made? What could you do to reduce the chance of that type of error?

31. A study was conducted of the effects of a special class designed to improve children's verbal skills. Each child took a verbal skills test twice, both before and after a three-week period in the class. Let Y be the second exam score minus the first exam score. Hence, if μ (the population mean for Y) is equal to 0, the class has no effect on the average. Test the null hypothesis of no effect against the alternative hypothesis that the effect is positive, if the scores on Y for a random sample of four children having learning problems were 3, 7, 3, 3. Interpret the P-value. (Note: The scores could tend to improve simply from the students feeling more comfortable with the testing process. A more appropriate design would also administer the exam twice to a control group that does not take the special class, comparing the changes for the experimental and control groups.)

32. A jury list contains the names of all individuals who may be called for jury duty. The proportion of the available jurors on the list who are women is .53. If a jury of size 12 is selected at random from the list of available jurors,
a) Find the probability that no women are selected.
b) Find the probability that exactly one woman is selected.
c) Find the expected value and standard deviation of the number of women selected.

33. Refer to the previous problem. Test the hypothesis that the selections are random, if no woman is selected out of a sample of size 12. Report the P-value, and interpret.

34. At current rates, the proportion of deaths of American females that are due to suicide is .01 (*Statistical Abstract of the United States, 1995*). For a random sample of ten deaths, find the probability that
a) None of them is due to suicide.
b) At most one is due to suicide.

35. Refer to the previous problem. The proportion of deaths due to suicide for American males is .02.
a) For 1000 American male deaths, find the mean and the standard deviation of the probability distribution of the number that were due to suicide.
b) Would it be surprising if none were due to suicide? Explain.

36. For each free throw, a basketball player has probability .80 of making the shot and .20 of missing it.
a) Find the probability that he makes ten free throws in a row.
b) He is fouled and takes two free throws. Find the probabilities in the distribution of the number of free throws that he makes.
c) What assumptions must you make for the calculations in (a) and (b) to be valid?

37. A football team has probability $\pi = .50$ of winning any particular game.
a) Find the probability that the team wins all or none of its six conference games. What assumptions does this calculation require?
b) If the team wins all six games, can you be quite confident that its actual probability of winning any single game differs from .50? Explain.

38. A person claiming to possess extrasensory perception (ESP) says she can guess more often than not the outcome of a flip of a balanced coin in another room, not visible to her.
a) Introduce appropriate notation, and state hypotheses for testing her claim.
b) Of ten coin flips, she guesses the correct result seven times. Find the P-value and interpret.

39. A weather forecaster states "The chance of rain is 50% on Saturday and 50% again on Sunday. So, there's a 100% chance of rain sometime over the weekend." If whether it rains on Saturday is independent of whether it rains on Sunday, find the actual probability of rain at least once during the weekend.

40. * Refer to Table 6.2. Of those subjects responding in one of the two extreme categories, let π denote the population proportion who make the extremely liberal response, with $1 - \pi$ being the population proportion making the extremely conservative response. Using small-sample methods, show that the P-value for testing that these probabilities are identical equals 1.0. Interpret.

41. In a given year, the probability that an American female adult dies in a motor vehicle accident equals .0001 (*Statistical Abstract of the United States, 1995*).
a) In a city having 1 million American female adults, find the mean and standard deviation of the number of deaths from motor vehicle accidents.
b) Based on the normal approximation to the binomial, find an interval within which the number of deaths has probability .95 of occurring.
c) The rate for American male adults is twice as large. Repeat (a) and (b) using the rate .0002, and compare results to those for females.

42. Refer to Problem 6.16. Find the P-value for testing H_0: $\pi = .5$ against H_a: $\pi \neq .5$ when five people are randomly selected and all of them prefer a particular candidate. Interpret. For these hypotheses, show that a sample of at least six people is needed to obtain P-value below .05.

43. A fraternal organization admits 80% of all applicants who satisfy certain requirements. Of four members of a minority group who recently applied for admission, all met the requirements but none was accepted.
a) Find the probability that none would be accepted if the same admissions standards were applied to the minority group, other things being equal.
b) Find the P-value for testing that the probability of admission for a minority group member equals .80, against the alternative that it is less than .80, and interpret.

44. A decision is planned in a test of H_0: $\mu = 0$ against H_a: $\mu > 0$, using $\alpha = .05$. If $\mu = 5$, the probability of a Type II error equals .17.
a) Explain the meaning of this last sentence.
b) If the test were conducted at the $\alpha = .01$ level, would the probability of a Type II error be less than, equal to, or greater than .17? Explain.
c) If $\mu = 10$, would the probability of a Type II error be less than, equal to, or greater than .17? Explain.

45. Refer to Example 6.11.
a) Show that the probability of Type II error is less than .001 at $\mu = 15$.
b) Show that the probability of Type II error equals .49 at $\mu = 5$.
c) Find the probability of Type II error at $\mu = 8$.
d) If $\alpha = .01$ instead of .05, verify that the probability of Type II error at $\mu = 10$ equals .16.
e) If $\alpha = .01$ but $n = 81$, verify that the probability of Type II error at $\mu = 10$ equals .004.

46. Refer to the hypotheses in Example 6.11. Suppose a sample of size 25 were taken (instead of 36) and suppose the estimated standard error of $\bar{Y}$ is $\hat{\sigma}_{\bar{Y}} = 3.6$. When $\alpha = .05$, compute the probabilities of a Type II error for the alternative μ values of (a) 5, (b) 10, (c) 15. Compare these to the probabilities of Type II error when $\hat{\sigma}_{\bar{Y}} = 3.0$, and use them to illustrate the effect of the sample size on P(Type II error).

47. Let π denote the proportion of schizophrenics who respond positively to a particular treatment. A test is conducted of H_0: $\pi = .5$ against H_a: $\pi > .5$, based on a sample of size 25, using $\alpha = .05$.
a) Find the region of sample proportion values for which H_0 is rejected, for a large-sample z test.
b) Suppose that, unknown to the researcher, the true value of π is .60. Find the probability that a Type II error occurs.
c) Explain why the probability of Type II error increases toward .95 as the true value of π moves down toward .5. (Assume n and α stay fixed.)

Concepts and Applications

48. Refer to the WWW data set (Problem 1.7).
a) Test whether the mean political ideology differs from 4.0. Report the P-value, and interpret.
b) Test whether the proportion favoring legalized abortion equals, or differs, from .50. Report the P-value, and interpret.

49. Refer to the data file you created in Problem 1.7. For variables chosen by your instructor, conduct inferential statistical analyses. Prepare a report, summarizing and interpreting your findings. In this report, also use graphical and numerical methods presented earlier in this text to describe the data and, if necessary, to check assumptions that you make for your analysis.

50. In response to the question "Do you think it should or should not be the government's responsibility to provide a job for everyone who wants one," the categories "Definitely should be," "Probably should be," "Probably should not be," "Definitely should not be" had response counts 242, 330, 333, 362 in the 1991 General Social Survey. Applying appropriate inferential methods, analyze these data.

51. An article in a sociology journal that deals with changes in religious beliefs over time states, "For these subjects, the difference in their responses on the scale of religiosity between age 16 and the current survey was significant ($P < .05$)."
a) Explain what it means for the result to be "significant."
b) Explain why it would have been more informative if the authors provided the actual P-value rather than simply indicating that it is below .05. What other information might they have provided?
c) Presumably the statement refers to a population mean μ of difference scores between the current survey and age 16. Based on the authors' statement, what can you conclude about μ?
d) Can you conclude that an important change in religiosity has occurred between age 16 and the time of the current survey? Why or why not?

52. A newly created random number generator is supposed to generate a sequence of digits such that each digit is equally likely to be any of 0, 1, 2, ..., 9. The first 20 numbers generated are

$$7, 7, 3, 0, 5, 6, 3, 2, 6, 1, 0, 9, 9, 4, 0, 8, 5, 0, 6, 2$$

As a check of whether the process works correctly, test whether the mean differs significantly from the value expected. Report the P-value and interpret.

53. In making a decision in a significance test, a researcher worries about the possibility of rejecting the null hypothesis when it is actually true. Explain how to control the probability of this type of error.

54. Consider the analogy between making a decision about a hypothesis in a significance test and making a decision about the innocence or guilt of a defendant in a criminal trial. Identify "H_0 true," "H_0 false" with "Defendant innocent," "Defendant guilty," and "Reject H_0," "Do not reject H_0" with "Convict defendant," "Acquit defendant."
a) Explain the difference between Type I and Type II errors in this trial setting. Which type of error would you consider more serious and hope to set at a very small level? Why?
b) In this setting, explain why decreasing the chance of Type I error increases the chance of Type II error.

55. Medical tests for diagnosing conditions such as cancer or HIV+ are fallible, just like decisions in significance tests. Identify (H_0 true, H_0 false) with (Disease absent, Disease present), and (Reject H_0, Do not reject H_0) with (Diagnostic test is positive, Diagnostic test is negative), where a positive diagnosis means that the test predicts that the disease is present. Explain the difference between Type I and Type II errors in this setting. Explain why decreasing the chance of Type I error increases the chance of Type II error.

56. An article in a political science journal states that "no significant difference was found between men and women in their voting rates ($P = .63$)." Can we conclude that the population voting rates are identical for men and women? Explain.

57. A research study conducts 60 significance tests. Of these, 3 are significant at the .05 level. The authors write a report stressing only the three cases in which they found "significant" results, not mentioning the other 57 tests they conducted that were "not significant." Explain what is misleading about this report.

58. Some medical journals have a policy of publishing results only of research that is statistically significant at some level.
a) Explain the dangers of this by describing what could happen if researchers at 20 different institutions independently, and without communication among each other, conducted similar studies and tested the same null hypothesis using $\alpha = .05$, when that hypothesis is actually true.
b) When medical stories in the mass media report supposed large dangers or benefits of certain agents (e.g., coffee drinking, fiber in cereal), later research often suggests that the effects are not nearly as large as first believed. Explain why.

Select the correct response(s) in Problems 6.59–6.62.

59. A 95% confidence interval for μ is (96, 110). Which of the following statements about significance tests for the same data is (are) correct?
a) In testing H_0: $\mu = 100$ against H_a: $\mu \neq 100$, $P > .05$.
b) In testing H_0: $\mu = 100$ against H_a: $\mu \neq 100$, $P < .05$.
c) In testing H_0: $\mu = \mu_0$ against H_a: $\mu \neq \mu_0$, $P > .05$ if μ_0 is any of the numbers inside the confidence interval.
d) In testing H_0: $\mu = \mu_0$ against H_a: $\mu \neq \mu_0$, $P > .05$ if μ_0 is any of the numbers outside the confidence interval.

60. The P-value for testing H_0: $\mu = 100$ against H_a: $\mu \neq 100$ is $P = .001$. This indicates that
a) There is strong evidence that $\mu = 100$.
b) There is strong evidence that $\mu \neq 100$.
c) There is strong evidence that $\mu > 100$.
d) There is strong evidence that $\mu < 100$.
e) If μ were equal to 100, it would be unusual to obtain data such as those observed.

61. Refer to the previous problem. Suppose the value of the test statistic was $z = 3.29$. Then
a) There is strong evidence that $\mu = 100$.
b) There is strong evidence that $\mu > 100$.
c) There is strong evidence that $\mu < 100$.

62. Refer to Problem 6.30. When we make a decision using $\alpha = .05$, this means that
a) If the plant is not exceeding the limit, but actually $\mu = 500$, there is only a 5% chance that we will conclude that they are exceeding the limit.
b) If the plant is exceeding the limit, there is only a 5% chance that we will conclude that they are not exceeding the limit.

c) The probability of getting the sample mean we observed would equal .05 if H_0 were true.
d) If we reject H_0, the probability that it is actually true is .05.
e) All of the above.

63. True or False: P(Type II error) = 1 − P(Type I error). Explain.

64. True or False: If we reject H_0 using $\alpha = .01$, then we also reject it using $\alpha = .05$. Explain.

65. An article in an anthropology journal reports $P = .043$ for testing H_0: $\mu = 0$ against H_a: $\mu \neq 0$. True or False: If the authors had instead reported a 95% confidence interval for μ, then the interval would have contained zero. Explain, and discuss what one could learn from the confidence interval but not from the test.

66. **a)** Define (i) P-value, (ii) α-level, (iii) Type II error, (iv) rejection region, (v) power.
b) Explain the difference between one-sided and two-sided hypotheses, and explain how this affects calculation of the P-value.

67. Explain why the terminology "do not reject H_0" is preferable to "accept H_0."

68. A random sample of size 40 has $\bar{Y} = 120$. The P-value for testing H_0: $\mu = 100$ against H_a: $\mu \neq 100$ is $P = .057$. Explain what is incorrect about each of the following interpretations of this P-value, and provide a proper interpretation.
a) The probability that the null hypothesis is correct equals .057.
b) The probability that $\bar{Y} = 120$ if H_0 is true equals .057.
c) If in fact $\mu \neq 100$, the probability equals .057 that the data would be at least as contradictory to H_0 as the observed data.
d) The probability of Type I error equals .057.
e) We can accept H_0 at the $\alpha = .05$ level.
f) We can reject H_0 at the $\alpha = .05$ level.
g) The value of the z test statistic is $z = 1.58$.

69. * The P-value for testing H_0: $\mu = 100$ against H_a: $\mu > 100$ is $P = .043$. Does a 95% confidence interval for μ contain 100? Explain.

70. Explain why one can interpret the P-value as the smallest α-level at which one can reject H_0; that is, P equals the smallest level at which the data are significant. Illustrate using the P-value of .057 from Problem 6.68.

71. * Refer to Example 6.2 and to the correspondence between results of confidence intervals and two-sided tests. Since $P = .52$ in that example, and since $1 - .52 = .48$, explain why the 48% confidence interval is the narrowest confidence interval centered about $\bar{Y}$ that contains $\mu_0 = 4.0$.

72. * We know the sample mean $\bar{Y}$ of n measurements. Show that if we know $(n - 1)$ of those measurements, then we can determine the remaining observation. In other words, given the value of $\bar{Y}$, the values of $(n - 1)$ observations determine the remaining one. In summarizing scores on a quantitative variable, there are $(n - 1)$ *degrees of freedom*, since only that many observations are independent.

73. * A researcher conducts a significance test every time she analyzes a new data set. Over time, she conducts 100 tests.
a) Suppose the null hypothesis is true in every case. What is the distribution of the number of times she rejects the null hypothesis at the .05 level?
b) Suppose she rejects the null hypothesis in seven of the tests. Is it plausible that the null hypothesis is correct in every case? Explain.

74. * Each year in Liverpool, New York, a public librarian estimates the mean number of times the books in that library have been checked out in the previous year. To do this, the librarian randomly samples computer records for 100 books and forms a 95% confidence interval for the mean. This has been done for 20 years.
a) Find the probability that all the confidence intervals contain the true means. (*Hint*: Use the binomial distribution.)
b) Find the probability that at least one confidence interval does not contain the true mean.
c) Find the mean and standard deviation of the probability distribution of the number of confidence intervals that do not contain the true mean.

75. * Refer to Problem 6.53. The process was recently modified because the outcome 0 tended to occur too frequently. Using these data, test whether the proportion of times that 0 would occur in the long run equals .1, the value expected for a proper random number generator, versus the alternative that it exceeds .1. Report the P-value, and interpret.

76. * Refer to Problem 6.37.
a) For what values of the number of wins X can you reject $H_0 : \pi = .50$ in favor of $H_a : \pi \neq .50$, using $\alpha = .05$?
b) For what values of X can you reject H_0 using $\alpha = .01$? (Note: For small samples with discrete data, it may not be possible to achieve very small P-values.)
c) * Suppose you test H_0 using $\alpha = .05$. What is the probability of making a Type I error? (Note: For discrete distributions, the actual chance of Type I error may be less than the intended chance. It is better simply to report the P-value. Unfortunately, this fact also adversely affects small-sample confidence intervals for proportions, making them conservative.)

77. Let β denote the probability of Type II error. For an $\alpha = .05$-level test of H_0: $\mu = 0$ against H_a: $\mu > 0$ based on $n = 30$ observations, $\beta = .36$ at $\mu = 4$. Select the correct response(s).
a) At $\mu = 5$, $\beta > .36$.
b) If $\alpha = .01$, then at $\mu = 4$, $\beta > .36$.
c) If $n = 50$, then at $\mu = 4$, $\beta > .36$.
d) The power of the test is .64 at $\mu = 4$.
e) This must be false, because necessarily $\alpha + \beta = 1$.

Bibliography

Cohen, J. (1988). *Statistical Power Analysis for the Behavioral Sciences*, 2nd ed. Hillsdale, NJ: Lawrence Erlbaum.

Mosteller, F., and Wallace, D. L. (1964). *Inference and Disputed Authorship: The Federalist*. Reading, MA: Addison-Wesley.

Chapter 7

Comparison of Two Groups

Comparing some characteristic of two groups is a fundamental analysis in the social sciences. One study might compare mean income for men and women having similar jobs and experience, another might compare Democrats and Republicans on the proportions who favor national health insurance. Comparisons of means are appropriate for quantitative variables (e.g., income) and comparisons of proportions are appropriate for qualitative variables (e.g., opinion favorable on some issue).

A quite common type of comparison looks at a mean or proportion parameter at two points in time. This type of study is said to be ***longitudinal***. For instance, one might compare subjects' opinions about national health insurance today to their opinions five years ago. By contrast, a study that uses a single survey to compare groups is called a ***cross-sectional*** study. One example is a current comparison of Democrats and Republicans on the proportion favoring health insurance.

Another possible comparison looks at a parameter for an experimental group and a control group. For instance, to analyze whether a SAT preparatory course makes any difference on students' performance on the SAT, a study might randomly split a class of students into two groups, one of which takes the course and one of which doesn't; after the course, both groups take the SAT, and mean scores are compared. Research studies in the social sciences, however, are rarely experimental in the same way as laboratory sciences. There is limited control over other factors that could affect the results.

Dependent and Independent Samples

Most of this chapter refers to comparisons using ***independent random samples*** from the groups. The choice of subjects for one sample does not depend on which subjects are in the other sample. Many comparisons of two groups result from dividing a larger sample into subsamples according to classification on some variable, such as gender or race. If the overall sample was randomly selected, then the subsamples constitute independent random samples from the corresponding subpopulations.

Dependent samples result when a natural matching occurs between each subject in one sample and a subject in the other sample. The most common instance of this happens when each sample has the same subjects, but measurements for the two samples occur at different times. For instance, a study that analyzes the extent to which a tutoring program improves mathematical understanding might administer a standard math achievement test to a sample of students both before and after they go through the program. In this type of situation, a subject's first response is likely to be related to the second response: The students who perform relatively well on one exam may tend to perform well on the second exam also. Data sets with dependent samples require different statistical methods than data sets with independent samples.

Bivariate Analyses with Response and Explanatory Variables

Two groups being compared constitute a ***binary*** or ***dichotomous*** variable—a variable having only two categories. For instance, in a comparison of the mean income of men and women, men and women are the two categories of the binary variable, gender. Methods for comparing two groups are special cases of ***bivariate*** statistical methods—an outcome variable of some type is analyzed within each of two categories of some other variable.

The outcome variable about which comparisons are made is called the ***response variable***. The variable that defines the groups is called the ***explanatory variable***. The analysis studies how the outcome on the response variable *depends on* or is *explained by* the value of the explanatory variable. In a comparison of men and women on income, income is the response variable and gender is the explanatory variable; income may depend on gender, not gender on income. A comparison of Democrats and Republicans on the proportion who favor national health insurance is a bivariate analysis of the response variable—opinion about national health insurance, and the explanatory variable—political party affiliation.

The response variable is also often called the ***dependent variable***, and the explanatory variable is also often called the ***independent variable***. The terminology *dependent variable* refers to the goal of investigating the degree to which the response on that variable *depends on* the group to which the subject belongs. We prefer not to use these terms, since *independent* and *dependent* are used for so many other things in statistical methods, such as labeling the types of samples being compared.

Section 7.1 presents methods for comparing means, and Section 7.2 presents methods for comparing proportions, both for independent samples. Section 7.3 presents small-sample methods, and Section 7.4 presents methods for dependent samples. Finally, Section 7.5 discusses alternative "nonparametric" methods that require fewer assumptions about the population distribution.

7.1 Quantitative Data: Comparing Two Means

Do women tend to spend more time on housework than men? If so, how much more? Based on data from the National Survey of Families and Households, a recent study (by S. South and G. Spitze, *American Sociological Review*, Vol. 59, 1994, pp. 327–347) reported the descriptive statistics in Table 7.1 for the hours spent in housework per week. We shall analyze these data in this section.

TABLE 7.1 Housework Hours, per Week, for a National Survey of Men and Women

		Housework Hours	
Sex	Sample Size	Mean	Standard Deviation
Men	4252	18.1	12.9
Women	6764	32.6	18.2

To compare two groups on a quantitative characteristic, we make inferences about their population means μ_1 and μ_2 and the difference between them. You may wish to review Sections 5.2 and 6.2 on methods for analyzing means in the one-sample case.

Confidence Interval for $\mu_2 - \mu_1$

A natural way to compare two population means is to estimate the difference between them, $\mu_2 - \mu_1$. We treat $\mu_2 - \mu_1$ as a parameter and estimate it from the sample data by the difference of sample means, $\bar{Y}_2 - \bar{Y}_1$.

Let n_1 denote the sample size for the sample from the first population and n_2 the sample size for the sample from the second population. When the sample is random and large, the sampling distribution of the sample mean is approximately normal about the true population mean. Not surprisingly, the estimator $\bar{Y}_2 - \bar{Y}_1$ has a sampling distribution that is approximately normal about $\mu_2 - \mu_1$, as portrayed in Figure 7.1.

Let $\sigma_{\bar{Y}_2 - \bar{Y}_1}$ denote the standard error of the sampling distribution of the estimated difference, $\bar{Y}_2 - \bar{Y}_1$. This describes the degree to which $\bar{Y}_2 - \bar{Y}_1$ would vary if we repeatedly took random samples of size n_1 and n_2. As usual for large samples, a confidence interval takes the estimate and adds and subtracts a tabulated z-score times the standard error. This confidence interval for $\mu_2 - \mu_1$ is

$$(\bar{Y}_2 - \bar{Y}_1) \pm z\sigma_{\bar{Y}_2 - \bar{Y}_1}$$

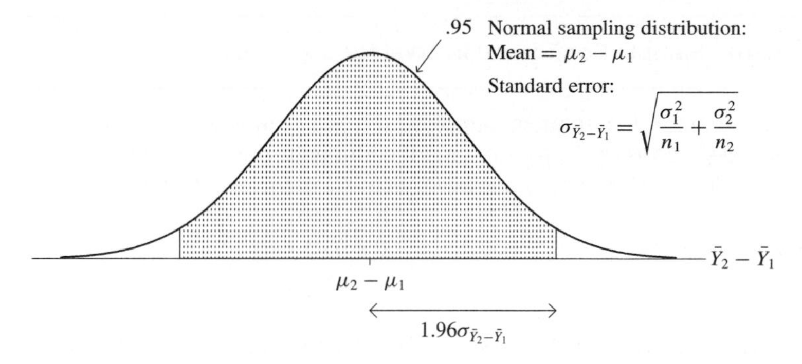

Figure 7.1 The Sampling Distribution of the Estimator $\bar{Y}_2 - \bar{Y}_1$ of the Difference of Means Is Approximately Normal About $\mu_2 - \mu_1$

The z-score depends on the confidence coefficient in the usual way (e.g., 1.96 for 95% confidence). The general rule given in the box enables us to estimate the standard error $\sigma_{\bar{Y}_2-\bar{Y}_1}$.

Variance of Sampling Distribution of Difference Between Two Estimates

When two estimates are formed from independent samples, the sampling distribution of their difference has variance equal to the sum of the variances of the sampling distributions of the separate estimates.

For instance, when $\bar{Y}_1$ and $\bar{Y}_2$ are sample means from two independent samples,

$$\sigma^2_{\bar{Y}_2-\bar{Y}_1} = \sigma^2_{\bar{Y}_1} + \sigma^2_{\bar{Y}_2}$$

Sampling error is associated with each estimate, and these errors add in determining the sampling error of the difference of the estimates.

Recall that the variance of the sampling distribution of a sample mean equals

$$\sigma^2_{\bar{Y}} = \frac{\sigma^2}{n}$$

where σ denotes the population standard deviation. Therefore, the difference between two sample means has standard error

$$\sigma_{\bar{Y}_2-\bar{Y}_1} = \sqrt{\sigma^2_{\bar{Y}_2-\bar{Y}_1}} = \sqrt{\sigma^2_{\bar{Y}_1} + \sigma^2_{\bar{Y}_2}} = \sqrt{\frac{\sigma_1^2}{n_1} + \frac{\sigma_2^2}{n_2}}$$

We estimate this by substituting the sample standard deviations for the unknown population values, which provides the result stated in the accompanying box.

Large-Sample Confidence Interval for $\mu_2 - \mu_1$

For independent random samples from two groups, denote the sample sizes by n_1 and n_2, the means by $\bar{Y}_1$ and $\bar{Y}_2$, and the standard deviations by s_1 and s_2. A large-sample confidence interval for $\mu_2 - \mu_1$ is $(\bar{Y}_2 - \bar{Y}_1) \pm z\hat{\sigma}_{\bar{Y}_2 - \bar{Y}_1}$, which equals

$$(\bar{Y}_2 - \bar{Y}_1) \pm z\sqrt{\frac{s_1^2}{n_1} + \frac{s_2^2}{n_2}}$$

The z-score is chosen to provide the desired confidence level. This interval is usually valid when both n_1 and n_2 are at least 20.

Example 7.1 Comparing Housework Hours of Men and Women

Now we return to the data in Table 7.1. Denote the population mean housework hours by μ_1 for men and μ_2 for women. The point estimate of $\mu_2 - \mu_1$, the difference between the mean for women and the mean for men, equals $\bar{Y}_2 - \bar{Y}_1 = 32.6 - 18.1 = 14.5$. The estimated standard error of this difference equals

$$\hat{\sigma}_{\bar{Y}_2 - \bar{Y}_1} = \sqrt{\frac{s_1^2}{n_1} + \frac{s_2^2}{n_2}} = \sqrt{\frac{(12.9)^2}{4252} + \frac{(18.2)^2}{6764}} = .30$$

The sample sizes are very large, so we use a high confidence coefficient. A 99% confidence interval for $\mu_2 - \mu_1$ is

$$(\bar{Y}_2 - \bar{Y}_1) \pm 2.58\hat{\sigma}_{\bar{Y}_2 - \bar{Y}_1} = 14.5 \pm 2.58(.30), \quad \text{or } 14.5 \pm .8, \quad \text{or } (13.7, 15.3)$$

We infer that the true difference falls between 13.7 and 15.3 hours. That is, the mean amount of weekly time that women spend on housework is between 13.7 and 15.3 hours more than men spend on it. □

Interpretations

Since the confidence interval in the last example contains only positive values, and since we took the difference between the mean for women and the mean for men, we can conclude that the mean is higher for women. A confidence interval for $\mu_2 - \mu_1$ that contains only positive numbers suggests that $\mu_2 - \mu_1$ is positive, meaning that μ_2 is larger than μ_1. A confidence interval for $\mu_2 - \mu_1$ that contains only negative numbers suggests that μ_2 is smaller than μ_1.

When the confidence interval contains 0, insufficient evidence exists to conclude which of μ_1 or μ_2 is larger. In this case, 0 is a plausible value for $\mu_2 - \mu_1$, meaning that it is plausible that $\mu_1 = \mu_2$. To illustrate, suppose the confidence interval for $\mu_2 - \mu_1$ equals $(-5.0, 8.0)$. Then the mean for women may be as much as 5 less or as much as 8 more than the mean for men.

The identification of which group is called 1 and which is called 2 is arbitrary. If we reversed the labels, each endpoint reverses sign. For instance, a confidence interval of (−5.0, 8.0) for $\mu_2 - \mu_1$ is equivalent to one of (−8.0, 5.0) for $\mu_1 - \mu_2$.

As in the one-sample case, larger sample sizes contribute to narrower confidence intervals. In addition, higher confidence coefficients yield wider confidence intervals. A 95% confidence interval for the difference of mean housework hours equals (13.9, 15.1), which is narrower than the 99% interval of (13.7, 15.3).

Significance Tests About $\mu_2 - \mu_1$

One can also compare population means μ_1 and μ_2 by a significance test of the hypothesis H_0: $\mu_1 = \mu_2$. For the difference of means parameter, this hypothesis is H_0: $\mu_2 - \mu_1 = 0$.

A large-samples ($n_1 \geq 20, n_2 \geq 20$) significance test of this hypothesis uses $\bar{Y}_2 - \bar{Y}_1$ and its standard error. The standard form for a z test statistic is

$$z = \frac{\text{Estimate of parameter} - \text{null hypothesis value of parameter}}{\text{Standard error of estimator}}$$

Treating the difference $\mu_2 - \mu_1$ as the parameter, we test that $\mu_2 - \mu_1 = 0$; that is, the null hypothesis value of the parameter $\mu_2 - \mu_1$ is 0. The estimated value of $\mu_2 - \mu_1$ is $\bar{Y}_2 - \bar{Y}_1$, and the estimated standard error of $\bar{Y}_2 - \bar{Y}_1$ is the same as in a confidence interval. The form of the large-samples z test statistic is

$$z = \frac{(\bar{Y}_2 - \bar{Y}_1) - 0}{\hat{\sigma}_{\bar{Y}_2 - \bar{Y}_1}} = \frac{\bar{Y}_2 - \bar{Y}_1}{\sqrt{\frac{s_1^2}{n_1} + \frac{s_2^2}{n_2}}}$$

The P-value for the test depends in the usual way on whether the alternative hypothesis is the two-sided one, H_a: $\mu_1 \neq \mu_2$ (i.e., $\mu_2 - \mu_1 \neq 0$), or one of the one-sided ones, H_a: $\mu_1 > \mu_2$ ($\mu_2 - \mu_1 < 0$) or H_a: $\mu_1 < \mu_2$ ($\mu_2 - \mu_1 > 0$). Most common is the two-sided alternative, in which case the P-value is the two-tail probability beyond the observed test statistic value.

Example 7.2 Test Comparing Mean Housework for Men and Women

We continue with the data from Example 7.1, now testing for a difference between the population mean housework hours, μ_1 for men and μ_2 for women. We test $H_0 : \mu_1 = \mu_2$ against $H_a : \mu_1 \neq \mu_2$. From Example 7.1, the estimate $\bar{Y}_2 - \bar{Y}_1$ equals $32.6 - 18.1 = 14.5$, and it has an estimated standard error of .30.

The test statistic equals

$$z = \frac{\bar{Y}_2 - \bar{Y}_1}{\hat{\sigma}_{\bar{Y}_2 - \bar{Y}_1}} = \frac{(32.6 - 18.1)}{.30} = 48.8$$

This is a huge value for a z statistic and gives a P-value that is essentially 0. We can conclude, with little risk of making an error, that the population means differ. Specifically, the sample means show that the difference takes the direction of a higher mean for women. □

Correspondence Between Confidence Intervals and Tests

The equivalence between two-sided tests of hypotheses and confidence intervals for means mentioned in Section 6.2 also applies in the two-sample case. For example, since the two-sided P-value in Example 7.2 is less than .01, we would reject the null hypothesis that $\mu_2 - \mu_1 = 0$ at the $\alpha = .01$ level. Similarly, a 99% confidence interval for $\mu_2 - \mu_1$ would not contain 0, the null hypothesis value. In fact, that interval equals (13.7, 15.3).

When a 99% confidence interval for the difference between two means does not contain 0, then 0 is not a plausible value for $\mu_2 - \mu_1$, at the $\alpha = .01$ level; that is, the P-value for testing H_0: $\mu_2 - \mu_1 = 0$ against H_a: $\mu_2 - \mu_1 \neq 0$ must be less than .01. Thus, H_0 would be rejected at the .01 level and, in fact, at any α-level above the P-value.

As in one-sample problems, confidence intervals are more informative than tests. The confidence interval tells us not only that the mean housework hours differs for men and women, but it shows us just how large that difference is likely to be, and in which direction.

In practice, significance tests are much more common for two-sample comparisons than for one-sample analyses such as those shown in Chapter 6. For example, it is usually artificial to test whether the population mean equals one particular value, such as we do in testing a hypothesis of form $H_0 : \mu = \mu_0$. However, it is often relevant to test whether a *difference* exists between two population means, such as we do in testing a hypothesis of the form $H_0 : \mu_1 = \mu_2$. For instance, we may not have any idea what to hypothesize for the mean income of men in a particular occupation, but we may want to know whether that mean (whatever its value) is the same as, larger than, or smaller than the mean income for women. Comparison methods focus on the nature and degree of differences between groups, not on the specification of exact values of characteristics of the separate distributions.

7.2 Qualitative Data: Comparing Two Proportions

This section discusses comparison of independent samples for a qualitative response variable. Confidence intervals and tests compare the proportions of subjects classified in a particular category of that response variable.

Let π_1 denote the proportion for the first population and π_2 the proportion for the second population. For independent random samples of sizes n_1 and n_2, let $\hat{\pi}_1$ and $\hat{\pi}_2$ denote the sample proportions. These estimate the population proportions. You may wish to review Sections 5.3 and 6.3 on methods for analyzing proportions in the one-sample case.

Example 7.3 Change over Time in Views About Women's Roles

Table 7.2 summarizes responses from General Social Surveys in 1982 and in 1994 to the question, "Do you agree or disagree with this statement? Women should take care of running their homes and leave running the country up to men." For 345 respondents in 1982, 122 subjects agreed. For a separate set of 1900 respondents in 1994, 268 agreed.

TABLE 7.2 Responses to the Statement, "Women should take care of running their homes and leave running the country up to men."

Year	Agree	Disagree	Total
1982	122	223	345
1994	268	1632	1900

Is there a change in opinion from 1982 to 1994? Let π_1 denote the population proportion who agreed with this statement in 1982, and let π_2 denote the population proportion in 1994. From Table 7.2, the sample proportions equal

$$\hat{\pi}_1 = \frac{122}{345} = .35, \quad \hat{\pi}_2 = \frac{268}{1900} = .14$$

Similarly, the sample proportions of disagree responses were $1 - \hat{\pi}_1 = .65$ in 1982 and $1 - \hat{\pi}_2 = .86$ in 1994. □

Confidence Interval for Difference of Proportions

We compare two population proportions using the difference between them, $\pi_2 - \pi_1$. The difference of sample proportions, $\hat{\pi}_2 - \hat{\pi}_1$, estimates $\pi_2 - \pi_1$. If n_1 and n_2 are relatively large, the estimator $\hat{\pi}_2 - \hat{\pi}_1$ has a sampling distribution that is approximately normal with mean equal to the true value $\pi_2 - \pi_1$, as portrayed in Figure 7.2.

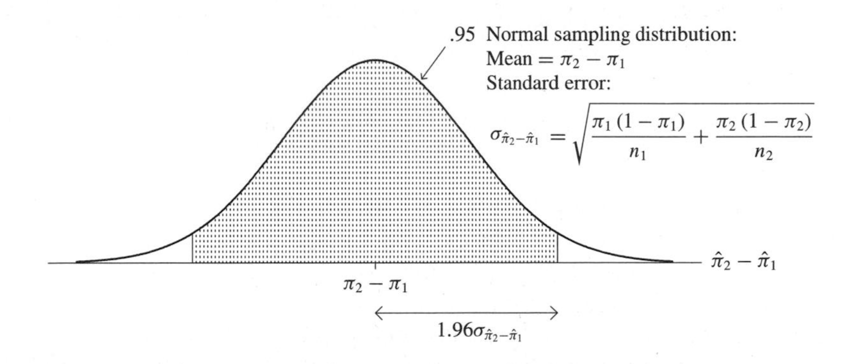

Figure 7.2 Large-Sample Normal Sampling Distribution of the Estimator $\hat{\pi}_2 - \hat{\pi}_1$ of the Difference of Proportions

Let $\sigma_{\hat{\pi}_2-\hat{\pi}_1}$ denote the standard error of the sampling distribution of the estimated difference, $\hat{\pi}_2 - \hat{\pi}_1$. As usual, the confidence interval takes the estimate and adds and subtracts a tabulated z score times the estimated standard error,

$$(\hat{\pi}_2 - \hat{\pi}_1) \pm z\hat{\sigma}_{\hat{\pi}_2-\hat{\pi}_1}$$

From the general rule in the box in Section 7.1, the variance of the difference of sample proportions equals the sum of the variances of the separate proportions. From Section 5.3, the estimated variance of the sampling distribution of a sample proportion equals

$$\hat{\sigma}^2_{\hat{\pi}} = \frac{\hat{\pi}(1-\hat{\pi})}{n}$$

Therefore, the difference between two proportions has estimated standard error equal to

$$\hat{\sigma}_{\hat{\pi}_2-\hat{\pi}_1} = \sqrt{\hat{\sigma}^2_{\hat{\pi}_2-\hat{\pi}_1}} = \sqrt{\hat{\sigma}^2_{\hat{\pi}_1} + \hat{\sigma}^2_{\hat{\pi}_2}} = \sqrt{\frac{\hat{\pi}_1(1-\hat{\pi}_1)}{n_1} + \frac{\hat{\pi}_2(1-\hat{\pi}_2)}{n_2}}$$

Large-Sample Confidence Interval for $\pi_2 - \pi_1$

A large-sample confidence interval for $\pi_2 - \pi_1$ is $(\hat{\pi}_2 - \hat{\pi}_1) \pm z\hat{\sigma}_{\hat{\pi}_2-\hat{\pi}_1}$, or

$$(\hat{\pi}_2 - \hat{\pi}_1) \pm z\sqrt{\frac{\hat{\pi}_1(1-\hat{\pi}_1)}{n_1} + \frac{\hat{\pi}_2(1-\hat{\pi}_2)}{n_2}}$$

The sample is large enough to use this formula if, for each sample, more than five observations fall in the category for which the proportion is estimated, and more than five observations do not fall in that category. Most studies easily satisfy this.

We illustrate with the data introduced in Example 7.3 on women's roles. The parameter of interest is the difference $\pi_2 - \pi_1$ between 1994 and 1982 in the proportion of agree responses. Since $\hat{\pi}_1 = .35$ and $\hat{\pi}_2 = .14$, the estimated difference equals $\hat{\pi}_2 - \hat{\pi}_1 = -.21$, a drop of .21 in the proportion who agree that women should take care of running their homes and leave running the country up to men.

To determine the accuracy of this estimate, we form a confidence interval for the true population difference. A 95% confidence interval for $\pi_2 - \pi_1$ is

$$(\hat{\pi}_2 - \hat{\pi}_1) \pm 1.96\sqrt{\frac{\hat{\pi}_1(1-\hat{\pi}_1)}{n_1} + \frac{\hat{\pi}_2(1-\hat{\pi}_2)}{n_2}}$$

which equals

$$(.14 - .35) \pm 1.96\sqrt{\frac{(.35)(.65)}{345} + \frac{(.14)(.86)}{1900}} = -.21 \pm 1.96(.0269)$$
$$= -.21 \pm .05 \quad \text{or} \quad (-.26, -.16)$$

Since both endpoints are negative, we conclude that $\pi_2 - \pi_1$ is negative; that is, π_2 is less than π_1. The proportion who believe that women's place is in the home has decreased between 1982 and 1994. Moreover, since the numbers in the confidence interval fall quite far from 0, the change in opinion is quite large in practical terms.

When a confidence interval for $\pi_2 - \pi_1$ contains entirely positive numbers, then we conclude that $\pi_2 - \pi_1$ is positive; that is, π_2 is larger than π_1. When the interval contains 0, insufficient evidence exists to conclude which of π_1 or π_2 is larger, and it is plausible that the true proportions are equal.

For instance, in 1993, 154 of 1042 subjects agreed with the statement about women's role, a proportion of .15. A 95% confidence interval for the difference between the proportions agreeing in 1994 and in 1993 is (−.03, .02). We infer that the true proportion in 1994 could be as much as .03 lower or as much as .02 higher than in 1993. Since this interval contains 0, 0 is a plausible value for the difference, which translates to possibly equal population proportions. The lower and upper bounds show that if a difference exists, it is small.

Significance Tests About $\pi_2 - \pi_1$

One can also compare population proportions π_1 and π_2 by a significance test of the hypothesis H_0: $\pi_1 = \pi_2$. For the difference of proportions parameter, this hypothesis is H_0: $\pi_2 - \pi_1 = 0$.

The test compares the estimate $\hat{\pi}_2 - \hat{\pi}_1$ of $\pi_2 - \pi_1$ to the null hypothesis value of 0. For large samples, the test statistic is

$$z = \frac{\text{Estimate} - \text{null hypothesis value}}{\text{Standard error}} = \frac{(\hat{\pi}_2 - \hat{\pi}_1) - 0}{\hat{\sigma}_{\hat{\pi}_2 - \hat{\pi}_1}}$$

Although this test statistic can use the standard error formula from the confidence interval for $\pi_2 - \pi_1$, it is preferable to use an alternative formula based on the H_0 assumption that $\pi_1 = \pi_2$. Let $\hat{\pi}$ denote the proportion of the *total* sample in the category of interest. This estimates the common value of π_1 and π_2, for testing that $\pi_1 = \pi_2$, and it is called the *pooled estimate*. The standard error formula for the test is

$$\hat{\sigma}_{\hat{\pi}_2 - \hat{\pi}_1} = \sqrt{\frac{\hat{\pi}(1 - \hat{\pi})}{n_1} + \frac{\hat{\pi}(1 - \hat{\pi})}{n_2}} = \sqrt{\hat{\pi}(1 - \hat{\pi})\left(\frac{1}{n_1} + \frac{1}{n_2}\right)}$$

To illustrate, in Example 7.3, $\hat{\pi}_1 = 122/345 = .35$ and $\hat{\pi}_2 = 268/1900 = .14$. The pooled estimate based on the entire sample equals (122 + 268)/(345 + 1900) = .174. The standard error estimate for the test equals

$$\hat{\sigma}_{\hat{\pi}_2 - \hat{\pi}_1} = \sqrt{\hat{\pi}(1 - \hat{\pi})\left(\frac{1}{n_1} + \frac{1}{n_2}\right)} = \sqrt{.174(.826)\left(\frac{1}{345} + \frac{1}{1900}\right)}$$

$$= \sqrt{.00049} = .022$$

slightly smaller than the standard error of .027 for the confidence interval. The test statistic for testing H_0: $\pi_1 = \pi_2$ equals

$$z = \frac{\hat{\pi}_2 - \hat{\pi}_1}{\hat{\sigma}_{\hat{\pi}_2 - \hat{\pi}_1}} = \frac{.14 - .35}{.022} = -9.6$$

For the two-sided alternative hypothesis H_a: $\pi_1 \neq \pi_2$, the P-value is the two-tail probability from the standard normal distribution. A z-score of -9.6 is enormous and has a P-value that is 0 to several decimal places. Extremely strong evidence exists of a change in opinion during this 12-year period, with opinions becoming more liberal about women's roles.

Contingency Tables and Conditional Probabilities

Table 7.2 is an example of a statistical type of table called a ***contingency table***, because it displays how outcomes of the response variable are *contingent* on the category of the explanatory variable. The table displays the combinations of possible outcomes on two categorical variables. Each row is a category of one variable and each column is a category of the other variable. In Table 7.2, one variable is a binary response variable (opinion about women's roles), and one variable is a binary variable defining the two groups compared (year: 1982 and 1994). The ***cells*** of the table contain frequency counts for the four possible combinations of outcomes. Table 7.2 contains observations for 2245 subjects, cross-classified according to the year at which they were interviewed and their response about women's proper role.

The proportions π_1 and π_2 estimated using the data in the contingency table are called ***conditional probabilities***. This term refers to probabilities for a response variable evaluated under two conditions, namely the two levels of the explanatory variable. For instance, under the condition that year = 1994 (i.e., "conditional on year = 1994"), the conditional probability of agreeing with the statement about women's roles is estimated to be 268/1900 = .14.

This section has presented methods for qualitative response categories having only two levels. Instead, the characteristic could have several categories. For example, the response categories might be (Agree, Disagree, Undecided). In this case, we could compare the two groups in terms of the conditional probabilities of observations in each of the three categories. Moreover, there could be more than two groups to compare, as we might include survey results from 1972 as well. Chapter 8 presents analyses of contingency tables that have more than two rows or columns.

7.3 Small-Sample Inference for Comparing Means and Proportions

The past two sections have presented large-sample methods for comparing proportions and comparing means. We now present methods that are preferable when at least one of the sample sizes is small.

We first present small-sample methods for comparing means. These are applicable if either sample size n_1 or n_2 is less than 20. Single-sample methods introduced in Section 6.5 for small n using the t distribution extend to the two-sample case. Like those methods, the two-sample methods assume that the population distributions are normal.

An *approximate* approach uses the same formulas as in Section 7.1 for large samples, but with the substitution of the t distribution for the normal distribution. For instance, the confidence interval has the form

$$(\bar{Y}_2 - \bar{Y}_1) \pm t\sqrt{\frac{s_1^2}{n_1} + \frac{s_2^2}{n_2}}$$

with a t-score in place of the z-score. Computation of the *df* value to determine the t-score is complex, however. If you use software to conduct the analysis, this is no obstacle, and this strategy is discussed later in this section in the subsection on "Approximate Small-Sample Inference Using Computer Software."

In practice, the most commonly used small-sample methods use similar formulas but with a simpler *df* expression for an *exact t* distribution that occurs under the additional assumption that the population standard deviations are equal, that is, $\sigma_1 = \sigma_2$. We present these in the next two subsections. Though it seems disagreeable to make an additional assumption, confidence intervals and two-sided tests are fairly robust against violations of this and the normality assumption, particularly when the sample sizes are similar and not extremely small.

Small-Sample Confidence Interval for $\mu_2 - \mu_1$

For small-sample methods that make the assumption that the two groups have the same variability, the common value σ of σ_1 and σ_2 is estimated by

$$\hat{\sigma} = \sqrt{\frac{\sum(Y_{1i} - \bar{Y}_1)^2 + \sum(Y_{2i} - \bar{Y}_2)^2}{n_1 + n_2 - 2}} = \sqrt{\frac{(n_1 - 1)s_1^2 + (n_2 - 1)s_2^2}{n_1 + n_2 - 2}}$$

In this formula, the term $\sum(Y_{1i} - \bar{Y}_1)^2$ is a sum of squared distances of measurements in the first sample from their mean $\bar{Y}_1$, and the sum of squares $\sum(Y_{2i} - \bar{Y}_2)^2$ refers to distances of measurements in the second sample from their mean. The estimate σ pools information from the two samples to provide a single estimate of variability. The degrees of freedom for this estimate equal $df = n_1 + n_2 - 2$. This equals the total number of observations $(n_1 + n_2)$ minus the number of parameters estimated in order to calculate $\hat{\sigma}$ (namely, the two means, μ_1 and μ_2, estimated by $\bar{Y}_1$ and $\bar{Y}_2$).

Using the single value $\hat{\sigma}$ to estimate σ_1 and σ_2, the estimated standard error of $\bar{Y}_2 - \bar{Y}_1$ simplifies to

$$\hat{\sigma}_{\bar{Y}_2 - \bar{Y}_1} = \sqrt{\frac{\hat{\sigma}^2}{n_1} + \frac{\hat{\sigma}^2}{n_2}} = \hat{\sigma}\sqrt{\frac{1}{n_1} + \frac{1}{n_2}}$$

The confidence interval for $\mu_2 - \mu_1$ has the form

$$(\bar{Y}_2 - \bar{Y}_1) \pm t\hat{\sigma}_{\bar{Y}_2 - \bar{Y}_1}$$

The t-score comes from the t table with $df = n_1 + n_2 - 2$, for the desired confidence level (e.g., use $t_{.025}$ for 95% confidence).

Example 7.4 Comparing Clinical Therapies with Small Samples

A clinical psychologist wants to choose between two therapies for treating severe cases of mental depression. She selects six patients who are similar in their depressive symptoms and in their socioeconomic backgrounds and general health status. She randomly selects three of the patients to receive therapy A, and the other three receive therapy B. She selects small samples for ethical reasons; if her experiment indicates that one therapy is superior, that therapy will be used on all her other patients having these symptoms.

After one month of intense treatment, the degree of improvement in each patient is measured by the change in score on a standardized scale for measuring severity of mental depression. The improvement scores are 10, 20, 30 for the patients receiving therapy A, and 30, 45, 45 for the patients receiving therapy B; we identify A as group 1 and B as group 2.

For these data, $\bar{Y}_1 = 20$, $\bar{Y}_2 = 40$, and

$$\begin{aligned}\sum(Y_{1i} - \bar{Y}_1)^2 &= (10-20)^2 + (20-20)^2 + (30-20)^2 = 200\\ \sum(Y_{2i} - \bar{Y}_2)^2 &= (30-40)^2 + (45-40)^2 + (45-40)^2 = 150\end{aligned}$$

On the assumption that both populations have the same standard deviation, the pooled estimate of that common standard deviation equals

$$\hat{\sigma} = \sqrt{\frac{\sum(Y_{1i} - \bar{Y}_1)^2 + \sum(Y_{2i} - \bar{Y}_2)^2}{n_1 + n_2 - 2}} = \sqrt{\frac{200 + 150}{3 + 3 - 2}} = \sqrt{\frac{350}{4}} = 9.35$$

Now, $\bar{Y}_2 - \bar{Y}_1 = 40 - 20 = 20$ has an estimated standard error of

$$\hat{\sigma}_{\bar{Y}_2 - \bar{Y}_1} = \hat{\sigma}\sqrt{\frac{1}{n_1} + \frac{1}{n_2}} = 9.35\sqrt{\frac{1}{3} + \frac{1}{3}} = 7.64$$

Let μ_1 and μ_2 denote the mean improvement scores for these therapies for the hypothetical populations that the samples represent. When $df = (n_1 + n_2 - 2) = 4$, the t-score for a 95% confidence interval for $(\mu_2 - \mu_1)$ is $t_{.025} = 2.776$. The interval is $(40 - 20) \pm 2.776(7.64)$, or 20 ± 21.2, or $(-1.2, 41.2)$. We conclude that the mean for therapy B could be as much as 1.2 less or as much as 41.2 higher than the mean for therapy A. Since the interval contains 0, it is plausible that the means are identical. On the other hand, therapy B could be considerably better. Not surprisingly, the small samples result in a very wide interval.

When the sample sizes are very small, it may be worth sacrificing a bit of confidence to achieve more precision. A 90% confidence interval is $20 \pm 2.132(7.64)$, or 20 ± 16.3, or (3.7, 36.3). At this confidence level, we conclude that therapy B is better. □

Small-Sample Test of $H_0 : \mu_1 = \mu_2$

Now, consider the null hypothesis, H_0: $\mu_1 = \mu_2$. The small-sample test statistic is

$$t = \frac{(\bar{Y}_2 - \bar{Y}_1)}{\hat{\sigma}_{\bar{Y}_2 - \bar{Y}_1}}$$

This uses the same standard error as the confidence interval just given, and the test statistic has the t distribution with $df = n_1 + n_2 - 2$.

To illustrate, we test H_0: $\mu_1 = \mu_2$ against H_a: $\mu_1 \neq \mu_2$ for the data in Example 7.4 comparing two therapies for mental depression. The test statistic equals

$$t = \frac{\bar{Y}_2 - \bar{Y}_1}{\hat{\sigma}_{\bar{Y}_2 - \bar{Y}_1}} = \frac{40 - 20}{7.64} = 2.62$$

This statistic has $df = n_1 + n_2 - 2 = 6 - 2 = 4$. From the t-table, this statistic has a two-sided P-value of $P < .10$ but $P > .05$ (from computer software, $P = .059$).

There is some evidence, particularly given the small sample sizes, of better success using therapy B. For such small samples, it is difficult to achieve small P-values unless sample effects are very large. A P-value exceeding .05 is consistent with the 95% confidence interval for the difference containing 0; $P < .10$ is consistent with the 90% confidence interval not containing 0.

Approximate Small-Sample Inference Using Computer Software

Table 7.3 illustrates the way computer software reports results of the two-sample t test. (Appendix A shows SAS code for this analysis.) For each group (therapy), the output lists the sample size, sample mean, sample standard deviation, and estimated standard error. The last part of the table shows the value of the t test statistic and its df value. The test statistic is the negative of the value we reported above, simply because the software took mean 1 minus mean 2, rather than mean 2 minus mean 1. The P-value is labeled Prob $> |T|$, representing the probability that the absolute value of t is at least as large as observed, which is the two-sided P-value. Some software (e.g., SPSS) also reports a 95% confidence interval for the difference of means.

Table 7.3 shows results of two tests for comparing means, differing in terms of whether they assume equal population variances. The t test just presented assumes that $\sigma_1 = \sigma_2$. The t statistic that software reports for the unequal variances case is the z statistic of Section 7.1 for comparing two means, $z = (\bar{Y}_2 - \bar{Y}_1)/\hat{\sigma}_{\bar{Y}_2 - \bar{Y}_1}$, with

$$\hat{\sigma}_{\bar{Y}_2 - \bar{Y}_1} = \sqrt{\frac{s_1^2}{n_1} + \frac{s_2^2}{n_2}}$$

TABLE 7.3 Output from Software for Performing Two-Sample t Test

THERAPY	N	Mean	Std Dev	Std Error
a	3	20.000	10.000	5.7735
b	3	40.000	8.660	5.0000

Variances	T	DF	Prob>|T|
Unequal	-2.6186	3.9	0.0605
Equal	-2.6186	4.0	0.0589

In using this alternative statistic with small samples, treat it as a t statistic rather than a z statistic. The t distribution is only the approximate sampling distribution, however, and its *df* formula is now less than $n_1 + n_2 - 2$. The *df* formula, called the *Welch–Satterthwaite approximation*, is complex and depends on s_1 and s_2 as well as n_1 and n_2. Software reports this approximate *df* value, and the associated two-sided P-value. When $n_1 = n_2$, the "equal variances" and "unequal variances" test statistics are identical, and they are usually similar if n_1 and n_2 are close.

Many texts and most software present a statistic denoted by F for testing that the population variances are equal. This test only behaves well if the populations are very close to normal, and statisticians do not recommend it for general use because of its lack of robustness. If the data show evidence of a potentially large difference in standard deviations (with, say, one standard deviation being at least double the other), it is wisest to use the approximate two-sample t test that does not require the equal variance assumption. In fact, if you use the computer to analyze your data, you don't have to worry about the complex computation of *df* for the unequal variance method, so you can simply rely on that result in all cases. It typically yields a conclusion different from the equal variance method only if the sample variances are quite different.

Fisher's Exact Test for Comparing Proportions*

The large-sample methods for comparing proportions presented in Section 7.2 are used when all the counts in the cells of contingency tables such as Table 7.2 exceed about five. For smaller sample sizes, the methods may not work well, because the sampling distribution of $\hat{\pi}_2 - \hat{\pi}_1$ may be insufficiently close to normality. For small samples, one can compare two proportions π_1 and π_2 using a method called ***Fisher's exact test***, due to the eminent statistician R. A. Fisher.

The calculations for this test are complex and beyond the scope of this text. The principle behind the test is straightforward (see Problem 7.51), however, and statistical software provides its P-value, as shown in the appendix. As usual, the P-value is the probability of a sample result even more extreme than the observed data, under the assumption that the null hypothesis is true. The smaller the P-value, the stronger the evidence that the two proportions are truly different. For the formula for computing probabilities for Fisher's exact test, see Agresti (1996, p. 39), Fleiss (1981, p. 25), or Sprent (1993, p. 214).

Example 7.5 Parental Influence on Sexual Orientation of Children

A recent study (S. Colombok and F. Tasker, *Developmental Psychology*, Vol. 32, 1996, pp. 3–11) investigated the sexual orientation of adults who had been raised as children in lesbian families. Twenty-five children of lesbian mothers and a control group of 20 children of heterosexual mothers were seen at age 10 and again at age about 24. At the later time, they were interviewed about their sexual identity, with possible responses "Bisexual/Lesbian/Gay" or "Heterosexual." Table 7.4 shows results, in the form of a printout (SAS) for conducting Fisher's exact test.

TABLE 7.4 Sexual Identity of Young Adults, by Family Type of Upbringing, with Computer Printout for Performing Fisher's Exact Test

```
              IDENTITY

    MOTHER    | B/L/G    | HETERO |  Total
    ----------+---------+--------+
    Lesbian   |       2 |     23 |    25
    ----------+---------+--------+
    Heterosx  |       0 |     20 |    20
    ----------+---------+--------+
    Total           2       43        45

   STATISTICS FOR TABLE OF MOTHER BY IDENTITY

   Statistic                          Prob
   -----------------------------------------------------
   Fisher's Exact Test (Left)         1.000
                       (Right)        0.303
                       (2-Tail)       0.495
```

Denote the population proportion identifying as bisexual/lesbian/gay by π_1 for those with lesbian mothers and by π_2 for those with heterosexual mothers. Then, $\hat{\pi}_1 = 2/25 = .08$ and $\hat{\pi}_2 = 0/20 = 0.0$. We test H_0: $\pi_1 = \pi_2$ against H_a: $\pi_1 > \pi_2$. Two of the four counts are very small, however, and large-sample methods are inappropriate. From the printout, Fisher's exact test provides a P-value of $P = .303$ for this one-sided alternative, based on the right-hand tail of the sampling distribution of $(\hat{\pi}_1 - \hat{\pi}_2)$. There is insufficient evidence to conclude that bisexual/gay/lesbian orientation is more likely for children raised by lesbians. The P-value for the two-sided alternative equals .495 and the P-value for H_a: $\pi_1 < \pi_2$ is 1.000. □

Small-Sample Estimation Comparing Two Proportions*

Unfortunately, no adequate solution exists for constructing confidence intervals comparing two proportions when the sample sizes are small. The calculations for available methods are complex and require highly specialized computer software. Moreover, the resulting confidence interval is "conservative." The procedure for constructing "95%" confidence intervals has actual confidence level of *at least* that level; the actual level may be 98%, for instance, and is unknown.

7.4 Comparing Dependent Samples

Dependent samples occur when each observation in sample 1 matches with an observation from sample 2. The data are often called ***matched pairs*** data, because of this matching. Dependent samples most commonly occur when each sample has the same subjects. This is an example of *repeated measurement* data, each subject being observed repeatedly on the response variable.

We first present dependent-samples analyses for comparing means for a quantitative variable, and then we present an analysis for comparing proportions for a qualitative variable.

Comparing Dependent Means Using Paired Differences

In Example 7.4, the random choice of subjects for each therapy would have been inadvisable if the subjects differ in ways thought to affect the experimental results. For instance, if the subjects selected for therapy A happened to have poor overall health quality and little potential for relief of depressive symptoms regardless of treatment, compared to those selected for therapy B, results might have an inadvertent bias in favor of treatment B. For moderate to large samples, these other factors tend to balance by virtue of the randomization. For very small samples, though, an imbalance could occur.

With this in mind, suppose the psychologist redesigned the experiment. She selects three pairs of subjects, such that the patients matched in any given pair are very similar in health and other factors. For each pair, one subject is randomly selected for each therapy. This matched-pairs plan is a simple example of a ***randomized block design***. Each pair of subjects forms a *block*, and within blocks subjects are randomly assigned to the treatments being compared. Table 7.5 shows the sample results.

For matched-pairs data, each sample has the same number of observations. Each observation in one sample pairs with an observation in the other sample. Then, for the ith pair, let

$$D_i = \text{Observation in sample 2} - \text{observation in sample 1}$$

The sample mean $\bar{D}$ of these difference scores estimates μ_D, the population mean difference. In fact, the parameter μ_D is identical to $\mu_2 - \mu_1$, the difference between the

TABLE 7.5 Matched Pairs for Comparing Improvement Scores for Two Depression Therapies

Pair	Therapy A	Therapy B	Difference $(B) - (A)$
1	10	30	+20
2	20	45	+25
3	30	45	+15

population means for the two groups; the mean of the differences equals the difference between the means.

> For matched-pairs data, the difference between the means of the two groups equals the mean of the difference scores.

We can base analyses about $\mu_2 - \mu_1$ on inferences about μ_D, using the single sample of difference scores. This simplifies the analysis, since it reduces a two-sample problem to a one-sample problem.

Let n denote the number of observations in each sample. This equals the number of difference scores $\{D_i\}$. The confidence interval for μ_D is

$$\bar{D} \pm t\frac{s_D}{\sqrt{n}}$$

Here, $\bar{D}$ is the sample mean of the difference scores $\{D_i\}$,

$$s_D = \sqrt{\frac{\sum(D_i - \bar{D})^2}{n - 1}}$$

is the sample standard deviation of the difference scores, and t is the t-score from the t table for the desired confidence level, having $df = n - 1$.

We calculate the standard deviation s_D of the difference scores the same way we calculate a standard deviation s for any set of n numbers. In fact, this confidence interval has the same form as the one presented in Section 6.5 for a single mean; we simply apply the formula to the single sample of n differences rather than the original two sets of observations. When n exceeds 30, z-scores can substitute for the t-scores.

For testing H_0: $\mu_1 = \mu_2$ against H_a: $\mu_1 \neq \mu_2$, we express the hypotheses as H_0: $\mu_D = 0$ and H_a: $\mu_D \neq 0$. The test statistic is

$$t = \frac{\bar{D}}{\hat{\sigma}_{\bar{D}}} = \frac{\bar{D}}{s_D/\sqrt{n}}$$

This compares the sample mean of the differences to the null hypothesis value of 0, measuring the difference in terms of the number of standard errors. The standard error is the same one used in the confidence interval. As usual, the P-value is the two-tail probability. Since this test uses the difference scores for the pairs of observations, it is called a ***paired-difference t test***.

Example 7.6 Comparing Therapies with Matched-Pairs (Repeated Measures) Data

We now analyze the matched-pairs data in Table 7.5 for the improvement scores for two therapies for treating depression. The three difference scores have a sample mean of $\bar{D} = 20$, which is the same as the difference between the sample means of 40 and 20 for the two therapies. The sample standard deviation equals

$$\begin{aligned} s_D &= \sqrt{\frac{\sum(D_i - \bar{D})^2}{n-1}} = \sqrt{\frac{(20-20)^2 + (25-20)^2 + (15-20)^2}{2}} \\ &= \sqrt{25} = 5.0 \end{aligned}$$

The standard error of $\bar{D}$ is $\hat{\sigma}_{\bar{D}} = s_D/\sqrt{n} = 5.0/\sqrt{3} = 2.89$.

For a 95% confidence interval for $\mu_D = \mu_2 - \mu_1$, Table B reports $t_{.025} = 4.303$ for $df = 2$. The confidence interval equals

$$\bar{D} \pm 4.303\hat{\sigma}_{\bar{D}} = 20 \pm 4.303(2.89), \quad \text{or } (7.6, 32.4)$$

Even though the sample has only three pairs, the interval suggests that therapy B is better than therapy A.

The t test statistic for a significance test of H_0: $\mu_D = 0$ against H_a: $\mu_D \neq 0$ is

$$t = \frac{\bar{D}}{s_D/\sqrt{n}} = \frac{20}{2.89} = 6.9$$

based on $df = 2$. The (two-tailed) P-value equals .020. There is considerable evidence that therapy B is superior for this psychologist. Table 7.6 shows how SPSS software reports these results for its paired-samples t test option. □

When used with small samples, paired difference inferences make the usual assumptions for t procedures. Specifically, they assume that the observations (the difference scores) are randomly selected from a population distribution that is normal. Confidence intervals and two-sided tests work quite well even if the normality assumption is violated, unless the sample size is very small and the true distribution is very highly skewed.

Independent Versus Dependent Samples

To contrast analyses, we used the same raw scores here as in the small-sample analysis for independent samples in the previous section. The analyses and results differ according to whether the samples are independent or dependent. For instance, the difference of means equals 20 in each case, but the standard error equals 7.64 when the

TABLE 7.6 SPSS Printout for Matched-Pairs Analysis Comparing Therapy A with Therapy B on Depression Scores

```
t-tests for Paired Samples

                  Number of
Variable          pairs            Mean         SD       SE of Mean
THERAPY A                         20.000      10.000       5.774
                     3
THERAPY B                         40.000       8.660       5.000

Paired Differences
   Mean          SD      SE of Mean     t-value     df    2-tail Sig
  -20.0000      5.00       2.887         -6.93       2       .020
    95% CI   (-32.421, -7.579)
```

samples are independent and 2.89 when the samples are dependent. In this case, greater accuracy is achieved with the dependent samples analysis.

The use of dependent samples can have certain benefits. First, known sources of potential bias are controlled. Using the same subjects in each sample, for instance, keeps many other factors fixed that could affect the analysis. Second, the standard error of $\bar{Y}_2 - \bar{Y}_1$ may be smaller with dependent samples, as we observed in this example. This is because the variability in the difference scores may be less than the variability in the original scores. This happens when the scores in the two samples are strongly associated, in the sense that the subjects who score high in sample 1 are the ones who tend to score high in sample 2, and the subjects who score low in sample 1 tend to score low in sample 2.

McNemar Test for Comparing Dependent Proportions*

We next present methods for comparing proportions, for dependent samples. We use the following example to illustrate methods.

Example 7.7 Opinions About Government Spending

Table 7.7 illustrates dependent-samples data for a qualitative response. A recent General Social Survey asked subjects their opinions regarding government spending on health and government spending on law enforcement; for each, should it increase, or should it decrease? Of 340 subjects, 317 favored increased spending on health and 23 favored decreased spending. For the same subjects, 306 favored increased spending on law enforcement, and 34 favored decreased spending. Let π_1 denote the population proportion who favor increased spending on health, and let π_2 denote the population proportion who favor increased spending on law enforcement. The sample estimates are $\hat{\pi}_1 = 317/340 = .932$ and $\hat{\pi}_2 = 306/340 = .900$.

The rows of Table 7.7 are the response categories for opinion about health spending. The columns are the same categories for opinion about spending on law enforcement. The counts in the margins of the table display the frequencies for the response outcomes for the two samples. The row marginal counts (317, 23) are the (increase, decrease) totals for spending on health, and the column marginal counts (306, 34) are the totals for spending on law enforcement. We will compare the proportion of responses favoring an increase in spending, for these two types of spending. The samples for making the comparson are dependent, because each sample has the same subjects.

If the proportions favoring increased spending were identical for health and law enforcement, the number of observations in the first row of the table would equal the number of observations in the first column. The first cell (the one containing 292 in Table 7.7) is common to both the first row and first column, so the other cell count in the first row would equal the other cell count in the first column. That is, the number of subjects favoring an increase in health spending but a decrease in law enforcement spending would equal the number of subjects favoring a decrease in health spending and an increase in law enforcement spending. One can test the hypothesis $H_0 : \pi_1 = \pi_2$ using the counts in those two cells. That is, if the null hypothesis is true, then of the subjects in the population who support increased spending on one but decreased spending on the other, a proportion of 1/2 support an increase on health and a decrease on law enforcement, and 1/2 support a decrease on health and an increase on law enforcement.

So, as in the matched-pairs test for a mean, we reduce the inference to one about a single parameter for one sample. Of the population having a positive opinion about one issue and a negative one about the other, we test whether half have a positive opinion about the first. In Table 7.7, of the 25 + 14 = 39 subjects who feel spending should be increased on one and decreased on the other, the sample proportion 25/39 = .64 believe spending should increase on health. Under the null hypothesis that the true proportion is .50, the standard error of the sample proportion is $\sqrt{(.50)(.50)/39} = .080$. From Section 6.3, the z statistic for testing that the true proportion equals .50 is

$$z = \frac{.64 - .50}{.080} = 1.76$$

The two-sided P-value equals .08. This provides slight evidence that support for increased spending is higher for health than for law enforcement. □

TABLE 7.7 Opinions About Government Spending on Health and on Law Enforcement

	Law Enforcement Spending		
Health Spending	Increase	Decrease	Total
Increase	292	25	317
Decrease	14	9	23
Total	306	34	340

A very simple form exists for this test statistic for comparing two dependent proportions. For a table of the form of Table 7.7, denote the cell counts in the two relevant cells by n_{12} for those who are in row 1 for the row variable and in column 2 for the column variable and by n_{21} for those who are in row 2 for the row variable and in column 1 for the column variable. The test statistic equals

$$z = \frac{n_{12} - n_{21}}{\sqrt{n_{12} + n_{21}}}$$

When $n_{12}+n_{21}$, the total number of differing responses on the row and column classification, exceeds about 20, this statistic has approximately a standard normal distribution when the null hypothesis is true. This test is often referred to as ***McNemar's test***. For smaller samples, one uses the binomial distribution to conduct the test.

To illustrate, for Table 7.7, $n_{12} = 25$, the number who favor increased spending on health and decreased spending on law enforcement, and $n_{21} = 14$, the number who favor the reverse. The test statistic equals

$$z = \frac{25 - 14}{\sqrt{25 + 14}} = 1.76$$

The two-sided P-value of .08 provides slight evidence against the null hypothesis that the proportion favoring increased spending is the same for these two types of spending. Specifically, based on the sample proportions, it appears that the population proportion favoring increased spending was higher for spending on health, although we are less than 95% confident of this conclusion.

Estimating Differences of Proportions*

A confidence interval for the difference of proportions is more informative than a significance test. Let $\hat{\pi}_{11}$ denote the sample proportion of cases in the cell in row 1 and column 1, $\hat{\pi}_{12}$ the proportion in row 1 and column 2, $\hat{\pi}_{21}$ the proportion in row 2 and column 1, and $\hat{\pi}_{22}$ the proportion in row 2 and column 2. In Table 7.7, for instance, $\hat{\pi}_{12} = 25/340$. For dependent samples, the estimated variance of the sample difference $\hat{\pi}_2 - \hat{\pi}_1$ equals

$$\frac{\hat{\pi}_1(1 - \hat{\pi}_1) + \hat{\pi}_2(1 - \hat{\pi}_2) - 2(\hat{\pi}_{11}\hat{\pi}_{22} - \hat{\pi}_{12}\hat{\pi}_{21})}{n}$$

The square root of the estimated variance is a standard error for confidence intervals for the true difference of proportions. To illustrate, for Table 7.7, the difference of sample proportions favoring increased spending equals .932 – .900 = .032. From substituting into the formula just given, its estimated variance equals

$$\frac{1}{340}\left\{(.932)(.068) + (.900)(.100) - 2\left[\left(\frac{292}{340}\right)\left(\frac{9}{340}\right) - \left(\frac{25}{340}\right)\left(\frac{14}{340}\right)\right]\right\}$$
$$= \frac{.114}{340} = .000335$$

The standard error is $\sqrt{.000335} = .0183$. A 95% confidence interval equals $.032 \pm 1.96(.0183)$, or $(-.004, .068)$. We conclude that the proportion favoring increased spending on health is between about .004 lower and .068 higher than the proportion favoring increased spending on law enforcement.

In summary, it is plausible that the population proportions are equal. If they differ, the difference is small, with the proportion favoring increased spending possibly being as much as about .07 higher for health than for law enforcement.

7.5 Nonparametric Statistics*

We have seen repeatedly that a variety of statistics have large-sample normal distributions, even if the data do not come from a normal distribution. In fact, nearly all estimators of population parameters have normal distributions, for large sample sizes. Small samples, though, often require additional assumptions. For instance, inferences for means using the t distribution assume that the data have a normal population distribution.

A body of methods exist that require relatively few assumptions about the population distribution. These methods are called ***nonparametric*** statistics. They contrast with the traditional (so-called *parametric*) methods that require assumptions such as normal populations. Nonparametric methods are useful, for instance, when the normality assumption required for methods using the t distribution is badly violated. They are primarily useful for small samples, when parametric methods commonly require extra assumptions such as normality.

When n_1 or n_2 is less than 20, the t distribution procedure for comparing means discussed in Section 7.3 requires that the population distributions be normal with identical standard deviations. Most nonparametric comparisons of groups assume identical shapes for the population distributions, but the shapes are not required to be normal. The most popular test of this type is called the *Wilcoxon* or *Mann-Whitney* test. It is an ordinal-level method, in the sense that it uses only the rankings of the observations. The combined sample of $n_1 + n_2$ measurements are ranked from 1 to $n_1 + n_2$, and the means of the ranks are computed for observations in each sample. The test statistic compares these mean ranks.

Nonparametric methods have the benefit of requiring fewer assumptions. In practice, social scientists do not use nonparametric methods nearly as much as parametric methods. Partly this reflects the large sample sizes for most studies. In addition, nonparametric methods for multivariate data sets are not as thoroughly developed as the parametric methods. Due to lack of space, nonparametric methods are beyond the scope of this text. For an introduction to the topic, see Sprent (1993).

Treating Ordinal Variables as Quantitative

Social scientists often use statistical methods for quantitative variables when the data themselves are only ordinal. They do this by assigning scores to the ordered categories.

Example 6.1 showed an example of this. One should realize, though, that different choices of scores may lead to different conclusions (see Problem 7.48). In assigning scores to ordered levels, one should conduct a sensitivity analysis. Try a few "sensible" score choices to check the dependence of the conclusions on that choice.

Alternatively, one can apply nonparametric methods to ordinal data. The reason is that nonparametric methods do not use quantitative scores, but rather rankings of the observations; in other words, they use ordinal features of the data. We shall see examples of such methods in the following chapter.

7.6 Chapter Summary

This chapter introduced methods for comparing two groups. Table 7.8 summarizes the methods for ***independent random samples***, when the sample sizes are large. For small samples, one uses the same formulas but with the t distribution instead of the standard normal.

- For quantitative data, we compared the ***means*** for the two groups. Large-sample confidence intervals and significance tests apply to the difference $\mu_2 - \mu_1$. The test analyzes whether 0 is a plausible value for this difference, in which case the

TABLE 7.8 Summary of Comparison Methods for Two Groups, for Large, Independent Random Samples

	Type of Response Variable	
	Qualitative	Quantitative
Estimation		
1. Parameter	$\pi_2 - \pi_1$	$\mu_2 - \mu_1$
2. Point estimate	$\hat{\pi}_2 - \hat{\pi}_1$	$\bar{Y}_2 - \bar{Y}_1$
3. Standard error	$\hat{\sigma}_{\hat{\pi}_2-\hat{\pi}_1} = \sqrt{\frac{\hat{\pi}_1(1-\hat{\pi}_1)}{n_1} + \frac{\hat{\pi}_2(1-\hat{\pi}_2)}{n_2}}$	$\hat{\sigma}_{\bar{Y}_2-\bar{Y}_1} = \sqrt{\frac{s_1^2}{n_1} + \frac{s_2^2}{n_2}}$
4. Confidence interval	$(\hat{\pi}_2 - \hat{\pi}_1) \pm z\hat{\sigma}_{\hat{\pi}_2-\hat{\pi}_1}$	$(\bar{Y}_2 - \bar{Y}_1) \pm z\hat{\sigma}_{\bar{Y}_2-\bar{Y}_1}$
Significance testing		
1. Assumptions	More than 5 observations in each category, for each group	$n_1, n_2 \geq 20$
2. Hypotheses	H_0: $\pi_1 = \pi_2$ $(\pi_2 - \pi_1 = 0)$ H_a: $\pi_1 \neq \pi_2$	H_0: $\mu_1 = \mu_2$ $(\mu_2 - \mu_1 = 0)$ H_a: $\mu_1 \neq \mu_2$
3. Test statistic	$z = \frac{\hat{\pi}_2-\hat{\pi}_1}{\hat{\sigma}_{\hat{\pi}_2-\hat{\pi}_1}}$	$z = \frac{\bar{Y}_2-\bar{Y}_1}{\hat{\sigma}_{\bar{Y}_2-\bar{Y}_1}}$
4. P-value	Two-tail probability from standard normal (Use one-tail for one-sided alternative)	

population means may be equal. If the confidence interval contains zero, it is plausible that the means are equal.

- For qualitative data, we compared the ***proportions*** of individuals in the two groups that were classified in a particular category. Confidence intervals and tests apply to the difference between the proportions, $\pi_2 - \pi_1$.
- Both for differences of proportions and differences of means, large-sample confidence intervals have the form

$$\text{Estimate } \pm z \text{ (standard error)}$$

that is,

$$(\hat{\pi}_2 - \hat{\pi}_1) \pm z\hat{\sigma}_{\hat{\pi}_2 - \hat{\pi}_1} \quad \text{and} \quad (\bar{Y}_2 - \bar{Y}_1) \pm z\hat{\sigma}_{\bar{Y}_2 - \bar{Y}_1}$$

The test statistic equals the estimated difference divided by the standard error.

- For small samples, tests and confidence intervals comparing means use the t distribution instead of the normal. They apply when either n_1 or n_2 is less than 20, and assume normal population distributions with equal standard deviations. The small-sample test for proportions is *Fisher's exact test*.
- For ***dependent*** samples, each observation in one sample is matched with an observation in the other sample. For quantitative data, we compare means by analyzing the mean of difference scores computed between the paired observations. The confidence interval and test procedures are simply the one-sample methods of Chapters 5 and 6 applied to the difference scores. The significance test is called a ***paired-difference*** test.
- For dependent samples with qualitative data, the significance test is called ***McNemar's test***. It is based on comparing the number of subjects who are in category 1 in the first sample and category 2 in the second sample to the number of subjects who are in category 2 in the first sample and category 1 in the second.

PROBLEMS

Practicing the Basics

1. In the 1982 General Social Survey, 350 subjects reported the time spent every day watching television. The sample mean was 4.1 hours, with standard deviation 3.3. In the 1994 General Social Survey, 1965 subjects reported a mean time spent watching television of 2.8 hours, with standard deviation 2.0.

a) Construct the 95% confidence interval for the difference between the means in 1982 and 1994. Interpret. Is it plausible that the mean was the same in both years?

b) Set up the hypotheses of a significance test to analyze whether the population means differ in 1982 and 1994.

c) For this test, construct the test statistic. Report and interpret the P-value.

d) Do you think the distribution of the response variable is approximately normal? Why? Does this affect the validity of your inference? Explain.

2. Refer to Table 6.2 and Example 6.7. Table 7.9 is a computer printout that summarizes the responses on the seven-point political ideology scale for the surveys in 1978 and in 1994.
a) Are these samples independent or dependent?
b) Construct a 95% confidence interval for the difference between the mean political ideology in 1994 and 1978. Interpret.
c) Show how to test the hypothesis that the population mean political ideology was equal in 1978 and 1994, against the alternative of a difference. Report the P-value, and interpret.
d) Show the correspondence between the results of the inferences in (b) and (c). Which inference is more informative? Explain.

TABLE 7.9

```
Variable: IDEOLOGY

YEAR          N        Mean      Std Dev     Std Error
------------------------------------------------------
1978         627       4.03         1.26        .0503
1994        2879       4.17         1.39        .0259

Variances      T           Prob>|T|
-------------------------------------
Unequal      2.474        0.0134
```

3. Refer to Problem 5.39 and Table 5.4. Construct a 99% confidence interval for the difference between the mean number of hours employed per week for men and for women. Interpret. Based on this interval, indicate what result would occur in a two-sample test comparing the means. Explain.
4. A recent study compared substance use, delinquency, psychological well-being, and social support among various family types, for a sample of urban African-American adolescent males (M. Zimmerman et al., *Child Development*, Vol. 66, 1995, pp. 1598–1613). The sample contained 108 subjects from single-mother households and 44 from households with both biological parents. Delinquency was measured with four dichotomous items asking the youths if they had ever (i) been in trouble with the police, (ii) been arrested, (iii) appeared before a juvenile court, or (iv) been incarcerated in the last year; the delinquency score was the number of items to which a youth answered yes. This score had a mean of 1.9 and standard deviation of 1.5 for single-mother households and a mean of 1.4 and standard deviation of 1.5 for households with both biological parents.
a) Construct a 99% confidence interval for the difference between the true means for the populations represented by this sample. Interpret.
b) Test the hypothesis of equal means against the alternative of unequal means. Report the P-value and interpret.
c) Explain the correspondence between the results of (a) and (b).
d) Do you think that delinquency is normally distributed for each household type? Explain, and indicate whether this has any effect on your inferential analyses.
5. Refer to the preceding exercise. The youths also responded to a battery of questions that provides a measure of perceived parental support. This measure had sample means of 46 ($s = 9$) for the single-mother households and 42 ($s = 10$) for the households with both bio-

logical parents. Consider the conclusion, "The mean parental support was 4 units higher for the single-mother households. If the true means were equal, a difference of this size could be expected only 2% of the time. For samples of this size, 95% of the time one would expect this difference to be within 3.4 of the true value." Explain how this conclusion refers to the results of **a)** a confidence interval, **b)** a test.

6. For 30 female assistant professors selected at random from arts and science departments in a large state university system, the mean academic year salary is $41,000. The mean salary for a random sample of 50 male assistant professors is $41,800. The sample standard deviations are $3000 for females and $3400 for males.
a) Construct a 95% confidence interval for the difference in mean salary between male and female assistant professors in this system. Interpret.
b) Perform a significance test of whether the true means are identical. Interpret the P-value.
c) What decision would you make in the test, using $\alpha = .05$? Explain how this decision relates to information displayed by the confidence interval.

7. Seegmiller et al. (*J. Psychology*, Vol. 102, 1979, pp. 215–224) studied the effect of tester sex on sex-role differentiation scores of a sample of preschool children. For the Occupational Preference Test, children were asked to give three choices of what they wanted to be when they grew up. Each occupation was rated on a scale from 1 (traditionally feminine) to 5 (traditionally masculine), and a child's score was the mean of the three selections.
a) The mean score was 4.7 for the 152 boys ($s = .7$), and the mean score was 3.1 for the 140 girls ($s = 1.3$). Construct a 95% confidence interval to summarize how much more masculine the boys' choices tended to be than the girls' choices. Interpret.
b) When the tester was male, the 50 girls had $\bar{Y} = 2.9$ and $s = 1.4$, whereas when the tester was female, the 90 girls had $\bar{Y} = 3.2$ and $s = 1.2$. Test the hypothesis that the true mean is the same for each tester sex, against the alternative that they differ. Report the P-value and interpret.
c) For the test in (b), what is your conclusion, using $\alpha = .05$? If you have made an error, is it a Type I error or a Type II error?

8. Families headed by low-wage workers may be more likely to be living in poverty now than two decades ago. According to a report by the Center on Budget and Policy Priorities, in 1977 7.7% of families with children in which the head of the household worked were below the poverty level; in 1993, this poverty rate was 11.4%. Report a point estimate to compare the statistics from 1977 and 1993, and interpret.

9. Chatterjee et al. (1995, p. 132) described two studies about the effect of condoms in reducing the spread of AIDS. One two-year Italian study followed heterosexual couples where one partner was infected with the HIV virus. Of 171 couples who always used condoms, 3 partners became infected with HIV, while of 55 couples who did not always use condoms, 8 partners became infected.
a) Estimate the infection rates for the two groups, and construct a 95% confidence interval to compare them. Interpret.
b) Test whether the rates are significantly different. Report the P-value and interpret.

10. For a random sample of 1600 Canadians taken in January, 880 people indicate approval of the prime minister's performance. A similar poll a month later of a separate random sample of 1600 Canadians has a favorable rating by 944 people. Let π_1 denote the true proportion in January that approve of the prime minister's performance, and let π_2 represent the true proportion in February.

a) Calculate point estimates of these proportions. Calculate the difference, and interpret.
b) Construct a 99% confidence interval for $\pi_2 - \pi_1$. Interpret.
c) Based on the interval in (b), explain why (i) there may have been no change in support, (ii) if an increase in support occurred, it may have been fairly important, (iii) if a decrease in support occurred, it was probably so small as to be substantively unimportant.

11. The National Health Interview Survey conducted of 42,000 adults by the U.S. National Center for Health Statistics in 1991 indicated that 25.6% of adults were current smokers. A similar study conducted in 1965 indicated that 42.4% were current smokers.
a) Construct a 95% confidence interval for the difference between the proportion of current smokers in 1965 and the proportion of current smokers in 1991. (Assume that the sample size was the same in 1965.) Interpret.
b) Using the standard error from the confidence interval, test that there was no change in the proportion of current smokers. Report the test statistic and P-value, and interpret.
c) What statistical factors cause the test statistic in (b) to be so large?

12. Refer to Example 7.3.
a) Find point estimates of the proportions of Americans who *disagree* with the statement about women's roles in 1982 and in 1994.
b) Find a 95% confidence interval for the change in the disagree proportion, and compare it to the interval found following Example 7.3 for the change in the agree proportion. Explain the correspondence.
c) Conduct a test for equality of the disagree proportions in 1982 and 1994. Compare to the result of the test for the agree proportions, and explain the correspondence.

13. The General Social Survey asks, "Do you favor or oppose the death penalty for persons convicted of murder?" In 1982, 154 were in favor and 165 opposed. In 1994, 2215 were in favor and 580 were opposed. Construct a 95% confidence interval for the difference between the proportions in favor of the death penalty in 1994 and in 1982. Interpret.

14. In a survey conducted in 1992 by the Wright State University School of Medicine, senior high school students were asked if they had ever used marijuana. Of the females sampled, 445 said yes and 675 said no; of the males sampled, 515 said yes and 641 said no. Compare and interpret these responses,
a) Using a 95% confidence interval.
b) Using a significance test.

15. A study using a probability sample from the 1984 National Health Interview Study (R. Coward et al., *The Gerontologist*, Vol. 35, 1995, pp. 24–34) reported that the percent of respondents with income below the poverty level was 11.3% for residents of metropolitan areas ($n = 7331$) and 20.6% for residents of nonmetropolitan areas ($n = 4235$). Construct and interpret a 99% confidence interval for the difference in poverty rates between nonmetropolitan and metropolitan areas. What is the effect of the very large sample sizes on the interval?

16. Table 7.10 shows results from a recent General Social Survey on two variables, gender and whether one believes in an afterlife.
a) Which of the two variables is more naturally treated as the response variable?
b) Let π_1 denote the population proportion of females who would respond "yes" to whether they believe in an afterlife, and let π_2 denote the population proportion for males. Construct a 95% confidence interval for $\pi_1 - \pi_2$, and interpret.
c) Is it plausible that $\pi_1 = \pi_2$? Explain.
d) Suppose, unknown to us, that $\pi_1 = .78$ and $\pi_2 = .71$. Does the confidence interval in

TABLE 7.10

	Belief in Afterlife		
Gender	Yes	No or Undecided	Total
Female	435	147	582
Male	375	134	509

(b) contain the value of the parameter it is designed to estimate? Explain.

e) Conduct a test of H_0: $\pi_1 = \pi_2$. Make a decision using $\alpha = .05$. If you have made an error, what type of error is it, Type I or Type II?

17. The 1974–1975 General Social Survey asked "Do you believe that women should take care of running their homes and leave running the country up to men?" For 1305 male respondents, 465 replied yes and 840 replied no. For 1566 female respondents, 555 replied yes and 1,011 replied no.

a) Find a 95% confidence interval for the difference between the proportion of males and the proportion of females who would respond yes. Interpret the interval.

b) Of 487 respondents having 8 years or less education, 330 replied yes. Of 902 respondents having at least 13 years of education, 153 replied yes. Find a 95% confidence interval for the difference between the proportion at the low education level and the proportion at the high education level who would respond yes. Interpret the interval.

c) Which variable, gender or educational level, seems to have had the greater influence on opinion? In other words, did opinion tend to differ more between men and women, or between the most and least educated?

18. *Newsweek* magazine (September 19, 1983) reported results of a poll to investigate whether there was a difference between the percentage of women and the percentage of men who approved of the way Ronald Reagan was handling his job as president. Of 511 women interviewed, 40% indicated approval. Of 507 men, 49% indicated approval.

a) Find a 90% confidence interval for the difference between the proportion of men and the proportion of women who approved of Reagan's performance. What assumption do you make to use this method?

b) Interpret the interval obtained in part (a). Can you conclude that there is a gender gap?

19. A study of bulimia among college women (J. Kern and T. Hastings, *Journal of Clinical Psychology*, Vol. 51, 1995, p. 499) considered the effect of childhood sexual abuse on various components of a Family Environment Scale. For a measure of family cohesion, the sample mean for the bulimic students was 2.0 for 13 sexually abused students ($s = 2.1$) and 4.8 for 17 nonabused students ($s = 3.2$).

a) Use small-sample methods to construct a 95% confidence interval for the difference in mean family cohesion for sexually abused students and nonabused students. Interpret. What assumptions have you made?

b) Find the P-value for testing whether the population means are equal, against a two-sided alternative. Interpret.

20. The property values for three homes selected at random in the Forest Ridge subdivision are (in thousands of dollars) 110, 120, 130. The property values for two homes selected at random in the Hermitage subdivision are 180, 200.

a) Use small-sample methods to construct a 90% confidence interval for the difference in mean property values between the Hermitage and Forest Ridge subdivisions. Interpret. What assumptions have you made?

b) Find the *P*-value for testing H_0: $\mu_1 = \mu_2$ against H_a: $\mu_1 \neq \mu_2$. Interpret. Explain the correspondence between results of the confidence interval and results of the test.
c) Using computer software, compare results of the t test that assumes equal population variances to that of the approximate t test that does not require that assumption. Interpret.

21. For the WWW data for University of Florida social science students, described in Problem 1.7, the responses on political ideology had a mean of 3.18 and standard deviation of 1.72 for the 51 nonvegetarian students and a mean of 2.22 and standard deviation of .67 for the 9 vegetarian students.
a) Show how to compare the means using a 90% confidence interval. Interpret.
b) When we use software to compare the means with a significance test, we obtain the printout

```
Variances          T          DF     Prob>|T|
---------------------------------------------
Unequal       2.9146        30.9       0.0066
Equal         1.6359        58.0       0.1073
```

Explain why the results of the two tests differ so much, and give your conclusion about whether the true means are equal.

22. Three editors of large-city newspapers are selected at random and asked to rate on a score of 0–100 the fairness of the news media in political reporting (0 corresponds to very unfair). The same question is posed to three editors of small-town papers. The large-city editors gave scores of 55, 90, 95, and the small-town editors gave scores of 40, 60, 80. Table 7.11 shows a computer printout for results of a two-sample comparison of means. Explain how to interpret results from this printout.

TABLE 7.11

```
Variable: FAIRNESS
PAPER         N        Mean        Std Dev       Std Error
-----------------------------------------------------------
large         3       80.00         21.794          12.583
small         3       60.00         20.000          11.547

Variances          T       DF     Prob>|T|
------------------------------------------
Unequal       1.1711      4.0       0.3071
Equal         1.1711      4.0       0.3066
```

23. Refer to Example 7.5. The young adults were also asked whether they had ever had a same-gender sexual relationship. Table 7.12 shows results. Use software to test whether the probability of this is higher for those raised by lesbian mothers. Interpret.

TABLE 7.12

	Same-Gender Relationship	
Mother	Yes	No
Lesbian	6	19
Heterosexual	0	20

24. Refer to the previous problem. Subjects were also asked whether they had ever had a same-gender sexual attraction. Table 7.13 shows a computer printout with results of the analysis. Interpret.

TABLE 7.13

MOTHER / Frequency	ATTRACT No	Yes	Total
Heterosx	16	4	20
Lesbian	16	9	25

Statistic		Prob
Fisher's Exact Test	(Left)	0.936
	(Right)	0.200
	(2-Tail)	0.327

25. Refer to Problem 6.12. Compare the proportions of blacks and whites who think that affirmative action has been good for the country:
a) Using a 95% confidence interval. Interpret.
b) Using a significance test. Report the P-value, and interpret.

26. Refer to Examples 7.4 and 7.6. Construct and interpret a 90% confidence interval for $\mu_2 - \mu_1$:
a) In the independent samples case.
b) In the dependent samples case.
c) Compare results of the two analyses.

27. A recent study compared personality characteristics between 49 adult children of alcoholics and a control group matched on age and gender (D. Baker and L. Stephenson, *Journal of Clinical Psychology*, Vol. 51, 1995, p. 694). On a measure of well-being, the 49 children of alcoholics had a mean of 26.1 ($s = 7.2$) and the 49 subjects in the control group had a mean of 28.8 ($s = 6.4$). The difference scores between the matched subjects from the two groups had a mean of 2.7 ($s = 9.7$).
a) Construct a 95% confidence interval for the difference between means (i) treating the samples as independent, (ii) treating the samples as dependent. Interpret.
b) Conduct a test of equality of the two means (i) treating the samples as independent, (ii) treating the samples as dependent. Report the P-values and interpret.
c) Compare results of the independent-samples and dependent-samples analyses. (Note: Matching on age and gender does not normally result in a "correlation" between responses in the two samples, so the results are similar.)

28. Refer to the previous problem. For the 29 pairs of women, the authors reported a mean of 24.8 on the well-being measure for the children of alcoholics, and a mean of 29.0 for the control group. They reported a t-score of 2.67 for the test comparing the means. Assuming that this is the result of a dependent-samples analysis, report the P-value, and interpret.

29. Refer to Problem 6.26. The changes in heart rate for 15 males had a mean of -1.22 with standard deviation 19.7. Show how to compare the mean changes for females and males, **a)** using a 95% confidence interval, **b)** using a significance test. Interpret in each case.

30. Each of a random sample of ten college freshmen takes a mathematics aptitude test both before and after undergoing an intensive training course designed to improve such test scores. Then, the scores for each student are paired, as shown in Table 7.14.

TABLE 7.14

Student	Before	After
1	60	70
2	73	80
3	42	40
4	88	94
5	66	79
6	77	86
7	90	93
8	63	71
9	55	70
10	96	97

a) Compare the mean scores after and before the training course by (i) finding the difference of the sample means, (ii) finding the mean of the difference scores. Compare.
b) Calculate and interpret the P-value for testing whether the mean change equals 0.
c) Compare the mean scores after and before the training course by constructing and interpreting a 90% confidence interval for the population mean difference.
d) Explain the correspondence between the result of the significance test and the result of the confidence interval. What assumptions does each inference make?

31. A paired-difference experiment dealing with response latencies for noise detection under two conditions used a sample of twelve 9-month-old children and reported a sample mean difference of 70.1 and standard deviation of 49.4 for the differences (J. Morgan and J. Saffran, *Child Development*, Vol. 66, 1995, pp. 911–936). In their discussion, the authors reported a t statistic of 4.9 having $P < .01$ for a two- sided alternative. Show how they constructed the t statistic, and confirm the P-value.

32. Refer to Example 7.7. Table 7.15 shows opinions about spending on the environment and on cities.
a) Find the sample proportion favoring increased spending, for each item.
b) Test whether the population proportions are equal. Report the P-value, and interpret.
c) Construct a 95% confidence interval for the difference of proportions. Interpret.

TABLE 7.15

Spending	Spending on Environment	
on Cities	Increase	Decrease
Increase	108	5
Decrease	157	24

33. Table 7.16 refers to a sample of juveniles convicted of a felony in Florida in 1987. Matched pairs were formed using criteria such as age and the number of prior offenses. For each pair, one subject was handled in the juvenile court and the other was transferred to the adult court. The response of interest was whether the juvenile was rearrested by the end of 1988.
a) Test the hypothesis that the true proportions rearrested were identical for the adult and juvenile court assignments. Use a two-sided alternative.
b) Find a 90% confidence interval for the difference between the true proportions of rearrest for the adult and juvenile court assignments. Interpret.

TABLE 7.16

Adult	Juvenile Court	
Court	Rearrest	No Rearrest
Rearrest	158	515
No Rearrest	290	1134

Source: Based on a study at the University of Florida by D. Bishop, C. Frazier, L. Lanza-Kaduce, and L. Winner. Thanks to Dr. Larry Winner for showing me these data.

34. A recent study used data from the Longitudinal Study of Aging to investigate how older people's health and social characteristics influence how far they live from their children (M. Silverstein, *Demography*, Vol. 32, 1995, p. 35). Consider Table 7.17, which shows whether an older subject lives with a child at a given time and then again four years later.
a) Test the hypothesis that the true proportion living with a child is the same at the two times. Report and interpret the P-value. Silverstein expected that as people aged and their health deteriorated, they would be more likely to live with children. Do these data support this belief?
b) Construct a 95% confidence interval for the change in the proportion living with a child. Interpret.

TABLE 7.17

First	Four Years Later	
Survey	Yes	No
Yes	423	138
No	217	2690

Concepts and Applications

35. Refer to the WWW data set (Problem 1.7). Using software, use a confidence interval and a test to compare males and females in terms of
a) The mean number of times reading a newspaper. Interpret.
b) Opinions about legalized abortion. Interpret.

36. For the WWW data set (Problem 1.7), compare Democrat and Republican students in terms of political ideology,

a) Using back-to-back stem and leaf plots.

b) Using numerical summaries.

c) Using inferential statistical methods. Interpret.

37. For the 60 students in the WWW data set (Problem 1.7), compare the mean weekly time spent watching TV to the mean weekly time in sports and other physical exercise, using both descriptive and inferential methods.

38. Refer to the data file created in Problem 1.7. For variables chosen by your instructor, conduct inferential statistical analyses, using methods of this chapter. Prepare a report that summarizes and interprets your findings. In this report, also use graphical and numerical methods presented earlier in this text to describe the data and, if necessary, to check assumptions you make for your analysis.

39. Two new drugs have been proposed for treating subjects who suffer from high blood pressure. A sample of ten such subjects were randomly allocated to two groups for an experiment comparing the drugs. The five subjects in one group took drug A, and the five subjects in the other group took drug B. Following three months of treatment, the drop in blood pressure was measured. The sample values were

Drug A: 0, 2, 2, 3, 3
Drug B: 3, 6, 6, 7, 8

Analyze these data, and interpret your results.

40. The results in Table 7.18 are from a recent study of physical attractiveness and subjective well-being (E. Diener et al., *Journal of Personality and Social Psychology*, Vol. 69, 1995, pp. 120–129). As part of the study, a sample of college students were rated by a panel on their physical attractiveness. The table presents the number of dates in the past three months for students rated in the top or bottom quartile of attractiveness. Analyze these data, and interpret.

TABLE 7.18

	No. Dates, Men			No. Dates, Women		
Attractiveness	Mean	Std. Dev.	n	Mean	Std. Dev.	n
More	9.7	10.0	35	17.8	14.2	33
Less	9.9	12.6	36	10.4	16.6	27

41. A study of the effects of race on professional football players' compensation (L. Kahn, *Industrial & Labor Relations Review*, Vol. 45, 1992, p. 295) reported mean salaries, in thousands of dollars, in the National Football League in 1989 of 328 for 544 whites ($s = 277$) and 315 for 819 blacks ($s = 248$). The mean number of times the player had been to the pro bowl was .24 for whites ($s = .87$) and .26 for blacks ($s = .82$). The proportions who had received college degrees were .39 for whites and .20 for blacks. Analyze these data. Prepare a short report, summarizing your analyses and conclusions.

42. For a sample of high school students in South Carolina, the percentage who reported they had had sexual intercourse was 65% for white males ($n = 1147$), 54% for white females ($n = 1314$), 89% for black males ($n = 612$), and 71% for black females ($n = 820$) (D.

Richter et al., *Journal of School Health*, Vol. 63, 1993, p. 91). Analyze these data, and interpret.

43. Refer to Problem 7.1. In 1982, 11 of the 350 subjects reported watching no television. In 1994, 74 of the 1965 subjects reported watching no television. Analyze these data, and interpret.

44. According to *Newsweek* magazine (July 31, 1995), the mean household income in 1994 was $40,708 for whites and $25,409 for blacks. The article did not provide sample sizes or standard deviations. Suppose it had reported a *standard error* of $200 for the mean for whites and $300 for the mean for blacks. Is this sufficient information to construct a 99% confidence interval comparing the true difference in means? If so, construct the interval and interpret; if not, explain why not.

45. In the United States, the estimated annual probability that a woman over the age of 35 dies of lung cancer equals .00130 for current smokers and .00012 for nonsmokers (M. Pagano and K. Gauvreau, *Principles of Biostatistics*, Belmont, CA: Duxbury Press, 1993, p. 134).
a) Calculate and interpret the difference between smokers and nonsmokers in the proportions who die of lung cancer in a given year. Interpret. Are the proportions similar, or quite different, for the two groups?
b) Another way to compare the two proportions is through their *ratio*. The ratio equals 1.0 when the proportions are equal, and ratios farther from 1.0 represent larger differences between the groups. Calculate the ratio of the proportions for smokers and nonsmokers, and interpret. According to this measure, are the proportions similar, or quite different, for the two groups? (Note: The ratio of proportions is also called the ***relative risk***. It is often more relevant than the difference of proportions when both proportions are close to zero.)

46. Refer to the previous exercise. According to the U.S. National Center for Health Statistics, in 1991 the probability that a male between the ages of 20 and 24 was a homicide victim was .00164 for blacks and .00015 for whites. Compare these rates using the difference of proportions and the relative risk, and interpret.

47. A poll by Louis Harris and Associates of 1249 Americans in July 1994 indicated that 36% believe in ghosts and 37% believe in astrology.
a) Is it valid to compare the proportions using inferential methods for independent samples? Explain.
b) Do you have enough information to compare them using methods for dependent samples? Explain.

48. Table 7.19 compares two hospitals on the outcomes of patient admissions for severe pneumonia. Although patient status is an ordinal variable, two researchers who analyze the data treat it as an interval variable. The first researcher assigns the scores (0, 5, 10) to the three categories. The second researcher, believing that the middle category is much closer to the third category than to the first, uses the scores (0, 9, 10). Each researcher calculates the means for the two institutions and identifies the institution with the higher mean as the one having more success in treating its patients.
a) Find the two means for the scoring system used by (i) the first researcher, (ii) the second researcher. Interpret.
b) Notice that the conclusion depends on the scoring system; hence, if one uses methods for quantitative variables with ordinal data, one should take care in selecting scores.

TABLE 7.19

	Patient Status		
	Died in Hospital	Released After Lengthy Stay	Released After Brief Stay
Hospital A	1	29	0
Hospital B	8	8	14

49. A recent General Social Survey asked respondents whether premarital sex was always wrong, and whether extramarital sex was always wrong (yes, no). Of the respondents, 144 said yes on both, 86 said no on both, 243 said no on premarital and yes on extramarital, and 2 said yes on premarital and no on extramarital. Analyze these data, and interpret.

50. A short questionnaire measuring anxiety is given to a random sample of two students taken from a large introductory statistics class. Anxiety is measured both before and after the students take a midterm exam in statistics. The scores are 91 before and 80 after for student 1 and 69 before and 60 after for student 2. Analyze these data.

51. A pool of six candidates for three managerial positions includes three females and three males. Table 7.20 shows the results.

a) Denote the three females by F_1, F_2, F_3 and the three males by M_1, M_2, M_3. Identify the 20 distinct samples of size three that can be chosen from these six individuals.

b) Let $\hat{\pi}_1$ denote the sample proportion of males selected and $\hat{\pi}_2$ the sample proportion of females. For Table 7.20, $\hat{\pi}_1 - \hat{\pi}_2 = (2/3) - (1/3) = 1/3$. Of the 20 possible samples, show that 10 have $\hat{\pi}_1 - \hat{\pi}_2 \geq 1/3$. Thus, if the three managers were randomly selected, the probability would equal $10/20 = .5$ of obtaining $\hat{\pi}_1 - \hat{\pi}_2 \geq 1/3$. In fact, this is the reasoning that provides the one-sided P-value for Fisher's exact test.

c) Find the P-value if all three selected are male. Interpret.

TABLE 7.20

	Chosen for Position	
Gender	Yes	No
Male	2	1
Female	1	2

52. Refer to Problem 7.10. Suppose the same 1600 people who were interviewed in January were interviewed again in February, with the results shown in Table 7.21.

TABLE 7.21

	February		
January	Approve	Disapprove	Total
Approve	794	86	880
Disapprove	150	570	720
Total	944	656	1600

a) Test for a difference between the proportions who approve in February and in January. Interpret.

b) Find a 95% confidence interval for the difference between the proportions who approve

in February and in January. Interpret. Can you conclude that the approval rating changed from January to February? Explain.
c) Compare results to those in Problem 7.10 for the independent samples. In particular, compare the standard error for the difference of proportions for the two analyses. How does this affect the width of the confidence interval? Which analysis provides more precise results, and why does this happen?

53. Provide an example of a situation in which it is more sensible to compare means using dependent samples than independent samples.

In Problems 7.54–7.56, select the correct response(s).

54. A 99% confidence interval for the difference $\pi_2 - \pi_1$ between the proportions of men and women in Florida who are alcoholics equals (.02, .09).
a) We are 99% confident that the proportion of alcoholics is between .02 and .09.
b) We are 99% confident that the proportion of men in Florida who are alcoholics is between .02 and .09 larger than the proportion of women in Florida who are.
c) At this confidence level, there is insufficient evidence to infer that the population proportions are different.
d) We are 99% confident that a minority of Florida residents are alcoholics.
e) Since the confidence interval does not contain 0, it is impossible that $\pi_1 = \pi_2$.

55. To compare the population mean annual incomes for Hispanics (μ_1) and for whites (μ_2) having jobs in construction, we construct a 95% confidence interval for $\mu_2 - \mu_1$.
a) If the confidence interval is (3000, 6000), then at this confidence level we conclude that the mean income for whites is higher than for Hispanics.
b) If the confidence interval is (−1000, 3000), then the corresponding $\alpha = .05$ level test of H_0: $\mu_1 = \mu_2$ against H_a: $\mu_1 \neq \mu_2$ rejects H_0.
c) If the confidence interval is (−1000, 3000), then it is plausible that $\mu_1 = \mu_2$.
d) If the confidence interval is (−1000, 3000), then we are 95% confident that the mean annual income for whites is between \$1000 less and \$3000 more than the mean annual income for Hispanics.

56. For small samples, the Wilcoxon test differs from parametric procedures (for means) in the sense that
a) The alternative hypothesis cannot be one-sided.
b) It is unnecessary to assume that the population distribution is normal.
c) Random sampling is not assumed.
d) It applies directly to ordinal as well as interval response variables.

57. * A test consists of 100 true–false questions. Joe did not study, so on each question, he randomly guesses the correct response.
a) Find the probability that he scores at least 70, thus passing the exam. (*Hint*: Use the sampling distribution for the proportion of correct responses.)
b) Jane studied a little and has a .60 chance of a correct response for each question. Find the probability that her score is nonetheless lower than Joe's. (*Hint*: Use the distribution of the difference of sample proportions.)
c) How do the answers to (a) and (b) depend on the number of questions? Explain.

58. Using their formulas, for independent samples show that $\hat{\sigma}_{\bar{Y}_2-\bar{Y}_1}$ is larger than $\hat{\sigma}_{\bar{Y}_1}$ or $\hat{\sigma}_{\bar{Y}_2}$ for either mean separately. Provide some intuition for why this happens. (Hint: Notice that if $\bar{Y}_1$ is above (below) μ_1 and $\bar{Y}_2$ is below (above) μ_2, then $\bar{Y}_2 - \bar{Y}_1$ is even further below (above) $\mu_2 - \mu_1$. Thus there is the potential for making a larger error in using $\bar{Y}_2 - \bar{Y}_1$ to estimate $\mu_2 - \mu_1$ than in using $\bar{Y}_1$ to estimate μ_1 or $\bar{Y}_2$ to estimate μ_2.)

59. * Let X_i denote the observation for the ith subject at time 1. Let Y_i denote the observation for the ith subject at time 2. Let $D_i = Y_i - X_i$.
a) Letting $\bar{X}$, $\bar{Y}$, and $\bar{D}$ denote the means of these observations, show that $\bar{D} = \bar{Y} - \bar{X}$.
b) Is the median difference (i.e., the median of the D_i values) equal to the difference between the medians of the Y_i and X_i values? Show this is true, or give a counterexample to show that it is false.

Bibliography

Agresti, A. (1996). *An Introduction to Categorical Data Analysis.* New York: Wiley.

Chatterjee, S., Handcock, M., and Simonoff, J. (1995). *A Casebook for a First Course in Statistics and Data Analysis.* New York: Wiley.

Fleiss, J. (1981). *Statistical Methods for Rates and Proportions*, 2nd ed. New York: Wiley.

Sprent, P. (1993). *Applied Nonparametric Statistical Methods*, 2nd ed. London: Chapman & Hall.

Wilcox, R. (1987). *New Statistical Procedures for the Social Sciences.* Hillsdale, NJ: Lawrence Erlbaum.

Chapter 8

Analyzing Association Between Categorical Variables

The previous chapter introduced bivariate methods in which the explanatory (independent) variable is a binary variable that defines the two groups compared and the response (dependent) variable is quantitative or qualitative. In the quantitative case, we compared two means, and in the qualitative case, we compared two proportions.

We now turn our attention to bivariate methods in which both variables are *qualitative* (categorical) with an arbitrary number of levels. The explanatory variable, no longer restricted to being binary, may have *several* categories. For instance, it could define several groups to be compared. The response variable may also have several categories; for instance, we may want to analyze the proportions of (liberal, moderate, conservative) responses.

Association

This chapter presents methods for detecting and describing *associations* between a pair of categorical variables. There is said to be an *association* between two variables if the distribution of the response variable changes in some way as the value of the explanatory variable changes. Certain values of one variable tend to go with certain values of the other.

For example, consider the categorical variables "religious affiliation," with categories (Protestant, Catholic, Other) and "ethnic group," with categories (Anglo-American, African-American, Mexican-American). In the United States, Anglo-Americans are more likely to be Protestant than are Mexican-Americans, who are overwhelmingly Catholic. African-Americans are even more likely to be Protestant. An association exists between religious affiliation and ethnic group, because the probabilities for the distribution of religious affiliation change as ethnic group changes. For instance, the probability of being Protestant is not the same for each ethnic group. On the other hand, if 60% of each ethnic group were Protestant, 30% of each group were Catholic, and 10% of each group were Other, then the probability of each response would be *independent* of ethnic group, and ethnic group and religious affiliation would not be associated.

The analysis of association is relevant whenever a study considers two or more variables. In Chapter 7, in comparing two means for a quantitative response variable or in comparing two proportions for a qualitative response variable, we analyzed whether an association exists between that response variable and the variable that defines the groups. An association exists if the true means or true proportions differ between the groups.

This chapter presents methods for analyzing association for both primary types of categorical data—nominal and ordinal scale. The methods of Chapter 7 for comparing proportions for two groups are special cases of ones considered here in which both variables have only two categories. The methods of this chapter help us answer such questions as "Does political conservatism vary by political party affiliation?" "Are adolescents in different social class levels equally likely to have experienced premarital sex?" "Is there regional variation in the attitudes of people about federal aid for the arts?"

Section 8.1 introduces terminology for categorical data analysis and defines *statistical independence*, a type of lack of association. Section 8.2 presents the *chi-squared test of independence* for determining whether two categorical variables are associated, and Section 8.3 follows up that test by a *residual analysis* that describes the nature of that association.

Besides determining whether two variables are associated, we should determine whether the association is strong enough to have practical importance. Section 8.4 shows how to summarize the strength of the association, using the *difference of proportions* and the *odds ratio*. Section 8.5 presents a specialized summary for ordinal variables, *gamma*. The chi-squared test works best for nominal variables, and Section 8.6 presents alternative inferences for determining whether two ordinal variables are associated. Section 8.7 introduces a way of summarizing association, the *proportional reduction in error* approach, that is helpful for all types of data, qualitative or quantitative.

8.1 Contingency Tables

Data for the analysis of categorical variables are displayed in ***contingency tables***. This type of table, introduced in Section 7.2 for comparing two proportions, displays the number of subjects observed at all the combinations of possible outcomes for the two variables.

Example 8.1 Data on Gender Gap

In recent years in the United States there has been much discussion about whether a "gender gap" exists in political beliefs. Do women and men tend to differ in their political thinking and voting behavior? To investigate this, we study Table 8.1, from the 1991 General Social Survey. The categorical variables are gender and party identification. Subjects indicated whether they identified more strongly with the Democratic or Republican party or as Independents.

Table 8.1 contains responses for 980 subjects, cross-classified by their gender and party identification. For instance, 279 subjects in the sample are Females who identify themselves as Democrats. Contingency tables are identified by their numbers of rows and columns. The number of rows is given first, followed by the number of columns. Table 8.1 is a 2×3 (read "2–by–3") contingency table, having two rows and three columns.

The row totals and the column totals of this table are called the ***marginal distributions***. The sample marginal distribution for party identification, for instance, is the set of marginal frequencies (444, 120, 416). □

Constructing a contingency table is the first step in investigating an association between two categorical variables. We now introduce methods for describing the frequencies in the table, as the next step.

TABLE 8.1 Party Identification and Gender

	Party Identification			
Gender	Democrat	Independent	Republican	Total
Females	279	73	225	577
Males	165	47	191	403
Total	444	120	416	980

Note: Data from 1991 General Social Survey.

Percentage Comparisons

To interpret Table 8.1, we need to study how party identification depends on gender. To compare females and males, we convert the frequencies to percentages within each row, as Table 8.2 shows. For example, of 577 females in the sample, 279 identify themselves as Democrat. This is a proportion of 279/577 = .483, or 48.3% in percentage terms. The percentage of males who identify themselves as Democrat equals 40.9% (165 out of 403). The row totals in Table 8.1 are the basis of these percentage calculations. The percentage comparisons suggest that, compared to Males, Females are more likely to identify as Democrats and less likely to identify as Republicans.

TABLE 8.2 Party Identification and Gender: Percentages Computed Within Rows of Table 8.1

	Party Identification				
Gender	Democrat	Independent	Republican	Total %	n
Females	48.3	12.7	39.0	100.0	577
Males	40.9	11.7	47.4	100.0	403

The two sets of percentages for females and males are called the ***conditional distributions*** on the response variable, party identification. They refer to the sample distribution of party identification, *conditional* on the level for gender. The females' conditional distribution on party identification is the set of percentages (48.3, 12.7, 39.0) for the responses (Democrat, Independent, Republican). The males' conditional distribution on party identification is the set of percentages (40.9, 11.7, 47.4). The percentages sum to 100 in each row. The proportions (.483, .127, .390) for females and (.409, .117, .474) for males are the genders' estimated ***conditional probabilities*** on party identification. Figure 8.1 portrays graphically the two conditional distributions.

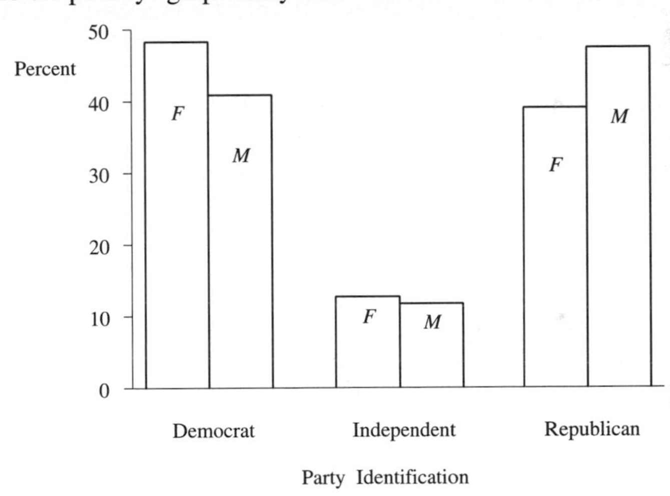

Figure 8.1 Portrayal of Conditional Distributions for Females and Males in Table 8.2

In a similar way, one could compute conditional distributions on gender for each party identification. Each column would then contain a percentage distribution. The first column would indicate, for example, that 62.8% of Democrats are females and 37.2% are males. There would probably be more interest, though, in describing how party identification depends on gender than how gender depends on party identification. Therefore, we have computed conditional distributions within rows only.

Most bivariate analyses identify one variable as a response variable and the other variable as an explanatory variable. Then, it is standard to form the conditional distribution for the response variable, within categories of the explanatory variable. In this example, party identification is a natural response variable, so Table 8.2 reports percentages within rows, which tells us the percentage of Democrats, Independents, and Republicans within each gender.

Another way to report percentages provides a single set computed over all cells in the table, using the total sample size as the base. To illustrate, out of the 980 subjects sampled for Table 8.1, 279 or 28.5% fall in the cell (Female, Democrat), 165 or 16.8% fall in the cell (Male, Democrat), and so forth. This percentage distribution is called the sample ***joint distribution***. It is useful for comparing the relative frequencies of occurrences for combinations of variable levels. When we distinguish between response and explanatory variables, though, conditional distributions (i.e., those computed within rows or within columns) are more informative than the joint distribution.

Guidelines in Forming Contingency Tables

Here are some guidelines when computing percentages in contingency tables. First, as just noted, compute percentages for the response variable within the categories of the explanatory variable. Suppose you construct the table so that the column variable is the response variable; that is, the response categories list across the top of the table. Then, you compute percentages within each row; divide each cell frequency by the row total and multiply the result by 100. The sum of the cell percentages within rows equals 100, apart from slight variations due to rounding.

Second, give the table a title that specifies the variables cross-classified and other relevant information for understanding the table. The two variables and their categories should be clearly labeled.

Third, include the total sample sizes on which the percentages or proportions are based. That way, readers can determine the cell frequencies, if they are not listed. In addition, they can then compute standard errors to analyze how close sample proportion estimates fall to the true values. For instance, an estimate of .40 with a sample of size 1000 has much less sampling error and deserves greater respect than an estimate of .40 with a sample size of 10.

Independence and Dependence

Whether an association exists in Table 8.1 is a matter of whether females and males differ in their conditional distributions on party identification. We answer the question "Is party identification associated with gender?" with reference to the concepts of statistical *independence* and *dependence*.

Statistical Independence and Dependence

Two categorical variables are ***statistically independent*** if the population conditional distributions on one of them are identical at each category of the other. The variables are ***statistically dependent*** if the conditional distributions are not identical.

In other words, two variables are statistically independent if the percentage of the population in any particular category of one variable is the same across all categories of the other variable. In Table 8.2, the two conditional distributions are not identical. For example, the percentage of Democrats is higher for females than for males. If those observations were the entire population, then the variables would be statistically dependent.

For simplicity, we usually use the term *independent* rather than *statistically independent*. Table 8.3 is a contingency table showing independence. The table contains the population data for two variables—party identification and ethnic group. The percentage of Democrats is the same for each ethnic group, 44%. Similarly. the percentage of Independents and the percentage of Republicans is the same for each ethnic group. The probability that a person has a particular party identification is the same for each ethnic group, and so party identification is independent of ethnic group.

TABLE 8.3 Population Cross-Classification Exhibiting Statistical Independence. The conditional distribution is the same in each row, (44%, 14%, 42%).

	Party Identification			
Ethnic Group	Democrat	Independent	Republican	Total
White	440 (44%)	140 (14%)	420 (42%)	1000 (100%)
Black	44 (44%)	14 (14%)	42 (42%)	100 (100%)
Hispanic	110 (44%)	35 (14%)	105 (42%)	250 (100%)

Statistical independence is a symmetric property between two variables. That is, if the conditional distributions within rows are identical, then so are the conditional distributions within columns. In Table 8.3, for example, you can check that the conditional distribution within each column equals (74%, 7%, 19%).

Example 8.2 Beliefs of College Freshmen

Every year, a large-scale poll of college freshmen conducted by the Higher Education Research Institute at UCLA asks their opinions on a variety of issues. For several years, the percentage of freshmen agreeing that abortion should be legal has been about 60% both for males and females. On the other hand, there has been a large difference between females and males regarding whether homosexual relations should be legally prohibited; in 1994, for instance, 45% of males agreed, but only 24% of females agreed. Thus, it appears that gender and opinion about abortion may be statistically independent, but gender and opinion about legalizing homosexual relations are statistically dependent. □

8.2 Chi-Squared Test of Independence

Table 8.1 contains sample data from the population of American adults. The definition of statistical independence refers to the population. Two variables are independent if the *population* conditional distributions on the response variable are identical. Since Table 8.1 refers to a sample rather than a population, it provides evidence but does not definitively answer whether party identification and gender are independent. Even if they are independent, we would not expect the *sample* conditional distributions to be identical. Because of sampling variability, each sample percentage typically differs from the true population percentage.

We next study whether it is plausible that, in the population, party identification and gender are independent. We determine whether the observed *sample* differences between females and males in their conditional distributions could be due to sampling variation. If the variables are truly independent, would sampled differences of this size be likely? Or are the observed differences in percentages so great that statistical independence in the population is implausible? A statistical test answers these questions.

The null hypothesis for the test states that the two categorical variables are statistically independent; that is, the hypotheses are

H_0: The variables are statistically independent
H_a: The variables are statistically dependent.

The test requires random samples or stratified random samples in which the categories of one variable are the strata. The sample size must be large, in a certain sense explained later in the section.

Expected Frequencies for Independence

The chi-squared test compares the observed frequencies in the cells of the contingency table with values expected from the null hypothesis of independence.

Notation for Observed and Expected Frequencies

Let f_o denote an ***observed*** frequency in a cell of the table.
Let f_e denote an ***expected*** frequency; this is the count expected in a cell if the variables were independent.

Table 8.4 shows the observed frequencies from Table 8.1, with the expected frequencies in parentheses. To illustrate, the cell in the upper left-hand corner has observed frequency $f_o = 279$ and expected frequency $f_e = 261.4$. The expected frequencies have the same row and column totals as the observed frequencies, but satisfy independence. The next box provides the rule for calculating them.

Calculation of Expected Frequency

The expected frequency f_e for a cell equals the product of the row and column totals for that cell, divided by the total sample size.

TABLE 8.4 Party Identification by Gender, with Expected Frequencies in Parentheses

	Party Identification			
Gender	Democrat	Independent	Republican	Total
Female	279 (261.4)	73 (70.7)	225 (244.9)	577
Male	165 (182.6)	47 (49.3)	191 (171.1)	403
Total	444	120	416	980

For instance, the cell in the upper left-hand corner refers to Females who identify as Democrats. Its expected frequency is $f_e = (577)(444)/980 = 261.4$, the product of the row total for Females and the column total for Democrats, divided by the overall sample size.

Let's see why this rule makes sense. In the entire sample, 444 out of 980 people (45.3%) identify as Democrats. If the variables were independent, we would expect 45.3% of males and 45.3% of females to identify as Democrats. For instance, 45.3% of the 577 Females should be classified in the Democrat category. The expected frequency for the cell is then

$$f_e = \left(\frac{444}{980}\right)577 = .453(577) = 261.4$$

Chi-Squared Test Statistic

The test statistic for the test of independence summarizes how close the expected frequencies fall to the observed frequencies. Symbolized by χ^2, it is called the ***chi-squared statistic***. It is the oldest test statistic in use today, having been introduced by the British statistician Karl Pearson in 1900. It equals

$$\chi^2 = \sum \frac{(f_o - f_e)^2}{f_e}$$

The summation is taken over all cells in the contingency table. For each cell, we square the difference between the observed and expected frequencies and then divide that square by the expected frequency. After calculating this term for every cell, we sum the terms to construct the χ^2 statistic.

When H_0 is true, f_o and f_e tend to be close for each cell, and χ^2 is relatively small. If H_0 is false, at least some f_o and f_e values tend not to be close, leading to large $(f_o - f_e)^2$ values and a large test statistic. The larger the χ^2 value, the greater the evidence against the null hypothesis of independence.

Substituting the f_o and f_e values from Table 8.4 into the formula for the χ^2 test statistic,

$$\begin{aligned}\chi^2 &= \sum \frac{(f_o - f_e)^2}{f_e} \\ &= \frac{(279 - 261.4)^2}{261.4} + \frac{(73 - 70.7)^2}{70.7} + \frac{(225 - 244.9)^2}{244.9} \\ &\quad + \frac{(165 - 182.6)^2}{182.6} + \frac{(47 - 49.3)^2}{49.3} + \frac{(191 - 171.1)^2}{171.1} \\ &= 1.19 + .07 + 1.62 + 1.70 + .11 + 2.31 = 7.0.\end{aligned}$$

The calculation is messy, but it is simple to obtain this statistic and its P-value using statistical software, as shown in the appendix. We next study how to interpret the magnitude of the χ^2 test statistic.

The Chi-Squared Distribution

To convert the χ^2 test statistic to a P-value, we use the sampling distribution of the χ^2 statistic. This distribution indicates how large χ^2 must be before strong evidence exists that H_0 is false.

For large sample sizes, the sampling distribution of χ^2 is the ***chi-squared probability distribution***. The name of the test and the symbol for the test statistic refer to the name of the sampling distribution. Figure 8.2 shows a sketch of several chi-squared distributions.

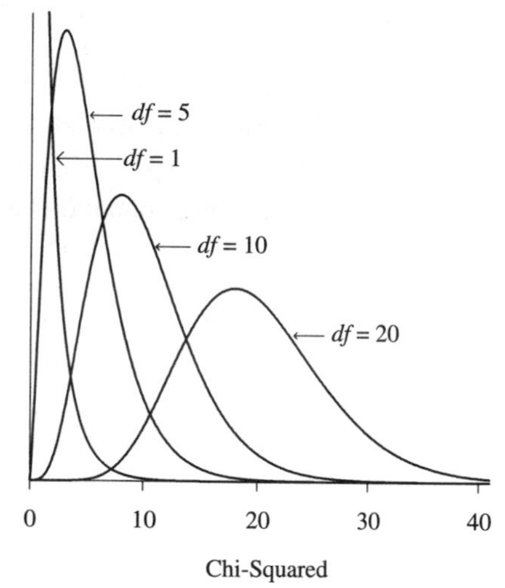

Figure 8.2 The Chi-Squared Distribution. The curve has larger mean and standard deviation as the degrees of freedom increase.

The main properties are as follows:

- It is concentrated on the positive part of the real line. The χ^2 test statistic cannot be negative, since it sums squared differences divided by positive expected frequencies. The minimum possible value, $\chi^2 = 0$, occurs when $f_o = f_e$ in each cell; that is, when the variables are completely independent in the sample.

- It is skewed to the right.
- The precise shape of the distribution depends on the ***degrees of freedom***. For testing independence in a table with r rows and c columns, the formula for the degrees of freedom is

$$df = (r - 1)(c - 1)$$

For a 2×3 table such as Table 8.4, for example, $r = 2$ and $c = 3$ and

$$df = (2 - 1)(3 - 1) = 1 \times 2 = 2$$

- The mean μ and standard deviation σ of the chi-squared distribution depend on the df value, with $\mu = df$ and $\sigma = \sqrt{2df}$. For a 2×3 table, for instance, $\mu = df = 2$ and $\sigma = \sqrt{2df} = \sqrt{4} = 2.0$. The mean and standard deviation both increase as df increases. Thus, the distribution tends to shift to the right and become more spread out for larger values of df. Since $df = (r - 1)(c - 1)$, larger numbers of rows and columns produce larger df values. Since larger tables have more terms in the summation for the χ^2 statistic, the test statistic values also tend to be larger. In addition, as df increases, the shape of the chi-squared curve more closely resembles a normal curve.
- The larger the χ^2 value, the stronger the evidence against the null hypothesis of independence. Values of χ^2 greater than the observed value are ones providing even greater evidence against H_0 than the observed data. Thus, the P-value equals the right-tail probability under the chi-squared curve above the observed χ^2 value; that is, it equals the probability, when H_0 is true, that χ^2 is at least as large as the observed value. Figure 8.3 depicts the P-value.
- Table C in the back of the text lists scores from the chi-squared distribution having various right-hand tail probabilities. These are χ^2 test statistic values that have P-values equal to those probabilities. For example, Table 8.4 has $df = 2$. For $df = 2$, the χ^2 value of 5.991 has $P = .05$, whereas $\chi^2 = 9.210$ has $P = .01$.

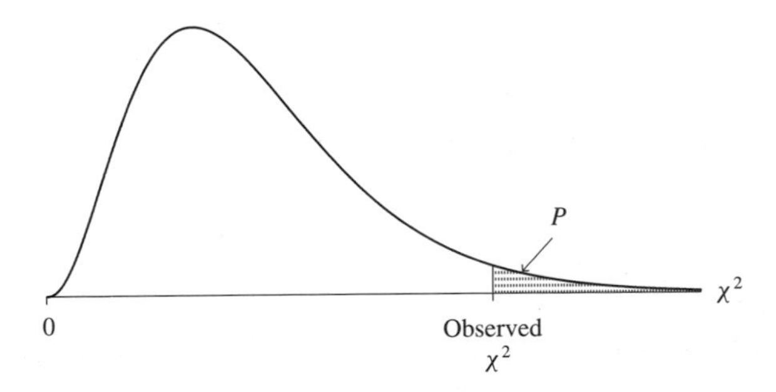

Figure 8.3 The P-Value for the Chi-Squared Test of Independence Is the Right-Hand Tail Probability, Above the Observed Value of the Test Statistic

Example 8.3 Chi-Squared and Gender Gap Data

The sample for the General Social Survey results from a rather complex multistage design. For this design, however, the characteristics of the data are similar to those of a random sample, so we will apply the chi-squared test to Table 8.4. We test the null hypothesis that party identification and gender are statistically independent. The alternative hypothesis is that they are statistically dependent.

Previously, we obtained a test statistic value of $\chi^2 = 7.0$ for these data. In Table C, for $df = 2$, 7.0 falls between 5.991 and 7.824, the chi-squared values having tail probabilities .05 and .02. Thus, we conclude that $.02 < P < .05$. In fact, computer software indicates that $P = .030$.

This small P-value provides moderately strong evidence against the null hypothesis. It seems likely that party identification and gender are associated, in the population. If the variables were independent, it would be unusual for a random sample to have this large a chi-squared statistic or one even larger. □

Sample Size Requirements

The chi-squared distribution is the sampling distribution of the χ^2 test statistic only if the sample size is large. A rough guideline for this requirement is that the expected frequency f_e should exceed 5 in each cell. Otherwise, the chi-squared distribution may poorly approximate the actual distribution of the χ^2 statistic.

The final part of Section 8.3 discusses a small-sample test of independence that applies when the sample size is not sufficiently large to use the chi-squared test. Table 8.5 summarizes the five elements of the chi-squared test.

TABLE 8.5 The Five Elements of the Chi-Squared Test of Independence

1. Assumptions:	Two categorical variables Random sample or stratified random sample $f_e \geq 5$ in all cells (otherwise, use small-sample exact test)
2. Hypotheses:	H_0: Statistical independence of variables H_a: Statistical dependence of variables
3. Test statistic:	$\chi^2 = \sum \frac{(f_o - f_e)^2}{f_e}$, where $f_e = \frac{\text{(Row total)(Column total)}}{\text{Total sample size}}$
4. P-value:	P = right-hand tail probability above observed χ^2 value, for chi-squared distribution with $df = (r-1)(c-1)$
5. Conclusion:	Report P-value If decision needed, reject H_0 at α-level if $P \leq \alpha$

Using Computer Software to Conduct Chi-Squared Tests

The χ^2 test statistic is the most computationally complex statistic we have encountered so far in this text. It is messy enough that we would normally not compute it by hand, or even with a calculator. Statistical software can easily calculate χ^2 and its P-value.

Table 8.6 illustrates the output one obtains in using software to conduct a chi-squared test of independence for Table 8.1. SAS lists the P-value under the heading *Prob*, and SPSS lists it under *Significance*. Most software also reports an alternative test statistic, called the *likelihood-ratio statistic*, which usually provides similar results. Chapter 15 introduces this statistic.

TABLE 8.6 Printout for Chi-Squared Test of Independence

```
GENDER      PARTY
Frequency|
Expected |democrat|indep   |repub   |  Total
---------+--------+--------+--------+
female   |   279  |    73  |   225  |   577
         | 261.42 | 70.653 | 244.93 |
---------+--------+--------+--------+
male     |   165  |    47  |   191  |   403
         | 182.58 | 49.347 | 171.07 |
---------+--------+--------+--------+
Total        444      120      416      980

Statistic                     DF      Value          Prob
---------------------------------------------------------
Chi-Square                     2      7.010         0.030
Likelihood Ratio Chi-Square    2      7.003         0.030
```

Interpretation of Degrees of Freedom

The *df* term in a chi-squared test has the following interpretation: given the row and column marginal frequencies, the observed frequencies in a rectangular block of size $(r - 1) \times (c - 1)$ within the contingency table determine the other cell frequencies.

To illustrate, in Table 8.4, suppose we know the two frequencies 279 and 73 in the upper left-hand part of the table; this is a block of size 1×2, shown in Table 8.7. Then, given the row and column totals, we can determine all the other cell counts. For instance, since 279 of the 444 Democrats are classified female, the other $444-279 = 165$ must be classified male. Since 73 of the 120 Independents are classified female, the other $120 - 73 = 47$ must be classified male. Also, since the total of the female row is 577, and since the first two cells contain 352 (i.e., $279 + 73$) subjects, then the remaining cell must have $577 - 352 = 225$ observations. From this and the fact that the last column has 416 observations, there must be $416 - 225 = 191$ observations in the second cell in that column.

Once the marginal frequencies are fixed in a contingency table, a block of only $(r - 1) \times (c - 1)$ cell entries is free to vary, since these cell entries determine the remaining ones. The degrees of freedom value equals the number of cells in this block, or $df = (r - 1)(c - 1)$. Similarly, a block of $(r - 1)(c - 1)$ expected frequencies determine the remaining ones, since they have the same row and column totals as the observed frequencies.

TABLE 8.7 Illustration of Degrees of Freedom; a Block of $(r - 1)(c - 1)$ Cell Counts Determine the Others

	Party Affiliation			
Gender	Democrat	Independent	Republican	Total
Female	279	73	—	577
Male	—	—	—	403
Total	444	120	416	980

Chi-Squared Tests and Treatment of Categories

In the chi-squared test of independence, the value of the χ^2 test statistic depends on the way the categories are defined for the two variables. If religion is classified as Protestant and non-Protestant, the conclusion of the test might be different than if the categories were Protestant, Jewish, Catholic, None, and Other. A report of the test result should identify the categories for the variables.

The chi-squared test treats the classifications as nominal. That is, the same value results for χ^2 if the rows or columns of the table are reordered in any way. If either classification is ordinal, the chi-squared test does not use the information on the ordering.

8.3 More on Testing Independence

Misuses of Chi-Squared

Unfortunately, the chi-squared test is frequently misused. A common misuse is to apply it when the expected frequencies are too small. Another misuse is to compute the χ^2 statistic for numbers other than the observed frequencies, such as percentages or rates. Another misuse is to apply it when separate rows or columns are dependent samples, such as when each row has the same subjects, in which case McNemar's test (Section 7.4) is more appropriate.

Although the chi-squared statistic treats the classifications as nominal, it is used in practice for any type of categorical data. When the variables are ordinal or interval, however, it is usually better to apply stronger statistical methods designed for the higher level of measurement. Section 8.6 presents a test of independence for ordinal classifications, and Section 9.5 presents one for interval data.

The chi-squared test of independence, like other significance tests, provides limited information. If the P-value is of moderate size (e.g., $P > .10$), then it is plausible that the variables are independent. In such a case, though, it is inappropriate to *accept* the null hypothesis. The sample size simply may not be large enough to have much power for detecting whatever effect truly exists. If the P-value is very small, on the other hand, strong evidence exists against the null hypothesis of independence. In that case, we can conclude that the variables are associated. The chi-squared statistic and that P-value tells us nothing, however, about the nature or strength of the association.

The next two sections introduce methods one can use following a chi-squared test to learn more about the association.

Residual Analysis

A test statistic and its P-value summarize the strength of evidence against the null hypothesis. If χ^2 is large for testing independence, then *somewhere* in the contingency table the data depart from what independence predicts. The test statistic does not indicate, however, whether all cells deviate greatly from independence or perhaps only one or two of the cells do so.

A cell-by-cell comparison of observed and expected frequencies reveals the nature of the evidence. The difference $(f_o - f_e)$ between an observed and expected cell frequency is called a ***residual***. For the first cell of Table 8.4, for instance, the residual equals $279 - 261.4 = 17.6$. The residual is positive when, as in this cell, the observed frequency f_o exceeds the value f_e that independence predicts; it is negative when the observed frequency is smaller than independence predicts.

How do we know whether a residual is large enough to indicate a significant departure from independence? To answer this, we use an adjusted form of the residual that behaves like a z-score.

Adjusted Residual

The ***adjusted residual*** for a cell equals

$$\frac{f_o - f_e}{\sqrt{f_e(1 - \text{row proportion})(1 - \text{column proportion})}}$$

The denominator is the standard error of the difference $(f_o - f_e)$, when the variables are truly independent. It uses the marginal proportions for the row and column in which the cell falls.

Suppose the null hypothesis of independence is true. Then an adjusted residual reports the number of standard errors that the observed count falls from the expected count, and it has a large-sample standard normal distribution. In that case, the adjusted residuals fluctuate around a mean of 0, with a standard deviation of 1. Thus, there is only about a 5% chance that any particular adjusted residual exceeds 2 in absolute value. A large adjusted residual provides evidence against independence in that cell; a value exceeding about 3 provides strong evidence.

Example 8.4 Adjusted Residuals for Gender Gap Data

Table 8.8 displays the adjusted residuals for testing independence in Table 8.1. For the first cell, for instance, $f_o = 279$ and $f_e = 261.4$. The first row and first column marginal proportions equal 577/980 = .589 and 444/980 = .453. Substituting into the formula, the adjusted residual for this cell equals

$$\frac{279 - 261.4}{\sqrt{[261.4(1 - .589)(1 - .453)]}} = 2.3$$

Since the adjusted residual exceeds 2.0, this cell shows a greater discrepancy between f_o and f_e than is expected if the variables were truly independent.

TABLE 8.8 Adjusted Residuals (in Parentheses) for Testing Independence in Table 8.1

	Party Identification		
Gender	Democrat	Independent	Republican
Females	279 (2.3)	73 (0.5)	225 (−2.6)
Males	165 (−2.3)	47 (−0.5)	191 (2.6)

Table 8.8 exhibits large positive residuals for female Democrats and male Republicans, which are the two cells in which f_o is much larger than f_e. This means there were significantly more female Democrats and male Republicans than the hypothesis of independence predicts. The table exhibits large negative residuals for female Republicans and male Democrats, the two cells in which f_o is much smaller than f_e. There were significantly fewer female Republicans and male Democrats than one would expect if party affiliation were independent of gender.

For each party, Table 8.8 contains only one nonredundant adjusted residual; the one for females is the negative of the one for males. The observed counts and the expected frequencies have the same row and column totals. Thus, in a given column, if $f_o > f_e$ in one cell, the reverse must happen in the other cell. The differences $f_o - f_e$ have the same magnitude but different sign in the two cells, implying the same pattern for their adjusted residuals. □

Along with the χ^2 statistic, software can easily provide adjusted residuals. See the appendix at the end of the text for details.

Chi-Squared and Difference of Proportions for 2×2 Tables

We now discuss the special case of 2×2 contingency tables. Tables of this size often display the responses of two groups on a binary variable. The response outcomes could be, for example, (employed, unemployed), (support president, do not support president), or (yes, no) on an opinion question such as whether abortion should be legal. For convenience, we label the two possible outcomes for that binary variable by the generic labels *success* and *failure*.

Let π_1 represent the proportion of successes in population 1, and let π_2 represent the proportion of successes in population 2. Then $(1-\pi_1)$ and $(1-\pi_2)$ are the proportions of failures for the two populations. Table 8.9 displays the notation. The rows are the groups to be compared and the columns are the response categories.

TABLE 8.9 2×2 Table for Comparing Populations 1 and 2 on a Binary Response Variable Having Outcomes *Success* and *Failure*

	Proportion Making Each Response		
Group	Success	Failure	Total
1	π_1	$1 - \pi_1$	1.0
2	π_2	$1 - \pi_2$	1.0

If the response variable is statistically independent of the populations considered, then $\pi_1 = \pi_2$; that is, the probability of success does not depend on whether a subject is from population 1 or population 2. The null hypothesis of independence in the chi-squared test corresponds to H_0: $\pi_1 = \pi_2$. In fact, the chi-squared test is equivalent to a test for equality of two population proportions, π_1 and π_2.

Section 7.2 presented a z test statistic for testing equality of two proportions, based on dividing the difference of sample proportions by its standard error; that is,

$$z = \frac{\hat{\pi}_2 - \hat{\pi}_1}{\hat{\sigma}_{\hat{\pi}_2 - \hat{\pi}_1}}$$

The chi-squared statistic for 2×2 tables is related to this z statistic by $\chi^2 = z^2$.

The chi-squared statistic for 2×2 tables has $df = 1$. Its P-value from the chi-squared distribution is the same as the P-value for the two-sided test with the z test statistic. This is because of a direct connection between the standard normal distribution and the chi-squared distribution with $df = 1$: squaring z-scores with certain tail probabilities yields chi-squared scores with $df = 1$ having the same tail probabilities. For instance, $z = 1.96$ is the z-score with a two-tail probability of .05. The square of this, $(1.96)^2 = 3.84$, is the chi-squared score for $df = 1$ with a P-value of .05. (You should check this in Table C.)

Example 8.5 Women's Roles Example Continued

Table 7.2 in Example 7.3 is a 2×2 contingency table dealing with comparing opinions in 1982 and 1994 on whether women should take care of running their homes and leave running the country up to men. Table 8.10 shows the data again. The test in Section 7.2 comparing the proportions who supported this statement in each year had a test statistic value of $z = -9.6$. The chi-squared statistic for this table equals 91.9. This equals the square of the z test statistic from the previous analysis. Both statistics show extremely strong evidence against the null hypothesis of equal population proportions. □

TABLE 8.10 Responses to the Statement, "Women should take care of running their homes and leave running the country up to men," with Adjusted Residuals in Parentheses

Year	Agree	Disagree	Total
1982	122 (9.6)	223 (−9.6)	345
1994	268 (−9.6)	1632 (9.6)	1900

Adjusted Residuals for 2×2 Tables

Let's follow up the test for Table 8.10 with a residual analysis. Table 8.10 also shows the adjusted residuals for the counts in that table. The residuals in the first column suggest that more subjects agreed with the statement in 1982 and fewer agreed in 1994 than one would expect if opinion were independent of the year of the survey. Notice that *every* adjusted residual equals either +9.6 or −9.6; the absolute value of the adjusted residual is 9.6 in every cell.

For chi-squared tests with 2×2 tables, $df = 1$. This means that only one nonredundant piece of information exists about whether an association exists. Once we calculate the adjusted residual for one cell, other adjusted residuals in the same row or column take the negative of that value. If an observed count exceeds the expected frequency in one cell, the observed count is smaller than the expected frequency in the other cell in that row or column. This is because the expected frequencies have the same row and column totals as the observed counts. In fact, in 2×2 tables, all four adjusted residuals have absolute value equal to the square root of the χ^2 test statistic. The adjusted residuals have the same size as the z test statistic for comparing the two proportions.

Small-Sample Test of Independence*

The chi-squared test, like one- and two-sample z tests for means and proportions, is a large-sample test. When the expected frequencies are small (i.e., any of them being less than 5), it should not be used. In that case, small-sample tests of independence are more appropriate than the chi-squared test.

For 2×2 contingency tables, a small-sample test of independence is ***Fisher's exact test***, discussed in Section 7.3. This test extends to tables of arbitrary size $r \times c$. For these larger tables, computations are so complex that a computer is necessary, and the task can even be difficult for a computer if r and c are large. The test requires specialized software, such as SAS (the EXACT option in PROC FREQ) or SPSS (the Exact module), as discussed in the appendix.

Fisher's exact test for 2×2 tables and its extension for $r \times c$ tables both utilize the *hypergeometric* distribution for the cell counts. When n is small, this distribution is highly discrete, and the related sampling distribution of the χ^2 test statistic is poorly approximated by the continuous chi-squared curve. Thus, P-values based on the small-sample exact test may be quite far from the crude approximations provided by referring the test statistic to the chi-squared distribution.

Example 8.6 Small-Sample Test Better than Chi-Squared

Table 8.11 is a 3×9 table having many zero entries and small counts. For it, $\chi^2 = 22.3$ with $df = 16$. The chi-squared approximation provides a P-value of $P = .13$. Because the cell counts are so small, the validity of this approximation is suspect. Software for the exact test provides an exact P-value of .001 (see the appendix), quite different from the result using the large-sample approximation. Contrary to what the chi-squared distribution suggests, very strong evidence exists against independence. □

TABLE 8.11 Example of 3×9 Table for Small-Sample Test of Independence

	Response Category								
Row	1	2	3	4	5	6	7	8	9
1	0	7	0	0	0	0	0	1	1
2	1	1	1	1	1	1	1	0	0
3	0	8	0	0	0	0	0	0	0

8.4 Measuring Association in 2×2 Tables

At least three questions are normally addressed in analyzing a contingency table.

- *How likely is it that at least the observed degree of association would occur in a sample, if the variables are truly independent in the population?* The chi-squared test addresses this question of whether evidence exists of an association.
- *How do the data depart from independence?* When two variables appear to be associated, the adjusted residuals highlight the cells in which the counts are significantly different from what independence predicts.
- *How strong is the association?* For this we use a statistic such as a difference of proportions, forming a confidence interval to estimate how strong the association is likely to be in the population.

TABLE 8.12 Cross-classification of Opinion About Legalized Abortion by Race, Showing No Association

	Opinion		
Race	Favor	Oppose	Total
White	360	240	600
Black	240	160	400
Total	600	400	1000

TABLE 8.13 Cross-classification of Opinion About Legalized Abortion by Race, Showing Maximum Association

	Opinion		
Race	Favor	Oppose	Total
White	600	0	600
Black	0	400	400
Total	600	400	1000

Analyzing the *strength* of the association reveals whether the association is an important one, or if it is statistically significant but weak and unimportant in practical terms. This section discusses this third matter of describing the strength of the association.

Measures of Association

Measure of Association

A ***measure of association*** is a statistic that summarizes the strength of the statistical dependence between two variables.

This section introduces two measures of association for contingency tables. The measures refer to 2×2 tables but are also useful for describing association in larger tables.

Let's first consider what is meant by *strong* versus *weak* association. Tables 8.12 and 8.13 show two hypothetical sets of frequencies for contingency tables relating race to opinion about legalized abortion.

Table 8.12, which exhibits statistical independence, represents the weakest possible association. Both whites and blacks have 60% in favor and 40% opposed to legalized abortion. Opinion is not associated with race. By contrast, Table 8.13 exhibits the strongest possible association. All whites favor legalized abortion, whereas all blacks oppose it. In this table, opinion is completely dependent on race. For these subjects, if we know their race, we know their opinion about legalized abortion.

A measure of association describes how similar a particular table is to the tables representing the strongest and weakest associations. It takes a range of values from one extreme to another as data range from the weakest to strongest association.

Difference of Proportions

As discussed in Sections 7.2 and 8.3, many 2×2 tables summarize the responses of two groups on a binary response variable. In such cases, an easily interpretable measure of association is the difference between the group proportions for a given response level. For example, we could measure the difference between the proportions of whites and blacks who favor legalized abortion. For Table 8.12, this difference is

$$\frac{360}{600} - \frac{240}{400} = .6 - .6 = 0$$

The population difference of proportions is 0 whenever the conditional distributions are identical, that is, when the variables are independent. The difference is 1 or −1 for the strongest possible association. For Table 8.13, for instance, the difference is

$$\frac{600}{600} - \frac{0}{400} = 1.0$$

which is the maximum possible absolute value for the difference.

Section 7.2 studied the difference of proportions and showed how to construct a confidence interval for a population difference. This measure falls between −1 and +1. The stronger the association, the larger the difference is in absolute value. The following contingency tables illustrate the increase in this measure as the degree of association increases:

Cell Counts:	25 25 / 25 25	30 20 / 20 30	35 15 / 15 35	40 10 / 10 40	45 5 / 5 45	50 0 / 0 50
Difference of Proportions:	0	.2	.4	.6	.8	1.0

For the second table, for instance, the proportion falling in the first column equals 30/(30 + 20) = .60 in row 1 and 20/(20 + 30) = .40 in row 2, for a difference of .60 − .40 = .20.

Chi-Squared and Association

A large value for χ^2 in the test of independence suggests that the variables are associated. It does not imply, however, that the variables have a strong association. This statistic measures how close the observed frequencies are to the frequencies expected if the variables were independent. It merely indicates, however, *how certain* we can be that the variables are dependent, not how strong that dependence is. For a given association, larger χ^2 values occur for larger sample sizes. Large χ^2 values can occur with weak associations, if the sample size is large.

For example, consider contingency Tables A, B, and C in Table 8.14. The association in each table is very weak—the conditional distribution for whites on opinion (49% favor, 51% oppose) is nearly identical to the conditional distribution for blacks (51% favor, 49% oppose). All three tables show exactly the same degree of association, with the difference between the proportions of blacks and whites who favor legalized abortion being .51 − .49 = .02 in each table.

TABLE 8.14 Cross-Classifications of Opinion on Legalized Abortion by Race, Showing Weak but Identical Associations

	A			B			C		
	Yes	No	Total	Yes	No	Total	Yes	No	Total
White	49	51	100	98	102	200	4,900	5,100	10,000
Black	51	49	100	102	98	200	5,100	4,900	10,000
	100	100	200	200	200	400	10,000	10,000	20, 000
	$\chi^2 = .08$			$\chi^2 = .16$			$\chi^2 = 8.0$		
	$P = .78$			$P = .69$			$P = .005$		

For the sample of size 200 in Table A, $\chi^2 = .08$, which has a P-value of $P = .78$. For the sample of size 400 in Table B, $\chi^2 = .16$, for which $P = .69$. So, when the cell counts double, χ^2 doubles. Similarly, for the sample size of 20,000 (100 times as large as $n = 200$) in Table C, $\chi^2 = 8.0$ (100 times as large as $\chi^2 = .08$); for that value, $P = .005$.

In summary, for a fixed percentage assignment to the cells of a contingency table, the value of χ^2 is directly proportional to the sample size—larger values occur with larger sample sizes. Like other test statistics, the larger the χ^2 statistic, the smaller the P-value and the stronger the evidence against the null hypothesis. However, a small P-value results from a weak association when the sample size is large, as Table C shows.

The Odds Ratio*

The difference of proportions is an easily interpretable measure of association. However, several other measures are available and are reported by statistical software. This subsection presents the most important one for categorical data analysis, the *odds ratio*.

For a 2×2 table, again suppose that the two rows represent two groups and the two columns represent the response outcomes *success* and *failure*, as in Table 8.9. The probability of success equals π_1 in row 1 and π_2 in row 2. Within a row, the ***odds*** of success are defined to be

$$\text{Odds} = \frac{\text{Probability of success}}{\text{Probability of failure}}$$

For instance, if the probability of success equals .75, then the probability of failure equals $1 - .75 = .25$, and the odds of success equal $.75/.25 = 3$. If the probability of success equals .50, then the odds of success equal $.50/.50 = 1.0$. If the probability of success equals .25, then the odds of success equal $.25/.75 = 1/3$. The odds are nonnegative, with value greater than 1.0 when a success is more likely than a failure. When odds = 3.0, a success is three times as likely as a failure; we expect about three successes for every failure. When odds = 1/3, a failure is three times as likely as a success; we expect about one success for every three failures.

The success probability is the function of the odds,

$$\text{Probability} = \frac{\text{Odds}}{\text{Odds} + 1}.$$

For instance, when odds = 3, the success probability equals $3/(3+1) = .75$.

The ratio of odds from the two rows is called the ***odds ratio***. For instance, if the odds in row 1 equal 4.5 and the odds in row 2 equal 3.0, then the odds ratio equals 4.5/3.0 = 1.5; the odds of success in row 1 then equal 1.5 times the odds of success in row 2. The odds ratio is denoted by the Greek letter θ (theta).

Example 8.7 Odds Ratio for Race of Murder Victims and Offenders

To illustrate the odds ratio, we analyze Table 8.15. For murders in the United States in 1993 having a single victim and single offender, Table 8.15 cross classifies the race of the victim by the race of the offender.

TABLE 8.15 Cross Classification of Race of Victim and Race of Offender

Race of	Race of Victim		
Offender	White	Black	Total
White	4686	304	4990
Black	849	5393	6242

Source: U.S. Department of Justice, *Crime in the United States, 1993.* (Washington, D.C.: U.S. Government Printing Office).

We treat race of victim as the response variable. For white offenders, the proportion of victims who were white equals $4686/4990 = .939$ and the proportion who were black equals $304/4990 = .061$. The odds of a white victim equaled $.939/.061 = 15.4$. This equals $(4686/4990)/(304/4990) = 4686/304$, so one can calculate the odds by the ratio of the counts in the two cells in row 1, without converting them to proportions.

The value 15.4 means that for white offenders, there were 15.4 white victims for every 1 black victim. For black offenders, the odds of a white victim equalled 849/5393 = .157. This means there were .157 white victims for every 1 black victim. Equivalently, since 5393/849 = 1/.157 = 6.4, black offenders had 6.4 black victims for every white victim.

For Table 8.15, the odds ratio equals

$$\theta = \frac{\text{Odds for white offenders}}{\text{Odds for black offenders}} = \frac{15.4}{.157} = 97.9$$

For white offenders, the odds of a white victim were about 100 times the odds of a white victim for black offenders. □

In summary,

Odds and Odds Ratio

The ***odds*** for a binary response equal the number of successes divided by the number of failures.
The ***odds ratio*** is a measure of association for 2×2 contingency tables that equals the odds in row 1 divided by the odds in row 2.

Properties of the Odds Ratio*

In Table 8.15, suppose we treat race of offender, rather than race of victim, as the response variable. When victims were white, the odds the race of offender was white equaled 4686/849 = 5.52. When victims were black, the odds the race of offender was white equaled 304/5393 = .056. The odds ratio equals 5.52/.056 = 97.9. For white victims, the odds of a white offender were about 100 times the odds of a white offender for black victims.

For each choice of the response variable, we found an odds ratio of 97.9. In fact,

- The odds ratio takes the same value regardless of the choice of response variable.

The same value occurs when we treat (1) the columns as the response variable and the rows as the explanatory variable or (2) the rows as the response variable and the columns as the explanatory variable. Since the odds ratio treats the variables symmetrically, it does not require identifying a response variable. This makes the odds ratio a natural measure when there is no obvious distinction between the variables, such as when they are both response variables.

- The odds ratio θ equals the ratio of the products of cell counts from diagonally opposite cells.

For Table 8.15, for instance,

$$\theta = \frac{(4686)(5393)}{(304)(849)} = 97.9$$

Because of this property, the odds ratio is also called the ***cross-product ratio***.

- The odds ratio can equal any nonnegative number.
- When the success probabilities are identical in the two rows of a 2×2 table (i.e., $\pi_1 = \pi_2$), then $\theta = 1$.

When $\pi_1 = \pi_2$, the odds are also equal. The odds of success do not depend on the row level of the table, and the variables are then independent, with $\theta = 1$. The value $\theta = 1$ for independence serves as a baseline for comparison. Odds ratios on each side of 1 reflect certain types of associations.

- When $\theta > 1$, the odds of success are higher in row 1 than in row 2.

For instance, when $\theta = 4$, the odds of success in row 1 are four times the odds of success in row 2. Subjects in row 1 are more likely to have successes than are subjects in row 2; that is, $\pi_1 > \pi_2$. To illustrate, suppose $\pi_1 = .8$ and $\pi_2 = .5$. The odds of success equal .8/.2 = 4.0 in row 1 and .5/.5 = 1.0 in row 2, and the odds ratio equals 4.0/1.0 = 4. This odds ratio value of 4 also occurs for many other combinations of probabilities, such as $\pi_1 = .5$ and $\pi_2 = .2$.

- When $\theta < 1$, the odds of success are lower in row 1 than in row 2.

In this case, $\pi_1 < \pi_2$. For instance, if $\pi_1 = .5$ and $\pi_2 = .8$, then the odds equal .5/.5 = 1.0 in row 1 and .8/.2 = 4.0 in row 2, and the odds ratio equals 1.0/4.0 = .25.

- Values of θ farther from 1.0 in a given direction represent stronger associations.

An odds ratio of 4 is farther from independence than an odds ratio of 2, and an odds ratio of .25 is farther from independence than an odds ratio of .50. The most extreme values for the odds ratio are 0 and ∞ (infinity). They occur when the probability of success equals 1 or 0 in one of the rows.

- Two values for θ represent the same strength of association, but in opposite directions, when one value is the inverse of the other.

For instance, $\theta = 4.0$ and $\theta = 1/4.0 = .25$ represent the same strength of association. When $\theta = .25$, the odds of success in row 1 are .25 times the odds of success in row 2. Equivalently, the odds of success in row 2 are 1/.25 = 4.0 times the odds of success in row 1. When the order of the rows is reversed or the order of the columns is reversed, the new value of θ is the inverse of the original value. This ordering of rows or columns is usually arbitrary, so whether we get 4.0 or .25 for the odds ratio is simply a matter of how we identify the rows and columns.

- The sampling distribution of the sample odds ratio $\hat{\theta}$ is highly skewed unless the sample size is extremely large, in which case the distribution is approximately normal. We defer details about inference for odds ratios to Chapter 15.

The Ratio of Proportions*

An odds ratio of 97.9 does *not* mean that π_1 is 97.9 times π_2. Instead, $\theta = 97.9$ means that the *odds* in row 1 equal 97.9 times the odds in row 2.

The ratio of proportions π_1/π_2 is itself a useful measure. Problem 7.45 introduced this measure, called the ***relative risk***. It is not used as much as the odds ratio, mainly because the odds ratio connects better to advanced methods for contingency tables presented in Chapter 15.

Odds Ratios for $r \times c$ Tables*

For contingency tables with more than two rows or more than two columns, the odds ratio describes patterns in any 2×2 subtable formed from a rectangular array of two rows combined with two columns. We illustrate using the data on party identification and gender from Table 8.1, shown again in Table 8.16.

TABLE 8.16 Party Identification and Gender

	Party Identification		
Gender	Democrat	Independent	Republican
Females	279	73	225
Males	165	47	191

Consider first the 2×2 subtable formed from the first two columns of this table. The sample odds ratio equals $(279 \times 47) \div (73 \times 165) = 1.09$. The odds that a female's response was Democrat rather than Independent equal 1.09 times the odds for males. Of those subjects who responded Democrat or Independent, females were slightly more likely than males to respond Democrat; however, this value is so close to 1.0 that this part of the table reveals essentially no association.

The sample odds ratio for the last two columns of this table equals $(73 \times 191) \div (225 \times 47) = 1.32$. The odds that a female's response was Independent rather than Republican equal 1.32 times the odds for males. Of those subjects who responded Independent or Republican, females were more likely than males to respond Independent.

Finally, for the 2×2 subtable formed from the first and last columns, the sample odds ratio equals $(279 \times 191) \div (225 \times 165) = 1.44$. The odds that a female's response was Democrat rather than Republican equal 1.44 times the odds for males. Of those subjects who responded Democrat or Republican, females were more likely than males to respond Democrat.

The odds ratio value of 1.44 for the first and last columns equals (1.09)(1.32), the product of the other two odds ratios. For 2×3 tables, $df = 2$, meaning that only two bits of nonredundant information exist about the association. Two of the odds ratios determine the third. Generally, in a $r \times c$ table, a basic set of $(r-1)(c-1)$ odds ratios determine all possible odds ratios formed from the table.

8.5 Association Between Ordinal Variables*

We now turn our attention to the analysis of contingency tables that cross-classify two ordinal variables. The categories of ordinal variables are ordered from high to low. Statistical analyses for ordinal data take this ordering into account. Specialized measures of association exist for ordinal data, and special methods of inference apply. This section introduces a popular ordinal measure of association, called *gamma*, and Section 8.6 discusses methods of inference for ordinal data.

Example 8.8 Job Satisfaction and Income

Table 8.17 is a contingency table with ordinal variables. These data, from the 1991 General Social Survey, refer to the relation between job satisfaction and income for a sample of black Americans. The next two sections show ways of analyzing these data using methods for ordinal data.

Let's first get a better feel for the data by studying the conditional distributions on job satisfaction. Table 8.17 shows these in parentheses. For instance, the conditional distribution (27.3, 59.1,13.6) displays the percentages in the job satisfaction categories for subjects with income below \$5000. Only 13.6% of them are very satisfied with their jobs, whereas 33.3% of the subjects at the highest income level are very satisfied. Conversely, a lower percentage (12.5%) of the high-income group are dissatisfied with their jobs compared to those in the lowest income group (27.3%). It seems that subjects with higher incomes tended to have greater job satisfaction. □

TABLE 8.17 Job Satisfaction and Income for a Sample of Black Americans

	Job Satisfaction			
Income	Dissatisfied	Moderately Satisfied	Very Satisfied	Total
< 5000	6 (27.3)	13 (59.1)	3 (13.6)	22 (100.0)
\$5000–25,000	9 (15.5)	37 (63.8)	12 (20.7)	58 (100.0)
> 25,000	3 (12.5)	13 (54.2)	8 (33.3)	24 (100.0)
Total	18	63	23	104

Source: General Social Survey, 1991.

Ordinal data exhibit two primary types of association—*positive* and *negative*. Positive association between variables X and Y results when subjects at the high end of the scale on X tend also to be high on Y, and those who are low on X tend to be low on Y. For example, a positive association might exist between income and job satisfaction; those with low incomes tend to have lower job satisfaction, and those with high incomes tend to have higher job satisfaction. Negative association occurs when subjects classified high on X tend to be classified low on Y, and those classified low on X tend to be high on Y. For example, a negative association might exist between religiosity and favorableness of attitude toward legalization of abortion; the more religious, the less the support tends to be for legalization.

Concordance and Discordance

Most ordinal measures of association are based on the information about the association provided by all the pairs of observations.

Concordant Pair, Discordant Pair

A pair of observations is ***concordant*** if the subject who is higher on one variable also is higher on the other variable.
A pair of observations is ***discordant*** if the subject who is higher on one variable is lower on the other.

We illustrate with Table 8.17. “Dissatisfied” is the low end and “very satisfied” is the high end of the scale on Y = job satisfaction, and “< $5000” is the low end and “> $25,000” is the high end of the scale on X = income. By convention, we construct contingency tables for ordinal variables so that the low end of the row variable is the first row and the low end of the column variable is the first column. (There is no standard, however, and other books or software may use a different convention.)

Consider a pair of subjects, one of whom is classified (< $5000, dissatisfied) on the two variables, and the other of whom is classified ($5,000–25,000, moderately satisfied). The first subject is one of the 6 classified in the upper left-hand corner of Table 8.17, and the second subject is one of the 37 classified in the middle cell in the table. This pair of subjects is concordant, since the second subject is higher than the first subject both in job satisfaction (moderately satisfied versus dissatisfied) and in income ($5,000–25,000 versus < $5000); that is, the subject who is higher on one variable is also higher on the other. Now, one can pair each of the 6 subjects classified (< $5000, dissatisfied) with each of the 37 subjects classified ($5,000–25,000, moderately satisfied). So, there are $6 \times 37 = 222$ concordant pairs of subjects from these two cells.

By contrast, each of 13 subjects in the cell (< $5000, moderately satisfied) forms a discordant pair when matched with each of the 9 subjects in the cell ($5,000–25,000, dissatisfied). The 13 subjects have lower income than the other 9 subjects, yet they have higher job satisfaction. All $13 \times 9 = 117$ of these pairs of subjects are discordant.

Concordant pairs of observations provide evidence of positive association since, for such a pair, the subject who is higher on one variable also is higher on the other. On the other hand, the more prevalent the discordant pairs, the more evidence there is of a negative association.

Notation for Numbers of Concordant and Discordant Pairs

Let C denote the total number of concordant pairs of observations, and let D denote the total number of discordant pairs of observations.

We calculate C and D using the cell counts in the contingency table. A general rule for the number of concordant pairs C is this: Start at the corner of the table for the low level for each variable (the cell in row 1 and column 1 for the convention we use in constructing the tables). Multiply that cell count by the count in every cell that is higher on both variables (those cells below and to the right in Table 8.17). Similarly, for every other cell, multiply the cell count by the counts in cells that are higher on both variables. (For the cells in the row or in the column at the highest level of a variable, such as row ">$25,000" or column "very satisfied" in Table 8.17, no observations are higher on both variables.) The number of concordant pairs is the sum of these products.

In Table 8.17, the 6 subjects in the first cell are concordant when matched with the (37+12+13+8) subjects below and to the right who are higher on each variable. Similarly, the 13 subjects in the second cell in the first row are concordant when matched with the (12 + 8) subjects who are higher on each variable, and so forth, so that

$$C = 6(37 + 12 + 13 + 8) + 13(12 + 8) + 9(13 + 8) + 37(8) = 1165$$

Table 8.18 illustrates this calculation of the total number of concordant pairs.

TABLE 8.18 Illustration of Calculation of Number of Concordant Pairs, C

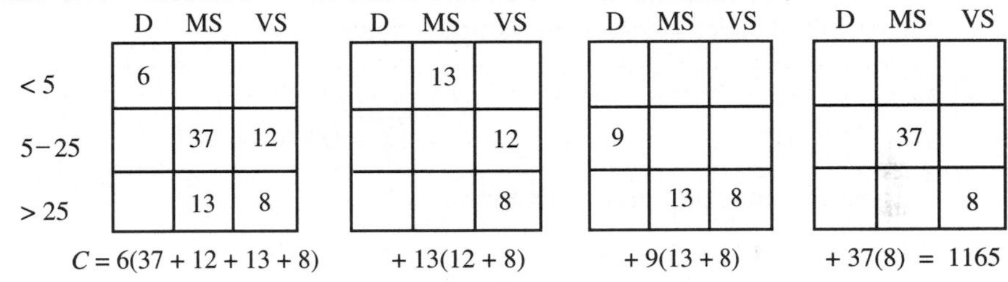

	D	MS	VS	D	MS	VS	D	MS	VS	D	MS	VS
< 5	6				13							
5–25		37	12			12	9				37	
> 25		13	8			8		13	8			8

$C = 6(37 + 12 + 13 + 8)$ $+ 13(12 + 8)$ $+ 9(13 + 8)$ $+ 37(8) = 1165$

To find the total number of discordant pairs D, we start at the corner of the table that is the high level of one variable and the low level of the other. For example, the 3 subjects in the cell (< $5000, very satisfied) form discordant pairs when combined with the (9 + 37 + 3 + 13) subjects below and to the left in the table who are higher on income but lower on job satisfaction. We multiply the count in each cell by the counts in all cells falling below and to the left, which are the ones higher on income but lower in job satisfaction. The total number of discordant pairs is

$$D = 3(9 + 37 + 3 + 13) + 13(9 + 3) + 12(3 + 13) + 37(3) = 645$$

Table 8.19 illustrates the calculation of the number of discordant pairs.

TABLE 8.19 Illustration of Calculation of Number of Discordant Pairs, D.

	D	MS	VS	D	MS	VS	D	MS	VS	D	MS	VS
< 5			3					13				
5–25	9	37				12	9				37	
> 25	3	13		3	13		3			3		

$D = 3(9 + 37 + 3 + 13)$ $+ 12(3 + 13)$ $+ 13(9 + 3)$ $+ 37(3) = 645$

In summary, Table 8.17 has $C = 1165$ and $D = 645$. More pairs show evidence of a positive association (i.e., concordant pairs) than show evidence of a negative association (discordant pairs).

Gamma

Most ordinal measures of association refer in some way to $C - D$. A positive difference for $C - D$ occurs when $C > D$ and indicates a positive association, since concordant pairs are more common than discordant pairs. A negative difference reflects a negative association.

Larger sample sizes have larger numbers of pairs with, typically, larger absolute differences in $C - D$. Therefore, we standardize this difference to make it easier to interpret. To do this, we divide $C - D$ by its maximum possible value, $C + D$, which is the total number of pairs that are either concordant or discordant. This gives the measure of association called ***gamma***. Its sample formula is

$$\hat{\gamma} = \frac{C - D}{C + D}$$

Here are some properties of gamma:

- The value of gamma falls between -1 and $+1$.
- The sign of gamma indicates whether the association is positive or negative.
- The magnitude of gamma indicates the strength of the association.

The larger the absolute value of gamma, the stronger the association. The value $+1$ represents the strongest positive association. This occurs when there are no discordant pairs ($D = 0$), so all the pairs reveal a positive association. Gamma equals -1 when $C = 0$, so all pairs reveal a negative association. Gamma equals 0 when $C = D$.

For Table 8.17, $C = 1165$ and $D = 645$, so that

$$\hat{\gamma} = \frac{1165 - 645}{1165 + 645} = .287$$

This sample of 104 subjects exhibits a positive association between income and job satisfaction. The higher the income, the higher the job satisfaction tends to be. However, the sample value is closer to 0 than to 1, so the association is not especially strong.

The calculation of C and D needed for gamma is rather messy. Most statistical software reports gamma, however, as discussed in the appendix.

Gamma as a Difference Between Proportions

Another interpretation for the magnitude of gamma follows from the expression

$$\hat{\gamma} = \frac{C - D}{C + D} = \frac{C}{C + D} - \frac{D}{C + D}$$

Now, $(C + D)$ is the total number of pairs that are concordant or discordant. The ratio $C/(C + D)$ is the proportion of those pairs that are concordant, $D/(C + D)$ is the proportion of the pairs that are discordant, and $\hat{\gamma}$ is the difference between the two proportions.

For example, suppose $\hat{\gamma} = .6$. Then, since .8 and .2 are the two proportions that sum to 1 and have a difference of $.8 - .2 = .6$, 80% of the pairs are concordant and 20% are discordant. Similarly, $\hat{\gamma} = -.333$ indicates that 1/3 of the pairs are concordant and 2/3 of the pairs are discordant, since $1/3 + 2/3 = 1$ and $1/3 - 2/3 = -.333$.

For Table 8.17, out of the $1165 + 645 = 1810$ pairs that are concordant or discordant, the proportion $1165/1810 = .644$ of the pairs are concordant, and the proportion $645/1810 = .356$ are discordant; $\hat{\gamma} = .287$ is the difference between these proportions.

Common Properties of Ordinal Measures

Gamma is one of many ordinal measures of association for contingency tables with ordered categories. Others are ***Kendall's tau-b*** and ***tau-c***, ***Spearman's rho-b*** and ***rho-c***, and ***Somers' d***. All these measures are similar in their basic purposes and characteristics. For lack of space, we do not define these other measures, but we will list some of their common properties.

- Ordinal measures of association take values between -1 and $+1$.
- If the variables are statistically independent, then the population values of ordinal measures equal 0.
- The stronger the relationship, the larger the absolute value of the measure. A table for which an ordinal measure takes a value of .6 or $-.6$ exhibits a stronger association than one for which the measure takes a value of .3 or $-.3$, for example. Values of 1.0 and -1.0 represent the strongest associations, but in opposite directions.
- With the exception of Somers' d, the ordinal measures of association named above do not distinguish between response and explanatory variables. They take the same value when variable Y is the response variable as when it is the explanatory variable.

Effect of Category Choice on Gamma

Gamma is the most popular ordinal measure of association, but it has one property that is disadvantageous. Suppose two variables that are inherently continuous are grouped into ordered categories. For distance, annual income was measured in Table 8.17 with categories (<\$5,000, \$5,000–25,000, over \$25,000), but it is inherently continuous. In such cases, the absolute value of gamma for the contingency table tends to be inflated above the value it would take for the continuous variables. Gamma equaled .287 for Table 8.17, but if we had measured income and job satisfaction more finely, the value might have been somewhat smaller, say .20. Gamma tends to be larger when a variable is measured using only two or three categories rather than several categories. For instance, if gamma = .20 for a 10×10 table, perhaps gamma = .30 if the data were more crudely measured in a 2×3 table. Thus, different researchers could reach somewhat different conclusions if they studied the same variables but used different categorizations. Ideally the value of the measure should not be so highly dependent on the choice of categories.

Kendall's tau-b is an ordinal measure that tends to be more stable than gamma under different categorizations. This measure is more complex to define than gamma, and we leave it to Problem 8.61.Most software for contingency tables supplies both gamma and Kendall's tau-*b*. You can simply take the value of Kendall's tau-*b* from a printout and interpret it in much the same way as gamma. It falls between −1 and +1, and it equals the approximate difference between the proportions of concordant and discordant pairs if the variables were measured more precisely as continuous variables.

So far, we have discussed the use of gamma only for description. The next section presents statistical inference, namely, confidence intervals and tests for ordinal data.

8.6 Inference for Ordinal Associations*

Section 8.2 introduced the chi-squared test of whether two categorical variables are independent. That test treats the variables as nominal, and other tests are usually more appropriate when the variables are ordinal. This section presents such a test. It also shows how to construct confidence intervals for ordinal measures of association such as gamma.

Common Test for Ordinal Measures

First, we present a test that utilizes the category orderings. As in the chi-squared test, the null hypothesis is that the variables are statistically independent. We express the test in terms of gamma, but a similar approach works with other ordinal measures.

Let γ denote the population value of gamma. The alternative hypothesis in the test usually takes the two-sided form $H_a: \gamma \neq 0$. It can also take a one-sided form, $H_a: \gamma > 0$ or $H_a: \gamma < 0$, when the sign of the association is predicted. For nominal variables, the notions of positive or negative association do not exist, so there is only one possible alternative hypothesis, H_a: Dependence.

For a sample value of gamma, $\hat{\gamma}$, let $\hat{\sigma}_{\hat{\gamma}}$ denote the standard error. This describes the extent of variation in sample gamma among samples of the given size. The test statistic has the usual z statistic form of dividing the difference between the sample estimate and the null hypothesis value by the standard error of the estimator. Specifically, it is

$$z = \frac{\hat{\gamma} - 0}{\hat{\sigma}_{\hat{\gamma}}}$$

The test statistic z takes the sample estimate $\hat{\gamma}$ of gamma, which summarizes the sample tendency toward positive or negative association, subtracts the value of 0 expected for gamma when the null hypothesis of independence is true, and divides by the standard error. The test statistic has approximately the standard normal distribution, when H_0 is true. The test is best applied to a large random sample. As a rough guideline, each of C and D should exceed about 50.

Unfortunately, the formula for the standard error $\hat{\sigma}_{\hat{\gamma}}$ is very complicated. However, standard errors are reported by most software having procedures for contingency tables, so it is straightforward to conduct the test.

Example 8.9 Tests for Job Satisfaction Data

To illustrate the ordinal test, we again use the data from Table 8.17 relating income and job satisfaction. Is there an association between job satisfaction and income? The chi-squared test of independence has $\chi^2 = 4.1$ with $df = 4$, for which the P-value equals .39. This test does not show any evidence of an association. The chi-squared test treats the variables as nominal, however, and ordinal-level methods may provide a different answer.

Table 8.20 illustrates a computer printout for the analysis of Table 8.17. From the printout, the sample value of gamma is $\hat{\gamma} = .287$. Its standard error of .151 is given under the heading ASE (the label is ASE1 in SPSS), where ASE denotes "asymptotic (large-sample) standard error." Thus, $\hat{\sigma}_{\hat{\gamma}} = .151$, and the test statistic equals

$$z = \frac{\hat{\gamma} - 0}{\hat{\sigma}_{\hat{\gamma}}} = \frac{.287 - 0}{.151} = 1.90$$

(Some software reports a standard error for tests based on a simpler formula that only applies under the null hypothesis of independence; an example is ASE0 in SPSS. It is also valid to form a z test statistic using that standard error, but it is inappropriate to use it in confidence intervals.)

TABLE 8.20 Part of a Computer Printout for Analyzing Table 8.17

```
Statistic                 DF      Value       Prob
-----------------------------------------------------
Chi-Square                 4      4.094      0.393

Statistic                         Value       ASE
-----------------------------------------------------
Gamma                             0.287      0.151
Kendall's Tau-b                   0.168      0.090
```

From the standard normal table, the P-value for the two-sided alternative equals .057. There is some, though not convincing, evidence of an association. Since the sample value of gamma was positive, it appears that a positive association may exist between income and job satisfaction, though independence is also plausible. The test for the one-sided alternative of a positive association has $P = .029$. □

Ordinal Tests versus Pearson Chi-Squared

The z test result for these data providing some evidence of an association may seem surprising. The chi-squared statistic of $\chi^2 = 4.1$ with $df = 4$ provides no evidence ($P = .39$). (Some of the cell counts are small, and it may be safer to do a small-sample exact test. Results are then similar to chi-squared, with $P = .43$.)

A test of independence based on an ordinal measure is usually preferred to the chi-squared test when both variables are ordinal, since χ^2 ignores the ordering of the categories. If a positive or negative trend exists for the relationship, ordinal measures are usually more powerful for detecting it than the χ^2 statistic, which is the same no matter how the levels are ordered. Unfortunately, the situation is not clear cut. It is possible for the chi-squared test to be more powerful even if the data are ordinal.

To explain this, we first note that the null hypothesis of independence is not equivalent to a value of 0 for population gamma. Although independence implies $\gamma = 0$, the converse is not true. Namely, γ may equal 0 even though the variables are not statistically independent. For example, Table 8.21 shows a relationship between two variables that does not have a single trend. Over the first two columns there is a positive relationship, since Y increases when X increases. Over the last two columns there is a negative relationship, as Y decreases when X increases. For the entire table, $C = 25(25+25) = 1250 = D$, so $\gamma = 0$. The proportion of concordant pairs equals the proportion of discordant pairs. However, there is not independence, because the conditional distribution on Y for the low level of X is completely different from the conditional distribution on Y for the high level of X.

TABLE 8.21 A Relationship for Which Ordinal Measures of Association Equal 0; the Variables Are Dependent Even Though Gamma Equals 0

		Level of Y			
		Very low	Low	High	Very high
Level of X	Low	25	0	0	25
	High	0	25	25	0

Thus, an ordinal measure of association may equal 0 when the variables are statistically dependent, but the dependence does not have an overall positive or overall negative trend. The chi-squared test is usually better than the ordinal test when the relationship does not have a single trend. In practice, most relationships with ordinal variables have primarily one trend, if any, so the ordinal test is usually better than the chi-squared test.

Confidence Intervals for Measures of Association

Significance tests help determine whether an association exists between two variables. As in other analyses, confidence intervals are more informative. A confidence interval for a measure of association helps us gauge the strength of that association in the population.

Confidence intervals for ordinal measures of association have the usual form of estimate plus and minus a z-score times the estimated standard error. For instance, confidence intervals for the population value γ equal

$$\hat{\gamma} \pm z\hat{\sigma}_{\hat{\gamma}}$$

For the data on income and job satisfaction, sample gamma of .287 has a standard error of .151. A 95% confidence interval for γ is

$$\hat{\gamma} \pm 1.96\hat{\sigma}_{\hat{\gamma}}, \quad \text{or } .287 \pm 1.96(.151), \quad \text{or } .287 \pm .296$$

which equals $(-.007, .583)$. We can be 95% confident that gamma is no less than $-.007$ and no greater than .583. It is plausible that no association exists between income and job satisfaction, but it is also plausible that a moderate positive association exists. We need a larger sample size to estimate this more precisely.

Our conclusion with this confidence interval is consistent with the test conducted earlier in this section. Namely, we concluded with the test that independence was plausible. The confidence interval gives us considerably more information, showing that a moderate positive association is also plausible.

Inference for Other Ordinal Measures

The methods presented here for gamma apply also to other ordinal measures of association. For instance, for a confidence interval for Kendall's tau-b, one takes the sample value and adds and subtracts a z-score times the standard error, which is available using software. From the printout in Table 8.20, a 95% confidence interval for population Kendall's tau-b is $.168 \pm 1.96(.090)$, or $(-.008, .344)$.

Test results are usually similar for any ordinal measure based on the difference between the numbers of concordant pairs and discordant pairs, such as gamma, Kendall's tau-b, or Somers' d. In all these cases, this procedure tests the null hypothesis of independence against the alternative that the measure is not equal to zero, and all these measures are nonzero in the same situations.

Whenever possible, it is better to choose the categories for ordinal variables finely rather than crudely. For instance, it is better to use four or five categories than only two. Variables with several categories tend to have smaller standard errors, for a given sample size. Thus, the finer the categorizations, the shorter the confidence interval for a population measure of association tends to be. In addition, the various ordinal measures tend to take more similar values when the measurement is finer. Finally, finer measurement makes it more valid to treat the data as quantitative and use the more powerful methods presented in the following chapter for quantitative variables.

Mixed Ordinal–Nominal Contingency Tables

One can also use ordinal measures for description and inference with cross-classifications of ordinal and nominal variables, when the nominal variable has only two categories. In that case, the sign of the measure indicates which level of the nominal variable is associated with higher responses on the ordinal variable. For instance, suppose gamma $= -.18$ for the association in a 2×3 table relating gender (rows female, male) to job satisfaction (columns dissatisfied, somewhat satisfied, very satisfied). Since the sign is negative, the "higher" level of gender (i.e., male) tends to occur with lower satisfaction. The association is weak, however.

An alternative approach treats the ordinal variable as quantitative by assigning scores to its levels. For instance, suppose the two nominal categories are groups to be compared on the ordinal response, such as females and males compared on job satisfaction. One can then use the methods of Section 7.1 to compare mean responses for the two groups.

In a mixed ordinal–nominal contingency table in which the nominal variable has more than two categories, it is inappropriate to use an ordinal measure such as gamma. One can treat both variables as nominal, but that usually entails a loss of information and power. It is usually better to treat the ordinal variable as quantitative, assigning scores. The methods of Chapter 12, which generalize comparisons of two means to several groups, are then appropriate.

Sometimes the choice of scores for an ordinal variable is straightforward. For categories (too little, about right, too much) for opinion about government spending on the environment, any set of equally spaced scores is sensible, such as (1, 2, 3) or (0, 5, 10). When the choice is unclear, such as with categories (dissatisfied, somewhat satisfied, very satisfied), it is a good idea to perform a sensitivity study. Choose two or three reasonable sets of potential scores, such as (0, 5, 10), (0, 6, 10), (0, 7, 10), and check whether the ultimate conclusions are similar for each. They usually are similar unless the data are quite unbalanced, with most observations falling in one response category. Section 15.5 presents a more modern approach that does not require assigning scores to ordinal response variables.

8.7 Measuring Association: Proportional Reduction in Error*

We end this chapter by introducing a concept that unifies measures of association for qualitative variables with measures for quantitative variables. For measures formulated with this concept, the strength of association is described by how well one can predict values on one variable based on their values on the other variable. For example, race and attitude toward capital punishment are highly associated if race is a good predictor of attitude; that is, if knowing their race, we can make much better predictions about people's attitudes toward capital punishment than if we did not know it.

The predictions refer to the response variable, using the explanatory variable as the predictor. The measures of association compare two summaries of prediction error. One summary refers to errors in using the explanatory variable to predict the response variable. This is compared to an analogous summary of the errors in predicting the response variable *without* using the explanatory variable. The overall measure is defined as the proportion fewer errors using information on the explanatory variable instead of ignoring it in making the predictions. The magnitude of the measure of association is interpreted as the ***proportional reduction in error***.

Elements of PRE Measures

Measures of association with proportional reduction in error (PRE) structure have four elements:

1. ***Rule 1: A rule for predicting a subject's value on the response variable, ignoring the value of the explanatory variable***
2. ***Rule 2: A rule for predicting a subject's value on the response variable, using the value of the explanatory variable for the subject***
3. ***A definition of prediction error:*** Each measure has a way of summarizing the prediction error using each rule. We denote the total prediction error using rule 1 by E_1 and using rule 2 by E_2.
4. ***The PRE definition of the measure:*** The difference between the prediction errors for the two rules is $E_1 - E_2$; that is, $E_1 - E_2$ is the reduction in error using rule 2 instead of rule 1. We standardize this reduction in error by dividing by E_1, so the result falls between 0 and 1. This yields the PRE measure of association,

$$\text{PRE} = \frac{E_1 - E_2}{E_1}$$

For example, for a particular sample, suppose we make 100 errors in predicting attitude toward capital punishment when the prediction rule does not use information on the explanatory variable, race. Suppose we make 70 errors when the rule uses information on the explanatory variable to predict attitude. Then $E_1 = 100$, $E_2 = 70$, the reduction in error is $E_1 - E_2 = 30$, and the proportional reduction in error equals

$$\text{PRE} = \frac{E_1 - E_2}{E_1} = \frac{100 - 70}{100} = .30$$

We make 30% fewer prediction errors in using the explanatory variable in making the predictions, so the degree of association is summarized by that number, .30.

The PRE value summarizes how much better the predictions are using rule 2 than using rule 1. The larger the PRE, the better the response variable can be predicted using the explanatory variable, and the stronger the association between the variables. The weakest possible association, $\text{PRE} = 0$, occurs when no reduction in error occurs (i.e., $E_1 = E_2$) by having the additional information about the subjects' values on the explanatory variable. In that case, the response variable is not associated with the explanatory variable. The strongest possible association, $\text{PRE} = 1$, occurs when $E_2 = 0$. In that case, one can predict a subject's value on the response variable perfectly from the value on the explanatory variable.

Gamma as a PRE Measure

Gamma is usually interpreted as the difference between the proportions of concordant and discordant pairs, but it also has a PRE interpretation. Since gamma is an ordinal measure, the rules refer to predicting orderings of values on Y based on orderings on X; that is, they predict pairs of observations as concordant or discordant. Rule 1 predicts which of a pair of observations is higher on variable Y without knowing which is higher on variable X. Rule 2 predicts which is higher on Y based on knowing which is higher on X.

Rule 1 randomly predicts for each pair whether it is concordant or discordant. With probability .5 concordance is predicted and with probability .5 discordance is predicted. With this rule, we expect an incorrect prediction half the time. For the $C + D$ predictions, the expected number of errors is $E_1 = (C + D)/2$.

Rule 2 predicts which subject in a pair is higher on Y based on knowing which subject is higher on X. For example, suppose that $C > D$. Then, if the first subject in a pair is higher on X than the second subject, the rule predicts that that subject also is higher on Y. In other words, it predicts that the pair is concordant. Rule 2 always predicts concordance if $C > D$ and always predicts discordance if $C < D$.

Now, if every pair is predicted to be concordant, a prediction error occurs whenever a pair is discordant. Hence, the number of errors using rule 2 is $E_2 = D$, the total number of discordant pairs. The proportional reduction in error is

$$\frac{E_1 - E_2}{E_1} = \frac{(C + D)/2 - D}{(C + D)/2} = \frac{C - D}{C + D} = \hat{\gamma}$$

For instance, $\hat{\gamma} = .4$ means that 40% fewer prediction errors occur in predicting whether each pair is concordant or discordant, when we know which type of pair occurs more frequently.

When $\hat{\gamma} < 0$, the absolute value of gamma has a PRE interpretation. In that case $C < D$, so we predict that every pair is discordant. For example, if $\hat{\gamma} = -.2$, then $|\hat{\gamma}| = .2$, and a 20% reduction in error occurs in predicting pair orderings by always predicting discordance.

PRE Measures for Nominal Variables

Two PRE measures were defined in 1954 for nominal variables by the prominent statistician–social scientists who introduced gamma, Leo Goodman and William Kruskal. The measures are referred to as ***tau*** and ***lambda*** and are introduced in Problems 8.58 and 8.60. These PRE measures are preferred to many other nominal measures of association that are not as readily interpretable. Most computer output for analyzing contingency tables prints a large number of measures. Many of these, such as the contingency coefficient and Cramer's V, are very difficult to interpret and are less informative than the PRE measures.

Unfortunately, when r or c exceed 2 in a contingency table for nominal variables, it is usually an oversimplification to describe the table with a single summary measure of association. Even for measures with PRE structure, the value the measure attains can be very highly dependent on the relative numbers of observations falling in different rows or columns. For instance, if two tables have the same conditional distributions on Y but differ dramatically in the relative numbers of observations in the various levels of X, most measures of association will differ for the two tables, perhaps considerably. An exception to this is the odds ratio. See Problem 8.59.

For ordinal data, it is often helpful to describe an overall positive or negative trend by a measure such as gamma. For nominal data, however, positive and negative trends do not make sense, and too many possible patterns of association exist to describe an $r \times c$ table well by a single number. Because of this and because we feel other topics are more important, we have not defined nominal measures such as Goodman and Kruskal's tau and lambda in the body of this chapter.

Historically, summary measures of association were more frequently used 20 or 30 years ago. It is now common instead to look at more detailed analyses of the data using recently developed modeling methods, rather than to try to summarize the data by a single number. One can get a better feel for the association by making percentage comparisons of conditional distributions, viewing the pattern of adjusted residuals in the cells of the table, constructing odds ratios in 2×2 subtables, or building models for the data such as those presented in Chapter 15. These methods become even more highly preferred to summary measures of association when one attempts multivariate, rather than bivariate, analyses of categorical data.

8.8 Chapter Summary

This chapter introduced ways of analyzing association between two categorical variables:

- By cross-classifying observations in a ***contingency table***. Each cell of the table counts the number of subjects having a particular combination of responses on the two variables.
- By *describing the counts* in contingency tables using percentage distributions, called ***conditional distributions***, across the categories of the response variable. If the population conditional distributions are identical, the two variables are ***statistically independent***—the probability of any particular response is the same for each level of the explanatory variable.
- By using ***chi-squared*** to ***test for independence*** between variables. The χ^2 test statistic compares each observed cell frequency f_o to the expected frequency f_e satisfying the null hypothesis, using

$$\chi^2 = \sum \frac{(f_o - f_e)^2}{f_e}$$

The test statistic has a large-sample chi-squared distribution. The ***degrees of freedom*** depend on the number of rows r and the number of columns c, through

$df = (r-1)(c-1)$. The P-value is the right-hand tail probability above the observed value of χ^2.

- By *describing the pattern of association* using ***adjusted residuals*** for the cells of the table. An adjusted residual reports the number of standard errors that the observed count falls from the expected count; a value larger than about 2 in absolute value indicates that that cell provides evidence of association.
- By *describing the strength of association*. For 2×2 tables, the ***difference of proportions*** is useful. The ***odds ratio***, the ratio of odds from the two rows, is another useful measure; each odds measures the proportion of successes divided by the proportion of failures. When there is independence, the difference of proportions equals 0 and the odds ratio equals 1. The stronger the association, the farther the measures fall from these baseline values.

This chapter also presented methods for analyzing association between two ordinal variables. ***Ordinal measures of association*** such as ***gamma*** use the numbers of ***concordant*** and ***discordant pairs***.

- A pair is ***concordant*** if the subject who is higher on X also is higher on Y, and ***discordant*** if the subject who is higher on X is lower on Y.
- Of the pairs that are concordant or discordant, ***gamma*** equals the difference between the proportions of the two types.
- Gamma falls between -1 and $+1$, with larger absolute values indicating stronger association.
- When the variables are independent, gamma equals 0.

The chi-squared test treats the data as nominal. When the variables are ordinal and one expects a positive or negative association trend, methods that treat the data as ordinal are normally more powerful. For instance, one can use a z test based on gamma to perform an ordinal test of independence, and one can construct a confidence interval for gamma to make inferences about the true strength of association.

Many measures of association for qualitative and quantitative data have ***proportional reduction in error*** structure. The association is strong if one can make much better predictions about the response variable based on knowing a subject's value for the explanatory variable.

The next chapter introduces similar methods for describing and making inferences about the association between two quantitative variables.

PROBLEMS

Practicing the Basics

1. A Gallup Poll in 1995 suggested that in the United States, about 15% of males and 15% of females believe that abortion should be illegal in all circumstances (*The Gallup Poll Monthly*, No. 354. Princeton, NJ: The Gallup Poll, March 1995, p. 30).

a) For males and for females, report the conditional distribution on whether abortion should be illegal in all circumstances, using categories (yes, no).
b) Based on these results, does statistical independence seem plausible between gender and opinion about whether abortion should be illegal in all circumstances? Explain.

2. Whether a woman becomes pregnant in the next year is a categorical variable with categories (yes, no), and whether she and her partner use contraceptives is another categorical variable with categories (yes, no). Would you expect these variables to be statistically independent, or associated? Explain.

3. For a sample of 100,000 subjects, Table 8.22 cross-classifies whether a subject is HIV+ by whether a diagnostic test for HIV+ status is positive (i.e., indicates that the subject is HIV+).
a) Construct the conditional distributions for the test result, given the true disease status.
b) For those who truly are HIV+, what percentage are diagnosed as HIV+ by the diagnostic test? For those who are not HIV+, what percentage get a negative test result? Does the diagnostic test appear to be a good one?
c) Construct the conditional distribution on disease status, for those who have a positive diagnostic test result. Of those subjects, what percentage truly are HIV+? Comment. (For discussion of the possibility of a high false positive rate even for a seemingly good diagnostic test, see Chatterjee et al., 1995, pp. 37–40).

TABLE 8.22

		Diagnostic Test	
		Positive	Negative
HIV+	Yes	475	25
Status	No	4975	94,525

4. In an article about crime in the United States, *Newsweek* magazine (January 10, 1994) quoted FBI statistics stating that of all blacks slain in 1992, 94% were slain by blacks, and of all whites slain in 1992, 83% were slain by whites. Let Y denote race of victim and X denote race of murderer.
a) Which conditional distributions do these statistics refer to, those of Y at given levels of X, or those of X at given levels of Y? Set up a table with race of murderer as rows and race of victim as columns, showing these conditional distributions.
b) Are X and Y independent or dependent? Explain.

5. A young child wonders what causes women to have babies. For each woman who lives on her block, she observes whether her hair is gray and whether she has young children, with the results shown in Table 8.23.
a) Construct the 2×2 contingency table that cross-classifies "gray hair" (yes, no) with "young children" (yes, no) for these nine women.
b) Treating "young children" as the response variable, obtain the conditional distributions for those women who have gray hair and for those who do not. Does there seem to be an association?
c) Noticing this association, the child concludes that not having gray hair is what causes women to have children. Use this example to explain why association does not necessarily imply causation. (Chapter 10 discusses the connection between association and causation.)

TABLE 8.23

Woman	Gray Hair	Young Children
Andrea	No	Yes
Mary	Yes	No
Linda	No	Yes
Jane	No	Yes
Maureen	Yes	No
Judy	Yes	No
Margo	No	Yes
Carol	Yes	No
Donna	No	Yes

6. How large a χ^2-value provides a P-value of .05 for testing independence for the following table dimensions?
   a) 2×2 **b)** 3×3 **c)** 2×5 **d)** 5×5 **e)** 3×9
7. Show that the contingency table in Table 8.24 has four degrees of freedom, by filling in the missing cell counts.

TABLE 8.24

10	20		60
30	40		100
50	80	70	

TABLE 8.25

		Cigarette Use	
		Yes	No
Alcohol Use	Yes	1449	500
	No	46	281

Source: Study conducted by Wright State University School of Medicine and United Health Services, Dayton, Ohio. Thanks to Professor Harry Khamis for providing these data.

8. Table 8.25 refers to a survey conducted in 1992 of senior high school students in Dayton, Ohio.
   a) Construct conditional distributions that treat cigarette smoking as the response variable. Interpret.
   b) Construct conditional distributions that treat alcohol use as the response variable. Interpret.
   c) Test whether cigarette use and alcohol use are statistically independent. Report the P-value, and interpret.
9. Table 8.26 shows an SPSS printout for some analyses for a data set taken from the 1991 General Social Survey, with variables race and party identification.
   a) Form the conditional distribution on party identification, for each race. Interpret.
   b) Report the expected frequency for the first cell, and show how SPSS obtained it.
   c) Test the hypothesis of independence between party identification and race. Report the test statistic, P-value, and interpret.
   d) Use the adjusted residuals to describe the evidence about association.

e) Summarize the association between race and whether one is a Democrat or a Republican, by computing the odds ratio or by comparing percentages. Interpret.

TABLE 8.26

```
                Count     PARTY
              Exp Val
              Adj Res                                    Row
                         democr   indep     repub      Total
RACE        black          103      15        11        129
                          58.4    15.8      54.8
                           8.5     -.2      -8.4

            white          341     105       405        851
                         385.6   104.2     361.2
                          -8.5      .2       8.4

          Column Total     444     120       416        980

      Chi-Square          Value       DF     Significance
-----------------       ---------    ----    ------------
Pearson                  79.431       2          .00000
Likelihood Ratio         90.331       2          .00000
```

TABLE 8.27

	Teenage Birth Control			
Premarital Sex	Strongly Disagree	Disagree	Agree	Strongly Agree
Always wrong	81 (7.6)	68 (3.1)	60 (−4.1)	38 (−4.8)
Almost always wrong	24 (2.3)	26 (1.8)	29 (−0.8)	14 (−2.8)
Wrong only sometimes	18 (−2.7)	41 (1.0)	74 (2.2)	42 (−1.0)
Not wrong at all	36 (−6.1)	57 (−4.6)	161 (2.4)	157 (6.8)

Source: 1991 General Social Survey.

10. Table 8.27 is from the 1991 General Social Survey. Subjects were asked their opinion about a man and woman having sex relations before marriage. They were also asked whether methods of birth control should be made available to teenagers between the ages of 14 and 16.
a) Software reports that $\chi^2 = 128.7$, based on $df = 9$. Interpret.
b) Table 8.27 also shows, in parentheses, the adjusted residuals. Use these to describe how the data deviate from independence in the four corner cells. Interpret.

11. Refer to Problem 8.8.
a) Describe the nature of the association using adjusted residuals.
b) Describe the strength of association using the difference between users and nonusers of alcohol in the proportions who have used cigarettes. Interpret.

c) Describe the strength of association using the difference between users and nonusers of cigarettes in the proportions who have used alcohol. Interpret.
d) Describe the strength of association using the odds ratio. Interpret.

12. Table 8.28 refers to 68,694 passengers in autos and light trucks involved in accidents in the state of Maine in 1991. The table classifies passengers by whether they were wearing a seat belt and by whether they were injured or killed. Test the hypothesis of independence. Interpret the P-value.

13. Refer to the previous problem. Show how to follow up the test descriptively with
a) An analysis using adjusted residuals.
b) The difference between two proportions.
c) The odds ratio.

14. Table 8.29 refers to a national study of 15- and 16-year-old adolescents. The groups are four combinations of gender and race. The event of interest is ever having sexual intercourse. Is there an association between group and the response? Interpret the results in the computer printout supplied with the table.

TABLE 8.28

		Injury	
		Yes	No
Seat	Yes	2409	35,383
Belt	No	3865	27,037

Source: Thanks to Dr. Cristanna Cook, Medical Care Development, Augusta, Maine, for supplying these data.

TABLE 8.29

	Intercourse	
Group	Yes	No
Black females	22	36
White females	26	149
Black males	29	23
White males	43	134

```
STATISTICS FOR TABLE OF GROUP BY INTERCOURSE

Statistic                        DF     Value     Prob
------------------------------------------------------
Chi-Square                        3    39.830    0.001
Likelihood Ratio Chi-Square       3    37.517    0.001
```

Source: S. P. Morgan and J. D. Teachman, *Journal of Marriage and Family*, Vol. 50 (1988), 929–936.

15. Refer to the previous exercise.
a) Describe the nature of the association, using conditional distributions.
b) Describe the nature of the association for females, using adjusted residuals for the *Yes* category in the first two rows.
c) Describe the strength of association, using the difference between the proportions of blacks and whites who have experienced sexual intercourse, separately for females and for males.
d) Describe the strength of association, using the difference between the proportions of males and females who have experienced sexual intercourse, separately for blacks and for whites.
e) Describe the strength of association using the odds ratio between gender and response, separately for each race.
f) Describe the strength of association using the odds ratio between race and response, separately for each gender.
g) Summarize what your analyses have revealed about these data.

16. Refer to the previous two exercises.
a) Analyze the association between race and response, given gender, by conducting two chi-squared tests, one for females using the first two rows of the table and one for males using the last two rows of the table. Interpret.
b) Analyze the overall association between gender and response by combining the first two rows, combining the last two rows, and conducting a chi-squared test for the resulting 2×2 table. Interpret. (Note: The construction in (a) and (b) shows a way of analyzing the table with three separate chi-squared statistics, each having $df = 1$, rather than a single statistic having $df = 3$. The sum of the three separate statistics having $df = 1$ is approximately equal to the overall statistic having $df = 3$. For discussion of ways of "partitioning" chi-squared, see Agresti, 1996, Sec. 2.4.6.)

17. Table 8.30 is from the 1991 General Social Survey. White subjects were asked: "If your party nominated a (Negro/Black) for President, would you vote for him if he were qualified for the job?" and "During the last few years, has anyone in your family brought a friend who was a (Negro/Black) home for dinner?"
a) Find the statistical significance of the association using the chi-squared test. Interpret the P-value.
b) Some expected frequencies are small, so the chi-squared test may not be valid. If you have software available, conduct a small-sample exact test, and interpret the P-value.
c) Follow up the test, describing the nature of the association.

18. Refer to the previous exercise. Using the yes and no categories of each response, estimate the odds ratio. Interpret.

19. A newspaper article preceding the 1994 World Cup semifinal match between Italy and Bulgaria stated that "Italy is favored 10-11 to beat Bulgaria, which is rated at 10-3 to reach the final." Suppose this means that the odds that Italy wins are $11/10 = 1.1$ and the odds that Bulgaria wins are $3/10 = .3$. Find the probability that each team wins, and comment.

20. According to data from the 1993 Substance Abuse and Mental Health Administration national household survey on drug abuse, for Americans aged 26–34, 59% had used marijuana at least once in their lifetime, and 26% had used cocaine at least once.
a) Find the odds of having used marijuana. Interpret.
b) Find the odds of having used cocaine. Interpret.

c) Find the odds ratio comparing marijuana use to cocaine use (i.e., this odds ratio refers to the 2×2 table having rows (marijuana, cocaine) and columns (used at least once, never used). Interpret.

21. When asked by the National Opinion Research Center "Is there any area right around here—that is, within a mile—where you would be afraid to walk alone at night?", in 1973 60% of females answered yes and 20% of males answered yes, whereas in 1994 60% of females answered yes and 30% of males answered yes.
a) Calculate the odds that a female responded yes in (i) 1973, (ii) 1994. Interpret.
b) Calculate the odds that a male responded yes (i) in 1973, (ii) in 1994.
c) Calculate the odds ratio between gender and response, in 1973 and in 1994. In which year is the association stronger?

22. For murders in the the United States in 1993 having a single victim and single offender, Table 8.31 cross classifies the sex of the victim by the sex of the offender. Calculate and interpret the odds ratio.

TABLE 8.30

Black For President	Home For Dinner		
	Yes	No	Don't Know
Yes	113	239	1
No	6	49	0
Don't Know	0	13	2

TABLE 8.31

Sex of Offender	Sex of Victim	
	Female	Male
Female	290	1004
Male	2765	7487

Source: U.S. Department of Justice, *Crime in the United States, 1993* (Washington, D.C.: U.S. Government Printing Office).

23. According to the U.S. Department of Justice, in 1994 the incarceration rate in the nation's prisons was 646 per 100,000 male residents, 45 per 100,000 female residents, 1471 per 100,000 black residents, and 207 per 100,000 white residents (*Bureau of Justice Statistics Bulletin: Prisoners in 1994*).
a) Find the odds ratio between gender and whether incarcerated. Interpret.
b) Find the odds ratio between race and whether incarcerated. Interpret.
c) According to the odds ratio, which has the stronger association with whether incarcerated, gender or race?

24. For college freshmen in 1994, the odds ratio between gender (male, female) and opinion about whether homosexual relations should be legally prohibited (yes, no) equaled 2.6 (Higher Education Research Institute, University of California at Los Angeles).
a) Explain what is wrong with the interpretation, "The probability of a yes response for males is 2.6 times the probability of a yes response for females." Give the correct interpretation.
b) The odds of a yes response equaled .82 for males. What is the probability of a yes response for males?
c) Based on the odds of .82 for males and the odds ratio of 2.6, find the probability of a yes response for females.

25. Consider the 3×3 table having entries, by row, of (4, 2, 1 / 2, 2, 2 / 1, 2, 4).
a) Is the chi-squared test of independence appropriate for these data? Explain.
b) Using software, conduct a valid test of independence. Interpret.

26. Refer to the previous exercise. Suppose the variables are ordinal, with levels (high, medium, low) for each classification.
a) Compute the numbers of concordant and discordant pairs.
b) Compute gamma, and interpret.
c) Show how to express gamma as a difference between two proportions.
d) Using software, conduct a test of independence using gamma, and compare the result to the previous exercise. Why is the P-value so much smaller here?

27. Table 8.32 refers to a study that assessed factors associated with women's attitudes toward mammography. The columns refer to their response to the question, "How likely is it that a mammogram could find a new case of breast cancer?" Use gamma to summarize the strength of association. Interpret the sign and magnitude.

TABLE 8.32

Mammography	Detection of Breast Cancer		
Experience	Not Likely	Somewhat Likely	Very Likely
Never	13	77	144
Over one year ago	4	16	54
Within the past year	1	12	91

Source: D. Hosmer and S. Lemeshow, *Applied Logistic Regression* (New York: Wiley 1989), p. 224.

28. A study on educational aspirations of high school students (S. Crysdale, *International Journal of Comparative Sociology*, Vol. 16, 1975, pp. 19–36) measured aspirations using the scale (some high school, high school graduate, some college, college graduate). For students whose family income was low, the counts in these categories were (9, 44, 13, 10); when family income was middle, the counts were (11, 52, 23, 22); when family income was high, the counts were (9, 41, 12, 27). Software provides the results shown in Table 8.33.

TABLE 8.33

```
Statistic                 DF    Value       Prob
------------------------------------------------
Chi-Square                 6    8.871       0.181

Statistic                       Value       ASE
------------------------------------------------
Gamma                           0.163       0.080
Kendall's Tau-b                 0.108       0.053
```

a) Report the value of an ordinal measure of association, and use it to summarize the association.
b) Test independence of educational aspirations and family income using the chi-squared test. Interpret, and explain the deficiency of this test for these data.
c) Find a 90% confidence interval for an ordinal measure. Interpret.
d) Conduct an alternative test of independence that takes category ordering into account. Interpret results, and compare to results of the chi-squared test.

29. Table 8.34 comes from a study of economic conditions and political attitudes of Cuban workers toward the Cuban revolution. A computer analysis reports:

```
Statistic                    Value          ASE
-----------------------------------------------
Gamma                       -0.481        0.117
Kendall's Tau-b             -0.238        0.059
```

a) Interpret one of these sample statistics.
b) Obtain a 95% confidence interval for the population value. Interpret.
c) Conduct an ordinal test of independence. Interpret.

TABLE 8.34

	Months Worked In Year Before Revolution		
Attitude	6 or Less	7–9	10 or More
Hostile	3	3	26
Indecisive	6	2	14
Favorable	54	14	65

Source: M. Zeitlin, *Revolutionary Politics and the Cuban Working Class* (Princeton, NJ: Princeton University Press, 1967), p. 55. Reprinted by permission of Princeton University Press.

30. Refer to Problem 8.10. The analysis there does not take into account the ordinality of the variables. Using software, summarize the strength of association by finding and interpreting gamma. Construct and interpret a 95% confidence interval for the population value of gamma.

31. Suppose $\hat{\gamma} = .30$ for the relationship between two ordinal variables.
a) Of the pairs that are concordant or discordant, what proportion are concordant? Discordant?
b) Is this a stronger or a weaker association than one having $\hat{\gamma} = -.70$? Explain.

32. Conduct the test of independence for Table 8.17 using Kendall's tau-*b* rather than gamma. Compare results to those obtained using gamma.

Concepts and Applications

33. Refer to the WWW data set (Problem 1.7). Using computer software, create and analyze descriptively and inferentially the contingency table relating
a) Political affiliation and opinion about abortion.
b) Religiosity and opinion about abortion.
c) Gender and belief in life after death.

34. Refer to the data file you created in Problem 1.7. For variables chosen by your instructor, conduct descriptive and inferential statistical analyses. Interpret and summarize your findings.

35. Several sociologists have reported that racial prejudice varies according to religious group. Examine this using Table 8.35, for white respondents to the 1980 General Social Survey. Religious preference has four categories: (i) Liberal Protestant, encompassing the theologically more liberal groups such as Presbyterians, Episcopalians, Methodists, and

TABLE 8.35

Religious Preference	Laws Against Marriage Favor	Oppose	Total
Liberal Protestant	103	187	290
Conservative Protestant	182	238	420
Catholic	80	286	366
None	16	74	90
Total	381	785	1166

Source: 1980 General Social Survey

TABLE 8.36

	Political Ideology				
Political Party	Very Liberal	Slightly Liberal	Moderate	Slightly Conservative	Very Conservative
Democratic	80	81	171	41	55
Republican	30	46	148	84	99

Source: 1991 General Social Survey

Congregationalists; (ii) Conservative Protestant, consisting of conservative and fundamentalist denominations and sects such as Baptists and Pentecostals; (iii) Catholic; and (iv) None, including those with no religious preference. (Jews and other religious preferences are not shown because of the small number of cases in the sample.) The indicator of racial prejudice is the survey question, "Do you think there should be laws against marriages between blacks and whites?" Analyze these data. Prepare a report, describing your analyses and providing interpretations of the data.

36. Analyze the data in Table 8.36, relating political ideology to political party. Prepare a report of your analyses, emphasizing interpretations of the results.

37. Table 8.37 classifies a sample of psychiatric patients by their diagnosis and by whether their treatment prescribed drugs. Analyze these data, providing interpretations of all your analyses.

TABLE 8.37

	Drugs	
Diagnosis	Yes	No
Schizophrenia	105	8
Affective disorder	12	2
Neurosis	18	19
Personality disorder	47	52
Special symptoms	0	13

Source: E. Helmes and G. C. Fekken, *Journal of Clinical Psychology*, Vol. 42 (1986), pp. 569–576. Copyright by Clinical Psychology Publishing Co., Inc., Brandon, VT.

TABLE 8.38

	Belief In Afterlife		
Gender	Yes	No	Undecided
Female	435	89	58
Male	275	84	50

38. Table 8.38, from the 1991 General Social Survey, refers to the association between belief in the afterlife and gender. Analyze these data.

39. The printout in Table 8.39 refer to a sample of voters from a recent presidential primary in Wisconsin. Democrats and Republicans were classified on political ideology. Explain how to interpret these analyses.

TABLE 8.39

```
PARTY        IDEOLOGY

Frequency|
Expected |Conserv |Moderate| Liberal|  Total
---------+--------+--------+--------+
Democrat |   100  |   156  |   143  |   399
         | 147.75 |  148.4 | 102.84 |
---------+--------+--------+--------+
Republ   |   127  |    72  |    15  |   214
         | 79.246 | 79.595 | 55.158 |
---------+--------+--------+--------+

Statistic                    DF     Value        Prob
------------------------------------------------------
Chi-Square                    2    90.242       0.001

Statistic                           Value         ASE
------------------------------------------------------
Gamma                               0.628       0.047
Kendall's Tau-b                     0.362       0.032
Lambda Asymmetric C|R               0.143       0.034
```

40. The sample in Table 8.17 consists of 104 black Americans. A similar table relating income and job satisfaction for white subjects in the 1991 General Social Survey has counts (18, 39, 36) in row 1, (43, 163, 158) in row 2, and (27, 97, 156) in row 3. Analyze these data.

41. Refer to Problem 8.8. A similar table relating alcohol use as rows to marijuana use as columns has cell counts (322, 5) in row 1 and (994, 955) in row 2. Analyze these data.

42. In a survey conducted in 1994 by Market Segment Research and Consulting, Inc., as part of the 1994 Ethnic Market Report, about 5000 respondents were asked "Which of the following issues facing your community today do you feel is the most important?" Table 8.40 shows the data. Describe the patterns of association that this table displays.

43. Shortly before a gubernatorial election, a random sample of 50 potential voters are asked the following questions:

    Do you consider yourself to be a Democrat (D), a Republican (R), or Independent (I)?

    If you were to vote today, would you vote for the Democratic candidate (D), the Republican (R), or would you be undecided (U) about how to vote?

    Do you plan on voting in the election? Yes (Y) or no (N)?

TABLE 8.40

Ethnic Group	Most Important Issue: Education	Drugs	Gang Violence and Crime	Child Abuse	AIDS	Health Care	Other
White	440	130	180	80	80	60	30
Hispanic	660	540	420	140	140	60	40
Black	410	210	200	30	90	30	30
Asian	460	40	250	30	80	10	60

Source: U.S. Department of Justice, *Sourcebook of Criminal Justice Statistics, 1994* (Washington, DC: U.S. Government Printing Office), p. 141.

TABLE 8.41

(D, U, N)	(R, R, Y)	(I, D, Y)	(I, U, N)	(R, U, N)
(I, D, N)	(R, R, Y)	(I, U, N)	(D, U, Y)	(D, R, N)
(I, D, N)	(D, D, Y)	(D, D, Y)	(I, D, Y)	(R, U, N)
(D, R, N)	(R, D, N)	(D, U, N)	(D, D, Y)	(R, R, Y)
(R, R, Y)	(D, D, N)	(D, D, Y)	(I, D, Y)	(R, R, N)
(D, D, Y)	(D, R, Y)	(I, U, N)	(D, D, N)	(D, D, Y)
(R, R, Y)	(R, R, Y)	(D, U, N)	(I, R, N)	(I, R, Y)
(R, R, Y)	(I, U, Y)	(D, D, Y)	(D, R, Y)	(D, D, N)
(D, D, Y)	(I, R, Y)	(R, R, Y)	(I, D, Y)	(R, R, N)
(R, R, Y)	(D, D, Y)	(I, D, Y)	(I, R, N)	(R, R, Y)

For each person interviewed, the answers to the three questions are entered in a data file. For example, the entry (D, U, N) represents a registered Democrat who is undecided and who does not expect to vote. Table 8.41 summarizes results of the 50 interviews. Using computer software, create a data file and conduct the following analyses:

a) Construct the 3×3 contingency table relating political party affiliation to intended vote (the first two variables listed). Report the conditional distributions on intended vote for each of the three political party affiliations. Are they very different?

b) Report the result of the test of the hypothesis that intended vote is independent of political party affiliation; provide the test statistic, the P-value, and interpret the result.

c) Construct the 2×3 contingency table relating expectation of voting to political party affiliation (the first and third variables listed). Report the conditional distributions on expectation of voting for each of the political party affiliations, and interpret. Test the null hypothesis of independence, and interpret.

d) Supplement the analyses in (a)–(c) to investigate the association more fully. Interpret.

44. Give an example of a contingency table for which the chi-squared test of independence should not be used, because of:

a) Sample size

b) Measurement scale

c) The rows of the table are dependent samples

45. True or false? Interchanging two rows in a contingency table has no effect on the chi-squared statistic.

46. True or false? The null hypothesis for the test of independence between two categorical variables is $H_0 : \chi^2 = 0$.

47. **a)** When the sample size is very large, we have not necessarily established an important result when we show a statistically significant association. Explain.
b) The remark in Section 8.4 about small P-values not necessarily referring to an important effect applies for any significance test. Explain why, illustrating with the observation in Example 6.7 that, when n is large, a small P-value may occur in testing H_0: $\mu = \mu_0$ when $\bar{Y}$ is not very different from μ_0 in practical terms.

48. True or false? If $\gamma = 0$ for two variables, then the variables are statistically independent. Explain.

49. True or false? Interchanging two rows in a contingency table has no effect on gamma.

50. The correct answer in Problem 8.45 implies that if the chi-squared statistic is used for a contingency table having ordered categories in both directions, then (select the correct response(s)):
a) The statistic actually treats the variables as nominal.
b) Information about the ordering is ignored.
c) The test is usually not as powerful for detecting association as a test statistic based on numbers of concordant and discordant pairs.
d) The statistic cannot differentiate between positive and negative associations.

51. Show how to get the chi-squared values for $df = 1$ in Table C from z-scores in the standard normal table (Table A). Illustrate for the chi-squared value that has P-value .01.

52. Each subject in a sample of 100 men and 100 women is asked to indicate which of the following factors (one or more) are responsible for increases in crime committed by teenagers: A – the increasing gap in income between the rich and poor, B – the increase in the percentage of single-parent families, C – insufficient time that parents spend with their children, D – criminal penalties given by courts are too lenient, E – increasing problems with drugs in society, F – increasing levels of violence shown on TV. To analyze whether responses differ by gender of respondent, we cross-classify the responses by gender, as Table 8.42 shows.
a) Is it valid to apply the chi-squared test of independence to these data? Explain.
b) Explain how this table actually provides information needed to cross-classify gender with each of six variables. Construct the contingency table relating gender to opinion about whether the increasing gap in income is responsible for increases in teenage crime, and analyze.

TABLE 8.42

Gender	A	B	C	D	E	F
Men	60	81	75	63	86	62
Women	75	87	86	46	82	83

TABLE 8.43

$n/2$	0
0	$n/2$

53. * Table 8.43 exhibits the maximum possible association between two binary variables for a sample of size n. Show that $\chi^2 = n$ for this table and, hence, that the maximum value of χ^2 for 2×2 tables is n.

54. Refer to the previous exercise. The ***phi-squared*** measure of association for 2×2 contingency tables has sample value

$$\hat{\phi}^2 = \frac{\chi^2}{n}$$

Explain why this measure falls between 0 and 1, with a population value of 0 corresponding to independence. (It is a special case, for 2×2 tables, of the *Goodman and Kruskal tau* measure introduced in Problem 8.58 and of the r^2 measure introduced in the next chapter.)

55. For 2×2 tables, gamma simplifies to a measure first proposed about 1900 by the statistician G. Udny Yule, who also introduced the odds ratio. In that special case, gamma is called ***Yule's* Q**.

a) Show that for a generic table with counts (a, b) in row 1 and (c, d) in row 2, the number of concordant pairs equals ad, the number of discordant pairs equals bc, and $Q = (ad - bc)/(ad + bc)$.

b) Show that the absolute value of gamma equals 1 for any 2×2 table in which one of the cell frequencies is 0.

56. You have data on two ordinal variables, degree of political conservatism (low, medium, high) and willingness to raise taxes for welfare programs (low, medium, high). If you compute gamma for these data, would you expect its value to be positive, or would you expect it to be negative? Explain.

57. * Construct a 3×3 table for each of the following conditions:

a) Gamma equals 1. (*Hint*: There should be no discordant pairs.)

b) Gamma equals -1.

c) Gamma equals 0.

58. * ***Goodman and Kruskal's tau***, also called the ***concentration measure***, is a PRE measure of association for nominal variables. We illustrate it for Table 8.35. Rule 1 for predicting attitude without using religious preference is based on the marginal distribution for that response variable. To illustrate, since the proportion $381/1166 = .327$ of the sample is classified *Favor*, we guess a classification of *Favor* for that same proportion of the sample. We predict *Oppose* for the remaining 785 people. For rule 2, in predicting attitudes, we now do it for the conditional distribution of the response variable within each level of religious affiliation. For example, we predict *Favor* for a random sample of 103 of 290 Liberal Protestants and predict *Oppose* for the other 187 Liberal Protestants.

a) Show that the number of errors expected in predicting attitude using rule 1 is $E_1 = 381(785/1166) + 785(381/1166) = 513.0$.

b) Show that the expected number of errors in predicting attitude using rule 2 is $E_2 = 490.4$.

c) Show that tau, the proportional reduction in error, equals .044. (Properties of this measure include (i) $0 \le \tau \le 1$, (ii) $\tau = 0$ corresponds to independence, (iii) $\tau = 1$ when for each category of the predictor variable, *all* the observations occur in one category of the response variable.)

59. Refer to the previous problem. A contingency table relating race (the rows) to opinion (the columns) has counts (490, 410) in row 1 and (10, 90) in row 2. A contingency table relating gender (the rows) to opinion has counts (350, 150) in row 1 and (150, 350) in row 2.

a) Compute the difference of proportions in each table. Which table seems to have the stronger association?

b) Repeat (a) using the odds ratio.

c) Now compare the tables using Goodman and Kruskal's tau. This reveals the dependence of this measure on the margins. It is easier to get a larger value for the second table, for which row totals are similar for each gender, than for the first table, for which row totals are quite different for the races. One should use tau to compare tables only if they have similar marginals.

d) For the first table, suppose the counts equal (50, 450) instead of (10, 90) in row 2. Show that the conditional distributions do not change, the difference of proportions does not change, the odds ratio does not change, but tau changes.

60. * Another PRE measure of association for nominal variables is called ***lambda***. For this measure, rule 1 predicts for each observation the *modal category* for the marginal distribution of the response variable. For Table 8.35, for example, rule 1 predicts Oppose always. Rule 2 predicts the modal category for the conditional distribution of the response variable, within each level of the predictor variable. For each row in Table 8.35, rule 2 predicts Oppose.

a) Show that $E_1 = 381$, $E_2 = 103 + 182 + 80 + 16 = 381$, and $\hat{\lambda} = (E_1 - E_2)/E_1 = 0$.
b) Explain why lambda equals 0 when the modal category of the response variable is the same for each level of the explanatory variable. (The main difference between lambda and tau is that lambda can equal 0 even when the variables are statistically dependent.)

61. * In cross-classifications of ordinal variables, not all pairs are concordant or discordant. Subjects in the same category of a variable are *tied* on that variable. Two observations that fall in the same row are tied on the row variable, and two observations that fall in the same column are tied on the column variable. Let T_x denote the number of pairs tied on X, and let T_y denote the number of pairs tied on Y. The sample value of ***Kendall's tau-b*** is

$$\hat{\tau}_b = \frac{C - D}{\sqrt{[n(n-1)/2 - T_x][n(n-1)/2 - T_y]}}$$

Show that $\hat{\gamma} = \hat{\tau}_b$ when $T_x = T_y = 0$. (The common value is then known as ***Kendall's tau***.)

Bibliography

Agresti, A. (1996). *An Introduction to Categorical Data Analysis*. New York: Wiley.

Chatterjee, S., Handcock, M., and Simonoff, J. (1995). *A Casebook for a First Course in Statistics and Data Analysis*. New York: Wiley.

Fleiss, J. L. (1981). *Statistical Methods for Rates and Proportions*. New York: Wiley.

Kendall, M. G., and Gibbons, J. D. (1990). *Rank Correlation Methods*, 5th ed. London: Edward Arnold.

Chapter 9

Linear Regression and Correlation

Table 9.1 shows recent data from *Statistical Abstract of the United States* for the 50 states and the District of Columbia on several variables:

- Murder rate: The number of murders per 100,000 people in the population.
- Violent crime rate: The number of murders, forcible rapes, robberies, and aggravated assaults per 100,000 people in the population.
- Percentage of the population with income below the poverty level.
- Percentage of the population living in metropolitan areas.
- Percentage of the population who are white.
- Percentage of the population who are high school graduates or higher.
- Percentage of families living below the poverty level.
- Percentage of families headed by a single parent (male householders with no wife present and with own children, or female householders with no husband present and with own children).

All these variables are quantitative. Crime rate and murder rate are natural response variables. One might treat the other variables as explanatory variables for these responses.

This chapter introduces methods for analyzing relationships between a pair of quantitative variables. We present three different, but related, aspects of such relationships:

TABLE 9.1 Statewide Data Used to Illustrate Regression Analyses

State	Violent Crime Rate	Murder Rate	Metropolitan Residents	Percent White	High School Graduates	Poverty Rate	Single Parent
AK	761	9.0	41.8	75.2	86.6	9.1	14.3
AL	780	11.6	67.4	73.5	66.9	17.4	11.5
AR	593	10.2	44.7	82.9	66.3	20.0	10.7
AZ	715	8.6	84.7	88.6	78.7	15.4	12.1
CA	1078	13.1	96.7	79.3	76.2	18.2	12.5
CO	567	5.8	81.8	92.5	84.4	9.9	12.1
CT	456	6.3	95.7	89.0	79.2	8.5	10.1
DE	686	5.0	82.7	79.4	77.5	10.2	11.4
FL	1206	8.9	93.0	83.5	74.4	17.8	10.6
GA	723	11.4	67.7	70.8	70.9	13.5	13.0
HI	261	3.8	74.7	40.9	80.1	8.0	9.1
IA	326	2.3	43.8	96.6	80.1	10.3	9.0
ID	282	2.9	30.0	96.7	79.7	13.1	9.5
IL	960	11.4	84.0	81.0	76.2	13.6	11.5
IN	489	7.5	71.6	90.6	75.6	12.2	10.8
KS	496	6.4	54.6	90.9	81.3	13.1	9.9
KY	463	6.6	48.5	91.8	64.6	20.4	10.6
LA	1062	20.3	75.0	66.7	68.3	26.4	14.9
MA	805	3.9	96.2	91.1	80.0	10.7	10.9
MD	998	12.7	92.8	68.9	78.4	9.7	12.0
ME	126	1.6	35.7	98.5	78.8	10.7	10.6
MI	792	9.8	82.7	83.1	76.8	15.4	13.0
MN	327	3.4	69.3	94.0	82.4	11.6	9.9
MO	744	11.3	68.3	87.6	73.9	16.1	10.9
MS	434	13.5	30.7	63.3	64.3	24.7	14.7
MT	178	3.0	24.0	92.6	81.0	14.9	10.8
NC	679	11.3	66.3	75.2	70.0	14.4	11.1
ND	82	1.7	41.6	94.2	76.7	11.2	8.4
NE	339	3.9	50.6	94.3	81.8	10.3	9.4
NH	138	2.0	59.4	98.0	82.2	9.9	9.2
NJ	627	5.3	100.0	80.8	76.7	10.9	9.6
NM	930	8.0	56.0	87.1	75.1	17.4	13.8
NV	875	10.4	84.8	86.7	78.8	9.8	12.4
NY	1074	13.3	91.7	77.2	74.8	16.4	12.7
OH	504	6.0	81.3	87.5	75.7	13.0	11.4
OK	635	8.4	60.1	82.5	74.6	19.9	11.1
OR	503	4.6	70.0	93.2	81.5	11.8	11.3
PA	418	6.8	84.8	88.7	74.7	13.2	9.6
RI	402	3.9	93.6	92.6	72.0	11.2	10.8
SC	1023	10.3	69.8	68.6	68.3	18.7	12.3
SD	208	3.4	32.6	90.2	77.1	14.2	9.4
TN	766	10.2	67.7	82.8	67.1	19.6	11.2
TX	762	11.9	83.9	85.1	72.1	17.4	11.8
UT	301	3.1	77.5	94.8	85.1	10.7	10.0
VA	372	8.3	77.5	77.1	75.2	9.7	10.3
VT	114	3.6	27.0	98.4	80.8	10.0	11.0
WA	515	5.2	83.0	89.4	83.8	12.1	11.7
WI	264	4.4	68.1	92.1	78.6	12.6	10.4
WV	208	6.9	41.8	96.3	66.0	22.2	9.4
WY	286	3.4	29.7	95.9	83.0	13.3	10.8
DC	2922	78.5	100.0	31.8	73.1	26.4	22.1

1. We investigate *whether an association exists* between the two variables by testing the hypothesis of statistical independence.
2. We study the *strength of their association* using a measure of association called the *correlation*.
3. We study the *form of the relationship*. Using the data, we estimate a formula that predicts a subject's score on the response variable from the score on the explanatory variable. For instance, a formula of this type predicts a state's murder rate from the percentage of its population living below the poverty level.

The analyses conducted in studying these three aspects of the relationship between two quantitative variables are collectively called a ***regression analysis***. Section 9.1 introduces the use of a straight line to describe the form of the relationship. Section 9.2 presents the method, called *least squares*, that estimates the best line for a particular data set. Section 9.3 describes the *linear regression model*, which takes into account variability of the data about the straight line. Section 9.4 defines the *Pearson correlation* for describing the strength of a linear relationship. Section 9.5 presents statistical inference for a regression analysis. The final section takes a closer look at assumptions and potential pitfalls in using regression.

9.1 Linear Relationships

The univariate methods of Chapters 5 and 6 summarized quantitative variables by their means. As noted in Chapters 7 and 8, most analyses involve more than one variable and distinguish between response and explanatory variables. This chapter presents methods for analyzing a pair of quantitative variables, one of which is a response variable and one of which is an explanatory variable.

Notation for Response and Explanatory Variables

Let Y denote the *response* variable and let X denote the *explanatory* variable.

We shall analyze how the response variable Y tends to change from one subset of the population to another, as defined by values of X. For categorical variables, we did this by comparing the conditional distributions of Y at the various categories of X, in a contingency table. For quantitative variables, a mathematical formula describes how the conditional distribution of Y varies according to the value of X. This type of formula describes how Y = murder rate varies among states for different levels of X = percent below the poverty level. For instance, does the murder rate tend to be higher for states that have higher poverty levels?

Linear Functions

A variety of different formulas might describe how Y relates to X. Any particular formula might provide a good description of the relationship or a poor one. This chapter introduces the simplest class of such formulas, *straight lines*. They are called ***linear functions***.

> **Linear Function**
>
> The formula $Y = \alpha + \beta X$ expresses the response variable Y as a ***linear function*** of the explanatory variable X. The formula maps out a straight-line graph with ***slope*** β (beta) and ***Y-intercept*** α (alpha).

Example 9.1 Example of a Linear Function

The formula $Y = 3 + 2X$ is a linear function. This has the form $Y = \alpha + \beta X$ with $\alpha = 3$ and $\beta = 2$; the Y-intercept equals 3 and the slope equals 2.

Each real number X, when substituted into the formula $Y = 3+2X$, yields a distinct value for Y. For instance, the value $X = 0$ has $Y = 3 + 2(0) = 3$; the value $X = 1$ has $Y = 3 + 2(1) = 5$. Figure 9.1 plots this function. The horizontal axis, called the ***X-axis***, lists the possible values of the explanatory variable, X; the vertical axis, called the ***Y-axis***, lists the possible values of Y. The axes intersect at the point where $X = 0$ and $Y = 0$, called the *origin*. □

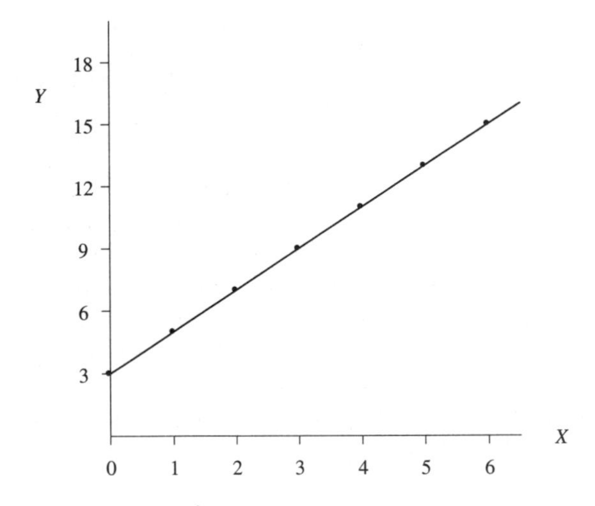

Figure 9.1 Graph of the Straight Line $Y = 3 + 2X$. The Y-intercept is 3 and the slope is 2.

Interpreting the *Y*-Intercept and Slope

At $X = 0$, the equation $Y = \alpha + \beta X$ simplifies to $Y = \alpha + \beta X = \alpha + \beta(0) = \alpha$. Thus, the constant α in this equation is the value of Y when $X = 0$. Now, points on the Y-axis have $X = 0$, so the line has height α at the point of its intersection with the Y-axis. Because of this, α is called the ***Y-intercept***. The straight line $Y = 3 + 2X$ intersects the Y-axis at $\alpha = 3$, as Figure 9.1 shows.

The ***slope*** β equals the change in Y for a one-unit increase in X. That is, for two X-values that differ by 1.0 (such as $X = 0$ and $X = 1$), the Y-values differ by β. For the line $Y = 3 + 2X$, at $X = 0$ and $X = 1$, the Y values are 3 and 5. These Y values differ by $5 - 3 = 2$, which is the value of β for this line. Similarly, two X-values that are 10 units apart differ by 10β in their Y-values. For example, when $X = 0$, $Y = 3$, and when $X = 10$, $Y = 3 + 2(10) = 23$, and $23 - 3 = 20 = 10\beta$. Figure 9.2 portrays the interpretation of the Y-intercept and slope of a straight line.

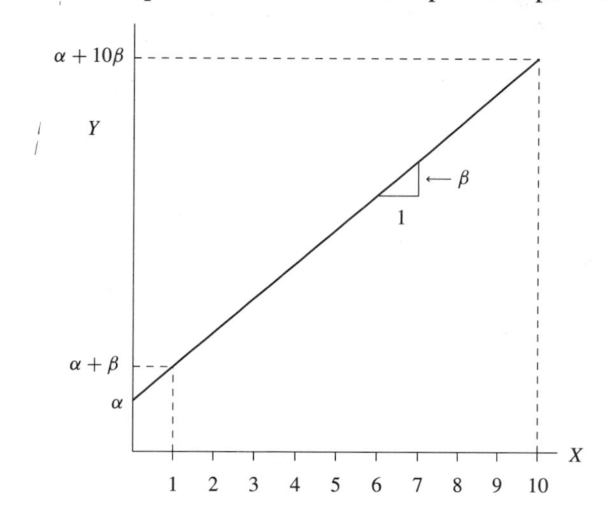

Figure 9.2 Graph of the Straight Line $Y = \alpha + \beta X$. The Y-intercept is α and the slope is β.

One can draw the straight line by finding any two separate pairs of (X, Y) values on the graph and then drawing the line through the points. For instance, we could use the points just discussed: $(X = 0, Y = 3)$ and $(X = 1, Y = 5)$. The point on the graph with $(X = 0, Y = 3)$ is three units up the Y-axis. To find the point with $(X = 1, Y = 5)$, we start at the origin $(X = 0, Y = 0)$ and move one unit to the right on the X-axis and five units upward parallel to the Y-axis (see Figure 9.1). After plotting the two points, drawing the straight line through the two points graphs the function $Y = 3 + 2X$.

The slope describes the rate of change of the Y-values as X increases. A one-unit increase in X corresponds to a change of β units in Y, no matter what the value of X. The larger the absolute value of β, the steeper the line. A line with $\beta = 4$, such as $Y = 3 + 4X$, climbs more quickly than one with $\beta = 2$.

Example 9.2 Straight Line for Violent Crime and Poverty

For the 50 states, consider the variables Y = violent crime rate (number of violent crimes per 100,000 population) and X = poverty rate (percentage of state residents living below the poverty level). The straight line $Y = 210 + 25X$ approximates the relation between these variables. The Y-intercept equals 210. This represents the violent crime rate at poverty rate $X = 0$ (unfortunately, there are no such states). The slope equals 25. When the poverty rate increases by one percent, the violent crime rate increases by about 25 (i.e., 25 crimes per 100,000 population). Similarly, if the poverty rate increases by 10%, the crime rate increases by about 10(25) = 250.

By contrast, if instead the explanatory variable X refers to the percentage of the population living in metropolitan areas, the straight line approximating the relationship is $Y = 26 + 8X$. The slope of 8 is smaller than the slope of 25 when poverty rate is the predictor. A 1% change in poverty rate has a greater effect on the violent crime rate than a 1% change in metropolitan residence. Figure 9.3 shows the lines relating the violent crime rate to poverty rate and metropolitan residence. □

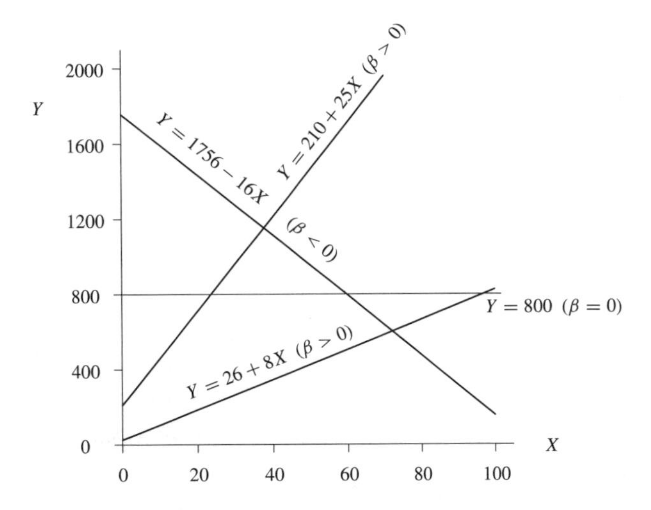

Figure 9.3 Graphs of Lines Showing Positive Relationships ($\beta > 0$), a Negative Relationship ($\beta < 0$), and Independence ($\beta = 0$)

If β is positive, then Y increases as X increases; that is, large values of Y occur with large values of X, and small values of Y occur with small values of X. The straight line then goes upward, like the lines just mentioned relating violent crime rate to poverty rate or metropolitan residence. When a relationship between two variables follows a straight line with $\beta > 0$, the relationship is said to be ***positive***.

If β is negative, then Y decreases as X increases. The straight line then goes downward, and the relationship is said to be ***negative***. For instance, the equation $Y = 1756 - 16X$ approximates the relationship between Y = violent crime rate and X = percentage of residents who are high school graduates. This shows a negative relationship with slope -16. For each increase of 1.0 in the percent who are high school graduates, the violent crime rate decreases by about 16. Figure 9.3 also shows this line.

When $\beta = 0$, the graph of a linear function is a horizontal line. The value of Y is constant and does not vary as X varies. If two variables are independent, with the value of Y not depending on the value of X, then a straight line with $\beta = 0$ represents their relationship. The line $Y = 800$ shown in Figure 9.3 is an example of a line with $\beta = 0$.

Models

The linear function is the simplest mathematical function and provides the simplest form for the relationship between two quantitative variables. The function $Y = \alpha + \beta X$ is a ***model***. That is, the formula provides a simple approximation for the true relationship between X and Y. For a given value of X, the model predicts a value for Y. The better these predictions tend to be, the better the model.

9.2 Least Squares Prediction Equation

Using sample data, one can estimate the linear model relating Y and X. The process treats α and β in the linear function $Y = \alpha + \beta X$ as unknown parameters and yields estimates of these parameters. The estimated linear function then provides predictions about Y at fixed values for X.

Scatter Diagram

The first step of model fitting is to look at the data. A plot of the data reveals whether a model with a straight line trend makes sense.

The values (X, Y) of the two variables for any particular subject form a point relative to the X and Y axes. To portray graphically the sample relationship, we plot the sample of n observations on X and Y as n points. This graphical plot is called a ***scatter diagram*** or ***scatterplot***.

Example 9.3 Scatter Diagram for Murder Rate and Poverty

We use the data from Table 9.1 on X = poverty rate and Y = murder rate throughout the chapter to illustrate aspects of regression analysis. To check whether Y is approximately linearly related to X, we first construct a scatter diagram for the 51 observations. Figure 9.4 shows this plot.

Each point in Figure 9.4 portrays the values of poverty rate and murder rate for a given state. For Maryland, for instance, the poverty rate is $X = 9.7$, and the murder rate is $Y = 12.7$. Its point $(X, Y) = (9.7, 12.7)$ has coordinate 9.7 for the X-axis and 12.7 for the Y-axis. We plot the point (9.7, 12.7) by moving from the origin 9.7 units in the X (horizontal) direction and 12.7 units in the Y (vertical) direction. This brings us to the spot labeled MD in Figure 9.4.

Figure 9.4 indicates that the trend of points seems to be approximated well by a straight line. Notice, though, that one point is far removed from the rest. This is the point for D.C., for which the murder rate was much higher than that of any state. This point lies far from the overall trend. Figure 9.4 also shows box plots for these variables. They reveal that D.C. is an extreme *outlier* on murder rate; in fact, it falls 6.5 standard deviations above the mean. We shall see that outliers can have a serious impact on the results of a regression analysis. □

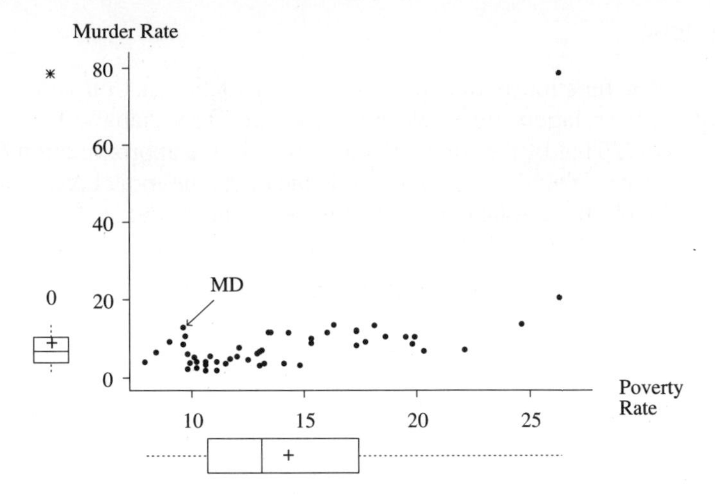

Figure 9.4 Scatter Diagram for Y = Murder Rate and X = Percentage of Residents Below the Poverty Level, for 50 States and D.C.

The scatter diagram provides a visual means of checking whether a relationship is approximately linear. When the relationship seems strongly nonlinear, it does not make sense to use a straight line to model the data. Figure 9.5 illustrates such a case. This figure shows a negative relationship over part of the range of X values, and a positive relationship over the rest. These cancel each other out using a linear model. For such data, one needs a different type of model, presented in Section 14.4.

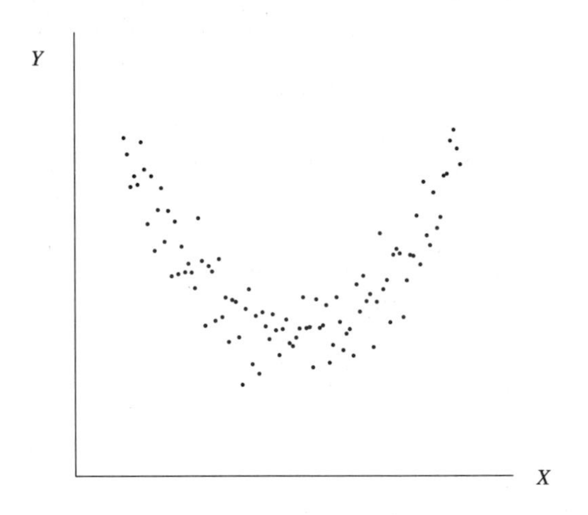

Figure 9.5 A Nonlinear Relationship, for Which It Is Inappropriate to Use a Straight Line Regression Model

Prediction Equation

When the scatter diagram suggests that the linear model $Y = \alpha + \beta X$ is realistic, we estimate this unknown line. The notation

$$\hat{Y} = a + bX$$

represents a *sample* equation, based on the data, that estimates the linear function $Y = \alpha + \beta X$. In the sample equation, the Y-intercept (a) estimates the unknown Y-intercept α of the linear model, and the slope (b) estimates the unknown slope β.

Substituting a particular value of X into the formula $a + bX$ provides a value, denoted by $\hat{Y}$, that predicts the value of Y at that value of X. The sample equation $\hat{Y} = a + bX$ is called the ***prediction equation***, since it provides predictions about the response variable Y for any value of X.

The prediction equation is the best straight line summarizing the trend of the points in the scatter diagram. It falls closest to those points, in a certain average sense discussed later in this section. The formulas for the coefficients of this prediction equation are

$$b = \frac{\sum(X - \bar{X})(Y - \bar{Y})}{\sum(X - \bar{X})^2}, \qquad a = \bar{Y} - b\bar{X}$$

If an observation has both X- and Y-values above their means, or both X- and Y-values below their means, then $(X - \bar{X})(Y - \bar{Y})$ is positive. The slope estimate b tends to be positive when most observations are like this, that is, when points with large X-values also tend to have large Y-values and points with small X-values tend to have small Y-values.

We shall not dwell on these formulas or even illustrate how to use them. The formula for b is complex and is messy to use even for a small to moderate sample size. Although a somewhat simpler computational formula exists, anyone who does any serious regression modeling these days uses a computer or a calculator that has these formulas programmed. For the extensions of this linear model studied in later chapters, it is impractical to calculate prediction equations even with a calculator but very simple to do so using statistical software. We strongly encourage that you use modern computing methods, rather than hand calculation, to conduct regression analyses. To use statistical software on a PC or other type of computer, you supply the data set and either select the desired methods from a menu or enter certain code specifying those methods (e.g., regression) and the options desired (e.g., scatter diagrams, prediction equation). The appendix at the end of the text provides details.

Example 9.4 Prediction Equation for Murder Rate and Poverty

Using statistical software for the 51 observations on murder rate and poverty rate from Table 9.1, we get the results shown in the sample printout in Table 9.2. Usually, software provides summary statistics such as the mean and standard deviation for each variable, as well as statistics for the regression analysis. For instance, murder rate has $\bar{Y} = 8.7$ and $s = 10.7$, indicating that it is probably highly skewed to the right. In fact, the box plot for murder rate in Figure 9.4 shows that the extreme outlying observation for D.C. contributes a lot to this outcome.

The estimates of α and β are listed under the heading "Parameter Estimate." The estimate of the Y-intercept is $a = -10.14$, listed opposite "INTERCEP." The estimate

TABLE 9.2 Part of Computer Printout for Fitting Linear Regression Model to 51 Observations (Including D.C.) on X = Percent in Poverty and Y = Murder Rate

```
Variable      Mean    Std Dev          Dependent Variable: MURDER
-----------------------------                    Parameter  Standard
MURDER       8.727     10.718          Variable   Estimate     Error
POVERTY     14.259      4.584
-----------------------------          INTERCEP   -10.1364    4.1206
                                       POVERTY      1.3230    0.2754
```

of the slope is $b = 1.32$, listed opposite the variable name of which it is the coefficient in the prediction equation, "POVERTY." Therefore, the prediction equation relating Y = murder rate to X = poverty rate is $\hat{Y} = a + bX = -10.14 + 1.32X$.

For this prediction equation, the slope is $b = 1.32$. Since the slope is positive, the sample relationship is positive. The larger the poverty rate, the larger the murder rate tends to be. The value of 1.32 for the slope indicates that, on the average, an increase of 1% in the poverty rate relates to an increase of 1.32 in the murder rate.

Similarly, an increase of 10% in the poverty rate corresponds to a 10(1.32) = 13.2-unit increase in the murder rate. If one state has a 12% poverty rate and another has a 22% poverty rate, for example, the predicted number of murders per 100,000 population is 13.2 higher in the second state than the first state. Since the mean murder rate is 8.7, it seems as if poverty rate is an important predictor of murder rate. This differential of 13 murders per 100,000 population translates to 130 per million or 1300 per 10 million population. If the two states each had populations of 10 million, the one with the higher poverty rate would be predicted to have 1300 more murders per year. □

Effect of Outliers on Prediction Equation

Figure 9.6 plots the prediction equation from Example 9.4 over the scatter diagram. The diagram shows that one observation, the one for D.C., is a serious outlier. This outlier appears to have a substantial effect on the results, since the line seems to be pulled up toward it and away from the center of the general trend of points.

Table 9.3 shows part of a computer printout from using the observations for the 50 states but not the one for D.C. The prediction equation equals $\hat{Y} = -.86 + .58X$. Figure 9.6 also shows this line, which passes more directly through the 50 points. The slope is .58, compared to 1.32 when the observation for D.C. is included. The one outlying observation has the effect of more than doubling the slope! An observation is called ***influential*** if removing it results in a large change in the prediction equation. Unless the sample size is large, an observation can have a strong influence on the slope if its X value is low or high compared to the rest of the data.

In summary, the line for the data set including the D.C. observation seems to distort the relationship for the other 50 states. It seems wiser to use the equation based on data

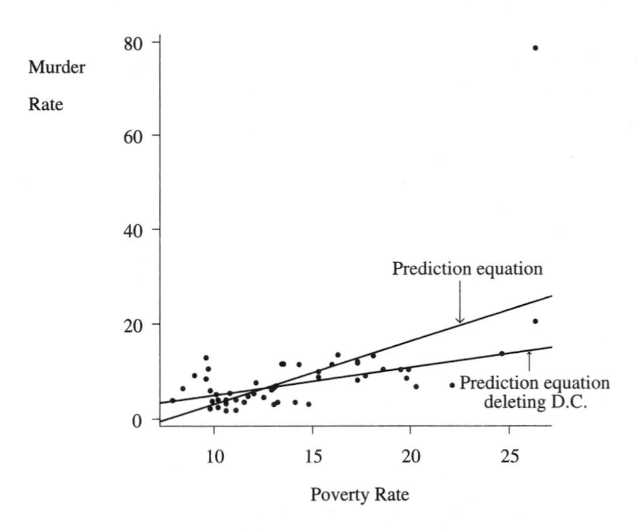

Figure 9.6 Prediction Equations Relating Murder Rate and Percentage in Poverty, with and without D.C. Observation

for the 50 states alone rather than to use a single equation both for the 50 states and D.C. This line for the 50 states better represents the overall trend for the states. In reporting these results, we would note that the murder rate for D.C. falls outside this trend, being much larger than this equation predicts.

Residuals

The prediction equation $\hat{Y} = -.86 + .58X$ provides predicted murder rates for states with various levels of X = poverty rate. For the sample data, a comparison of the predicted values to the actual murder rates checks the goodness of the prediction equation.

For example, consider Massachusetts, for which $X = 10.7$ and $Y = 3.9$. The predicted murder rate ($\hat{Y}$) at $X = 10.7$ is $\hat{Y} = -.86 + .58X = -.86 + .58(10.7) = 5.4$. The prediction error is the difference between the actual Y value of 3.9 and the predicted value of 5.4, or $Y - \hat{Y} = 3.9 - 5.4 = -1.5$. Thus, the prediction equation overestimates the murder rate by 1.5. Similarly, for Louisiana, $X = 26.4$ and $\hat{Y} = -.86 + .58(26.4) = 14.6$. The actual murder rate is $Y = 20.3$, so the prediction is too low. The prediction error is $Y - \hat{Y} = 20.3 - 14.6 = 5.7$.

The prediction errors are called ***residuals***. For a given observation, the response variable equals Y and the predicted value equals $\hat{Y} = a + bX$. The difference $Y - \hat{Y}$ is the residual for that observation.

Residual

The difference between observed and predicted values of the response variable, $Y - \hat{Y}$, is called a ***residual***.

TABLE 9.3 Part of Computer Printout for Fitting Linear Model to 50 States (Excluding D.C.) on X = Percent in Poverty and Y = Murder Rate

Source	DF	Sum of Squares	Mean Square	Variable	Parameter Estimate
Model	1	307.342	307.34	INTERCEP	-0.8567
Error	48	470.406	9.80	POVERTY	0.5842
Total	49	777.749			
			Root MSE 3.1305		

Obs	Dep Var MURDER	Predict Value	Residual
1	9.0000	4.4599	4.5401
2	11.6000	9.3091	2.2909
3	10.2000	10.8281	-0.6281
4	8.6000	8.1406	0.4594
5	13.1000	9.7765	3.3235
6	5.8000	4.9273	0.8727

Table 9.3 shows the murder rates, the predicted values, and the residuals for some of the states. A positive residual results when the observed value Y is larger than the predicted value $\hat{Y}$, and $Y - \hat{Y} > 0$. A negative residual results when the observed value is smaller than the predicted value. The smaller the absolute value of the residual, the better is the prediction, since the predicted value is closer to the observed value.

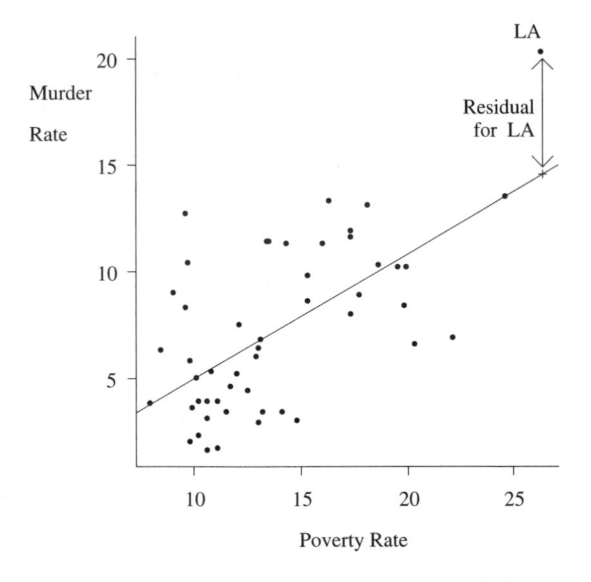

Figure 9.7 Prediction Equation and Residuals

Graphically in the scatter diagram, the residual for an observation is the vertical distance between the point for the observation and the prediction line. Figure 9.7 illustrates this for the murder rate data for the 50 states. For example, the observation for Louisiana is the point with (X, Y) coordinates (26.4, 20.3). The prediction is represented by the point (26.4, 14.6) on the prediction line corresponding to substituting $X = 26.4$ into the prediction equation $\hat{Y} = -.86 + .58X$. The residual is the difference between the observed and predicted points, which is the vertical distance $Y - \hat{Y} = 20.3 - 14.6 = 5.7$.

Method of Least Squares

Each observation in a sample has a residual. Some residuals are positive, some are negative, and their average equals 0. Just as deviations of observations from their mean $\bar{Y}$ satisfy $\sum(Y - \bar{Y}) = 0$, so is the prediction equation defined so that $\sum(Y - \hat{Y}) = 0$.

If the prediction line falls close to the points in the scatter diagram, the residuals tend to be small. We summarize the size of the residuals by the sum of their squared values. This quantity, denoted by SSE, equals

$$\text{SSE} = \sum(Y - \hat{Y})^2$$

In other words, the residual is computed for every observation in the sample; each residual is squared, and then SSE is the sum of these squares. The symbol SSE is an abbreviation for the ***sum of squared errors***. This terminology refers to the use of the residual as a measure of prediction error from using $\hat{Y}$ to predict Y. The measure SSE is also called the ***residual sum of squares***. It describes the variation of the observed points around the prediction line.

The better the prediction equation, the smaller the residuals tend to be and, hence, the smaller the summary measure SSE tends to be. Any particular equation has a corresponding set of residuals and a value of SSE. The prediction equation specified by the usual formulas for estimates a and b of α and β is the one with the smallest value of SSE out of all possible linear prediction equations.

Method of Least Squares, Least Squares Estimates

The ***method of least squares*** provides the prediction equation $\hat{Y} = a + bX$ having the minimal value of $\text{SSE} = \sum(Y - \hat{Y})^2$. The ***least squares estimates*** a and b are the values determining the prediction equation for which the sum of squared errors SSE is a minimum.

The best prediction equation, according to the least squares criterion, is the one with the smallest sum of squared residuals. Because of this criterion, the prediction line $\hat{Y} = a + bX$ is called the ***least squares line***. If we square the residuals (such as those in Table 9.3) for the least squares line $\hat{Y} = -.86 + .58X$ and then sum them, we get

$$\text{SSE} = \sum(Y - \hat{Y})^2 = (4.54)^2 + (2.29)^2 + \cdots = 470.4$$

This value of SSE is smaller than the value of SSE for *any* other straight line predictor, such as $\hat{Y} = -.88 + .60X$. In this sense, the observed points fall closer to this line than to *any* other line. Software for regression lists the value of SSE. Table 9.3 reports it in the "Sum of Squares" column, in the row labeled "Error." In some software, such as SPSS, this is labeled as "Residual" in the sum of squares column.

9.3 The Linear Regression Model

For the model $Y = \alpha + \beta X$, each value of X corresponds to a single value of Y. Such a model is said to be ***deterministic***. It is unrealistic in social science research, since *variability* occurs in Y-values among subjects with the same X-value.

For example, let X = number of years of education and Y = annual income. Consider the subjects having $X = 12$ years of education. They do not all have the same income, since income is not completely dependent upon education. Instead, there is a probability distribution of annual income for individuals with $X = 12$. This distribution refers to the variability in the Y values at a *fixed* value of X, so it is a ***conditional distribution***. Similarly, a separate conditional distribution applies for those with $X = 13$ years of education, and others apply for those with each possible value of X. The conditional distributions of income differ at the different levels of education. The mean of the conditional distribution probably is higher at higher levels of education.

A ***probabilistic*** model for the relationship between X and Y is one that allows for variability in the values of Y at each value of X. We now show how a linear function is the basis for a probabilistic model.

Linear Regression Function

A probabilistic model uses $\alpha + \beta X$ to represent the *mean* of Y, rather than Y itself, as a function of X. For a given value of X, $\alpha + \beta X$ represents the mean of the conditional distribution of Y for subjects having that value of X.

Expected Value of *Y*

Let $E(Y)$ denote the mean of a conditional distribution of Y. The symbol E represents *expected value*, which is another term for the *mean*.

We now use the form of equation

$$E(Y) = \alpha + \beta X$$

to model the relationship between X and the mean of the conditional distribution of Y. For instance, let Y = violent crime rate and X = poverty rate for the 50 states. The probabilistic model $E(Y) = 210 + 25X$ is more plausible than the deterministic model $Y = 210 + 25X$. At $X = 10$, for instance, $E(Y) = 210 + 25(10) = 460$. For states having 10% of their residents below the poverty level, the model predicts that the mean violent crime rate is 460. Different states having $X = 10$ would have different violent crime rates, but the average response would be about 460.

An equation of the form $E(Y) = \alpha + \beta X$ that relates values of X to the mean of the conditional distribution of Y is called a *regression function.*

Regression Function

A ***regression function*** is a mathematical function that describes how the mean of the response variable changes according to the value of an explanatory variable.

The function $E(Y) = \alpha + \beta X$ is called a *linear* regression function, since it uses a straight line for the relationship between the mean of Y and the values of X. The Y-intercept α and the slope β are called the ***regression coefficients*** for the linear regression function.

In practice, the parameters of the linear regression function are unknown. Least squares provides the sample prediction equation $\hat{Y} = a + bX$. At a fixed value of X, $\hat{Y} = a + bX$ estimates the mean of Y for all subjects in the population having that value of X.

Describing Variation About the Regression Line

For each fixed value of X, there is a conditional distribution of Y-values. The mean of that distribution equals $E(Y) = \alpha + \beta X$. The linear regression model has an additional parameter σ describing the standard deviation of that distribution. That is, σ measures the variability of the Y values for all subjects having the same X-value. We refer to σ as the ***conditional standard deviation***.

Example 9.5 Income Variation for Education Predictor

Suppose the linear regression model $E(Y) = -5 + 3X$ with $\sigma = 13$ describes the relationship between Y = annual income, in thousands of dollars, and X = number of years of education. According to this model, for individuals with X years of education, their incomes have a mean of $E(Y) = -5 + 3X$ and a standard deviation of 13.

For instance, those having a high school education ($X = 12$) have a mean income of $E(Y) = -5+3(12) = 31$ and a standard deviation of 13. If the conditional distribution of income is approximately bell-shaped, then about 95% of the incomes fall between \$5000 and \$57,000. These are the values that are two standard deviations from the mean; that is, $31 - 2(13) = 5$ and $31 + 2(13) = 57$. Similarly, those with a college education ($X = 16$) have a mean annual income of $E(Y) = -5+3(16) = 43$ thousand dollars, with about 95% of the incomes falling between \$17,000 and \$69,000.

The slope of $\beta = 3$ implies that mean income increases 3 thousand dollars for each year increase in education. Figure 9.8 pictures this linear regression model with $\alpha = -5$, $\beta = 3$, and $\sigma = 13$. That figure plots the regression function $E(Y) = -5+3X$ and the conditional income distributions at $X = 8$, 12, and 16 years. □

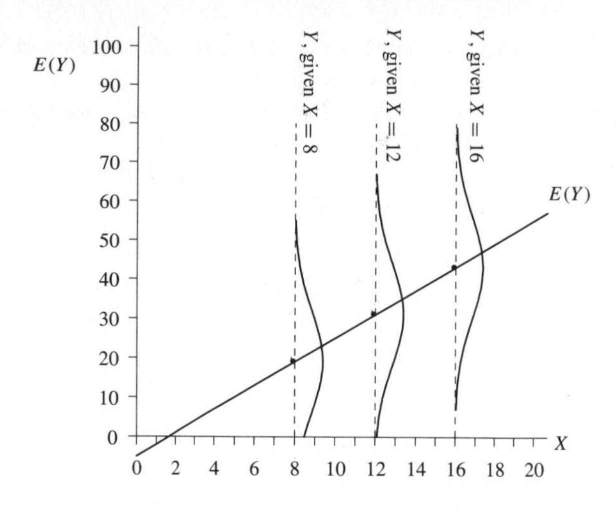

Figure 9.8 The Regression Model $E(Y) = -5 + 3X$, with $\sigma = 13$, Relating Y = Income (in Thousands of Dollars) to X = Education (in Years)

In Figure 9.8, each conditional distribution is normal, and each has the same conditional standard deviation, $\sigma = 13$. In practice, of course, the conditional distributions would not be exactly normal, and the standard deviation need not be the same for each. In fact, *any particular model never holds exactly in practice*. It is merely a simple approximation for reality. When we have sample data, we can check whether a particular model is realistic by checking the fit of the prediction equation on the scatter diagram.

Estimating Conditional Variation

This section has assumed that the mean of Y is related to X by a *linear* regression equation and that it, therefore, makes sense to use a straight line to estimate that equation. The optimality of the least squares estimates is based partly on the further assumption that the standard deviation σ of the conditional distribution of Y is identical at the various values of X.

The conditional standard deviation σ refers to the variability of Y values about the conditional population mean $E(Y) = \alpha + \beta X$ for all subjects with the same value of X. To estimate σ, we use the numerical value for $\text{SSE} = \sum(Y - \hat{Y})^2$, which measures sample variability about the least squares line. The estimate is

$$\hat{\sigma} = \sqrt{\frac{\text{SSE}}{n-2}} = \sqrt{\frac{\sum(Y - \hat{Y})^2}{n-2}}$$

Example 9.6 Estimating Variability for Murder Rates

We return to the analysis of Y = murder rate and X = poverty rate, for the 50 states. From Section 9.2 and Table 9.3, the prediction equation is $\hat{Y} = -.86 + .58X$, and the sum of squared errors in using X to predict Y equals $\text{SSE} = 470.4$. The estimate of the conditional standard deviation is

$$\hat{\sigma} = \sqrt{\frac{\text{SSE}}{n-2}} = \sqrt{\frac{470.4}{50-2}} = 3.1$$

The model predicts that at any fixed value X of poverty rate, the murder rates vary around a mean of $-.86 + .58X$ with a standard deviation of 3.1.

At $X = 20$, for instance, the conditional distribution of murder rates has an estimated mean of $-.86 + .58(20) = 10.7$ and standard deviation of 3.1. For an approximately bell-shaped distribution, about 95% of the observations fall within two standard deviations of the mean. Then, for a particular value of X, most murder rates fall within about $2\hat{\sigma} = 2(3.1) = 6.2$ of the true mean murder rate at that value of X. □

Chapter 14 shows how to check the constant variation assumption, through analysis of the residuals. A quick check with the scatter diagram of whether the variability around the fitted line is fairly constant for various values of X is also a way to evaluate whether the assumption is reasonable. If it is, $\hat{\sigma}$ describes the variability of the observations about the regression line, at any fixed X value. Otherwise, $\hat{\sigma}$ provides a measure of *average* variability about the line.

Conditional Versus Marginal Variation

From Sections 3.3 and 5.1, a sample point estimate of the standard deviation of a variable Y is given by

$$s = \sqrt{\frac{\sum(Y - \bar{Y})^2}{n - 1}}$$

This is the standard deviation of the *marginal* distribution of Y, since the formula makes no reference to any other variable X. To emphasize that the standard deviation of the marginal distribution of Y depends on values of Y alone, the remainder of the text denotes it by s_Y in a sample and σ_Y in a population. It differs from the standard deviation of the *conditional* distribution of Y, for a fixed value of X, which we denote by $\hat{\sigma}$ for the sample and σ for the population.

For the data on murder rates for the 50 states, the marginal standard deviation is $s_Y = 4.0$ (It is 10.7 if D.C. is included!), whereas the conditional standard deviation is $\hat{\sigma} = 3.1$. Typically, there is less spread in Y at a fixed value of X than there is totaled over all such values. Thus, the conditional standard deviation of Y is usually less than the marginal standard deviation of Y.

For another example, the *marginal* distribution of college GPAs (Y) at a particular university may primarily fall between about 1.0 and 4.0, perhaps having a standard deviation of $\sigma_Y = .60$. For those students with a high school GPA (X) of 3.8, however, the *conditional* distribution of college GPAs might cluster between 2.7 and 4.0, perhaps having $\sigma = .25$. Figure 9.9 illustrates the distinction between the marginal and conditional distributions and their standard deviations, the conditional distribution typically being less disperse.

The term $(n - 2)$ in the denominator of $\hat{\sigma}$ is called the ***degrees of freedom*** (df) for the estimate. In general, when a regression equation has p unknown parameters, then $df = n - p$. The linear model $E(Y) = \alpha + \beta X$ has two parameters (α and β), so $df = n - 2$. The problem studied in Chapter 5 of estimating a single mean μ corresponds to estimating the parameter in the simpler regression model, $E(Y) = \mu$,

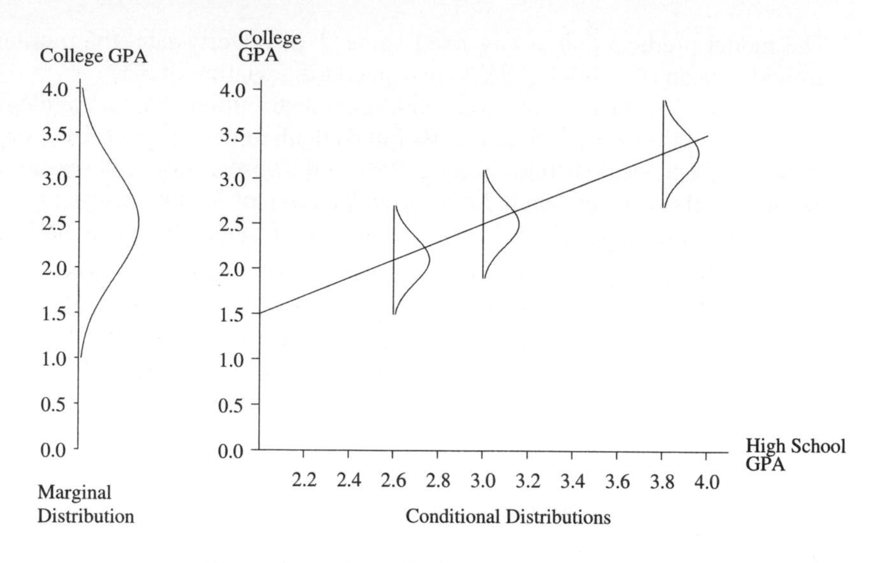

Figure 9.9 Marginal and Conditional Distributions

with a single parameter. Thus, the estimate s_Y of the standard deviation of the marginal distribution has $df = n - 1$.

Table 9.3 lists SSE = 470.4 and its degrees of freedom, $df = n - 2 = 50 - 2 = 48$. The ratio of these, $\hat{\sigma}^2 = 9.80$, is listed on the printout and called the "Mean Square Error." This is often abbreviated by MSE. Its square root is the estimate of the conditional standard deviation of Y, $\hat{\sigma} = \sqrt{9.80} = 3.13$, which is listed under the heading "ROOT MSE."

9.4 Measuring Linear Association—The Correlation

The linear regression model uses a straight-line prediction equation to describe the relationship between two quantitative variables. This section introduces two measures of the strength of linear association between the variables.

The Slope and Strength of Association

The slope b of the prediction equation tells us the *direction* of the association; its sign indicates whether the prediction line slopes upward or downward as X increases. That is, it indicates whether the association is positive or negative. The slope does not, however, directly tell us the strength of the association. The reason for this is that the numerical value of the slope is intrinsically linked to the units of measurement of the variables.

For example, consider the prediction equation $\hat{Y} = -.86 + .58X$ for Y = murder rate and X = poverty rate for the 50 states. Suppose we measure murder rate as the number of murders per 1,000,000 population instead of 100,000 population. A one-unit increase in the poverty rate corresponds to a $b = .58$ increase in the predicted number of murders per 100,000 people, which is equivalent to a 5.8 increase in the predicted number of murders per 1,000,000 people. Thus, if Y = number of murders per 1,000,000 population, the slope of the prediction equation is 5.8 instead of .58. The strength of the association is the same in each case, since the variables and data are the same; only the units of measurement for the response variable differed.

In summary, the slope b of the prediction equation depends on the units of measurement. It doesn't directly indicate whether the association is strong or weak, since we can make b as large or as small as we like by an appropriate choice of units.

The slope *is* useful for comparing effects of two predictors having the same units. For instance, the prediction equation for the 50 states relating the murder rate to percentage living in metropolitan areas is $3.28 + .06X$. A one-unit increase in the percentage living in metropolitan areas corresponds to a .06 increase in the murder rate, whereas a one-unit increase in the percentage below the poverty level corresponds to a .58 increase in the murder rate. A change of 1% in poverty rate has a much greater effect on murder rate than a change of 1% in metropolitan residence.

The measures of association we now study do not vary according to the units of measurement. Like the measures of association presented in the previous chapter for categorical data, their magnitudes indicate the strength of association.

The Pearson Correlation

The measure of association for quantitative variables known as the ***Pearson correlation***, or simply the ***correlation***, is a *standardized* version of the slope. It is a type of slope for which the value, unlike that of b, does not depend on the units of measurement. The standardization adjusts the slope b for the fact that the marginal distributions of X and Y have standard deviations that depend on the units of measurement for X and Y. The correlation is the value the slope assumes if the measurement units for the two variables are such that their standard deviations are equal.

Let s_X and s_Y denote the sample standard deviations of X and Y,

$$s_X = \sqrt{\frac{\sum(X - \bar{X})^2}{n - 1}} \quad \text{and} \quad s_Y = \sqrt{\frac{\sum(Y - \bar{Y})^2}{n - 1}}$$

Pearson Correlation

The ***Pearson correlation***, denoted by r, is related to the slope b of the prediction equation $\hat{Y} = a + bX$ by

$$r = \left(\frac{s_X}{s_Y}\right) b$$

Multiplying the slope by the ratio of the sample standard deviations provides a standardized measure. When the sample dispersions are equal ($s_X = s_Y$), the correlation is

$$E = \sum (\text{observed } Y \text{ value} - \text{predicted } Y \text{ value})^2$$

For rule 1, the predicted values all equal $\bar{Y}$. The total prediction error equals

$$E_1 = \sum (Y - \bar{Y})^2$$

This is called the ***total sum of squares*** of the Y-values about their mean. We denote this by TSS. For rule 2, the predicted values are the $\hat{Y}$ values from the prediction equation. The total prediction error equals

$$E_2 = \sum \left(Y - \hat{Y}\right)^2$$

We previously denoted this by SSE, the ***sum of squared errors***.

When a strong linear relationship exists between X and Y, the prediction equation provides predictions ($\hat{Y}$) that are much better than $\bar{Y}$, in the sense that the sum of squared prediction errors is substantially less. Figure 9.13 shows graphical representations of the two predictors and their prediction errors. For rule 1, the same prediction ($\bar{Y}$) applies for the value of Y, regardless of the value of X. For rule 2 the prediction changes as X changes, and the prediction errors tend to be smaller.

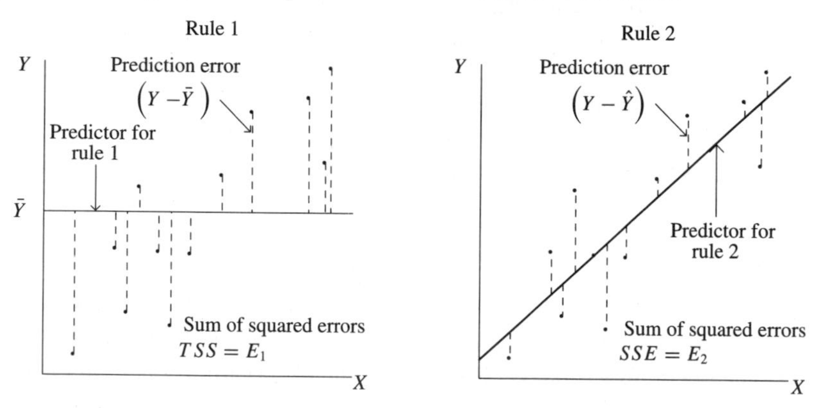

Figure 9.13 Graphical Representation of rule 1 and Total Sum of Squares $E_1 = TSS = \sum(Y - \bar{Y})^2$, rule 2 and Residual Sum of Squares $E_2 = SSE = \sum(Y - \hat{Y})^2$

Definition of Measure: The proportional reduction in error from using the linear prediction equation instead of $\bar{Y}$ to predict Y is called the ***coefficient of determination***. It is denoted by r^2. The PRE formula for r^2 is

$$r^2 = \frac{E_1 - E_2}{E_1} = \frac{\text{TSS} - \text{SSE}}{\text{TSS}} = \frac{\sum(Y - \bar{Y})^2 - \sum(Y - \hat{Y})^2}{\sum(Y - \bar{Y})^2}$$

Properties of Coefficient of Determination

We use the notation r^2 for this measure because, in fact, it equals the square of the Pearson correlation r. If we know the correlation, it is simple to calculate the coefficient of determination. The PRE formula above is useful for interpreting r^2, but it is not needed for its calculation.

The coefficient of determination, like the Pearson correlation, measures the strength of *linear* association between X and Y. We emphasize *linear* because r^2 compares the linear prediction equation to $\bar{Y}$ for making predictions.

Example 9.9 r^2 for Murder and Poverty Rates

The correlation between poverty rate and murder rate for the 50 states is $r = .629$. The coefficient of determination is, therefore, $r^2 = (.629)^2 = .395$. For predicting murder rate, the linear prediction equation $\hat{Y} = -.86 + .58X$ has 39.5% less error than $\bar{Y}$.

One can also calculate r^2 directly from the definition. Software for regression routinely provides tables of sums of squares, such as shown in part of Table 9.3. From it, the sum of squared errors using the prediction equation is $\text{SSE} = \sum(Y - \hat{Y})^2 = 470.4$; the total sum of squares is $\text{TSS} = \sum(Y - \bar{Y})^2 = 777.7$ (Alternatively, if we already know $s_Y = 3.98$, then $\sum(Y - \bar{Y})^2 = (n-1)s_Y^2 = 49(3.98)^2$). Thus,

$$r^2 = \frac{\text{TSS} - \text{SSE}}{\text{TSS}} = \frac{777.7 - 470.4}{777.7} = \frac{307.3}{777.7} = .395$$

Normally, it is unnecessary to perform this computation, since most software reports r or r^2 or both. □

The properties of the coefficient of determination r^2 follow directly from those of the Pearson correlation r.

- Since $-1 \le r \le 1$, r^2 falls between 0 and 1.
- The minimum possible value for SSE is 0, in which case $r^2 = \text{TSS}/\text{TSS} = 1$. For $\text{SSE} = 0$, all sample points must fall exactly on the prediction line. In that case, there is no prediction error using X to predict Y. This condition corresponds to $r = \pm 1$.
- If SSE = TSS, then $r^2 = 0$, in which case the slope $b = 0$ also. Now, SSE can be no larger than TSS, since the least squares line provides the minimum sum of squared errors about a line. When the least squares slope $b = 0$, the Y-intercept a equals $\bar{Y}$ (since $a = \bar{Y} - b\bar{X}$, which equals $\bar{Y}$ when $b = 0$), so that $\hat{Y} = \bar{Y}$ for all X. The two prediction rules are then identical, so that SSE = TSS and $r^2 = 0$.
- The closer r^2 is to 1, the stronger the linear association, in the sense that the more effective the least squares line $\hat{Y} = a + bX$ is relative to $\bar{Y}$ in predicting the response variable.
- r^2 does not depend on the units of measurement.
- r^2 takes the same value when X predicts Y as when Y predicts X.

Conditional Versus Marginal Variability

To summarize, the Pearson correlation r falls between -1 and $+1$. It indicates the direction of the association, positive or negative, through its sign. It is a standardized slope, indicating what the slope equals for units of measurement such that X and Y are equally disperse. A one standard deviation change in X corresponds to an r standard deviation change in Y. The square of the correlation has a proportional reduction in error interpretation related to the reduction in error from predicting Y using $\hat{Y} = a + bX$ rather than $\bar{Y}$.

The total sum of squares, TSS = $\sum(Y - \bar{Y})^2$, summarizes the *variability* of the observations on Y, since this quantity divided by $n - 1$ is the sample variance s_Y^2 of the Y-values. Similarly, SSE = $\sum(Y - \hat{Y})^2$ summarizes the variability around the prediction equation, which refers to variability for the conditional distributions. When $r^2 = .39$, the variability in Y using X to make the predictions (via the prediction equation) is 39% less than the overall variability of the Y values. Thus, the r^2 result is often expressed as "X explains 39% of the variability in Y" or "39% of the variance in Y is explained by its linear relationship with X." Roughly speaking, the variance of the conditional distribution of Y for a given X is 39% smaller than the variance of the marginal distribution of Y.

When two variables X and Y are strongly associated, the variation in the conditional distributions is considerably less than the variation in the marginal distribution. For instance, the estimated conditional standard deviation $\hat{\sigma}$ is then much smaller than the sample marginal standard deviation s_Y of Y. Figure 9.9 illustrated this.

9.5 Inferences for the Slope and Correlation

Sections 9.1–9.3 showed how a linear function can represent the *form* of relationships between quantitative variables. Section 9.4 used the Pearson correlation and its square to describe the *strength* of the association. These parts of a regression analysis are descriptive. We now present inferential methods for the parameters in regression models.

A test of whether the two quantitative variables are statistically independent has the same purpose as the chi-squared test for qualitative variables. More usefully, one can construct a confidence interval for the slope β of the regression equation or for the population Pearson correlation. These inferences enable us to judge whether the variables are associated and to estimate the direction and strength of the association.

Assumptions for Statistical Inference

The statistical inferences presented in this section make the following assumptions:

- The mean of Y is related to X by the linear equation $E(Y) = \alpha + \beta X$.
- The conditional standard deviation is identical at each X-value.
- The conditional distribution of Y at each value of X is normal.
- The sample is selected randomly.

The first assumption states that the linear regression model is valid. The second assumption is one under which the least squares estimates are optimal. The third assumption is needed for the standardized version of the sample slope or correlation to have a t sampling distribution. In practice, of course, none of these assumptions is ever satisfied exactly. For large samples, the normality assumption is unimportant. If the second assumption is violated, other estimates may be more efficient than least squares, but the methods we discuss here are still approximately valid. The first and last assumptions are very important. If the true relationship deviates greatly from a straight line, for instance, it does not make sense to use a slope or a correlation to describe it.

Test of Independence

Under these assumptions, suppose the mean of Y is identical at each X-value. In other words, the normal conditional distribution of Y is the same at each X-value. Then, the two quantitative variables are statistically independent. For the linear regression model $E(Y) = \alpha + \beta X$, this means that the slope β of the regression line equals 0 (see Figure 9.14). The null hypothesis that the variables are statistically independent is $H_0 : \beta = 0$.

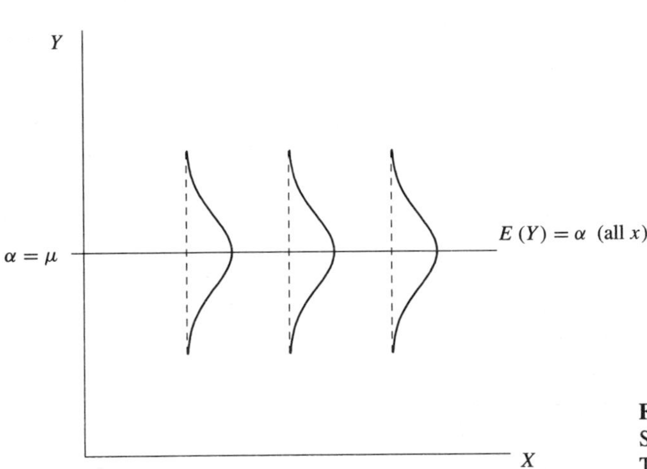

Figure 9.14 X and Y Are Statistically Independent When the True Slope Equals 0

We normally test independence against the two-sided alternative hypothesis H_a: $\beta \neq 0$. Occasionally we use a one-sided alternative, H_a: $\beta > 0$ or H_a: $\beta < 0$, to predict the direction of the association. The test statistic in either case equals

$$t = \frac{b}{\hat{\sigma}_b}$$

where $\hat{\sigma}_b$ denotes the standard error of the sample slope b. The form of the test statistic is the usual one for a t or z test. We take the estimate b of the parameter β, subtract the null hypothesis value of the parameter ($\beta = 0$), and divide by the standard error of the estimate b. Under the assumptions, this test statistic has the t sampling distribution with $df = n - 2$.

The formula for the estimated standard error of b is

$$\hat{\sigma}_b = \frac{\hat{\sigma}}{\sqrt{\sum(X - \bar{X})^2}} = \frac{\hat{\sigma}}{s_{\hat{X}}\sqrt{n-1}}$$

This depends on the point estimate $\hat{\sigma}$ of the standard deviation of the conditional distributions of Y. From Section 9.3, this estimate is

$$\hat{\sigma} = \sqrt{\frac{\text{SSE}}{n-2}}$$

called the ***root mean square error*** by most software.

The smaller the standard deviation of the conditional distributions, the more precisely b estimates β. A small $\hat{\sigma}$ occurs when the data points show little variability about the prediction equation. Also, the standard error of b is inversely related to $\sum(X - \bar{X})^2$, the sum of squares of the observed X-values about their mean. This sum increases, and hence b estimates β more precisely, as the sample size n increases and when the X-values are more highly spread out.

As in the two-sided t test for a mean, the P-value for the alternative $H_a : \beta \neq 0$ is the two-tail probability from the t distribution (Table B). The degrees of freedom for the t sampling distribution are the same as the df of the estimate $\hat{\sigma}$, namely, the denominator $n - 2$. When $df > 30$, recall that the t distribution is very similar to the standard normal distribution. We can then treat the t test statistic as a z statistic and approximate the P-value using the normal probability table (Table A). Computer software provides the appropriate P-value for any value of df.

Example 9.10 Regression for Selling Price of Homes

What factors affect the selling price of a house? Table 9.4 shows 93 observations on home sales in Gainesville, Florida, in January 1996. Variables listed are selling price (in thousands of dollars), size of house (in thousands of square feet), number of bedrooms, number of bathrooms, and whether the house is newly built. For now, we use only the data on Y = selling price and X = size of house.

Since these 93 observations come from one city alone, we cannot use them to make inferences about the relationship between X and Y in general. The houses sold are not even a random sample of homes in Gainesville. Nevertheless, we treat them as a random sample of a conceptual population of home sales in this market in order to analyze how these variables seem to be related.

Figure 9.15 shows a scatter diagram of selling price and size of house, with a box plot for each variable. The diagram displays a strong positive trend, and the model $E(Y) = \alpha + \beta X$ seems appropriate. A couple of the points are severe outliers on each variable, however, and one of them falls rather far from the overall trend; we discuss this abnormality in Section 14.5, which introduces an alternative model for these data that does not assume constant variability around the regression line.

TABLE 9.4 Selling Prices and Related Factors for a Sample of Sales of Homes in Gainesville, Florida

P	S	Be	Ba	New	P	S	Be	Ba	New	P	S	Be	Ba	New
48.5	1.10	3	1	0	76.0	1.66	3	2	0	108.0	1.79	4	2	1
55.0	1.01	3	2	0	81.8	1.33	3	2	0	107.5	1.85	3	2	0
68.0	1.45	3	2	0	84.5	1.34	3	2	0	109.9	2.06	4	2	1
137.0	2.40	3	3	0	83.5	1.40	3	2	0	110.0	1.76	4	2	0
309.4	3.30	4	3	1	86.0	1.15	2	2	1	120.0	1.62	3	2	1
17.5	.40	1	1	0	86.9	1.58	3	2	1	115.0	1.80	4	2	1
19.6	1.28	3	1	0	86.9	1.58	3	2	1	113.4	1.98	3	2	0
24.5	.74	3	1	0	86.9	1.58	3	2	1	114.9	1.57	3	2	0
34.8	.78	2	1	0	87.9	1.71	3	2	0	115.0	2.19	3	2	0
32.0	.97	3	1	0	88.1	2.10	3	2	0	115.0	2.07	4	2	0
28.0	.84	3	1	0	85.9	1.27	3	2	0	117.9	1.99	4	2	0
49.9	1.08	2	2	0	89.5	1.34	3	2	0	110.0	1.55	3	2	0
59.9	.99	2	1	0	87.4	1.25	3	2	0	115.0	1.67	3	2	0
61.5	1.01	3	2	0	87.9	1.68	3	2	0	124.0	2.40	4	2	0
60.0	1.34	3	2	0	88.0	1.55	3	2	0	129.9	1.79	4	2	1
65.9	1.22	3	1	0	90.0	1.55	3	2	0	124.0	1.89	3	2	0
67.9	1.28	3	2	0	96.0	1.36	3	2	1	128.0	1.88	3	2	1
68.9	1.29	3	2	0	99.9	1.51	3	2	1	132.4	2.00	4	2	1
69.9	1.52	3	2	0	95.5	1.54	3	2	1	139.3	2.05	4	2	1
70.5	1.25	3	2	0	98.5	1.51	3	2	0	139.3	2.00	4	2	1
72.9	1.28	3	2	0	100.1	1.85	3	2	0	139.7	2.03	3	2	1
72.5	1.28	3	1	0	99.9	1.62	4	2	1	142.0	2.12	3	3	0
72.0	1.36	3	2	0	101.9	1.40	3	2	1	141.3	2.08	4	2	1
71.0	1.20	3	2	0	101.9	1.92	4	2	0	147.5	2.19	4	2	0
76.0	1.46	3	2	0	102.3	1.42	3	2	1	142.5	2.40	4	2	0
72.9	1.56	4	2	0	110.8	1.56	3	2	1	148.0	2.40	5	2	0
73.0	1.22	3	2	0	105.0	1.43	3	2	1	149.0	3.05	4	2	0
70.0	1.40	2	2	0	97.9	2.00	3	2	0	150.0	2.04	3	3	0
76.0	1.15	2	2	0	106.3	1.45	3	2	1	172.9	2.25	4	2	1
69.0	1.74	3	2	0	106.5	1.65	3	2	0	190.0	2.57	4	3	1
75.5	1.62	3	2	0	116.0	1.72	4	2	1	280.0	3.85	4	3	0

Note: P = selling price (thousands of dollars), S = size (thousands of square feet), Be = number of bedrooms, Ba = number of bathrooms, New = 1 if new and 0 if not. Data provided by Jane Myers, Coldwell-Banker Realty.

Table 9.5 shows part of a SAS computer printout for a regression analysis. The least squares estimates of α and β are $a = -25.2$ and $b = 75.6$, so the prediction equation is $\hat{Y} = -25.2 + 75.6X$. The predicted selling price increases by $b = 75.6$ thousand dollars for an increase in size of 1 thousand square feet; that is, it increases about $76 per square foot. Figure 9.15 also superimposes the prediction equation over the scatter diagram.

Table 9.5 reports that the standard error of the slope estimate is $\hat{\sigma}_b = 3.865$. This is listed under "Standard Error." This value estimates the variability in sample slope values that would result from repeatedly selecting random samples of 93 house sales in Gainesville and calculating prediction equations.

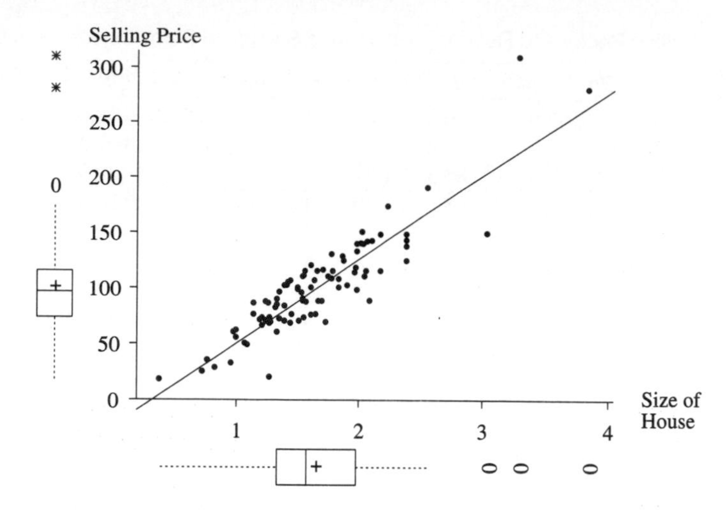

Figure 9.15 Scatter Diagram and Prediction Equation for Y = Selling Price of House (in Thousands of Dollars) and X = Size of House (in Thousands of Square Feet)

To test independence, $H_0 : \beta = 0$, we form the t test statistic,

$$t = \frac{b}{\hat{\sigma}_b} = \frac{75.61}{3.865} = 19.6$$

This statistic is listed in Table 9.5 under "T for H0: Parameter = 0." Since the sample size equals $n = 93$, the degrees of freedom for this statistic are $df = n - 2 = 91$. This t statistic has sampling distribution that is essentially identical to the standard normal distribution. Clearly, this is an extremely large test statistic.

The P-value, listed in Table 9.5 under the heading PROB > $|T|$, is .0001 to four decimal places. This refers to the two-sided alternative $H_a : \beta \neq 0$; that is, it is the two-tailed probability of a t statistic at least as large in absolute value as the absolute value of the observed t, $|t| = 19.6$, if H_0 were true. For the one-sided alternative $H_a : \beta > 0$, this level is halved, since P is then the right-hand tail probability of $t > 19.6$.

Table 9.6 shows part of a SPSS printout for the same analysis. For SPSS, "B" denotes the estimated regression coefficient, and "(Constant)" is the label for the Y-intercept. "SE B" denotes the standard error of b, "Sig T" denotes the two-sided P-value for the t test for a coefficient, and "Beta" denotes the estimated standardized regression coefficient (this is the Pearson correlation for the regression model of this chapter; it is not to be confused with the population slope, β, which is unknown).

Both the SAS and SPSS printouts also contain a standard error and t test for the Y-intercept. This information is usually not of interest, since rarely is there any reason to test the hypothesis that a Y-intercept equals 0. In fact, for this example, the Y-intercept does not have any interpretation, since houses of size 0 do not exist.

TABLE 9.5 Part of a SAS Computer Printout for Regression Analysis of Selling Price and Size of House

Variable	N	Mean	Std Dev
PRICE	93	99.533	44.184
SIZE	93	1.650	0.525

Source	DF	Sum of Squares	Mean Square	Root MSE	R-square
Model	1	145097.464	145097.464	19.473	0.8079
Error	91	34508.402	379.213		
Total	92	179605.867			

Variable	DF	Parameter Estimate	Standard Error	T for H0: Parameter=0	Prob > |T|
INTERCEP	1	-25.1936	6.6884	-3.767	0.0003
SIZE	1	75.6068	3.8652	19.561	0.0001

TABLE 9.6 Part of a SPSS Computer Printout for Regression Analysis of Selling Price and Size of House

	DF	Sum of Squares	Mean Square
Regression	1	145097.464	145097.464
Residual	91	34508.402	379.213

R Square	.8079
Standard Error	19.4734

Variable	B	SE B	Beta	T	Sig T
SIZE	75.6068	3.8652	.8988	19.561	.0000
(Constant)	-25.1936	6.6884		-3.767	.0003

In summary, we can feel highly confident in concluding that size has a positive effect on price. On the average, selling price increases as size of house increases. This is no big surprise. Indeed, we would be shocked if these variables were independent, and a test of independence is irrelevant, for practical purposes, for these data. As is usually the case, estimation is more informative than significance testing. □

Confidence Interval for the Slope

A small P-value for $H_0 : \beta = 0$ in a test of independence suggests that the regression line has a nonzero slope. However, we should be more concerned with the size of the slope than in knowing merely that it is not 0. If the absolute value of the slope is small, in relation to the units of measurement for X and Y, the association could be statistically significant but practically unimportant. It is more informative, therefore, to construct a confidence interval for the true slope β.

A confidence interval for the slope has the formula

$$b \pm t\hat{\sigma}_b$$

The t-score is the value from Table B, with $df = n - 2$, for the desired confidence coefficient. As usual, use the t-score with half the error probability in each tail, for instance $t_{.025}$ for a 95% confidence interval. This interval is similar in form to the confidence interval for a mean (Section 6.5). We take the estimate b and add and subtract some t multiple of the standard error of the estimate.

Example 9.11 Estimating the Slope for Housing Data

For the data on X = size of house and Y = selling price, $b = 75.6$ and $\hat{\sigma}_b = 3.87$. The parameter β refers to the change in the mean selling price for a thousand square foot increase in size (i.e., a one-unit increase in X). For a 95% confidence interval, we use the $t_{.025}$ value for $df = n - 2 = 91$, which is $t_{.025} = 1.99$. (It is permissible to use the z-score if your software does not supply t scores or confidence intervals.) The interval is

$$\begin{aligned} b \pm t_{.025}\hat{\sigma}_b &= 75.6 + 1.99(3.87) \\ &= 75.6 \pm 7.7 \quad \text{or} \quad (67.9, 83.3) \end{aligned}$$

We can be 95% confident that β lies between 67.9 and 83.3. The mean selling price increases by between \$67,900 and \$83,300 for a thousand-square-foot increase in house size; equivalently, this is \$67.90 to \$83.30 per square foot. □

A confidence interval for β may not be meaningful if a one-unit increase in X is relatively small or large, in practical terms. One should make inferences about the change in $E(Y)$ for an increase in X that is a more relevant portion of the actual range of X values. To obtain endpoints for the confidence interval for a constant multiple of the slope (e.g., $.1\beta$, the change in the mean of Y for an increase of .1 unit in X), multiply the endpoints of the interval for β by the same constant.

For Table 9.4, X = size of house has $\bar{X} = 1.65$ and $s_X = .53$. A change of 1.0 thousand square feet in size is very large. Let us estimate the effect of a 100-square-foot increase in area. Now, 100 square feet equals .1 thousand square feet, so we refer to the effect of a .1 unit change in X. The change in the mean of Y is $.1\beta$.

A confidence interval for $.1\beta$ has endpoints that are .1 times the endpoints of the interval for β. Thus, a 95% confidence interval for $.1\beta$ has endpoints .1(67.9) = 6.79 and .1(83.3) = 8.33. We infer that the mean selling price increases by at least \$6790 and at most \$8330, for a 100-square-foot increase in house size.

To illustrate, we compare the selling prices of homes having $X = 1.7$ and $X = 1.6$ thousand square feet of size. The estimated difference in selling price is $.1b =$

7.56 thousand dollars. Assuming that the linear regression model is appropriate, we conclude that the mean is between \$6790 and \$8330 higher for houses of 1700 square feet than for houses of 1600 square feet.

Reading the Computer Printout

Let's take a closer look at the SAS computer printout in Table 9.5. It contains a considerable amount of additional information we have not yet discussed. For instance, the middle part of the printout reports a sum of squares table. From it, the sum of squared errors (SSE) is 34,508.4. The estimated conditional standard deviation of Y for fixed X is

$$\hat{\sigma} = \sqrt{\text{SSE}/(n-2)} = \sqrt{34{,}508.4/91} = 19.5$$

This is labeled "Root MSE," for square root of the mean square error. It describes the estimated variability in selling prices, for any fixed value for size of house.

The sum of squares table also reports the total sum of squares, $\text{TSS} = \sum(Y-\bar{Y})^2 = 179,605.9$. From this value and SSE,

$$r^2 = \frac{\text{TSS} - \text{SSE}}{\text{TSS}} = \frac{179{,}605.9 - 34{,}508.4}{179{,}605.9} = .808$$

This is the proportional reduction in error in using house size to predict selling price, and is listed on the printout under the heading R-SQUARE. Since the slope of the prediction equation is positive, the Pearson correlation is the positive square root of this value, or .899. A strong positive association exists between these variables.

In Table 9.5, the total sum of squares TSS partitions into two parts, the sum of squared errors, $\text{SSE} = 34{,}508.4$, and the difference between TSS and SSE, $\text{TSS} - \text{SSE} = 145{,}097.5$. This difference is the numerator of the r^2 measure and is called the ***model sum of squares*** or sometimes the ***regression sum of squares*** or ***explained sum of squares***. It represents the amount of the total variation TSS in Y that is explained by X in using the least squares line. The ratio of the model sum of squares to the total sum of squares equals r^2.

The table of sums of squares has an associated list of degrees of freedom values. The degrees of freedom for the total sum of squares $\text{TSS} = \sum(Y - \bar{Y})^2$ is $n - 1 = 92$, since TSS refers to variability in the *marginal* distribution of Y, which has sample variance $s_Y^2 = \text{TSS}/(n-1)$. The degrees of freedom for SSE equals $n-2 = 91$, since it refers to variability in the *conditional* distribution of Y, which has variance estimate $\hat{\sigma}^2 = \text{SSE}/(n-2)$ for a model having two parameters. The model sum of squares has *df* equal to the number of explanatory variables in the regression model, in this case 1. The sum of *df* for the model sum of squares and *df* for the sum of squared errors equals $df = n - 1$ for the total sum of squares.

In the SPSS printout of Table 9.6, the "Regression" sum of squares is the model sum of squares and the "Residual" sum of squares is SSE. The term labeled "Standard Error" is the root mean square error, the square root of SSE divided by its *df*.

Inference for the Pearson Correlation*

The Pearson correlation r equals 0 in the same situations in which the slope b of the least squares line equals 0. Let ρ (rho) denote the value of the Pearson correlation for the regression equation $E(Y) = \alpha + \beta X$; that is, ρ is the population value of r. Then, $\rho = 0$ precisely when $\beta = 0$. Thus, a test of H_0: $\rho = 0$ using the sample value r is equivalent to the t test of $H_0 : \beta = 0$ using the sample value b.

The test statistic for testing $H_0 : \rho = 0$ is

$$t = \frac{r}{\sqrt{(1 - r^2)/(n - 2)}}$$

This provides the same value as the test statistic $t = b/\hat{\sigma}_b$. Use either statistic to test the null hypothesis of independence, since each has the same t sampling distribution with $df = n - 2$ and yields the same P-value. For example, the Pearson correlation of .899 for the house price data in Example 9.10 leads to

$$t = \frac{r}{\sqrt{(1 - r^2)/(n - 2)}} = \frac{.899}{\sqrt{(1 - .808)/91}} = 19.6$$

This is the same t-value as in Example 9.10 for testing $H_0 : \beta = 0$.

For a set of variables, software for regression analyses reports their correlations in a ***correlation matrix***. This matrix is a square table listing the variables as the rows and again as the columns. It reports the correlation for each pair, and usually also the two-sided P-value for testing the significance of the correlation.

Table 9.7 illustrates the way software reports the correlation matrix for four of the variables from Table 9.1, deleting the observation for D.C. The correlation between each pair of variables appears twice. For instance, the correlation of .3688 between violent crime rate and percentage in poverty occurs both in the row for "VIOLENT" and column for "POVERTY" and in the row for "POVERTY" and column for "VIOLENT." The P-value for testing that that true correlation equals zero against the two-sided alternative is .0084, listed underneath the correlation. The correlations on the diagonal running from the upper left-hand corner to the lower right-hand corner of the table all equal 1.000, indicating simply that the correlation between a variable and itself is 1.0. For instance, if we know the value of Y, then we can predict the value of Y perfectly.

Although tests are simple, constructing a confidence interval is more complicated for ρ than for the slope β. The reason is that the sampling distribution of r is not symmetric except when $\rho = 0$. The lack of symmetry is caused by the restricted range $[-1, 1]$ for r values. If ρ is close to 1.0, for instance, then the sample r cannot fall much above ρ, but it can fall well below ρ, and the sampling distribution of r is skewed to the left. Problem 9.46 shows how to construct confidence intervals for correlations.

TABLE 9.7 Computer Printout of Correlation Matrix for Variables from Table 9.1

```
Pearson Correlations / Prob > |R| under Ho: Rho=0

            VIOLENT     MURDER      POVERTY     SINGLE
VIOLENT     1.00000     0.78147     0.36875     0.64868
            0.0         0.0001      0.0084      0.0001

MURDER      0.78147     1.00000     0.62862     0.72807
            0.0001      0.0         0.0001      0.0001

POVERTY     0.36875     0.62862     1.00000     0.43031
            0.0084      0.0001      0.0         0.0018

SINGLE      0.64868     0.72807     0.43031     1.00000
            0.0001      0.0001      0.0018      0.0
```

9.6 Model Assumptions and Violations

We end this chapter by reconsidering the assumptions underlying linear regression analysis. In particular, we discuss the effects of violating these assumptions and the effects of "influential" observations. Finally, we show an alternate way to express the model.

Comments About Assumptions

The linear regression model assumes that the relationship between X and the mean of Y follows a straight line. The actual form is unknown, and it is unlikely to be *exactly* linear. Nevertheless, a linear function often provides a decent approximation for the actual form. Figure 9.16 illustrates a straight line falling close to an actual curvilinear relationship. If the assumption of linearity is badly violated, as with a U-shaped relationship, results and conclusions using the linear model may be very misleading. For this reason, you should always construct a scatter diagram to check this fundamental assumption.

For instance, the inferences discussed in the previous section are appropriate for detecting positive or negative linear associations. If the true relationship were U-shaped (such as in Figure 9.5), the variables would be statistically dependent, since the mean of Y would change according to the value of X. The t test of independence might not detect it, though, since the slope b of the least squares line would be close to 0. In other words, a small P-value would probably not occur for H_0: $\beta = 0$ even though an association exists. In summary, $\beta = 0$ need not correspond to independence if the assumption of a linear regression model is violated.

The least squares line and the measures of association r and r^2 are valid descriptive statistics no matter what the shape of the conditional distribution of Y-values for each X-value. However, the statistical inferences in Section 9.5 make the additional

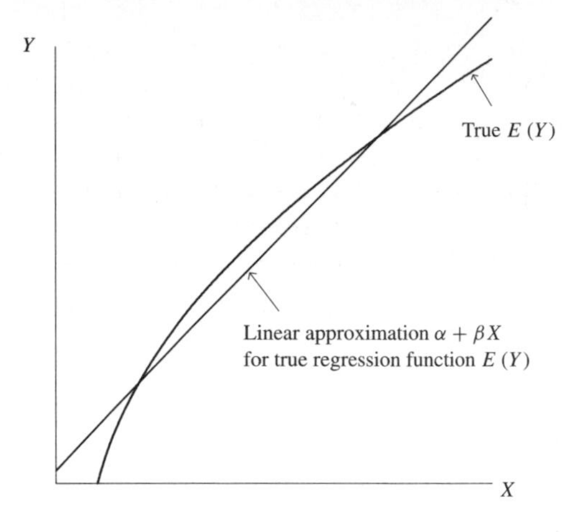

Figure 9.16 A Linear Regression Equation as an Approximation for Nonlinear Relationship

assumptions that the conditional distributions of Y are (1) normal, with (2) identical standard deviation σ for each X-value. These assumptions are also never *exactly* satisfied in practice. However, the closer reality falls to this ideal model, the more appropriate are the confidence interval and test procedures introduced in that section. For large samples, the normality assumption is not crucial, since an extended Central Limit Theorem implies that sample slopes and correlations have approximately normal sampling distributions.

Chapter 14 discusses ways of checking the assumptions of regression modeling and making modifications to the analysis, if necessary.

Extrapolation Is Dangerous

It is dangerous to apply a prediction equation to values of X outside the range of observed values of X. The relationship might not be linear outside that range. We might get poor or even absurd predictions by extrapolating beyond the observed range.

To illustrate, the prediction equation $\hat{Y} = -.86 + .58X$ in Section 9.2 relating X = poverty rate to Y = murder rate was based on sample poverty rates between 8.0 and 26.4. It is not valid to extrapolate much below or above this range. For instance, the predicted murder rate for a poverty rate of $X = 0\%$ is $\hat{Y} = -.86$. This is an impossible value for murder rate, which cannot be negative.

Influential Observations

The least squares method has a long history and is the standard way to fit prediction equations to data. A disadvantage of least squares, however, is that individual observations can unduly influence the fit of the model. A single observation can have a large effect on the fit if its X value is unusually large or unusually small and if it falls quite far from the trend that the rest of the data follow.

Figure 9.17 illustrates this. The figure plots observations for several African and Asian nations, taken from Table 9.13 in Problem 9.17, on Y = crude birth rate (number of births per 1000 population size) and X = number of televisions per 100 people. We added to the figure an observation on these variables for the United States, which is the outlier that is much lower than the other countries in birth rate but much higher on number of televisions. Figure 9.17 shows the prediction equations both without and with the U.S. observation. The prediction equation changes from $\hat{Y} = 29.8 - .024X$ to $\hat{Y} = 31.2 - .195X$. Adding only a single point to the data set causes the prediction line to tilt dramatically downward.

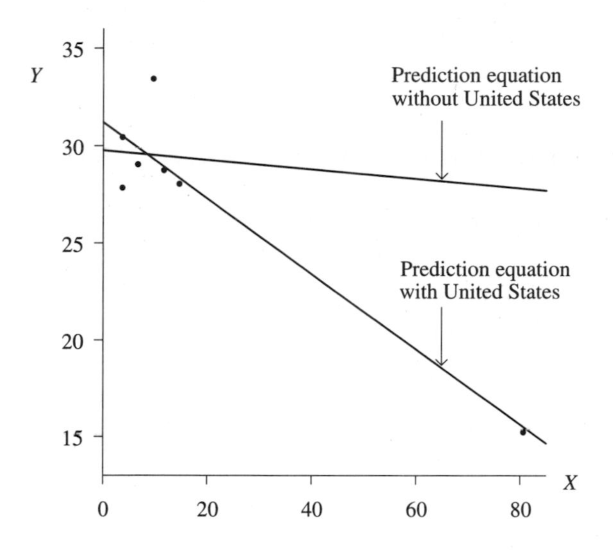

Figure 9.17 Prediction Equations for Y = Birth Rate and X = Television Ownership, with and without Observation for United States

Section 9.2 showed a not-so-extreme version of this. The slope of the prediction equation more than doubled when we included the observation for D.C. in the data set.

When a scatter diagram reveals a severe outlier, the reasons for it should be investigated. An observation may have been incorrectly recorded. If the observation is correct, perhaps that observation is fundamentally different from the others in some way, such as the U.S. observation in Figure 9.17. It may suggest an additional predictor for the model. It is often worthwhile to refit the model without one or two extreme outliers to see if those observations have a large effect on the parameter estimates, as we did in Example 9.4 with the D.C. observation for the murder rate data.

Observations that have a large influence on the model parameter estimates can also have a large impact on the correlation. For instance, for the data in Figure 9.17, the correlation is $-.935$ when the outlier is included and $-.051$ when it is deleted from the data set. One point can make quite a difference, especially when the sample size is small.

Factors Influencing the Correlation

Besides being influenced by outliers, the sample correlation depends on the range of X-values sampled. When a sample has a much narrower range of variation in X than the population, for example, the sample correlation tends to underestimate drastically (in absolute value) the population correlation.

Figure 9.18 shows a scatter diagram of 500 points that is very regular and has a correlation of $r = .705$. Suppose, instead, we had only sampled the middle half of the points, roughly between X values of 43 and 57. Then the correlation equals only $r = .330$, considerably lower. For another example, consider the relation between housing price and size of house, portrayed in Figure 9.15. The correlation equals .899. If we sampled only those sales in which house size is between 1.3 and 2.0 thousand feet, which include 48 of the 93 observations, the correlation decreases to .564.

The correlation is most appropriate as a summary measure of association when the sample (X, Y)-values are a random sample of the population. This way, there is a representative sample of the X variation as well as the Y variation.

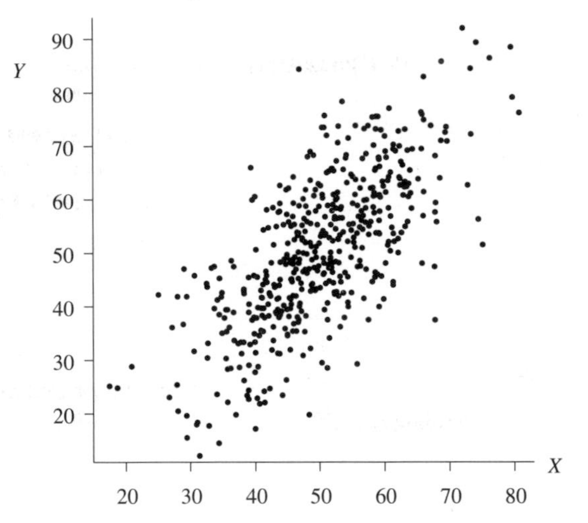

Figure 9.18 The Correlation is Affected by the Range of X Values; Correlation Decreases from .705 to .330 Using Only Points with X Between 43 and 57

Example 9.12 Does the SAT Predict College GPA?

In studying the association between score on a college entrance exam (such as the SAT) and later performance in college (such as GPA at end of second year), the strength of the correlation depends on the variability in SAT scores in the sample. If we study the association only for students at Harvard University, the correlation will probably not be strong, because the sample SAT scores will be concentrated very narrowly at the upper end of the scale. By contrast, if we randomly sampled from the population of all high school students who take the SAT and placed those students in the Harvard environment, students with poor SAT scores would tend to have low GPAs at Harvard, and we would then observe a much stronger correlation. □

Other aspects of regression, such as fitting a prediction equation to the data and making inferences about the slope, remain valid when we randomly sample Y within a restricted range of X-values. We simply limit our predictions to that range. The slope of the prediction equation is not affected by a restriction in the range of X. For Figure 9.18, for instance, the sample slope equals .97 for the full data and .96 for the restricted middle set. The correlation makes most sense, however, when both X and Y are random, rather than only Y.

Regression Model with Error Terms*

We now show an alternative way of writing the linear regression equation. Recall that at a fixed value of X, the values of Y fluctuate around their mean, $E(Y) = \alpha + \beta X$. Any specific observation on Y may fall above that mean (i.e., above the regression line) or below that mean (below the regression line). The standard deviation component σ of the model summarizes the typical sizes of the deviations from the mean.

An alternative formulation for the model expresses each observation on Y, rather than the mean $E(Y)$ of the values, in terms of X. One does this by including a term for the deviation of the observation from the mean. This approach models an observation on Y as

$$Y = \alpha + \beta X + \varepsilon$$

The term denoted by ε (the Greek letter, epsilon) represents the deviation of the observation from the mean, $\alpha + \beta X$. Each observation has its own value for ε.

If ε is positive, then $\alpha + \beta X + \varepsilon$ is larger than $\alpha + \beta X$, and the observation falls above the mean. See Figure 9.19. If ε is negative, the observation falls below the mean. When $\varepsilon = 0$, the observation falls exactly at the mean. The mean of the ε-values is 0.

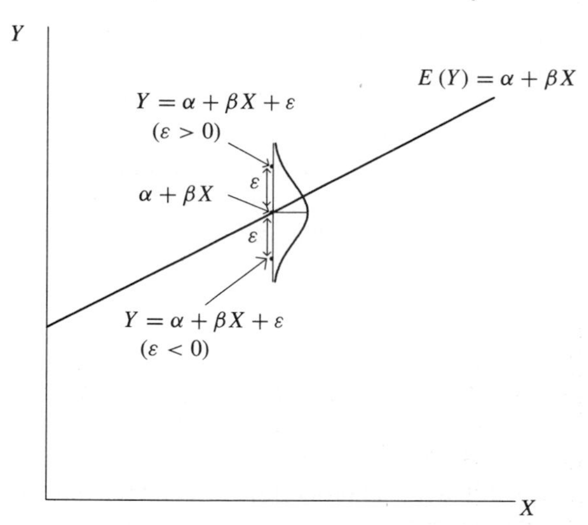

Figure 9.19 Positive and Negative ε-Values Correspond to Observations Above and Below the Conditional Mean

The ε term in the equation $Y = \alpha + \beta X + \varepsilon$ occurs because observations with the same X-values do not all have the same Y-values; that is, the variables are not perfectly correlated. The model $Y = \alpha + \beta X + \varepsilon$ is a *probabilistic* one. For each X, variability in the Y-values corresponds to variability in ε. The ε term is called the ***error term***, since it represents the error that results from using the mean value $(\alpha + \beta X)$ of Y at a certain value of X for the prediction of the individual observation on Y.

For a particular sample of n observations, we do not know the n values for ε, just like we do not know the parameter values and the true mean $\alpha + \beta X$. For the sample data and their prediction equation, let e be such that

$$Y = a + bX + e$$

That is, $Y = \hat{Y} + e$, so that $e = Y - \hat{Y}$. Then e is simply the *residual*, the difference between the observed and predicted values of Y. Since $Y = \alpha + \beta X + \varepsilon$, the residual e estimates ε. We can interpret ε as a ***population residual***. Thus, ε is the difference between the observation Y and the mean $\alpha + \beta X$ of all possible observations on Y at that value of X. Graphically, ε is the vertical distance between the observed point and the true regression line.

One can write the regression model either as $E(Y) = \alpha + \beta X$ or as $Y = \alpha + \beta X + \varepsilon$. Although we use the first equation, we introduced this alternative one since some books and articles express the regression model in this way.

Models and Reality

We emphasize again that the regression model *approximates* the true relationship. No sensible researcher expects a relationship to be exactly linear, with exactly normal conditional distributions at each X and with exactly the same standard deviation of Y-values at each X-value. However, by definition, models merely approximate reality.

If the model seems too simple to be adequate, the scatter diagram or other diagnostics may suggest improvement by using more general models introduced in the remainder of this text. Such models can be fitted, rechecked, and perhaps modified further. Model building is an iterative process. Its goals are to find a realistic model that is adequate for describing the relationship and making predictions but that is still simple enough to be easily interpreted. Chapters 11–15 present ways of extending the model so that it applies to situations in which the assumptions of this chapter are too simplistic.

9.7 Chapter Summary

Chapters 7–9 have dealt with the detection and description of *association between two variables*. Chapter 7 showed how to compare means or proportions for two groups. The explanatory variable is a binary variable defining the two groups; we compare means when the response variable is quantitative and proportions when the response variable is qualitative. When the variables are statistically independent, the means or proportions are identical for the two groups.

Chapter 8 dealt with *association between two qualitative variables.* Measures of association such as the difference of proportions, the odds ratio, and gamma describe the strength of association. The chi-squared statistic for nominal data or a z statistic based on sample gamma for ordinal data tests the hypothesis of independence.

This chapter dealt with *association between quantitative variables.* A new element studied here was a regression model to describe the *form* of the relationship between the explanatory variable X and the mean $E(Y)$ of the response variable. The major aspects of the analysis are as follows:

- The ***linear regression equation*** $E(Y) = \alpha + \beta X$ describes the *form* of the relationship. This regression model is appropriate when a straight line approximates the relationship between X and Y.
- A ***scatter diagram*** plots the data and checks whether the relationship is approximately linear. If it is, the ***method of least squares*** provides estimates of the Y-intercept α and the slope β for a linear prediction equation $\hat{Y} = a + bX$.
- The ***Pearson correlation r*** and its square, the ***coefficient of determination*** r^2, describe the *strength* of the linear association between two quantitative variables. The Pearson correlation is a standardized slope, having the same sign but falling between -1 and $+1$. Its square has a proportional reduction in error (PRE) interpretation; it describes the reduction in variability about the prediction equation compared to the variability about the sample mean of Y.
- For inference about the *existence* of a relationship between the variables, a t test using the slope or correlation tests the ***null hypothesis of independence***, namely, that the population slope and correlation equal 0. More informatively, one can construct confidence intervals for the unknown parameters.

Table 9.8 summarizes the methods studied in the past three chapters.

TABLE 9.8 Summary of Tests of Independence and Measures of Association

	Measurement Levels Of Variables		
	Nominal	Ordinal	Interval
Null hypothesis	H_0: Independence	H_0: Independence	H_0: Independence ($\beta = 0$)
Test statistic	$\chi^2 = \sum \frac{(f_o - f_e)^2}{f_e}$	$z = \frac{\hat{\gamma}}{\hat{\sigma}_{\hat{\gamma}}}$	$t = \frac{b}{\hat{\sigma}_b} = \frac{r}{\sqrt{\frac{1-r^2}{n-2}}}$, $df = n - 2$
Measure of association	$\hat{\pi}_2 - \hat{\pi}_1$	$\hat{\gamma} = \frac{C-D}{C+D}$	$r = b\left(\frac{s_X}{s_Y}\right)$
	Odds ratio		$r^2 = \frac{E_1 - E_2}{E_1} = \frac{\text{TSS}-\text{SSE}}{\text{TSS}}$

Chapter 11 introduces the ***multiple regression*** model, a generalization that permits *several* explanatory variables in the model. Chapter 12 shows how to include qualitative predictors in a regression model, and Chapter 13 includes both qualitative and quantitative predictors. Chapter 14 introduces models for more complex relationships, such as nonlinear ones. Finally, Chapter 15 presents regression models for qualitative response variables. Before discussing these multivariate models, however, we introduce in the next chapter some new concepts that help us to understand and interpret multivariate relationships.

PROBLEMS

Practicing the Basics

1. For the following variables in a regression analysis, which variable more naturally plays the role of X (explanatory variable) and which plays the role of Y (response variable)?
   a) College grade point average (GPA) and high school GPA.
   b) Number of children and mother's education level.
   c) Annual income and number of years of education.
   d) Annual income and assessed value of home.
2. Sketch plots of the following lines, for values of X between 0 and 10:
   a) $Y = 7 + .5X$
   b) $Y = 7 + X$
   c) $Y = 7 - X$
   d) $Y = 7 - .5X$
   e) $Y = 7$
   f) $Y = X$
3. For the data on the 50 states in Table 9.1 on Y = violent crime rate and X = poverty rate, the prediction equation is $\hat{Y} = 209.9 + 25.5X$.
   a) Sketch a plot of the prediction equation for X between 0 and 100.
   b) Interpret the Y-intercept and the slope.
   c) Find the predicted violent crime rate for Massachusetts, which has $X = 10.7$ and $Y = 805$.
   d) Find the residual for the Massachusetts prediction. Interpret.
   e) Two states differ by 10.0 in their poverty rates. Find the difference in their predicted violent crime rates.
   f) The state poverty rates range from 8.0 (for Hawaii) to 24.7 (for Mississippi). Over this range, find the range of predicted values for violent crime rate.
   g) What is the sign of the Pearson correlation between these variables? Why?
4. A college admissions officer claims that the prediction equation $\hat{Y} = .5 + 7.0X$ approximates the relationship between Y = college GPA and X = high school GPA (both measured on a four-point scale) for students at that college.
   a) Is this equation realistic? Why or why not?
   b) Suppose that the prediction equation is actually $\hat{Y} = .5 + .7X$. Interpret the slope.
   c) Using the prediction equation in (b), find the predicted GPA for a student having a high school GPA of (i) 3.0, (ii) 4.0.
   d) Suppose the prediction equation is $\hat{Y} = X$. Identify the Y-intercept and slope, and interpret their values.
5. A recent study of mail survey response rate patterns of the elderly found a prediction equation relating X = age and Y = percentage of subjects responding of $\hat{Y} = 90.2 - .6X$, for ages between about 60 and 90 (D. Kaldenberg et al., *Public Opinion Quarterly*, Vol. 58, 1994, p. 68).

a) Interpret the slope.
b) Find the predicted response rate for a (i) 60 year old, (ii) 90 year old.
c) Find the difference in predicted response rates for two age groups that are ten years apart.

6. Refer to Problem 9.24. For those counties, Table 9.9 shows part of the printout for the regression analysis relating Y = median income (thousand of dollars) to X = percent of residents with at least a high school education.
a) Report the prediction equation, and interpret the slope.
b) Is the Y-intercept meaningful? Explain.
c) Find the predicted median income for a county with 70% of the residents having at least a high school education.
d) County A has 10% more of its residents than county B with at least a high school education. Find their difference in predicted median incomes.
e) Find the Pearson correlation. Interpret using (i) the sign, (ii) the magnitude, (iii) the standardized slope.
f) Find the coefficient of determination. Explain its PRE interpretation.

TABLE 9.9

Variable	Mean	Std Dev
INCOME	24.51	4.69
EDUCATIO	69.49	8.86

Variable	Parameter Estimate
INTERCEP	-4.63
EDUCATIO	0.42

7. A high school student analyzes whether a relationship exists between X = number of books read for pleasure in the previous year and Y = daily average number of hours spent watching television. For her three best friends, the observations are as shown in Table 9.10.
a) Construct a scatter diagram. From inspection of the diagram, state the prediction equation, and interpret. (Note: You can do this without using the least squares formulas.)
b) Report the sample correlation between X and Y, and interpret.

TABLE 9.10

x	y
0	5
5	3
10	1

8. For the WWW data set described in Problem 1.7, the sample correlation between Y = political ideology (scored 1 to 7) and X = mumber of times a week reading a newspaper is $r = -.066$.
a) Interpret the sign of the correlation.
b) Interpret the square of the correlation. Would you conclude that the sample association is strong, or weak?
c) When Y is predicted using X = religiosity (how often attend religious services, scored 0, 1, 2, 3), the sample correlation is $r = .580$. Which of these two explanatory variables seems to have a stronger linear relationship with Y? Explain.

TABLE 9.14

R Square	.2973				
Standard Error	8.821				

Variable	B	SE B	Beta	T	Sig T
WOMEN-EC	-0.277	0.089	-0.545	-3.119	0.0048
(Constant)	36.324	4.700		7.729	0.0001

19. Refer to the previous two exercises. Using software, obtain the correlation matrix for these data. Which pairs of variables are highly correlated? Describe the nature of those correlations, and explain how your software handled the missing values. (For a particular analysis, most software deletes observations for which data are missing on at least one variable used in the analysis. Better strategies exist; see, for instance, R. Little and D. Rubin, *Sociological Methods and Research*, Vol. 18, 1989, pp. 292–326.)

20. For a random sample of U.S. counties, data are obtained on X = percentage of the population aged over 50 and Y = per student expenditure on education. Table 9.15 is part of the computer printout for the analysis.
a) What was the sample size for this study?
b) Fill in the blanks in Table 9.15.

TABLE 9.15

Source	DF	Sum of Squares	Mean Square	Root MSE	R-square
Model	1	------	------	-----	-----
Error	--	400000	------		
Total	49	500000			

Variable	Parameter Estimate	Standard Error	T for H0: Parameter=0	Prob > |T|
INTERCEP	1300.0	6.334	205.2	.0000
AGE	-5.0	1.443	-----	-----

21. For Table 9.1, use software to analyze the data on violent crime rate and percent single parent families.
a) Construct a scatter diagram. What does it show?
b) One point is quite far removed from the others, having a much higher value on both variables than the rest of the sample, but it fits in well with the linear trend exhibited by the rest of the points. Show that the correlation changes from .839 to .649 when you delete this observation. Why does it drop so dramatically?

Concepts and Applications

22. Refer to the WWW data set (Problem 1.7). Using software, conduct regression analyses relating (i) Y = political ideology and X = religiosity, (ii) Y = high school GPA and X = hours of TV watching. Prepare a report,
a) Using graphical ways of portraying the individual variables and their relationship.

b) Interpreting descriptive statistics for summarizing the individual variables and their relationship.
c) Summarizing and interpreting results of inferential analyses.
d) Checking effects of possibly influential outliers.

23. Refer to the data file you created in Problem 1.7. For variables chosen by your instructor, conduct a regression and correlation analysis. Report both descriptive and inferential statistical analyses, interpreting and summarizing your findings.

24. Table 9.16 shows recent data from all 67 Florida counties on crime rate (number of crimes per 1000 residents), median income (in thousands of dollars), percentage of residents with at least a high school education (of those aged at least 25), and the percentage of the county's residents living in an urban environment. Using crime rate as the response vari-

TABLE 9.16

County	Crime Rate	Median Income	High School	Percent Urban	County	Crime Rate	Median Income	High School	Percent Urban
ALACHUA	104	22.1	82.7	73.2	LAFAYETTE	0	20.7	58.2	0.0
BAKER	20	25.8	64.1	21.5	LAKE	42	23.4	70.6	43.2
BAY	64	24.7	74.7	85.0	LEE	59	28.4	76.9	86.1
BRADFORD	50	24.6	65.0	23.2	LEON	107	27.3	84.9	82.5
BREVARD	64	30.5	82.3	91.9	LEVY	45	18.8	62.8	0.0
BROWARD	94	30.6	76.8	98.9	LIBERTY	8	22.3	56.7	0.0
CALHOUN	8	18.6	55.9	0.0	MADISON	26	18.2	56.5	20.3
CHARLOTTE	35	25.7	75.7	80.2	MANATEE	79	26.0	75.6	88.7
CITRUS	27	21.3	68.6	31.0	MARION	64	22.5	69.6	39.6
CLAY	41	34.9	81.2	65.8	MARTIN	53	31.8	79.7	83.2
COLLIER	55	34.0	79.0	77.6	MONROE	89	29.4	79.7	73.2
COLUMBIA	69	22.0	69.0	31.1	NASSAU	42	30.2	71.2	44.9
DADE	128	26.9	65.0	98.8	OKALOOSA	37	27.9	83.8	84.0
DESOTO	69	21.0	54.5	44.6	OKEECH.	51	21.4	59.1	30.1
DIXIE	49	15.4	57.7	0.0	ORANGE	93	30.3	78.8	93.1
DUVAL	97	28.5	76.9	98.8	OSCEOLA	78	27.3	73.7	66.4
ESCAMBIA	70	25.2	76.2	85.9	PALM B.	90	32.5	78.8	94.7
FLAGLER	34	28.6	78.7	63.1	PASCO	42	21.5	66.9	67.4
FRANKLIN	37	17.2	59.5	30.2	PINELLAS	70	26.3	78.1	99.6
GADSDEN	52	20.0	59.9	28.8	POLK	84	25.2	68.0	70.3
GILCHRIST	15	20.6	63.0	0.0	PUTNAM	83	20.2	64.3	15.7
GLADES	62	20.7	57.4	0.0	SANTA R.	43	27.6	79.9	57.2
GULF	19	21.9	66.4	35.2	SARASOTA	58	29.9	71.7	92.1
HAMILTON	6	18.7	58.4	0.0	SEMINOLE	56	35.6	78.5	44.4
HARDEE	57	22.1	54.8	16.7	ST JOHNS	54	29.9	81.3	93.2
HENDRY	47	24.9	56.6	44.7	ST LUCIE	58	27.7	84.6	92.8
HERNANDO	44	22.7	70.5	61.3	SUMTER	37	19.6	64.3	19.3
HIGHLANDS	56	21.1	68.2	24.8	SUWANEE	37	19.8	63.8	23.6
HILLSBOR.	110	28.5	75.6	89.2	TAYLOR	76	21.4	62.1	41.8
HOLMES	5	17.2	57.1	16.8	UNION	6	22.8	67.7	0.0
INDIAN R.	58	29.0	76.5	83.0	VOLUSIA	62	24.8	75.4	83.9
JACKSON	32	19.5	61.6	21.7	WAKULLA	29	25.0	71.6	0.0
JEFFERSON	36	21.8	64.1	22.3	WALTON	18	21.9	66.5	20.9
					WASHINGTON	21	18.3	60.9	22.9

Source: Dr. Larry Winner, University of Florida.

able and percent urban as the predictor, analyze these data. In your report, provide interpretations of all the analyses.

25. Refer to the previous exercise. Using income as the response variable and percentage of high school graduates as the explanatory variable, analyze these data. Prepare a report, and explain carefully the interpretations of all your analyses.

26. Refer to Table 9.1. Analyze the relationship between violent crime rate and percentage having at least a high school education. Write a report showing your analyses, providing interpretations, and summarizing your findings.

27. Refer to Table 9.1. Analyze the relationship between violent crime rate and percentage of single-parent families. Write a report showing your analyses, providing interpretations, and summarizing your findings.

28. Repeat the previous exercise, using murder rate as the response variable.

29. Refer to Table 9.13. Analyze the relationship between newspaper circulation and gross national product. Tell why you conducted each part of the analysis and explain how to interpret the results.

30. Recently the General Social Survey has asked subjects to rate various groups using the "feeling thermometer." Ratings between 50 and 100 mean you feel favorable and warm toward the group, whereas ratings between 0 and 50 mean that you don't feel favorable. It also asks subjects to rate themselves on political ideology, using scores 1 = extremely liberal, 2 = liberal, 3 = slightly liberal, 4 = moderate, 5 = slightly conservative, 6 = conservative, 7 = extremely conservative, and to describe their religious attendance, using the categories (never, less than once a year, once or twice a year, several times a year, about once a month, 2-3 times a month, nearly every week, every week, several times a week). Table 9.17 shows data for ten of the subjects in a recent survey, where the feelings thermometer refers to feelings about liberals and using religion scores that are the category numbers.

a) Analyze Table 9.17. Tell why you conducted each analysis, and explain how to interpret the results.

d) Suppose the feelings response for the first subject had incorrectly been recorded as 90 instead of 10. How would this have affected results of your analyses?

TABLE 9.17

Feelings	Ideology	Religion	Feelings	Ideology	Religion
10	7	9	60	4	8
70	3	5	40	5	6
50	4	7	80	3	7
30	6	4	50	4	6
90	4	3	40	6	8

31. In an article in *USA Today* (December 28, 1984), sociologists N. Glenn and B. A. Shelton are quoted as showing a strong link between residential mobility and divorce rates. In Table 9.18, divorce rate is the annual number of divorces and annulments per 1000 population, and mobility rate is the percentage of people living in a different house from five years ago. Analyze these data.

32. Describe a situation in which it is inappropriate to use the Pearson correlation to measure the association between two quantitative variables.

TABLE 9.18

Region	Mobility Rate	Divorce Rate
New England	41	4.0
Middle Atlantic	37	3.4
East North Central	44	5.1
West North Central	46	4.6
South Atlantic	47	5.6
East South Central	44	6.0
West South Central	50	6.5
Mountain	57	7.6
Pacific	56	5.9

33. Annual income, in dollars, is the response variable in a regression analysis. For a British version of the report on the analysis, all responses are converted to British pounds sterling (1 pound equals about 1.5 dollars, as of 1997).
a) How, if at all, does the slope of the prediction equation change?
b) How, if at all, does the correlation change?

34. The variables Y = annual income (thousands of dollars), X_1 = number of years of education, and X_2 = number of years experience in job are measured for all the employees having city-funded jobs, in Knoxville, Tennessee. The following prediction equations and correlations apply.
i. $\hat{Y} = 10 + 1.0X_1$, $r = .30$.
ii. $\hat{Y} = 14 + .4X_2$, $r = .60$.
The correlation is $-.40$ between X_1 and X_2. Which of the following statements are true?
a) The strongest sample association is between Y and X_2.
b) The weakest sample association is between X_1 and X_2.
c) The prediction equation using X_2 to predict X_1 has negative slope.
d) A standard deviation increase in education corresponds to a predicted increase of .3 standard deviations in income.
e) There is a 30% reduction in error in using education, instead of $\bar{Y}$, to predict income.
f) Each additional year on the job corresponds to a $400 increase in predicted income.
g) When X_1 is the predictor of Y, the sum of squared residuals (SSE) is larger than when X_2 is the predictor of Y.
h) The predicted mean income for employees having 20 years of experience is $4000 higher than the predicted mean income for employees having 10 years of experience.
i) If $\hat{\sigma} = 8$ for the model using X_1 to predict Y, then it is not unusual to observe an income of $70,000 for an employee who has 10 years of education.
j) It is possible that $s_Y = 12.0$ and $s_{X_1} = 3.6$.
k) It is possible that $\bar{Y} = 20$ and $\bar{X}_1 = 13$.

Select the best response(s) in Problems 7.35–7.37.

35. One can interpret $r = .3$ as follows:
a) A 30% reduction in error occurs in using X to predict Y.
b) A 9% reduction in error occurs in using X to predict Y compared to using $\bar{Y}$ to predict Y.

c) 9% of the time $\hat{Y} = Y$.
d) Y changes .3 unit for every one-unit increase in X.
e) When X predicts Y, the average residual is .3.
f) X changes .3 standard deviations when Y changes one standard deviation.

36. The correlation is inappropriate as a measure of association between two quantitative variables:
a) When different people measure the variables using different units.
b) When the relationship is highly nonlinear.
c) When the data points fall exactly on a straight line.
d) When the slope of the prediction equation is 0 using nearly all the data, but a couple of outliers are extremely high on Y at the high end of the X scale.
e) When Y tends to decrease as X increases.
f) When we have data for the entire population rather than a sample.
g) When the sample has a much narrower range of X-values than does the population.

37. The slope of the least squares prediction equation and the Pearson correlation coefficient are similar in the sense that
a) They do not depend on the units of measurement.
b) They both must fall between -1 and $+1$.
c) They both have the same sign.
d) They both equal 1 when there is the strongest association.
e) Their squares both have PRE interpretations.
f) They have the same t statistic value for testing H_0: Independence.
g) They both can be strongly affected by severe outliers.

38. Describe the assumptions (a) in using the regression equation $E(Y) = \alpha + \beta X$ to represent the relationship between two variables and (b) in making inferences about that equation using the least squares prediction equation. Which assumptions are most critical?

39. Refer to the previous exercise. In view of these assumptions, indicate why such a model would or would not be good in the following situations.
a) $X =$ time, $Y =$ percentage unemployed workers in the United States. (*Hint*: Does this continually tend to increase or decrease?)
b) $X =$ income, $Y =$ charitable contributions within the previous year. (*Hint*: Would poor people show as much variation as wealthy people?)
c) $X =$ age, $Y =$ annual medical expenses. (*Hint*: Suppose expenses tend to be relatively high for the newborn and for the elderly.)
d) $X =$ per capita income, $Y =$ life expectancy, for nations. (*Hint*: The increasing trend eventually levels off.)

40. For a class of 100 students, the teacher takes the 10 students who perform poorest on the midterm exam and enrolls them in a special tutoring program. The overall class mean is 70 both on the midterm and final, but the mean for the specially tutored students increases from 50 to 60. Can we conclude that the tutoring program was successful? Explain.

41. Refer to Problem 9.24. For these counties, the correlation between high school education rate and income equals .79. Suppose we also have data at the individual level as well as aggregated for a county. Sketch a scatter diagram to show that at the individual level, the correlation could be much weaker. (*Hint*: Show that lots of variability could exist for individuals, yet the summary values for counties could fall close to a straight line.) Because of this result, it is misleading to extend results from the aggregate level to individuals. Making predictions about individuals based on the behavior of aggregate groups is known as the ***ecological fallacy*** (See W. S. Robinson, *American Sociological Review*, Vol. 15, 1950, p. 351).

42. Explain why the correlation between X = number of years of education and Y = annual income is likely to be smaller if we use a random sample of adults who have a college degree than if we use a random sample of all adults.

43. Explain carefully the interpretations of the standard deviations (a) s_Y, (b) s_X, (c) $\hat{\sigma}$, (d) $\hat{\sigma}_b$.

44. *A report summarizing the results of a study on the relationship between scores for students on a verbal aptitude test X and a mathematics aptitude test Y states that $\bar{X} = 480$, $\bar{Y} = 500$, $s_X = 80$, $s_Y = 120$, and $r = .60$.
a) Using the formulas for the correlation and for the least squares estimates, find the prediction equation.
b) Find the prediction equation for predicting verbal test result using math test result.

45. *Observations on both X and Y are standardized, having estimated means of 0 and standard deviations of 1 (see Section 4.2). Show that the prediction equation has the form $\hat{Y} = rX$, where r is the sample correlation between X and Y; that is, for the standardized variables, the Y-intercept equals 0 and the slope is the same as the correlation.

46. *A confidence interval for a population correlation ρ requires a mathematical transformation of r for which the sampling distribution is approximately normal. This transformation is $T(r) = (1/2)\log[(1+r)/(1-r)]$, where *log* denotes the natural (base-*e*) logarithm. The transformation of the population value ρ is denoted by $T(\rho)$. The variable $T(r)$ is approximately normally distributed about $T(\rho)$ with standard error $\sigma_T = 1/\sqrt{n-3}$. A confidence interval for $T(\rho)$ is $T(r) \pm z\sigma_T$. Once we get the endpoints of the interval for $T(\rho)$, we substitute each endpoint for T in the inverse transformation $\rho = (e^{2T} - 1)/(e^{2T} + 1)$, where e denotes the exponential function (the inverse of the natural log function). These two values form the endpoints of the confidence interval for ρ.
a) For the correlation of .899 for housing price and size in Table 9.4, show that $T(r) = 1.47$. Show that the standard error of $T(r)$ is .1054.
b) Show that a 95% confidence interval for $T(\rho)$ is (1.26, 1.67).
c) Show that the corresponding confidence interval for ρ is (.85, .93). (Unless $r = 0$, the confidence interval for ρ is not symmetric about the point estimate r, because of the nonsymmetry of the sampling distribution of r.)
d) A confidence interval for the population value ρ^2 of the coefficient of determination follows directly by squaring the limits of the confidence interval for ρ. Find and interpret this confidence interval.
e) If the confidence interval for ρ includes 0, explain why the lower endpoint of the confidence interval for ρ^2 is also 0, and the upper endpoint is the larger of the squared endpoints of the confidence interval for ρ.

47. *Refer to the previous exercise and to Problem 9.16. Find and interpret 95% confidence intervals for the population Pearson correlation and the population coefficient of determination.

48. *Refer to Problem 9.46. Let ρ_1 and ρ_2 denote the population correlation values between two variables for two separate populations. Let r_1 and r_2 denote sample values for independent random samples from the populations. To test the null hypothesis H_0: $\rho_1 = \rho_2$, the test statistic is

$$z = \frac{T_2 - T_1}{\sigma_{T_2 - T_1}}$$

where T_1 and T_2 are the transformed values of r_1 and r_2 and

$$\sigma_{T_2 - T_1} = \sqrt{\frac{1}{n_1 - 3} + \frac{1}{n_2 - 3}}$$

If H_0 is true, this test statistic has approximately the standard normal distribution. In Table 9.4, the correlation between housing price and size of home is $r_1 = .932$ for the 28 new homes and $r_2 = .915$ for the 65 older homes. Find the P-value for testing H_0: $\rho_1 = \rho_2$ against H_a: $\rho_1 \neq \rho_2$. Interpret.

49. *Show that substituting $X = \bar{X}$ into the prediction equation $\hat{Y} = a + bX$ yields the predicted Y-value of $\hat{Y} = \bar{Y}$. (*Hint*: The least squares formula for the Y-intercept is $a = \bar{Y} - b\bar{X}$.) Show that this means that the least squares prediction equation passes through the point with coordinates $(\bar{X}, \bar{Y})$, the center of gravity of the data.

50. *Alternative formulas for defining the Pearson correlation use the data in formulas similar to the one for b:

$$r = \frac{\sum(X - \bar{X})(Y - \bar{Y})}{\sqrt{\left[\sum(X - \bar{X})^2\right]\left[\sum(Y - \bar{Y})^2\right]}} = \frac{1}{n-1}\sum\left(\frac{X - \bar{X}}{s_X}\right)\left(\frac{Y - \bar{Y}}{s_Y}\right)$$

Roughly, the correlation is the average cross-product of the z-score for X times the z-score for Y. Using this formula, explain why (a) the correlation has the same value when X predicts Y as when Y predicts X, (b) the correlation does not depend on the units of measurement. (Note: For the population, the correlation is often defined as

$$\frac{\text{Covariance of } X \text{ and } Y}{(\text{Standard deviation of } X)(\text{Standard deviation of } Y)}$$

where the ***covariance*** between X and Y is the average of the cross-products $(X - \mu_X)$ $(Y - \mu_Y)$ about the population means.)

51. *The values of Y are multiplied by a constant c. From their formulas, show that the standard deviation s_Y and the least squares slope b are also then multiplied by c. Thus, show that $r = bs_X/s_Y$ remains the same, so that r does not depend on the units of measurement.

52. *Suppose that the linear regression equation $E(Y) = \alpha + \beta X$ with normality and constant standard deviation σ is truly appropriate for the relationship between Y and X. Then, the interval of numbers

$$\hat{Y} \pm t\hat{\sigma}\sqrt{1 + \frac{1}{n} + \frac{(X - \bar{X})^2}{\sum(X - \bar{X})^2}}$$

predicts where a new observation on Y will fall at that value of X. This interval is called a ***prediction interval*** for Y. To make an inference about the *mean* of Y (rather than a single value of Y) at that value of X, one can use the ***confidence interval***

$$\hat{Y} \pm t\hat{\sigma}\sqrt{\frac{1}{n} + \frac{(X - \bar{X})^2}{\sum(X - \bar{X})^2}}$$

The t-value in these intervals is based on $df = n - 2$. Most software has options for calculating these formulas (e.g., the options CLI and CLM in PROC REG in SAS). Refer to the housing data in Table 9.4, at house size $X = 2.0$.

a) Show that $\hat{Y} = 126.0$ and a 95% prediction interval is (87.0, 165.0).

b) Show that a 95% confidence interval for the mean selling price is (121.2, 130.8).

c) Explain intuitively why a prediction interval for a single observation is much wider than a confidence interval for the mean.

d) Results using these formulas are typically overly optimistic, because the model assumptions never hold exactly. Explain how prediction intervals would likely be in error if, in fact, the variability in housing prices tends to increase as house size increases.

53. *Refer to Problem 9.16 and the previous exercise.

a) Construct a 95% confidence interval for the mean number of children for mothers having $X = 16$.

b) Explain why the prediction interval is probably inappropriate. (*Hint*: Is Y approximately normal?)

54. *Refer to Example 9.8 on regression toward the mean. Does this imply that, over time, variation in height is decreasing until, eventually, everyone has the same height? Explain.

55. *To implement least squares, one can find the formulas for the a and b estimates that minimize SSE = $\sum(Y - \hat{Y})^2 = \sum[Y - (a + bX)]^2$ using calculus, by taking the derivative of this function with respect to a, taking the derivative with respect to b, setting the two derivatives equal to 0, and solving the two linear equations simultaneously for a and b. Take the derivative with respect to a and solve for a, showing that $a = \bar{Y} - b\bar{X}$.

Bibliography

Chatterjee, S., Handcock, M., and Simonoff, J. (1995). *A Casebook for a First Course in Statistics and Data Analysis*. New York: Wiley.

Draper, N. R. and Smith, H. (1981). *Applied Regression Analysis*, 2nd ed. New York: Wiley.

Weisberg, S. (1985). *Applied Linear Regression*, 2nd ed. New York: Wiley.

Younger, M. S. (1985). *A First Course in Linear Regression*, 2nd ed. Boston: Duxbury Press.

Chapter 10

Introduction to Multivariate Relationships

The previous three chapters introduced methods for analyzing the association between two variables. In most social science research, these analyses are but the first step. Subsequent steps use multivariate methods to analyze, at the same time, other variables that might influence the relationship between the original two. For instance, Example 8.3 showed that gender influences party identification, with women more likely than men to be Democrats. To analyze why this is so, one might look at whether differences between men and women in political views could explain the association. Perhaps women tend to be more liberal than men, and being liberal tends to be associated with being Democrat.

Several types of research questions require adding variables to the analysis. These questions often involve notions of *causal* connections among the variables. Does being female cause one to be more likely to be a Democrat, or is perhaps political ideology or income or educational level or religion the causal factor behind party identification? If we study conservatives or other subjects having a particular political ideology, for instance, is it still true that women are more likely than men to be Democrats?

Section 10.1 discusses the relationship of data analysis to the notions of causation and outlines methods for testing causal assumptions. Section 10.2 introduces the fundamental tool of *statistical control*, which involves studying the association between two variables while removing the influence of other variables. Section 10.3 presents various types of multivariate relationships that statistical control can reveal. The final section discusses issues dealing with inferential statistical methods for multivariate relationships.

10.1 Association and Causality

Causality is central to the scientific endeavor. Most people are familiar with this concept, at least in an informal sense. We know, for instance, that being exposed to a virus can cause the flu. But few understand how to test rigorously a possible causal relationship. It can be particularly tricky to do so for social science variables, where relationships are probabilistic rather than deterministic. For instance, what causes juvenile delinquency? Being poor? Coming from a single-parent home? A lack of moral and religious training? Genetic factors? A combination of these and other factors? We now look at some guidelines that help us assess a hypothesis of the form, "X causes Y."

In causal relationships, there's usually an asymmetry, with one variable having an influence on the other, but not vice versa. An arrow drawn between two variables X and Y, pointing to the response variable, denotes a causal association between the variables. Thus,

$$X \rightarrow Y$$

specifies that X is an explanatory variable having a causal influence on Y. For example, suppose we suspect that being a Boy Scout has a causal effect on being a juvenile delinquent, scouts being less likely to be delinquents. We are hypothesizing that $S \rightarrow D$, where S (for Scouting) and D (for Delinquency) denote the binary variables "whether a Boy Scout (yes, no)" and "whether a juvenile delinquent (yes, no)."

If we suspect that one variable is causally related to another, how do we analyze whether it actually is? A relationship must satisfy three criteria to be considered a causal one. These criteria are

- Association between the variables
- An appropriate time order
- The elimination of alternative explanations

We discuss these criteria below. If all three are met, then the evidence supports the hypothesized causal relationship. If one or more criteria are not met, then we can conclude there is no causal relationship. We can never *prove* that one variable is a cause of another, since causation is hypothesized by the researcher but not actually observed. We can disprove causal hypotheses, however, by showing that empirical evidence contradicts them by not satisfying at least one of these criteria.

Association

The first criterion for causality is ***association***. We must show that X and Y are statistically dependent to support the hypothesis that X causes Y. If $X \rightarrow Y$, then as X

changes, the distribution of Y should change in some way. If scouting causes lower delinquency rates, for example, then the population proportion of delinquents should be higher for nonscouts than for scouts. For sample data, a statistical test (such as chi-squared for qualitative data or a test for the regression slope for quantitative data) or a confidence interval for a measure of association analyzes whether this criterion is satisfied.

Time Order

The second criterion for causation is that the two variables have the appropriate ***time order***, with the cause preceding the effect. Sometimes this is just a matter of logic. For instance, race, age, and gender exist prior to current attitudes or achievements, so any causal association must treat them as causes rather than effects.

In other cases, the causal direction is not as obvious. The example of scouting and delinquency is one such case. It is possible and logically reasonable that scouting reduces delinquency tendencies. On the other hand, it is just as plausible that delinquent boys avoid scouting but nondelinquent boys do not. Thus, the time order is not clear, and both possibilities, $S \rightarrow D$ and $D \rightarrow S$, are plausible. Just showing that an association exists does not solve this dilemma, since a lower proportion of delinquents among scout members is consistent with both explanations.

Some research can use an experimental design to fix the time order. For instance, does a new drug have a beneficial effect in treating a disease? One might randomly assign subjects suffering from the disease to receive either the drug or a placebo. Then, to analyze whether the drug assignment may have a causal influence on the response outcome, one would observe whether the proportion successfully treated was significantly higher for the drug treatment group. The outcome for a subject (success or not) is observed after the treatment, so the time order is certain.

Elimination of Alternative Explanation

Suppose an association exists between two variables, and suppose they have the proper time order to satisfy a potential casual relation. It is a common mistake to deduce causality from association and time order alone. There may be an ***alternative explanation***. Airline pilots turn on the "fasten seat belt" sign just before their planes encounter turbulence, but this does not imply that turning on the sign causes turbulence.

This third criterion of a lack of alternative explanation is the most difficult to achieve. Finding an alternative, plausible explanation for an association is responsible for rejecting most hypotheses of causal relationships. The main types of alternative explanations to consider are the following:

- The relationship may be ***spurious***, as defined in Section 10.2, both variables being dependent on a third variable.
- The relationship may be ***conditional*** on the values of another variable, perhaps disappearing at certain levels of that variable.

- The relationship may work through another variable that intervenes between the two, as explained in Section 10.3.
- The observed relationship may be a result of sampling error.

The first three types of alternatives are explained in this chapter, and they form the focus of inquiry in most research. The fourth, the possibility that the observed relationship results from sampling variation and does not represent a true relationship in the population, is dealt with by inferential statistics and by replication of research studies.

Association, Causality, and Anecdotal Evidence

The association between smoking and lung cancer is one that is now regarded as having a causal link. The association is moderately strong, there is the proper time order (lung cancer following a period of smoking), and no alternative explanation has been found to explain the relationship. In addition, the causal link has been bolstered by biological theories that explain how smoking could cause lung cancer.

Sometimes you hear people give anecdotal evidence to attempt to disprove causal relationships. "My Uncle Reg is 85 years old, he still smokes two packs of cigarettes a day, and he's as healthy as a horse." An association does not need to be perfect, however, to be causal. Not all people who smoke two packs of cigarettes a day will get lung cancer, but a much higher percentage of them will do so than people who are nonsmokers. Perhaps Uncle Reg is still in fine health, but that should not encourage us to tempt the fates by smoking two packs a day.

Anecdotal evidence is not enough to disprove causality unless it can deflate one of the three criteria for causality. In other words, it must show a lack of any association, or show an improper time order, or provide an alternative explanation for the association.

10.2 Controlling for Other Variables

A fundamental component to evaluating whether X could cause Y is the study of whether their association remains when we remove the effects of other variables on this association. In a multivariate analysis, a variable is said to be ***controlled*** when its influence on the other variables is removed.

In a laboratory experiment, one can control variables that could confound the results by holding their values constant while studying the relationships among the other variables. For instance, an experiment in chemistry or physics might control temperature and atmospheric pressure by holding them constant in a laboratory environment during the course of the experiment. A lab experiment investigating the effect of different doses of a carcinogen on mice might control the age and diet of the mice.

Statistical Control in Social Research

Unlike laboratory sciences, social science research is usually *observational* rather than *experimental*. We cannot fix values of variables we might like to control, such as intelligence, before obtaining data on the explanatory and response variables of interest. But we can approximate an experimental type of control by grouping together observations with similar or equal values on the control variables. In controlling education, for instance, we might group the sample results into those subjects with less than a high school education, those with a high school education but no college education, those with some college education, and those with at least one college degree. This is ***statistical control***, rather than experimental control.

The following example illustrates statistical control in a social science setting, holding a key variable constant.

Example 10.1 Causal Effect of Height on Math Achievement?

Do tall students tend to do better in math than short students? We might think so looking at a random sample of students from Lake Wobegon school district who take a math achievement test. They have a correlation of .81 between height and math test score. Taller students tend to have higher scores.

Could being tall be a causal influence on scoring well on the math achievement score? Perhaps some alternative explanation exists for this association. It might be explained by the sample having students of various ages. As age increases, both height and test score would tend to increase. Older students tend to be taller, and older students tend to have stronger math knowledge.

We can remove the effects of age from the association by *statistical control*, studying the association between height and test score for students of the same age. That is, we control for age by analyzing the association separately at each age level. Then, variation in age cannot jointly cause both variation in height and test score.

In fact, the achievement test was administered to students from grades 2, 5, and 8 at Lake Wobegon, so the sample contained considerable variability in the students' ages. Figure 10.1 shows a computer printout of a scatterplot of the observations, with labels indicating the grade for each student. The overall pattern of points shows a strong positive correlation, with higher math scores at higher heights. View the points within a fixed grade level, however, and you see random variation, with no particular pattern of increase or decrease. The correlation is close to zero within each grade level.

Thus, for students of (about) the same age, an association does not exist between height and math test score. Height is not a causal influence for test score, since the association disappears when age is held constant. □

In summary, we control a variable by holding its value constant. In other words, we study the relationship between X and Y for cases with equal (or similar) values of that variable. The variable controlled is called a ***control variable***. In holding the control variable constant, we remove the part of the association between X and Y that is caused by its variation.

The scatterplot in Figure 10.1 describes the association between two quantitative variables, controlling for a third variable. To study the association between two qualitative variables, while controlling for a third variable, we form contingency tables relating those variables separately for subjects in each category of that control variable.

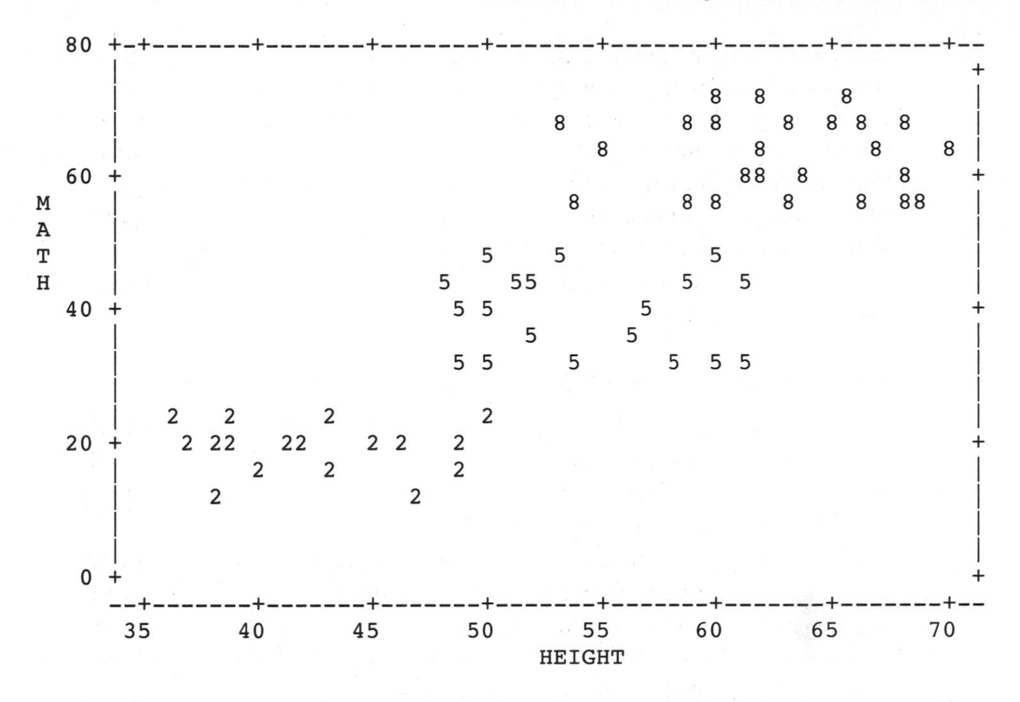

Figure 10.1 Computer Printout Showing Relationship Between Height and Math Achievement Test Score, with Observations Labeled by Grade Level

The separate tables that display the relationships within the fixed levels of the control variable are called ***partial tables***.

Example 10.2 Statistical Control for Scouting–Delinquency

Table 10.1 shows a hypothetical table relating scouting to delinquency. The percentage of delinquents among scout members is lower than among nonscouts. We refer to this table as ***bivariate***, since it contains data only on *two* variables. All other variables are ignored; none is controlled.

TABLE 10.1 Contingency Table Relating Scouting and Delinquency

		Delinquency		
		Yes	No	Total
Boy Scout	Yes	36 (9%)	364 (91%)	400
	No	60 (15%)	340 (85%)	400

In considering possible explanations for the association, we could control for variables such as church attendance or social class. Perhaps boys who attend church are more likely than nonattenders to be scouts, and perhaps boys who attend church are less likely to be delinquent. Then the difference in delinquency rates between scouts and nonscouts might be due to variation in church attendance.

To control for church attendance, we examine the association between scouting and delinquency within partial tables formed by various levels of church attendance. Table 10.2 shows partial tables for three levels defined as follows: Low = no more than once a year, Medium = more than once a year but less than once a week, High = at least once a week. Adding these three partial tables together produces the bivariate table (Table 10.1), which ignores rather than controls church attendance. For instance, the number of Boy Scouts who are delinquents is $36 = 10 + 18 + 8$.

TABLE 10.2 Contingency Table Relating Scouting and Delinquency, Controlling for Church Attendance

		Church Attendance					
		Low		Medium		High	
Delinquency		Yes	No	Yes	No	Yes	No
Scout	Yes	10 (20%)	40 (80%)	18 (12%)	132 (88%)	8 (4%)	192 (96%)
	No	40 (20%)	160 (80%)	18 (12%)	132 (88%)	2 (4%)	48 (96%)

In each partial table, the percentage of delinquents is the same for scouts as for nonscouts. When we control for church attendance, no association exists between scouting and delinquency. These data provide an alternative explanation for the association between scouting and delinquency, as discussed below, making us skeptical of any causal links. □

Spurious Associations

An association between two variables is said to be ***spurious*** if both variables are dependent on a third variable, and their association disappears when it is controlled. Such an association is an effect of the relationship of those variables to the control variable, rather than indicating a causal connection. No association exists between them in the partial tables that hold the control variable constant and thus remove its effect.

Example 10.3 Examples of Spurious Associations

The association between height and mathematics achievement test score disappears at fixed levels of age. That association is spurious, with age being a common cause of both height and math achievement.

Table 10.1 displayed an association between scouting and delinquency. Controlling for church attendance, the partial tables in Table 10.2 showed no association. This is also consistent with spuriousness. Table 10.2 shows that as church attendance increases, the percentage of delinquents decreases (compare percentages across the partial tables) and the percentage of scout members increases. By the nature of these two associations, it is not surprising that scouts exhibit lower overall delinquency rates than nonscouts in Table 10.1. □

In practice, even if an association between two variables is truly spurious, *sample* partial tables would not look like those in Table 10.2. Because of sampling error, they would not show a perfect lack of association. Moreover, few relationships are *completely* spurious. There may be *some* causal connection between two variables, even within each partial table, but not as strong as the bivariate table suggests.

When we add additional explanatory variables to an analysis, we will use subscripts to distinguish between them. For instance, for the three-variable relationships discussed in this chapter, we denote the response variable by Y and the explanatory or control variables by X_1 and X_2.

Figure 10.2 graphically depicts a spurious association between X_1 and Y. They are associated only because they both depend on a common cause, X_2. The figure illustrates this for Y = math test score, X_1 = height, and X_2 = age. As X_2 changes, it produces changes simultaneously in X_1 and Y, so that X_1 and Y appear to be statistically dependent. In reality they are related, but only because of their common dependence on the third variable. They are not *causally* related, because the variable X_2 provides an *alternative explanation* for their association. Thus, showing that the association between two variables is spurious disproves the hypothesis of a causal connection between them.

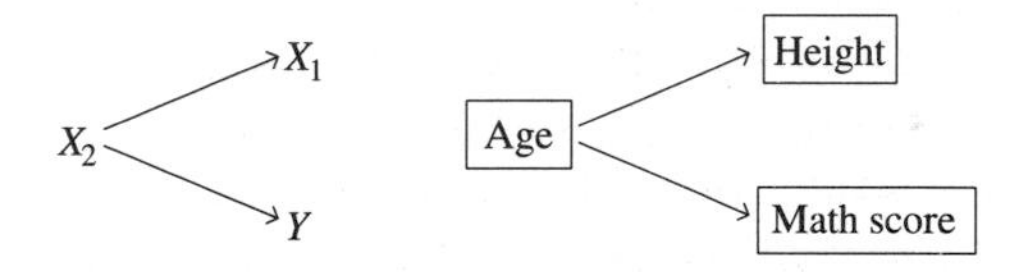

Figure 10.2 Graphical Depiction of a Spurious Association Between X_1 and Y. The association disappears when we control for X_2, which causally affects both X_1 and Y.

For instance, an alternative explanation exists for the association between scouting and delinquency, namely, that both these variables depend on church attendance. Youngsters who attend church are less likely to be delinquents and more likely to be scouts. For a fixed level of church attendance, scouting has no association with delinquency. Since the association can be explained by a dependence on church attendance, no causal link exists between scouting and delinquency.

It is not always obvious which variables require control in a study of an association. A strong background in the theory and previous work in a field of study helps a researcher to know which variables to control and makes detection of spurious associations more likely.

Statistical Control for Quantitative Variables

Tables 10.1 and 10.2 illustrate statistical control for an association between two qualitative (categorical) variables. Concepts such as statistical control and spuriousness apply to all variables, quantitative as well as qualitative. For instance, they apply to regression analyses, as Example 10.1 showed, and to comparisons of means for quantitative variables.

Example 10.4 Home Computers and Math Achievement

Table 10.3 shows a bivariate association between the quantitative response variable, math achievement test score, and the qualitative explanatory variable, home computer ownership (yes, no). Students having home computers have higher mean test scores. Is there a causal link, whereby ownership of a home computer improves the math test score? We must consider the alternative explanation that another variable is causally related to both of these.

TABLE 10.3 Relationship Between Math Test Score and Home Computer Ownership

		Mean Test Score	n
Home Computer	Yes	82.5	500
	No	73.8	500

Social class or a related variable such as parental education or income is often a prime candidate for control in social science research. Perhaps ownership of a home computer is more likely for students of middle-class or higher backgrounds than for students from a lower-class background. Therefore, in comparing test scores for students on the basis of home computer ownership, we may be comparing groups with quite different social class composition.

Social class may itself be a factor in school performance, with higher test scores tending to occur for students from higher social classes. The association between test score and home computer ownership may simply reflect higher social class being associated both with higher test scores and with greater chance of computer ownership. To study the effect of home computer ownership on test score, one should remove the effect of social class by controlling it.

Table 10.4 shows a potential result of controlling social class of family. That table suggests that home computers do not help math test scores. When we control for social class, essentially no association exists between home computer ownership and the math

TABLE 10.4 Relationship Between Math Test Score and Home Computer Ownership, Controlling for Social Class

Social Class	Home Computer	Mean Test Score	n
Lower	Yes	69.5	25
	No	68.7	275
Middle	Yes	80.7	300
	No	79.3	200
Upper	Yes	87.5	175
	No	86.8	25

test score. The mean score is about the same (and perhaps equal, within the limits of sampling error) for those who have home computers and for those who do not, at each level of social class.

Table 10.4 also indicates that social class has an effect both on test score and on computer ownership. Higher social class corresponds to higher mean test score and to higher probability of a home computer. This evidence is consistent with a spurious relationship between test score and computer ownership. Figure 10.3 depicts such a relationship. □

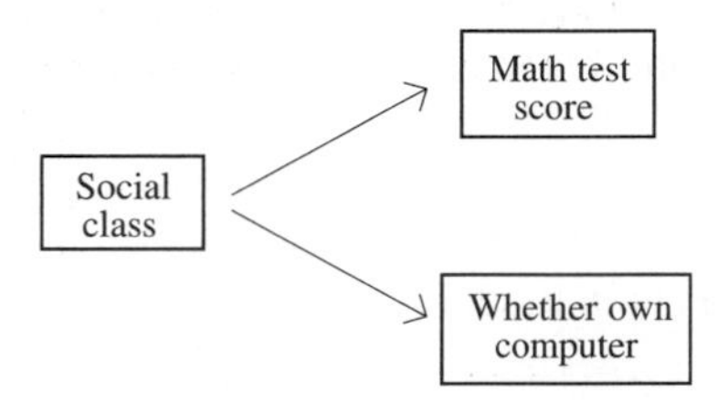

Figure 10.3 Graphical Depiction of Spurious Association Between Home Computer Ownership and Math Test Score

The partial tables in Table 10.4 do not show *exactly* no difference in means, but all differences are close enough to 0 to be substantively unimportant. Even if the true relationship is spurious, samples do not show a complete lack of association because of random sampling variations.

Finally, several causal explanations can be possible for a data set. The association in Table 10.3 might also be explained by the joint dependence of math test score and home computer ownership on the degree of parental emphasis on education. Or the students having home computers may have been ones encouraged to pursue computer-related studies because of high math test scores in previous years.

10.3 Types of Multivariate Relationships

Section 10.2 showed that an association between Y and X_1 may change dramatically when we control for another variable, X_2. For instance, the X_1–Y association is *spurious* when X_2 is a causal influence of both X_1 and Y and explains their association, the X_1–Y association disappearing at fixed levels of X_2. This section describes other types of multivariate relationships that commonly occur in social science research.

Chain Relationships

Spurious associations are not the only ones for which the association disappears when we control for a third variable. Another common way is with a chain of causation, in which X_1 affects X_2, which in turn affects Y. Figure 10.4 depicts the chain. Here, X_1 is an *indirect*, rather than direct, cause of Y. Variable X_2 in Figure 10.4 is called an

intervening variable. To explain how X_1 affects Y, we need to identify the intervening variable.

$$X_1 \longrightarrow X_2 \longrightarrow Y$$

Figure 10.4 A Chain Relationship, in Which X_1 Indirectly Affects Y Through the Intervening Variable X_2

If the hypothesized causal chain is correct, a bivariate association exists between X_1 and Y, but that association disappears when we control for X_2. For instance, suppose that race is related to frequency of juvenile arrests, with black juveniles more likely to be arrested than white juveniles. This may primarily be an effect of different socioeconomic status levels of blacks and whites. It is plausible, for instance, that (1) race affects family income, with blacks tending to have lower family incomes than whites and (2) being poor increases the chance of being arrested as a juvenile. Figure 10.5 depicts this causal chain model.

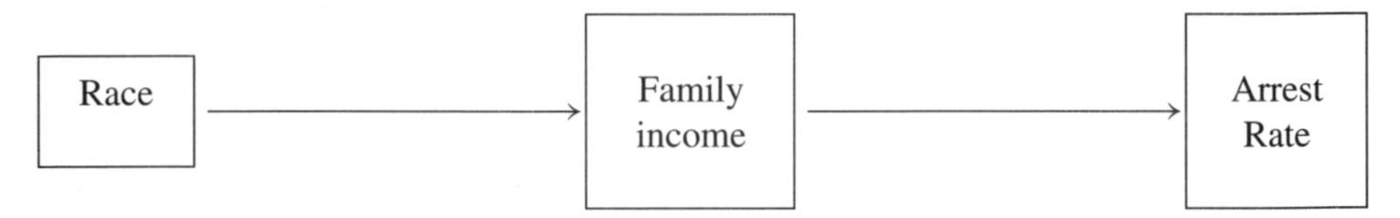

Figure 10.5 Example of a Chain Relationship

To support this model, one must show that the differential arrest rate between whites and blacks disappears after controlling for family income. That is, within fixed levels of family income, the intervening variable, no significant difference should occur in the arrest rates of blacks and whites. If this happens, race does not directly affect arrest rate, but it is an indirect cause through the intervening variable, family income.

For both spurious relationships and chain relationships, the association between X_1 and Y disappears when we control for X_2. The difference between the two is in the causal order among the variables and in the interpretation of the relationships. For a spurious association, X_2 is causally prior to both X_1 and Y, whereas in a chain association X_2 intervenes between the two. Refer to Figures 10.2 and 10.4.

To illustrate, suppose that X_1 = father's education is positively associated with Y = son's income at age 40, but this association disappears after controlling for X_2 = son's education. Specifically, the regression analysis of son's income on father's education has a Pearson correlation of .50; however, for the regression analysis conducted separately at fixed levels of son's education, either for those with X_2 = high school education or less, or those with X_2 = some college but no degree, or those with X_2 = at least a college degree, the correlation does not differ significantly from zero. In this case, father's education probably affects son's income mainly through its effects on other factors, such as son's education. Thus, it is natural to treat this as a chain relationship,

$$\text{Father's education} \longrightarrow \text{Son's education} \longrightarrow \text{Son's income}$$

For the relationship to be spurious, X_2 = son's education would need to be causally prior to both father's education and son's income, an improbable situation. It usually helps to draw arrow diagrams of variables' relationships to help guide data analysis and interpretations.

Multiple Causes

Response variables in the social world usually have more than one cause. For instance, a variety of factors could have causal influences on juvenile delinquency. Figure 10.6 depicts a common situation. This figure indicates that X_1 and X_2 are separate causes of Y. We say that Y has ***multiple causes***.

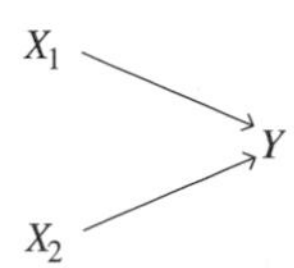

Figure 10.6 Graphical Depiction of Multiple Causes of Y

Sometimes two variables X_1 and X_2 that are separate causes of Y are themselves statistically independent. That is, they are *independent causes*. For instance, X_1 = gender and X_2 = race are essentially statistically independent. If they both have effects on juvenile delinquency, with delinquency rates varying both according to gender and race, they are likely to be independent causes.

In the social sciences, independent causes are not common. Most variables are associated. Both being poor and being from a single-parent family may cause delinquency, but those factors are themselves probably associated. Because of complex association linkages, when we control a variable X_2, the X_1–Y association usually changes somewhat. Often the association decreases somewhat, though usually it does not completely disappear as in a spurious or chain relationship.

For instance, the original X_1–Y association might be only partially a chain, and also partially a valid causal association. Perhaps X_1 has direct effects on Y and also indirect effects through a variable X_2, as shown in Figure 10.7. For instance, perhaps being from a single-parent family has direct effects on delinquency but also indirect effects through being more likely to be poor.

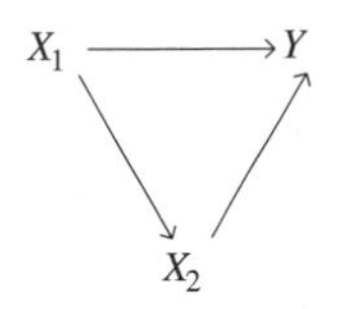

Figure 10.7 Graphical Depiction of Direct and Indirect Effects of X_1 on Y

Most response variables in social research have many causes, both direct and indirect, so that three-variable models are hardly ever adequate. However, once you understand the basic concepts presented in this chapter, you can better handle their extensions to more than three variables.

Suppressor Variables

So far, we have discussed examples in which an association disappears or changes somewhat when we control for another variable. By contrast, sometimes no association is apparent between two variables until a third variable is controlled. That control variable is called a ***suppressor variable***.

Example 10.5 Age Suppresses the Association Between Education and Income
Is educational level positively related with income? We would expect so, income tending to increase with education. Table 10.5 shows such a relationship, measured as binary variables, controlling for age. In each partial table, the percentage of subjects at the high level of income is higher when education is high than when education is low. In fact, it does seem that these variables are positively related.

TABLE 10.5 Partial Tables Relating Education and Income, Controlling for Age

		Age = Low			Age = High		
	Income	High	Low	% High	High	Low	% High
Education	High	125	225	35.7%	125	25	83.3%
	Low	25	125	16.7%	225	125	64.3%

Suppose now that we ignore age, adding these two partial tables together. The bivariate table for education and income is the first panel of Table 10.6. Every count equals 250. Both when education is high and when education is low, the percentage having a high income is 50%. For the bivariate table, no association exists between education and income.

A look at the other two bivariate tables in Table 10.6 reveals how this could happen. Age is positively associated with income but negatively associated with education. Older subjects tend to have higher income, but they tend to have lower education. Thus, when we ignore rather than control age, we give an inadvertent boost to the relative numbers of people at high incomes with low educational levels and at low incomes with high educational levels. □

The potential existence of suppressor variables shows the importance of studying partial associations even when bivariate analyses reveal a lack of association. One should control for a third variable if it may be related to both the explanatory and response variables, even though no bivariate association is evident. The first panel of Table 10.6 illustrates a *spurious nonassociation*, since both education and income depend on the third variable, age.

TABLE 10.6 Bivariate Tables Relating Education, Income, and Age

	Income			Income			Education	
Education	High	Low	Age	High	Low	Age	High	Low
High	250	250	High	350	150	High	150	350
Low	250	250	Low	150	350	Low	350	150

Statistical Interaction

Often the strength of association between two variables changes according to the value of a third variable. In other words, the association appears under certain conditions, and it disappears or changes in intensity or direction when other conditions happen. Those conditions are the values of a third variable.

For example, a data set discussed later in this text studies how the relationship between Y = annual income (in thousands of dollars) and X = number of years of education varies according to the level of race. According to recent census data, the relationships can be approximated by the linear regression equations, $E(Y) = -12.6 + 2.6X$ for whites and $E(Y) = -3.3 + 1.4X$ for blacks. On the average, income for whites increases by 2.6 thousand dollars for every year of education, whereas for blacks it increases by 1.4 thousand dollars for every year of education. The effect of education on income appears to be nearly twice as large for whites as for blacks. That is, the association between income and education varies according to the values of a control variable, race of subject.

When the true association between a variable Y and another variable X_1 changes at different levels of a control variable X_2, the three-variable relationship is said to exhibit *statistical interaction*.

Statistical Interaction

Statistical interaction exists when the true association between two variables changes as the value of a third variable changes.

One can assess whether evidence of interaction exists in the sample by comparing the X_1–Y associations among the levels of X_2. When the sample association is similar at each level of X_2, there is no evidence of interaction. The sample degree of interaction is worth noting when the variability in associations is large, for instance, when the association is positive at one level of X_2 and negative at another, or strong at one level and weak or nonexistent at another.

Example 10.6 Statistical Interaction with Death Penalty Verdicts

Table 10.7 refers to all court trials about homicides in Florida between 1976 and 1987. The table portrays the relationship between race and punishment. It summarizes the proportion of defendants who received the death penalty, for each combination of defendant's race and victim's race. Let us consider the effect of X_1 = defendant's race on Y = death penalty verdict, controlling for X_2 = victim's race.

For each partial table, we summarize the association by the difference of proportions. Let π_1 denote the proportion of white defendants receiving the death penalty, and let π_2 denote the proportion of black defendants receiving the death penalty. When the victim was white, $\pi_1 = .049$ and $\pi_2 = .126$. There is a fairly substantial difference, one proportion being less than half as large as the other, and $\pi_1 - \pi_2 = -.077$. When

TABLE 10.7 Death Penalty Verdict by Defendant's Race, Controlling for Victim's Race (Top of Table) and Ignoring Victim's Race (Bottom of Table)

Victim's Race	Defendant's Race	Death Penalty Yes	Death Penalty No	Proportion Yes
White	White	227	4418	.049
	Black	92	639	.126
Black	White	9	255	.034
	Black	36	4392	.008
Total	White	236	4673	.048
	Black	128	5031	.025

Source: M. L. Radelet and G. L. Pierce, *Florida Law Review*, Vol. 43 (1991), pp. 1–34.

the victim was black, $\pi_1 = .034$ and $\pi_2 = .008$. Both proportions are small, and the difference is small, with $\pi_1 - \pi_2 = .026$.

In summary, the association has opposite direction in the two cases. With white victims, $\pi_1 - \pi_2$ is negative, with the death penalty being much more likely for black defendants than for white defendants; with black victims, $\pi_1 - \pi_2$ is positive, the death penalty being less likely for black defendants than for white defendants. The effect of defendant's race on the death penalty verdict differs dramatically according to the level of victim's race. Since the results are quite different for the two partial tables, these data show evidence of statistical interaction.

Combining the partial tables for white victims and black victims yields the lower panel of Table 10.7, which ignores rather than controls victim's race. This table suggests that the proportion receiving the death penalty is slightly larger for whites than for blacks. This is misleading, since a strong difference occurs in the opposite direction when victims are white. When interaction exists, describe the partial tables separately, since the association changes according to the value of the third variable. □

Figure 10.8 depicts a three-variable relationship having statistical interaction. The effect of variable X_2 is on the relationship between X_1 and Y.

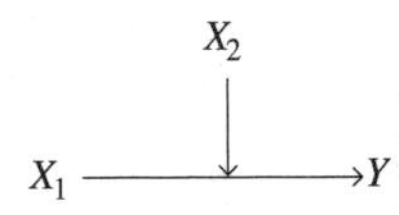

Figure 10.8 Graphical Depiction of Statistical Interaction

Association and Interaction

A lack of interaction does *not* mean that X_2 has no association with X_1 or Y. It is possible for no statistical interaction to occur even when all the variables are associated. For instance, Tables 10.2, 10.4, and 10.5 showed no interaction—in each case the association was similar in each partial table. However, in each case the control variable was associated with each variable. In Table 10.5, for instance, age is negatively associated with education and positively associated with income.

In summary, for spurious and chain relationships, the association disappears under the appropriate control. For multiple causes, the association may change under a control but does not disappear. When there is a suppressor variable, the association only appears under the control. When there is statistical interaction, the association has different strengths and/or directions at different values of the control variable.

This does not exhaust the possible association structures for three variables. It is even possible that, after controlling for a variable, each partial association has the opposite direction as the bivariate association. This is called ***Simpson's paradox*** and is illustrated in Problems 10.8, 10.17, and 10.33.

In summary, to master data analysis, one must know when to use controls, which variables to control, and how to interpret the results. This is a difficult, but vital, element of research technique. For further discussions of these issues, see Babbie (1995) and Rosenberg (1968).

10.4 Inferential Issues in Statistical Control

The first three sections of this chapter have ignored inferential matters, to avoid confusing them with the new concepts presented. We now discuss some inferential issues that arise in controlling variables.

Effects of Sample Size

In controlling for X_2, we study the X_1–Y association at fixed levels of X_2. The sample size at a fixed level of X_2 may be much smaller than in the full data set. Even if no reduction in association occurs relative to the full data, standard errors of parameter estimators tend to be larger. Thus, confidence intervals for those parameters at fixed levels of X_2 tend to be wider, and test statistic values tend to be smaller.

For qualitative data, for example, one could compute the χ^2 statistic within a particular partial table to test whether the variables are independent at that level of X_2. This χ^2 value may be small relative to the χ^2 value for the bivariate X_1-Y table. This could be due partly to a decrease in the degree of association, but it could also reflect the reduction in sample size. Section 8.4 showed that larger sample sizes tend to produce larger χ^2-values, for a particular degree of association.

Hence, a test statistic may become nonsignificant under a control simply because of the reduction in sample size. It is, therefore, important to compare the value of a measure of association (such as the difference of proportions or odds ratio) from the bivariate table to those from the partial tables even if the values from the partial tables are not statistically significant. The measure of association values in the partial tables need not necessarily be smaller than in the overall table, even if the χ^2 values are smaller.

Effects of Categorization

For simplicity, the control variables in the examples of the past few sections had only a few categories. In practice, however, one should avoid overly crude categorizations of quantitative control variables. The greater the number of control levels, the more nearly constant X_2 is forced to be within each partial table. Unless the control variable *naturally* has only two levels (e.g., sex), it is usually best to use at least three or four partial tables.

On the other hand, it's best not to use so many partial tables that each one has a very small sample size. When a study controls at once several variables, the sample sizes in some partial tables are very small. The separate results may have large standard errors, making it difficult to conduct precise inferences within the partial tables.

In addition, when the number of partial tables is large, it is difficult to assimilate the results, because many comparisons are necessary. Fortunately, the advanced model-building methods presented in the rest of the text allow us to assess patterns of partial association and interaction without necessarily performing separate analyses at the various combinations of levels of the control variables.

Comparing and Pooling Measures

After controlling for a variable, it is often useful to compare certain parameters from different levels of the control variable. One can construct confidence intervals for a difference between two parameter values in the same way as Chapter 7 showed for a difference of proportions or a difference of means. Suppose that the two sample estimates are based on independent random samples, with standard errors denoted by $\hat{\sigma}_1$ and $\hat{\sigma}_2$. If the estimates have approximately normal sampling distributions, then a confidence interval for the difference between the parameters is

$$(\text{Estimate}_2 - \text{Estimate}_1) \pm z\sqrt{\hat{\sigma}_1^2 + \hat{\sigma}_2^2}$$

If the interval does not include 0, the evidence suggests that the parameter values differ.

The term ***partial association*** refers to the strength of the association between two variables after controlling the influence of another. When the association is similar in the partial analyses, one can combine the values into a summary measure of partial association. The rest of the text shows how to do this in various situations, using models that handle all the variables at once.

10.5 Chapter Summary

It's almost always necessary to use more than two variables for more than a cursory statistical analysis of relationships. To demonstrate a causal relationship, we must show ***association*** between the variables, proper ***time order***, and ***eliminate alternative explanations*** for the association.

To consider alternative explanations, we introduce ***control variables*** into the analysis. We perform statistical control by analyzing the association between Y and X_1

while keeping the value of the control variable X_2 essentially constant. This helps us to detect

- ***Spuriousness***, in which X_2 jointly affects both Y and X_1.
- ***Chain relationships***, in which X_2 is an ***intervening variable***, so that X_1 affects Y indirectly through its effects on X_2.
- ***Suppressor variables***, in which the X_1–Y association appears only after controlling for X_2.
- ***Statistical interaction***, in which X_2 affects the X_1–Y association, the nature or strength of that association varying according to the fixed value for X_2.

Table 10.8 summarizes some possible three-variable relationships. The remainder of this text presents statistical methods for multivariate relationships.

TABLE 10.8 Some Three-Variable Relationships

Graph	Name of Relationship	What Happens After Controlling for X_2
$X_2 \nearrow X_1$, $X_2 \searrow Y$	Spurious X_1–Y association	Association between X_1 and Y disappears.
$X_1 \longrightarrow X_2 \longrightarrow Y$	Chain relationship; X_2 intervenes; X_1 indirectly causes Y	Association between X_1 and Y disappears.
$X_2 \downarrow$ ($X_1 \longrightarrow Y$)	Interaction	Association between X_1 and Y varies according to level of X_2.
$X_2 \searrow Y$, $X_1 \nearrow Y$	Multiple causes	Association between X_1 and Y does not change.
$X_1 \longrightarrow Y$, $X_1 \searrow X_2 \nearrow Y$	Both direct and indirect effects of X_1 on Y	Association between X_1 and Y changes, but does not disappear.

PROBLEMS

Practicing the Basics

1. For each of the three criteria for a causal relationship, describe a relationship between two variables that is not causal because that criterion would be violated.
2. An association exists between college GPA and whether one has ever used marijuana.
   a) Explain how the direction of a causal arrow might go in either direction.
   b) Explain how a third variable might be responsible for the association.
3. Give an example of two variables for which the time order for a causal connection is ambiguous, and explain why.
4. Explain clearly what it means to *control* for a variable. Describe a situation in which an association would change dramatically after controlling a certain variable.

5. Explain what is meant by a *spurious* association.
a) Illustrate using the variables X_1 = shoe size, X_2 = age, and Y = number of books one has ever read, for a random sample of children from schools in Winnipeg, Canada.
b) Illustrate using the variables X_1 = shoe size, X_2 = gender, and Y = annual income, for a random sample of adults. Draw a scatter diagram to illustrate this case if, overall, men tend to have larger shoes and higher income, on the average, than females.
6. Refer to Problem 9.17 (Table 9.13). Those data exhibit a strong negative correlation between birth rate and per capita television ownership. While there *could* potentially be a causal relationship, there are other plausible explanations for this association. Using another variable from that exercise, provide an explanation for which this association is spurious.
7. Table 10.9 relates occupational level (white collar, blue collar) and political party choice, controlling for income.
a) Construct the bivariate table between occupational level and political party, ignoring income. Is there an association? If so, describe it.
b) Do the partial tables display an association? Interpret them.
c) Using the nature of the association between income and each of the other variables, explain why the bivariate table has such different association than the partial tables.
d) Do the partial tables exhibit statistical interaction? Explain.
e) Construct a chain diagram that might explain the relationships for these data, identifying the intervening variable.
f) Show that the data are also consistent with a spurious association, and draw the corresponding diagram.
g) Refer to parts (e) and (f). Which diagram seems more appropriate? Why?

TABLE 10.9

		High Income		Medium High Income		Medium Low Income		Low Income	
		White Collar	Blue Collar	White Collar	Blue Collar	White Collar	Blue Collar	White Collar	Blue Collar
Party	Democrat	45	5	100	25	75	300	45	405
	Republican	405	45	300	75	25	100	5	45

8. In murder trials in 20 Florida counties during 1976 and 1977, the death penalty was given in 19 out of 151 cases in which a white killed a white, in 0 out of 9 cases in which a white killed a black, in 11 out of 63 cases in which a black killed a white, and in 6 out of 103 cases in which a black killed a black (M. Radelet, *American Sociological Review*, Vol. 46, 1981, pp. 918–927).
a) Construct partial tables to study the association between defendant's race and the death penalty verdict, controlling for victim's race.
b) Describe the association in the partial tables, using the difference of proportions or the odds ratio.
c) Construct the bivariate table, ignoring rather than controlling victim's race. Describe the association, and compare to (b).
d) *Simpson's paradox* states that the associations in partial tables can all have a different direction than the association in the bivariate table. Show that these data satisfy Simpson's paradox.

9. The news media report that a study has found that children who eat breakfast get better math grades than those who do not eat breakfast. This result was based on a simple bivariate association, with X = whether eat breakfast (yes, no) and Y = grade in last math course taken. How might this result be spurious, and how could you check for that possibility?

10. For a particular Big Ten university, the mean income for male faculty is $8000 higher than the mean income for female faculty. Explain how this difference could disappear:
a) Controlling for number of years since received highest degree, if male professors tend to be older and more experienced, relatively few female professors having been hired until recent years.
b) Controlling for college of employment, if relatively more female faculty are in low-salary colleges (Arts and Science, Education) and relatively more male faculty are in high-salary colleges (Medical School, Law, Engineering).
c) In (a) and (b), explain how the variables might satisfy a chain relationship.

11. Refer to the variables in Table 9.4. The number of bedrooms has a moderately strong positive correlation with selling price ($r = .59$). Controlling for size of home, however, this association diminishes greatly. Explain how this could happen, illustrating with a diagram showing potential direct and indirect effects of number of bedrooms on selling price.

12. Refer to the variables in Table 9.16. Perhaps surprisingly, a moderate positive correlation exists between crime rate and percent who are at least high school graduates ($r = .468$). Percentage living in metropolitan areas is also strongly positively correlated both with crime rate ($r = .678$) and with high school graduation rate ($r = .791$).
a) Explain why the association between crime rate and high school graduation rate could disappear, or even change direction, when we control for percentage living in metropolitan areas.
b) Which type of relationship is more plausible for these variables, a spurious relationship or a chain relationship? Explain.

13. Give an example of three variables for which the effect of X_1 on Y would be
a) Spurious, disappearing when X_2 is controlled.
b) Part of a chain relationship, disappearing when an intervening variable X_2 is controlled.
c) Weakened, but not eliminated, when X_2 is controlled.
d) Unaffected by controlling X_2.
e) Different at different levels of X_2 (i.e., showing interaction).

14. Opposition to the legal availability of abortion is stronger among the very religious than the nonreligious, and it is also stronger among those with conservative sexual attitudes than those with more permissive attitudes.
a) Draw a three-variable diagram of how these variables might be related, treating abortion attitude as the response variable. (Note: More than one diagram is plausible.)
b) Explain how you would test your diagram, showing potential sample data that are consistent with it.

15. Table 10.10 lists the mean salary, in thousands of dollars, of full-time instructional faculty on nine-month contracts in United States institutions of higher education in 1993–1994, by gender and rank.
a) Suppose that gender is the explanatory variable. Identify the response variable and the control variable.
b) Describe the bivariate relationship between gender and salary.
c) Describe the relationship between gender and salary, controlling for academic rank.

d) A hypothesis of interest for these variables is "Controlling for academic rank, annual salary and gender are independent." Draw a causal diagram that is consistent with this hypothesis. Refer to your interpretation in part (c), and comment on whether the hypothesis in fact seems plausible.
e) The overall difference of 9.5 thousand dollars between mean income of men and women is larger than the difference for each academic rank. Explain how this could be.

TABLE 10.10

	Academic Rank				
Gender	Professor	Associate	Assistant	Instructor	Overall
Men	61.9	46.2	38.8	29.8	49.6
Women	54.7	43.2	36.2	28.1	40.1

Source: U.S. Department of Education, National Center for Education Statistics, *Digest of Education Statistics, 1995*, p. 240.

16. Refer to Table 10.11. Form the overall table, summing over social class. Compare the proportions of Catholics and Protestants favoring the death penalty, both ignoring and controlling social class. What type of relationship do these data satisfy?

TABLE 10.11

		Social Class			
		Working Class		Middle Class	
	Death Penalty	Favor	Oppose	Favor	Oppose
Religion	Catholic	240	160	30	70
	Protestant	80	20	200	200

17. Refer to Table 15.4 on death penalty verdict, defendant's race, and victims' race.
a) Construct the bivariate table relating defendant's race to death penalty verdict, ignoring victims' race. Calculate the conditional distribution on the death penalty verdict, for each race of defendant.
b) Construct partial tables relating death penalty verdict to defendant's race, treating victims' race as a control variable. For each partial table, describe the association by computing the difference between the proportions of white and black defendants who received the death penalty. Interpret.
c) Notice that the association changes direction when we control for victims' race; that is, in the bivariate table relatively more whites received the death penalty, but in each partial table relatively more blacks received the death penalty. (This is Simpson's paradox; see Problem 10.8d.) By describing how victims' race is associated with each of these variables, explain why the partial association differs as it does from the bivariate association.
d) For these variables, indicate whether each of the following diagrams seems to provide a reasonable model. Give your reasoning. (Here, P = death penalty, D = defendant's race, V = victims' race.)

i. $V \begin{smallmatrix} \nearrow P \\ \searrow D \end{smallmatrix}$ ii. $D \longrightarrow V \longrightarrow P$ iii. $\begin{smallmatrix} D \searrow \\ V \nearrow \end{smallmatrix} P$ iv. $\begin{smallmatrix} D & \longrightarrow & P \\ & \searrow V \nearrow & \end{smallmatrix}$

e) Is it plausible that victims' race may be a suppressor variable? Why or why not?

18. For the cell counts in Table 10.12 relating Y = exam score (1 = below median, 2 = above median) to gender, controlling for subject of exam (Math, Verbal),
a) Is subject of exam a suppressor variable? Explain.
b) Is there interaction? Explain.

TABLE 10.12

	Math		Verbal	
Gender	$Y = 1$	$Y = 2$	$Y = 1$	$Y = 2$
Females	100	50	50	100
Males	50	100	100	50

19. Refer to Problem 8.43 in Chapter 8.
a) Treat expectation of voting as the control variable, and construct the partial tables relating political party affiliation to present voting intention.
b) Show how to combine the partial tables to obtain the bivariate table constructed in Problem 8.43.
c) From constructing conditional distributions in the partial tables in part (a), indicate whether sample evidence of statistical interaction seems to exist.

20. Refer to Table 15.17 on AZT use and AIDS symptoms.
a) Using the difference of proportions or the odds ratio, describe the association for (i) whites, (ii) blacks.
b) Does there seem to be strong evidence of statistical interaction? Explain.

21. A study of the association between whether a smoker (yes, no) and whether have had some form of cancer (yes, no) has odds ratio 1.1 for subjects of age less than 30, 2.4 for subjects of age 30 to 50, and 4.3 for subjects of age over 50.
a) Identify the response variable, explanatory variable, and control variable, and interpret the associations.
b) Does the study show evidence of interaction? Explain.

22. Refer to Table 8.29. Treat the four groups as a cross-classification of the two variables, race and gender. Treating race as a control variable, is there evidence of severe statistical interaction for these data? Explain.

23. A study of students at Oregon State University found an association between frequency of church attendance and favorability toward the legalization of marijuana. Both variables were measured in ordered categories. When gender of student was controlled, the resulting gamma measures for the two partial tables were as follows:
Males: gamma $= -.287$, standard error $= .081$
Females: gamma $= -.581$, standard error $= .091$
a) Interpret these results.
b) These results show a slight degree of______, since the association is somewhat stronger for females than males.

24. A study of the relationship between student's high school GPA and mother's employment (yes, no) suspects an interaction with gender of student. Controlling gender of student, Table 10.13 contains the mean GPA for each combination of gender and mother's employment.
a) Describe the relationship between mother's employment and GPA for females and for males. Does this sample show evidence of statistical interaction? Explain.

b) A journal article written about the study states, "Having a mother who is employed outside the home seems to have positive effects on daughter's achievement in high school, but no substantive effect on son's achievement." Explain how Table 10.13 suggests this interpretation.

TABLE 10.13

Gender	Mother Employed	Mother Not Employed
Females	2.94	2.71
Males	2.72	2.74

Concepts and Applications

25. Refer to the WWW data set (Problem 1.7). Construct partial tables relating opinion about abortion to opinion about life after death, controlling for attendance at religious services, measured using the two categories, (Never or occasionally, Most weeks or every week). Prepare a report (a) posing and interpreting a possible arrow diagram, before you analyze the data, for relationships among the variables, (b) interpreting the sample associations in the bivariate table and the partial tables, (c) revising, if necessary, your arrow diagram based on the evidence in the sample data.

26. Refer to Problem 1.7. Repeat the previous problem for three variables chosen by your instructor.

27. Refer to Problem 1.7. Are there any pairs of these variables for which you expect the association to disappear under control for a third variable? Explain.

28. Using the most recent General Social Survey, construct a contingency table relating gender and party affiliation ("PARTYID"). Is there still a gender gap? Control for political ideology ("POLVIEWS") by forming partial tables for the most conservative and the most liberal subjects. Does the association seem to persist for these subjects?

29. Table 10.14 is a contingency table that shows counts on self-esteem, race, and cumulative GPA, for a sample of females. Analyze the type of multivariate relationship that these variables seem to satisfy. Prepare a report, explaining the analyses you conducted to reach your conclusion and providing interpretations.

TABLE 10.14

	Black		White	
Cumulative GPA	High Esteem	Low Esteem	High Esteem	Low Esteem
High	13	22	22	32
Low	24	23	3	17

Source: D. Demo and K. Parker, *Journal of Social Psychology*, Vol. 127, 1987, pp. 345–355.

30. Table 10.15 shows the relation between attitude toward a nuclear freeze and church attendance, controlling for educational levels of the respondents. Assessing association by the difference of proportions, analyze these data using methods of this chapter, and explain why it is misleading to report only the results from the bivariate table.

TABLE 10.15

		Education			
		12 years or less		More than 12 years	
Attendance		Seldom or never attend	Often attend church	Seldom or never attend	Often attend church
Attitude	Favor	82	96	140	164
	Oppose	66	117	80	55

31. In 1980, SAT total scores (verbal plus math) had a mean of 890. In 1984 the mean increased to 897, for an increase of 7. Table 10.16 shows the means and their changes, controlling for race. For each year, the table also shows the percentage representation in the sample of each race.
a) In each year, using the means for each race and the percentage representations, verify the overall means.
b) For the overall sample, the change in the mean from 1980 to 1984 (+7) was smaller than the change for either whites (+8) or nonwhites (+15). Explain this apparent paradox.

TABLE 10.16

	Mean SAT			Percentage Representation	
Race	1980	1984	Change	1980	1984
White	924	932	+8	72.2	69.5
Nonwhite	802	817	+15	27.8	30.5
All	890	897	+7	100.0	100.0

Source: H. Wainer, *Journal of Educational Statistics*, Vol. 11, (1986), pp. 239–244.

32. Table 10.17 reports the median family income (in dollars) in 1992 for white and black families. The median for white families is $17,748 higher than the median for black families. For each of the three types of families, however, the median for white families is about $8000 higher than the median for black families. Explain what could cause this discrepancy.

TABLE 10.17

	Race		Difference
Type of Family	Black	White	of Medians
Married couple	34,196	42,738	8,542
Male householder	20,678	29,671	8,993
Female householder	11,956	20,130	8,174
Overall	21,161	38,909	17,748

Source: U.S. Department of Commerce, Bureau of the Census, *Money Income of Households, Families, and Persons in the United States: 1992*, *Current Population Reports*, P60–184.

33. Table 10.18 shows the mean number of children in Canadian families, classified by whether the family was English speaking or French speaking and by whether the family was in Quebec or in another province. Let Y = number of children in family, X_1 = primary language of family, and X_2 = province (Quebec, others).
a) Describe the association between Y and X_1, based on the overall means in this table.
b) Describe the association between Y and X_1, controlling for X_2.
c) Explain how it is possible that for each level of province the mean is higher for French speaking families, yet overall the mean is higher for English speaking families. (This illustrates *Simpson's paradox*; see Problem 10.8d).

TABLE 10.18

Province	English	French
Quebec	1.64	1.80
Other	1.97	2.14
Overall	1.95	1.85

34. Religious affiliation is associated with attitudes about racial intermarriage, with white Catholics tending to be less racially prejudiced than Protestants, members of theologically liberal Protestant groups tending to be less prejudiced than those from the more conservative denominations, and Jews and the nonreligious being least prejudiced (see Problem 8.35). Describe some variables than you might control to see if they can explain, at least partially, this association.

35. A research study funded by Wobegon Springs Mineral Water, Inc., discovers that the probability that a newborn child has a birth defect is lower for families that regularly buy bottled water than for families that do not. Is this association likely to reflect a causal link between drinking bottled water and a reduction in birth defects? Why or why not?

36. The percentage of women who get breast cancer is higher now than at the beginning of this century. Suppose that cancer incidence tends to increase with age, and suppose that women tend to live longer lives now than earlier in this century. Explain why a comparison of breast cancer rates now with the beginning of this century could show different results from these if we control for the age of the woman.

37. Refer to Table 9.1. A regression analysis reveals a moderately strong negative correlation (−.677) between percent of white residents and violent crime rate. Another variable measured in Table 9.1 is percent of families headed by a single parent. This is positively correlated with violent crime rate (.839) and negatively correlated with percent white residents (−.657).
a) Controlling for single-parent rate, would you expect the association between percentage of white residents and violent crime rate to be stronger, or weaker, or the same as in the bivariate relationship. Explain.
b) Explain how these variables could plausibly have a chain relationship.

38. In the United States, median age of residents is lowest in Utah. At each age level, the death rate from heart disease is higher in Utah than in Colorado; yet overall, the death rate from heart disease is lower in Utah than Colorado. Are there any contradictions here, or is this possible? Explain.

39. For lower-level managerial employees of a fast-food chain, the prediction equation relating Y = annual income (thousands of dollars) to X_1 = number of years experience on the

job equals $\hat{Y} = 14.2 + 1.1X_1$ for $X_2 =$ males and $\hat{Y} = 14.2 + .4X_1$ for $X_2 =$ females. These equations show evidence of (select one, and explain)
a) Spurious relationship
b) Chain relationship
c) Statistical interaction

40. Statistical interaction refers to which of the following?
a) Association exists between two variables.
b) The degree of association between two variables varies greatly over the partial levels of a control variable.
c) The partial association is the same at each level of the control variable, but it is different from the overall bivariate association, ignoring the control variable.
d) For a collection of three variables, each pair of variables is associated.
e) All of the above.

41. Consider the relationship between $Y =$ political party preference (Democrat, Republican) and $X_1 =$ race (Black, White) and $X_2 =$ gender. There is an association between Y and both X_1 and X_2, with the Democrat preference being more likely for blacks than whites and for women than men. Select the correct response(s).
a) X_1 and X_2 are probably independent causes of Y.
b) The association between Y and X_1 is probably spurious, controlling for X_2.
c) Since both variables affect Y, there is probably interaction.
d) The variables probably satisfy a chain relationship.
e) Race is probably a suppressor variable.
f) None of the above.

Bibliography

Babbie, E. (1995). *The Practice of Social Research*, 7th ed. Belmont, CA: Wadsworth.
Crosson, C. (1994). *Tainted Truth: The Manipulation of Fact in America*. New York: Simon & Schuster.
Davis, J. (1985). *The Logic of Causal Order*. Beverly Hills, CA: Sage.
Hirschi, T., and Selvin, H. (1973). *Principles of Survey Analysis*. New York: The Free Press.
Rosenberg, M. (1968). *The Logic of Survey Analysis*. New York: Basic Books.
Zeisel, H. (1985). *Say It with Figures*. New York: Harper & Row.

Chapter 11

Multiple Regression and Correlation

Chapter 9 introduced a regression model for the relationship between two quantitative variables. Multivariate relationships require more complex models, containing several explanatory variables. Some of these may be predictors of theoretical interest, and some may be control variables.

To predict Y = college GPA, for example, it is sensible to use several predictors in the same model, such as X_1 = high school GPA, X_2 = math score on college entrance exam, X_3 = verbal score on college entrance exam, and X_4 = rating by high school guidance counselor. This chapter presents a model for the relationship between a response variable Y and a collection of explanatory variables. The model provides better predictions of Y than does a model with a single explanatory variable.

This multivariable model also can analyze *partial* relationships between two variables, controlling for other variables. This is important because an effect of a predictor may change considerably when controls are introduced. The previous chapter showed that partial associations, controlling certain variables, can be quite different from bivariate associations in which the other variables are ignored rather than controlled. Thus, this model provides information not available with simple models that analyze only two variables at a time.

The first section of the chapter extends the linear regression model to a ***multiple regression model*** that can have any number of explanatory variables. Section 11.2 introduces an example that illustrates aspects of the model throughout the chapter. Section 11.3 defines some multivariate measures of association. These include a *multiple correlation* and a r^2-like proportional reduction in error measure that describe association between a response variable Y and a set of explanatory variables. Section 11.4 presents inference procedures for the multiple regression model. Section 11.5 shows how to allow *statistical interaction* in the model, and Section 11.6 presents a test of whether a complex model provides a better fit than a simpler model. Finally, Section 11.7 and 11.8 introduce *partial correlations* and *standardized regression coefficients* that summarize the partial association between the response variable and each explanatory variable.

11.1 The Multiple Regression Model

Chapter 9 modeled the relationship between the explanatory variable X and the mean of the response variable Y by the linear equation $E(Y) = \alpha + \beta X$. We refer to this model containing a *single* predictor as a ***bivariate model***, since it contains only two variables.

The Basic Model

In a multivariable study, denote the explanatory variables by $X_1, X_2, X_3, \ldots, X_k$, where k denotes the number of predictors. The bivariate model generalizes to

$$E(Y) = \alpha + \beta_1 X_1 + \beta_2 X_2 + \cdots + \beta_k X_k$$

In this equation, $\alpha, \beta_1, \ldots, \beta_k$ are parameters discussed below. This equation is called a ***multiple regression*** equation.

The multiple regression model is more difficult to portray graphically than the bivariate model. For $k = 2$ explanatory variables, the X_1 and X_2 axes are perpendicular but lie in a horizontal plane and the Y axis is vertical and perpendicular to both the X_1 and X_2 axes. The equation $E(Y) = \alpha + \beta_1 X_1 + \beta_2 X_2$ traces a plane (a flat surface) cutting through three-dimensional space, as Figure 11.1 portrays. For $k > 2$, the equation traces a surface in $(k + 1)$-dimensional space. This makes mathematical sense for any k, but it is difficult to visualize for $k > 2$.

The simplest way to interpret the model treats all but one explanatory variable as control variables and fixes them at particular levels. This leaves a linear equation describing the relationship between the mean of Y and the remaining explanatory variable. One can plot this straight line, and its slope is the same for every combination of values of the other variables.

Example 11.1 Do Higher Levels of Education Cause Higher Crime Rates?

Problem 9.24 in Chapter 9 contains recent measurements on several variables for the 67 counties in the state of Florida. For that data set, let Y = crime rate (annual number of crimes per 1000 population), X_1 = education (percentage of adult residents having

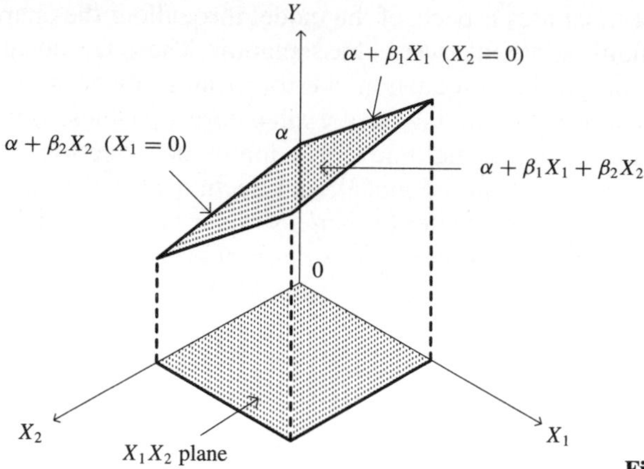

Figure 11.1 Graphical Depiction of a Multiple Regression Equation

at least a high school education), and X_2 = urbanization (percentage living in an urban environment).

The bivariate relationship between crime rate and education is approximated by $E(Y) = -51.3 + 1.5X_1$. Surprisingly, the association is moderately *positive*, the Pearson correlation being $r = .47$. As the percentage of county residents having at least a high school education increases, so does the crime rate.

A closer look at the data reveals strong positive associations between crime rate and urbanization ($r = .68$) and between education and urbanization ($r = .79$). This suggests that the association between crime rate and education may be spurious. Perhaps urbanization is a common causal factor. See Figure 11.2. As urbanization increases, both crime rate and education increase, resulting in a positive correlation between crime rate and education.

The relation between crime rate and both predictors considered together is approximated by the multiple regression model

$$E(Y) = 58.9 - .6X_1 + .7X_2$$

One can find the expected crime rate for counties having any particular values of X_1 and X_2 by substituting those values into the regression equation. For instance, the expected crime rate for a county at the mean levels of education ($\bar{X}_1 = 70$) and urbanization ($\bar{X}_2 = 50$) is $E(Y) = 58.9 - .6X(70) + .7(50) = 52$ annual crimes per 1000 population.

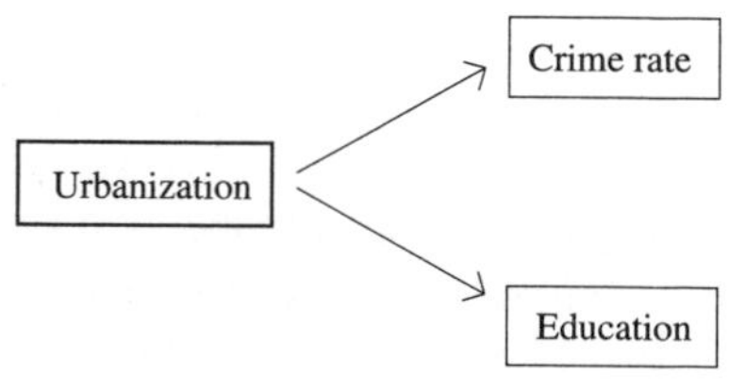

Figure 11.2 Spurious Positive Association Between Crime Rate and Education

Let's study the relationship between Y and X_1, controlling for X_2. We first set X_2 at its mean level of 50. Then, the partial relationship between crime rate and education is

$$E(Y) = 58.9 - .6X_1 + .7(50) = 58.9 - .6X_1 + 35.0 = 93.9 - .6X_1$$

Figure 11.3 plots this line. Controlling for X_2 by fixing it at 50, the relationship between crime rate and education is now negative, rather than positive. The slope decreased and changed sign from 1.5 in the bivariate relationship to $-.6$. At a fixed level of urbanization, a negative relationship exists between education and crime rate.

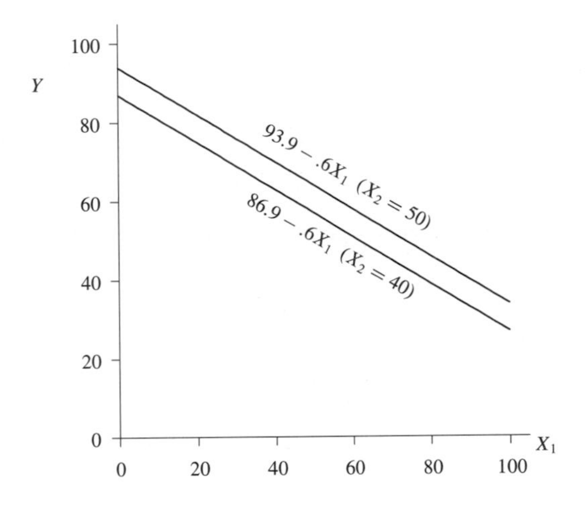

Figure 11.3 The Partial Relationship Between Y and X_1 for the Multiple Regression Equation $E(Y) = 58.9 - .6X_1 + .7X_2$

Next we fix X_2 at a different level, say $X_2 = 40$ instead of 50. Then, you can check that $E(Y) = 86.9 - .6X_1$. Thus, decreasing X_2 by 10 units shifts the partial line relating Y to X_1 downward by $10\beta_2 = 7.0$ units (see Figure 11.3). The slope of $-.6$ for the relationship remains the same, so the line is parallel to the original one. Setting X_2 at a variety of values yields a collection of parallel lines, each having slope $\beta_1 = -.6$.

Similarly, setting X_1 at a variety of values yields a collection of parallel lines, each having slope .7, relating the mean of Y to X_2. In other words, controlling for education, the slope of the partial relationship between crime rate and urbanization is $\beta_2 = .7$.

In summary, education has an overall positive effect on crime rate, but a negative effect controlling for urbanization. The partial association has the opposite direction from the bivariate association. This is called ***Simpson's paradox***. Figure 11.4 illustrates how this happens. It shows the scatter diagram relating crime rate to education, portraying the overall positive association between these variables. On the diagram, we have circled the 19 counties that are highest in urbanization. That subset of points for which urbanization is nearly constant has a negative trend between crime rate and education. The high positive association between education and urbanization is reflected by the fact that most of the highlighted observations that are highest on urbanization also have high values on education. □

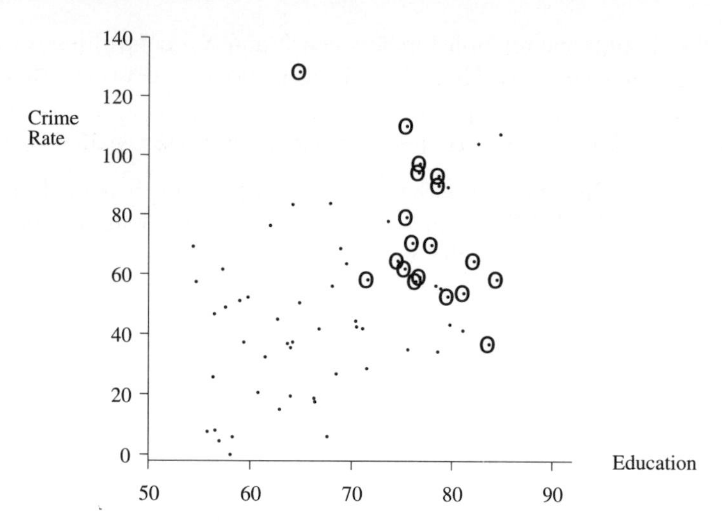

Figure 11.4 Scatter Diagram Relating Crime Rate and Education. Circled points are the counties highest on Urbanization.

Interpretation of Regression Coefficients

We have seen that for a fixed value of X_2, the equation $E(Y) = \alpha + \beta_1 X_1 + \beta_2 X_2$ is a linear equation in X_1 with slope β_1. Thus, controlling for X_2, a linear relationship exists between $E(Y)$ and X_1 with slope β_1. The Y-intercept depends on the fixed value of X_2. In addition, the slope β_1 of the controlled relationship between Y and X_1 is the same for each fixed value of X_2. Thus, if X_1 is increased 1 unit with X_2 held constant, the change in the mean of Y is β_1.

The parameter β_1 measures the partial effect of X_1 on Y, that is, the effect on Y of a one-unit increase in X_1, holding X_2 constant. The partial effect of X_2 on Y, holding X_1 constant, has slope β_2. Similarly, for the multiple regression model with several predictors, the coefficient β_i of an arbitrary X_i describes the change in the mean of Y for a one-unit increase in X_i, controlling for the other variables in the model.

The parameter α in the multiple regression equation is the mean of Y when each explanatory variable equals 0. The parameters $\beta_1, \beta_2, \ldots, \beta_k$, which are the slopes of the partial relationships, are called ***partial regression coefficients***. The adjective *partial* distinguishes these parameters from the regression coefficients for the *bivariate* relationships between Y and each explanatory variable. The coefficient β in the bivariate model $E(Y) = \alpha + \beta X$ describes the slope between $E(Y)$ and X, *ignoring* rather than *controlling* effects of the other explanatory variables. Since the bivariate model omits other explanatory variables, that model cannot control their influences by fixing them at constant levels.

The partial slope for X_1 in a multiple regression model need not differ from its slope in the bivariate model. For instance, when $k = 2$, they are equal if the correlation between X_1 and X_2 equals 0. When X_1 and X_2 are independent causes of Y, the effect of X_1 on Y does not change when we control for X_2.

This multiple regression model assumes that the slope of the partial relationship between Y and each X_i is identical for *all* combinations of values of the other explanatory variables. This means that the model is appropriate when there is *no statistical interaction*, in the sense of Section 10.3. If the true partial slope between Y and X_1 is very different at $X_2 = 50$ than at $X_2 = 40$, for example, we need a more complex model of a type presented in Section 11.5.

Prediction Equation and Residuals

Once we propose the model $E(Y) = \alpha + \beta_1 X_1 + \cdots + \beta_k X_k$ for the effects of a set of predictors on a response variable, the next step is to estimate the model parameters using sample data.

Notation for Prediction Equation

Let $\{a, b_1, b_2, \ldots, b_k\}$ denote point estimates of $\{\alpha, \beta_1, \beta_2, \ldots, \beta_k\}$. Then $\hat{Y} = a + b_1 X_1 + \cdots + b_k X_k$ denotes the prediction equation that estimates the multiple regression equation $E(Y) = \alpha + \beta_1 X_1 + \cdots + \beta_k X_k$.

When the model contains at least two explanatory variables, calculating the prediction equation is extremely cumbersome and it is almost imperative to use computer software. The calculation formulas use matrix algebra and are not shown in this text.

Using the prediction equation, software provides predicted values of Y for each subject by substituting the X-values into the equation. Like the bivariate model, the multiple regression model has ***residuals*** that measure prediction errors. For a subject with predicted response $\hat{Y}$ and observed response Y, the residual is the difference $Y - \hat{Y}$. The next section shows an example.

The ***sum of squared errors*** (SSE),

$$\text{SSE} = \sum \left(Y - \hat{Y}\right)^2$$

summarizes the closeness of fit of the prediction equation to the response data. The formula for this prediction error summary is identical to SSE in Chapter 9. The only difference is that the predictions of Y result from using several explanatory variables simultaneously, instead of just a single predictor.

The parameter estimates in the prediction equation satisfy the ***least squares*** criterion: The prediction equation has the *smallest* SSE value of all possible equations of form $\hat{Y} = a + b_1 X_1 + \cdots + b_k X_k$.

11.2 Example with Multiple Regression Computer Output

We illustrate the methods of this chapter with the data introduced in the following example. To aid the data analysis, we present some computer outputs that are typical of those provided by software for multiple regression analyses.

Example 11.2 Multiple Regression for Mental Health

A study in Alachua County, Florida, investigated the relationship between certain mental health indices and several explanatory variables. Primary interest focused on an index of mental impairment, which incorporates various dimensions of psychiatric symptoms, including aspects of anxiety and depression. This measure, which is the response variable Y, ranged from 17 to 41 in the sample; higher scores indicate greater psychiatric impairment.

The two explanatory variables used here are X_1 = life events score and X_2 = socioeconomic status (SES). The life events variable is a composite measure of both the number and severity of major life events the subject experienced within the past three years. These events range from severe personal disruptions such as a death in the family, a jail sentence, or an extramarital affair, to less severe events such as getting a new job, the birth of a child, moving within the same city, or having a child marry. This measure, developed by E. Paykel et al. (*Archives of General Psychiatry*, Vol. 75, 1971, pp. 340–347), ranged from 3 to 97 in the sample. A high X_1 score represents a greater number and/or greater severity of these life events.

The second explanatory variable, X_2 = SES, is a composite index based on occupation, income, and education. Measured on a standard scale, it ranged from 0 to 100; the higher the score, the higher the status. These three variables are, at best, crude approximations for the traits that are truly of interest. However, much of social science research deals with the construction, use, and improvement of such indices for traits that are not directly measureable.

Table 11.1 presents the scores on the three variables for a random sample of 40 adults in the county. [These data are based on a larger survey reported by Holzer (1977). The authors thank Dr. Holzer for permission to use the study as the basis of this example.] Table 11.2 summarizes the sample means and standard deviations of the three variables. □

Plots of Bivariate and Partial Relationships

As in the bivariate case, plots of the data provide an informal check of whether the relationships are linear. Most software has the option of constructing scatter diagrams on a single page for each pair of the variables. Figure 11.5 shows the plots for the variables from Table 11.1. This type of plot is called a ***scatterplot matrix***. Like a correlation matrix, it shows each pair of variables twice. In one plot, a variable is on the Y-axis and in one it is on the X-axis. For instance, mental impairment is on the Y-axis for the plots

TABLE 11.1 Scores on Y = Mental Impairment, X_1 = Life Events Index, and X_2 = Socioeconomic Status

Y	X_1	X_2	Y	X_1	X_2	Y	X_1	X_2
17	46	84	26	50	40	30	44	53
19	39	97	26	48	52	31	35	38
20	27	24	26	45	61	31	95	29
20	3	85	27	21	45	31	63	53
20	10	15	27	55	88	31	42	7
21	44	55	27	45	56	32	38	32
21	37	78	27	60	70	33	45	55
22	35	91	28	97	89	34	70	58
22	78	60	28	37	50	34	57	16
23	32	74	28	30	90	34	40	29
24	33	67	28	13	56	41	49	3
24	18	39	28	40	56	41	89	75
25	81	87	29	5	40			
26	22	95	30	59	72			

TABLE 11.2 Estimated Means and Standard Deviations of Y = Mental Impairment, X_1 = Life Events, and X_2 = Socioeconomic Status

Variable	Mean	Standard Deviation
Impairment (Y)	27.30	5.46
Life Events (X_1)	44.42	22.62
SES (X_2)	56.60	25.28

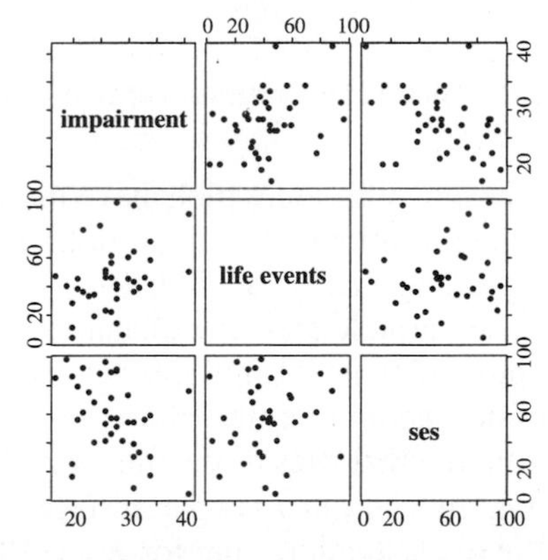

Figure 11.5 Scatter Diagrams for Pairs of Variables from Table 11.1

in the first row of Figure 11.5. These plots show no evidence of nonlinearity, and linear models seem appropriate. The plots suggest that life events has a mild positive effect and SES has a mild negative effect on mental impairment.

The multiple regression model states, in addition, that each predictor has a linear effect with common slope, controlling for the other. To check this, some software plots Y versus each predictor, for subsets of points that are nearly constant on the other predictor. In such plots, it is usually not possible to keep control variables exactly constant, or else not enough points are left to show an informative plot. So, software takes regions of values within which the control variable does not vary much. This type of plot is called a ***conditioning plot*** or ***coplot***.

Figure 11.6 shows a conditioning plot for the relation between mental impairment and life events, controlling for SES. The bars above the plot show the range of SES values for each plot. The three bars below the dotted line refer to the three plots on the bottom. For instance, the first plot on the bottom refers to SES values between about 3 and 40. These plots have small sample sizes, but linearity seems plausible for the partial effect of life events on mental impairment.

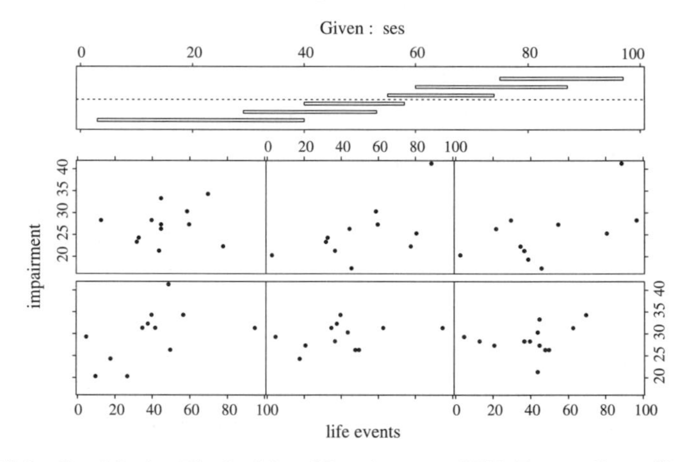

Figure 11.6 Conditioning Plot for Mental Impairment and Life Events, Controlling for SES

For small samples, conditioning plots have the disadvantage that each panel contains relatively few observations. A more informative single picture is provided by the ***partial regression plot***, also called an ***added-variable plot***. It displays the relationship between the response variable and an explanatory variable X_i after removing the effects of the other predictors in the multiple regression model. It does this by plotting the residuals from the model that uses those other variables to predict Y against the residuals from the model that uses those other variables to predict X_i.

To illustrate, Figure 11.7 shows a partial regression plot for Y = mental impairment and X_1 = life events, controlling for X_2 = SES. It plots the residuals from the model $E(Y) = \alpha + \beta X_2$ against the residuals from the model $E(X_1) = \alpha + \beta X_2$. For these residuals, the effect of X_2 is removed. The least squares slope for the points in this plot is necessarily the same as the estimated partial slope in the multiple regression model. Figure 11.7 suggests that the partial effect of life events is approximately linear. The partial regression plot for SES (not given here) shows that its partial effect is also approximately linear. It is easy to obtain such plots with standard software. (See the appendix.)

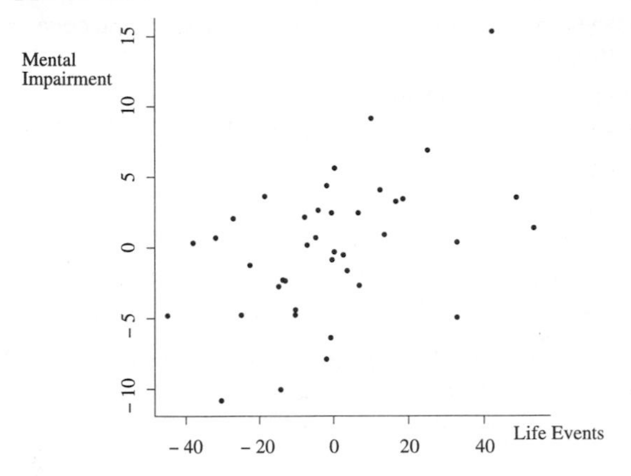

Figure 11.7 Partial Regression Plot for Mental Impairment and Life Events, Controlling for SES

Sample Computer Printouts

Tables 11.3 and 11.4 are computer printouts for the bivariate relationships between Y and X_1 and between Y and X_2. The prediction equations are

$$\hat{Y} = 23.31 + .090X_1 \text{ and } \hat{Y} = 32.17 - .086X_2$$

In the sample, mental impairment is positively related to life events, since the coefficient of X_1 (.090) is positive. The greater the number and severity of life events in the previous three years, the higher the index of psychiatric symptoms (i.e., the poorer the

TABLE 11.3 Regression Analysis for Y = Mental Impairment and X_1 = Life Events Score

Source	DF	Sum of Squares	Mean Square	F Value	Prob>F
Model	1	161.048	161.048	6.112	0.0180
Error	38	1001.352	26.351		
Total	39	1162.400			

Root MSE 5.13336 R-square 0.1385

Variable	Parameter Estimate	Standard Error	T for H0: Parameter=0	Prob > |T|
INTERCEP	23.3095	1.8068	12.901	0.0001
LIFE	0.0898	0.0363	2.472	0.0180

mental health) tends to be. Also, mental impairment is negatively related to socioeconomic status. The greater the SES level, the lower the index of psychiatric symptoms tends to be. The r^2 coefficient of determination values between mental impairment and the explanatory variables are modest, $r^2_{YX_1} = .14$ and $r^2_{YX_2} = 16$.

TABLE 11.4 Regression Analysis for Y = Mental Impairment and X_2 = Socioeconomic Status

Source	DF	Sum of Squares	Mean Square	F Value	Prob>F
Model	1	184.654	184.654	7.177	0.0109
Error	38	977.746	25.730		
Total	39	1162.400			

Root MSE 5.072 R-square 0.1589

Variable	Parameter Estimate	Standard Error	T for H0: Parameter=0	Prob > |T|
INTERCEP	32.1720	1.9876	16.186	0.0001
SES	-0.0861	0.0321	-2.679	0.0109

Table 11.5 shows part of a SAS computer printout for the multiple regression model $E(Y) = \alpha + \beta_1 X_1 + \beta_2 X_2$. (Table 11.8 later in the chapter shows the SPSS printout.) Much of what that printout lists is unfamiliar at this stage but is explained in this chapter. From the "Parameter Estimate" column, the prediction equation is

$$\hat{Y} = a + b_1 X_1 + b_2 X_2 = 28.230 + .103X_1 - .097X_2$$

Controlling for SES, the sample relationship between mental impairment and life events is positive, since the coefficient of life events ($b_1 = .103$) is positive. There is a .1-unit increase in the estimated mean of mental impairment for every 1-unit increase in the life events score, controlling for SES. Since $b_2 = -.097$, a negative association exists between mental impairment and SES, controlling for life events. For example, over the 100-unit range of potential SES values (from a minimum of 0 to a maximum of 100), the estimated mean mental impairment changes by $100(-.097) = -9.7$. Since mental impairment ranges only from 17 to 41 with a standard deviation of 5.5, a decrease of 9.7 points in the mean is noteworthy.

From Table 11.1, the first subject in the sample had $Y = 17$, $X_1 = 46$, and $X_2 = 84$. This subject's predicted mental impairment is $\hat{Y} = 28.230 + .103(46) - .097(84) = 24.8$. The prediction error (residual) is $Y - \hat{Y} = 17 - 24.8 = -7.8$. Table 11.5 also lists the predictions and residuals for the first four subjects. From the "Sum of Squares" column, the sum of squared errors (residuals) for this prediction equation is

$$\text{SSE} = \sum \left(Y_i - \hat{Y}_i\right)^2 = (-7.8)^2 + (-3.8)^2 + \cdots = 768.2$$

TABLE 11.5 Multiple Regression Analysis for Y = Mental Impairment, X_1 = Life Events Score, and X_2 = Socioeconomic Status

Source	DF	Sum of Squares	Mean Square	F Value	Prob>F
Model	2	394.238	197.119	9.495	0.0005
Error	37	768.162	20.761		
Total	39	1162.400			

Root MSE	4.556	R-square	0.3392
Dep Mean	27.300	Adj R-sq	0.3034

Variable	Parameter Estimate	Standard Error	T for H0: Parameter=0	Prob > |T|
INTERCEP	28.2298	2.1742	12.984	0.0001
LIFE	0.1033	0.0325	3.177	0.0030
SES	-0.0975	0.0291	-3.351	0.0019

Pearson Correlation Coefficients

	IMPAIR	LIFE	SES
IMPAIR	1.0000	0.3722	-0.3986
LIFE	0.3722	1.0000	0.1233
SES	-0.3986	0.1233	1.0000

Obs	Dep Var IMPAIR	Predict Value	Residual
1	17.0000	24.7918	-7.7918
2	19.0000	22.8018	-3.8018
3	20.0000	28.6784	-8.6784
4	20.0000	20.2542	-0.2542

Some software, such as SPSS, labels this the "Residual" sum of squares. Table 11.5 also displays a correlation matrix, listing the sample Pearson correlation between each pair of variables.

Table 11.6 illustrates how one might report results from the computer printout for the multiple regression model. It shows standard errors in parentheses below the parameter estimates. That table also summarizes the results for the two bivariate models. The partial slopes are similar to the slopes for the bivariate models. In each case, the introduction of the second predictor does little to alter the effect of the other one. This suggests that these predictors may be nearly independent in their sample effects on Y. In fact, the sample correlation between X_1 and X_2 is very weak ($r_{X_1 X_2} = .123$). The next section shows how to measure the joint association of the explanatory variables with the response variable, and shows how to interpret the R^2 value listed for the multiple regression model.

TABLE 11.6 Summary of Regression Models for Mental Impairment

	Predictors in Regression Model		
Effect	Multiple	Life Events	SES
Intercept	28.230	23.309	32.172
Life events	0.103	.090	—
	(.032)	(.036)	
SES	−0.097	—	−0.086
	(.029)		(.032)
R^2	.339	.138	.159
(n)	(40)	(40)	(40)

11.3 Multiple Correlation and R^2

The Pearson correlation r and the coefficient of determination r^2 describe strength of association for bivariate relationships. This section presents analogous measures for the multiple regression model. They describe the strength of association between Y and the set of explanatory variables acting together as predictors in the model.

Multiple Correlation Coefficient

Intuitively, the explanatory variables collectively are strongly related to Y if the observed values of Y correlate highly with the $\hat{Y}$ predicted values. The Pearson correlation between the observed and predicted values summarizes this association.

Multiple Correlation

The ***multiple correlation*** for a regression model is the Pearson correlation between the observed Y values and the predicted $\hat{Y}$ values.

For each subject, the prediction equation provides a predicted value $\hat{Y}$. So, each subject has a Y-value and a $\hat{Y}$-value. Table 11.5, for example, lists these under the headings "Dep Var IMPAIR" (for dependent variable) and "Predict Value." The sample Pearson correlation computed between the Y- and $\hat{Y}$-values is the multiple correlation. It is denoted by R.

The predicted values $\hat{Y}$ cannot correlate negatively with Y. These predictions are at least as good as the sample mean $\bar{Y}$, which is the prediction when all $b_i = 0$, and $\bar{Y}$ has zero correlation with Y. So, R always falls between 0 and 1. In this respect, the correlation between Y and $\hat{Y}$ differs from the correlation between Y and a predictor X, which falls between -1 and $+1$. The larger the multiple correlation R, the better the predictions of Y by the set of explanatory variables.

Coefficient of Multiple Determination

Another measure uses the *proportional reduction in error* concept, generalizing r^2, the coefficient of determination for bivariate models. This measure summarizes the relative improvement in predictions using the prediction equation instead of $\bar{Y}$. For the multiple regression model to be useful for prediction, it should provide improved predictions relative not only to $\bar{Y}$ but also to the separate bivariate models for Y and each explanatory variable.

An analog of r^2 for the multiple regression model has the following elements:

Rule 1 (Predict the response variable Y without using $X_1, \ldots, X_k$): The best predictor is then the sample mean, $\bar{Y}$.

Rule 2 (Predict the response variable Y using $X_1, \ldots, X_k$): The best predictor is the prediction equation $\hat{Y} = a + b_1X_1 + b_2X_2 + \cdots + b_kX_k$. Substituting a subject's observed values of $X_1, \ldots, X_k$ into this equation yields that subject's predicted value, $\hat{Y}$.

Prediction Errors: The prediction error for a subject is the difference between the observed and predicted values of Y. With rule 1, the error is $Y - \bar{Y}$; with rule 2, it is the residual $Y - \hat{Y}$. In either case, the sample summary of the total prediction error is the sum of the squared prediction errors. For rule 1, this is $\text{TSS} = \sum(Y - \bar{Y})^2$. For rule 2, it is $\text{SSE} = \sum(Y - \hat{Y})^2$, the sum of squared errors using the prediction equation.

Definition of Measure: The proportional reduction in error from using the prediction equation $\hat{Y} = a + b_1X_1 + \cdots + b_kX_k$ instead of $\bar{Y}$ to predict Y is called the ***coefficient of multiple determination***.

Coefficient of Multiple Determination

$$R^2 = \frac{\text{TSS} - \text{SSE}}{\text{TSS}} = \frac{\sum(Y - \bar{Y})^2 - \sum(Y - \hat{Y})^2}{\sum(Y - \bar{Y})^2}$$

This measures the proportion of the total variation in Y that is explained by the simultaneous predictive power of all the explanatory variables, through the multiple regression model. The uppercase notation R^2 distinguishes this PRE measure from the

PRE measure r^2 for the bivariate model. Their formulas are identical, and r^2 is the special case of R^2 applied to a regression equation with $k = 1$ explanatory variable.

Example 11.3 Multiple Correlation and R^2 for Mental Impairment

For the data on Y = mental impairment, X_1 = life events, and X_2 = SES introduced in Example 11.2, the prediction equation is $\hat{Y} = 28.23 + .103X_1 - .097X_2$. Table 11.5 showed the computer output for this model. From the "Sum of Squares" column, the total sum of squares is TSS $= \sum(Y - \bar{Y})^2 = 1162.4$, and the sum of squared errors from using the prediction equation to predict Y is SSE $= \sum(Y - \hat{Y})^2 = 768.2$.

The coefficient of multiple determination is

$$R^2 = \frac{\text{TSS} - \text{SSE}}{\text{TSS}} = \frac{1162.4 - 768.2}{1162.4} = .34$$

This statistic appears in Table 11.5 under the heading "R-square." Most software also reports an adjusted version of R^2 that is a less biased estimate of the population value. Problem 11.51 defines this measure, and Table 11.5 reports its value of .30 under the heading "Adj R-sq."

Since $R^2 = .34$, using life events and SES together to predict mental impairment provides a 34% reduction in the prediction error relative to using only $\bar{Y}$. Since $r_{YX_1}^2 = .14$ and $r_{YX_2}^2 = .16$, the multiple regression model provides a substantially larger reduction in error than either bivariate model. It is more useful than those models for predictive purposes.

The multiple correlation between mental impairment and the two explanatory variables is $R = +\sqrt{.34} = .58$. This equals the Pearson correlation between the observed Y- and predicted $\hat{Y}$-values for the model. Notice that $R > .37 = |r_{YX_1}|$ and $R > .40 = |r_{YX_2}|$. □

Properties of R and R^2

The properties of R^2 are similar to those of r^2 for bivariate models.

- R^2 falls between 0 and 1. Predictions can be no worse overall with $\hat{Y}$ instead of $\bar{Y}$ as a predictor, since $\hat{Y}$ uses additional information (the values of $X_1, \ldots, X_k$). Therefore, SSE can be no larger than TSS, and R^2 must fall between 0 and 1.
- The larger the value of R^2, the better the set of explanatory variables $(X_1, \ldots, X_k)$ collectively predict Y.
- $R^2 = 1$ only when all the residuals are 0, that is, when all $Y = \hat{Y}$, so that SSE $= 0$. In that case, the prediction equation passes through all the data points.
- $R^2 = 0$ when the predictions do not vary as any of the X-values vary. In that case, $b_1 = b_2 = \cdots = b_k = 0$, and $\hat{Y}$ is identical to $\bar{Y}$, since the explanatory variables do not add any predictive power. When this happens, the Pearson correlation between Y and each explanatory variable equals 0.

- R^2 cannot decrease when we add an explanatory variable to the model. It is impossible to explain *less* variation in Y by adding explanatory variables to a regression model. For example, the worst that can happen when X_2 is added to a model already containing X_1 is that it is useless for predictive purposes, in which case $b_2 = 0$. Then, the predictions are the same as when X_1 alone is an explanatory variable, so SSE is unchanged. Therefore, SSE cannot increase when we add a variable to the model.
- As a consequence of the previous property, R^2 for the multiple regression model is at least as large as $r^2_{YX_1}, r^2_{YX_2}, \ldots, r^2_{YX_k}$ for the separate bivariate models. That is, R^2 for the multiple regression model is at least as large as for each bivariate model, Y as a linear function of X_1, Y as a linear function of X_2, and so forth.
- The multiple correlation R equals the positive square root of the coefficient of multiple determination, $R = +\sqrt{R^2}$.

Properties of the multiple correlation R follow directly from the ones for R^2. For instance, if R_c denotes the multiple correlation for a "complete" model and R_r denotes the multiple correlation for a "reduced" model having only some of the explanatory variables from the first model, then $R_c \geq R_r$. The multiple correlation for the model $E(Y) = \alpha + \beta_1 X_1 + \beta_2 X_2 + \beta_3 X_3$ is at least as large as the multiple correlation for the model $E(Y) = \alpha + \beta_1 X_1 + \beta_2 X_2$.

The numerator of R^2, TSS − SSE, represents the amount of variation in Y explained by the multiple regression model. This difference, which equals $\sum(\hat{Y} - \bar{Y})^2$, is called the ***model sum of squares***. For instance, Table 11.5 lists TSS − SSE = 394.2 next to "Model" in the "Sum of Squares" column. (Some software, such as SPSS and Minitab, labels this the "Regression" sum of squares.) Thus, the total sum of squares TSS of the Y-values about $\bar{Y}$ partitions into the variation explained by the model (model sum of squares) plus the variation not explained by the model (SSE).

Multicollinearity

When we add explanatory variables to a multiple regression model, R^2 cannot decrease. On the other hand, once certain important predictors are in the model, the addition of other variables often provides only a small boost in R^2. These other variables may have small associations with Y, given the variables already in the model. This often happens in social science research when the explanatory variables are highly correlated, no one having much unique explanatory power. Section 14.3 discusses this condition, called ***multicollinearity***.

Figure 11.8, which portrays the portion of the total variability in Y explained by each of three predictors, shows a common occurrence. The size of the set for a predictor in this figure represents the size of its r^2-value in predicting Y. The amount a set for a predictor overlaps with the set for another predictor represents its association with that predictor. The part of the set for a predictor that does not overlap with other sets represents the part of the variability in Y explained uniquely by that predictor.

In Figure 11.8, all three predictors have moderate associations with Y, and together they explain considerable variation. Once X_1 and X_2 are in the model, however, X_3 explains little unique variation in Y, because of its strong correlations with X_1 and X_2. Because of this overlap, R^2 increases very little when X_3 is added to a model already containing X_1 and X_2.

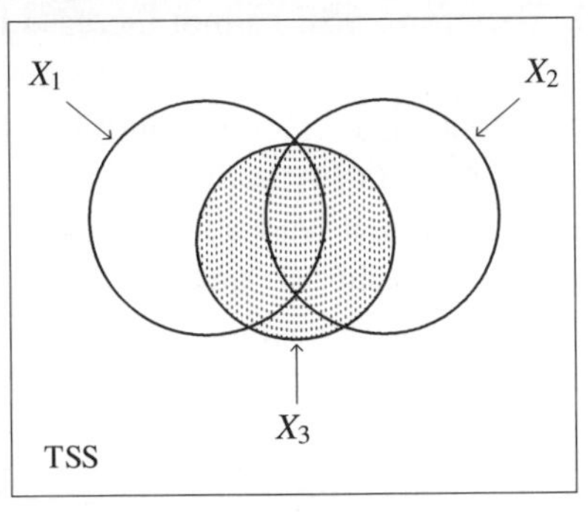

Figure 11.8 R^2 Does Not Increase Much When X_3 Is Added to the Model Already Containing X_1 and X_2

For predictive purposes, we gain little by adding explanatory variables to a model that are strongly correlated with ones already in the model, since R^2 does not increase much. Ideally, for prediction purposes one should model Y using explanatory variables having weak correlations with each other but strong correlations with Y. In practice, this is not always possible, especially if we want to include certain variables in the model for theoretical reasons.

11.4 Inferences for Multiple Regression Coefficients

The basic multiple regression model assumes that the relationship between the mean of the response variable Y and the explanatory variables $X_1, \ldots, X_k$ follows the equation

$$E(Y) = \alpha + \beta_1 X_1 + \cdots + \beta_k X_k$$

For particular values of $X_1, \ldots, X_k$, $\alpha + \beta_1 X_1 + \cdots + \beta_k X_k$ is the mean of Y for the population having those values of $X_1, \ldots, X_k$.

Inferences about the parameters of the multiple regression model require additional elements to this basic model. These are

- The population distribution of Y is normal, for each combination of values of $X_1, \ldots, X_k$.
- The standard deviation, σ, of the conditional distribution of responses on Y is the same at each combination of values of $X_1, \ldots, X_k$.
- The sample is randomly selected.

These assumptions are necessary for the true sampling distributions to equal those quoted in this section. In practice, the assumptions are never satisfied perfectly. The more closely they are satisfied, however, particularly randomness, the more likely the inferences are to be correct.

We present two types of inferences. The first of these is a global test of independence. It checks whether *any* of the explanatory variables are statistically related to Y. The second studies the partial regression coefficients individually, to assess which explanatory variables have significant partial effects on Y and to estimate the sizes of those effects.

Collective Influence of the Explanatory Variables

The first matter is whether the explanatory variables collectively have a statistically significant effect on the response variable Y. We check this by testing the null hypothesis

$$H_0 : \beta_1 = \beta_2 = \cdots = \beta_k = 0$$

This states that the mean of Y does not depend on the values of $X_1, \ldots, X_k$. That is, if all the partial regression coefficients equal 0, then Y is statistically independent of all k explanatory variables.

The alternative hypothesis is

$$H_a : \text{At least one } \beta_i \neq 0$$

This states that *at least one* explanatory variable is related to Y, controlling for the others. The test judges whether using $X_1, \cdots, X_k$ together to predict Y, with the prediction equation $\hat{Y} = a + b_1X_1 + \cdots + b_kX_k$, is better than simply using $\bar{Y}$.

Let **P** (uppercase rho) denote the population multiple correlation coefficient. These hypotheses about $\{\beta_i\}$ are equivalent to

$$H_0 : \mathbf{P} = 0 \quad H_a : \mathbf{P} > 0$$

The equivalence occurs because the multiple correlation equals 0 only in those situations in which all the partial regression coefficients equal 0. The hypotheses are also equivalent to H_0: $\mathbf{P}^2 = 0$ and H_a: $\mathbf{P}^2 > 0$, where $\mathbf{P}^2$ is the population analog of R^2.

The test statistic equals

$$F = \frac{R^2/k}{(1 - R^2)/[n - (k + 1)]}$$

The sampling distribution of this statistic is the ***F distribution***. We next study this distribution and its properties.

F Distribution

The symbol for the F statistic and its distribution honors the most eminent statistician in history, R. A. Fisher, who discovered the F distribution in 1922. Like the chi-squared

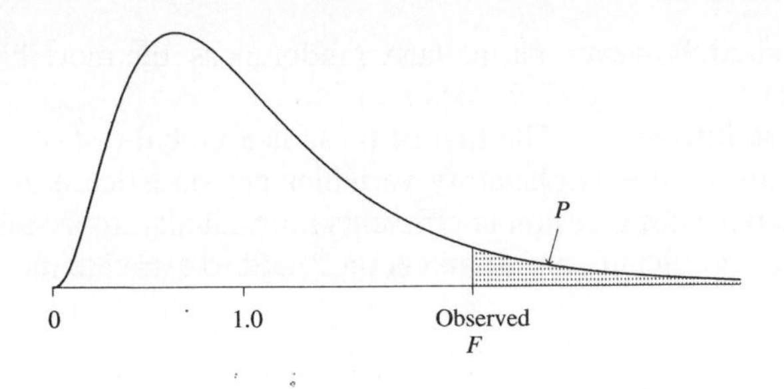

Figure 11.9 The F Distribution and the P-Value for F Tests

distribution, the F distribution can assume only nonnegative values and it is somewhat skewed to the right. Figure 11.9 illustrates.

The shape of the F distribution is determined by two degrees of freedom terms, denoted by df_1 and df_2. The first of these, $df_1 = k$, is the divisor of the numerator term (R^2) in the F statistic. It equals the number of explanatory variables in the multiple regression model. The second term, $df_2 = n-(k+1)$, is the divisor of the denominator term $(1 - R^2)$ term in the F statistic. It equals the sample size n minus the number of parameters in the multiple regression model; that number of parameters equals $k + 1$, representing the k beta terms and the alpha term. The mean of the F distribution equals $df_2/(df_2 - 2)$. This approaches 1.0 as n, and thus df_2, increases.

The larger the sample R^2 value, the larger the ratio $R^2/(1 - R^2)$, and the larger the F test statistic becomes. Thus, larger values of the F test statistic provide stronger evidence against H_0. The P-value is the probability the F statistic is at least as large as the observed F value, if H_0 were true. This is the right-tail probability under the F distribution beyond the observed F-value, as Figure 11.9 shows. Table D at the end of the text lists the F scores having P-values of .05, .01, and .001, for various combinations of df_1 and df_2. This table allows us to determine whether $P > .05$, $.01 < P < .05$, $.001 < P < .01$, or $P < .001$. Software for regression reports the actual P-value.

Example 11.4 F Test for Mental Health Data

The model introduced in Example 11.2 for mental impairment refers to $n = 40$ observations and $k = 2$ explanatory variables, namely, life events and SES. For this model, the null hypothesis H_0: $\beta_1 = \beta_2 = 0$, or H_0: $\mathbf{P} = 0$, states that mental impairment is statistically independent of life events and SES.

From Example 11.3, the coefficient of multiple determination for this model is $R^2 = .34$. The F test statistic value is therefore

$$F = \frac{R^2/k}{(1 - R^2)/[n - (k + 1)]} = \frac{.34/2}{.66/[40 - (2 + 1)]} = 9.5$$

The two degrees of freedom terms for the F distribution are $df_1 = k = 2$ and $df_2 = n - (k + 1) = 40 - 3 = 37$, the two divisors of this statistic.

From Table D, the F-value with right-hand tail probability of .001, when $df_1 = 2$ and $df_2 = 37$, falls between 8.77 and 8.25. Since the observed F statistic of 9.5 falls above these two, it is farther out in the tail and has smaller tail probability than .001.

Thus, the P-value is $P < .001$. From the printout (Table 11.5), the actual P-value is $P = .0005$. This value appears under the heading PROB > F, which is next to F VALUE, the F test statistic.

This extremely small P-value provides strong evidence against H_0 and suggests that at least one explanatory variable is related to mental impairment. We obtain significantly better predictions of Y using the multiple prediction equation than by using $\bar{Y}$. □

Normally, unless the sample size is small and the associations are weak, this test has a small P-value. If we choose variables wisely for a study, at least one of them should have *some* explanatory power. When the P-value is small, we can conclude only that at least one explanatory variable affects the response variable, controlling for the other explanatory variables. More narrowly focused analyses judge *which* partial effects are nonzero and estimate the sizes of those effects.

Inferences for Individual Regression Coefficients

For the bivariate model, $E(Y) = \alpha + \beta X$, the t test of H_0: $\beta = 0$ (Section 9.5) evaluates whether Y and X are statistically dependent, *ignoring* other variables. By contrast, the test presented next evaluates whether variables are dependent, *controlling* for other variables.

The test for the partial effect of an explanatory variable X_i on Y has null hypothesis H_0: $\beta_i = 0$, where β_i is the coefficient of X_i in the multiple regression model. If $\beta_i = 0$, the mean of Y is identical for all values of X_i, controlling for the other explanatory variables in the model. The alternative hypothesis normally is two sided, H_a: $\beta_i \neq 0$; the one-sided form H_a: $\beta_i > 0$ or H_a: $\beta_i < 0$ predicts the direction of the partial association.

The test statistic is

$$t = \frac{b_i}{\hat{\sigma}_{b_i}}$$

where $\hat{\sigma}_{b_i}$ denotes the standard error of the estimate b_i of β_i. As usual, the t test statistic takes the best estimate (b_i) of the parameter (β_i), subtracts the null hypothesis value of the parameter (0), and divides by the standard error ($\hat{\sigma}_{b_i}$) of the point estimate. The formula for $\hat{\sigma}_{b_i}$ is complex, but software for multiple regression provides its value.

If H_0 is true and the model assumptions hold, the t statistic has the t distribution with $df = n - (k + 1)$. The df value for this test is the same as df_2 in the F test of H_0: $\beta_1 = \cdots = \beta_k = 0$ just studied.

It is more informative to estimate the size of a partial regression coefficient than to test simply whether it is zero. A confidence interval for β_i equals

$$b_i \pm t\hat{\sigma}_{b_i}$$

where the t score comes from the t table, with $df = n - (k + 1)$, for the desired confidence coefficient. The confidence interval refers to the change in the mean of Y for a one-unit increase in X_i, controlling for the other variables in the model.

Example 11.5 Inference for Separate Predictors of Mental Impairment
Consider the parameter β_1 in the multiple regression model

$$E(Y) = \alpha + \beta_1 X_1 + \beta_2 X_2$$

for Y = mental impairment, X_1 = life events, and X_2 = SES. The null hypothesis H_0: $\beta_1 = 0$ states that, controlling for SES, mental impairment is statistically independent of life events. If it is, the full multiple regression model reduces to the simple bivariate model $E(Y) = \alpha + \beta_2 X_2$. If H_0 is false, then $\beta_1 \neq 0$ and the full model provides a better fit than the bivariate model having X_2 alone as an explanatory variable.

From Table 11.5, the point estimate of β_1 is $b_1 = .103$. The estimated standard error of b_1 is listed under the heading "Standard Error" in that table, namely, $\hat{\sigma}_{b_1} = .0325$. The test statistic equals

$$t = \frac{b_1}{\hat{\sigma}_{b_1}} = \frac{.103}{.0325} = 3.18$$

This appears under the heading "T for H0: Parameter = 0." The statistic has $df = n-(k+1) = 40-3 = 37$. The P-value appears under "Prob > |T|." For this example, it is $P = .003$, the probability that the t statistic is at least 3.18 in absolute value. There is strong evidence that mental impairment is related to life events, controlling for SES.

A 95% confidence interval for β_1 uses $t_{.025} = 2.026$, the t-value for $df = 37$ having a probability of .05/2 = .025 in each tail. This interval equals

$$b_1 \pm t_{.025}\hat{\sigma}_{b_1} = .103 \pm 2.026(.0325) = .103 \pm .066$$

or (.04, .17). Controlling for SES, we are 95% confident that the change in mean mental impairment per one-unit increase in life events is between .04 and .17. The interval does not contain 0, in agreement with rejecting H_0: $\beta_1 = 0$ in favor of H_a: $\beta_1 \neq 0$ at the $\alpha = .05$ level.

Since this interval contains only positive numbers, the relationship between mental impairment and life events is positive, controlling for SES. The interval is relatively wide, however, in the sense that an increase of 100 units in life events corresponds to anywhere from a $100(.04) = 4$ to a $100(.17) = 17$ unit increase in mean mental impairment. This is due primarily to the small sample size. □

A note of caution: When multicollinearity exists, possibly none of the individual partial effects have small P-values, even if R^2 is large and a large F statistic occurs in the global test for the βs. Any particular variable may explain uniquely little of the variation in Y, even though together the variables explain a lot of the variation. Section 14.3 discusses reasons for this.

Variability and Mean Squares*

The precision of the least squares estimates relates to the size of the conditional standard deviation σ that measures variability of Y at fixed values of the predictors. The smaller the variability in the values of Y about the regression equation, the smaller the standard errors become. The estimate of σ^2 is

$$\hat{\sigma}^2 = \frac{\sum\left(Y - \hat{Y}\right)^2}{n - (k + 1)} = \frac{\text{SSE}}{df}$$

The degrees of freedom for the estimate equals the sample size n minus the number of parameters in the regression model. This general model with k explanatory variables has $(k + 1)$ parameters, the k β-values and the α term, so $df = n - (k + 1)$. This df value is also df for t inferences for regression coefficients, and it is df_2 for the F test for the collective effect of the predictors. When a model has only $k = 1$ explanatory variable, df reduces to $n - 2$, the term in the denominator of $\hat{\sigma}^2$ in Section 9.3.

From Table 11.5 for the multiple regression model with the mental impairment data, $\text{SSE} = 768.2$, $n = 40$, and $k = 2$, so $df = n - 3 = 37$ and

$$\hat{\sigma}^2 = \frac{\text{SSE}}{df} = \frac{768.2}{37} = 20.8$$

This estimate of the conditional variance is called the ***mean square error***, abbreviated by MSE. It appears in the computer printout in the "Mean Square" column, in the row labeled "Error." The estimate of the conditional standard deviation is $\hat{\sigma} = \sqrt{20.8} = 4.6$, which appears under the heading "Root MSE" in the printout. To the extent that the conditional distributions are approximately bell-shaped, most mental health scores fall within about ten units (two standard deviations) of the regression equation.

An alternative formula for the F test statistic for testing $H_0 : \beta_1 = \cdots = \beta_k$ about the collective influence of the predictors uses this and another variance estimate. The second estimate uses the model sum of squares, (TSS − SSE). The degrees of freedom for that sum of squares equals k, the number of explanatory variables in the model, which is df_1 for the F test. The variance estimate equals the ratio of the model sum of squares to its degrees of freedom and is called the ***model mean square*** (or, in some software, the ***regression mean square***). This estimate tends to *overestimate* σ^2 when H_0 is false, and is an unbiased estimate only when H_0 is true. The F test statistic equals the ratio of these two estimates; that is, F equals the model mean square divided by the mean square error. If H_0 is false, the F ratio tends to be relatively large, since the model mean square tends to overestimate σ^2.

On the printout (Table 11.5) for this example, the mean square error equals $\hat{\sigma}^2 = 20.8$, and the model mean square equals

$$\frac{\text{Model SS}}{df_1} = \frac{394.2}{2} = 197.1$$

The F test statistic equals

$$F = \frac{\text{Model mean square}}{\text{Mean square error}} = \frac{197.1}{20.8} = 9.5$$

which is also the value from the formula based on R^2.

A statistical test of this type, based on comparing two variance estimates, is called an ***analysis of variance*** procedure. Chapter 12 presents the principles underlying such tests.

Relationship Between F and t Statistics*

Some regression software lists F-values instead of t-values for the tests of the individual regression coefficients. The two statistics are related and have the same P-values. The square of the t statistic for testing that a partial regression coefficient equals 0 is an F test statistic having the F distribution with $df_1 = 1$ and $df_2 = n - (k + 1)$.

To illustrate, in Example 11.5, one could test H_0: $\beta_1 = 0$ against H_a: $\beta_1 \neq 0$ using $F = t^2 = 3.18^2 = 10.1$. This has the F distribution with $df_1 = 1$ and $df_2 = 37$. The P-value for this F value is .002, the same as Table 11.5 reports for the two-sided t test.

In general, if a statistic has the t distribution with d degrees of freedom, then the square of that statistic has the F distribution with $df_1 = 1$ and $df_2 = d$. A disadvantage of the F approach is that it lacks information about the direction of the association. It cannot be used for one-sided alternative hypotheses.

11.5 Modeling Interaction

In the multiple regression model

$$E(Y) = \alpha + \beta_1 X_1 + \beta_2 X_2 + \cdots + \beta_k X_k$$

the partial relationship between Y and each X_i is linear. Also, the slope β_i of that relationship is identical for all values of the other explanatory variables. That is, the change in the mean of Y for a 1-unit increase in X_i equals β_i, no matter what the fixed values of those other variables. This implies a parallelism of lines relating the two variables, at various values of the other variables, as Figure 11.3 illustrated.

This model is sometimes too simple to be adequate. Often, for instance, there is ***statistical interaction***. Section 10.3 introduced this term. It refers to relationships between two variables that change according to the value of a control variable. The multiple regression model studied so far assumes a lack of interaction, in the sense that the slope of the relationship between Y and each X_i is the same at all levels of the other Xs. Interaction occurs when this does not hold.

Interaction

For quantitative variables, ***statistical interaction*** exists when the slope of the relationship between the response variable and an explanatory variable changes as the levels of the other variables change.

For example, suppose the relationship between X_1 and the mean of Y is $E(Y) = 2+5X_1$ when $X_2 = 0$, it is $E(Y) = 4+15X_1$ when $X_2 = 50$, and it is $E(Y) = 6+25X_1$ when $X_2 = 100$. The slope for the partial effect of X_1 changes markedly as the fixed value for X_2 changes. There is interaction between X_1 and X_2 in their effects on Y.

Cross-product Terms

The most common approach for modeling interaction introduces ***cross-product terms*** of the explanatory variables into the multiple regression model. For instance, for two explanatory variables, the model allowing interaction has the form

$$E(Y) = \alpha + \beta_1 X_1 + \beta_2 X_2 + \beta_3 X_1 X_2$$

This is a special case of the multiple regression model with three explanatory variables, in which X_3 is an artificial variable created as the cross-product $X_3 = X_1 X_2$ of the two primary explanatory variables.

Let's see how Y is related to X_1, controlling for X_2, in this model. We rewrite the equation in terms of X_1 as

$$E(Y) = (\alpha + \beta_2 X_2) + (\beta_1 + \beta_3 X_2)X_1 = \alpha' + \beta' X_1$$

where

$$\alpha' = \alpha + \beta_2 X_2 \qquad \text{and} \qquad \beta' = \beta_1 + \beta_3 X_2$$

So, for fixed X_2, the mean of Y changes linearly as a function of X_1. The Y-intercept of the relationship is $\alpha' = (\alpha + \beta_2 X_2)$, which depends on the value of the controlled variable. The slope of the relationship is $\beta' = (\beta_1 + \beta_3 X_2)$, which also depends on the value of X_2. As X_2 changes, the slope for the effect of X_1 changes.

For this model, the mean of Y is a linear function of X_1, but the slope of the line depends on X_2. Similarly, the mean of Y is a linear function of X_2, but the slope of the relationship varies according to the value of X_1.

Example 11.6 Interaction Model for Mental Impairment

The data set on Y = mental impairment, X_1 = life events, and X_2 = SES has only 40 observations. More complex models require larger sample sizes to estimate the parameters well. A sample of size 40 is a bit small to estimate four parameters very precisely, but we apply the interaction model to these data for illustrative purposes. We create a third explanatory variable X_3 that is the cross-product of X_1 and X_2 for the 40 individuals. For example, the X_3 scores are $46(84) = 3864$ for the first subject, $39(97) = 3783$ for the second, and so forth. Software makes it easy to create this variable without doing the calculations yourself.

Table 11.7 shows part of the printout for this model. The prediction equation equals

$$\hat{Y} = 26.037 + .156X_1 - .060X_2 - .00087X_1 X_2$$

TABLE 11.7 Interaction Model for Y = Mental Impairment, X_1 = Life Events, and X_2 = SES

Source	DF	Sum of Squares	Mean Square	F Value	Prob>F
Model	3	403.631	134.544	6.383	0.0014
Error	36	758.769	21.077		
Total	39	1162.400			

Root MSE	4.591	R-square	0.347
Dep Mean	27.300	Adj R-sq	0.293

Variable	Parameter Estimate	Standard Error	T for H0: Parameter=0	Prob > |T|
INTERCEP	26.036649	3.948826	6.594	0.0001
LIFE	0.155865	0.085338	1.826	0.0761
SES	-0.060493	0.062675	-0.965	0.3409
LIFE*SES	-0.000866	0.001297	-0.668	0.5087

Figure 11.10 portrays the relationship between mental impairment and life events for a few distinct SES values. For an SES score of $X_2 = 0$, the estimated relationship between $E(Y)$ and X_1 is

$$\hat{Y} = 26.037 + .156X_1 - .060(0) - .00087X_1(0) = 26.037 + .156X_1$$

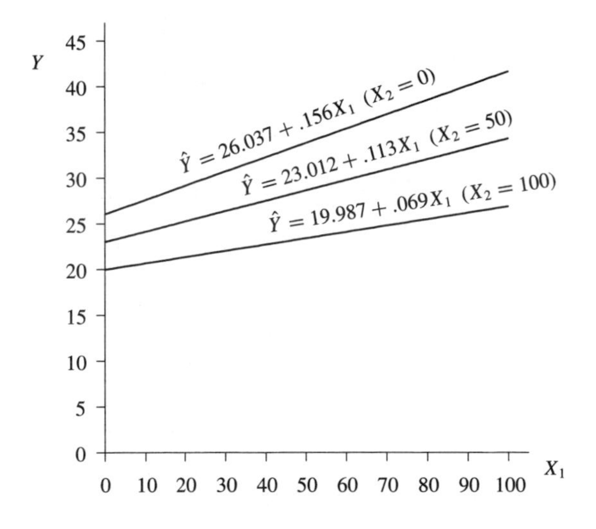

Figure 11.10 Interaction Relationship Between Y and X_1, Controlling for X_2

When $X_2 = 50$, the relationship is

$$\hat{Y} = 26.037 + .156X_1 - .060(50) - .00087(50)X_1 = 23.012 + .113X_1$$

When $X_2 = 100$, the prediction equation is

$$Y = 19.987 + .069X_1$$

The higher the value of $X_2 = \text{SES}$, the smaller the slope between mental impairment and life events; thus, the smaller is the effect of life events on mental impairment. This suggests that subjects who possess greater resources, in the form of higher SES, are better able to withstand the mental stress of potentially traumatic life events. □

Testing for Interaction

For two explanatory variables, the model allowing interaction is

$$E(Y) = \alpha + \beta_1 X_1 + \beta_2 X_2 + \beta_3 X_1 X_2$$

The simpler model assuming a lack of interaction is the special case $\beta_3 = 0$. The hypothesis of no interaction is H_0: $\beta_3 = 0$. As usual, the t statistic divides the estimate of the parameter (β_3) by its standard error.

From the printout in Table 11.7, this t statistic equals $t = -.00087/.0013 = -.67$. The P-value for testing H_0: $\beta_3 = 0$ against H_a: $\beta_3 \neq 0$ equals $P = .51$. Little evidence exists of interaction. The variation in the slope of the relationship between mental impairment and life events for various SES levels could be due to sampling variability. The sample size here is small, however, and standard errors suffer accordingly, making it difficult to estimate effects precisely. Studies based on larger sample sizes (e.g., Holzer 1977) have shown that interaction of the type seen in this example does exist for these variables.

Notice in Table 11.7 that neither the test of H_0: $\beta_1 = 0$ or of H_0: $\beta_2 = 0$ have small P-values. Yet, the tests of both H_0: $\beta_1 = 0$ and H_0: $\beta_2 = 0$ are highly significant for the "no interaction" model $E(Y) = \alpha + \beta_1 X_1 + \beta_2 X_2$; from Table 11.5, the P-values are .003 and .002. This loss of significance occurs because $X_3 = X_1 X_2$ is quite strongly correlated with X_1 and X_2, with $r_{X_1 X_3} = .779$ and $r_{X_2 X_3} = .646$. These substantial correlations are not surprising, since $X_3 = X_1 X_2$ is completely determined by X_1 and X_2.

Since considerable overlap occurs in the variation in Y that is explained by X_1 and by $X_1 X_2$, and also by X_2 and $X_1 X_2$, the *partial* variability explained by each is relatively small. For example, much of the predictive power contained in X_1 is also contained in X_2 and $X_1 X_2$. The *unique* contribution of X_1 (or X_2) to the model is relatively small, and nonsignificant, when X_2 (or X_1) and $X_1 X_2$ are in the model.

When strong evidence of interaction does not occur, it is best to drop the interaction term from the model before testing hypotheses about partial effects such as H_0: $\beta_1 = 0$

or H_0: $\beta_2 = 0$. On the other hand, if the evidence of interaction is strong, it no longer makes sense to test these other hypotheses; if there is interaction, then the effect of each variable exists and differs according to the level of the other variable.

Generalizations and Limitations

When the number of explanatory variables exceeds two, one can model interaction by introducing cross-products for each pair of explanatory variables. For example, for three explanatory variables, the interaction model is

$$E(Y) = \alpha + \beta_1 X_1 + \beta_2 X_2 + \beta_3 X_3 + \beta_4 X_1 X_2 + \beta_5 X_1 X_3 + \beta_6 X_2 X_3$$

This is a special case of multiple regression with six explanatory variables, identifying $X_4 = X_1 X_2$, $X_5 = X_1 X_3$, and $X_6 = X_2 X_3$. These are called ***second-order*** interaction terms, to distinguish them from *higher-order* interaction terms, such as $X_1 X_2 X_3$, used in even more complex models. Significance tests can judge which, if any, of the cross-product terms are needed in the model.

When interaction exists and the model contains cross-product terms, it is more difficult to summarize simply the relationships. One approach is to sketch a collection of lines such as those in Figure 11.10 to describe graphically how the relationship between two variables changes according to the values of control variables. Alternatively, one could divide the data into groups according to the value on a control variable (e.g., high on X_2, medium on X_2, low on X_2) and report the correlation or slope between Y and X_1 within each subset as a means of describing the interaction. The coplot graphic described in Section 11.2 is also useful for this purpose.

11.6 Comparing Regression Models

When the number k of explanatory variables in the multiple regression model

$$E(Y) = \alpha + \beta_1 X_1 + \cdots + \beta_k X_k$$

increases, the model becomes more difficult to interpret and some variables may become redundant. This is particularly true when some explanatory variables are cross-products of others, to allow for interaction. Not all predictors may be needed in the model. We next present a test of whether the model fits significantly better than a simpler model containing only a subset of the predictors.

Complete and Reduced Models

We refer to the full model with all k predictors as the ***complete model***. The second model, called the ***reduced model***, has

$$E(Y) = \alpha + \beta_1 X_1 + \cdots + \beta_g X_g$$

with $g < k$. The complete model contains all g of these predictors besides an additional $(k - g)$ variables. The first g variables $(X_1, \ldots, X_g)$ are identical in both models. The extra variables in the complete model are $X_{g+1}, \ldots, X_k$. The reduced model is said to be *nested* within the complete model, being a special case of it.

The complete and reduced models are identical if the partial regression coefficients $\beta_{g+1}, \beta_{g+2}, \ldots, \beta_k$ for the extra variables all equal 0. In that case, none of the extra predictors increases the explained variability in Y, in the population of interest. Testing whether the complete model is identical to the reduced model is equivalent to testing

$$H_0 : \beta_{g+1} = \cdots = \beta_k = 0$$

The alternative hypothesis is that at least one of these extra parameters is not 0, in which case the complete model is better than the reduced model.

For instance, a model with three explanatory variables and all the second-order interaction terms is

$$E(Y) = \alpha + \beta_1 X_1 + \beta_2 X_2 + \beta_3 X_3 + \beta_4 X_1 X_2 + \beta_5 X_1 X_3 + \beta_6 X_2 X_3$$

The test comparing this complete model to the reduced model

$$E(Y) = \alpha + \beta_1 X_1 + \beta_2 X_2 + \beta_3 X_3$$

with no interaction terms has H_0: $\beta_4 = \beta_5 = \beta_6 = 0$. The complete model has $k = 6$ explanatory variables, and the reduced model has $g = 3$ explanatory variables.

Comparing Models by Comparing SSE or R^2 Values

The test statistic for comparing two regression models is based on comparing the sums of squared errors for the two models. Denote $\text{SSE} = \sum(Y - \hat{Y})^2$ for the reduced model by SSE_r and for the complete model by SSE_c. Now, $\text{SSE}_r \geq \text{SSE}_c$, since the reduced model has only some of the terms from the complete model and its predictions tend to be poorer. Even if the null hypothesis were true, the estimates of $\beta_{g+1}, \ldots, \beta_k$ and the difference $(\text{SSE}_r - \text{SSE}_c)$ are not 0. Some reduction in error occurs from fitting the extra terms because of sampling variability.

Since the complete model has k explanatory variables, it has $n - (k + 1)$ degrees of freedom for SSE_c. Similarly, SSE_r for the reduced model, which contains g explanatory variables, has $df = n - (g + 1)$. The addition of the extra $(k - g)$ terms in the complete model, in other words, reduces the error df by $(k - g)$. The reduction in error, $\text{SSE}_r - \text{SSE}_c$, that results from adding the extra variables is based on $(k - g)$ degrees of freedom.

The test statistic for testing H_0: $\beta_{g+1} = \cdots = \beta_k = 0$ refers to these SSE values, and an equivalent statistic refers to the R^2 values, R_c^2 and R_r^2, for the complete and reduced models. The statistic equals

$$F = \frac{(\text{SSE}_r - \text{SSE}_c)/(k - g)}{\text{SSE}_c/[n - (k + 1)]} = \frac{(R_c^2 - R_r^2)/(k - g)}{(1 - R_c^2)/[n - (k + 1)]}$$

The numerator divides the reduction in error by its degrees of freedom; the denominator divides the error for the complete model by its degrees of freedom. A relatively large reduction in error, compared to what is expected if H_0 were true, yields a large F test statistic and small P-value.

The sampling distribution for the F test statistic is the F distribution with $df_1 = k-g$ and $df_2 = n-(k+1)$. The P-value is the right-hand tail probability, representing the probability the F statistic is at least as large as the observed F-value, when H_0 is true.

Example 11.7 Comparing Models for Mental Impairment

For the mental impairment data, a comparison of the complete model

$$E(Y) = \alpha + \beta_1 X_1 + \beta_2 X_2 + \beta_3 X_1 X_2$$

to the reduced model

$$E(Y) = \alpha + \beta_1 X_1 + \beta_2 X_2$$

analyzes whether statistical interaction exists. The complete model has just one additional term, and the null hypothesis is H_0: $\beta_3 = 0$. It has $k = 3$ predictors, the reduced model has $g = 2$ predictors, and $k - g = 1$.

The sum of squared errors for the complete model is $\text{SSE}_c = 758.8$ (Table 11.7), while for the reduced model it is $\text{SSE}_r = 768.2$ (Table 11.5). The difference

$$\text{SSE}_r - \text{SSE}_c = 768.2 - 758.8 = 9.4$$

is based on $df_1 = k - g = 3 - 2 = 1$. Since the sample size is $n = 40$, $df_2 = n - (k + 1) = 40 - (3 + 1) = 36$, the df for SSE in Table 11.7. The F test statistic equals

$$F = \frac{(\text{SSE}_r - \text{SSE}_c)/(k-g)}{\text{SSE}_c/[n-(k+1)]} = \frac{9.4/1}{758.8/36} = .45$$

Equivalently, using $R_r^2 = .339$ and $R_c^2 = .347$,

$$F = \frac{\left(R_c^2 - R_r^2\right)/(k-g)}{\left(1 - R_c^2\right)/[n-(k+1)]} = \frac{(.347 - .339)/1}{(1 - .347)/36} = .45$$

From software, the P-value from the F distribution with $df_1 = 1$ and $df_2 = 36$ is $P = .51$. There is little evidence that the complete model is better. The null hypothesis that the reduced model is adequate seems plausible.

When the null hypothesis contains a single parameter, one can simply conduct a t test. In fact, from the previous section, the t statistic equals

$$t = \frac{b_3}{\hat{\sigma}_{b_3}} = \frac{-.00087}{.0013} = -.67$$

and also has P-value of $P = .51$ for $H_a : \beta_3 \neq 0$. We get the same result with the t test as with the F test for complete and reduced models. In fact, the F test statistic equals the square of the t statistic. (Refer to the final subsection in Section 11.4.) □

The t test method is limited to testing one parameter at a time. The F test can test a number of regression parameters together to analyze whether any of them are nonzero, such as in the global F test of $H_0 : \beta_1 = \cdots = \beta_k$ or the test comparing a complete model with k predictors to a reduced model with g predictors. F tests are equivalent to t tests only when the null hypothesis contains a single parameter.

11.7 Partial Correlation*

Multiple regression models describe the relationship between two variables, controlling for other variables of interest. We next study measures that describe the strength of the association between two variables, controlling for other variables. For example, we measure the association between mental impairment and life events, controlling for SES. We might ask, "Controlling for SES, what proportion of the variation in mental impairment does life events explain?"

These measures describe the partial association between Y and a particular predictor, whereas the multiple correlation and R^2 describe the association between Y and the entire set of predictors in the model. The *partial correlation* relates directly to the bivariate Pearson correlations between each pair of variables. For a single control variable, it is defined as follows:

Partial Correlation

The sample ***partial correlation*** between Y and X_2, controlling for X_1, is

$$r_{YX_2 \cdot X_1} = \frac{r_{YX_2} - r_{YX_1} r_{X_1X_2}}{\sqrt{\left(1 - r_{YX_1}^2\right)\left(1 - r_{X_1X_2}^2\right)}}$$

In the symbol $r_{YX_2 \cdot X_1}$, the variable to the right of the dot represents the controlled variable. The analogous formula for $r_{YX_1 \cdot X_2}$, controlling X_2, is

$$r_{YX_1 \cdot X_2} = \frac{r_{YX_1} - r_{YX_2} r_{X_1X_2}}{\sqrt{\left(1 - r_{YX_2}^2\right)\left(1 - r_{X_1X_2}^2\right)}}$$

Since one variable is controlled, the partial correlations $r_{YX_1 \cdot X_2}$ and $r_{YX_2 \cdot X_1}$ are called ***first-order partial correlations***.

Example 11.8 Partial Correlation Between Education and Crime Rate

For the data described in Example 11.1 on Y = crime rate, X_1 = education, and X_2 = urbanization for the counties of Florida, the pairwise Pearson correlations equal $r_{YX_1} = .468$, $r_{YX_2} = .678$, and $r_{X_1X_2} = .791$. Is the positive correlation between crime rate and education explained by their joint dependence on urbanization? This is plausible if the association disappears when we control for urbanization.

The partial correlation between crime rate and education, controlling for urbanization, equals

$$r_{YX_1 \cdot X_2} = \frac{r_{YX_1} - r_{YX_2} r_{X_1X_2}}{\sqrt{(1 - r_{YX_2}^2)(1 - r_{X_1X_2}^2)}} = \frac{.468 - .678(.791)}{\sqrt{(1 - .678^2)(1 - .791^2)}} = -.152$$

Not surprisingly, $r_{YX_1 \cdot X_2}$ is much smaller than r_{YX_1}, and it even changes direction (Simpson's paradox). The relationship between crime rate and education may well be spurious, reflecting their joint dependence on urbanization. □

Interpreting Partial Correlations

We now present interpretations of partial correlations and their squares. We discuss them in terms of $r_{YX_1 \cdot X_2}$, but analogous interpretations apply to $r_{YX_2 \cdot X_1}$.

- Under the assumptions in Section 11.4 for conducting inference, $r_{YX_1 \cdot X_2}$ estimates the Pearson correlation between Y and X_1 at every *fixed* value of X_2. If we could control X_2 by considering a subpopulation of members all having the same value on X_2, then $r_{YX_1 \cdot X_2}$ approximates the correlation between Y and X_1 for that subpopulation. (In fact, the sample partial correlation is identical to the Pearson correlation computed for the points in the *partial regression plot* introduced in Section 11.2.)
- Like the Pearson correlation, $r_{YX_1 \cdot X_2}$ falls between -1 and $+1$.
- The partial correlation $r_{YX_1 \cdot X_2}$ has the same sign as the partial slope (b_1) for the effect of X_1 in the prediction equation $\hat{Y} = a + b_1 X_1 + b_2 X_2$. This happens because the same variable (X_2) is controlled in the model as in the correlation. An analogous property holds in bivariate regression, the Pearson correlation having the same sign as the slope.
- The larger the absolute value of $r_{YX_1 \cdot X_2}$, the stronger the association between Y and X_1, controlling for X_2.
- The value of the partial correlation does not depend on the units of measurement of the variables.
- The square of the partial correlation has a proportional reduction in error (PRE) interpretation.

The PRE property states that $r_{YX_2 \cdot X_1}^2$ is the proportion of variation in Y explained by X_2, controlling for X_1. This squared measure describes the effect of removing from consideration the portion of the total sum of squares (TSS) in Y that is explained by X_1, and then finding the proportion of the remaining unexplained variation in Y that is explained by X_2.

Squared Partial Correlation

The square of the partial correlation $r_{YX_2 \cdot X_1}$ represents the proportion of the variation in Y that is explained by X_2, out of that left unexplained by X_1, or

$$r^2_{YX_2 \cdot X_1} = \frac{R^2 - r^2_{YX_1}}{1 - r^2_{YX_1}} = \frac{\text{Partial proportion explained uniquely by } X_2}{\text{Proportion unexplained by } X_1}$$

Now, $r^2_{YX_1}$ is the proportion of the variation in Y explained by X_1, and the remaining proportion $(1 - r^2_{YX_1})$ represents the variation left unexplained. When X_2 is added to the model, it accounts for some additional variation. The total proportion of the variation in Y accounted for by X_1 and X_2 jointly is R^2, the coefficient of multiple determination for the model with both X_1 and X_2 as explanatory variables. So, $R^2 - r^2_{YX_1}$ is the additional proportion of the variability in Y explained by X_2, after the effects of X_1 have been removed or controlled. Figure 11.11 illustrates. The maximum this difference could be is $1 - r^2_{YX_1}$, the proportion of variation yet to be explained after accounting for the influence of X_1. The additional explained variation $R^2 - r^2_{YX_1}$ divided by this maximum possible difference is a measure that has a maximum possible value of 1. In fact, this ratio equals the squared partial correlation between Y and X_2, controlling for X_1.

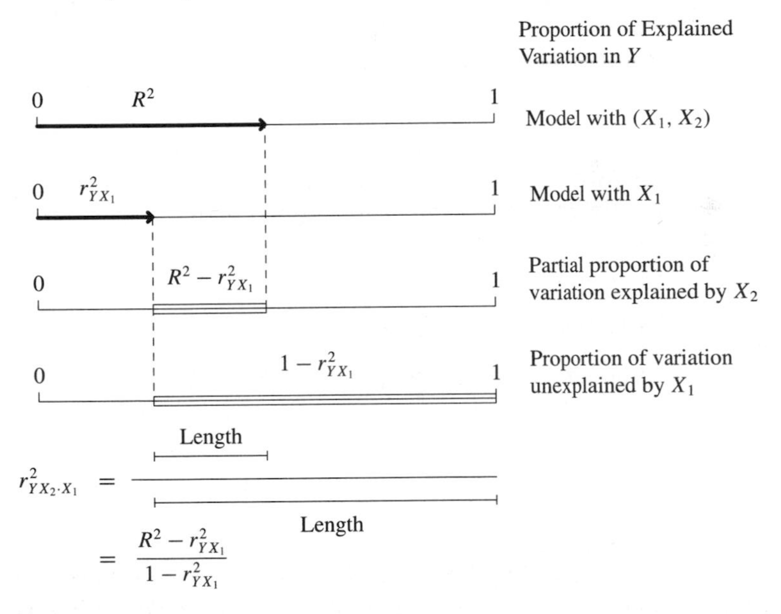

Figure 11.11 Representation of $r^2_{YX_2 \cdot X_1}$ as the Proportion of Variability That Can Be Explained by X_2, of that Left Unexplained by X_1

Figure 11.11 represents this relationship by the ratio of the partial contribution of X_2 beyond that of X_1, namely, $R^2 - r^2_{YX_1}$, divided by the proportion $(1 - r^2_{YX_1})$ left unexplained by X_1. Similarly, the square of $r_{YX_1 \cdot X_2}$ equals

$$r^2_{YX_1 \cdot X_2} = \frac{R^2 - r^2_{YX_2}}{1 - r^2_{YX_2}}$$

the proportion of variation in Y explained by X_1, out of that part unexplained by X_2.

Example 11.9 Partial Correlation of Life Events with Mental Impairment

From Table 11.5 for the mental impairment data, $r_{YX_1} = .372$, $r_{YX_2} = -.399$, and $r_{X_1X_2} = .123$. From the definition, the partial correlation between Y = mental impairment and X_1 = life events, controlling for X_2 = SES, is

$$r_{YX_1 \cdot X_2} = \frac{r_{YX_1} - r_{YX_2} r_{X_1X_2}}{\sqrt{\left(1 - r^2_{YX_2}\right)\left(1 - r^2_{X_1X_2}\right)}} = \frac{.372 - (-.399)(.123)}{\sqrt{[1 - (-.399)^2](1 - .123^2)}} = .463$$

The partial correlation, like the Pearson correlation between mental impairment and life events, is moderately positive.

The PRE partial measure equals $r^2_{YX_1 \cdot X_2} = (.463)^2 = .21$. Controlling for SES, 21% of the variation in mental impairment is explained by life events. Alternatively, since $R^2 = .339$ (Table 11.5),

$$r^2_{YX_1 \cdot X_2} = \frac{R^2 - r^2_{YX_2}}{1 - r^2_{YX_2}} = \frac{.339 - (-.399)^2}{1 - (-.399)^2} = .21$$

From $r^2_{YX_1 \cdot X_2} = .21$, we can recover the partial correlation, $r_{YX_1 \cdot X_2} = +\sqrt{.21} = .46$. We take the positive square root, since $r_{YX_1 \cdot X_2}$ has the same sign as the coefficient b_1 of X_1, controlling for X_2, in the prediction equation for the multiple regression model $(b_1 = .103)$. □

Higher-Order Partial Correlations

Similar expressions apply when the number of control variables exceeds one. For example, the square of the partial correlation between Y and X_3, controlling for X_1 and X_2, is

$$r^2_{YX_3 \cdot X_1, X_2} = \frac{R^2_{Y(X_1,X_2,X_3)} - R^2_{Y(X_1,X_2)}}{1 - R^2_{Y(X_1,X_2)}}$$

Here, $R^2_{Y(X_1,X_2,X_3)}$ denotes the value of R^2 for the multiple regression model with three predictors, and $R^2_{Y(X_1,X_2)}$ is the value of R^2 for the model with only the two variables treated as controls. The difference $R^2_{Y(X_1,X_2,X_3)} - R^2_{Y(X_1,X_2)}$ is the increase in the proportion of explained variance from adding X_3 to the model. The denominator $1 - R^2_{Y(X_1,X_2)}$

is the proportion of the variation left unexplained when X_1 and X_2 are the only predictors in the model.

The partial correlation $r_{YX_3 \cdot X_1, X_2}$ is called a ***second-order partial correlation***, since it controls two variables. It has the same sign as b_3 in the prediction equation $\hat{Y} = a + b_1 X_1 + b_2 X_2 + b_3 X_3$.

Inference for Partial Correlations

Controlling for a certain set of variables, the slope β_i of the partial relationship between Y and X_i is 0 in the same situations in which the partial correlation between Y and X_i is 0. An alternative formula for the t test for an individual partial effect uses the partial correlation.

Denote the population partial correlation by $\rho_{YX_i \cdot —}$, where the horizontal bar after the dot indicates that all other explanatory variables in the model are control variables. For example, with five explanatory variables, $\rho_{YX_2 \cdot —}$ represents $\rho_{YX_2 \cdot X_1, X_3, X_4, X_5}$. The test of H_0: $\beta_i = 0$ is equivalent to a test of H_0: $\rho_{YX_i \cdot —} = 0$. The equivalent t test statistic is

$$t = \frac{r_{YX_i \cdot —}}{\sqrt{(1 - r^2_{YX_i \cdot —})/[n - (k + 1)]}}$$

where $r_{YX_i \cdot —}$ is the sample estimate.

This statistic has the t distribution with $df = n - (k + 1)$. It equals $t = b_i / \hat{\sigma}_{b_i}$ and, hence, has the same P-value. When no variables are controlled (i.e., the number of explanatory variables is $k = 1$), the t statistic reduces to

$$t = \frac{r}{\sqrt{(1 - r^2)/(n - 2)}}$$

the statistic for testing that the population Pearson correlation equals 0 (Section 9.5).

We illustrate by testing H_0: $\rho_{YX_1 \cdot X_2} = 0$ about the true partial correlation between mental impairment and life events, controlling for SES. From Example 11.9, $r_{YX_1 \cdot X_2} = .463$. There are $k = 2$ explanatory variables and $n = 40$ observations. The test statistic equals

$$t = \frac{r_{YX_1 \cdot X_2}}{\sqrt{(1 - r^2_{YX_1 \cdot X_2})/[n - (k + 1)]}} = \frac{.463}{\sqrt{[1 - (.463)^2]/37}} = 3.18$$

This equals the test statistic for H_0: $\beta_1 = 0$ in Example 11.5. Thus, the P-value is also the same, $P = .003$.

Confidence intervals for partial correlations are more complex. See Problem 11.53.

11.8 Standardized Regression Coefficients*

The sizes of partial regression coefficients depend on the units of measurement for the variables. To compare the relative effects of different explanatory variables, it is appropriate to compare their coefficients only if the variables have the same units. Otherwise, *standardized* versions of the regression coefficients provide more meaningful comparisons.

Standardized Regression Coefficient

The ith ***standardized regression coefficient***, denoted by β_i^*, represents the change in the mean of Y, in Y standard deviations, for a one standard deviation increase in X_i, controlling for the other explanatory variables.

The Standardization Mechanism

The standardized regression coefficient β_i^* is the value the coefficient of X_i assumes in multiple regression when the units are such that Y and X_i have equal standard deviations. We standardize the estimated partial regression coefficients $\{b_1, b_2, \ldots, b_k\}$ by adjusting for the differing standard deviation of Y and each X_i. Let s_Y denote the sample standard deviation of Y, and let $s_{X_1}, s_{X_2}, \ldots, s_{X_k}$ denote the sample standard deviations of the explanatory variables. The standardization multiplies b_i by the ratio of s_{X_i} to s_Y.

If b_i is the estimate of the partial regression coefficient β_i, then

$$b_i^* = b_i\left(\frac{s_{X_i}}{s_Y}\right)$$

is the estimate of the standardized regression coefficient, β_i^*.

The value b_i^* estimates the partial effect of X_i in standard units. These coefficients measure the relative weights attached to the various explanatory variables in contributing to the mean of the response variable. The larger the absolute value of β_i^*, the greater is the effect on Y of a standard deviation change in X_i, controlling for the other variables. If $|\beta_2^*| > |\beta_1^*|$, for example, then a change in X_2 has a greater partial effect on Y than does a comparable change in X_1, measured in standard deviations.

Example 11.10 Standardized Coefficients for Mental Impairment

The prediction equation relating mental impairment to life events and SES is

$$\hat{Y} = 28.23 + .103X_1 - .097X_2$$

From Table 11.2, the sample standard deviations are $s_Y = 5.5$, $s_{X_1} = 22.6$, and $s_{X_2} = 25.3$. Since the unstandardized coefficient of X_1 is $b_1 = .103$, the estimated standardized coefficient is

$$b_1^* = b_1\left(\frac{s_{X_1}}{s_Y}\right) = .103\left(\frac{22.6}{5.5}\right) = .43$$

TABLE 11.8 SPSS Printout for Multiple Regression Model with Mental Impairment Data

```
Multiple R              .582
R Square                .339
Adjusted R Square       .303
Standard Error         4.556

                DF   Sum of Squares      Mean Square
Regression       2        394.238           197.119
Residual        37        768.162            20.761

     F =   9.495      Signif F = .0005

Variable             B          SE B         Beta          T      Sig T
LIFE             .1033         .0325        .4279      3.177      .0030
SES             -.0975         .0291       -.4513     -3.351      .0019
(Constant)     28.2298        2.1742                  12.984      .0000
```

Similarly, since $b_2 = -.097$, the standardized value equals

$$b_2^* = b_2 \left(\frac{s_{X_2}}{s_Y} \right) = -.097 \left(\frac{25.3}{5.5} \right) = -.45$$

The estimated change in the mean of Y for a standard deviation increase in X_1, controlling for X_2, has similar magnitude as the estimated change for a standard deviation increase in X_2, controlling for X_1. However the partial effect of X_1 is positive, whereas the partial effect of X_2 is negative.

Table 11.8 shows the way SPSS reports results of the multiple regression model. It reports the estimated standardized regression coefficients under the heading BETA, reflecting the alternative name ***beta weights*** for these coefficients. □

Properties of Standardized Regression Coefficients

For bivariate regression, standardizing the regression coefficient yields the Pearson correlation. For the multiple regression model, though, the standardized partial regression coefficient is not the same as the partial correlation. Moreover, unlike the partial correlation, it is not symmetric. It is related to the partial correlation (Problem 11.57), however, and it usually takes similar value.

Unlike partial or Pearson correlations, b_i^* does not necessarily fall between -1 and $+1$. A value $|b_i^*| > 1$ occasionally occurs when X_i is highly correlated with the set of other explanatory variables in the model. In such cases, the standard errors are usually large and the estimates are unreliable.

Since a standardized regression coefficient is a multiple of the unstandardized coefficient, one equals 0 when the other does. The test of H_0: $\beta_i^* = 0$ is equivalent to the t test of H_0: $\beta_i = 0$. Thus, it is unnecessary to study separate inference procedures for these coefficients.

Standardized Form of Prediction Equation*

Prediction equations have an expression in terms of the estimated standardized regression coefficients. In this equation, the variables appear in standardized form.

> **Notation for Standardized Variables**
>
> Let $z_Y, z_{X_1}, \ldots, z_{X_k}$ denote the standardized versions of the variables $Y, X_1, \ldots, X_k$. For instance, the z-score $z_Y = (Y - \bar{Y})/s_Y$ represents the number of standard deviations that an observation on Y falls from its mean.

For each subject, one can transform the scores on $Y, X_1, \ldots, X_k$ to z-scores for the variables, $z_Y, z_{X_1}, \ldots, z_{X_k}$. If a subject's score on X_1 is such that $z_{X_1} = (X_1 - \bar{X}_1)/s_{X_1} = 2.0$, for instance, then that subject falls two standard deviations above the mean $\bar{X}_1$ of that variable.

Let $\hat{z}_Y = (\hat{Y} - \bar{Y})/s_Y$ denote the predicted z-score for the response variable. For the standardized variables and the estimated standardized regression coefficients, the prediction equation is

$$\hat{z}_Y = b_1^* z_{X_1} + b_2^* z_{X_2} + \cdots + b_k^* z_{X_k}$$

This equation predicts how far an observation on Y falls from its mean, in standard deviation units, based on how far the explanatory variables fall from their means, in standard deviation units. No intercept term appears. When the z-scores for all the X variables equal 0 (i.e., when the X-values equal their means), then $\hat{Y} = \bar{Y}$, so that

$$\hat{z}_Y = \frac{\hat{Y} - \bar{Y}}{s_Y} = 0$$

Example 11.11 Standardized Prediction Equation for Mental Impairment

The estimated standardized regression coefficients for the life events and SES predictors of mental impairment are $b_1^* = .43$ and $b_2^* = -.45$. The standardized prediction equation relating the z-score on mental impairment to the z-scores on life events and SES is

$$\hat{z}_Y = .43 z_{X_1} - .45 z_{X_2}$$

Consider a subject who is two standard deviations above the mean on life events but two standard deviations below the mean on SES. This subject has a predicted z-score on mental impairment of

$$\hat{z}_Y = .43(2) - .45(-2) = 1.8$$

The predicted mental impairment for that subject is 1.8 standard deviations above the mean. If the distribution of mental impairment is approximately normal, this subject would be predicted to show some mental health problems, since only about 4% of the scores in a normal distribution fall at least 1.8 standard deviations above their mean. □

Cautions in Comparing Standardized Regression Coefficients

To assess which predictor in a multiple regression model has the greatest impact on the response variable, it is tempting to compare their standardized regression coefficients. Make such comparisons only with caution. In some cases, the observed differences in the b_i^* may simply reflect sampling error. In particular, when multicollinearity exists, the standard errors are high and the estimated standardized coefficients may be very unstable.

Keep in mind also that the effects are partial ones, depending on which other variables are in the model. An explanatory variable that seems important in one system of variables may seem unimportant when other variables are controlled. For example, it is possible that $|b_2^*| > |b_1^*|$ in a model with two explanatory variables, yet when a third explanatory variable is added to the model, $|b_2^*| < |b_1^*|$.

It is unnecessary to standardize to compare the effect of the same variable for two groups, such as in comparing the results of separate regressions for females and males, since the units of measurement are the same in each group. In fact, it is usually unwise to standardize in this case, because the standardized coefficients are more susceptible than the unstandardized coefficients to differences in the standard deviations. For instance, Section 9.6 showed that the Pearson correlation depends strongly on the range of X-values sampled. Two groups that have the same value for an estimated regression coefficient have different standardized coefficients if the ratio s_X/s_Y differs for the two groups.

Finally, if X_i is highly correlated with the set of other explanatory variables, it is artificial to conceive of X_i changing and the other predictors remaining fixed in value. As an extreme example, suppose $Y =$ height, $X_1 =$ length of left leg, and $X_2 =$ length of right leg. The Pearson correlation between X_1 and X_2 is extremely close to 1. It does not make much sense to imagine how Y changes as X_1 changes while X_2 is controlled. Sometimes it is more relevant to compare uncontrolled effects of X_i on Y, using Pearson correlations, than to compare partial effects.

11.9 Chapter Summary

This chapter generalized the bivariate regression model to include additional explanatory variables. The ***multiple regression equation*** relating a response variable Y to a set of k explanatory variables is

$$E(Y) = \alpha + \beta_1 X_1 + \beta_2 X_2 + \cdots + \beta_k X_k$$

- The $\{\beta_i\}$ are ***partial regression coefficients***. The value β_i is the change in the mean of Y for a one-unit change in X_i, controlling for the other variables in the model.
- The ***multiple correlation*** R and its square, the ***coefficient of multiple determination*** R^2, describe the degree of predictability of the response variable Y by the collective set of explanatory variables.

- The multiple correlation equals the Pearson correlation between the observed and predicted Y-values. It falls between 0 and 1.

- $R^2 = (\text{TSS} - \text{SSE})/\text{TSS}$ is the ***proportional reduction in error*** from predicting Y using the prediction equation $\hat{Y} = a + b_1X_1 + \cdots + b_kX_k$ instead of $\bar{Y}$.

- The ***partial correlation***, such as $r_{YX_1 \cdot X_2}$, describes the association between two variables, controlling for others. It falls between -1 and $+1$.

- The squared partial correlation between Y and X_i represents the proportion of the variation in Y that can be explained by X_i, out of that part left unexplained by a set of control variables.

- An F ***statistic*** tests $H_0 : \beta_1 = \ldots = \beta_k$, that the response variable is independent of all the predictors. A small P-value suggests that at least one predictor affects the response.

- Individual t tests and confidence intervals for β_i analyze separate partial effects of each predictor, controlling for the other variables in the model.

- One can permit ***interaction*** by introducing cross-products of explanatory variables into the model.

- One can ***compare regression models***, a *complete* model and a simpler *reduced* model, using an F test based on comparing the SSE values or R^2 values.

- ***Standardized regression coefficients*** do not depend on the units of measurement. The estimated standardized coefficient b_i^* describes the change in Y, in Y standard deviation units, for a one standard deviation increase in X_i, controlling for the other explanatory variables.

The multiple regression model

$$E(Y) = \alpha + \beta_1X_1 + \beta_2X_2$$

has $k = 2$ explanatory variables. Fixing X_2, a straight line describes the relation between Y and X_1. Its slope β_1 is the change in the mean of Y for a one-unit change in X_1, controlling for X_2. The prediction equation is $\hat{Y} = a + b_1X_1 + b_2X_2$. The multiple correlation R is at least as large as the Pearson correlations between Y and each predictor. The squared partial correlation $r^2_{YX_2 \cdot X_1}$ is the proportion of the variation of Y that is explained by X_2, out of that part of the variation left unexplained by X_1. The estimated standardized regression coefficient $b_1^* = b_1(s_{X_1}/s_Y)$ refers to the effect of a standard deviation change in X_1, controlling for X_2.

Table 11.9 summarizes the basic properties and inference methods for these measures and those introduced in Chapter 9 for bivariate regression.

TABLE 11.9 Summary of Bivariate and Multiple Regression

	BIVARIATE REGRESSION	MULTIPLE REGRESSION	
Model	$E(Y) = \alpha + \beta X$	$E(Y) = \alpha + \beta_1 X_1 + \cdots + \beta_k X_k$	
Prediction equation	$\hat{Y} = a + bX$	$\hat{Y} = a + b_1 X_1 + \cdots + b_k X_k$	
		Simultaneous effect of $X_1 \ldots, X_k$	Partial effect of one X_i
Properties of measures	b = Slope r = Pearson correlation, standardized slope, $-1 \le r \le 1$, r has the same sign as b r^2 = Coefficient of determination, PRE measure, $0 \le r^2 \le 1$	R = Multiple correlation, $0 \le R \le 1$ R^2 = Coefficient of multiple determination, PRE measure, $0 \le R^2 \le 1$	b_i = Partial slope b_i^* = Standardized regression coefficient $r_{YX_i \cdot \text{—}}$ = Partial correlation, $-1 \le r_{YX_i \cdot \text{—}} \le 1$, same sign as b_i and b_i^*, $r^2_{YX_i \cdot \text{—}}$ is PRE measure
Tests of no association	H_0: $\beta = 0$ or H_0: $\rho = 0$, Y not associated with X	H_0: $\beta_1 = \cdots = \beta_k = 0$ or H_0: $\mathbf{P} = 0$, Y not associated with $X_1, \ldots X_k$	H_0: $\beta_i = 0$, or H_0: $\rho_{YX_i \cdot \text{—}} = 0$, Y not associated with X_i, controlling for other X variables
Test statistic	$t = \frac{b}{\hat{\sigma}_b} = \frac{r}{\sqrt{\frac{1-r^2}{n-2}}}$ $df = n - 2$	$F = \frac{\text{Regression mean square}}{\text{Error mean square}}$ $= \frac{R^2/k}{(1-R^2)/[n-(k+1)]}$, $df_1 = k$, $df_2 = n - (k+1)$	$t = \frac{b_i}{\hat{\sigma}_{b_i}} = \frac{r_{YX_i \cdot \text{—}}}{\sqrt{\frac{1-r^2_{YX_i \cdot \text{—}}}{n-(k+1)}}}$ $df = n - (k+1)$

The model studied in this chapter is still somewhat restrictive in the sense that all the predictors are quantitative. The next chapter shows how to include qualitative predictors in the model.

PROBLEMS

Practicing the Basics

1. In Table 9.16 in Problem 9.24 regarding Florida counties, refer to the variables Y = crime rate (number per 1000 residents), X_1 = median income (thousands of dollars), and X_2 = percent in urban environment.
   a) Figure 11.12 shows a SAS scatter diagram relating Y to X_1. Predict the sign that the estimated effect of X_1 has in the prediction equation $\hat{Y} = a + bX_1$. Explain.
   b) Figure 11.13 shows a SAS partial regression plot relating Y to X_1, controlling for X_2. Predict the sign that the estimated effect of X_1 has in the prediction equation $\hat{Y} = a + b_1 X_1 + b_2 X_2$. Explain.
   c) Table 11.10 shows part of a SAS printout for the bivariate and multiple regression models. Report the prediction equation relating Y to X_1, and interpret the slope.

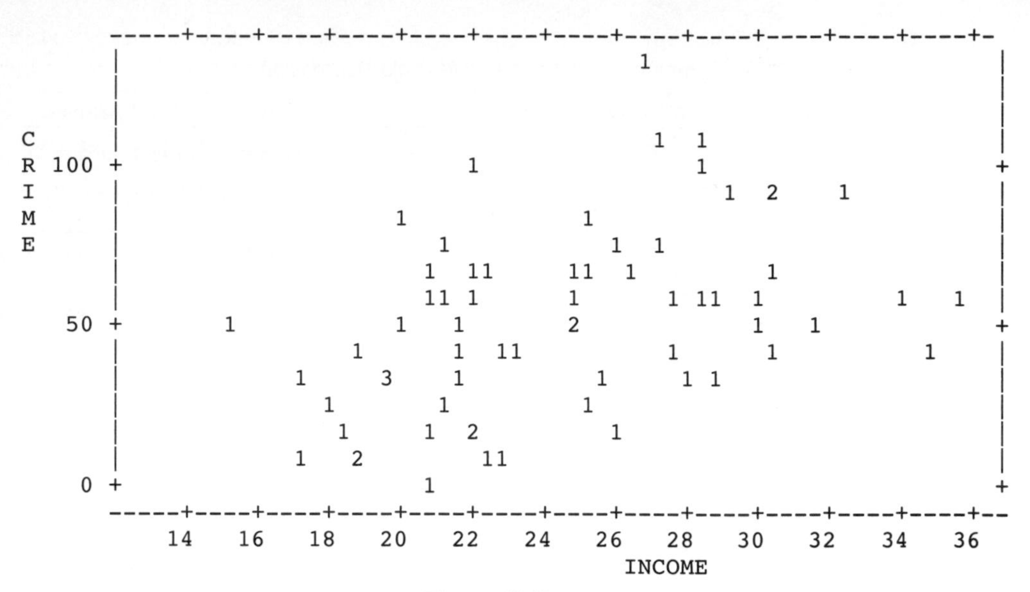

Figure 11.12

d) Report the prediction equation relating Y to both X_1 and X_2. Interpret the coefficient of X_1, and compare to (c).
e) The Pearson correlations are $r_{YX_1} = .43$, $r_{YX_2} = .68$, $r_{X_1X_2} = .73$. Use these to explain why the X_1 effect seems so different in (c) and (d).
f) Report the prediction equations relating crime rate to income at urbanization levels of (i) 0, (ii) 50, (iii) 100. Interpret.

TABLE 11.10

Variable	Parameter Estimate	Standard Error	T for H0: Parameter=0	Prob > |T|
INTERCEP	-11.526	16.834	-0.685	0.4960
INCOME	2.609	0.675	3.866	0.0003

Variable	Parameter Estimate	Standard Error	T for H0: Parameter=0	Prob > |T|
INTERCEP	40.261	16.365	2.460	0.0166
INCOME	-0.809	0.805	-1.005	0.3189
URBAN	0.646	0.111	5.811	0.0001

2. For students at Walden University, the relationship between Y = college GPA (with range 0–4.0) and X_1 = high school GPA (range 0–4.0) and X_2 = college board score (range 200–800) satisfies $E(Y) = .20 + .50X_1 + .002X_2$.
a) Find the mean college GPA for students having (i) high school GPA = 4.0 and college board score = 800, (ii) $X_1 = 3.0$ and $X_2 = 300$.
b) Show that the relationship between Y and X_1 for those students with $X_2 = 500$ is $E(Y) = 1.2 + .5X_1$.

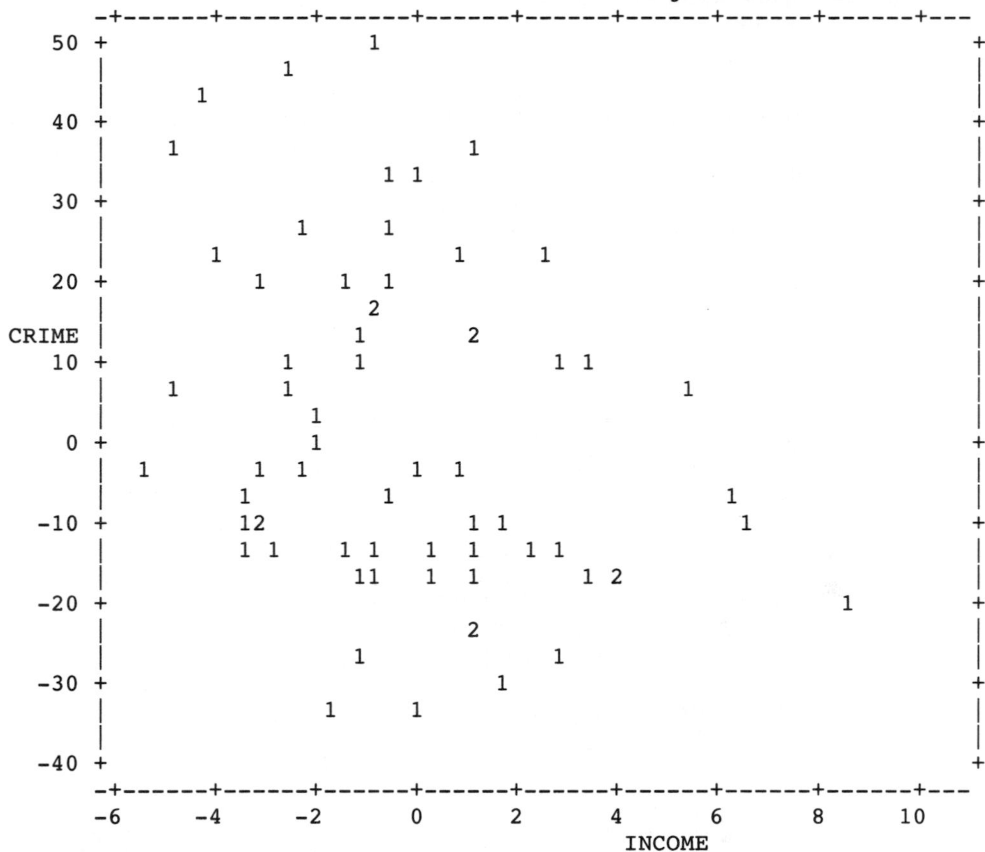

Figure 11.13

c) Show that when $X_2 = 600$, $E(Y) = 1.4 + .5X_1$. Thus, increasing X_2 by 100 shifts the line relating Y to X_1 upward by $100\beta_2 = .2$ units.
d) Show that setting X_1 at a variety of values yields a collection of parallel lines, each having slope .002, relating the mean of Y to X_2.
e) Since $\beta_1 = .50$ is larger than $\beta_2 = .002$, does this imply that X_1 has the greater partial effect on Y? Explain.

3. Refer to the data in Table 9.13 in Problem 9.17. Let Y = crude birth rate, X_1 = women's economic activity, and X_2 = GNP. The least squares equation is $\hat{Y} = 34.53 - .13X_1 - .64X_2$.
a) Interpret the estimated regression coefficients.
b) Plot on a single graph the relationship between Y and X_1 when $X_2 = 0$, $X_2 = 10$, and $X_2 = 20$. Interpret the results.
c) The bivariate prediction equation with X_1 is $\hat{Y} = 37.65 - .31X_1$. The Pearson correlations are $r_{YX_1} = -.58$, $r_{YX_2} = -.72$, and $r_{X_1X_2} = .58$. Explain why the coefficient of X_1

in the bivariate equation is so different from its value in the multiple predictor equation.
d) $R^2 = .56$. Interpret.

4. Refer to Example 11.1. Using computer software for the data in Problem 9.24 for those three variables:
a) Construct box plots for each variable and scatter diagrams and partial regression plots between Y and each of X_1 and X_2. If available with your software, also construct coplots. Interpret these plots.
b) Find prediction equations for the bivariate models, using education to predict crime rate and using urbanization to predict crime rate. Interpret.
c) Construct the correlation matrix for these three variables. Interpret.
d) Find the prediction equation for the multiple regression model. Interpret.
e) Find R^2 for the multiple regression model, and show that it is not much larger than r^2 for the model using urbanization alone as the predictor. Interpret.

5. Table 11.11 shows a SAS printout from fitting the multiple regression model to the data from Table 9.1, excluding D.C., on Y = violent crime rate, X_1 = poverty rate, and X_2 = percent living in metropolitan areas.

TABLE 11.11

Source	DF	Sum of Squares	Mean Square	F Value	Prob>F
Model	2	2448368.07	1224184.04	31.249	0.0001
Error	47	1841257.15	39175.68		
Total	49	4289625.22			

Root MSE 197.928 R-square 0.5708

Variable	Parameter Estimate	Standard Error	T for H0: Parameter=0	Prob > |T|	Standard -ized Estimate
INTERCEP	-498.683	140.988	-3.537	0.0009	0.000
POVERTY	32.622	6.677	4.885	0.0001	0.473
METRO	9.112	1.321	6.900	0.0001	0.668

Pearson Correlation Coefficients

	VIOLENT	POVERTY	METRO
VIOLENT	1.00000	0.36875	0.59396
POVERTY	0.36875	1.00000	-0.15562
METRO	0.59396	-0.15562	1.00000

a) Report the prediction equation, and interpret the estimated regression coefficients.
b) Find the predicted violent crime rate for Massachusetts. Find the residual, and interpret.
c) Interpret the fit by showing the prediction equation relating Y and X_1 for states with (i) $X_2 = 0$, (ii) $X_2 = 50$, (iii) $X_2 = 100$. Interpret.
d) Interpret the correlation matrix.
e) Report R^2 and the multiple correlation, and interpret.
f) Find the partial correlation between violent crime rate and poverty, controlling for metropolitan rate. Interpret.
g) Refer to (f). Interpret the squared partial correlation.

h) Show how to construct the F statistic for testing H_0: $\beta_1 = \beta_2 = 0$, report its df values and P-value, and interpret.
i) Show how to construct the t statistic for testing H_0: $\beta_1 = 0$, report its df and its P-value for H_a: $\beta_1 \neq 0$, and interpret.
j) Construct a 95% confidence interval for β_1, and interpret.

6. Repeat the previous exercise using software to fit the model with murder rate in Table 9.1 as the response variable.

7. Refer to Problem 11.5. With X_3 = percentage of single-parent families also in the model, Table 11.12 shows results.
a) Report the prediction equation and interpret the coefficients.
b) Report R^2, and interpret.
c) Report the test statistic for H_0: $\beta_1 = \beta_2 = \beta_3 = 0$, find its df values and P-value, and interpret.
d) Find the t statistic for H_0: $\beta_1 = 0$ and its df and P-value for H_a: $\beta_1 \neq 0$, and interpret.
e) Construct a 95% confidence interval for β_1, and interpret.
f) Without X_3 in the model, the coefficient of X_1 is 32.62. Why do you think it changes so much after X_3 is added?
g) Since Table 9.1 provides data for all states, what relevance, if any, do the inferences have in this exercise and Problem 11.5?

TABLE 11.12

Variable	Coefficient	Std. Error
Intercept	−1197.538	
Poverty	18.283	(6.136)
Metropolitan	7.712	(1.109)
Single-parent	89.401	(17.836)
F	39.9	
R^2	.722	
n	50	

TABLE 11.13

	Sum of Squares
Regression	31.8
Residual	199.3
Variable	B
MEDUC	−.24
FSES	.02
(Constant)	5.25

8. Table 11.13 is part of a SPSS printout for fitting a regression model to the relationship between Y = number of children in family, X_1 = mother's educational level (MEDUC) in years, and X_2 = father's socioeconomic status (FSES), for a random sample of 49 college students at Texas A&M University.
a) Write the prediction equation. Interpret parameter estimates.
b) For the first subject in the sample, $X_1 = 12$, $X_2 = 61$, and $Y = 5$. Find the predicted value of Y and the residual.
c) Report SSE. Use it to explain the least squares property of this prediction equation.
d) Find the multiple correlation. Interpret.
e) Is it possible that $r_{YX_1 \cdot X_2} = .40$? Explain.
f) Can you tell from this printout whether r_{YX_1} is positive or negative? Explain.

9. Refer to Problem 9.30 on feelings toward liberals, political ideology, and religious attendance. The sample size is small, but for illustrative purposes Table 11.14 shows results of fitting the multiple regression model with feelings toward liberals as the response, using the category numbers as scores for religion. Standard errors are shown in parentheses.
a) Report the prediction equation and interpret the estimates.

b) Report the predicted value and residual for the first observation, for which ideology = 7, religion = 9, and feelings = 10.
c) Report, and explain how to interpret, R^2.
d) Tables of this form often put * by an effect having $P < .05$, ** by an effect having $P < .01$, and *** by an effect having $P < .001$. Show how this was determined for the ideology effect, and discuss the disadvantage of summarizing in this manner.
e) Explain how the F value was obtained, report its df values, and explain how to interpret its result.
f) The estimated standardized regression coefficients are $-.79$ for ideology and $-.23$ for religion. Interpret.

TABLE 11.14

Variable	Coefficient
Intercept	135.31
Ideology	−14.07
	(3.16)**
Religion	−2.95
	(2.26)
F	13.93**
R²	.799
Adj. R²	.742
(n)	(10)

10. Refer to Table 11.5. Test the null hypothesis H_0: $\beta_2 = 0$ that mental impairment is independent of SES, controlling for life events. Report the test statistic, and report and interpret the P-value for (a) H_a: $\beta_2 \neq 0$, (b) H_a: $\beta_2 < 0$.

11. Use software with Table 9.4 to conduct a multiple regression analysis of Y = selling price of home, X_1 = size of home, X_2 = number of bedrooms, X_3 = number of bathrooms.
a) Use graphics to display the effects of the predictors. Interpret.
b) Show that the prediction equation is $\hat{Y} = -53.4 + 62.4X_1 + 1.64X_2 + 22.9X_3$. Interpret the estimates, and find the predicted selling price for a home with $X_1 = 2$, $X_2 = 4$, and $X_3 = 2$.
c) Inspect the correlation matrix, and report the variables having the (i) strongest association, (ii) weakest association.
d) Report R^2, and interpret.
e) Show how to calculate the F statistic for testing the effect of the three predictors, report its df values and its P-value, and interpret.
f) Show how to calculate the t statistic for testing $H_0 : \beta_2 = 0$, report its P-value for $H_a : \beta_2 > 0$, and interpret. Why do you think this effect is not significant?
g) Fit the simpler model without size of home as a predictor. Now again test the partial effect of number of bedrooms, and interpret.
h) Interpret the fit in (g) by showing the prediction equation relating Y and X_2 for homes with (i) one bathroom, (ii) two bathrooms.
i) Construct a 95% confidence interval for the coefficient of X_2 in the model in (g), and interpret.
j) Find the partial correlation between price and number of bedrooms, controlling for number of bathrooms. Compare it to the Pearson correlation, and interpret.

TABLE 11.15

Variable	N	Mean	Std Dev
BIRTHS	23	22.117	10.469
ECON	23	47.826	19.872
LITER	23	77.696	17.665

Pearson Correlation Coefficients / Prob > |R| under Ho: Rho=0

	BIRTHS	ECON	LITER
BIRTHS	1.00000 0.0	-0.61181 0.0019	-0.81872 0.0001
ECON	-0.61181 0.0019	1.00000 0.0	0.42056 0.0457
LITER	-0.81872 0.0001	0.42056 0.0457	1.00000 0.0

Source	DF	Sum of Squares	Mean Square	F Value	Prob>F
Model	2	1825.969	912.985	31.191	0.0001
Error	20	585.424	29.271		
C Total	22	2411.393			

Root MSE 5.410 R-square 0.7572

Variable	Parameter Estimate	Standard Error	T for H0: Parameter=0	Prob > |T|
INTERCEP	61.713	5.2453	11.765	0.0001
ECON	-0.171	0.0640	-2.676	0.0145
LITER	-0.404	0.0720	-5.616	0.0001

Variable	Standardized Estimate
ECON	-0.325
LITER	-0.682

k) Show how to calculate the estimated standardized regression coefficients for the model in (g), and interpret.
l) Write the prediction equation using standardized variables. Interpret.

12. Refer to the previous exercise and the model [in (g)] having numbers of bedrooms and bathrooms as predictors.
a) Fit the model allowing an interaction between these two predictors.
b) Interpret the fit by showing the prediction equation relating Y and X_2 for homes with (i) two bathrooms, (ii) three bathrooms.
c) Use a t test to analyze the significance of the interaction term. Interpret.

d) Use an F test for complete and reduced models to analyze the significance of the interaction term. Interpret, and compare results to c).

13. Refer to Problem 11.5. Add an interaction term.
a) Report the prediction equation.
b) As the percentage living in metropolitan areas increases, does the effect of poverty rate tend to increase or decrease? Explain.
c) Show how to interpret the prediction equation graphically.
d) Describe the substantive effect of adding the interaction term by comparing R^2 values for the model with and without this term.

14. Refer to Table 9.13 and Problem 9.17. Table 11.15 shows a SAS printout from fitting the model $E(Y) = \alpha + \beta_1 X_1 + \beta_2 X_2$ to Y = birth rate (BIRTHS), X_1 = women's economic activity (ECON), and X_2 = literacy rate (LITER). (We deleted observations for Germany, South Africa, and Vietnam.)
a) Report the value of each of the following:
(i) r_{YX_1} (ii) r_{YX_2} (iii) R^2
(iv) TSS (v) SSE (vi) MSE
(vii) $\hat{\sigma}$ (viii) s_Y (ix) $\hat{\sigma}_{b_1}$
(x) t for H_0: $\beta_1 = 0$
(xi) P for H_0: $\beta_1 = 0$ against H_a: $\beta_1 \neq 0$
(xii) P for H_0: $\beta_1 = 0$ against H_a: $\beta_1 < 0$
(xiii) F for H_0: $\beta_1 = \beta_2 = 0$
(xiv) P for H_0: $\beta_1 = \beta_2 = 0$
(xv) df for $\hat{\sigma}$ (xvi) P for H_0: $\rho_{YX_1} = 0$.
b) Report the prediction equation, and carefully interpret the three estimated regression coefficients.
c) Interpret the Pearson correlations r_{YX_1} and r_{YX_2}.
d) Show how to calculate R^2, and interpret its value.
e) Calculate the multiple correlation, and interpret.
f) Though inference may not be relevant for these data, show how to construct the F statistic for testing H_0: $\beta_1 = \beta_2 = 0$, report its df values and P-value, and interpret.
g) Show how to construct the t statistic for testing H_0: $\beta_1 = 0$, report its df and P-value for H_a: $\beta_1 \neq 0$, and interpret.

15. Refer to the previous exercise.
a) Find the partial correlation between Y and X_1, controlling for X_2. Interpret both the partial correlation and its square.
b) Show how to calculate the estimate of the conditional standard deviation, and interpret its value.
c) Show how to calculate the estimated standardized regression coefficient for X_1, and interpret its value.
d) Write the prediction equation using standardized variables. Interpret.
e) Find the predicted z-score for a country that is one standard deviation above the mean on both predictors. Interpret.

16. For a random sample of 66 state precincts, data are available on

Y = Percentage of adult residents who are registered to vote
X_1 = Percentage of adult residents owning homes
X_2 = Percentage of adult residents who are nonwhite

X_3 = Median family income (thousands of dollars)
X_4 = Median age of residents
X_5 = Percentage of residents who have lived in the precinct at least ten years

Table 11.16 shows a portion of the printout used to analyze the data.

TABLE 11.16

	DF	Sum of Squares	Mean Square	F Value	Prob > F	R-Square

Model	---	----	----	----	----	Root MSE
Error	---	2940.0	----			----
Total	---	3753.3				

Variable	Parameter Estimate	Standard Error	T For H0: Parameter = 0	Prob > |T|
Intercept	70.0000			
X1	0.1000	.0450	----	----
X2	-0.1500	.0750	----	----
X3	0.1000	.2000	----	----
X4	-0.0400	.0500	----	----
X5	0.1200	.0500	----	----

a) Fill in all the missing values in the printout.
b) State the prediction equation and interpret the coefficient of X_1.
c) Do you think it is necessary to include all five explanatory variables in the model? Explain.
d) Interpret the "R-Square" value.
e) To what test does the "F Value" refer? Interpret the result of that test.
f) To what test does the t-value opposite X1 refer? Report and interpret the P-value.
g) Interpret the value listed under "Root MSE."
h) Find a 95% confidence interval for the change in the mean of Y for a 1-unit increase in the percentage of adults owning homes, controlling for the other variables. Interpret.
i) Find a 95% confidence interval for the change in the mean of Y for a 50-unit increase in the percentage of adults owning homes, controlling for the other variables. Interpret.

17. *Refer to the data in Table 9.4 on Y = selling price of home, X_1 = size of home, and X_2 = whether the house is new (1 = yes, 0 = no). Chapters 12 and 13 show that one can incorporate qualitative predictors such as X_2 in regression models, and this exercise provides a preview. Table 11.17 shows part of the printout for the model using these predictors.
a) Report the prediction equation. By setting $X_2 = 0$ and then 1, construct the two separate lines for older and for new homes. Note that the model implies that the slope effect of size on selling price is the same for each.
b) Since X_2 takes only the values 0 and 1, explain why the coefficient of X_2 estimates the difference of mean selling prices between new and older homes, controlling for house size.
c) Test the significance of the partial effect of whether a house is new. Report the P-value, and interpret.

d) Construct a 95% confidence interval for the partial effect of whether a house is new, and interpret.

TABLE 11.17

Variable	Parameter Estimate	Standard Error	T for H0: Parameter=0	Prob > |T|
INTERCEP	-26.089	5.977	-4.365	0.0001
SIZE	72.575	3.508	20.690	0.0001
NEW	19.587	3.995	4.903	0.0001

18. * Refer to the previous exercise. Table 13.16 (Chapter 13) is a SPSS printout showing the effect of adding an interaction cross-product variable (NEWSIZE) to the model.
a) Report the prediction equation. Interpret the fit by reporting the prediction equation between selling price and size of house separately for new homes ($X_2 = 1$) and for old homes ($X_2 = 0$). Interpret. (This fit is equivalent to fitting lines separately to the data for new homes and for old homes.)
b) Interpret the fit by reporting the difference between the predicted selling prices for new and old homes for houses with X_1 equal to (i) 1.5, (ii) 2.0, (iii) 2.5.
c) Test the significance of the interaction term. Report the P-value, and interpret.
d) Figure 13.14 (Chapter 13) shows a plot of these data, identifying the points by whether the home is new (1) or not (0). When the new home with largest price is removed from the data set and the model is re-fitted, Table 13.17 shows results. Again test for interaction, and note the large impact a potential outlier can have.

19. A study analyzes relationships among Y = percentage vote for Democratic candidate, X_1 = percentage of registered voters who are Democrats, and X_2 = percentage of registered voters who vote in the election, for several congressional elections in 1996. The researchers expect interaction, since they expect a higher slope between Y and X_1 at larger values of X_2 than at smaller values. They obtain the prediction equation $\hat{Y} = 20 + .30X_1 + .05X_2 + .005X_1X_2$. Does this equation support the direction of their prediction? Explain.

20. A multiple regression analysis investigates the relationship between Y = college GPA and several explanatory variables, using a random sample of 195 students at Slippery Rock University. First, high school GPA and total SAT score are entered into the model. The sum of squared errors is SSE = 20. Next, parents' education and parents' income are added, to determine if they have an effect, controlling for high school GPA and SAT. For this expanded model SSE = 19. Test whether this complete model is significantly better than the one containing only high school GPA and SAT. Report and interpret the P-value.

21. Refer to Examples 11.1 and 11.8. Explain why the partial correlation between crime rate and high school graduation rate is so different from the Pearson correlation. (This is an example of *Simpson's paradox*; see Problem 10.8.)

22. For a group of 100 children of ages varying from 3 to 15, the Pearson correlation between vocabulary score on an achievement test and height of child is .65. The Pearson correlation between vocabulary score and age for this sample is .85, and the Pearson correlation between height and age is .75.
a) Show that the partial correlation between vocabulary and height, controlling for age, is .036. Interpret.
b) Test whether this partial correlation is significantly nonzero. Interpret.

c) Is it plausible that the relationship between height and vocabulary is spurious, in the sense that it is due to their joint dependence on age? Explain.

23. A multiple regression model describes the relationship among a collection of cities between Y = murder rate (number of murders per 100,000 residents) and

X_1 = Number of police officers (per 100,000 residents)

X_2 = Median length of prison sentence given to convicted murderers (in years)

X_3 = Median income of residents of city (in thousands of dollars)

X_4 = Unemployment rate in city

These variables are measured in 1996 for a random sample of thirty cities with population size exceeding 35,000. For these cities, the least squares equation is $\hat{Y} = 30 - .02X_1 - .1X_2 - 1.2X_3 + .8X_4$, and $\bar{Y} = 15$, $\bar{X}_1 = 100$, $\bar{X}_2 = 15$, $\bar{X}_3 = 13$, $\bar{X}_4 = 7.8$, $s_Y = 8$, $s_{X_1} = 30$, $s_{X_2} = 10$, $s_{X_3} = 2$, $s_{X_4} = 2$.
a) Can you tell from the coefficients of the prediction equation which explanatory variable has the greatest partial effect on Y? Explain.
b) Find the standardized regression coefficients and interpret their values.
c) Write the prediction equation using standardized variables. Find the predicted z-score on murder rate for a city that is one standard deviation above the mean on X_1, X_2, and X_3, and one standard deviation below the mean on X_4. Interpret.

24. Refer to Problem 11.5. Report the estimated standardized regression coefficients, and interpret, and express the prediction equation using standardized variables.

Concepts and Applications

25. Refer to the WWW data set (Problem 1.7). Using software, conduct a regression analysis using (i) Y = political ideology and using predictors number of times per week of newspaper reading and religiosity, (ii) Y = college GPA and predictors high school GPA and number of weekly hours of physical exercise. Prepare a report, summarizing your graphical analyses, bivariate models and interpretations, multiple regression models and interpretations, inferences, checks of effects of outliers, and overall summary of the relationships.

26. Refer to the data file you created in Problem 1.7. For variables chosen by your instructor, fit a multiple regression model and conduct descriptive and inferential statistical analyses. Interpret and summarize your findings.

27. Refer to data for the 50 states in Table 9.1. Using Y = violent crime rate, X_1 = percentage of single-parent families, and X_2 = metropolitan rate, analyze these data using regression. Provide interpretations for all your analyses, and provide a paragraph summary of your conclusions at the end of your report.

28. Repeat the previous exercise using murder rate as the response variable.

29. Refer to Problem 11.27. Repeat this problem, adding X_3 = percentage white as an explanatory variable.

30. Refer to Problem 11.27. Repeat this problem, including the observation for D.C. Describe the effect on the various analyses of this observation.

31. Refer to Problem 9.17 and Table 9.13. Construct a multiple regression model containing two explanatory variables that provide good predictions for birth rate. How did you select this model? (*Hint*: One way is based on entries in the correlation matrix.)

32. For Example 11.2, Table 11.18 shows the result of adding religious attendance as a predictor, measured as the approximate number of times the subject attends a religious service over the course of a year. Write a short report, interpreting the information from this table.

TABLE 11.18

Variable	Coefficient
Intercept	27.422
Life events	.0935
	(.0313)**
SES	−.0958
	(.0256)***
Religious attendance	−.0370
	(.0219)
R^2	.358
(n)	(40)

33. Refer to the variables in Problem 9.24. By constructing coplots, do you find any pairs of predictors that show evidence of interaction?

34. Describe a situation in which you would expect interaction. Describe the likely nature of the interaction; for example, as X_2 increases, would the slope of the relationship between Y and X_1 tend to increase, or decrease?

35. For a linear model with two explanatory variables X_1 and X_2, which of the following must be incorrect? Why?

a) $r_{YX_1} = .01$, $r_{YX_2} = -.2$, $R = .75$
b) $r_{YX_1} = .01$, $r_{YX_2} = -.75$, $R = .2$
c) $r_{YX_1} = .4$, $r_{YX_2} = .4$, $R = .4$

36. Table 11.19 shows results of fitting various regression models to data on Y = college GPA, X_1 = high school GPA, X_2 = mathematics entrance exam score, and X_3 = verbal entrance exam score.
Indicate which of the following statements are false. Give a reason for your answer.

TABLE 11.19

	Model		
Estimates	$E(Y) = \alpha + \beta X_1$	$E(Y) = \alpha + \beta_1 X_1 + \beta_2 X_2$	$E(Y) = \alpha + \beta_1 X_1 + \beta_2 X_2 + \beta_3 X_3$
Coefficient of X_1	.450	.400	.340
Coefficient of X_2		.003	.002
Coefficient of X_3			.002
R^2	.25	.34	.38

a) The correlation between Y and X_1 is positive.
b) A one-unit increase in X_1 corresponds to a change of .45 in the estimated mean of Y, controlling for X_2 and X_3.
c) The value of SSE increases as we add additional variables to the model.

d) It follows from the sizes of the estimates for the third model that X_1 has the strongest partial effect on Y.
e) The value of $r^2_{YX_3}$ is .40.
f) The partial correlation $r_{YX_1 \cdot X_2}$ is positive.
g) The partial correlation $r_{YX_1 \cdot X_3}$ could be negative.
h) Controlling for X_1, a 100-unit increase in X_2 corresponds to a predicted increase of .3 in college GPA.
i) For the first model, the estimated standardized regression coefficient equals .50.

37. In regression analysis, which of the following statements must be false? Why?
a) For the model $E(Y) = \alpha + \beta_1 X_1$, Y is significantly related to X_1 at the .05 level, but when X_2 is added to the model, Y is not significantly related to X_1 at the .05 level.
b) The estimated coefficient of X_1 is positive in the bivariate model, but negative in the multiple regression model.
c) When the model is refitted after Y is multiplied by 10, R^2, r_{YX_1}, $r_{YX_1 \cdot X_2}$, b_1^*, the F statistics and t statistics do not change.
d) $r_{YX_2 \cdot X_1}$ cannot exceed r_{YX_2}.
e) The F statistic for testing that all the regression coefficients equal 0 has $P < .05$, but none of the individual t tests have $P < .05$.
f) If you compute the standardized regression coefficient for a bivariate model, you always get the Pearson correlation.
g) $r^2_{YX_1} = r^2_{YX_2} = .6$ and $R^2 = .6$.
h) $r^2_{YX_1} = r^2_{YX_2} = .6$ and $R^2 = 1.2$.
i) The Pearson correlation between Y and $\hat{Y}$ equals $-.10$.
j) If X_3 is added to a model already containing X_1 and X_2, then if the prediction equation has $b_3 = 0$, R^2 stays the same.
k) For every F test, there is an equivalent test using the t distribution.

For Problems 11.38–11.42, select the correct answer(s) and indicate why the other responses are inappropriate.

38. If $\hat{Y} = 2 + 3X_1 + 5X_2 - 8X_3$, then controlling for X_2 and X_3, the predicted mean change in Y when X_1 is increased from 10 to 20 equals
a) 3 **b)** 30 **c)** .3 **d)** Cannot be given—depends on specific values of X_2 and X_3.

39. If $\hat{Y} = 2 + 3X_1 + 5X_2 - 8X_3$,
a) The strongest Pearson correlation is between Y and X_3.
b) The variable with the strongest partial influence on Y is X_2.
c) The variable with the strongest partial influence on Y is X_3, but one cannot tell from this equation which pair has the strongest Pearson correlation.
d) None of the above.

40. If $\hat{Y} = 2 + 3X_1 + 5X_2 - 8X_3$,
a) $r_{YX_3} < 0$
b) $r_{YX_3 \cdot X_1} < 0$
c) $r_{YX_3 \cdot X_1, X_2} < 0$
d) Insufficient information to answer.
e) Answers (a), (b), and (c) are all correct.

41. If $\hat{Y} = 2 + 3X_1 + 5X_2 - 8X_3$, and H_0: $\beta_3 = 0$ is rejected at the .05 level, then
a) H_0: $\rho_{YX_3 \cdot X_1, X_2} = 0$ is rejected at the .05 level.

b) H_0: $\rho_{YX_3} = 0$ is rejected at the .05 level.
c) $r_{YX_3 \cdot X_1, X_2} > 0$

42. The F test for comparing a complete model to a reduced model
a) Can be used to test the significance of a single regression parameter in a multiple regression model.
b) Can be used to test H_0: $\beta_1 = \cdots = \beta_k = 0$ in a multiple regression equation.
c) Can be used to test H_0: No second-order interaction, in the model

$$E(Y) = \alpha + \beta_1 X_1 + \beta_2 X_2 + \beta_3 X_3 + \beta_4 X_1 X_2 + \beta_5 X_1 X_3 + \beta_6 X_2 X_3$$

d) Can be used to test whether the model $E(Y) = \alpha + \beta_1 X_1 + \beta_2 X_2$ gives a significantly better fit than the model $E(Y) = \alpha + \beta_1 X_1 + \beta_2 X_3$.

43. Explain the difference in the purposes of the Pearson correlation, the multiple correlation, and the partial correlation.

44. Let Y = height, X_1 = length of right leg, X_2 = length of left leg. Describe what you expect for the relative sizes of the three pairwise Pearson correlations, R, and $r_{YX_2 \cdot X_1}$.

45. Give an example of three variables for which you expect $\beta \neq 0$ in the model $E(Y) = \alpha + \beta X_1$ but $\beta_1 = 0$ in the model $E(Y) = \alpha + \beta_1 X_1 + \beta_2 X_2$.

46. For the models $E(Y) = \alpha + \beta X$ and $E(Y) = \alpha + \beta_1 X_1 + \beta_2 X_2$, express null hypotheses in terms of correlations that are equivalent to the following:
a) H_0: $\beta = 0$
b) H_0: $\beta_1 = \beta_2 = 0$
c) H_0: $\beta_2 = 0$

47. * Whenever X_1 and X_2 are uncorrelated, then R^2 for the model $E(Y) = \alpha + \beta_1 X_1 + \beta_2 X_2$ satisfies $R^2 = r^2_{YX_1} + r^2_{YX_2}$. In this case, draw a picture that portrays the variability in Y, the part of that variability explained by each of X_1 and X_2, and the total variability explained by both of them together.

48. * Which of the following sets of correlations would you expect to yield the highest R^2 value? Why?
a) $r_{YX_1} = .4, \quad r_{YX_2} = .4, \quad r_{X_1X_2} = .0$
b) $r_{YX_1} = .4, \quad r_{YX_2} = .4, \quad r_{X_1X_2} = .5$
c) $r_{YX_1} = .4, \quad r_{YX_2} = .4, \quad r_{X_1X_2} = 1.0$

49. Suppose the Pearson correlation between Y and X_1 equals the multiple correlation between Y and X_1 and X_2. What does this imply about the partial correlation $r_{YX_2 \cdot X_1}$? Interpret.

50. * Let SSE_1 represent the sum of squared errors for model $E(Y) = \alpha + \beta X_1$, let SSE_{12} represent the sum of squared errors for model $E(Y) = \alpha + \beta_1 X_1 + \beta_2 X_2$, and so forth. Software reports four types of sums of squares in multiple regression models. For our purposes, the most important are the **Type I** (sometimes called *sequential*) sum of squares and the **Type III** (sometimes called *partial*) sum of squares. For a particular variable, the Type I sum of squares represents the variability explained by that variable, controlling for variables previously entered into the model, whereas the Type III sum of squares represents the variability explained by that variable, controlling for all other variables in the model. For the model with three predictors, the model sum of squares partitions into the Type I sum of squares for X_1 (i.e., the variability explained by X_1 in the bivariate model), the Type I sum of squares for X_2 controlling for X_1, and the Type I sum of squares for X_3 controlling for X_1 and X_2; by contrast, the Type III sum of squares for X_1 is the variability explained by X_1, controlling for X_2 and X_3.

a) Report the Type I and Type III sums of squares for each predictor in the model for mental impairment in Example 11.2.
b) For any multiple regression model, explain why the Type I sum of squares for X_1 is the model sum of squares for the bivariate model with X_1 as the predictor, whereas the Type I sum of squares for X_2 equals $\text{SSE}_1 - \text{SSE}_{12}$.
c) Explain why the Type I sum of squares for the last variable entered into a model is the same as the Type III sum of squares for that variable.
d) Using a diagram similar to that used in Figure 11.11, except letting the scale take on a maximum value of TSS instead of 1, show that

$$r^2_{YX_2 \cdot X_1} = \frac{\text{SSE}_1 - \text{SSE}_{12}}{\text{SSE}_1}$$

(*Hint*: SSE corresponds to unexplained variation, so SSE_1 and SSE_{12} correspond in length to $1 - r^2_{YX_1}$ and $1 - R^2$, respectively, in Figure 11.11.)

51. * The sample value of R^2 tends to be slightly biased upwards (i.e., it tends to overestimate the population value $\mathbf{P}^2$), because the sample data fall closer to the sample prediction equation than to the true population regression equation. This bias is greater if n is small or the number of predictors k is large. A somewhat better estimate is ***adjusted*** R^2,

$$R^2_{\text{adj}} = 1 - \frac{\hat{\sigma}^2}{s_Y^2} = R^2 - \left[\frac{k}{n - (k+1)}\right](1 - R^2)$$

where $\hat{\sigma}^2$ is the estimated conditional variance (i.e., the mean square error) and s_Y^2 is the sample variance of Y.
a) Suppose $R^2 = .339$ for a model with $k = 2$ explanatory variables (such as in Table 11.5). Calculate R^2_{adj} for the following sample sizes: 10, 40 (as in the text example), 100, and 1000. Show that R^2_{adj} approaches R^2 in value as n increases.
b) Show that R^2_{adj} is negative when $R^2 < k/(n-1)$. This is undesireable, and it is equated to 0 in such cases.
c) Explain why, unlike R^2, R^2_{adj} could decrease when we add an explanatory variable to a model.

52. * Let $R^2_{Y(X_1,\ldots X_k)}$ denote R^2 for the multiple regression model with k explanatory variables. Explain why the square of the partial correlation between Y and X_k, controlling for $X_1, X_2, \ldots, X_{k-1}$,equals

$$r_{YX_k \cdot X_1,\ldots,X_{k-1}} = \frac{R^2_{Y(X_1,\ldots,X_k)} - R^2_{Y(X_1,\ldots,X_{k-1})}}{1 - R_{Y(X_1,\ldots,X_{k-1})}}$$

53. * For large samples, a confidence interval for the population partial correlation $\rho_{YX_i \cdot}$— uses the transformation shown in Problem 9.46 for the Pearson correlation. The standard deviation of the transformed statistic equals $\sigma_T = 1/\sqrt{n - 3 - c}$, where c is the number of controlled variables. For the sample partial correlation, we obtain the transformed statistic T and form the confidence interval $T \pm z\sigma_T$. We then perform the inverse transformation with these endpoints to get the corresponding endpoints of the confidence interval for $\rho_{YX_i \cdot}$—. Construct a 95% confidence interval for the partial correlation between mental impairment and life events, controlling for SES, for the data in Table 11.1. Interpret.

54. * The numerator $R^2 - r^2_{YX_1}$ of the squared partial correlation $r^2_{YX_2 \cdot X_1}$ gives the increase in the proportion of explained variation from adding X_2 to the model. This increment, denoted by $r^2_{Y(X_2 \cdot X_1)}$, is called the squared ***semipartial*** (or ***part***) correlation. One can use squared semipartial correlations to partition the variation in the response variable. For instance, for three explanatory variables,

$$
\begin{aligned}
R^2_{Y(X_1,X_2,X_3)} &= r^2_{YX_1} + (R^2_{Y(X_1,X_2)} - r^2_{YX_1}) + (R^2_{Y(X_1,X_2,X_3)} - R^2_{Y(X_1,X_2)}) \\
&= r^2_{YX_1} + r^2_{Y(X_2 \cdot X_1)} + r^2_{Y(X_3 \cdot X_1,X_2)}.
\end{aligned}
$$

The total variation in Y explained by X_1, X_2, and X_3 together partitions into: (i) the proportion explained by X_1 (i.e., $r^2_{YX_1}$), (ii) the proportion explained by X_2 beyond that explained by X_1 (i.e., $r^2_{Y(X_2 \cdot X_1)}$), and (iii) the proportion explained by X_3 beyond that explained by X_1 and X_2 (i.e, $r^2_{Y(X_3 \cdot X_1,X_2)}$). These correlations have the same ordering as the t statistics for testing partial effects, and some researchers use them as indices of importance of the predictors.

a) In Example 11.2, show that $r^2_{Y(X_2 \cdot X_1)} = .20$ and $r^2_{Y(X_1 \cdot X_2)} = 18$. Interpret. Is it wise to conclude which variable is more important?

b) Show that the squared semipartial correlation $r^2_{Y(X_2 \cdot X_1)}$ cannot be larger than the squared partial correlation $r^2_{YX_2 \cdot X_1}$.

55. * The least squares prediction equation provides predicted values $\hat{Y}$ with the strongest possible correlation with Y, out of all possible prediction equations of that form. That is, the least squares equation yields the best prediction of Y in the sense that it represents the linear reduction of $X_1, \ldots, X_k$ to the single variable that is most strongly correlated with Y. Based on this property, explain why the multiple correlation cannot decrease when one adds a variable to a multiple regression model. (*Hint*: The prediction equation for the simpler model is a special case of a prediction equation for the full model that has coefficient 0 for the added variable.)

56. * **a)** Show that the square of the t statistic of Section 9.5 for testing $H_0 : \rho = 0$ about the population Pearson correlation is a special case of the F statistic of Section 11.4 for testing $H_0 : \beta_1 = \ldots = \beta_k = 0$ (equivalently, population multiple correlation $\mathbf{P} = 0$).

b) Show that the F test statistic of Section 11.4 for testing $H_0 : \beta_1 = \ldots = \beta_k = 0$ is itself a special case of the F test statistic of Section 11.6 that uses R^2 values to compare complete and reduced models. (*Hint*: the reduced model has $g = 0$ explanatory variables.)

57. * Let $\bar{b}^*_i$ denote the estimated standardized regression coefficient when X_i is treated as the *response* variable and Y as an *explanatory* variable, controlling for the same set of other variables. Then, $\bar{b}^*_i$ need not equal b^*_i. The partial correlation between Y and X_i, which *is* symmetric in the order of the two variables, satisfies

$$r^2_{YX_i \cdot —} = b^*_i \bar{b}^*_i$$

a) From this formula, explain why the partial correlation must fall between b^*_i and $\bar{b}^*_i$. (Note: When $a = \sqrt{bc}$, a is said to be the *geometric average* of b and c.)

b) Even though b^*_i does not necessarily fall between -1 and $+1$, explain why $b^*_i \bar{b}^*_i$ cannot exceed 1.

Bibliography

Draper, N. R. and Smith, H. (1981). *Applied Regression Analysis*, 2nd ed. New York: Wiley.

Holzer, C. E., III (1977). *The Impact of Life Events on Psychiatric Symptomatology*. Ph.D. dissertation, University of Florida, Gainesville.

Neter, J., Kutner, M.H., Nachtsheim, C. J., and Wasserman, W. (1996). *Applied Linear Regression Models*, 3rd ed. Homewood, IL.: Richard D. Irwin.

Weisberg, S. (1985). *Applied Linear Regression*, 2nd ed. New York: Wiley.

Younger, M. S. (1985). *A First Course in Linear Regression*, 2nd ed. Boston: Duxbury Press.

Chapter 12

Comparing Groups: Analysis of Variance Methods

Chapter 7 presented methods for comparing means of two groups. We next extend those methods to the comparison of means for *several* groups and to the comparison of means while controlling for another variable.

Chapter 8 presented methods of analyzing association between two qualitative variables. Chapters 9 and 11 presented regression methods that analyze association between quantitative variables. The methods of the current chapter refer to a quantitative response variable and a qualitative explanatory variable. The categories of the qualitative explanatory variable are the different groups compared. The comparison refers to means of the quantitative response variable. For example, in comparing men and women on their mean annual income, we are analyzing a quantitative response variable, annual income, and a qualitative explanatory variable, gender.

The method of comparing means of the various groups is called the *analysis of variance*. Section 12.1 introduces the basic test, which uses the F distribution. For estimation, Section 12.2 shows how to construct confidence intervals comparing pairs of group means. Section 12.3 shows that the method for comparing means is a special case of a multiple regression analysis. Qualitative explanatory variables appear in the model as a set of *dummy variables*. *Two-way analysis of variance*, introduced in Sections 12.4 and 12.5, corresponds to multiple regression with two qualitative explanatory variables, such as in modeling annual income using both gender and race. Section 12.6 presents methods for comparing means with dependent samples resulting from repeated measurements on the same sample of subjects. Section 12.7 extends these analyses to two explanatory variables. The final section discusses assumptions for analysis of variance methods.

12.1 Comparing Several Means: One-Way Analysis of Variance

For quantitative response variables, one of the most common analyses refers to comparing the mean responses of several groups. Section 7.3 introduced a method using the t distribution for comparing means of *two* groups. This section introduces an analogous method, called ***analysis of variance***, for simultaneously comparing means of *several* groups. R. A. Fisher developed this method in the 1920s, primarily for data from agricultural experiments. The heart of this analysis is a significance test, using his F distribution, for detecting evidence of differences among the population means.

Assumptions and Notation

In the following formulas, g denotes the number of groups. The means of the response variable for the g populations are $\mu_1, \mu_2, \ldots, \mu_g$. The analysis of variance is an F test of H_0: $\mu_1 = \mu_2 = \cdots = \mu_g$ against the alternative hypothesis H_a: at least two means are unequal.

We denote the group sample sizes by $n_1, n_2, \ldots, n_g$ and the total sample size by $N = n_1 + n_2 + \cdots + n_g$. The sample means are $\bar{Y}_1, \bar{Y}_2, \ldots, \bar{Y}_g$, and the sample standard deviations are $s_1, s_2, \ldots, s_g$.

The assumptions for the test are as follows:

- The population distributions on the response variable for the g groups are normal.
- The standard deviations of the population distributions for the g groups are equal. We denote the common value by σ.
- Independent random samples are selected from the g populations.

Figure 12.1 portrays the first two assumptions.

Example 12.1 Political Ideology by Party Identification

Table 12.1 summarizes observations on political ideology for three groups, based on data from a recent General Social Survey. The three groups are the (Democrat, Independent, Republican) categories of the explanatory variable, party identification (ID). Political ideology, the response variable, is measured on a seven-point scale, ranging from extremely liberal to extremely conservative. For each party ID, Table 12.1 shows the number of subjects who made each response. For instance, of 340 Democrats, 11 responded extremely liberal, 50 responded liberal, and so forth.

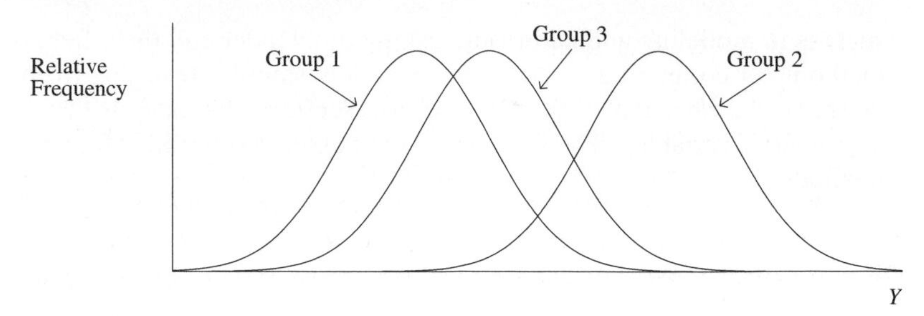

Figure 12.1 Assumptions About Population Distributions: Normal with Equal Standard Deviations, σ

Since Table 12.1 displays the data as counts in a contingency table, one might be tempted to use the methods of Chapter 8 for qualitative data. The chi-squared test treats both variables as nominal, however, whereas political ideology is ordinal. That test is not directed toward detecting whether responses have a higher or lower mean in some groups than others. Likewise, the ordinal analyses presented in that chapter are inappropriate, since they require both variables to be ordinal, whereas the groups (categories of party ID) are nominal.

TABLE 12.1 Political Ideology by Party Identification

	Political Ideology							Sample		Standard
Group (Party)	1	2	3	4	5	6	7	Size	Mean	Deviation
Democrat	11	50	60	139	35	39	6	340	3.82	1.32
Independent	8	33	47	142	37	40	6	313	3.99	1.27
Republican	2	19	30	99	65	61	14	290	4.53	1.28

Note: 1 = extremely liberal, 2 = liberal, 3 = slightly liberal, 4 = moderate, 5 = slightly conservative, 6 = conservative, 7 = extremely conservative.

Section 15.6 presents a model for an ordinal response with qualitative or quantitative predictors. When an ordinal response has several categories, however, it is common in practice to assign scores to its levels and treat it as a quantitative variable. This is a reasonable strategy when primary interest focuses on an overall measure of central tendency such as the mean rather than the proportions in particular categories. For Table 12.1, for instance, inquiry would probably focus on how liberal or conservative the responses tend to be for each group, in some average sense, rather than on the proportions falling in specific categories. We analyze these data by assigning the scores (1, 2, 3, 4, 5, 6, 7) to the levels of political ideology and comparing means. The higher the mean score, the more conservative the group's responses tended to be.

For these scores, Table 12.1 also shows the mean and standard deviation for each group. The sample means all fall relatively close to 4.0, the score for moderate ideology. We shall test whether the populations that these three samples represent have equal means. The null hypothesis is H_0: $\mu_1 = \mu_2 = \mu_3$, where μ_i is the population mean for party ID i. □

Variability Between and Within Groups

If the inferential method we shall present refers to population means, why is it called analysis of *variance*? The reason is that the test statistic uses two estimates of variance. It compares the variability of the sample means $\bar{Y}_i$ about the overall mean $\bar{Y}$ to the variability of the sample observations about their separate means—the observations from the first group about $\bar{Y}_1$, the observations from the second group about $\bar{Y}_2$, and so forth. To illustrate, suppose the sample observations from three groups are as shown in Figure 12.2a. It is clear that the means of the populations these samples represent are unequal. The basis for this conclusion is that the variability *between* sample means is large and the variability of the observations *within* each sample is small.

By contrast, in Figure 12.2b, the variability *within* the groups is also relatively large; it is not clear whether the population means differ, even though the sample means are the same as in Figure 12.2a. The greater the variability between sample means and the smaller the variability within each group of sample observations, the stronger the evidence that the null hypothesis of equal means is false.

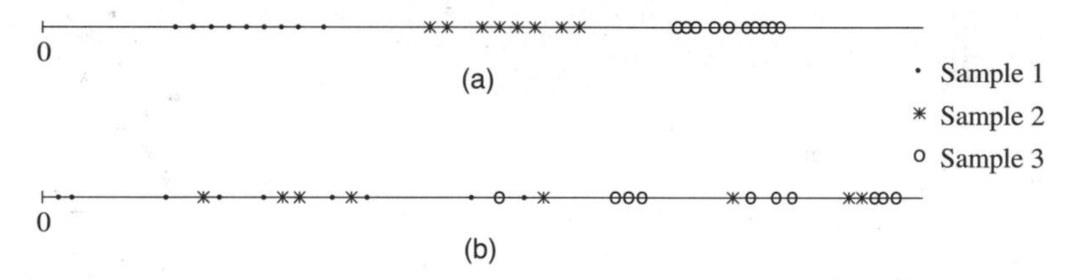

Figure 12.2 Two Sets of Samples: Variability within groups is larger in second set.

The test statistic is the ratio of two estimates of the population variance, σ^2, of the measurements in the groups. One of these uses the variability *within* each sample. This estimate is called the ***within-groups estimate*** of the variance, or simply the ***within estimate*** for short. It has good properties as an estimate, regardless of whether H_0: $\mu_1 = \mu_2 = \cdots = \mu_g$ is true.

The second estimate uses the variability *between* each sample mean $\bar{Y}_i$ and the overall sample mean $\bar{Y}$. It is called the ***between-groups estimate*** of the variance, or simply the ***between estimate*** for short. If H_0 is true, the between estimate tends to be good, about the same value as the within estimate, apart from sampling error. If H_0 is false, however, the between estimate tends to overestimate σ^2.

Within Estimate of Variance

The within estimate of the variance pools together the sums of squares of the observations about their means and divides by a degrees of freedom term. Now, for the n_1 observations from the first group, $\sum(Y - \bar{Y}_1)^2$ is the sum of squares of the observations about their mean. This sum of squares has $n_1 - 1$ degrees of freedom, the denominator of this sum for forming the sample variance s_1^2 for group 1. Similarly, for the n_2 observations from the second group, $\sum(Y - \bar{Y}_2)^2$ is the sum of squares of the observations about their sample mean, with $n_2 - 1$ degrees of freedom. The sum of these sum of squares terms for the g samples is called the ***within sum of squares***, since the sums of squares are calculated *within* each sample. Denote this sum by WSS.

The within sum of squares WSS has degrees of freedom equal to the sum of the *df* values of the component parts: $(n_1 - 1) + (n_2 - 1) + \cdots + (n_g - 1) = (n_1 + n_2 + \cdots + n_g) - g = N - g$. The ratio

$$\hat{\sigma}^2 = \frac{\text{WSS}}{df} = \frac{\text{WSS}}{N - g}$$

is the within estimate of the population variance σ^2 of the g groups.

This estimate summarizes information about variability from the separate samples. The estimate of σ^2 using only the first group is

$$s_1^2 = \frac{\sum(Y - \bar{Y}_1)^2}{n_1 - 1}$$

In Table 12.2, for example, this is the square of the reported standard deviation, $s_1 = 1.32$. Similarly, the sample variance for the second group is $s_2^2 = \sum(Y - \bar{Y}_2)^2/(n_2 - 1)$, and so forth for the remaining groups. Under the assumption that the population variances are identical, these terms all estimate the same parameter, σ^2. The numerator and denominator of the within estimate $\hat{\sigma}^2$ pool the information from these g estimates by adding their numerators and adding their denominators. The resulting estimate relates to the separate sample variances by

$$\hat{\sigma}^2 = \frac{(n_1 - 1)s_1^2 + (n_2 - 1)s_2^2 + \ldots + (n_g - 1)s_g^2}{N - g}$$

This estimate is a weighted average of the separate sample variances, with greater weight given to larger samples. It has good properties (e.g., unbiasedness, efficient), regardless of the values of $\mu_1, \mu_2, \ldots, \mu_g$.

We illustrate with the political ideology data in Table 12.1. We could calculate WSS by calculating the within sum of squares for each sample from the raw data, but it is simpler to use the formula just given based on the sample variances, or better yet, use computer software. The sample sizes for the $g = 3$ groups are $n_1 = 340$, $n_2 = 313$, $n_3 = 290$, for a total sample size of $N = 943$. Using the sample standard deviations from Table 12.1,

$$\hat{\sigma}^2 = \frac{(340 - 1)(1.32)^2 + (313 - 1)(1.27)^2 + (290 - 1)(1.28)^2}{943 - 3} = \frac{1570.8}{940} = 1.67$$

In summary, the within sum of squares equals WSS = 1570.8, with $df = 940$, providing a within-groups variance estimate of 1.67. The standard deviation estimate is $\hat{\sigma} = \sqrt{1.67} = 1.29$, a value that summarizes the three sample standard deviations from Table 12.1.

Between Estimate of Variance

The estimate of σ^2 based on variability between each sample mean and the overall mean equals

$$\frac{\sum_i n_i(\bar{Y}_i - \bar{Y})^2}{g - 1}$$

(Problem 12.51 motivates this formula.) The numerator of this estimate is called the ***between sum of squares*** and is denoted by BSS. It represents the sum of squares of the differences between each $\bar{Y}_i$ and the overall mean $\bar{Y}$, where each squared difference is weighted by the sample size upon which it is based. The ratio of BSS to its degrees of freedom, $g - 1$, is the between groups estimate of the variance σ^2.

This second estimate is unbiased only if H_0 is true. When the population means are unequal, the $\bar{Y}_i$ values tend to be more disperse than if the population means are equal. Thus, when H_0: $\mu_1 = \cdots = \mu_g$ is false, the between sum of squares tends to be inflated and the between estimate tends to be considerably larger than the within estimate $\hat{\sigma}^2$.

For the sample means in Table 12.1, the overall mean $\bar{Y} = 4.10$. The between sum of squares equals

$$\begin{aligned} \text{BSS} &= \sum_i n_i \left(\bar{Y}_i - \bar{Y}\right)^2 = n_1 \left(\bar{Y}_1 - \bar{Y}\right)^2 + n_2 \left(\bar{Y}_2 - \bar{Y}\right)^2 + n_3 \left(\bar{Y}_3 - \bar{Y}\right)^2 \\ &= 340(3.82 - 4.10)^2 + 313(3.99 - 4.10)^2 + 290(4.53 - 4.10)^2 \\ &= 85.38 \end{aligned}$$

This sum of squares has $df = g - 1 =$ number of groups $- 1 = 3 - 1 = 2$. The between estimate of the variance is

$$\frac{\text{BSS}}{g - 1} = \frac{85.38}{2} = 42.69$$

F Test Statistic

The F test statistic for H_0: $\mu_1 = \cdots = \mu_g$ is the ratio of the between estimate to the within estimate,

$$F = \frac{\text{Between estimate}}{\text{Within estimate}} = \frac{\text{BSS}/(g - 1)}{\text{WSS}/(N - g)}$$

This is called the ***analysis of variance F statistic***, or ***ANOVA F statistic*** for short. When H_0 is false, the between estimate tends to overestimate σ^2, so it tends to be larger than the within estimate. Then, the F test statistic tends to be considerably larger than 1.0.

When H_0 is true, this F test statistic has the F sampling distribution with $df_1 = g - 1$ and $df_2 = N - g$. The two df terms are the denominators of the between estimate and the within estimate. As in the F test for the parameters of a multiple regression model (Section 11.4), the P-value is the probability that the F test statistic is at least as large as the observed F value; that is, it the right-hand tail probability. The larger the F test statistic, the smaller the P-value.

For Table 12.1, the between-groups estimate of σ^2 equals 42.69 and the within-groups estimate equals 1.67. The F test statistic equals

$$F = \frac{\text{BSS}/(g-1)}{\text{WSS}/(N-g)} = \frac{85.38/2}{1570.84/940} = \frac{42.69}{1.67} = 25.6$$

In other words, the between estimate is over 25 times the within estimate. This statistic has $df_1 = g - 1 = 2$ and $df_2 = N - g = 940$. From Table D, the P-value is $P < .001$. This provides extremely strong evidence against H_0: $\mu_1 = \mu_2 = \mu_3$. We conclude that a difference exists among the true mean political ideology values for the three parties.

ANOVA Table

Computer software displays the results of ANOVA F tests in a table similar to the one used to summarize sums of squares in regression analysis. This table is called an ***ANOVA table***. Table 12.2 shows the basic format, illustrating for the test just conducted. The sums of squares divided by their degrees of freedom are called ***mean squares***. The two mean squares are the between-groups and within-groups estimates of the population variance σ^2.

In the "Between" row of the ANOVA table, BSS divided by df_1 is the between mean square BSS/df_1 = 85.38/2 = 42.69. In the "Within" row, WSS divided by df_2 is the within mean square WSS/df_2 = 1570.84/940 = 1.67. The ratio of the two mean squares is the F test statistic, $F = 42.69/1.67 = 25.6$, which has P-value $P < .0001$.

TABLE 12.2 ANOVA Table for Result of F Test for Table 12.1

Source	Sum of Squares	DF	Mean Square	F	Prob > F
Between	85.38	2	42.69	25.6	$P < .0001$
Within	1570.84	940	1.67		
Total	1656.22	942			

The sum of BSS and WSS is called the ***total sum of squares***, denoted by TSS. In fact, this equals

$$TSS = \sum (Y - \bar{Y})^2 = BSS + WSS$$

the sum of squares of the combined sample of N observations about the overall mean, $\bar{Y}$.

The analysis of variance test partitions the total variability about the overall mean, TSS, into two independent parts. One part, BSS, is the portion of the total explained by the differences among the group means. This is also called the *group sum of squares*, and most software replaces the "Between" label in Table 12.2 by the name of the group variable (e.g., PARTY ID). The other part, WSS, is the portion of the total that cannot be explained by the differences among the groups. It represents the variability that remains after classifying the observations into separate groups. The within sum of squares is also called the *error sum of squares*, and most software replaces the "Within" label in

Table 12.2 by "Error." Section 12.3 explains the analogy between these sums of squares and the model and error sums of squares in regression analysis.

The assumption that the population distributions are normal with identical standard deviations is a stringent one that is never satisfied exactly in practice. The last section of this chapter discusses the effects of violating these assumptions.

12.2 Multiple Comparisons of Means

The analysis of variance test of H_0: $\mu_1 = \mu_2 = \cdots = \mu_g$ is a global test of independence of the response and explanatory variables, just like the test of H_0: $\beta_1 = \cdots = \beta_k = 0$ for multiple regression models or chi-squared tests for contingency tables. The conclusion of the test does not specify which means are different or how different they are. More detailed inferences are necessary to determine the nature of the differences. Even if the P-value is not small for that test, it still is informative to estimate just how large differences in means could plausibly be.

Confidence Intervals Comparing Means

In practice, it is more informative to estimate the population means than merely to test whether they are all equal. We could, for example, construct a confidence interval for each mean or for each difference between a pair of means.

- A confidence interval for μ_i is

$$\bar{Y}_i \pm t \frac{\hat{\sigma}}{\sqrt{n_i}}$$

  In this formula, $\hat{\sigma} = \sqrt{\text{WSS}/(N-g)}$ is the square root of the within estimate of σ^2 used in the denominator of the ANOVA F test statistic, and the t-value from the t table for the chosen confidence level is based on the degrees of freedom for that estimate, $df = N - g$.
- A confidence interval for $\mu_i - \mu_j$ is

$$(\bar{Y}_i - \bar{Y}_j) \pm t\hat{\sigma}\sqrt{\frac{1}{n_i} + \frac{1}{n_j}}$$

  In this formula, as in the previous one, $df = N - g$ for the tabled t-value. Evidence exists of a difference between μ_i and μ_j when the interval does not contain 0. For the special case of $g = 2$ groups, $df = N - g = (n_1 + n_2 - 2)$, and this confidence interval is identical to the one introduced in Section 7.3 for $\mu_2 - \mu_1$.

Example 12.2 Comparing Mean Ideology of Democrats and Republicans

For Table 12.1, let's compare population mean ideology of Democrats (group 1) and Republicans (group 3). From Table 12.1, $\bar{Y}_1 = 3.82$ for $n_1 = 340$ Democrats and

$\bar{Y}_3 = 4.53$ for $n_3 = 290$ Republicans. From Table 12.2, the within estimate of the standard deviation is $\hat{\sigma} = \sqrt{1.67} = 1.29$, with $df = 940$. For a 95% confidence interval with $df = 940$, the t-score coincides with the z-score, 1.96. The interval for $\mu_3 - \mu_1$ is

$$(\bar{Y}_3 - \bar{Y}_1) \pm t\hat{\sigma}\sqrt{\frac{1}{n_1} + \frac{1}{n_3}} = (4.53 - 3.82) \pm 1.96(1.29)\sqrt{\frac{1}{340} + \frac{1}{290}}$$

$$= .72 \pm .20 \quad \text{or} \quad (.51, .92)$$

We infer that population mean ideology was between .5 and .9 units higher for Republicans than for Democrats. Since the interval contains only positive numbers, we conclude that $\mu_3 - \mu_1 > 0$; that is, μ_3 exceeds μ_1. On the average, Republicans were more conservative than Democrats. □

Dangers of Forming Large Numbers of Confidence Intervals

When the number of groups g is relatively large, the number of pairwise comparisons, which equals $g(g - 1)/2$, can be very large. Some pairs of means may appear to be different *even if all of the population means are equal.* When $g = 10$, for example, one can compare $g(g-1)/2 = 45$ pairs of means. Suppose we form a 95% confidence interval for the difference between each pair. Then, the error probability of .05 applies for each comparison. For the 45 comparisons, on the average $45(.05) = 2.25$ of the confidence intervals would not contain the true differences of means.

For 95% confidence intervals, the error probability of .05 is the probability that any particular confidence interval does not contain the true difference in population means. When we form a large number of confidence intervals, the probability that *at least* one confidence interval is in error is considerably larger than the error probability for any particular interval. The larger the number of groups being compared, the greater is the chance of at least one incorrect inference. In particular, suppose all the true means are equal. If we simply scan the sample means and select the largest and smallest sample means for a confidence interval, the probability may be quite large that the interval will not contain 0.

When we construct confidence intervals comparing all pairs of means, the probability that at least one interval is in error, not containing the *true* difference in means, is called the ***multiple comparison error rate***. We next show how to control this error rate at an admissibly small level.

Bonferroni Multiple Comparisons

When we plan a large number of pairwise comparisons, methods are available that control the probability that *all* intervals contain the true differences. Such intervals are called ***simultaneous confidence intervals***, since all intervals contain the true parameters *simultaneously* with an overall fixed probability. Procedures yielding a set of simultaneous confidence intervals are called ***multiple comparison*** techniques.

For example, when there are $g = 10$ means, the set of 95% simultaneous confidence intervals has probability .95 that *all* 45 of the intervals contain the pairwise differences $\mu_i - \mu_j$. Equivalently, the probability that *at least one* interval is in error is controlled at the level .05. This is the multiple comparison error rate.

There are several multiple comparison techniques. We present ***Bonferroni confidence intervals***, since this approach is simple and applies to a wide variety of situations. The Bonferroni approach uses the same formulas for confidence intervals introduced at the beginning of this section. However, it uses a more stringent confidence level for each interval, to ensure that the overall confidence level is acceptably high.

To illustrate, suppose we want a multiple comparison error rate of .10, that is, a probability of .90 that all confidence intervals are simultaneously correct. If we plan four comparisons of means, then the Bonferroni method uses error probability .10/4 = .025 for each one; that is, a 97.5% confidence level for each interval. This is a simple but somewhat conservative approach, actually ensuring that the overall error rate is *at most* .10 and that the overall confidence coefficient is *at least* .90. The method is based on an inequality in probability theory, due to C. E. Bonferroni, which states that the probability at least one of a set of events occurs can be no greater than the sum of the separate probabilities of the events. For instance, if the probability of an error for each of four confidence intervals equals $\alpha/4$, then the probability that at least one interval is in error is no greater than $(\alpha/4 + \alpha/4 + \alpha/4 + \alpha/4) = \alpha$.

Example 12.3 Bonferroni Intervals for Political Ideology Comparisons

We refer again to Table 12.1, for which $\hat{\sigma} = 1.29$. For the $g = 3$ groups, we compare $g(g-1)/2 = 3(2)/2 = 3$ pairs of means, μ_1 with μ_2, μ_1 with μ_3, and μ_2 with μ_3. We construct simultaneous confidence intervals having overall confidence level at least .95. For a multiple comparison error rate of .05 with three comparisons, we use error probability .05/3 = .0167 for each interval. These use the t-score with two-tail probability .0167, or single-tail probability .0083. For the large df value here (df = 940), this is the same as the z-score with this tail probability, which (from Table A) equals 2.39.

For instance, the interval for $\mu_2 - \mu_1$, the difference between the population mean ideology of Independents and Democrats, is

$$\begin{aligned}(\bar{Y}_2 - \bar{Y}_1) \pm t\hat{\sigma}\sqrt{\frac{1}{n_1} + \frac{1}{n_2}} &= (3.99 - 3.82) \pm 2.39(1.29)\sqrt{\frac{1}{340} + \frac{1}{313}} \\ &= .18 \pm .24 \quad \text{or} \quad (-.07, .42)\end{aligned}$$

Since this interval contains 0, it is plausible that $\mu_1 = \mu_2$. We construct the intervals for the other two pairs of means in a similar way. Table 12.3 displays them. All three confidence intervals hold with overall confidence coefficient at least .95. The probability that one or more of them does not contain the true difference is at most .05.

The starred intervals, for $\mu_3 - \mu_1$ and $\mu_3 - \mu_2$, do not contain 0. They show significant evidence of a difference between the true means, namely between Republicans

TABLE 12.3 Bonferroni and Tukey Methods for Multiple Pairwise Comparisons of Mean Political Ideology for Three Party ID Groups

Groups	Difference Of Means $\mu_i - \mu_j$	Estimated Difference $\bar{Y}_i - \bar{Y}_j$	Bonferroni 95% CI	Tukey 95% CI
(Independent, Democrat)	$\mu_2 - \mu_1$	.18	(−.07, .42)	(−.06, .41)
(Republican, Democrat)	$\mu_3 - \mu_1$	.72	(.47, .96)*	(.47, .96)*
(Republican, Independent)	$\mu_3 - \mu_2$	.54	(.29, .79)*	(.29, .79)*

Note: An asterisk * indicates a significant difference.

and each of the other groups. We summarize the three pairwise comparisons by the diagram:

D I R

____

where we ordered the party IDs (D, I, R) by their sample means. A line does not connect Democrats and Republicans or Independents and Republicans. Sufficient evidence exists that Republicans were more conservative than the other two groups. If only the difference between Democrats and Republicans had been significant, the diagram would be

D I R

____

        ____

The Bonferroni simultaneous 95% confidence intervals are wider than the separate 95% confidence intervals described at the beginning of this section. For instance, the ordinary interval comparing Republicans and Democrats is (.51, .92), whereas the Bonferroni interval is (.47, .96). This is because the multiple comparison approach uses a higher confidence level for each separate interval to ensure achieving the simultaneous confidence level. □

Other Multiple Comparison Methods

Many other methods exist for multiple comparisons of means, and most software reports a variety of them. Other popular methods are ***Tukey's method*** and ***Scheffé's method***. Scheffé's method is useful when, in addition to the standard pairwise intervals, one needs more complex comparisons such as between an average of one set of means and an average of another set.

The Tukey method is excellent for comparisons consisting of all the pairs of means. Its intervals are usually slightly narrower than the Bonferroni intervals. The Tukey method is more complex, using a probability distribution (the *Studentized range*) not discussed in this text, so we do not present its formula. However, most software can report intervals for that approach. For instance, Table 12.3 shows them for the political ideology data. The Tukey method is a good, general purpose method to follow-up ANOVA F tests.

12.3 Performing ANOVA by Regression Modeling

Chapter 11 used multiple regression models to analyze the relationship between the mean of a quantitative response variable and a collection of *quantitative* explanatory variables. The analysis of variance (ANOVA) analyzes the relationship between the mean of a quantitative response variable and a *qualitative* explanatory variable, the categories of which are the groups compared.

Regression with Dummy Variables

In fact, ANOVA is also a special case of multiple regression. Artificial explanatory variables in a regression model can represent the categories of the qualitative variable. Each variable takes only two values, 0 and 1, and indicates whether an observation falls in a particular group.

Suppose there are three groups, as in the political ideology example of the previous two sections. The first artificial variable, denoted by Z_1, equals 1 for observations from the first group and equals 0 otherwise. The second, denoted by Z_2, equals 1 for observations from the second group and equals 0 otherwise. That is, $Z_1 = 1$ and $Z_2 = 0$ for observations from the first group, and $Z_1 = 0$ and $Z_2 = 1$ for observations from the second group. When $Z_1 = Z_2 = 0$, the observation is from the third group. No more than one Z variable can be nonzero for a particular observation. The artificial variables Z_1 and Z_2 are called ***dummy variables***.

Table 12.4 summarizes the dummy variables for the three groups. It is unnecessary and redundant to create a dummy variable for the last (third) group, since values of 0 for Z_1 and Z_2 identify observations from it.

The dummy variables identify the group that an observation represents. For example, the combination ($Z_1 = 0$, $Z_2 = 0$) occurs for all subjects in the third group. The dummy variables denote classification, not magnitude, of an observation on the qualitative predictor.

TABLE 12.4 The Two Dummy Variables for Three Groups

Group	Z_1	Z_2
1	1	0
2	0	1
3	0	0

Now, with the dummy variables just defined we use the multiple regression equation

$$E(Y) = \alpha + \beta_1 Z_1 + \beta_2 Z_2$$

For observations from the third group, $Z_1 = Z_2 = 0$, and the equation reduces to

$$E(Y) = \alpha + \beta_1(0) + \beta_2(0) = \alpha$$

The parameter α represents the population mean μ_3 of the response variable for the last group. For observations from the first group, $Z_1 = 1$ and $Z_2 = 0$, so

$$E(Y) = \alpha + \beta_1(1) + \beta_2(0) = \alpha + \beta_1$$

equals the population mean μ_1 for that group. Similarly, $\alpha + \beta_2$ equals the population mean μ_2 for the second group (let $Z_1 = 0$ and $Z_2 = 1$).

Since $\alpha + \beta_1$ equals the mean μ_1 and α equals μ_3, β_1 represents the difference $\mu_1 - \mu_3$. Similarly, $\beta_2 = \mu_2 - \mu_3$. So, the β coefficients of the dummy variables represent differences between the mean of each group and the mean of the group not having its own dummy variable. Table 12.5 summarizes the parameters of the regression model and their correspondence with the three population means.

TABLE 12.5 Interpretation of Coefficients of Dummy Variables in Model $E(Y) = \alpha + \beta_1 Z_1 + \beta_2 Z_2$

Category	Z_1	Z_2	Mean of Y	Interpretation of β
1	1	0	$\mu_1 = \alpha + \beta_1$	$\beta_1 = \mu_1 - \mu_3$
2	0	1	$\mu_2 = \alpha + \beta_2$	$\beta_2 = \mu_2 - \mu_3$
3	0	0	$\mu_3 = \alpha$	

Dummy variable coding works because it allows the population means to take arbitrary values, with no assumed distances between groups. A coding such as $Z = 1$ for group 1, $Z = 2$ for group 2, and $Z = 3$ for group 3 would not work, because the model $E(Y) = \alpha + \beta Z$ would then assume an ordering as well as equal distances between groups. It treats the qualitative variable as if it were quantitative, which is improper. Whereas it takes only one term in a regression model to represent the linear effect of a quantitative explanatory variable, it requires $g - 1$ terms to represent the g categories of a qualitative variable.

Regression for Comparing Means

For three groups, the null hypothesis in the ANOVA F test is H_0: $\mu_1 = \mu_2 = \mu_3$. If H_0 is true, then $\mu_1 - \mu_3 = 0$ and $\mu_2 - \mu_3 = 0$. Since $\mu_1 - \mu_3 = \beta_1$ and $\mu_2 - \mu_3 = \beta_2$ in the regression model $E(Y) = \alpha + \beta_1 Z_1 + \beta_2 Z_2$ with dummy variables, the ANOVA null hypothesis H_0: $\mu_1 = \mu_2 = \mu_3$ is equivalent to H_0: $\beta_1 = \beta_2 = 0$ in that model. If all β-values in the model equal 0, then the mean of the response variable equals α

for each group. By setting up dummy variables, then, we can perform the ANOVA test using the F test of H_0: $\beta_1 = \beta_2 = 0$ for this multiple regression model.

For sample data, the prediction equation $\hat{Y} = a + b_1 Z_1 + b_2 Z_2$ estimates the regression equation. The estimates a, b_1, and b_2 in this equation relate to the sample means in the same manner that the regression parameters relate to the population means. Just as $\alpha = \mu_3$, so does $a = \bar{Y}_3$. Similarly, $b_1 = \bar{Y}_1 - \bar{Y}_3$ and $b_2 = \bar{Y}_2 - \bar{Y}_3$.

The assumption from regression analysis that the conditional distributions of Y about the regression equation are normal with constant standard deviation implies here that the population distributions for the groups are normal, with the same standard deviation for each group. These are precisely the assumptions for the ANOVA F test.

Example 12.4 Regression Analysis of Political Ideology and Party ID

For Table 12.1, the qualitative group variable ("Party ID") has three categories. The regression model for the ANOVA procedure with Y = political ideology is

$$E(Y) = \alpha + \beta_1 Z_1 + \beta_2 Z_2$$

The dummy variables satisfy only $Z_1 = 1$ for Democrats, only $Z_2 = 1$ for Independents, and $Z_1 = Z_2 = 0$ for Republicans. Table 12.6 shows a portion of a printout for fitting this regression model using SAS (PROC GLM).

TABLE 12.6 Printout for Regression Model $E(Y) = \alpha + \beta_1 Z_1 + \beta_2 Z_2$ with Data on Political Ideology and Party ID

```
Dependent Variable: IDEOLOGY
                         Sum of         Mean
Source        DF        Squares       Square       F Value      Pr > F
Model          2         85.382       42.691        25.55       0.0001
Error        940       1570.837        1.671
Total        942       1656.218

                                     T for H0:     Pr > |T|     Std Error of
Parameter           Estimate       Parameter=0                    Estimate
INTERCEPT              4.534          59.73         0.0001         0.0759
PARTY          1      -0.717          -6.94         0.0001         0.1033
               2      -0.541          -5.13         0.0001         0.1054
               3       0.000            .             .              .
```

The prediction equation is $\hat{Y} = 4.53 - .72Z_1 - .54Z_2$. No dummy variable estimate appears in Table 12.6 for Party 3 (Republicans), since it is redundant to include a dummy variable for the last group. The estimated mean ideology for Republicans is 4.53 (let $Z_1 = Z_2 = 0$), simply the corresponding sample mean.

Notice the similarity between the top part of Table 12.6 and the ANOVA table in Table 12.2. The "between sum of squares" in ANOVA is the "model sum of squares" in the regression model, or the variability explained by differences among the means. The "within sum of squares" is the sum of squared errors (SSE), the variability within the groups unexplained by including parameters in the model to account for the differences between the means. The sum of squared errors divided by its degrees of freedom is the mean square error (MSE), which is the within estimate $\hat{\sigma}^2 = 1.67$ of the variance of the observations within the groups. The model mean square is the between estimate.

The ratio of the model mean square to the mean square error is the F test statistic ($F = 25.6$), with $df_1 = 2$ and $df_2 = 940$, for testing H_0: $\beta_1 = \beta_2 = 0$. This hypothesis is equivalent to the hypothesis that the population means are identical for the three party IDs. The regression approach provides the same F statistic as ANOVA in Section 12.1.

□

Often the groups compared are the cells of a cross-classification of two or more qualitative variables. For example, the categories (white men, white women, black men, black women) result from the cross-classification of race and gender. The procedure for comparing the mean of a quantitative response variable across categories of one qualitative variable while controlling for another one is called a ***two-way analysis of variance***.

This chapter has so far studied only ***one-way ANOVA***, that is, a comparison across categories of one qualitative variable. This comparison ignores other variables. The computations for the two-way (or higher-order) analysis are more complicated. We will conduct such analyses by defining dummy variables and using computer software for the analysis. The rest of this chapter deals with these more complex methods, with special emphasis on two-way ANOVA.

12.4 Two-Way ANOVA

One-way ANOVA analyzes the relationship between the mean of the response variable and the categories of a single qualitative variable. Chapters 10 and 11 showed that such *bivariate* analyses are usually not as informative as multivariate analyses that can control for other variables. The rest of this chapter discusses such analyses for qualitative explanatory and control variables.

When the explanatory variables are qualitative, one can compare the mean of the response variable among the categories of any one of them, controlling for the other(s). To illustrate, we now analyze mean political ideology by categories of party ID and gender. We compare six means defined for the $2 \times 3 = 6$ combinations of categories of the two qualitative predictors, as shown in Table 12.7. The notation μ_{ij} represents the population mean political ideology for gender i and party j. For example, μ_{23} denotes mean political ideology for row 2 and column 3, males who are Republicans.

For this two-way classification of means, one comparison refers to the mean political ideology for the three party IDs, controlling for gender. For females, one compares the means μ_{11}, μ_{12}, and μ_{13} for the three party IDs; for males, one compares the means μ_{21}, μ_{22}, and μ_{23} for the three party IDs. A second comparison refers to the mean po-

TABLE 12.7 A Two-Way Classification of Population Mean Political Ideology by Party Identification and Gender

	Party Identification		
Gender	Democrat	Independent	Republican
Female	μ_{11}	μ_{12}	μ_{13}
Male	μ_{21}	μ_{22}	μ_{23}

litical ideology for males and females, controlling for party ID, by comparing means within each column of the table.

Interaction

Sections 10.3 and 11.5 showed that the study of *interaction* is important whenever we analyze multivariate relationships. An absence of interaction means that the relationship between the response variable and an explanatory variable does not change for different categories of a control variable.

We now illustrate interaction and its absence for two-way classifications of means. Suppose, first, that no interaction exists between gender and party ID in their effects on political ideology. Then, the difference between each pair of party IDs in population mean political ideology is the same for males and females. Also, the difference between females and males in population mean political ideology is the same for each party ID.

Table 12.8 shows a set of population means satisfying a lack of interaction. The difference between males and females in mean political ideology is 1.0 for each party. Similarly, the difference between each pair of parties in mean political ideology is the same for each gender. The difference between Republicans and Democrats, for example, equals 2.0 both for females and for males. Figure 12.3 plots the means for the party ID categories, within each gender. The ordering of categories on the horizontal axis is unimportant, since qualitative variables are nominal. The absence of interaction is indicated by the parallel sequences of points.

TABLE 12.8 Population Means for a Two-Way Classification with No Interaction

Gender	Democrat	Independent	Republican
Female	3.0	3.5	5.0
Male	4.0	4.5	6.0

Table 12.9 and Figure 12.4 show a set of means displaying interaction. The difference between females and males in mean political ideology is -2 for Democrats, 0 for Independents, and $+2$ for Republicans. Here, the difference in means between females and males depends on the party ID. Similarly, the party ID effect on ideology differs for females and males; for females, Republicans are the most conservative, whereas for males, Democrats are the most conservative.

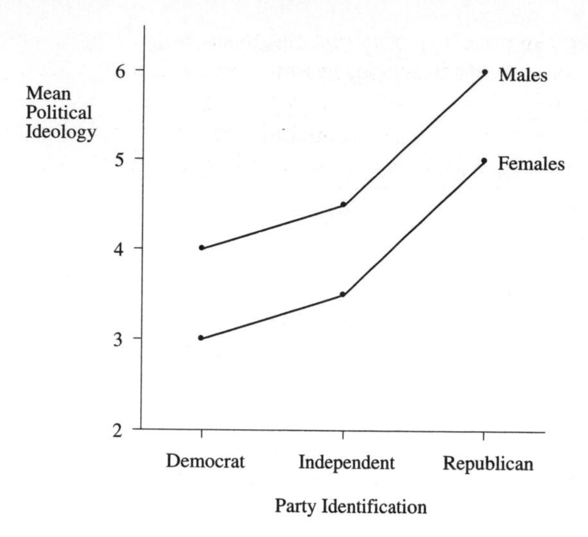

Figure 12.3 Mean Political Ideology, by Party Identification and Gender, Showing No Interaction

TABLE 12.9 Population Means for a Two-Way Classification with Interaction

Gender	Democrat	Independent	Republican
Females	3.0	4.0	5.0
Males	5.0	4.0	3.0

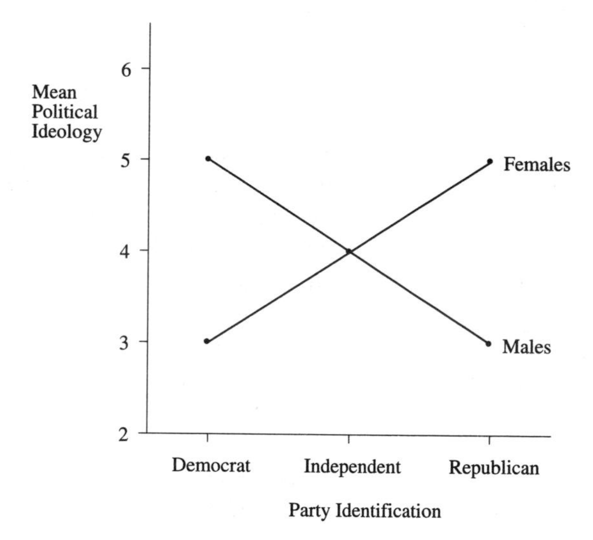

Figure 12.4 Mean Political Ideology, by Party Identification and Gender, Displaying Interaction

In Table 12.9, suppose the numbers of males and females are equal, for each party ID. Then the overall mean political ideology, ignoring gender, is 4.0 for each party. The overall difference in means between any two party IDs equals 0. A one-way analysis of mean political ideology by party ID would conclude that that predictor has no effect. However, a two-way analysis would detect the interaction and the differing effects for males and females.

Hypotheses in Two-Way ANOVA

The null hypothesis in one-way ANOVA states that the population mean for the response variable is identical in each category of the explanatory variable. Two-way ANOVA has three possible null hypotheses. One null hypothesis states that the population means are identical across categories of the first qualitative variable, controlling for the second one. For example, we could test that mean political ideology is identical for the three party IDs, controlling for gender. Table 12.10a displays a set of population means satisfying this null hypothesis.

TABLE 12.10 Population Mean Political Ideology Satisfying Main Effect Null Hypotheses: (a) No Effect of Party Identification, (b) No Effect of Gender

		Party Identification		
Table	Gender	Democrat	Independent	Republican
(a)	Female	3.0	3.0	3.0
	Malc	5.0	5.0	5.0
(b)	Female	3.0	4.0	5.0
	Male	3.0	4.0	5.0

A second null hypothesis states that the population means are identical across categories of the second qualitative variable, controlling for the first. For example, we could test whether mean political ideology is identical for the two genders, controlling for party ID. Table 12.10b displays a set of population means satisfying this null hypothesis. The effects of individual predictors tested in the first two null hypotheses are called ***main effects***.

The third null hypothesis states that there is a lack of interaction in the effects of the qualitative variables on the response; that is, the difference between means for two categories of one predictor is the same for each category of the other predictor, as in Table 12.8.

It is not meaningful to test the main effects hypotheses when interaction exists. A small P-value in the test of no interaction suggests that each qualitative predictor has an effect on the response, but the size of effect varies according to the category of the other one. For that reason, two-way ANOVA first tests the hypothesis of no interaction. If the evidence of interaction is not strong (i.e., if the P-value is not small), one then tests the main efffect hypotheses. On the other hand, if important evidence of interaction exists, one compares the means across categories of one classification separately within categories of the other. For instance, within each row one could compare the means for the column variable using confidence intervals.

The F tests of the three primary hypotheses in two-way ANOVA assume that the population distribution for each cell of the cross-classification is normal, and that the standard deviations are identical for each cell. The formulas for the test statistics are complicated except when the sample sizes in all cells are equal. We omit the computational details. Instead, the next section shows how to conduct the analyses using multiple regression models with a set of dummy variables for each qualitative predictor.

12.5 Two-Way ANOVA and Regression

One can compare means within two-way classifications of qualitative explanatory variables using a multiple regression model, by creating a set of dummy variables for the categories of each predictor. To illustrate, we analyze the relationship between political ideology and the qualitative predictors, party identification and gender. We refer to Democrat, Independent, and Republican as categories 1, 2, and 3 of party ID and to females and males as categories 1 and 2 of gender. Table 12.11 shows the data, again in the form of a contingency table. The table also shows the sample means and standard deviations of political ideology, based on scores (1, 2, 3, 4, 5, 6, 7), for the six combinations of gender and party ID.

TABLE 12.11 Political Ideology by Party Identification and Gender

		Political Ideology							Sample		Std.
Party	Gender	1	2	3	4	5	6	7	Size	Mean	Dev.
Democrat	Female	5	30	35	98	20	24	3	215	3.85	1.26
	Male	6	20	25	41	15	15	3	125	3.77	1.43
Independent	Female	4	17	27	83	16	17	5	169	3.95	1.24
	Male	4	16	20	59	21	23	1	144	4.04	1.30
Republican	Female	2	10	17	63	32	33	5	162	4.43	1.26
	Male	0	9	13	36	33	28	9	128	4.66	1.31

We use the symbol P for dummy variables for party ID and G as a dummy variable for gender; that is,

$$P_1 = \begin{cases} 1 & \text{if subject is Democrat} \\ 0 & \text{otherwise} \end{cases}$$

$$P_2 = \begin{cases} 1 & \text{if subject is Independent} \\ 0 & \text{otherwise} \end{cases}$$

Both P_1 and P_2 equal 0 when the subject is Republican. Also

$$G = \begin{cases} 1 & \text{if subject is female} \\ 0 & \text{if subject is male} \end{cases}$$

It is redundant to include dummy variables for the final categories.

Model Without Interaction

The regression model assuming that no interaction exists is

$$E(Y) = \alpha + \beta_1 P_1 + \beta_2 P_2 + \beta_3 G$$

To find the correspondence between the means μ_{ij} and the regression parameters, we substitute the various combinations of values for the dummy variables. To illustrate, for Republicans ($P_1 = P_2 = 0$) who are female ($G = 1$), the mean political ideology is

$$\mu_{13} = \alpha + \beta_1(0) + \beta_2(0) + \beta_3(1) = \alpha + \beta_3$$

Tables 12.12 and 12.13 show the means for the six combinations of party ID and gender. The difference in means between females and males equals β_3 for each party ID. That is, the coefficient β_3 of the dummy variable G for gender equals the difference between females and males in mean political ideology, controlling for party ID. Hence, the null hypothesis of no difference between females and males in the mean, controlling for party ID, is $H_0 : \beta_3 = 0$.

TABLE 12.12 Population Means of Political Ideology for the Two-Way Classification of Party ID and Gender, with No Interaction

	Party	Dummy Variables			Mean of Y
Gender	Identification	P_1	P_2	G	$\alpha + \beta_1 P_1 + \beta_2 P_2 + \beta_3 G$
Female	Democrat	1	0	1	$\mu_{11} = \alpha + \beta_1 + \beta_3$
	Independent	0	1	1	$\mu_{12} = \alpha + \beta_2 + \beta_3$
	Republican	0	0	1	$\mu_{13} = \alpha + \beta_3$
Male	Democrat	1	0	0	$\mu_{21} = \alpha + \beta_1$
	Independent	0	1	0	$\mu_{22} = \alpha + \beta_2$
	Republican	0	0	0	$\mu_{23} = \alpha$

TABLE 12.13 Population Means from Table 12.12, with Sample Means of Ideology

Gender	Democrat	Independent	Republican
Female	$\alpha + \beta_1 + \beta_3$	$\alpha + \beta_2 + \beta_3$	$\alpha + \beta_3$
	$\bar{Y}_{11} = 3.85$	$\bar{Y}_{12} = 3.95$	$\bar{Y}_{13} = 4.43$
Male	$\alpha + \beta_1$	$\alpha + \beta_2$	α
	$\bar{Y}_{21} = 3.77$	$\bar{Y}_{22} = 4.04$	$\bar{Y}_{23} = 4.66$

The β_1 term is the difference between the means for Democrats and Republicans, and β_2 is the difference between the means for Independents and Republicans, controlling for gender. The interpretations are similar to the β-values for the regression model for one-way ANOVA, except that here we also control for gender. The null hypothesis of no differences among the parties in mean political ideology, controlling for gender, is $H_0 : \beta_1 = \beta_2 = 0$.

One can test these two main effect hypotheses with the F test of Section 11.6 for complete and reduced regression models. For example, the test of H_0: $\beta_1 = \beta_2 = 0$ compares the complete model

$$E(Y) = \alpha + \beta_1 P_1 + \beta_2 P_2 + \beta_3 G$$

to the reduced model,

$$E(Y) = \alpha + \beta_3 G$$

This is the basis of the F tests that software provides in ANOVA tables.

Example 12.5 Two-Way ANOVA for Political Ideology, with No Interaction

For Table 12.11, Table 12.13 shows the sample mean political ideology for the six combinations of party ID and gender. The means show no obvious evidence of interaction; for each gender the sample mean conservatism increases as one moves from Democrat to Republican. Our analyses here assume a lack of interaction. We conduct the test of no interaction in the next subsection.

Most statistical software reports a table similar to Table 12.14 (based on PROC GLM in SAS) for summarizing the analyses. The sums of squares for "Party ID" and for "Gender" in the ANOVA table describe how much of the variability in political ideology is explained by each set of terms in the model. For example, the party ID sum of squares is the amount of the variation accounted for by introducing the terms $\beta_1 P_1 + \beta_2 P_2$ into the model, once the other terms are already there. It represents the difference between the sums of squared errors (SSE) when these terms are omitted and when they are included. There is a difference of two parameters in the two models, so this sum of squares has $df = 2$. As usual, the df value for SSE equals the total sample size minus the number of parameters in the regression model. For the complete model, this is $943 - 4 = 939$.

The mean square error (MSE) equals SSE divided by its df value. This estimates the variance of the population within each cell. For this model, it equals

$$\hat{\sigma}^2 = \frac{\text{SSE}}{df} = \frac{1569.53}{939} = 1.67$$

We test each null hypothesis by comparing this to another estimate of the variance that tends to be inflated when H_0 is not true. The variance estimates, listed in Table 12.14 under "Mean Square," divide each sum of squares by its df value. The F test statistics are the ratios of these estimates to $\hat{\sigma}^2$. The degrees of freedom for the F statistics are $df_1 = df$ for the numerator estimate, and $df_2 = df$ for $\hat{\sigma}^2$ (939, in this case). As usual, the P-value is the right-hand tail probability.

The null hypothesis of no difference in mean political ideology for the three party IDs, controlling for gender, is equivalent to H_0: $\beta_1 = \beta_2 = 0$ for the regression model. The F test statistic is the ratio

$$F = \frac{\text{Party ID mean square}}{\text{Mean square error}} = \frac{42.13}{1.67} = 25.2$$

TABLE 12.14 ANOVA Table for Two-Way Analysis of Mean Political Ideology by Party Identification and Gender, Assuming No Interaction

```
Dependent Variable: IDEOLOGY
                         Sum of        Mean
Source            DF    Squares       Square     F Value      Pr > F
Model              3     86.693       28.898      17.29       0.0001
Error            939   1569.525        1.671
Total            942   1656.218

Source        DF    Type III SS    Mean Square    F Value     Pr > F
PARTY          2       84.2516        42.1258      25.20     0.0001
GENDER         1        1.3110         1.3110       0.78     0.3760

                               T for H0:     Pr > |T|     Std Error of
Parameter          Estimate   Parameter=0                  Estimate
INTERCEPT            4.5768      51.02        0.0001        0.0897
PARTY         1     -0.7112      -6.87        0.0001        0.1035
              2     -0.5423      -5.15        0.0001        0.1054
              3      0.0000        .            .             .
GENDER        1     -0.0758      -0.89        0.3760        0.0856
              2      0.0000        .            .             .
```

based on $df_1 = 2$ and $df_2 = 939$. The P-value is $P = .0001$. Very strong evidence exists of a difference in mean political ideology among the three party IDs, controlling for gender.

The null hypothesis of no difference in mean political ideology between females and males, controlling for party ID, is H_0: $\beta_3 = 0$ for the regression model. The F test statistic, based on $df_1 = 1$ and $df_2 = 939$, is

$$F = \frac{\text{Gender mean square}}{\text{Mean square error}} = \frac{1.31}{1.67} = .78$$

The P-value is $P = .38$. There is negligible evidence that mean political ideology varies by gender, within each party ID.

The prediction equation for the no interaction model is

$$\hat{Y} = 4.58 - .71P_1 - .54P_2 - .08G$$

The coefficient of G, which is $-.08$, is the estimated difference between females and males in mean political ideology, for each party ID. The test of the gender main effect indicated that this difference is not statistically significant. The coefficient of P_1, which is $-.71$, is the estimated difference between Democrats and Republicans in mean political ideology, for each gender. The coefficient $-.54$ of P_2 is the estimated difference between Independents and Republicans, for each gender. The estimated difference between Democrats and Independents is $(-.71) - (-.54) = -.17$, for each gender.

Substituting dummy variable values into the prediction equation yields predicted cell means that satisfy the no interaction model. For instance, for female Republicans, $P_1 = P_2 = 0$ and $G = 1$, so $\hat{Y} = 4.58 - .71(0) + .54(0) - .08(1) = 4.50$. □

Model with Interaction

The model considered so far is inadequate when there is interaction. Section 11.5 showed that cross-product terms in a multiple regression model can represent interaction. Here, we take cross-products of dummy variables to obtain a regression model that includes interaction effects.

The interaction model for the two-way classification of party ID and gender is

$$E(Y) = \alpha + \beta_1 P_1 + \beta_2 P_2 + \beta_3 G + \beta_4 (P_1 G) + \beta_5 (P_2 G)$$

The last two terms use cross-products for the interaction. It is not necessary to take cross-products of dummy variables from categories of the same qualitative predictor, such as $P_1 P_2$. This is because no more than one dummy variable for a given predictor can be nonzero for any observation, since an observation cannot fall in more than one category. Hence, all such cross-products would equal 0.

Table 12.15 is a printout of an ANOVA table for the model that allows interaction. The sum of squares for interaction, shown in the row with the product label "PARTY * GENDER," is the amount of variability explained by the two interaction terms; that is, the difference between SSE without and with these terms in the model. The interaction mean square is an estimate of σ^2 based on

$$\frac{\text{Interaction SS}}{df} = \frac{3.64}{2} = 1.82$$

TABLE 12.15 ANOVA Table for Two-Way Analysis of Mean Political Ideology by Party Identification and Gender, Allowing Interaction

Dependent Variable: IDEOLOGY

Source	DF	Sum of Squares	Mean Square	F Value	Pr > F
Model	5	90.332	18.066	10.81	0.0001
Error	937	1565.886	1.671		
Total	942	1656.218			

Source	DF	Type III SS	Mean Square	F Value	Pr > F
PARTY	2	87.795	43.898	26.27	0.0001
GENDER	1	1.488	1.488	0.89	0.3456
PARTY*GENDER	2	3.640	1.820	1.09	0.3370

We test the hypothesis of no interaction (H_0: $\beta_4 = \beta_5 = 0$) using

$$F = \frac{\text{Interaction mean square}}{\text{Mean square error}} = \frac{1.82}{1.67} = 1.09$$

From the F distribution with $df_1 = 2$ and $df_2 = 937$, the P-value is $P = .34$.

There is not much evidence of interaction, so we can delete the cross-product terms from the model. We are justified in using the simpler model discussed previously. Since an absence of interaction is plausible, the main effect tests presented in Table 12.14 for party ID and gender are valid.

Partial Sums of Squares

The sums of squares for party, gender, and their interaction in Tables 12.14 and 12.15 are called ***partial sums of squares***. They are labeled ***Type III SS*** on SAS printouts. (See Problem 11.50 in the previous chapter.) They represent the variability in Y explained by those terms, once the other terms are already in the model. This equals the difference between the SSE values for the model without those terms and the model with them.

If the predictors are independent, such as when the same number of subjects occur at each combination of party ID and gender, then these sums of squares explain completely separate portions of the variability in Y. They then sum to the model sum of squares, which equals TSS − SSE and which represents the variability explained by all the terms together. This often happens in designed experiments, for which one can assign a fixed number of subjects to each cell of a classification. For survey research and observational data, however, this rarely happens. Gender is somewhat associated with party ID, for instance. Because of this, the partial sum of squares for party ID overlaps somewhat with the partial sum of squares for gender, and the Type III sums of squares listed in Tables 12.14 and 12.15 do not add up exactly to the model sum of squares.

Moreover, when the qualitative predictors are associated, the partial sum of squares explained by a predictor depends on whether the interaction terms are in the model. The partial sums of squares for party and for gender differ slightly between Tables 12.14 and 12.15. When there is not significant evidence of interaction, it is best to use the model without interaction terms in assessing the effects of the predictors and in conducting follow-up inference such as confidence intervals for the effects.

Multiple Comparisons Following Two-Way ANOVA

If any of the three null hypotheses have small P-values, we next investigate what effects cause this. Usually, interest focuses on estimating differences between means of pairs of categories of one of the qualitative variables, controlling for the other one.

Suppose the null hypothesis of no interaction seems plausible. Then, the difference in means between two categories for one predictor is the same at each category of the other. Thus, one can construct a single set of comparisons, rather than a separate set at each category of the other variable. In Example 12.5, one estimates the differences in the mean political ideology between each pair of party IDs, controlling for gender, a total of three comparisons. One can do this using ordinary confidence intervals for regression parameters; the form is the usual one of estimate plus and minus a t-score times the standard error, with df for t being the df value for the mean square error.

For instance, for the model assuming no interaction, the estimate of β_1 is $-.71$. This is the estimated difference between Democrats and Republicans in mean political ideology, controlling for gender. The standard error of this estimate, reported in Table 12.14, is .104. A 95% confidence interval is $-.71 \pm 1.96(.104)$, or $(-.91, -.51)$. Democrats are less conservative, on the average, for each gender.

The Bonferroni approach introduced in Section 12.2 for one-way ANOVA extends to higher-way ANOVA. A comparison of all three pairs of party IDs using a multiple comparison error rate of .05 uses error probability .05/3 = .0167 in determining the t score for each interval. For these data, we obtain very similar intervals to those shown in Table 12.3 following the one-way ANOVA.

When a practically significant degree of interaction exists, it is not appropriate to make summary comparisons of categories of one qualitative variable, controlling for the other. Instead, one usually compares the pairs of rows separately within each column or compares the pairs of columns separately within each row.

Factorial ANOVA

The methods of two-way ANOVA extend to models with several predictors. Qualitative explanatory variables in ANOVA are often called ***factors***. A multi-factor ANOVA in which one has observations from all the combinations of the factors is called ***factorial ANOVA***.

For instance, with three factors, ***three-way ANOVA*** considers possible interactions as well as main effects for those factors. For factors denoted by A, B, and C, the full model contains a main effect for each factor, A×B, A×C, and B×C two-factor interactions, and the A×B×C three-factor interaction. This model is equivalent to a regression model that has a set of dummy variables for each factor, cross-products of pairs of dummy variables for the two-factor interactions, and three-way products of dummy variables from all three factors for the three-factor interaction.

One first tests the three-factor interaction. If the P-value is small, one must compare pairs of categories for one variable at each combination of categories of the other two. Otherwise, one drops this term from the model and then tests the two-factor interactions. Suppose, for instance, that the P-value is small for the A×B interaction but not for the others. Then one can test the C main effect and make pairwise comparisons of means for various pairs of categories of C. One must make comparisons of means from categories of A separately at each category of B, and comparisons of means from categories of B separately at each category of A.

12.6 Repeated Measures ANOVA

The methods presented so far assume that the samples in the groups being compared are *independent*. Each group has a separate sample of subjects. In many studies, however, each group has the same subjects. The samples are then *dependent*, and the analysis must take this into account.

Example 12.6 Positive and Negative Influences on Children

In the 1990 General Social Survey, subjects were asked to respond to the following: "Children are exposed to many influences in their daily lives. What kind of influence does each of the following have on children? 1. Movies, 2. Programs on network television, 3. Rock music." The possible responses were (very negative, negative, neutral, positive, very positive). Table 12.16 shows responses for 12 of the sampled subjects, using scores $(-2, -1, 0, 1, 2)$ for the possible responses. □

TABLE 12.16 Opinions of Subjects About Three Influences on Children

	Influence		
Subject	Movies	TV	Rock
1	−1	0	−1
2	1	0	0
3	0	1	−2
4	2	0	1
5	0	−1	−1
6	−2	−2	−2
7	−1	−1	0
8	0	1	−1
9	−1	−1	−1
10	1	0	1
11	1	1	−1
12	−1	−1	−2
Mean	−.08	−.25	−.75

One-Way ANOVA with Repeated Measurement

For Table 12.16, the basic problem is like that in one-way ANOVA: comparing means for several groups. Are the population mean responses identical for the three influences? If they seem to be unequal, which mean is largest, which is smallest, and how large are the differences? One-way ANOVA is inappropriate for these data because the three samples for the categories of influence are not independent. The same subjects appear in each sample.

If we regarded the rows, as well as the columns, of Table 12.16 as different groups, then the data layout would resemble a two-way ANOVA. Each cell of the table is a combination of a subject with an influence. We could then construct a regression model that expresses the mean response as a function of 2 dummy variables for the 3 influences and 11 dummy variables for the 12 subjects. The test for a difference in mean response among the three influences is then the main effect test for the column variable in the two-way ANOVA. In fact, this is the appropriate test for data of this sort.

Table 12.16 is part of a much larger sample of over 1000 respondents, but for illustrative purposes we show the analysis for this small sample. Table 12.17 shows a

TABLE 12.17 ANOVA Table for Repeated Measures ANOVA of Opinion Response by Influence Type

Source	DF	Sum of Squares	Mean Square	F Value	Pr > F
Model	13	27.861	2.143	3.79	0.0029
Error	22	12.444	0.566		
Total	35	40.306			

Source	DF	Anova SS	Mean Square	F Value	Pr > F
INFLUENC	2	2.889	1.444	2.55	0.1006
SUBJECT	11	24.972	2.270	4.01	0.0027

printout of the ANOVA table for a two-way ANOVA (similar to that provided by SAS, using PROC ANOVA). Consider the null hypothesis of equal means for the three influences. The F test statistic for the influence main effect is the mean square for influence divided by the mean square error, or $F = 1.44/.57 = 2.55$. The df values for the mean squares are $df_1 = 2$ and $df_2 = 22$. The P-value equals $P = .101$. The evidence against the null hypothesis is not strong, but one can't expect much with only 12 subjects.

As usual, we learn more from estimating parameters. Table 12.16 shows the sample means for the three influences. Since the sample size is small, we loosen the confidence stringency a bit so that the intervals are not overly wide. Simultaneous 90% Bonferroni confidence intervals use error probability .10/3 = .0333 for each interval. The error $df = 22$, and the t-score with probability .0333/2 = .0167 in each tail is $t = 2.27$. Each group has 12 observations, and the square root of the mean square error equals $\hat{\sigma} = \sqrt{.566} = .75$, so the plus and minus part for each confidence interval is

$$t\hat{\sigma}\sqrt{\frac{1}{n_i} + \frac{1}{n_j}} = 2.27(.75)\sqrt{\frac{1}{12} + \frac{1}{12}} = .70$$

For instance, the confidence interval for the difference between the mean response on movies and on rock music is $(-0.08) - (-0.75) \pm .70$, or $(-.03, 1.37)$. It is plausible that the means are equal, but also plausible that the mean for movies is much more in the positive direction than the mean for rock music. Table 12.18 shows the set of Bonferroni comparisons. It is also plausible that the mean for TV is much more in the

TABLE 12.18 Bonferroni Simultaneous 90% Confidence Intervals for Pairwise Comparisons of Mean Responses for Three Influences

Influences	Difference of Means	Confidence Interval
Movies, TV	.17	(−.53, .87)
Movies, Rock	.67	(−.03, 1.37)
TV, Rock	.50	(−.20, 1.20)

positive direction than the mean for rock music. Confidence intervals can convey useful information even if the overall test statistic is not significant.

Fixed Effects and Random Effects

Let Y denote the response for the opinion about the various influences. The regression model for the previous analysis is

$$E(Y) = \alpha + \beta_1 I_1 + \beta_2 I_2 + \gamma_1 S_1 + \gamma_2 S_2 + \ldots + \gamma_{11} S_{11}$$

Here, I_1 is a dummy variable for the first influence (i.e., $I_1 = 1$ for a response on movies, 0 otherwise) and I_2 is a dummy variable for the second influence ($I_2 = 1$ for a response on TV, 0 otherwise); when $I_1 = I_2 = 0$, the response refers to the third medium, rock music. Similarly, S_1 is a dummy variable for subject 1, equaling 1 for that subject's three responses and 0 otherwise, and so on for the other 10 subject dummy variables. We use gamma instead of beta for the coefficients of these terms simply for convenience, so the index of the parameter agrees with the index of the dummy variable. As usual, each factor has one fewer dummy variable than its number of categories.

This analysis corresponds to a two-way ANOVA with only one observation in each of 36 cells. For such a design, we cannot include interaction terms and a test for interaction. The cross-products of the two influence dummy variables with the 11 subject dummy variables would take 22 terms. The model would then have as many parameters as observations, and all the variability would be explained. That is, then SSE = 0 and its $df = 0$, making construction of the F statistic impossible.

Suppose an analysis of this type has only two groups. That is, there are two dependent samples, with the same subjects in each group. Section 7.4 presented this case. Inference uses the t distribution with the difference scores between the two samples. For testing equality of means for the two groups, the F statistic from the ANOVA is simply the square of the t statistic from that matched-pairs t test.

A short-hand way of writing the regression model for the repeated measurement ANOVA is

$$E(Y) = \alpha + \beta_j + \gamma_i$$

Here β_j denotes the effect for influence j and γ_i is the effect for subject i, where the parameter for the final category of each variable equals 0. This expresses the expected response in the cell in row i and column j additively in terms of a row main effect and a column main effect. Testing equality of mean responses for the three influences corresponds to testing $H_0 : \beta_1 = \beta_2 = 0$.

In this model, the main focus is estimating the influence parameters $\{\beta_j\}$, not the subject parameters $\{\gamma_i\}$. The subject parameters depend on which subjects are chosen for the sample. The subject effect is called a ***random effect***, since the categories of that factor represent a random sample of all the possible ones.

By contrast, the factors that define the groups compared in this chapter, such as party identification, gender, influence type, and so forth, are called ***fixed effects***. The analyses refer to *all* the categories of interest, rather than a random sample of them, and provide inferences about differences among the means for those categories. Models studied in earlier sections of this chapter contain only fixed effects.

When the classification variables are a mixture of random and fixed effects, such as in the model discussed above, the model is called a ***mixed*** model. For more complex mixed models than this one, the test statistics differ for some tests from their form when all classification variables are fixed effects. The next section discusses an important case of this type.

Studies with repeated measurement designs have the advantage that they can improve precision of estimation. Having the same subjects in each group helps to eliminate extraneous sources of error. For instance, other variables that affect the response have the same values for each group, so differences between group means cannot reflect differences between groups on those variables. Controlling for subjects or other possibly confounding factors by keeping them fixed in each row of the data matrix is referred to as ***blocking***. For a detailed exposition of the linkage of analysis of variance procedures with experimental designs, see Howell (1992), Kirk (1982), and Winer et al. (1991).

12.7 Two-Way ANOVA with Repeated Measures on a Factor

Repeated measurement data sets often have more than one fixed effect. A second fixed effect may refer to groups to be compared on the repeated response. Most commonly those groups have independent samples. The following example illustrates:

Example 12.7 Comparing Three Treatments for Anorexia

For a sample of 72 young girls suffering from anorexia, Table 12.19 shows their weights before and after an experimental period. The girls were randomly assigned to receive one of three therapies during this period. One group, a control group, received the standard therapy. The study analyzed whether one treatment is better than the others, the girls tending to gain more weight under that treatment.

Figure 12.5 shows box plots, graphically describing the response distributions before and after the experimental period for each treatment. Table 12.20 shows the summary sample means. The three treatments have similar distributions originally. There is some evidence apparent of a greater average weight gain for the family therapy group, though there are a few low outlying weight values. □

Repeated Measures on One of Two Fixed Effects

Tables 12.19 and 12.20 refer to two fixed effects. One of them, "Treatment," has categories (CB = cognitive behavioral, FT = family therapy, C = control). It defines three groups of girls, represented by three independent samples. The second, "Time," consists of the two times for observations, (before, after). Each time has the same subjects, so the samples at its levels are dependent. Time is an example of a ***within-subjects factor***, since comparisons of its categories use repeated measurements within samples. Treatment is a ***between-subjects factor***, since comparisons of its categories use different subjects from each category.

TABLE 12.19 Weights of Anorexic Girls, in Pounds, Before and After Treatment Period

Cognitive Behavioral		Family Therapy		Control	
Weight Before	Weight After	Weight Before	Weight After	Weight Before	Weight After
80.5	82.2	83.8	95.2	80.7	80.2
84.9	85.6	83.3	94.3	89.4	80.1
81.5	81.4	86.0	91.5	91.8	86.4
82.6	81.9	82.5	91.9	74.0	86.3
79.9	76.4	86.7	100.3	78.1	76.1
88.7	103.6	79.6	76.7	88.3	78.1
94.9	98.4	76.9	76.8	87.3	75.1
76.3	93.4	94.2	101.6	75.1	86.7
81.0	73.4	73.4	94.9	80.6	73.5
80.5	82.1	80.5	75.2	78.4	84.6
85.0	96.7	81.6	77.8	77.6	77.4
89.2	95.3	82.1	95.5	88.7	79.5
81.3	82.4	77.6	90.7	81.3	89.6
76.5	72.5	83.5	92.5	78.1	81.4
70.0	90.9	89.9	93.8	70.5	81.8
80.4	71.3	86.0	91.7	77.3	77.3
83.3	85.4	87.3	98.0	85.2	84.2
83.0	81.6			86.0	75.4
87.7	89.1			84.1	79.5
84.2	83.9			79.7	73.0
86.4	82.7			85.5	88.3
76.5	75.7			84.4	84.7
80.2	82.6			79.6	81.4
87.8	100.4			77.5	81.2
83.3	85.2			72.3	88.2
79.7	83.6			89.0	78.8
84.5	84.6				
80.8	96.2				
87.4	86.7				

Source: Thanks to Prof. Brian Everitt, Institute of Psychiatry, London, for supplying these data.

TABLE 12.20 Sample Mean Weight, by Treatment and Time of Measurement

	Time	
Treatment	Before	After
Cognitive Behavioral (CB)	82.7	85.7
Family Therapy (FT)	83.2	90.5
Control (C)	81.6	81.1

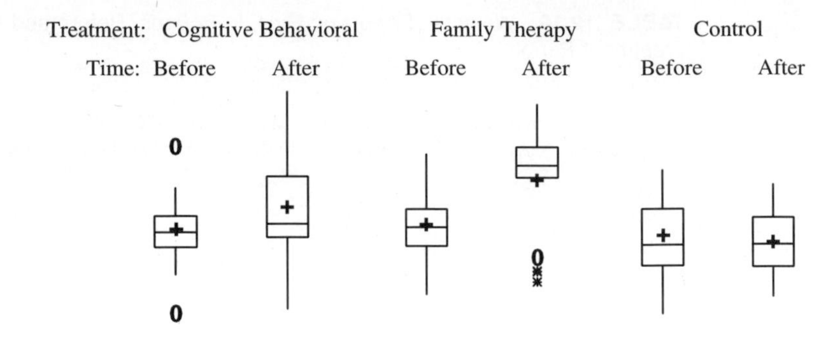

Figure 12.5 Box Plots for Weights of Anorexic Girls, by Treatment and Time of Measurement

Although the two factors (treatment and time) are fixed effects, the analysis differs from ordinary two-way ANOVA. This is because the repeated measurements on the within-subjects factor creates a third effect, a random effect for subjects. Each subject is measured at every category of the within-subjects factor (time), and subjects are said to be ***crossed*** with that factor. Each subject occurs at only one category of the between-subjects factor (treatment), and subjects are said to be ***nested*** within that factor.

As in ordinary two-way ANOVA, one can test each main effect as well as the interaction between them. However, the ordinary error term is partitioned into two parts. One part is based on the variability among mean scores of subjects; it forms an error term for testing the between-subjects factor. The other part is based on how the pattern of within-subject scores varies among subjects; it forms an error term for any test involving the within-subjects factor, both its main effect and its interaction with the other fixed effect. Figure 12.6 shows the partitioning of the total sum of squares for a repeated measures ANOVA with these data. Software automatically performs this partitioning and creates F statistics in the proper way for testing the main effects and interaction.

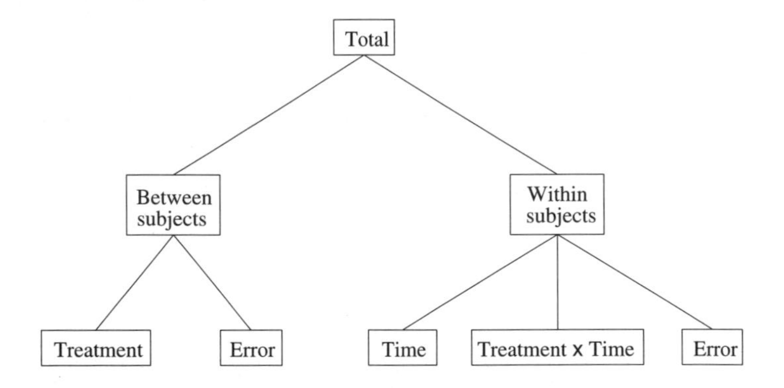

Figure 12.6 Partitioning of Variability in Two-Way ANOVA with Treatment and Time Factors and Repeated Measures on Time. Tests involving the within-subjects factor (time) use a separate error term.

TABLE 12.21 Computer Printout for Two-Way Analysis of Variance of Table 12.19 with Fixed Effects Treatment and Time and Repeated Measures on Time

```
                Repeated Measures Analysis of Variance
            Tests of Hypotheses for Between Subjects Effects

Source           DF        Anova SS   Mean Square   F Value      Pr > F
TREATMNT          2          644.23        322.12      6.20      0.0033
Error            69         3584.03         51.94

       Univariate Tests of Hypotheses for Within Subject Effects

Source: TIME
   DF       Anova SS     Mean Square      F Value     Pr > F
    1    275.0069444     275.0069444         9.70     0.0027

Source: TIME*TREATMNT
   DF       Anova SS     Mean Square      F Value     Pr > F
    2    307.3218334     153.6609167         5.42     0.0065

Source: Error(TIME)
   DF       Anova SS     Mean Square       69    1955.3712221     28.3387134
```

TABLE 12.22 ANOVA Table for Two-Way Analysis of Variance of Table 12.19 with Fixed Effects Treatment and Time and Repeated Measures on Time

Source	Sum of Squares	DF	Mean Square	F-Value	Prob > F
Between Subjects					
Treatment	644.2	2	322.1	6.2	.003
Error	3584.0	69	51.9		
Within Subjects					
Time	275.0	1	275.0	9.7	.003
Time × Treatment	307.3	2	153.7	5.4	.006
Error	1955.4	69	28.3		

Table 12.21 shows a SAS printout for the analysis for the anorexia data (Table 12.37 in Problem 12.30 shows how SPSS reports the results), and Table 12.22 shows how we might use this information to set up an ANOVA table. The sum of squares for a main effect is based on the variability among the means at the various categories for that variable. Each sum of squares summarizes the variation its terms explain in the regression model. This is the reduction in SSE when we add dummy variables to the regression model for that effect. Since treatment has three categories, it has two dummy variables in the regression model, and its sum of squares has $df = 2$. Since time has two levels, it has one dummy variable, and its sum of squares has $df = 1$. The interaction between these effects has two terms in the model, based on the cross-product of the two

dummy variables for treatment with the dummy variable for time, so it has $df = 2$. The appendix shows how to use software to conduct the analysis.

The error term for the between-subjects part of the table uses the variability among subjects' means within each group. It has $df = 69$, based on 28 dummy variables for the 29 subjects receiving therapy CB, 16 dummy variables for the 17 subjects receiving FT, and 25 dummy variables for the 26 subjects in group C $(28 + 16 + 25 = 69)$. The remaining variability, not accounted for by this error term or by the main effects and interaction terms, is the error sum of squares for testing the within-subjects effects. The total df, as in all ANOVAs, equals the number of measurements minus 1; here, 72 girls are measured twice, so total $df = 2(72) - 1 = 143$.

The ANOVA table indicates that the interaction is highly significant, with P-value equal to .006. The difference between means for the two times differs according to the treatment, and the difference between means for a pair of treatments varies according to the time. Because this is significant, it does not make sense to test the main effects. We need, instead, to conduct a more focused analysis that determines the nature of the interaction.

Follow-up Confidence Intervals

Table 12.20 showed the sample means for the six combinations of the two factors. The evidence of interaction is clear. The sample means for the three treatments are similar at the initial time. This is not surprising, if subjects were randomly allocated to the three groups at that time. At the second time, by contrast, the mean for the control group is similar to the initial time, but the mean is somewhat larger for the other two treatments than their initial means, particularly for the FT group.

To make within-subjects comparisons of means at the two times, for each treatment, the appropriate root mean square error for confidence intervals is the one from the within-subjects analysis. From Tables 12.21 and 12.22, this equals $\sqrt{28.3} = 5.3$, with $df = 69$. We illustrate by constructing a 95% confidence interval comparing the two means for family therapy (FT), which 17 girls received at each time. The t-score for 95% confidence when $df = 69$ equals 1.99, so the confidence interval has plus and minus part equal to this t-score times the root mean square error times the square root factor involving the inverse of each sample size (17 for each time). This interval equals

$$(90.5 - 83.2) \pm 1.99(5.3)\sqrt{\frac{1}{17} + \frac{1}{17}}, \quad \text{or } 7.3 \pm 3.6, \quad \text{or } (3.6, 10.9)$$

For this therapy, we conclude that the mean weight is between 3.6 and 10.9 pounds higher following the treatment period. Similarly, a 95% confidence interval comparing the two means equals (.2, 5.8) for the CB therapy and (−3.4, 2.5) for the control group. There is evidence of an increase, albeit a small one, for the CB therapy, but no evidence of change for the control group.

To make between-subjects comparisons of treatments, for each time, one cannot use the root mean square error from the between-subjects analysis. The reason is that these separate comparisons involve both the treatment main effect and the interaction, and these two sources of variation have different error terms in the repeated measures ANOVA. At a particular time, however, the subjects in the three treatments are independent samples. Thus, we can compare three means at a given time using a one-way ANOVA F test or using confidence intervals for those data alone.

For instance, for the 72 observations at time = after, the F test statistic for the one-way ANOVA comparing the three means is 8.6, based on $df_1 = 2$ and $df_2 = 69$, for which the P-value is .0004. At the follow-up time, very strong evidence exists of a difference among the treatment means. For this one-way ANOVA, the root MSE equals $\hat{\sigma} = 7.3$. The 95% confidence interval for the difference of means between the FT and the CB treatments, based on the 17 + 29 observations for the two groups, equals

$$(90.5 - 85.7) \pm 1.99(7.3)\sqrt{\frac{1}{17} + \frac{1}{29}}, \quad \text{or } 4.8 \pm 4.4, \quad \text{or } (.4, 9.2)$$

At the follow-up time, we conclude that the mean weight is between .4 and 9.2 pounds higher with treatment FT than with treatment CB. It is plausible that the true means are essentially equal, but if they differ, the advantage could be quite noticeable for the family therapy. Table 12.23 shows the intervals for each pair of treatments.

TABLE 12.23 95% Confidence Intervals Comparing Treatment Means After Treatment Period

Treatments Compared	Difference of Sample Means	Confidence Interval	Bonferroni Interval
FT - CB	4.8	(.4, 9.2)	(−.7, 10.3)
FT - C	9.4	(4.9, 13.9)	(3.8, 15.0)
CB - C	4.6	(.7, 8.5)	(−.2, 9.4)

As usual, the Bonferroni multiple comparison method controls the overall error rate for several comparisons. For instance, suppose we use three confidence intervals to compare treatments at time = after, and three intervals to compare times within the treatments. To ensure at least 90% confidence for the entire set, since .10/6 = .0167, we use a 98.33% confidence interval for each individual comparison. These intervals are wider than the ones just reported, since they use a t-score of 2.45 instead of 1.99. Table 12.23 shows them for the pairwise comparisons of treatments at time = after. With this more conservative approach, only the difference between the FT and C treatments is "significant," the interval not containing 0.

In summary, there is evidence that mean weight increases during the experimental period for both noncontrol treatments, and there is marginal evidence that the mean is higher after treatment for treatment FT than treatment CB. At this stage, further interest may relate to whether the change in means, between time = after and time = before, differed for the two noncontrol treatments. That is, do the difference scores for treatment FT have a significantly higher mean that the difference scores for treatment CB?

The difference scores have a mean of (90.5 − 83.2) = 7.3 for the FT treatment and (85.7 − 82.7) = 3.0 for the CB treatment, and we used these as the basis of separate confidence intervals for the mean change, above. Since the two groups are independent samples, the variance of the difference of these means is the sum of the variances. Thus, a 95% confidence interval for the difference between the mean changes in weight is

$$(7.3 - 3.0) \pm 1.99(5.3)\sqrt{\frac{1}{17} + \frac{1}{17} + \frac{1}{29} + \frac{1}{29}}, \quad \text{or } 4.3 \pm 4.6, \quad \text{or } (-.3, 8.8)$$

Although the mean change could be considerably larger for the FT treatment, it is also plausible that the mean changes could be identical.

More Complex Repeated Measures Analyses*

In some studies with two factors, repeated measures occur on both factors. For instance, the same subects may be measured for each treatment at each of several times. Then, subjects (a random effect) are crossed with both factors (fixed effects), and an observation occurs for every subject at every combination of factor levels. As in ordinary two-way ANOVA, the effects of interest refer to the fixed effects—their main effects and interaction. The complicating factor is that each test requires a separate mean square error, but software can easily conduct this analysis.

The analysis described in tbis section extends to more complex designs. Suppose, for instance, that a study has three factors, A, B, and C, with repeated measures across the levels of factor C. That is, subjects are crossed with C but nested within combinations of levels of A and B. The between-subjects effects, namely, the A and B main effects and the A×B interaction, are tested with a mean square error based on variability between subjects. All effects involving the within-subjects factor C, namely the C main effect, the A×C interaction, the B×C interaction, and the A×B×C interaction, are tested with a separate mean square error. The appendix shows how to conduct this analysis using software.

In Example 12.7 the repeated measurements occur at two times. When the within-subjects factor has more than two categories, such as measurement at several times, the repeated measures ANOVA is more complex. In particular, the results depend on assumptions about the correlation structure of the repeated measurements.

The traditional repeated measures ANOVA assumes ***sphericity***. This is satisfied if the correlation is identical between responses for each pair of categories of the within-subjects factor, a condition called ***compound symmetry***. If this assumption is badly violated, the P-value reported for the within-subjects effect tends to be too small. Most software provides a formal significance test (Mauchly's test) of the sphericity assumption. When the data strongly contradict that assumption, an approximate test for the within-subjects effect adjusts the degrees of freedom downward for the usual F test statistic for that effect, using an adjustment due to Greenhouse and Geisser. The technical details for these tests and adjustments are beyond the scope of this text, but standard software reports these results. Tests of the between-subjects effects are not affected by violation of the sphericity assumption, so no adjustment is needed for that F test.

An alternative approach to testing the within-subjects effects makes fewer assumptions. It treats the repeated responses as a multivariate vector of responses and uses standard methods for multivariate response data. These methods, such as *Wilks' lambda* or *Pillai's trace test*, are special cases of tests referred to as MANOVA, or *multivariate analysis of variance*. Although having fewer assumptions, this approach is less powerful than the traditional repeated measures ANOVA when the assumptions for that ANOVA are not badly violated, because the analysis requires estimating a larger number of parameters.

Recently developed methods (incorporated in software such as PROC MIXED in SAS) allow more varied types of modeling of the correlation structure for the repeated responses, providing options other than sphericity. The *autoregressive* option, for instance, allows observations that are closer together in time to be more highly correlated. The issues involved in choosing an analysis are complex. You should seek a statistician's guidance before attempting to analyze such data.

12.8 Violations of ANOVA Assumptions

Each ANOVA method presented in this chapter assumes that the cells defined by the cross-classification of the factors have population distributions that are normal with identical standard deviations. This is a stringent assumption that, like ordinary regression assumptions, is never exactly satisfied in practice.

Moderate departures from normality of the populations can be tolerated, in the sense that the F sampling distribution still provides a good approximation to the actual sampling distribution of the test statistics. This is particularly true for larger sample sizes, since the sampling distributions then have weaker dependence on the form of the population distribution.

Moderate departures from equal standard deviations can also be tolerated. When the sample sizes are identical for the various groups being compared, the F test is robust to violations of this assumption. Significance tests exist for checking whether several population standard deviations are equal. Some books recommend performing this test, and then conducting the ANOVA only if the null hypothesis of equal standard deviations is not rejected. This test of homogeneity of variability is *not* robust to violations of the normality assumption, however. It performs very poorly when the true distribution is not normal. Most statisticians believe that this test is inappropriate for determining whether to use ANOVA.

Unless the sample size is small, one can construct histograms for each sample distribution to check for extreme deviations from normality. Misleading results may occur in the F tests if the population distributions are highly skewed and the sample size is small, or if there are relatively large differences among the population standard deviations (say, the largest sample standard deviation is several times as large as the smallest one) and the sample sizes are unequal. Section 14.5 discusses alternative regression approaches for such gross violations.

Confidence intervals, like tests, are not highly dependent on the normality assumption. When the standard deviations are quite different, with the ratio of the largest to smallest exceeding about 2, it is preferable to use formulas for intervals based on separate standard deviations for the groups rather than a single pooled value. For large samples, for instance, one can use the confidence interval from Section 7.1 based on z-scores and separate standard deviations, since it does not assume equal standard deviations or normal population distributions.

As in other inferences, the quality of the sample is crucial. In one-way ANOVA, for instance, conclusions may be invalid if the observations in the separate groups compared are not independent random samples. ANOVA procedures are not robust to violations of sampling assumptions.

Consider, for instance, the validity of the assumptions for one-way ANOVA for the data in Table 12.1 on political ideology classified by party identification. The sample standard deviations are similar for the three party ID groups. Also, the sample sizes in Table 12.1 are large, so the normality assumption is not crucial. The full GSS sample was randomly obtained, so we may regard the three samples classified by party ID as independent random samples. ANOVA is suitable for these data.

The Kruskal-Wallis Test: A Nonparametric Approach

The ***Kruskal-Wallis test*** is an alternative to one-way ANOVA for comparing several groups without the normality assumption. This is a nonparametric procedure. It uses only the ordinal information in the data, since its formula is based on ranking the observations and comparing mean ranks for the various groups. It is particularly useful for small samples in which the effects of severe departures from normality may be influential. We shall not discuss this test here, but the result is similar to that of a test for coefficients of dummy variables for a qualitative predictor in a model for an ordinal response presented in Section 15.6. In practice, it is more informative to use a modeling approach, since the model parameter estimates help us describe sizes of effects. In addition, the modeling strategy adapts better to multivariate analyses.

Nonparametric tests also exist for more complex analyses. For instance, ***Friedman's test*** is an alternative to the F test of Section 12.6 comparing groups when the same subjects occur in each. An advantage of the parametric methods of this chapter is that they more easily generalize to multivariate modeling and to estimation of effects, which are more important than significance testing. Lehmann (1975) is a good source of details about nonparametric procedures.

12.9 Chapter Summary

This chapter presented ***analysis of variance*** (ANOVA) methods for comparing several groups according to their means on a quantitative response variable.

- One-way ANOVA methods compare means of categories of a single qualitative variable.
- Two-way ANOVA methods compare means across categories of one qualitative variable, controlling for another.
- ***Multiple comparison*** methods provide confidence intervals for the difference between each pair of means, while controlling the overall error probability.
- The ***Bonferroni*** multiple comparison method does this by splitting the desired error probability into as many parts as needed for the separate comparisons; for instance, to guarantee an overall error probability no greater than .05 in forming five confidence intervals, split it into five parts with error probability .01 each.
- Analysis of variance methods are special cases of multiple regression analyses. One uses ***dummy variables*** as explanatory variables in the model to represent the qualitative variables that define the groups being compared. Each dummy variable equals 1 for a particular category and 0 otherwise, and there is one fewer dummy variable than the number of categories.

Ordinary ANOVA methods, such as one-way and two-way ANOVA, compare groups based on *independent* random samples from those groups. For some studies, observations in different samples refer to the same subjects, rather than independent samples, such as when subjects are measured before and after some experimental period. Different methods apply for ***repeated measures ANOVA***. These methods correspond to regression models with ***random effects***, which represent the effects of the random sample of subjects who are measured repeatedly. Such methods treat *within-subjects* effects differently from *between-subjects* effects.

TABLE 12.24 ANOVA Tests for Comparing Several Groups on a Response Variable

Element of Test	One-Way ANOVA	Two-Way ANOVA	Repeated Measures ANOVA
1. Samples	Independent	Independent	Dependent
2. Hypotheses	H_0: Identical means H_a: At least two means not equal	H_0: Identical row means H_0 : Identical column means H_0: No Interaction	H_0: Identical means H_a: At least two means not equal
3. Test statistic	$F = \frac{\text{Between estimate}}{\text{Within estimate}}$	$F = \frac{\text{Effect mean square}}{\text{Mean square error}}$	$F = \frac{\text{Effect mean square}}{\text{Mean square error}}$
	F distribution $df_1 = g - 1$ $df_2 = N - g$	F distribution $df_1 = df$ for effect $df_2 = df$ for error	F distribution $df_1 = df$ for effect $df_2 = df$ for error
4. P-value	Right-hand tail prob.	Right-hand tail prob.	Right-hand tail prob.

Chapters 9 and 11 presented models for a quantitative response variable when the explanatory variables are also *quantitative*. This chapter has modeled a quantitative response variable as a function of *qualitative* explanatory variables, through the use of dummy variables. Models of the next chapter include both quantitative and qualitative explanatory variables. Table 12.24 summarizes the statistical tests discussed in Sections 12.1, 12.4, and 12.6 for comparing several groups on a quantitative response variable.

PROBLEMS

Practicing the Basics

1. Studies of the degree of residential segregation between blacks and whites use the *segregation index*, defined as the percentage of nonwhites who would have to change the block on which they live in order to produce a fully nonsegregated city—one in which the percentage of nonwhites living in each block is the same for all blocks in the city. This index can assume values ranging from 0 to 100, with higher values indicating greater segregation. (The national average was 65 in 1990.) Table 12.25 shows the index for a sample of cities in 1994, classified by region. In no sense is this a random sample, but use it to illustrate mechanics of one-way ANOVA. The overall mean is 72.75, and $\bar{Y}_1 = 81$, $\bar{Y}_2 = 88$, $\bar{Y}_3 = 71$, $\bar{Y}_4 = 51$, with $s_1 = 4.69$, $s_2 = 2.58$, $s_3 = 4.83$, $s_4 = 12.46$. Consider the null hypothesis $H_0 : \mu_1 = \mu_2 = \mu_3 = \mu_4$, where μ_i is the mean for all cities in region i.

TABLE 12.25

Northeast		North Central		South		West	
Buffalo, NY	84	Gary, IN	91	New Orleans, LA	73	Anchorage, AK	38
Newark, NJ	83	Detroit, MI	89	Orlando, FL	64	Dallas, TX	64
Philadelphia, PA	83	Chicago, IL	87	Miami, FL	75	Anaheim, CA	43
Pittsburgh, PA	74	Milwaukee, WI	85	Atlanta, GA	72	Houston, TX	59

a) For the four cities in the Northeast, show that the sum of squares of scores about their mean is 66.
b) The within sum of squares is 622.0. Show that its degrees of freedom equal 12, and that the within-groups estimate of the variance is 51.8.
c) Show that the between sum of squares equals 3107.0, based on 3 degrees of freedom, and the between-groups estimate of variance equals 1035.67.
d) Show that the F test statistic equals 20.0, and show that the P-value is $P < .001$. What do you conclude about the mean indices?
e) Find the plus and minus part that pertains to each comparison using the Bonferroni method for simultaneous 94% confidence intervals. Compare this to each difference of means to determine which pairs are significantly different. Construct a diagram to summarize.

2. Refer to the previous exercise.
a) State the assumptions for the analysis, and show that at least two of them may be violated. What is the impact?

b) Construct the regression model with dummy variables that applies for this ANOVA. What does the estimate of the coefficient of Z_1 equal?

3. A consumer protection group compares three different types of front bumpers for a brand of automobile. A test is conducted by driving an automobile into a brick wall at 15 miles per hour. The response is the amount of damage to the car, as measured by the repair costs, in hundreds of dollars. Due to the potentially large costs, the study conducts only two tests with each bumper type. Table 12.26 shows the results.
a) Report the sample means for the three bumpers.
b) Find the within-groups sum of squares and the associated variance estimate.
c) Find the between-groups sum of squares and its associated variance estimate.
d) Test the hypothesis that the mean repair costs are the same for the three types of bumpers. Report the test statistic, df values, P-value, and interpret.
e) Construct an ANOVA table for displaying the results of this analysis.
f) Construct a 95% confidence interval for the difference between the true means for bumpers A and B. Interpret.
g) Construct simultaneous 94% confidence intervals for the differences in mean repair costs for each pair of bumpers. Interpret the results, and provide a diagram that indicates which types of bumpers, if any, are judged to be different in mean repair cost.
h) Set up the regression model with dummy variables for these data. Show the correspondence between the hypothesis for the means and the hypothesis for the regression parameters. Provide the prediction equation.

TABLE 12.26

Bumper A	Bumper B	Bumper C
1	2	11
3	4	15

TABLE 12.27

Group A	Group B	Group C
4	1	9
6	5	10
8		5

4. Table 12.27 shows scores on the first quiz (maximum score 10 points) in a beginning French course. Students in the course are grouped as follows:
Group A: Never studied foreign language before, but have good English skills
Group B: Never studied foreign language before; have poor English skills
Group C: Studied other foreign language
a) Treating the students in this course as a random sample of all ninth-grade students taking this beginning course, show all steps of a statistical test of whether the three population means differ on this quiz. Use Table 12.28, based on using SPSS to perform the analyses, to provide results of calculations. Interpret the result.
b) Use simultaneous 85% confidence intervals to compare pairs of means. Interpret each interval. Construct a diagram to indicate the means, if any, that are significantly different.

5. The General Social Survey asks respondents to rate various groups using the "feeling thermometer." Ratings between 50 and 100 mean you feel favorable and warm toward the group, whereas ratings between 0 and 50 mean that you don't feel favorable. When asked to rate liberals, the mean response was 56.6 during 1983–87 ($n = 1351$, $s = 20.7$) and 57.8 for an independent sample during 1988–91 ($n = 2228$, $s = 25.0$). Report the ANOVA F statistic for comparing the mean responses. (You can either construct F or use the connection $F = t^2$ for the test statistic from the two-sample t test.)

TABLE 12.28

Source	D.F.	Sum of Squares	Mean Square	F Ratio	F Prob.
Between Groups	2	30.000	15.000	2.5000	.1768
Within Groups	5	30.000	6.00		
Total	7	60.00			

Group	Count	Mean	Standard Deviation	Standard Error
Grp 1	3	6.0000	2.0000	1.1547
Grp 2	2	3.0000	2.8284	2.0000
Grp 3	3	8.0000	2.6458	1.5275

6. Refer to Table 12.19.
   a) Using software, conduct a one-way ANOVA for the 72 observations at time = after. Verify that the F test statistic comparing the three treatments at that time equals 8.65, based on $df_1 = 2$ and $df_2 = 69$, for which $P = .0004$.
   b) Show how to obtain the same results by conducting a regression analysis.
7. Table 12.29 is a contingency table summarizing responses on political ideology in the 1991 General Social Survey by race and gender. Table 12.30 shows results of using SAS software to conduct an analysis of variance comparing the four groups, using the category labels for the ideology scores.
   a) Show all steps of a one-way ANOVA to test the hypothesis of equal population means. Report the P-value and interpret.
   b) The printout does not show the actual Bonferroni confidence intervals, but it shows the t-score used for simultaneous 95% intervals. Show how to construct the one comparing the first two groups.

TABLE 12.29

		Political Ideology						
Race	Gender	1	2	3	4	5	6	7
White	Female	9	51	63	216	57	63	11
	Male	9	35	50	127	67	55	14
Black	Female	3	7	16	30	11	11	2
	Male	1	8	9	11	3	11	0

8. Refer to the previous exercise, using the black sample only.
   a) Conduct an ANOVA comparing the two groups. Interpret.
   b) How does your analysis relate to the methods of Chapter 7 for comparing means?
   c) How, if at all, would the results differ if you used ideology scores (i) $(-3, -2, -1, 0, 1, 2, 3)$, (ii) $(10, 20, 30, 40, 50, 60, 70)$?
9. Use software with the data in Table 9.4.
   a) Conduct an ANOVA to test equality of the mean selling prices for homes with one, two, and three bathrooms. Interpret.
   b) Explain the difference between conducting a test of independence of selling price and number of bathrooms using the ANOVA test in (a) and using a regression t test for the coefficient of the number of bathrooms in a regression model.

TABLE 12.30

```
Analysis Variable : IDEOLOGY
 N           Mean      Std Dev
             GROUP=1
470          4.051      1.269
             GROUP=2
357          4.202      1.371
             GROUP=3
 80          4.000      1.350
             GROUP=4
 43          3.930      1.518

                         Sum of       Mean
Source          DF      Squares      Square    F Value    Pr > F
Model            3        6.852      2.284      1.30      0.2739
Error          946     1665.044      1.760
Total          949     1671.896

   Bonferroni T tests for variable: IDEOLOGY
Alpha= 0.05  Confidence= 0.95  df= 946  MSE= 1.760089
             Critical Value of T= 2.64382
```

c) Show how to conduct a regression analysis that is equivalent to the ANOVA in (a).
d) Use a multiple comparison method to construct simultaneous 85% confidence intervals comparing the means. Interpret. How do these intervals compare to ordinary 95% confidence intervals for individual comparisons?

10. Refer to Table 9.4. In that data set, whether a house is new is a dummy variable. Using software, put this as the sole predictor of selling price in a regression analysis.
a) Conduct the t test for the effect of this variable in the regression analysis (i.e., test H_0: $\beta = 0$). Interpret.
b) Conduct the F test for the analysis of variance comparing the mean selling prices of new and existing homes.
c) Explain the connection between the value of t in (a) and the value of F in (b).

11. A psychologist compares the mean amount of time of REM sleep for subjects under three conditions. She uses three groups of subjects, with four subjects in each group. Table 12.31 shows a SAS printout for the analysis.
a) Report and interpret all steps of the ANOVA F test.
b) Explain how to obtain the plus and minus part of 13.96 for the 95% Bonferroni confidence intervals.
c) Set up dummy variables for a regression model so that an F test for the regression parameters is equivalent to the ANOVA test. Express the null hypothesis both in terms of population means and regression parameters.
d) Report the prediction equation obtained in fitting this regression equation.

12. For g groups with large sample sizes, we plan to compare simultaneously all pairs of population means. We want the probability to equal at least .80 that the entire set of confidence intervals contain the true differences.
a) If $g = 10$, which tabled t-score should we use for each interval?
b) Which t-score should we use if $g = 5$?

TABLE 12.31

```
                 Analysis of Variance Procedure

Dependent Variable: TIME
                              Sum of      Mean
Source              DF       Squares     Square      F Value     Pr > F
Model                2        72.000     36.000        0.79      0.4813
Error                9       408.000     45.333
Total               11       480.000

      Bonferroni T tests for variable: TIME
          Alpha= 0.05  df= 9  MSE= 45.33333
              Critical Value of T= 2.93
          Minimum Significant Difference= 13.965

 Means with the same letter are not significantly different.
 Bon Grouping               Mean        N  GROUP
           A              18.000        4  3
           A
           A              15.000        4  2
           A
           A              12.000        4  1
```

c) Describe how the t- (or z-) score depends on the number of groups, and explain the implication regarding width of the intervals.

13. A geographer compares residential lot sizes in four quadrants of a city. To do this, he randomly samples 300 records from a city file on home residences and records the lot sizes (in thousands of square feet) by quadrant. The ANOVA table (Table 12.32) refers to a comparison of mean lot sizes for the northeast (NE), northwest (NW), southwest (SW), and southeast (SE) quadrants of the city.

TABLE 12.32

Source	Sum of Squares	DF	Mean Square	F	Prob > F
Quadrant	2700	3	—	—	—
Error	1480	296	—		
Total	4180	299			

a) Find the F test statistic for H_0: $\mu_1 = \mu_2 = \mu_3 = \mu_4$, and report the P-value. Interpret.
b) Find the square root of the mean square in the "Error" row of the table, and interpret its value.
c) The sample mean lot sizes for the NE, NW, SW, and SE quadrants are 8, 15, 11, and 9, with $n_1 = 100$, $n_2 = 100$, $n_3 = 50$, $n_4 = 50$. Illustrate simultaneous 94% confidence intervals for the pairwise differences in means by constructing the interval comparing the NE and NW quadrants. Would separate 94% confidence intervals for the differences between each pair of means be wider, or narrower? Explain.

d) Set up dummy variables for a regression model with which one could conduct the ANOVA F test. State the regression model and interpret the parameters.

14. A study compares the mean level of contributions to political campaigns in Pennsylvania by registered Democrats, registered Republicans, and unaffiliated voters.
a) Write a regression equation for this analysis, and interpret the parameters in the model.
b) Explain how one can use the regression model to test the null hypothesis of equal mean contributions for the three groups.

15. Use computer software to reproduce the one-way ANOVA results reported in Table 12.2 for Example 12.1.

16. According to the U.S. Department of Labor, the mean hourly wage in 1994 was \$15.71 for a college graduate and \$9.92 for a high school graduate. In 1979, the means (in 1994 dollars) were \$15.52 for college graduates and \$11.23 for high school graduates.
a) Construct a 2×2 table to display the results.
b) Compare the differences between 1994 and 1979 (i) for college graduates, (ii) for high school graduates. Is there interaction? Explain.
c) Show a set of four means that would display an absence of interaction.

17. Refer to Table 7.18 in Problem 7.40, which summarizes the mean number of dates in the past three months by gender and by level of physical attractiveness. Do these data appear to show interaction? Explain.

18. A recent regression analysis of college faculty salaries in 1984 (M. Bellas, *American Sociological Review*, Vol. 59, 1994, p. 807) included a large number of predictors, including a dummy variable for gender (male = 1) and a dummy variable for race (nonwhite = 1). For annual income measured in thousands of dollars, the estimated coefficients were .76 for gender and .62 for race.
a) Interpret these coefficients.
b) At particular settings of the other predictors, the estimated mean salary for white females was 30.2 thousand. Find the estimated means for the other three groups.
c) Show how to use the coefficients to find the estimated difference in mean income between nonwhite male faculty and white female faculty.

19. Refer to the prediction equation $\hat{Y} = 4.58 - .71P_1 - .54P_2 - .08G$ for the no interaction model in Example 12.5.
a) Using it, find the estimated means for each of the six cells, and show that they satisfy a lack of interaction.
b) Using results for this model, construct and interpret 95% Bonferroni confidence intervals comparing pairs of party IDs. Compare results to those in Table 12.3 for the one-way ANOVA. Which pairs are significantly different?

20. Use software with Table 12.11, analyzed in Section 12.5.
a) Fit the no interaction model, and verify the results given there.
b) Fit the interaction model. Compare SSE for this model to SSE for the no interaction model. Show how the difference between these values relates to the partial sum of squares for testing interaction.
c) Using the prediction equation for the interaction model, find the six estimated cell means, and compare them to the sample means. (Note: The model uses six parameters to summarize six means, so it has a perfect fit.)

21. Consider the regression model $E(Y) = \alpha + \beta_1 G + \beta_2 R + \beta_3 (GR)$, where Y = income (thousands of dollars), G = gender ($G = 1$ for men and $G = 0$ for women), and R = race ($R = 1$ for whites and $R = 0$ for blacks).

a) Suppose that, in the population, $\beta_3 = 0$. Interpret β_1 and β_2.
b) The prediction equation for a certain sample is $\hat{Y} = 16+2G+3R+8GR$. By finding the four predicted means for this equation, show that the coefficient 8 of the interaction term is the amount by which the mean for one of the four groups must increase or decrease for the interaction to disappear.

22. Refer to Problem 12.7. Table 12.33 shows the result of using SAS to conduct a two-way ANOVA, with both gender and race as predictors. The first panel of the table shows the result of the model with an interaction term, and the other three panels show results after dropping that term.
a) Test the hypothesis of no interaction. Report the F test statistic, the P-value, and interpret.
b) Test the hypothesis for the race main effect, controlling for gender. Report the F test statistic, the P-value, and interpret.
c) Test the hypothesis for the gender main effect, controlling for race. Report the F test statistic, the P-value, and interpret.
d) Construct a 95% confidence interval for the gender main effect, and interpret.
e) Defining dummy variables, explain how the tests in (a)–(c) relate to regression models.

TABLE 12.33

Source	DF	Type III SS	Mean Square	F Value	Pr > F
RACE	1	2.557	2.557	1.45	0.2284
GENDER	1	0.161	0.161	0.09	0.7626
RACE*GENDER	1	1.194	1.194	0.68	0.4104

Source	DF	Sum of Squares	Mean Square	F Value	Pr > F
Model	2	5.658	2.829	1.61	0.2009
Error	947	1666.238	1.759		
Total	949	1671.896			

Source	DF	Type III SS	Mean Square	F Value	Pr > F
RACE	1	1.812	1.812	1.03	0.3104
GENDER	1	3.545	3.545	2.01	0.1561

Parameter		Estimate	T for H0: Parameter=0	Pr > |T|	Std Error of Estimate
INTERCEPT		4.1865	61.80	0.0001	0.0677
RACE	b	-0.1303	-1.01	0.3104	0.1284
	w	0.0000	.	.	.
GENDER	f	-0.1239	-1.42	0.1561	0.0873
	m	0.0000	.	.	.

23. Table 12.34 shows results of an ANOVA on Y = depression index and the predictors gender and marital status (married, never married, divorced).
a) State the regression model for this analysis.
b) State the sample size and fill in the blanks in the ANOVA table.
c) Interpret the result of the F test for no interaction.

24. The 25 women faculty in the humanities division of a college have a mean salary of \$46,000, whereas the five in the science division have a mean salary of \$60,000. On the

TABLE 12.34

Source	Sum of Squares	DF	Mean Square	F	Prob > F
Gender	100	—	—	—	—
Marital status	200	—	—	—	—
Interaction	100	—	—	—	—
Error	___	—	—		
Total	4000	205			

other hand, the 20 men in the humanities division have a mean salary of \$45,000, and the 30 men in the science division have a mean salary of \$59,000.

a) Construct a table of sample mean incomes for the 2×2 cross-classification of gender and division of college. Find the overall means for men and women. Interpret.

b) Discuss how the results of a one-way comparison of mean incomes by gender would differ from the results of a two-way comparison of mean incomes by gender, controlling for division of college. (Note: This reversal of which gender has the higher mean salary, according to whether one controls division of college, illustrates *Simpson's paradox*. See Problem 10.8d.)

25. Refer to Example 12.6 and Table 12.16.

a) Using software, conduct the repeated measures analyses of Section 12.6.

b) Suppose one scored the influence categories (1, 2, 3, 4, 5). Would this have any effect on the test statistic? Explain.

c) Suppose one used scores $(-3, -2, 0, 2, 3)$. What would this assume about the response categories? Repeat the analyses using these scores. Are the conclusions sensitive to the choice of scores?

26. Recently the General Social Survey asked respondents, "Compared with ten years ago, would you say that American children today are (1) much better off, (2) better off, (3) about the same, (4) worse off, or (5) much worse off." Table 12.35 shows responses for ten of the subjects on three issues: quality of their education, safety of the neighborhoods they live in, and getting health care when they need it.

TABLE 12.35

	Status of Children		
Subject	Education	Neighborhood	Health Care
1	4	4	3
2	2	4	2
3	3	3	4
4	1	2	1
5	3	4	3
6	2	5	4
7	1	4	2
8	3	3	3
9	4	5	3
10	2	4	2

a) Test the hypothesis that the true means are equal. Report the P-value, and interpret.
b) Use a multiple comparison method to construct simultaneous 94% confidence intervals for differences between pairs of means.
c) For each of the following, indicate whether it is a fixed effect, random effect, or response variable: (i) opinion, (ii) issue, (iii) subject.

27. Refer to the previous exercise. The first five respondents were female, and the last five were male. Analyze these data using both gender and issue as factors.
a) Identify the between-subjects and within-subjects factors.
b) Test for interaction. Interpret.
c) Test the main effect of gender. Interpret.
d) Test the main effect of issue. Interpret.
e) Use 95% confidence intervals to compare the means for the three issues. Interpret, and summarize your findings from these analyses.

28. Refer to Problem 12.5. When these subjects were asked to rate conservatives, the mean responses were 61.9 during 1983–87 and 60.7 during 1988–91. Explain why a two-way ANOVA using time (1983–87, 1988–91) and group rated (Liberal, Conservative) as factors would require methods for repeated measures. Identify the within-subjects and between-subjects factors. Would you expect the test of no interaction to yield a small P-value? Explain.

29. Upjohn, a pharmaceutical company, conducted a randomized clinical trial comparing an active hypnotic drug with a placebo for patients suffering from insomnia. The outcome is patient response to the question, "How quickly did you fall asleep after going to bed?" Patients suffering from insomnia were randomly assigned to receive the active drug or a placebo. The study measured patients' responses at the start and at the conclusion of a two-week treatment period. The study analyzed whether the active drug helps subjects with insomnia problems. However, patients taking placebo may also tend to fall asleep more quickly at the conclusion of the study period, because of the *placebo effect*—thinking they are taking a beneficial drug may psychologically have a positive effect. Is the improvement with the active drug better than with the placebo? Table 12.36 shows a SAS printout from an analysis of these data. "RESP1" refers to the response at the initial occasion, and "RESP2" the response at the follow-up occasion. The first part of the table shows results of one-way ANOVAs comparing the two groups separately at each time. (For the raw data, see A. Agresti, *Statistics in Medicine*, Vol. 8, 1989, p. 1209.) The sample means are 50.0 initially and 27.8 at follow up for the 120 subjects taking active drug, and 50.3 initially and 37.4 at follow up for the 119 subjects taking placebo.
a) For the 239 observations at the initial occasion, report and interpret the F test statistic for the one-way ANOVA comparing the two means.
b) For the 239 observations at the follow-up occasion, report and interpret the one-way ANOVA F test statistic. For this one-way ANOVA, show that the 95% confidence interval for the difference of means equals (4.2, 15.0). Interpret.
c) Now treat both occasions at once, with repeated-measures ANOVA. Report and interpret the F statistic for testing interaction.
d) Does it make sense to test main effects? If not, why not? If so, do so and interpret.
e) Using the sample means, interpret the interaction descriptively.
f) To make within-subjects comparisons of occasion means, for each treatment, show that a 95% confidence interval equals (18.0, 26.4) for the active drug and (8.7, 17.2) for placebo. Interpret.

TABLE 12.36

Analysis of Variance Procedure

Dependent Variable: RESP1

Source	DF	Sum of Squares	Mean Square	F Value	Pr > F
Model	1	8.418	8.418	0.02	0.9014
Error	237	129711.457	547.306		
Total	238	129719.874			

Dependent Variable: RESP2

Source	DF	Sum of Squares	Mean Square	F Value	Pr > F
Model	1	5508.633	5508.633	12.06	0.0006
Error	237	108282.999	456.890		
Total	238	113791.632			

Repeated Measures Analysis of Variance

Tests of Hypotheses for Between Subjects Effects

Source	DF	Anova SS	Mean Square	F Value	Pr > F
GROUP	1	2973.865	2973.865	4.08	0.0444
Error	237	172591.093	728.232		

Univariate Tests of Hypotheses for Within Subject Effects

Source: OCCASION

DF	Anova SS	Mean Square	F Value	Pr > F
1	36815.951883	36815.951883	133.41	0.0001

Source: OCCASION*GROUP

DF	Anova SS	Mean Square	F Value	Pr > F
1	2543.185897	2543.185897	9.22	0.0027

Source: Error(OCCASION)

DF	Anova SS	Mean Square
237	65403.362220	275.963554

TABLE 12.37

Tests of Between-Subjects Effects.

Source of Variation	SS	DF	MS	F	Sig of F
WITHIN + RESIDUAL	3584.03	69	51.94		
TREATMNT	644.23	2	322.12	6.20	.003

Tests involving 'TIME' Within-Subject Effect.

Source of Variation	SS	DF	MS	F	Sig of F
WITHIN + RESIDUAL	1955.37	69	28.34		
TIME	366.04	1	366.04	12.92	.001
TREATMNT BY TIME	307.32	2	153.66	5.42	.006

30. Table 12.37 shows results of using SPSS in Example 12.7 with Table 12.19.
a) Explain how to use the information in this table to conduct the test of no interaction between treatment and time.
b) Explain how to determine the *df* values for treatment, time, and their interaction.

31. Using software, conduct the repeated measures ANOVA in Example 12.7 with Table 12.19.

Concepts and Applications

32. Refer to the WWW data set (Problem 1.7), with response variable the number of weekly hours engaged in sports and other physical exercise. Using software, conduct an analysis of variance and follow-up estimation, and prepare a report summarizing your analyses and interpretations using
a) Gender as a predictor.
b) Gender and whether a vegetarian as predictors.

33. Refer to the data file created in Problem 1.7. For variables chosen by your instructor, use ANOVA methods and related inferential statistical analyses. Interpret and summarize your findings.

34. The General Social Survey has frequently asked respondents how they would rate various countries on a scale from −5 to +5, where −5 indicates the country is disliked very much and +5 indicates the country is liked very much. Table 12.38 shows counts of the responses regarding ratings of Russia for surveys from various years. Using software, analyze these data. Prepare a report providing your analyses and showing interpretations and conclusions.

TABLE 12.38

	Opinion About Russia									
Years	−5	−4	−3	−2	−1	+1	+2	+3	+4	+5
1982	100	22	30	15	46	28	11	21	3	26
1993	95	25	43	49	86	299	141	159	43	74
1994	42	12	25	28	38	121	80	64	24	38

35. A random sample of 26 female students at a major university were surveyed about their attitudes toward abortion. Each received a score on abortion attitude according to how many from a list of eight possible reasons for abortion she would accept as a legitimate reason for a woman to seek abortion. Thus, the higher the score, the more favorable the attitude toward abortion as an option in a variety of circumstances. The students were classified as "fundamentalist" or "nonfundamentalist" in their religious beliefs. They were also classified according to their church attendance frequency, "frequent" (more than once a month) or "infrequent." Table 12.39 displays the 26 abortion attitude scores, classified by religion and frequency of church attendance. Using software, analyze the data. Discuss your findings in a short report, indicating the models fitted, hypotheses tested, parameters estimated, and interpretations that follow from your analyses.

36. Table 12.40, based on data from the 1989 General Social Survey, is a contingency table summarizing responses of 29 subjects regarding government spending on the environ-

TABLE 12.39

		Religion	
		Fundamentalist	Nonfundamentalist
Church Attendance	Frequent	0, 3, 4, 0, 3 2, 0, 1, 1	2, 5, 1, 2 3, 3
	Infrequent	4, 3, 4	6, 8, 6, 4 6, 3, 7, 4

TABLE 12.40

	Cities								
	1			2			3		
Law Enforcement Environment	1	2	3	1	2	3	1	2	3
1	3	1	0	6	3	0	5	3	1
2	1	0	0	2	1	0	1	1	0
3	1	0	0	0	0	0	0	0	0

ment, assistance to big cities, and law enforcement. The response scale was 1 = too little, 2 = about right, 3 = too much. Analyze these data, using repeated measures ANOVA to compare the mean responses for the three types of spending.

37. Refer to the WWW data set in Problem 1.7. Use repeated measures analyses to model the weekly number of hours of recreation in terms of type of activity (levels S = sports and physical exercise, T = TV watching) and gender. Interpret results of all analyses, and summarize.

38. An experiment used four randomly selected groups of five individuals each. The overall sample mean was 60.
a) What did the data look like if the one-way ANOVA for comparing the means had test statistic $F = 0$?
b) What did the data look like if $F = \infty$?

39. Explain carefully the difference between a probability of Type I error of .05 for a single comparison of two means and a multiple comparison error rate of .05 for comparing all pairs of means.

40. In multiple comparisons following a one-way ANOVA, is it possible that group A is not significantly different from group B and group B is not significantly different from group C, yet group A is significantly different from group C? Explain.

41. Construct a numerical example of means for a two-way classification under the following conditions:
a) Main effects are present only for the row variable.
b) Main effects are present for each variable, but there is no interaction.
c) Interaction effects are present.
d) No effects of any type are present.

42. The null hypothesis of equality of means for a factor is rejected in a two-way ANOVA. Does this imply that the hypothesis will be rejected in a one-way ANOVA F test, if the data are collapsed over the levels of the second variable? Explain.

43. For a two-way classification of means by factors A and B, at each level of B the means are equal for the levels of A. Does this imply that the overall means are equal at the various levels of A, ignoring B? Explain the implications, in terms of how results may differ between two-way ANOVA and one-way ANOVA.

Select the correct response(s) in Problems 12.44–12.47.

44. Analysis of variance and regression are similar in the sense that

a) They both assume a quantitative response variable.

b) They both have F tests for testing that the response variable is statistically independent of the explanatory variable(s).

c) For inferential purposes, they both assume that the response variable Y is normally distributed with the same standard deviation at all combinations of levels of the explanatory variable(s).

d) They both provide ways of partitioning the variation in Y into "explained" and "unexplained" components.

45. One-way ANOVA provides relatively more evidence that H_0: $\mu_1 = \cdots = \mu_g$ is false

a) The smaller the "between" variation and the larger the "within" variation.

b) The smaller the "between" variation and the smaller the "within" variation.

c) The larger the "between" variation and the smaller the "within" variation.

d) The larger the "between" variation and the larger the "within" variation.

46. For four means, a multiple comparison method provides simultaneous 95% confidence intervals for the differences between the six pairs. Then

a) For each confidence interval, there is a .95 chance that it contains the true difference.

b) The probability that all six confidence intervals are correct is .70.

c) The probability that all six confidence intervals are correct is .95.

d) The probability that all six confidence intervals are correct is $(.95)^6$.

e) The probability is .05 that at least one confidence interval does not contain the true difference.

f) The confidence intervals are wider than separate 95% confidence intervals for each difference.

47. Interaction terms are needed in a two-way ANOVA model when

a) Each pair of variables is associated.

b) Both explanatory variables have significant effects in the model without interaction terms.

c) The difference in means between two categories of one explanatory variable varies greatly among the categories of the other explanatory variable.

d) The mean square for interaction is huge compared to the mean square error.

48. * You know the sample mean, standard deviation, and sample size for each of three groups. Can you conduct an ANOVA F test, or would you need more information?

49. * You form a 95% confidence interval in five different situations.

a) Assuming that the results of the intervals are statistically independent, find the probability that *all* five intervals contain the parameters they are designed to estimate. Find the probability that at least one interval is in error. (*Hint*: Use the binomial distribution.)

b) What confidence coefficient should you use for each interval so that the probability that all five intervals contain the parameters equals exactly .95? Compare this to the confidence coefficient for each interval used in the Bonferroni method.

50. * Show how to construct a regression model for the analysis of mean income over a three-way classification of gender (male, female), race (white, black), and type of job (assembly-line worker, janitorial). Interpret the parameters in the model. Assume that there is no interaction.

51. * This exercise motivates the formula for the between estimate of the variance in one-way ANOVA. Suppose the sample sizes all equal n and the population means all equal μ. The sampling distribution of each $\bar{Y}_i$ then has mean μ and variance σ^2/n. The sample mean of the $\bar{Y}_i$ values, when the sample sizes are equal, is simply $\bar{Y}$.
a) Treating $\bar{Y}_1, \bar{Y}_2, \ldots, \bar{Y}_g$ as g observations having the sample mean $\bar{Y}$, explain why $\sum(\bar{Y}_i - \bar{Y})^2/(g-1)$ estimates the variance σ^2/n of the distribution of the $\bar{Y}_i$-values.
b) Using (a), explain why $\sum n(\bar{Y}_i - \bar{Y})^2/(g-1)$ estimates σ^2. For the unequal sample size case, replacing n by n_i yields the between estimate.

52. The ***Scheffé multiple comparison*** method for comparing g groups with overall error rate α provides interval for $\mu_i - \mu_j$ of

$$(\bar{Y}_i - \bar{Y}_j) \pm \hat{\sigma}\sqrt{(g-1)F_\alpha\left(\frac{1}{n_i} + \frac{1}{n_j}\right)}$$

where F_α denotes the value from the F distribution with $df_1 = g - 1$ and $df_2 = N - g$ having right-tail probability α. Apply this method to Example 12.3, interpret results, and compare the intervals to the ones for the Bonferroni method.

Bibliography

Howell, D. C. (1992) *Statistical Methods for Psychology*, 3rd ed. Belmont, CA: Duxbury.

Kirk, R. E. (1995) *Experimental Design: Procedures for the Behavioral Sciences*, 3rd ed. Pacific Grove, CA.: Brooks/Cole.

Lehmann, E. L. (1975). *Nonparametrics: Statistical Methods Based on Ranks*. San Francisco: Holden-Day.

Neter, J., Kutner, M. H., Nachtsheim, C. J., and Wasserman, W. (1996). *Applied Linear Statistical Models*, 4th ed. Homewood, IL.: Richard D. Irwin.

Winer, B. J., Brown, D. R., and Michels, K. M. (1991). *Statistical Principles in Experimental Design*, 3rd ed. New York: McGraw-Hill.

Chapter 13

Combining Regression and ANOVA: Analysis of Covariance

Chapters 9 and 11 introduced regression analysis as a method for describing the relationship between a quantitative response variable and *quantitative* explanatory variables. Chapter 12 showed that regression can also model the relationship between a quantitative response variable and *qualitative* explanatory variables, as in analysis of variance with dummy variables. We now show that regression can handle simultaneously both quantitative and qualitative explanatory variables. The model combines elements of standard regression analysis, for which the predictors are quantitative, and analysis of variance, for which the predictors are qualitative.

One could, for example, model mental impairment as a function of a life events index, socioeconomic status (quantitative predictors), marital status, and gender (qualitative predictors). The model takes the ordinary multiple regression form, but with dummy variables for each qualitative predictor.

Controlling for a Covariate

One-way ANOVA compares the mean of the response variable for several groups, treated as categories of a qualitative explanatory variable. In many applications, it is natural to do this while simultaneously controlling for another quantitative variable. For example, in comparing mean income for men and women, we might control for possibly differing levels of job experience between men and women. The quantitative control variable is called a ***covariate***. The use of regression for this type of comparison is called ***analysis of covariance***, and is one of the many statistical contributions of the brilliant British statistician, R. A. Fisher.

ANOVA ignores, rather than controls, quantitative covariates. Typically, a covariate is correlated with Y and also associated with the qualitative predictor, in the sense that some groups tend to have higher values than others on the covariate. When this happens, the results of analysis of covariance differ from the results of analysis of variance. For instance, job experience is usually positively correlated with income; thus, if men tend to have higher levels of experience than women at a particular job, the results of a comparison of mean income for men and women will depend on whether we control for experience.

For simplicity, this chapter illustrates concepts using only two explanatory variables, one qualitative and one quantitative. The first section describes two different emphases, according to whether the quantitative or the qualitative predictor is the control variable. Sections 13.2 and 13.3 show that regression models with dummy variables provide the basis for the analyses. Section 13.4 presents the primary tests for interaction and for the main effects. Section 13.5 presents adjusted means on the response variable for the categories of the qualitative predictor which, unlike ordinary means, control for the quantitative covariate.

13.1 Comparing Means and Comparing Regression Lines

Denote the response variable by Y, the quantitative explanatory variable by X, and the qualitative explanatory variable by Z. When the qualitative predictor has two categories, Z is simply a dummy variable; when it has several categories, Z symbolizes a set of dummy variables.

The interpretations differ for the quantitative and qualitative predictor effects. For the quantitative predictor X, the analysis has a regression flavor. It refers to the regression between Y and X within each category of the qualitative variable Z, treating Z as a control variable. For the qualitative predictor Z, the analysis has an ANOVA flavor. It refers to comparing the means of Y for the groups defined by the various categories of the qualitative variable Z, treating X as the control variable. The next two subsections illustrate this distinction.

Comparing Regression Lines

We first illustrate the case in which X is the predictor of interest, Z being a control. Refer to Table 9.4 (Section 9.5) on Y = selling price of home, using the quantitative predictor X = size of home and the qualitative predictor Z = whether a house is new (1 = yes, 0 = no). Studying the effect of X on Y while controlling for Z is equivalent to analyzing the regression between X and Y separately for new and older homes. We could find the best-fitting straight line for each set of points, one line for new homes and a separate line for older homes. We could then compare the relationships for new and older homes by comparing characteristics of the lines, for instance, whether they climb with similar or different slopes.

Whenever we study three or more variables simultaneously, *interaction* plays a central role. In this context, no interaction means that the true slope of the relationship between X and the mean of Y is the same for each category of Z. For instance, the true slope of the line relating selling price and size of home is the same for new and older homes. Equality of slopes implies that the regression lines are parallel (see Figure 13.1a). When the Y-intercepts are also equal, the regression lines coincide (see Figure 13.1b).

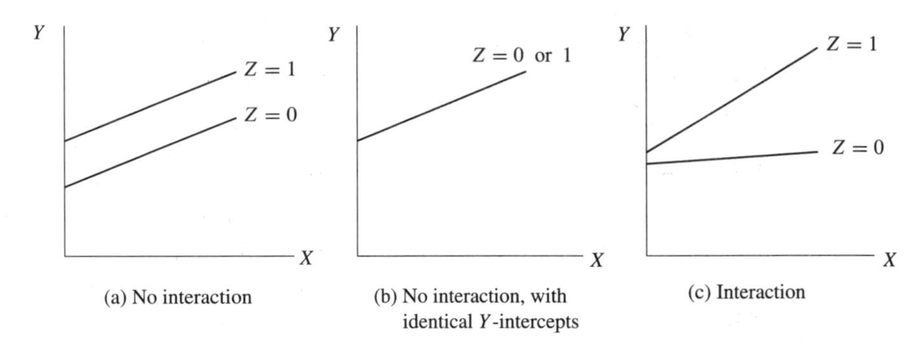

Figure 13.1 Regression Lines Between Quantitative Variables Y and X, Within Categories of a Qualitative Variable Z

If the slopes of the regression lines are not the same, then there is interaction. If the increase in selling price as a function of size of home differed for new and existing homes, for example, then the two regression lines would not be parallel (see Figure 13.1c). A comparison of the regression equations first checks for interaction by testing whether the best-fitting straight lines have significantly different slopes. If interaction exists, then separate lines with different slopes describe the system. If the hypothesis of no interaction seems plausible, then we can use lines having identical slopes.

The relationship between X and Y while controlling for a qualitative variable Z may differ in substantial ways from the relationship in the absence of control. For instance, the association between X and Y could disappear when we control for Z. Figure 13.2 displays a set of points having an overall positive relationship between Y and X when the control variable is ignored. Within each category of the control variable, however, the regression line relating Y to X is horizontal. The overall positive trend is due to the tendency for the categories with high (low) scores on Y to have high (low) scores on X also.

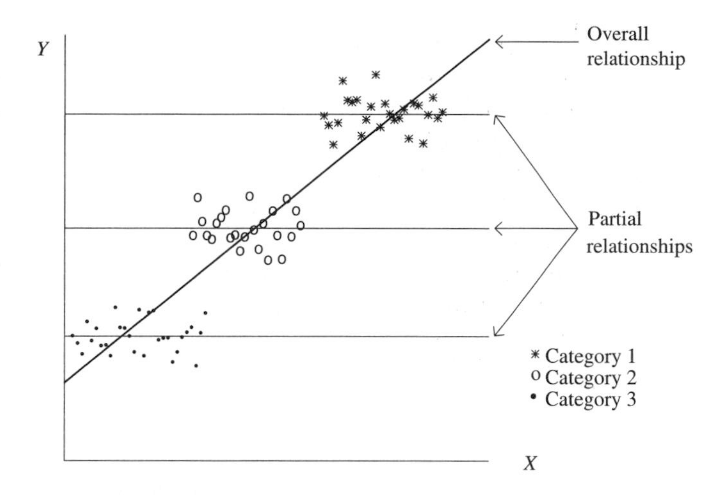

Figure 13.2 An Association Between X and Y That Disappears After Controlling for a Qualitative Variable

Comparing Means on Y, Controlling for X

When the qualitative variable Z is the predictor of interest, the goal is to compare the mean of Y for the groups defined by the categories of Z. Ignoring X, one could do this using one-way ANOVA methods. Instead of ignoring X, however, one can treat it as a control variable.

To illustrate, consider the relationship between Y = annual income and Z = gender for managerial employees of a chain of fast-food restaurants. From a two-sample comparison of men and women, mean annual income is higher for men than for women. In this company, annual income of managers tends to increase with X = number of years of experience. In addition, only recently have women received managerial appointments, so on the average they have less experience than the men. In summary, men tend to have greater experience, and greater experience tends to correlate with higher income. Perhaps this is why the overall mean annual income is higher for men. A chain relationship may exist, with gender affecting experience, which itself affects income. The difference in mean incomes between men and women could disappear when we control for experience.

To study whether this difference in mean incomes can be explained by differences in experience levels of men and women, we compare mean incomes for men and women having equal levels of experience. If there is no interaction, then the regression line between income and experience for the male employees is parallel to the one for the female employees. In that case, the difference between mean incomes for men and women is identical for all fixed values of X = number of years of experience. Figure 13.3a illustrates this, where the mean income for men is the same amount higher than the mean income for women at every experience level X. If the same regression line

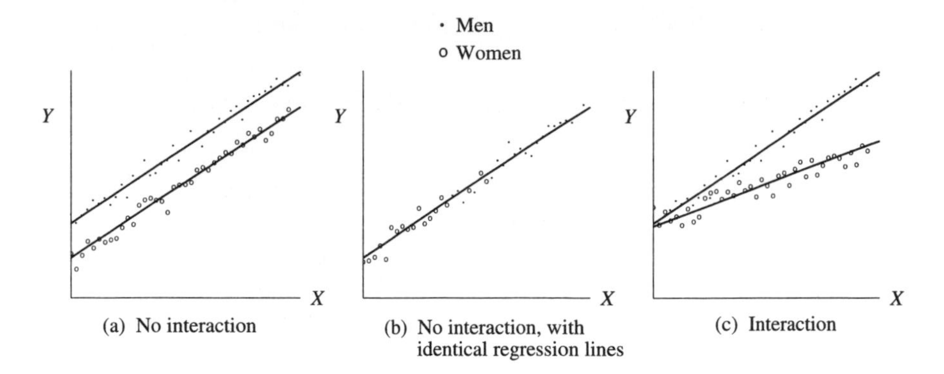

Figure 13.3 Y = Income by X = Number of Years Experience and Z = Gender

applies to each gender, as in Figure 13.3b, the mean income for each gender is identical at each level of experience. In that case, no difference occurs between male and female incomes, controlling for experience.

If interaction exists, then the regression line for men is not parallel to the regression line for women. In that case, the difference between the mean incomes of men and women varies by level of experience. For the lines depicted in Figure 13.3c, for example, the mean income for men is larger than the mean income for women at all experience levels, but the difference between the means increases as X increases. Thus, the discrepancy in mean incomes is relatively larger for individuals with more experience.

The steps in the analysis parallel those described earlier. We first check for interaction. If the differences between the group means can be assumed to be the same at all X-values, then we can estimate those differences or test whether they all equal 0.

If the groups have considerably different distributions on X, then the results of the analysis of covariance may differ considerably from the results of the analysis of variance, which ignores rather than controls X. For example, Figure 13.3b depicts a situation in which the sample mean income for men is greater than that for women. The two-sample comparison of means H_0: $\mu_1 = \mu_2$ would probably have a small P-value. However, the reason for the difference in this case is that men have more experience. In fact, the same regression line fits the relationship between income and experience for both genders. It appears that the mean incomes are equal, controlling for experience. Figure 13.3a, on the other hand, depicts a situation in which men have higher mean incomes, both ignoring and controlling for experience.

13.2 Analysis of Covariance Model

Multiple regression models can represent the relationship between the mean of Y and a quantitative predictor X and qualitative variable Z. This section presents the model without interaction, for which regression lines between Y and X are parallel within categories of Z.

Quantitative and Dummy Explanatory Variables

Ignoring X, the model relating Y to a qualitative predictor Z is the one-way ANOVA model. From Section 12.3, if Z has g categories, that model is

$$E(Y) = \alpha + \beta_1 Z_1 + \beta_2 Z_2 + \cdots + \beta_{g-1} Z_{g-1}$$

where $Z_1, \ldots, Z_{g-1}$ are dummy variables: $Z_1 = 1$ for observations from the first category and $Z_1 = 0$ otherwise; $Z_2 = 1$ for observations from the second category and $Z_2 = 0$ otherwise; and so forth. For observations from the last (gth) category, all dummy variables equal 0. Thus, $Z_1, \ldots, Z_{g-1}$ indicate the category of the qualitative variable for an observation. The β_i-values are differences in means of Y between the various categories and the last one.

We include the quantitative covariate X in the model in the usual linear manner. This gives the ***analysis of covariance*** model,

$$E(Y) = \alpha + \beta X + \beta_1 Z_1 + \cdots + \beta_{g-1} Z_{g-1}$$

where β (without a subscript) describes the partial effect of X on the mean of Y for the g groups. Successively setting each Z equal to 1 with the other Z-values equal to 0 provides the regression lines relating X to the mean of Y for each category of the qualitative variable.

Example 13.1 Regression of Income on Education and Racial–Ethnic Group

For a sample of adult Americans aged over 25, Table 13.1 shows data on Y = annual income (thousands of dollars), X = number of years of education (where 12 = high school graduate, 16 = college graduate), and Z = racial–ethnic group (black, Hispanic, white). The data exhibit patterns of a much larger sample recently taken by the U.S. Bureau of the Census. The sample contains $n_1 = 16$ blacks, $n_2 = 14$ Hispanics, and $n_3 = 50$ whites, for a total sample size of $N = 80$.

Table 13.2 shows mean income and education values for these subjects. Although the mean incomes differ among the three groups, these differences could result from the differing educational levels of the groups. For instance, though white subjects had higher mean incomes than blacks or Hispanics, they also had higher mean education. Perhaps the differences would disappear if we could control for education, making comparisons among the racial–ethnic groups at fixed levels of education.

Let's consider the regression model

$$E(Y) = \alpha + \beta X + \beta_1 Z_1 + \beta_2 Z_2$$

where

Y = income;
X = number of years of education;
Z_1 = 1 if subject is black, $Z_1 = 0$ otherwise;
Z_2 = 1 if subject is Hispanic, $Z_2 = 0$ otherwise;
$Z_1 = Z_2 = 0$ if subject is white.

TABLE 13.1 Observations on Y = Annual Income (in Thousands of Dollars) and X = Number of Years of Education, for Three Racial–Ethnic Groups

Black		Hispanic		White		White		White	
Y	X	Y	X	Y	X	Y	X	Y	X
8	10	16	16	15	14	31	16	25	16
9	7	8	11	24	14	12	10	25	14
13	9	10	10	20	7	25	13	11	11
8	11	29	16	42	18	16	10	13	12
17	14	15	12	25	10	17	16	23	16
11	12	13	10	19	12	26	18	11	9
21	16	10	8	15	12	12	12	12	9
21	16	20	12	38	16	11	14	32	14
8	9	16	10	24	16	10	13	14	12
10	10	11	11	18	11	15	14	16	12
33	16	10	10	20	11	12	13	19	14
13	12	28	14	22	12	60	18	22	12
10	10	16	12	15	10	11	10	11	12
15	15	15	11	30	15	41	16	9	10
10	10			12	9	9	12	12	12
15	19			44	17	13	12	28	20
				23	16	52	14		

TABLE 13.2 Mean Income and Education, by Racial–Ethnic Group

	Black	Hispanic	White	Overall
Mean income	$\bar{Y}_1 = 13.9$	$\bar{Y}_2 = 15.5$	$\bar{Y}_3 = 21.2$	$\bar{Y} = 18.8$
Mean education	$\bar{X}_1 = 12.2$	$\bar{X}_2 = 11.6$	$\bar{X}_3 = 13.1$	$\bar{X} = 12.7$
Sample size	$n_1 = 16$	$n_2 = 14$	$n_3 = 50$	$N = 80$

Table 13.3 shows part of the printout from using software (PROC GLM in SAS) to fit the model. The prediction equation is

$$\hat{Y} = -7.8 + 2.2X - 5.4Z_1 - 2.5Z_2$$

Substituting dummy variable values yields equations for the three groups. For blacks, for instance, the prediction equation results from letting $Z_1 = 1$ and $Z_2 = 0$; then,

$$\hat{Y} = -7.8 + 2.2X - 5.4(1) - 2.5(0) = -13.3 + 2.2X$$

The prediction equations for the other two racial–ethnic groups are

$$\begin{aligned} \hat{Y} &= -10.3 + 2.2X \quad \text{(Hispanics)} \\ \hat{Y} &= -7.8 + 2.2X \quad \text{(whites)} \end{aligned}$$

TABLE 13.3 Printout for Fitting Model Assuming No Interaction to Data from Table 13.1 on Y = Income, X = Education, and Z = Racial–Ethnic Status

```
Dependent Variable: INCOME
                          Sum of              Mean
  Source        DF        Squares            Square     F Value     Pr > F
  Model          3      3899.4255         1299.8085      21.75      0.0001
  Error         76      4541.0620           59.7508
  Total         79      8440.4875

          R-Square    0.4620            Root MSE    7.7299

                                   T for H0:      Pr > |T|   Std Error of
Parameter          Estimate      Parameter=0                   Estimate
INTERCEPT           -7.8317          -1.86         0.0665        4.2061
EDUC                 2.2158           7.16         0.0001        0.3096
RACE          b     -5.4372          -2.43         0.0174        2.2365
              h     -2.4669          -1.04         0.3036        2.3816
              w      0.0000            .             .             .
```

Figure 13.4 is a scatter diagram showing the relationship between Y and X for the three groups. That diagram also plots the prediction equations for the three groups. They are parallel, since they each have the same slope, 2.2. This is the coefficient of X in the overall prediction equation. □

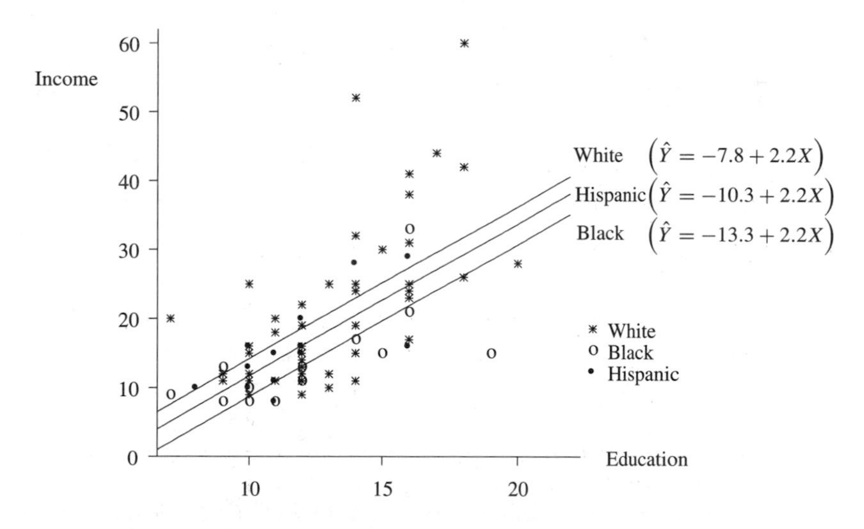

Figure 13.4 Plot of Prediction Equation for Model Assuming No Interaction

Interpretation of Parameters

We return now to the general case and discuss parameter interpretation. With $Z_1 = 1$ with the other dummies equal to 0, the relationship between X and the mean of Y within the first category is

$$\begin{aligned} E(Y) &= \alpha + \beta X + \beta_1(1) + \beta_2(0) + \beta_3(0) + \cdots + \beta_{g-1}(0) \\ &= (\alpha + \beta_1) + \beta X \end{aligned}$$

For the first category of the qualitative predictor, the mean of Y is linearly related to X with slope β and Y-intercept $\alpha + \beta_1$.

Similarly, in the second category, $E(Y) = (\alpha + \beta_2) + \beta X$. Generally, in the ith category, the relationship between X and the mean of Y is a straight line with slope β and Y-intercept $\alpha + \beta_i$. All the Z_i-values equal 0 in the last category, and that regression equation reduces to $E(Y) = \alpha + \beta X$.

Notice that the coefficient β of X is the slope of the regression lines between Y and X for all categories. The equation $E(Y) = \alpha + \beta X$ is the regression equation for the final category. The coefficients of the dummy variables tell us how the Y-intercept changes for the other categories. For instance, β_1 is the difference between the Y-intercept for the first and the final categories. In fact, since the regression lines are parallel, β_1 is the vertical distance between those two regression lines *at any fixed value of* X. That is, controlling for X, β_1 is the difference between the means of Y for the first and last categories. Table 13.4 displays the separate equations and the parameter interpretations for the case of $g = 3$ categories. Analogous results hold for sample data and their prediction equations.

TABLE 13.4 Summary of Equations and Parameters for Analysis of Covariance Model with No Interaction, When Qualitative Predictor Has $g = 3$ Categories

Category	Y-Intercept	Slope	Mean at Fixed X, $E(Y)$	Difference From Mean of Category 3, Controlling for X
1	$\alpha + \beta_1$	β	$(\alpha + \beta_1) + \beta X$	β_1
2	$\alpha + \beta_2$	β	$(\alpha + \beta_2) + \beta X$	β_2
3	α	β	$\alpha + \beta X$	0

We illustrate with the prediction equation $\hat{Y} = -7.8 + 2.2X - 5.4Z_1 - 2.5Z_2$, where Z_1 is a dummy variable for blacks and Z_2 is one for Hispanics. The coefficient -5.4 of Z_1 is the vertical distance between the lines for blacks and whites. The estimated mean annual income is \$5400 lower for blacks than for whites, at each fixed level of education. Similarly, the coefficient -2.5 of Z_2 represents the estimated difference ($-$\$2500) between mean income for Hispanics and whites, controlling for education.

In summary, the coefficients of the dummy variables estimate differences in means between each category and the final category, which does not have its own dummy variable. One can use these to estimate differences between any two categories. To illustrate, the difference between the means of Y in the first and second categories, at a fixed

X-value, is

$$[(\alpha + \beta_1) + \beta X] - [(\alpha + \beta_2) + \beta X] = \beta_1 - \beta_2$$

Figure 13.5 graphically displays the model for $g = 3$ categories.

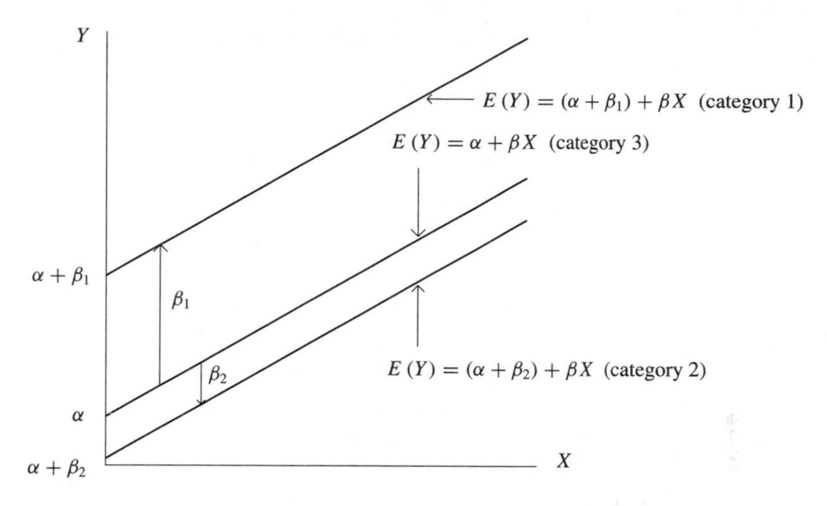

Figure 13.5 Graphic Portrayal of a Model with No Interaction, When the Qualitative Predictor Has Three Categories

13.3 Analysis of Covariance, Permitting Interaction

The analysis of covariance model that allows no interaction has the same slope for the relationship between X and Y in each category of Z. A generalized model, permitting interaction, allows different slopes.

Modeling Interaction with Cross-product Terms

We model interaction, as usual, by taking cross-products of the explanatory variables. It is not necessary to take cross-products of different dummy variables from the same predictor, such as $Z_1 Z_2$; only one Z can be nonzero at once, so such products always equal 0.

Example 13.2 Regression of Income on Education and Racial–Ethnic Group, Permitting Interaction

We illustrate this more general form of model by fitting it to Table 13.1. The model allows interaction between education and racial–ethnic group in their effects on income. The model has dummy variables Z_1 and Z_2 for two of the three groups and their cross-products XZ_1 and XZ_2 with the education predictor. Computer software provides the results shown in Table 13.5.

From Table 13.5, the overall prediction equation is

$$\hat{Y} = -12.9 + 2.6X + 9.7Z_1 + 4.6Z_2 - 1.2(XZ_1) - .6(XZ_2)$$

The prediction equation with both dummy variables equal to zero ($Z_1 = Z_2 = 0$) refers to the final racial–ethnic category, namely, whites. For that group,

$$\hat{Y} = -12.9 + 2.6X + 9.7(0) + 4.6(0) - 1.2X(0) - .6X(0) = -12.9 + 2.6X$$

TABLE 13.5 Printout for Fitting Interaction Model to Data from Table 13.1 on Income, Education, and Racial–Ethnic Status

Source	DF	Sum of Squares	Mean Square	F Value	Pr > F
Model	5	4072.385	814.477	13.80	0.0001
Error	74	4368.103	59.028		
Total	79	8440.488			

R-Square 0.4825 Root MSE 7.683

Parameter		Estimate	T for H0: Parameter=0	Pr > |T|	Std Error of Estimate
INTERCEPT		-12.9344	-2.46	0.0161	5.249
EDUC		2.6048	6.65	0.0001	0.391
RACE	b	9.6666	1.06	0.2940	9.146
	h	4.6320	0.38	0.7039	12.140
	w	0.0000	.	.	.
EDUC*RACE	b	-1.2053	-1.70	0.0933	0.709
	h	-0.5604	-0.56	0.5781	1.003
	w	0.0000	.	.	.

Permitting interaction, the prediction equation in the other categories may have a different slope and intercept. For the first category (blacks), $Z_1 = 1$ and $Z_2 = 0$, and the prediction equation simplifies to

$$\hat{Y} = -3.3 + 1.4X$$

Finally, for Hispanics ($Z_1 = 0$, $Z_2 = 1$), and the prediction equation is

$$\hat{Y} = -8.3 + 2.0X$$

The coefficient of Z_1 (i.e., 9.7) again describes the difference between the Y intercepts for the first and the last category. However, this is the difference *only* at $X = 0$, since the equations for the two categories have different slopes. Since the coefficient of X (i.e., 2.6) represents the slope for the last (third) category, the coefficient of (XZ_1)

(i.e., -1.2) represents the *difference in slopes* between the first and last categories. The two lines are parallel only when that coefficient equals 0. Similarly, for the second category, the coefficient of Z_2 is the difference between the Y-intercepts for the second and the last category, and the coefficient of (XZ_2) is the difference between their slopes. Table 13.6 summarizes the interpretations of the estimated parameters in the model.

TABLE 13.6 Summary of Prediction Equation Allowing Interaction, for Y = Income, X = Education, and Z = Racial–Ethnic Status

Category	Y-Intercept	Slope	$\hat{Y} = -12.9 + 2.6X + 9.7Z_1 + 4.6Z_2 - 1.2(XZ_1) - .6(XZ_2)$	Difference From Category 3 of Y-Intercept	Difference From Category 3 of Slope
1 (Black)	$-12.9 + 9.7$	$2.6 - 1.2$	$(-12.9 + 9.7) + (2.6 - 1.2)X$	9.7	-1.2
2 (Hispanic)	$-12.9 + 4.6$	$2.6 - .6$	$(-12.9 + 4.6) + (2.6 - .6)X$	4.6	$-.6$
3 (White)	-12.9	2.6	$-12.9 + 2.6X$	0	0

Figure 13.6 plots these three prediction equations. The sample slopes are all positive. Over nearly all the range of education values measured, whites have the highest estimated mean income, and blacks have the lowest. □

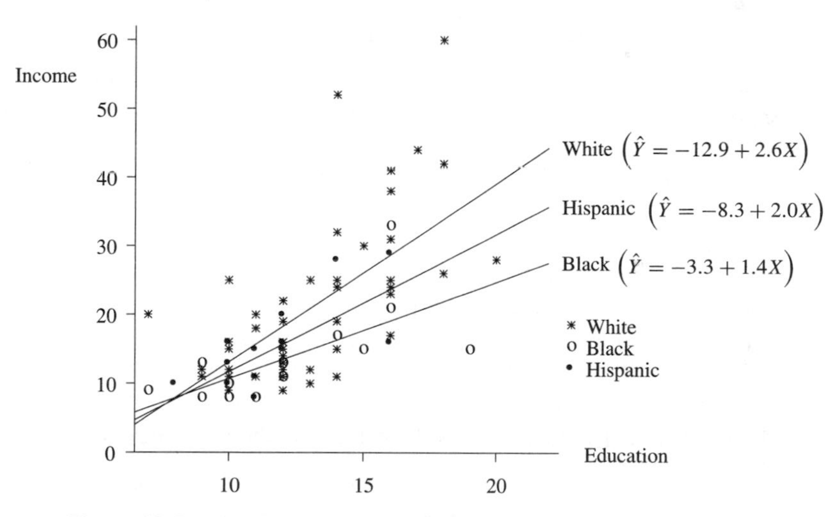

Figure 13.6 Plot of Prediction Equations for Model with Interaction Terms

In essence, the interaction model corresponds to fitting separate lines for each group. The prediction equations are the same as if one did a separate regression analysis for the data from each group.

Summary of Interpretations

In the model permitting interaction, the coefficients of the cross-product terms represent the differences between the slopes for the various categories and the slope for the final category of the qualitative variable. When all those coefficients equal 0, the slopes are all equal. In that case, the model reduces to the no interaction model studied in the previous section.

The coefficients of the dummy variables alone represent the differences between the Y-intercepts for the various categories and the Y-intercept for the final category. When the lines are not parallel, these coefficients are the vertical distances between the various lines and the line for the last category *only* at $X = 0$.

When interaction exists, the difference between means of Y for two groups varies as a function of X. For example, the difference between the estimated mean of Y for whites and Hispanics at a particular X-value is

$$(-12.9 + 2.6X) - (-8.3 + 2.0X) = -4.6 + .6X$$

This depends on the value of X. As X changes, the difference in the estimated means changes as a linear function of X. In this case, the difference between the estimated means is larger at higher levels of education. Figure 13.6 shows that the difference between the mean incomes of whites and blacks also gets larger at higher education levels.

Comparing R^2 for Different Models

To summarize how much better the model permitting interaction fits, we can check the increase in the coefficient of multiple determination, R^2. This measure has the usual interpretation. It is the proportional reduction in error from using the prediction equation to predict the response Y instead of using the overall mean $\bar{Y}$.

From the printout for the no-interaction model (Table 13.3), $R^2 = .462$. For it, SSE is the sum of squared prediction errors from predicting income using three separate but parallel lines. From the printout for the interaction model (Table 13.5), $R^2 = .482$. For it, SSE is the sum of squared errors from predicting income using three separate lines with different slopes. There is little to be gained by fitting the more complex model, as R^2 does not increase much.

Regression with Multiple Qualitative and Quantitative Predictors

Analysis of covariance models are simply regression models that include both quantitative and qualitative explanatory variables. For simplicity, our examples use only a single variable of each type. It is straightforward to generalize the model to include additional predictors of either type.

To introduce additional quantitative variables, simply add a βX term for each one. To introduce another qualitative variable, add a set of dummy variables for its categories. To permit interaction, introduce cross-product terms: Cross-products of X terms for interaction between quantitative predictors, cross-products between an X and a set of dummy variables for interaction between a quantitative and qualitative predictor, and cross-products between dummy variables for interaction between two qualitative predictors.

When there are several quantitative and qualitative predictors, the number of potential models is quite large. For each variable, we must consider both its main effect as well as its interaction with other variables. Moreover, some variables may overlap considerably in the variation they explain in the response variable. Chapter 14 discusses issues to consider in selecting a model.

13.4 Inference for Analysis of Covariance Models

We now present statistical inference for analysis of covariance models. This section presents significance tests and the following section presents estimation methods. As in other multivariable models, we first test the hypothesis of no interaction. This test analyzes whether the regression lines between Y and X are parallel for the different categories of the qualitative variable. If a lack of interaction is plausible, then further analyses of the main effects treat the regression lines as parallel.

One can test hypotheses about model parameters using the F test comparing complete and reduced regression models, introduced in Section 11.6. For instance, the test of no interaction compares the complete model containing the interaction terms to the reduced model deleting them. The test statistic for comparing models is

$$F = \frac{(\text{SSE}_r - \text{SSE}_c)/df_1}{\text{SSE}_c/df_2} = \frac{(R_c^2 - R_r^2)/df_1}{(1 - R_c^2)/df_2}$$

where SSE_r and SSE_c are the sums of squared errors for the reduced and complete models, df_1 equals the difference between the number of terms in the two models, and df_2 is the df value for the mean square for the complete model; equivalently, one can use the second expression for the R^2 values of the complete and reduced models. Software presents results of such tests in an ANOVA table.

Test of No Interaction

For a qualitative variable, the model permitting interaction has cross products of its dummy variables with the quantitative predictor, X. If there is no interaction, all coefficients of those cross-product terms equal 0. Hence, the null hypothesis of no interaction is that those parameters equal 0. The model under the null hypothesis is the reduced model without the cross-product terms, which has a common slope for all lines. Figure 13.7 depicts the two hypotheses for this test. This test has a small P-value if the addition of the cross-product interaction terms provides a significant improvement in the fit.

Example 13.3 Testing Interaction of Education and Racial–Ethnic Group on Income

For Table 13.1, we now test the hypothesis of no interaction between education and racial–ethnic group, in their effects on income. The complete model contains two in-

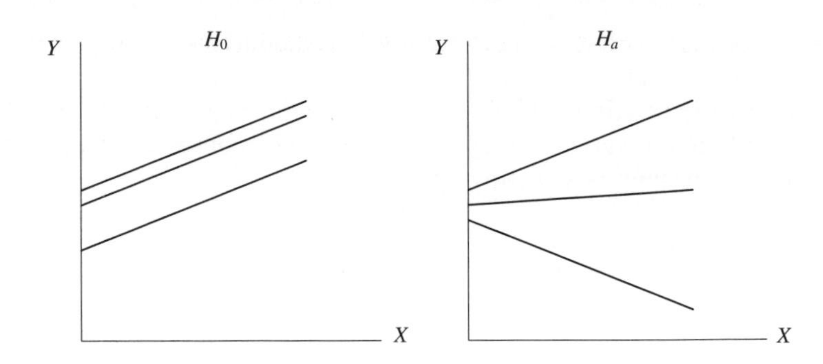

Figure 13.7 Graphical Representation of Null and Alternative Hypotheses in a Test of No Interaction, When the Qualititative Variable Has Three Categories

teraction terms, and Table 13.5 showed its printout. The model has a sum of squared errors of $\text{SSE}_c = 4368.1$, with $df = 74$.

The reduced model for the null hypothesis of no interaction is

$$E(Y) = \alpha + \beta X + \beta_1 Z_1 + \beta_2 Z_2$$

From Table 13.3, its sum of squared errors equals $\text{SSE}_r = 4541.1$. The difference in the number of terms in the two models is $df_1 = 2$, the number of interaction terms. The F test statistic for testing H_0: No interaction, based on $df_1 = 2$ and $df_2 = 74$, equals

$$F = \frac{(\text{SSE}_r - \text{SSE}_c)/df_1}{\text{SSE}_c/df_2} = \frac{(4541.1 - 4368.1)/2}{4368.1/74} = \frac{173.0/2}{4368.1/74} = \frac{86.5}{59.0} = 1.5$$

Software reports a P-value of .24. There is not much evidence of interaction, so we are justified in using the simpler model without cross-product terms. □

Table 13.7 shows how software summarizes sums of squares explained by various sets of terms in the model with interaction terms. The variability explained by the interaction terms, 173.0, simply equals the difference between the SSE values without and with those terms in the model. These sums of squares are ***partial sums of squares***, also called ***Type III sums of squares*** by some software (see Problem 11.50). They represent the variability explained by a variable after the other variables are already in the model.

TABLE 13.7 Computer Printout of Partial Sums of Squares Explained by Education, Racial–Ethnic Group, and Their Interaction, in the Analysis of Covariance Model Permitting Interaction

Source	DF	Type III SS	Mean Square	F Value	Pr > F
EDUC	1	1593.377	1593.377	26.99	0.0001
RACE	2	66.830	33.415	0.57	0.5702
EDUC*RACE	2	172.959	86.480	1.47	0.2377

Test for Main Effect of Qualitative Variable, Controlling for X

Possibly the model can be even further simplified. We next test the main effect for the qualitative predictor, racial–ethnic group. The null hypothesis states that each racial–ethnic group has the same regression line between X and Y; equivalently, each group has the same mean on Y, controlling for X.

For a qualitative predictor with three categories, this test compares the complete model

$$E(Y) = \alpha + \beta X + \beta_1 Z_1 + \beta_2 Z_2$$

to the reduced model

$$E(Y) = \alpha + \beta X$$

lacking effects of the qualitative predictor. The complete model, the no interaction model, represents three different but parallel regression lines between income and education, one for each racial–ethnic group. The reduced model states that the same regression line applies for all three groups. Figure 13.8 depicts this test. The P-value is small if the complete model with separate parallel lines provides a significantly better fit to the data than the reduced model of a common line.

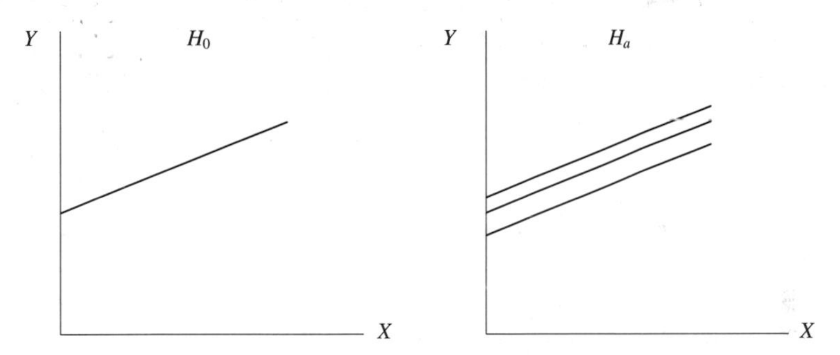

Figure 13.8 Graphical Representation of Null and Alternative Hypotheses in a Test of Equivalence of Regression Lines, When the Qualitative Variable Has Three Categories (Test Assumes No Interaction)

The null hypothesis for this test is

$$H_0 : \beta_1 = \beta_2 = 0$$

for the no interaction model. This hypothesis states that for each i, the difference β_i between the mean of Y in category i and in the last category is 0, controlling for X. This implies that the corresponding difference of means equals 0 between any two categories, controlling for X; that is, the mean of Y is the same for all groups, controlling for X. This resembles the null hypothesis in one-way ANOVA, except that this test compares the means while controlling for X.

Example 13.4 Testing Effect of Racial–Ethnic Group on Income, Controlling for Education

From Table 13.3, the sum of squared errors for the complete model is $\text{SSE}_c = 4541.1$, with $df_2 = 76$. Since the complete model here was the reduced model in the test of no interaction, SSE_c takes the same value as SSE_r in the previous test.

If H_0: $\beta_1 = \beta_2 = 0$ is true, the reduced model $E(Y) = \alpha + \beta X$ is appropriate, meaning that the same line represents the relationship between income and education for all three racial–ethnic groups. Table 13.8 shows part of its printout. The sum of squared errors is $\text{SSE}_r = 4906.2$. The difference between the number of terms in the complete and reduced models is $df_1 = 2$, the number of dummy variables. The F statistic for testing H_0: $\beta_1 = \beta_2 = 0$ is

$$F = \frac{(\text{SSE}_r - \text{SSE}_c)/df_1}{\text{SSE}_c/df_2} = \frac{(4906.2 - 4541.1)/2}{4541.1/76} = \frac{182.6}{59.8} = 3.1$$

based on $df_1 = 2$ and $df_2 = 76$.

TABLE 13.8 Printout for Fitting Model $E(Y) = \alpha + \beta X$ to Data from Table 13.1 on Y = Income and X = Education

Source	DF	Sum of Squares	Mean Square	F Value	Pr > F
Model	1	3534.279	3534.279	56.19	0.0001
Error	78	4906.208	62.900		
Total	79	8440.487			

Parameter	Estimate	T for H0: Parameter=0	Pr > |T|	Std Error of Estimate
INTERCEPT	-10.795	-2.67	0.0092	4.0417
EDUC	2.330	7.50	0.0001	0.3108

Table 13.9 shows how software reports the results of tests for the no interaction model. The P-value for the test of no effect of racial–ethnic group equals $P = .053$. There is some evidence that the regressions of Y on X are different for at least two of the racial–ethnic groups. The sample sizes for two of the three groups are very small, and if the estimates resulted from larger samples, the P-value would have been much smaller. □

Comparison with Results of ANOVA

The test just described compares the mean of Y among the categories of a qualitative variable, controlling for X. This is a different test than one-way ANOVA, because of the control for X. The ANOVA test compares the coefficients of the dummy variables for the simpler model that does not have the βX term.

TABLE 13.9 Printout of Partial Sums of Squares Explained by Education and Racial–Ethnic Group, in the Analysis of Covariance Model Without Interaction

Source	DF	Type III SS	Mean Square	F Value	Pr > F
EDUC	1	3061.308	3061.308	51.23	0.0001
RACE	2	365.146	182.573	3.06	0.0529

Table 13.10 shows part of the printout for fitting the regression model for one-way ANOVA to these data. The P-value equals .018. Somewhat stronger evidence exists of a difference among the means when we ignore, rather than control, education.

TABLE 13.10 ANOVA Printout for Fitting Model $E(Y) = \alpha + \beta_1 Z_1 + \beta_2 Z_2$ Analyzing Income by Racial–Ethnic Group

Source	DF	Sum of Squares	Mean Square	F Value	Pr > F
Model	2	838.117	419.059	4.24	0.0178
Error	77	7602.370	98.732		
Total	79	8440.487			

Parameter		Estimate	T for H0: Parameter=0	Pr > |T|	Std Error of Estimate
INTERCEPT		21.240	15.12	0.0001	1.4052
RACE	b	-7.365	-2.58	0.0118	2.8540
	h	-5.740	-1.91	0.0598	3.0045
	w	0.000	.	.	.

Chapter 12 showed that ANOVA methods partition the total sum of squares $\sum(Y-\bar{Y})^2$ of the response variable about its overall sample mean into component parts. The analysis of covariance, by contrast, partitions the term $\sum(X - \bar{X})(Y - \bar{Y})$ into component parts (Snedecor and Cochran, 1967). This term measures the *covariation* between X and Y (Problem 9.50). That is, it measures the way in which X and Y vary jointly around their means $\bar{X}$ and $\bar{Y}$. The cross-product $(X - \bar{X})(Y - \bar{Y})$ for an observation on X and Y summarizes the amount by which those observations are both above or both below their means (in which case it is positive), or one above and the other below their means (in which case the cross-product is negative). The name *analysis of covariance* originates from this representation of the method.

Test for Effect of X, Controlling for Qualitative Variable

Assuming no interaction, we have tested for the effect of the qualitative variable, controlling for the quantitative covariate. Alternatively, one could test for the effect of the quantitative variable, controlling for the qualitative variable. This test refers to the coefficient of X; that is, it tests $H_0 : \beta = 0$ in the model

$$E(Y) = \alpha + \beta X + \beta_1 Z_1 + \beta_2 Z_2$$

with no interaction. This hypothesis states that the straight line relating X to the mean of Y has slope 0 for each category of the qualitative variable.

One can perform this test using the ordinary t test for a regression parameter. From Table 13.3, the estimated slope of 2.22 for this model has a standard error of .310. The test statistic equals $t = 2.22/.310 = 7.2$, which has a P-value of .0001. The evidence is very strong of a positive slope.

Equivalently, the square of this t statistic equals the F statistic for comparing complete and reduced regression models. This is the value of 51.2 reported for the effect of education in Table 13.9. The complete model, containing the effect of X, is the analysis of covariance model with no interaction. The reduced model, which is this model in which the coefficient β of X equals 0, is simply the model for one-way ANOVA. Thus, testing the effect of X corresponds to comparing the models for analysis of variance and analysis of covariance.

Table 13.11 summarizes the hypotheses, sums of squared errors, and R^2 values for the models in the tests of this section. In bivariate models, education is a good predictor of income, considerably better than racial–ethnic group alone. There is some further reduction in error from using three parallel lines, rather than one line, to predict income. This is summarized by the second test in the table, referring to the effect of the qualitative predictor, controlling for the quantitative predictor. There is also a significant effect of the quantitative predictor, controlling for the qualitative one, as summarized by the third test. However, a small and insignificant reduction occurs by allowing different slopes for the three lines, the first test listed in the table.

13.5 Adjusted Means

As usual, we learn more by estimating parameters than by testing hypotheses. This section shows how to summarize and compare means on the response variable for the various categories of the qualitative predictor, while controlling for the quantitative covariate. These analyses are meaningful when a model assuming no interaction is adequate.

Adjusting Response Means, Controlling for the Covariate

One-way ANOVA and its follow-up confidence intervals compare the mean of the response variable Y for the groups formed by the categories of the qualitative predictor. An analysis of covariance model compares these means while controlling for a quantitative covariate X. This control takes into account the groups' differing means on that covariate. For this model, one can summarize the effect of controlling the covariate by reporting the values expected for the means on Y if those groups all had the same

TABLE 13.11 Results of Comparisons of Four Models

	Anal. Cov. Interaction	Anal. Cov. No Interaction	Bivariate Regression	One-Way ANOVA
	$E(Y) = \alpha + \beta X + \beta_1 Z_1 + \beta_2 Z_2 + \gamma_1(XZ_1) + \gamma_2(XZ_2)$	$E(Y) = \alpha + \beta X + \beta_1 Z_1 + \beta_2 Z_2$	$E(Y) = \alpha + \beta X$	$E(Y) = \alpha + \beta_1 Z_1 + \beta_2 Z_2$
SSE	4368.1	4541.1	4906.2	7602.4
R^2	.482	.462	.419	.099
H_0: No interaction $F = 1.5$, $P = .24$	Complete model	Reduced model	—	—
H_0: $\beta_1 = \beta_2 = 0$ (Equal adjusted means) $F = 3.1$, $P = .053$	—	Complete model	Reduced model	—
H_0: $\beta = 0$ (Zero slopes) $F = 51.2$, $P = .0001$	—	Complete model	—	Reduced model

mean on X. These values are called *adjusted means*, since they adjust for the groups' differing distributions on X.

Adjusted Mean

The ***adjusted mean*** of Y for a particular group is the regression function for that group evaluated at the overall mean of the X-values for all the groups. It represents the expected value for Y at the mean of X for the combined population.

Figure 13.9 illustrates the adjusted means. Since adjusted means are relevant when the no interaction model applies, this figure portrays that model. Let μ_X denote the mean of X for the combined population. The adjusted mean of Y for a particular group equals that group's regression function evaluated at μ_X.

The ***sample adjusted mean*** of Y for a particular group is the prediction equation for that group evaluated at $\bar{X}$, the overall sample mean of the X-values for all the groups. This represents the value expected for the group's sample mean of Y if the sample mean of X for the group had equaled the overall sample mean. Some software (such as SAS) calls these ***least squares means***.

Notation for Adjusted Means

Denote the sample adjusted means for g groups by $\bar{Y}_1', \bar{Y}_2', \ldots, \bar{Y}_g'$; that is, $\bar{Y}_i'$ is the value of the prediction equation for group i evaluated at $\bar{X}$.

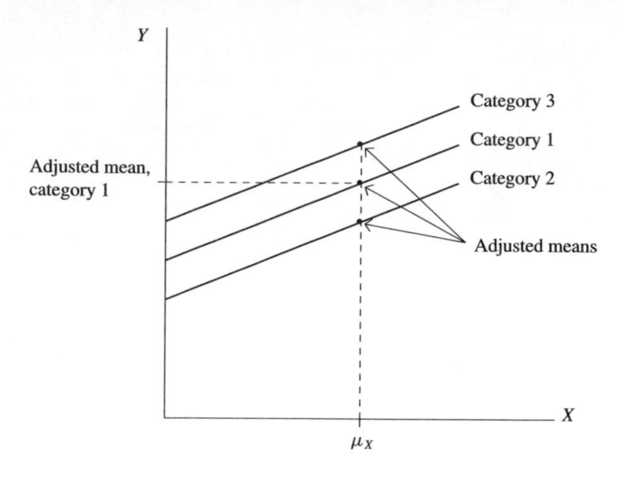

Figure 13.9 Population Adjusted Means, When a Qualitative Explanatory Variable Has Three Categories

Example 13.5 Adjusted Mean Incomes, Controlling for Education

We illustrate with the data in Table 13.1, calculating adjusted mean incomes for the three racial–ethnic groups, controlling for education. Table 13.12 lists the prediction equations for the three racial–ethnic groups for the model assuming no interaction, as well as the unadjusted and adjusted mean incomes.

TABLE 13.12 Sample Unadjusted Means, Adjusted Means, and Prediction Equations for Model Assuming No Interaction

Group	Prediction Equation	Mean of X	Mean of Y	Adjusted Mean of Y
Blacks	$\hat{Y} = -13.27 + 2.22X$	12.2	13.9	14.8
Hispanics	$\hat{Y} = -10.30 + 2.22X$	11.6	15.5	17.8
Whites	$\hat{Y} = -7.83 + 2.22X$	13.1	21.2	20.3

For blacks, for instance, the prediction equation is

$$\hat{Y} = -13.27 + 2.22X$$

From Table 13.2, the overall mean education for the combined sample of 80 observations is $\bar{X} = 12.7$. Hence, the sample adjusted mean income for blacks, controlling for education, is

$$\bar{Y}_1' = -13.27 + 2.22\bar{X} = -13.27 + 2.22(12.7) = 14.8$$

Similarly, the sample adjusted means for Hispanics and whites are

$$\bar{Y}_2' = -10.30 + 2.22(12.7) = 17.8$$
$$\bar{Y}_3' = -7.83 + 2.22(12.7) = 20.3$$

The adjustment process adjusts the ordinary sample means upward or downward according to whether mean education for the group is below or above average. For whites, for instance, the adjusted mean income of 20.3 is smaller than the unadjusted mean of 21.2. The reason is that the mean education for whites ($\bar{X}_3 = 13.1$) is larger than the mean education for the combined sample ($\bar{X} = 12.7$). Since a positive relationship exists between income and education, the model predicts that whites would have a lower mean income if their mean education were lower (equal to $\bar{X} = 12.7$). □

Comparing Adjusted Means

The coefficients of the dummy variables in the analysis of covariance model

$$E(Y) = \alpha + \beta X + \beta_1 Z_1 + \beta_2 Z_2$$

refer to differences between the true adjusted means. For example, β_1 is the coefficient of the dummy variable Z_1 for the first group. Thus, β_1 equals the difference between the adjusted mean for the first group ($\alpha + \beta_1 + \beta\mu_X$) and the adjusted mean for the gth group ($\alpha + \beta\mu_X$).

Inferences about the differences between the true adjusted means are based on the estimates of the dummy variable coefficients. These estimates are simply the differences in the sample adjusted means. The coefficient of Z_1 in the prediction equation equals $\bar{Y}'_1 - \bar{Y}'_g$, the coefficient of Z_2 equals $\bar{Y}'_2 - \bar{Y}'_g$, and so forth.

To illustrate, the estimated difference between adjusted mean incomes of blacks and whites is $\bar{Y}'_1 - \bar{Y}'_3 = 14.8 - 20.3 = -5.4$ (i.e., $-\$5400$). This is precisely the coefficient of the dummy variable Z_1 for blacks in the prediction equation for analysis of covariance (refer to Table 13.3). Similarly, the estimated difference between the adjusted means of Hispanics and whites equals $\bar{Y}'_2 - \bar{Y}'_3 = -2.5$, which is the coefficient of Z_2. Figure 13.10 depicts the sample adjusted means. The vertical distances between the lines represent the differences between these adjusted means.

Since the groups had different means on the covariate (education), the adjusted means and their differences differ from the unadjusted means and differences. For example, from Table 13.12, $\bar{Y}_2 - \bar{Y}_3 = 15.5 - 21.2 = -5.7$ is the unadjusted difference in mean income between Hispanics and whites, somewhat larger than the adjusted difference of -2.5. If whites and Hispanics had the same average educational level, the model predicts that the difference in their mean incomes would be less than half as large as it actually is in this sample.

Graphical Interpretation

Figure 13.11 depicts the relationship between the adjusted and unadjusted means. The figure depicts the line for the first category, but the same reasoning applies to any category. The prediction equation for the analysis of covariance model predicts a value of $\bar{Y}_i$ at the X-value of $X = \bar{X}_i$ for the ith category. In particular, the prediction line for the first category passes through the point with coordinates ($\bar{X}_1$, $\bar{Y}_1$). In other words, the *unadjusted mean* $\bar{Y}_1$ for the first category is the value of the prediction equation for

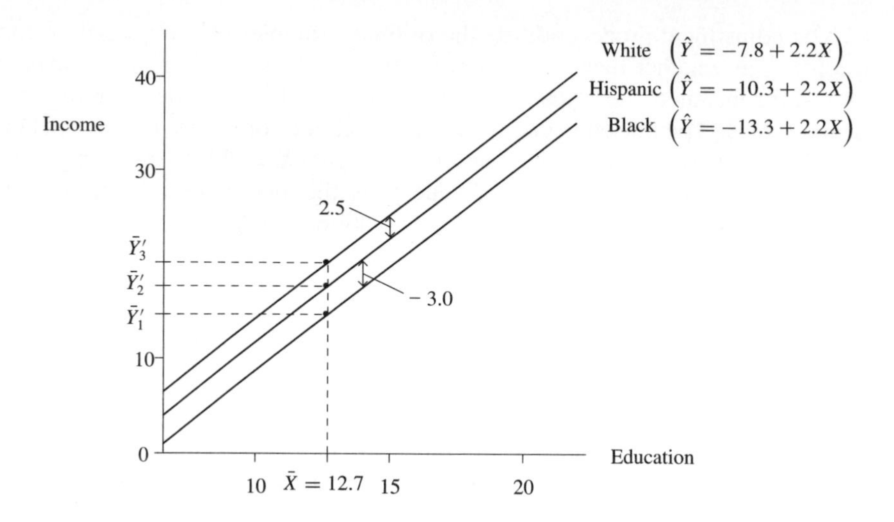

Figure 13.10 Sample Adjusted Means for Data on Income and Racial–Ethnic Group, Controlling for Education

that category evaluated at the X-value of $\bar{X}_1$, the mean of the X-values *for that category alone* [see the point $(\bar{X}_1, \bar{Y}_1)$ in Figure 13.11].

Now, the *adjusted mean* $\bar{Y}_1'$ for the first category is the value of that prediction equation evaluated at the *overall mean* $\bar{X}$ for the combined sample. Hence, the prediction line for that category also passes through the point $(\bar{X}, \bar{Y}_1')$, as shown in Figure 13.11. The difference between the X-coordinates for the points $(\bar{X}, \bar{Y}_1')$ and $(\bar{X}_1, \bar{Y}_1)$ is the horizontal distance $\bar{X} - \bar{X}_1$, indicated in the figure. Denoting the slope of the sample prediction equation by b, the vertical change in the line over the X distance of $\bar{X} - \bar{X}_1$ is $b(\bar{X} - \bar{X}_1)$. That is, the change in the Y direction for a $(\bar{X} - \bar{X}_1)$ unit change in the X-direction is $b(\bar{X} - \bar{X}_1)$. This vertical change is precisely the difference $\bar{Y}_1' - \bar{Y}_1$ between the Y coordinates of the points $(\bar{X}, \bar{Y}_1')$ and $(\bar{X}_1, \bar{Y}_1)$.

We have reasoned that the difference between the adjusted and unadjusted means for the first group equals

$$\bar{Y}_1' - \bar{Y}_1 = b(\bar{X} - \bar{X}_1)$$

Equivalently,

$$\bar{Y}_1' = \bar{Y}_1 + b(\bar{X} - \bar{X}_1)$$

The process of controlling for X takes the sample mean for Y and adjusts it by multiplying the difference $(\bar{X} - \bar{X}_1)$ in the means of X by the slope b of the prediction equation. The result estimates the value that $\bar{Y}$ would take for the first group *if* that group had the same mean on X as does the combined sample.

Similar reasoning applies to other groups. For the ith group, the difference between the adjusted and unadjusted means is

$$\bar{Y}_i' - \bar{Y}_i = b(\bar{X} - \bar{X}_i)$$

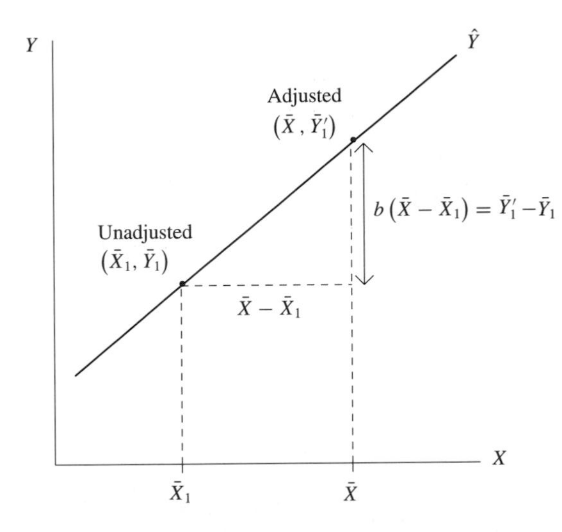

Figure 13.11 Graphical Depiction of Sample Adjusted and Unadjusted Means

If the slope b is positive, then the sample mean $\bar{Y}_i$ is adjusted upward if $\bar{X} > \bar{X}_i$ (i.e., if $\bar{X} - \bar{X}_i > 0$), as in Figure 13.11. This predicts that the mean of Y would have been larger had the distribution of X-values for that group had as large a mean as the combined samples. If the ith group is above average on X (i.e., $\bar{X}_i > \bar{X}$), then $b(\bar{X} - \bar{X}_i)$ is negative when $b > 0$, so the mean is adjusted downward. Figure 13.12 depicts the adjustment process on income for the three racial–ethnic groups. The mean is adjusted down for whites and up for blacks and Hispanics.

The difference between a group's adjusted and unadjusted means depends directly on the difference between $\bar{X}$ for the combined sample and $\bar{X}_i$ for that group. The adjusted means are similar to the unadjusted means if the $\bar{X}_i$-values are close to the overall $\bar{X}$, or if the slope b of the prediction equations is small.

Multiple Comparisons of Adjusted Means

Following an analysis of variance, the Bonferroni procedure compares all pairs of means simultaneously with a fixed overall confidence coefficient. This method extends directly to simultaneous comparison of *adjusted means*. The coefficients of the dummy variables in the prediction equation are the estimated differences between adjusted means. Software for multiple regression reports standard errors for these estimates. One can form t confidence intervals using these estimates and their standard errors, with df based on the sum of squared errors for the model.

Example 13.6 Confidence Intervals for Comparing Adjusted Mean Incomes

We construct simultaneous 95% confidence intervals for differences between the three pairs of adjusted mean incomes. With the Bonferroni approach, the error probability for each interval is $.05/3 = .0167$. The t-score with single-tail probability $.0167/2 = .0083$ and $df = 76$ (which is df for SSE for the no-interaction model) is 2.45.

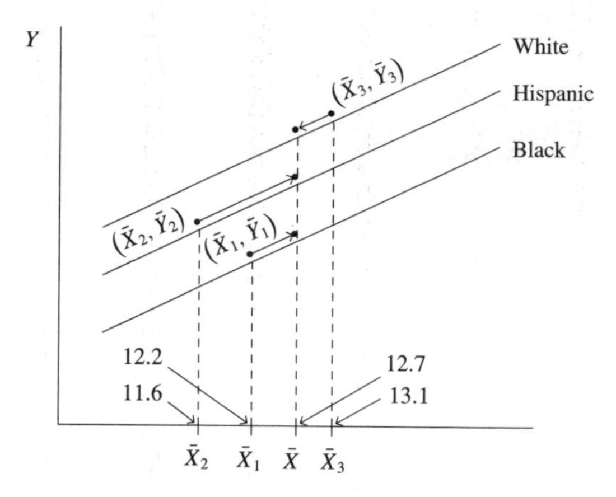

Figure 13.12 Adjustment Process for Income by Racial–Ethnic Group, Controlling for Education

The estimated difference between adjusted mean incomes of Hispanics and whites is the coefficient of the dummy variable Z_2 for Hispanics in the prediction equation. From Table 13.3, this equals -2.47 and has a standard error of 2.382. The Bonferroni confidence interval equals

$$-2.47 \pm 2.45(2.382), \quad \text{or } (-8.3, 3.4)$$

Controlling for education, the difference in mean incomes for Hispanics and whites is estimated to fall between $-\$8300$ and \$3400. Since the interval contains 0, it is plausible that the true adjusted mean incomes are equal for these two groups. The sample contained only 14 Hispanics, so the interval is wide.

Similarly, from Table 13.3, the confidence interval comparing blacks and whites is $-5.44 \pm 2.45(2.236)$, or $(-10.9, .04)$. To get the standard error for the comparison of blacks and Hispanics, we could fit the model with one of these categories as the baseline category lacking a dummy variable; or, we could use the general expression

$$\hat{\sigma}_{b_1 - b_2} = \sqrt{\hat{\sigma}^2_{b_1} + \hat{\sigma}^2_{b_2} - 2\text{Cov}(b_1, b_2)}$$

where $\text{Cov}(b_1, b_2)$ is taken from the *covariance matrix* of the parameter estimates, which software can provide. For these data, the standard error for this confidence interval equals 2.835. Table 13.13 summarizes the comparisons. We can be 95% confident that all three of these intervals simultaneously contain the differences in true adjusted means. None of the intervals show a significant difference, which is not surprising because the F test of the group effect in the previous section had a P-value of .053. Nonetheless, the intervals show that the adjusted means could be quite a bit smaller for blacks or Hispanics than for whites. More precise estimation requires a larger sample. □

TABLE 13.13 Bonferroni Multiple Comparisons of Differences in Adjusted Mean Income by Racial–Ethnic Group, Controlling for Education

Racial–Ethnic Group	Estimated Difference in Adjusted Means	Simultaneous 95% Confidence Intervals
Blacks, whites	$\bar{Y}_1' - \bar{Y}_3' = -5.4$	$(-10.9, .04)$
Hispanics, whites	$\bar{Y}_2' - \bar{Y}_3' = -2.5$	$(-8.3, 3.4)$
Blacks, Hispanics	$\bar{Y}_1' - \bar{Y}_2' = -3.0$	$(-9.9, 4.0)$

For the analysis of covariance model with no interaction, testing for identical adjusted means is equivalent to testing whether the regression lines are identical. That is, two parallel lines are identical if their corresponding adjusted means are equal. Therefore, just as confidence intervals reveal which pairs of adjusted means are significantly different, so do they reveal which pairs of lines are significantly different. For example, if the interval for the difference between a pair of adjusted means does not contain 0 (i.e., if the two population adjusted means are judged to be unequal), then the regression lines between Y and X for those two categories are judged to have different intercepts.

A Caution

Adjusted means are useful for comparing several groups by adjusting for differences in the means of a covariate X. Use them with caution, however, when the means on X are substantially different. The control process is a hypothetical one that infers what would happen *if* all groups had the same mean for X. If large differences exist among the groups in their means on X, the results of this control may be purely speculative. We must assume (1) that it makes sense to conceive of adjusting the groups on this covariate and (2) that the relationship between Y and X would continue to have the same linear form within each category as the X mean shifts for each category.

To illustrate the potential danger, recall the relationship between Y = annual income and X = experience and Z = gender shown in Figure 13.3b. The same line fits the relationship between Y and X for each gender, so it is plausible that the adjusted means are equal. However, nearly all the women have less experience than the men. The conclusion that the mean incomes are equal, controlling for experience, assumes that the regression line shown also applies to women with more experience than those in the sample and to men with less experience. If it does not, then the conclusion is incorrect.

Figure 13.13 portrays a situation in which the conclusion would be misleading. The dotted lines show the relationship for each group between Y and X over the X region not observed. At each fixed X value, a difference persists between the means of Y.

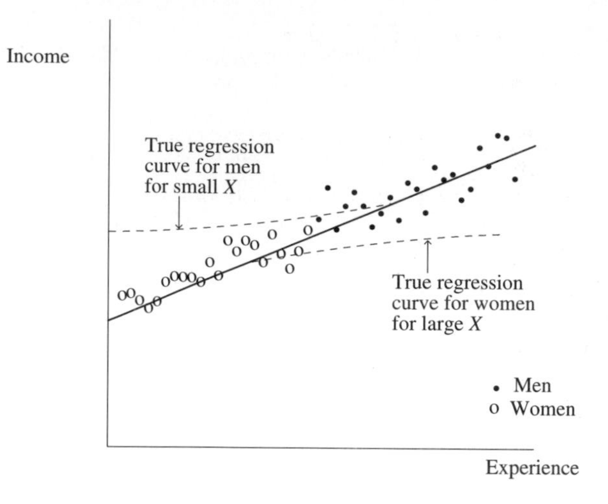

Figure 13.13 A Situation in Which Adjusted Means Are Misleading, Comparing Mean Incomes for Men and Women While Controlling for Experience

Whenever we form adjusted means, we should check the degree to which the distributions differ on the mean of X. Excessively large differences may mean that the conclusions need qualification. On the other hand, if relatively small differences exist among the $\bar{X}_i$, then controlling for X has little effect. The results of the comparisons of adjusted means are then similar to the results of the comparisons of unadjusted means in an analysis of variance.

13.6 Chapter Summary

This chapter showed that multiple regression can describe the relationship between a quantitative response variable and both quantitative and qualitative explanatory variables.

- The ***analysis of covariance*** model enables us to compare the mean of a response variable Y among various groups defined by categories of a qualitative predictor, while controlling for a quantitative *covariate* X.
- There is ***no interaction*** if the slope of the line relating X to the mean of Y is the same for each group. The model then provides a set of parallel lines.
- ***Adjusted means*** summarize the means on Y for the groups while controlling for X. They represent the model's prediction for the means of Y at the overall mean of the X values. They take into account the variation among groups on X, by predicting the mean of Y when all groups have means on X equal to its overall mean, $\bar{X}$.

Adjusted means are meaningful only when there is no interaction. One can test the hypothesis of no interaction, as well as the hypothesis of equal adjusted means (or, equivalently, identical regression lines), using an F test comparing complete and reduced regression models.

PROBLEMS

Practicing the Basics

1. The regression equation relating Y = education (number of years completed) to race (Z = 1 for whites, $Z = 0$ for nonwhites) in a certain country is $E(Y) = 11+2Z$. The regression equation relating education to race and to father's education (X) is $E(Y) = 3+.8X-.6Z$.
a) Find the mean education for whites, the mean education for nonwhites, and the difference between them (ignoring father's education).
b) Plot the relationship between X and the mean of Y for whites and for nonwhites.
c) Find the difference between the mean education of whites and nonwhites, controlling for father's education.
d) Find the mean education for whites and for nonwhites, when father's education equals 12 years.
2. A regression analysis for the 100th Congress, ending in 1988, predicted the proportion of each representative's votes on abortion issues that took the "pro-choice" position (R. Tatalovich and D. Schier, *American Politics Quarterly*, Vol. 21, 1993, p. 125). The prediction equation was

$$\hat{Y} = .350 + .011\text{ID} + .094\text{REL} + .005\text{NW} + .005\text{INC} + .063\text{GEN} - .167\text{PAR}$$

where REL = religion = 1 for non-Catholics, GEN = gender = 1 for women, PAR = party = 1 for Democrats, ID = ideology is the member's ADA score (ranging from 0 at most conservative to 100 at most liberal), NW = nonwhite is the percentage nonwhite of the member's district, and INC = income is the median family income of the member's district.
a) Interpret the effect of percentage nonwhite.
b) Interpret the effect of gender.
c) Interpret the coefficient for party. Does this imply that, ignoring the controls, Democrats are less likely than Republicans to take a pro-choice position?
d) Using standardized variables, the prediction equation is

$$\hat{z}_Y = .83\text{ID} + .21\text{REL} + .18\text{NW} + .05\text{INC} + .03\text{GEN} - .18\text{PAR}$$

Comment on the relative sizes of the partial effects. Interpret the coefficient of ideology.
3. Based on a national survey, Table 13.14 shows results of a prediction equation (ignoring some nonsignificant variables) reported for the response variable, Y = alcohol consumption, measured as the number of alcoholic drinks the subject drank during the past month (D. Umberson and M. Chen, *American Sociological Review*, Vol. 59, 1994, p. 152).
a) Setting up dummy variables F for whether father died in the past three years, G for gender, and (M_1, M_2, M_3) for the four categories of marital status, and letting X = alcohol consumption three years ago, report the prediction equation.
b) Find the predicted alcohol consumption for a divorced male whose father died in the previous three years and whose consumption three years previously was ten drinks per month.
c) Interpret the coefficient of gender.
d) The sample size was 1417. Test the null hypothesis that gender has no effect on the response, controlling for the other predictors. Interpret.

TABLE 13.14

Explanatory Variable	Estimate	Std. Error
Death of father (0 = no)	9.8	2.9
Gender (0 = male)	−5.3	1.6
Marital status (0 = married)		
Divorced, separated	7.0	2.0
Widowed	2.0	3.6
Never married	1.2	2.4
Alcohol consumption, three years ago	.501	.023
Intercept	8.3	

e) Construct a 95% confidence interval describing the true effect of gender, controlling for the other predictors. Interpret.

f) Marital status has three estimates. Dividing the coefficient of the divorced dummy variable by its standard error yields a t statistic. What hypothesis does it test? What would one need to do to test the effect of marital status on the response, controlling for the other variables?

4. Refer to Table 9.4 in Chapter 9. Table 13.15 shows a printout for modeling Y = selling price in terms of X = size of home and Z = whether the home is new ($Z = 1$, yes; $Z = 0$, no).

TABLE 13.15

```
Dependent Variable: PRICE
                       Sum of        Mean
Source        DF      Squares       Square       F Value      Prob>F

Model          2   152372.211    76186.105      251.775      0.0001
Error         90    27233.656      302.596
Total         92   179605.867

        Root MSE     17.3953       R-square     0.8484

               Parameter     Standard     T for H0:
Variable        Estimate        Error    Parameter=0     Prob > |T|
INTERCEP         -26.089       5.9775         -4.365         0.0001
SIZE              72.575       3.5077         20.690         0.0001
NEW               19.587       3.9948          4.903         0.0001
```

a) Specify the regression model that assumes no interaction. Interpret the parameters.

b) Report and interpret the prediction equation, and form separate equations relating Y to X for new and for older homes.

c) Find the predicted selling price for a home of 3000 square feet that is (i) new, (ii) older.

d) The mean selling prices were 123.41 for new homes and 89.25 for older ones. The mean sizes were 1.79 for new homes and 1.59 for older ones. Based on this information and the positive slope between selling price and size, predict whether the adjusted mean selling price for older homes would be larger or smaller than the unadjusted mean. Explain by sketching a plot.

5. Refer to the previous exercise. Table 13.16 shows a SPSS printout from fitting the model allowing interaction, where NEWSIZE refers to the cross-product term.

TABLE 13.16

```
R Square               .8675
Adjusted R Square      .8630
Standard Error       16.3509

Analysis of Variance
                    DF        Sum of Squares        Mean Square
Regression           3          155811.60588        51937.20196
Residual            89           23794.26079          267.35124

F =        194.26579          Signif F =   .0000

------------------ Variables in the Equation ------------------
Variable              B         SE B        Beta          T       Sig T
NEW             -31.826       14.818       -.332     -2.148       .0344
SIZE             66.604        3.694        .792     18.033       .0000
NEWSIZE          29.392        8.195        .571      3.587       .0005
(Constant)      -16.600        6.210                 -2.673       .0089
```

a) Interpret the parameter estimates. Report the lines relating Y to X for (i) new homes, (ii) older homes.
b) For the interaction model, find the predicted selling price for a home of 3000 square feet that is (i) new, (ii) older. Compare results to those in Problem 13.4(c).
c) For the interaction model, find the predicted selling price for a home of 1500 square feet that is (i) new, (ii) older. Comparing results to (b), explain how the difference in predicted prices between new and older homes changes as size of home increases.
d) Test for interaction between size of home and whether it is new. Interpret. How would you summarize these data?

6. Refer to the previous problem. Figure 13.14 shows a SAS scatter diagram for the relationship between selling price and size of home, identifying the points by a 1 when the home is new and a 0 when it is not. Because of the crudeness of scale, many of the points are hidden in this plot, but note that the observation with the highest selling price is a new home that is somewhat removed from the general trend of points. Table 13.17 shows a SAS printout for the interaction model after removing this single observation. For this adjusted data set, $R^2 = .845$ for the no interaction model and $R^2 = .848$ for the interaction model. Now, again answer the four parts of the previous exercise, and note what a large impact one observation can have on the conclusions.

7. Refer to Table 13.1. Not reported there were observations for ten Asian Americans. Their (X, Y) values follow:

Subject	1	2	3	4	5	6	7	8	9	10
Education	16	14	12	18	13	12	16	16	14	10
Income	35	21	12	28	16	19	29	41	18	10

a) Fit the analysis of covariance model using all four groups, assuming no interaction. Interpret the parameter estimates.

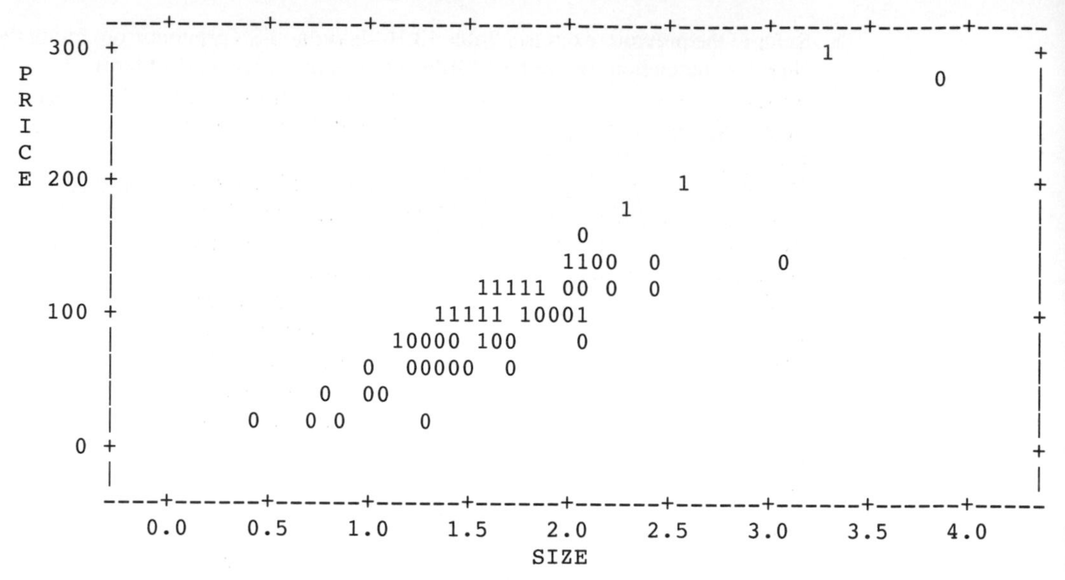

Figure 13.14

TABLE 13.17

Variable	Parameter Estimate	Standard Error	T for H0: Parameter=0	Prob > |T|
INTERCEP	-16.600	5.794	-2.865	0.0052
NEW	8.999	17.553	0.513	0.6095
SIZE	66.604	3.446	19.328	0.0001
NEW_SIZE	4.973	10.015	0.497	0.6208

b) Report the separate prediction equations for the four groups, and sketch a plot of the fitted model.
c) Find the predicted income for an Asian American with 16 years of education.
d) Find the estimated difference between mean income for blacks and whites, controlling for education. Compare the estimate to the value obtained without the Asian-American data in the dataset.
e) Using the data for all four groups, fit the interaction model. Interpret the parameter estimates.
f) Report the separate prediction lines for the four groups. Compare the lines to the ones for the no interaction model. Comment.
g) Fit the model relating Y to X using only the data for Asian Americans. Compare to the line for Asian Americans from the interaction model. (In fact, the fit of the interaction model is necessarily identical to conducting a separate regression analysis for each group.)
h) Explain how the estimated difference in mean incomes between blacks and whites changes as education changes.
i) Report R^2 for this model, and compare to R^2 for the simpler model without interaction. Does this model seem to provide much stronger predictive power?

j) Test the hypothesis of no interaction. Report the test statistic and P-value, and interpret.
k) Assuming no interaction, test the hypothesis that income is independent of ethnic–racial group, controlling for education. Report the test statistic and P-value, and interpret.
l) Assuming no interaction, test the hypothesis that income is independent of education, controlling for ethnic–racial group. Report the test statistic and P-value, and interpret.
m) Report the adjusted mean for blacks. Show how it compares to the unadjusted mean, and explain why the adjustment increased its value.
n) Estimate the difference between adjusted means for blacks and whites. Explain how to find this value from the coefficients of the prediction equation.
o) Construct a confidence interval for the difference between true adjusted means of blacks and whites. Use the appropriate confidence coefficient for forming simultaneous 94% confidence intervals for all six pairs of differences. Interpret.

TABLE 13.18

Source	DF	Sum of Squares	Mean Square	Parameter		Estimate
Model	1	7896.652	7896.652			
Error	38	713.248	18.770	INTERCEPT		-3.7226
Total	39	8609.900		REGISTER		0.7330

Source	DF	Sum of Squares	Mean Square	Parameter		Estimate
Model	2	5619.300	2809.650	INTERCEPT		26.600
Error	37	2990.600	80.827	RACE	a	25.700
Total	39	8609.900			b	4.400
					ma	0.000

Source	DF	Sum of Squares	Mean Square	Parameter		Estimate
				INTERCEPT		-2.7786
Model	3	7936.734	2645.578	REGISTER		0.7400
Error	36	673.166	18.699	RACE	a	-1.3106
Total	39	8609.900			b	-2.8522
					ma	0.0000

Source	DF	Sum of Squares	Mean Square	Parameter		Estimate
				INTERCEPT		-8.245
Model	5	7990.523	1598.105	REGISTER		0.878
Error	34	619.377	18.217	RACE	a	6.974
Total	39	8609.900			b	9.804
					ma	0.000
				REGISTER*RACE	a	-0.175
					b	-0.283
					ma	0.000

8. The printouts in Table 13.18 show results of using SAS to fit four models to data from a study of the relationship between Y = percentage of adults voting, X = percentage

of adults registered to vote, and racial–ethnic representation, for precincts in the state of Texas for a gubernatorial election. Racial–ethnic representation of a precinct is the group (Anglo, black, or Mexican American) having the strongest representation in a precinct. A random sample of 40 precincts is selected after a particular election and classified by racial–ethnic representation. **a)** Report the prediction equation for the analysis of covariance model assuming no interaction. Interpret the parameter estimates. Report the three racial–ethnic prediction equations.
b) Report the prediction equation for the analysis of covariance model allowing interaction. Interpret the parameter estimates. Report the three racial–ethnic prediction equations, and compare them to the equations for the no interaction model. Interpret the nature of the predicted interaction.
c) Test whether the regression lines relating Y and X for the three categories have the same slope. Report the test statistic and P-value, and interpret.
d) Report the prediction equation between Y and X, ignoring racial–ethnic representation. Interpret.
e) For the model assuming no interaction, test whether the regression lines between Y and X for the three groups are identical. Report the test statistic and P-value, and interpret.
f) For the model assuming no interaction, test whether the mean voting percentages are equal for the three categories of racial-ethnic representation, controlling for percentage registered. Report the test statistic and P-value, and interpret.
g) Report the test statistic and P-value for testing the null hypothesis that percentage voting and percentage registered are independent, controlling for racial–ethnic representation. Interpret.

9. Refer to the previous exercise. The means of percentage registered for the three categories are $\bar{X}_1 = 76.2$, $\bar{X}_2 = 49.5$, $\bar{X}_3 = 39.7$, with an overall mean of $\bar{X} = 60.4$.
a) Find the adjusted means on percentage voting, and interpret.
b) Compare the adjusted mean for Anglos to the unadjusted mean of 52.3, and interpret.
c) Sketch a plot of the no interaction model for these data, and identify on it the unadjusted and adjusted means.
d) Using the appropriate model, test the null hypothesis that the true unadjusted means on percentage voting are equal. Report the test statistic and P-value, and interpret. Compare the result of this test to the one in Problem 13.8(f).
e) Describe how to expand the no interaction model to include the additional qualitative variable, location of precinct, with categories urban, suburban, small town, and rural.
f) Describe how to expand the no interaction model to include the additional quantitative explanatory variable, percentage of residents in precinct who are homeowners.

10. An analysis of covariance model is fitted to annual income (thousands of dollars), using predictors age and marital status. Table 13.19 shows the sample mean incomes and the adjusted means for the model. How could the adjusted means be so different from the unadjusted means? Draw a sketch to help explain.

11. Refer to Problem 13.1. The overall mean of father's education is 12 years. Find the adjusted mean educational levels for whites and nonwhites, controlling for father's education, and compare them to the unadjusted means.

Concepts and Applications

12. Refer to the WWW data set (Problem 1.7). Using computer software, conduct and interpret an analysis of covariance using Y = political ideology. Prepare a report, presenting graphical, descriptive, and inferential analyses, using the predictors

TABLE 13.19

Group	Mean Age	Mean Income	Adjusted Mean Income
Married	44	40	30
Divorced	35	30	30
Single	26	20	30

a) religiosity and whether a vegetarian.

b) college GPA, number of times a week reading a newspaper, opinion about abortion, and whether a vegetarian.

13. Repeat the previous exercise, using Y = college GPA with predictors high school GPA, gender, and religiosity.

14. Refer to the data file created in Problem 1.7. For variables chosen by your instructor, use regression analysis as the basis of descriptive and inferential statistical analyses. Summarize your findings in a report in which you describe and interpret the fitted models and the related analyses.

15. Table 13.20 shows results of fitting a regression model to data from 1969 on salaries (in dollars) of about 35,000 college professors. Four predictors are qualitative (binary), with dummy variable defined in parentheses. The table shows estimates of parameters for each predictor, with standard errors in parentheses. Write a short report, interpreting the effects of the predictors.

TABLE 13.20

Variable	Estimate (Std. Error)
Years of Experience	162.7
	(2.0)
Terminal degree	264.0
(Ph.D. = 1)	(40.1)
Degree quality	616.8
	(34.5)
Research productivity	616.5
	(7.7)
Marital status	850.3
(married = 1)	(51.2)
Race	−310.3
(nonwhite = 1)	(107.0)
Gender	−1863.7
(female = 1)	(62.1)
Intercept	10, 033.2

Source: N. Langton and J. Pfeffer, *American Sociological Review*, Vol. 59 (1994), p. 236.

16. * Table 13.21 is a SPSS printout based on General Social Survey data combined for the years 1977, 1978, and 1980. The response variable is an index of attitudes toward premarital, extramarital, and homosexual sex. Higher scores represent more permissive attitudes. The qualitative explanatory variables are race (0 for whites, 1 for blacks), gender (0 for males, 1 for females), region (0 for South, 1 for non-South), and religion ($R_1 = 1$ for liberal Protestant sect, $R_2 = 1$ for conservative Protestant, $R_3 = 1$ for fundamentalist Protestant sect, $R_4 = 1$ for Catholic, $R_5 = 1$ for Jewish; no religious affiliation when $R_1 = \cdots = R_5 = 0$). The quantitative explanatory variables are age, education (number of years), attendance at church (higher values represent more frequent attendance), and a variable for which higher values represent greater intolerance of freedom of speech for atheists and communists.

a) State the regression model for this analysis, and give the prediction equation.

b) Based on the regression parameter estimates, give a profile of a person you would expect to be (i) least permissive, (ii) most permissive, with respect to sexual attitudes.

c) Summarize the main conclusions that you make from studying this printout.

TABLE 13.21

Analysis of Variance	DF	Sum of Squares	Mean Square	F
Regression	12.	2583.326	215.277	54.098
Residual	1092.	4345.534	3.979	

Multiple R	0.611
R Square	0.373
Adjusted R Square	0.366
Standard Error	1.995

Variables in the Equation

Variable	B	SE B	BETA	T	SIG T
RACE	0.993	0.2040	0.125	4.869	.000
AGE	-0.029	0.0042	-0.189	-6.957	.000
SEX	-0.289	0.1230	-0.058	-2.353	.019
EDUC	0.073	0.0223	0.092	3.281	.001
REGION	0.617	0.1401	0.115	4.403	.000
ATTEND	-0.286	0.0255	-0.304	-11.217	.000
R1	-0.296	0.2826	-0.049	-1.048	.295
R2	-0.605	0.2782	-0.113	-2.174	.030
R3	-1.187	0.3438	-0.128	-3.454	.001
R4	-0.127	0.2856	0.023	0.446	.656
R5	0.521	0.4417	0.034	1.179	.238
FREESPCH	-0.465	0.0581	-0.227	-8.011	.000
(Constant)	9.373				

17. A researcher is interested in factors associated with fertility in a Latin American city. Of particular interest is whether migrants from other cities or migrants from rural areas differ from natives of the city in their completed family sizes. The groups to be compared are urban natives, urban migrants, and rural migrants. Fertility is defined to be a woman's total number of live births. Since fertility is negatively related to educational level, and since education might differ among the three groups, it is decided to control that variable. Table 13.22 shows data for a random sample of married women above age 45. Analyze these data. In your report, provide graphical presentations as well as interpretations for all your analyses, and present a summary of the main results.

TABLE 13.22

Urban Natives		Urban Migrants		Rural Migrants	
Education	Fertility	Education	Fertility	Education	Fertility
0	7	0	7	0	4
0	5	0	6	0	6
1	5	0	7	0	10
1	4	1	5	0	8
2	7	2	2	1	7
4	4	2	6	2	8
5	4	3	3	3	5
7	3	4	6	3	6
8	5	7	4	4	7
8	2	7	4	5	7
8	3	8	4	6	8
9	3	11	3	6	6
10	4	11	3	7	4
11	3	11	4	7	5
12	3	12	4	8	4
		12	2	8	6
		12	3	8	5
				9	3
				9	7
				10	4

18. Refer to Table 9.1, not including the observation for D.C. Let Z be a dummy variable for whether a state is in the South, with $Z = 1$ for AL, AR, FL, GA, KY, LA, MD, MS, NC, OK, SC, TN, TX, VA, WV.
a) Analyze the relationship between Y = violent crime rate and the predictors X = poverty rate and Z.
b) Add percentage white as a predictor. Find a model that describes the data well, and interpret.
c) Repeat the analysis with D.C. in the data set, setting $Z = 1$ for it. Do the conclusions change in any way?

19. Refer to the previous exercise. Repeat using Y = murder rate.

20. Figure 13.4 exhibits at least one possibly influential outlier. Remove the observation with the highest income and reconduct the analyses. Did this one observation have any influence on the results?

21. You have two groups, and you want to compare their regressions of Y on X, in order to test the hypothesis that the true slopes are identical for the two groups. Explain how you can do this using regression modeling.

22. Let Y = death rate and X = average age of residents, measured for each county in Massachusetts and in Florida. Draw a hypothetical scatter diagram, identifying points for each state, when
a) The mean death rate is higher in Florida than in Massachusetts when X is ignored, but lower when it is controlled.
b) The mean death rate is higher in Florida than in Massachusetts both when X is ignored and when it is controlled.

23. Draw a scatter diagram of X and Y with sets of points representing two groups such that H_0: equal means on Y would be rejected in a one-way ANOVA, but would not be rejected in an analysis of covariance.

24. Give an example of a situation in which you expect interaction between a quantitative variable and a qualitative variable in their effects on a quantitative response variable.

In Problems 13.25–13.26, select the correct response(s).

25. In the model $E(Y) = \alpha + \beta_1 X + \beta_2 Z$, where Z is a dummy variable,
a) The qualitative predictor has two categories.
b) One line has slope β_1 and the other has slope β_2.
c) β_2 is the difference between the mean of Y for the second and first categories of the qualitative variable.
d) β_2 is the difference between the adjusted mean of Y (controlling for X) for the second and first categories of the qualitative variable.

26. In the United States, the mean annual income for blacks (μ_1) is smaller than for whites (μ_2), the mean number of years of education is smaller for blacks than for whites, and annual income is positively related to number of years of education. Assuming that there is no interaction, the difference in the mean annual income between whites and blacks, controlling for education, is
a) Less than $\mu_2 - \mu_1$
b) Greater than $\mu_2 - \mu_1$
c) Possibly equal to $\mu_2 - \mu_1$

27. Summarize the differences in purpose of the following:
a) A regression analysis for two quantitative variables
b) A one-way analysis of variance
c) A two-way analysis of variance
d) An analysis of covariance

Bibliography

Blalock, H. M. (1979). *Social Statistics*, rev. 2nd ed. New York: McGraw-Hill.

Neter, J., Kutner, M. H., Nachtsheim, C. J., and Wasserman, W. (1996). *Applied Linear Statistical Models*, 4th ed. Homewood, IL.: Richard D. Irwin.

Snedecor, G. W. and Cochran, W. G. (1967). *Statistical Methods*, 6th ed. Ames: Iowa State University Press.

Chapter 14

Model Building with Multiple Regression

The purpose of this chapter is twofold. First, we introduce basic tools for building regression models and evaluating the effects on their fit of highly correlated predictors or unusual observations. Section 14.1 discusses criteria for selecting a regression model by deciding which of a possibly large collection of variables to include in the model. Section 14.2 introduces methods for checking regression assumptions and evaluating the influence of individual observations. Section 14.3 discusses effects of multicollinearity on the model-building process. Second, we examine methods for modeling variables that do not satisfy the assumptions of earlier chapters. Sections 14.4 and 14.6 introduce models for nonlinear relationships, and Section 14.5 shows how to model response variables having distributions other than the normal.

14.1 Model Selection Procedures

Social research studies usually have several explanatory variables, any of which could be included in a regression model for the response variable. For example, for modeling mental impairment, potential predictors include income, educational attainment, life events, social and environmental stress, marital status, gender, age, self-assessment of health, number of jobs held in previous five years, number of relatives who live nearby, number of close friends, membership in social organizations, frequency of church attendance, and so forth.

Usually, the model includes some explanatory variables for theoretical reasons. Others may be included for exploratory purposes, perhaps to see if they explain much variation in the response variable. The model might also include terms to allow for interactions. In such situations, it can be difficult to decide which variables to use and which to exclude.

One possible strategy may be obvious: include every potentially useful predictor in the model and then delete those terms not making significant partial contributions at some preassigned α-level. Unfortunately, this usually is inadequate. Because of correlations among the explanatory variables, any one variable may have little unique predictive power, especially when the number included is large. It is conceivable that few, if any, explanatory variables would make significant *partial* contributions, given that all of the other explanatory variables are in the model. In Example 11.6 and Table 11.7, for instance, none of X_1, X_2, and X_1X_2 had small P-values in the tests for the partial effects, even though X_1 and X_2 were highly significant when X_1X_2 was not included in the model.

Selecting Explanatory Variables for a Model

Here are two basic guidelines for selecting explanatory variables for a regression model. First, include enough of them to make the model useful for theoretical and predictive purposes. Include those variables having theoretical importance, and try to obtain good predictive power as measured by a reasonably high R^2.

Second, as a counterbalance to the first goal, keep the model simple. The presence of unnecessary or redundant variables makes the model more difficult to interpret, presents more parameters to be estimated, may result in inflated standard errors of the estimates of these parameters, and may make it impossible to assess the partial contributions of variables that are important theoretically. To avoid multicollinearity (Section 14.3), it is helpful for the explanatory variables to be correlated with the response variable but not highly correlated among themselves.

Related to this second goal, it is best not to build complex models if the data set is small. If you have only 25 observations, you won't be able to untangle the complexity of relationships among 10 variables. Even with large data sets, it is difficult to build "believable" models containing more than about 5–10 explanatory variables, and with small to moderate sample sizes (say, 100 or less) it is safest to use only a few predictors.

Keeping these thoughts in mind, no unique or optimal approach exists for selecting predictors. For k potential predictors, since each can be either included or omitted (two possibilities for each variable), there are 2^k potential subsets. For $k = 2$, for example, there are $2^k = 2^2 = 4$ possible models: one with both X_1 and X_2, one with X_1 alone, one with X_2 alone, and one with neither variable. The set of potential models is too large to evaluate practically if k is even moderate; if $k = 7$ there are $2^7 = 128$ potential models.

Most software contains automated variable selection procedures that scan the explanatory variables to choose a subset for the model. These routines construct a model by sequentially entering or removing variables, one at a time according to some criterion. This takes much less time than fitting and comparing *all* 2^k possible regression models. For any particular sample and set of variables, however, different procedures may select different subsets of variables, and we shall see that this approach is not a panacea.

Among the most popular automated variable selection methods are *backward elimination*, *forward selection*, and *stepwise regression*. We first describe backward elimination.

Backward Elimination

Backward elimination begins by placing all of the predictors under consideration in the model. It deletes one at a time from the model until reaching a point where the remaining variables all make significant partial contributions to predicting Y. For most software, the variable deleted at each stage is the one that results in the smallest decrease in R^2. Equivalently, it is the one explaining the least variation in Y, controlling for the other predictors in the model at that stage.

Specifically, here's the sequence of steps for backward elimination: The initial model contains all potential explanatory variables. If all those variables make significant partial contributions at some fixed α-level, according to the usual t test or F test, then that model is the final one. Otherwise, the explanatory variable making the smallest contribution, controlling for the other variables in the model, is removed. Next, for the model with that variable removed, the partial contributions of the variables remaining in the model are reassessed, controlling for the other variables still in the model. If they are all significant, that model is the final model. Otherwise, the variable making the smallest contribution is removed. The process continues until each remaining predictor explains a significant partial amount of the variability in Y.

Example 14.1 Selecting Predictors of Home Selling Price

We refer to the 93 observations in Table 9.4 on selling price of home, size of home, number of bedrooms, number of bathrooms, and whether the home is new. We use backward elimination with these variables as potential predictors, requiring a variable to reach significance at the $\alpha = .05$ level for inclusion in the model.

Table 14.1 shows the first stage of the process, fitting the model containing all the predictors. The reported F statistic is the square of the t statistic, which divides the estimate by its standard error. The variable making the least partial contribution to the model is BEDROOMS. Its P-value ($P = .49$) is the largest, and R^2 decreases least by dropping it from the model (from .869 to .868). Although number of bedrooms is, by itself, a good predictor of selling price, the other predictors together explain most of the same variability in selling price. Once those variables are in the model, number of

TABLE 14.1 Model Fit at Initial Stage of Backward Elimination for Predicting Home Selling Price

Variable	Parameter Estimate	Standard Error	Sum of Squares	F	Prob>F
INTERCEP	-41.795	12.104	3191.08	11.92	0.0009
SIZE	64.761	5.630	35418.89	132.33	0.0001
BEDROOMS	-2.766	3.960	130.55	0.49	0.4868
BATHS	19.203	5.650	3091.84	11.55	0.0010
NEW	18.984	3.873	6431.68	24.03	0.0001

bedrooms is essentially redundant, explaining uniquely only 130.55 of the total sum of squares of 179,605.87.

Table 14.2 shows the second stage, refitting the model after dropping BEDROOMS as a predictor. Each variable now makes a significant contribution, controlling for the others in the model. Thus, this is the final model. Backward elimination provides the prediction equation

$$\hat{Y} = -48.0 + 62.3S + 20.1B + 18.4N$$

Other things being equal, an extra thousand square feet of size increases the selling price by about 62 thousand dollars, an extra bathroom increases the predicted selling price by about 20 thousand, and having a new home increases it by about 18 thousand.

TABLE 14.2 Model Fit at Second Stage of Backward Elimination for Predicting Home Selling Price

Variable	Parameter Estimate	Standard Error	Sum of Squares	F	Prob>F
INTERCEP	-47.992	8.209	9096.07	34.18	0.0001
SIZE	62.263	4.335	54898.18	206.30	0.0001
BATHS	20.072	5.495	3550.12	13.34	0.0004
NEW	18.371	3.761	6349.27	23.86	0.0001

If we had included interactions in the original model, we would have ended up with a different final model. However, the model given here (1) has the advantage of simplicity, (2) itself has very good predictive power ($R^2 = .868$), and (3) it is not a good idea to fit overly complex models with such a modest sample size ($n = 93$). □

Forward Selection and Stepwise Regression Procedures

Whereas backward elimination begins with all the potential explanatory variables in the model, ***forward selection*** begins with none of them. It adds one variable at a time to the model until reaching a point where no remaining variable not yet in the model makes

TABLE 14.3 Steps of Forward Selection for Predicting Home Selling Price

Step	Variables in Model	P-Value for New Term	R^2	C_p
0	None	—	.000	
1	S	.001	.808	39.9
2	S, N	.001	.848	14.8
3	S, N, BA	.001	.868	3.5
4	S, N, BA, BE	.487	.869	5.0

a significant partial contribution to predicting Y. At each step, the variable added is the one providing the largest boost in R^2.

To illustrate, consider again the data on selling prices of homes, with predictors S, BE, BA, and N. Table 14.3 depicts the process. The variable most highly correlated with selling price is S, so it is added first. Once S is in the model, N provides the greatest boost to R^2, and it is significant ($P = .001$), so it is the second variable added. Once both S and N are in the model, BA provides the greatest boost to R^2 and it is significant ($P = .001$), so it is added next. At this stage, BE is the only variable left for inclusion, but it does not make a significant contribution ($P = .49$), so the final model does not include it. Forward selection reaches the same final model as backward elimination.

Once forward selection provides a final model, not all the predictors appearing in it need necessarily be significantly related to Y. The variability in Y explained by a variable entered at an early stage may overlap with the variability explained by variables added later, so it may no longer be significant. Figure 14.1 illustrates. The figure portrays the portion of the total variability in Y explained by each of three predictors. Variable X_1 explains a similar amount of variability, by itself, as X_2 or X_3. However, X_2 and X_3 between them explain much of the same variation that X_1 does. Once X_2 and X_3 are in the model, the unique variability explained by X_1 is minor.

Stepwise regression is a modification of forward selection that drops variables from the model if they lose their significance as other variables are added. The approach is the same as forward selection except that at each step, after entering the new variable,

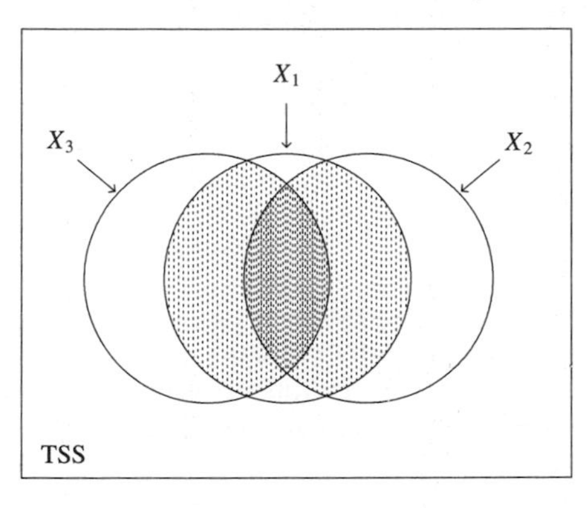

Figure 14.1 Variability in Y Explained by X_1, X_2, and X_3. Shaded portion is amount explained by X_1 that is also explained by X_2 and X_3

the procedure drops from the model any variables no longer making significant partial contributions. A variable entered into the model at some stage may eventually be eliminated because of its overlap with variables entered at later stages.

For the home sales data, stepwise regression behaves the same as forward selection. At each stage, each variable in the model makes a significant contribution, so no variables are dropped. For these variables, backward elimination, forward selection, and backward elimination all agree. This does not always happen, as Problem 14.2 illustrates.

Limitations and Abuses of Automatic Selection Procedures

It seems appealing to have a procedure that automatically selects variables according to established criteria. But any variable selection method should be used with caution and should not substitute for careful thought. There is no guarantee that the final model chosen will be sensible.

For instance, when it is not known whether explanatory variables interact in their effects on a response variable, one might specify all the pairwise interactions as well as the main effects as the potential explanatory variables for the selection procedure. In this case, it is inappropriate to remove a main effect from a model that contains an interaction composed of that variable. Yet, most software does not have this safeguard.

To illustrate, we used backward elimination with the home sales data, including the four predictors from above as well as their six cross-product interaction terms. Using PROC REG in SAS, the original model with all ten terms had $R^2 = .912$; the final model had $R^2 = .909$, using only four interaction terms, $S \times BE$, $S \times BA$, $S \times N$, $BE \times N$. It is inappropriate, however, to use these interactions as predictors without the main effects.

By contrast, using S-Plus software, which does not permit dropping a main effect if an interaction with that variable is in the model, the final model included these four interaction terms plus their main effects. This model is very complex. It is much more difficult to interpret than the one we obtained above using only the three main effects, and it provides only a relatively small gain in R^2 over the value of .868 attained with those effects. Although the software routine tells us it is a better model, this is so only in terms of statistical significance and not necessarily in terms of practical interpretation.

Also, a variable selection procedure may exclude an important predictor that really should be in the model according to other criteria. For instance, using backward elimination with the four predictors of selling price and their interactions, the variable S was the first removed. In other words, when all the other main effects and the interaction terms were in the model, S explained a tiny and insignificant part of the variation in selling price. Nevertheless, it is by far the best single predictor of selling price, having an $R^2 = .808$ by itself. (Refer to the first line of the forward selection process in Table 14.3.) Since size of home is such an important determinant of selling price, it seems sensible that any final model should include it as a predictor.

One can modify variable selection routines to include a certain set of explanatory variables in the model, regardless of whether they make significant partial contributions to the response variable. For instance, one can conduct backward elimination subject to the constraint that the model include the specified terms. This ensures that the final model includes the variables of primary interest as well as other variables that are useful for predictive purposes. Most software has this option.

Although P-values in selection procedures provide a guide for making decisions about adding or dropping variables, they are not the true P-values for the tests conducted. We add or drop a variable at each stage according to a maximum or minimum significance, but the sampling distribution of the maximum or minimum of a set of t or F statistics differs from the sampling distribution for the statistic for an a priori chosen test. For instance, suppose we add variables in forward selection according to whether the P-value is less than .05. Even if none of the potential predictors truly affect Y, the probability is considerably larger than .05 that at least one of the separate test statistics provides a P-value below .05 (Problem 14.60). At least one variable that is not really important may look impressive simply due to chance, because of sampling error.

Similarly, once we choose a final model with a selection procedure, any inferences conducted with that model are highly approximate. In particular, P-values are likely to appear smaller than they should be and confidence intervals are likely to be too narrow, since the model was chosen that most closely reflects the data, in some sense.

There is a basic difference between *explanatory* and *exploratory* modes of model selection. In ***explanatory research***, one has a theoretical model that one tests with multiple regression. One might test whether a hypothesized spurious association disappears when other variables are controlled, for example. In such research, automated selection procedures are usually not appropriate, because theory determines which variables are in the model.

In ***exploratory research***, by contrast, the goal is not to examine theoretically specified relationships, but simply to find a good set of predictors. One attempts to maximize R^2, without concern about theoretical explanations. Thus, educational researchers might use a variable selection procedure to search for a set of test scores and other factors that predict well how students perform in college. They should be cautious about giving causal interpretations to the effects of the different variables. For example, possibly the "best" predictor of students' success in college is whether their parents owned a personal computer when they lived at home.

In summary, automated variable selection procedures are no substitute for careful theory construction in guiding the formulation of models. For most scientific research, we do not recommend these procedures.

The C_p Statistic*

Most software routines for selecting variables report a summary statistic that describes how well each model fits compared to the full model with all the predictors. This statistic is denoted by C_p, where p denotes the number of parameters in the regression model (including the Y-intercept). For a given number of parameters p in the model, smaller values of C_p indicate a better fit. For the full model, necessarily $C_p = p$. A simpler model than the full one that has C_p close to p provides essentially as good a fit, apart from sampling error. Models having values of C_p considerably larger than p do not fit as well.

In using the C_p index to select a model, one attempts to use the smallest number of predictors necessary to give a value of C_p close to p. For that number of predictors, one selects the model with the minimum value of C_p.

To illustrate, Table 14.3 shows the C_p statistic for the subsets of variables sequentially selected using the forward selection procedure. The full model has four predictors and $p = 5$ parameters (including the Y-intercept), so it has $C_p = 5.0$. The model with all predictors except BE has $p = 4$ parameters and has $C_p = 3.5$. Since C_p is close to p, this model seems to fit essentially as well as the full model, apart from sampling error. The simpler models listed in the table have C_p considerably larger than p, and provide poorer fits.

14.2 Regression Diagnostics

Once we have selected a set of predictors for a model, how do we know that that model fits the data reasonably well? This section introduces diagnostics that indicate (1) when model assumptions are grossly violated and (2) when certain observations are highly influential in affecting the model fit or inference about model parameters.

Inference methods with regression models are founded on several assumptions:

- The true regression function has the form used in the model (e.g., linear).
- The conditional distribution of Y is normal.
- The conditional distribution of Y has constant standard deviation throughout the range of values of the explanatory variables. This condition is called ***homoscedasticity***.
- The observations on Y are statistically independent, such as in random sampling.

Significance tests exist for checking these assumptions. In practice, they are never perfectly fulfilled, but the regression model can still be useful. It is usually adequate to check that none of the assumptions is grossly violated.

Examination of the Residuals

Several checks of assumptions use the residuals, $Y - \hat{Y}$. They represent the deviations of the observations from the prediction equation values.

We first discuss the normality assumption. If the observations are normally distributed about the true regression equation with constant conditional standard deviation σ, then the residuals should be approximately normally distributed. To check this, plot the residuals about their mean value 0, using a histogram or stem and leaf plot. They should have approximately a bell shape about 0.

If an observation has a residual that is several $\hat{\sigma}$ below or above 0, it is an outlier and should be checked. If an outlier represents a measurement error, it could cause a major bias in the prediction equation. Even if it is not an error, it should be investigated because it represents an observation that is not typical of the sample data. One should try to ascertain reasons for the peculiarity. Often the outliers differ from the other observations on some variable not included in the model, and once that variable is added, they cease to be outliers.

Example 14.2 Residuals for Modeling Housing Price

For the housing data of Table 9.4 with Y = selling price, variable selection procedures in Example 14.1 suggested the model having predictors S = size of home, BA = number of bathrooms, and N = whether new. The prediction equation is

$$\hat{Y} = -48.0 + 62.3S + 20.1BA + 18.4N$$

and the estimated conditional standard deviation is $\hat{\sigma} = 16.3$.

Figure 14.2 is a stem and leaf plot of the residuals for this fit. No severe nonnormality seems to be indicated, since the residuals are roughly bell-shaped about 0, falling between -35 and $+28$. However, one extreme outlier on selling price has a residual of 73, which is 4.5 standard deviations above the mean of 0. This occurs for the observation having the highest selling price, \$309,400. The selling price was 73 thousand dollars higher than the predicted selling price for a new home of size 3300 square feet with three bathrooms.

Since a severe outlier on Y can substantially affect the fit, especially when the values of the explanatory variables are not near their mean, we refitted the model without this observation. The R^2 value changes from .868 to .870, and the prediction equation

```
-3 | 5
-3 | 3  2  1
-2 | 9  5
-2 | 2  2  2  0
-1 | 9  8  8  7  6  6
-1 | 4  4  4  1  1  1  0
-0 | 9  9  9  8  7  7  6  5  5
-0 | 4  4  3  2  2  2  1  1  0  0
 0 | 0  0  1  1  1  1  1  1  1  2  3  3  3  4  4  4  4  4  4
 1 | 1  2  2  2  4  4  4
 1 | 5  7  8  9  9
 2 | 1  1  1  2
 2 | 5  6  8
 3 |
 3 |
 4 |
 4 |
 5 |
 5 |
 6 |
 6 |
 7 | 3
```

Figure 14.2 Stem and Leaf Plot of Residuals for Multiple Regression Model Fitted to Housing Price Data

changes to

$$\hat{Y} = -38.2 + 57.8S + 18.6BA + 16.5N$$

The parameter estimates are slightly smaller, but so are their standard errors, and each predictor still has a very small P-value for the partial effect. This extreme outlier would have had a more substantial effect if the sample size had been smaller. □

Plotting Residuals Against Explanatory Variables

The normality assumption is not as important as the assumption that the model provides a good approximation for the true relationship between the predictors and the mean of Y. If the model assumes a linear effect but the effect is actually strongly nonlinear, the conclusions will be faulty.

For bivariate models, the scatter diagram provides a simple check on the form of the relationship. For multiple regression, it is also useful to construct a scatter diagram plotting each explanatory variable against the response variable. This displays only the *bivariate* relationships, however, whereas the model refers to the *partial* effect of each predictor, with the others held constant. The *conditioning plot* and *partial regression plot* introduced in Section 11.2 provides some information about this. The conditioning plot reveals whether a predictor has a linear effect within regions of values of another predictor. The partial regression plot removes the effects of the other variables by plotting the residuals from the model using those other variables to predict Y against the residuals from the model using those other variables to predict the given predictor. This provides a single summary picture of the partial relationship.

For multiple regression models, plots of the residuals against the predicted values $\hat{Y}$ or against each explanatory variable also help us check for potential problems. If the residuals appear to fluctuate randomly about 0 with no obvious trend or change in variation as the values of a particular X_i increase, then no violation of assumptions is indicated. The pattern should be roughly like Figure 14.3a. A scattering of the residuals as in Figure 14.3c suggests that Y is actually nonlinearly related to X_i, since Y tends to be below $\hat{Y}$ for very small and very large X_i-values (giving negative residuals) and above $\hat{Y}$ for medium-sized X_i-values (giving positive residuals). Sections 14.4 and 14.6 show how to address nonlinearity.

A fairly common occurrence is that residuals tend to become dramatically more (or less) variable as X_i increases, as in Figure 14.3b. This indicates that the conditional standard deviation of Y is not constant. For nonnegative responses, the variability often increases as the mean increases. If the change in variability is severe, then a method other than ordinary least squares provides a more appropriate fit with more valid standard errors. Section 14.5 presents a model for which the standard deviation increases as the mean does.

Figure 14.4 is a residual plot for the model relating selling price of home to size, number of bathrooms, and whether new. It plots the residuals against size. There is some suggestion of less variability at the lower size values. However, if we disregard

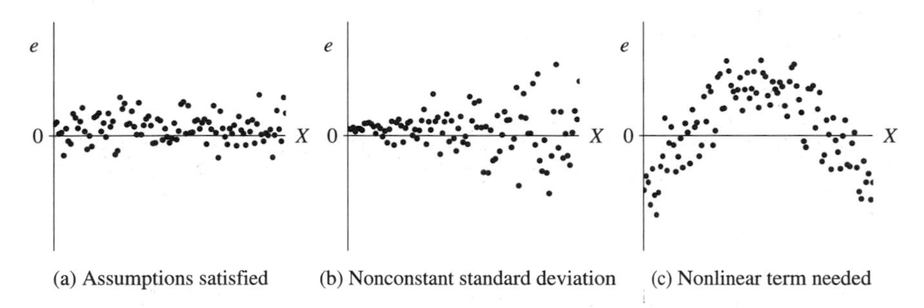

Figure 14.3 Possible Patterns for Residuals (e), Plotted Against an Explanatory Variable X

the outlier with the residual of 73, the points show a relatively regular trend with no strong irregularities.

In practice, residual patterns are rarely as neat as the ones in Figure 14.3. Be careful not to let a few outliers or ordinary sampling variability influence too strongly your reading of a pattern from the plot. Also, the plots described here just scratch the surface of the graphical tools now available for diagnosing potential problems. Fox (1991) describes a variety of modern graphical displays.

Time Series Data

Some social research studies collect observations sequentially over time. For economic variables, in particular, the observations may occur daily, monthly, or yearly. The observations are then probably not independent. Sampling subjects randomly from some population ensures that one observation is not statistically dependent on another. However, neighboring observations from a time sequence are usually related in some way. For example, if nationwide unemployment is relatively low in January 1997, it is probably also relatively low in February 1997.

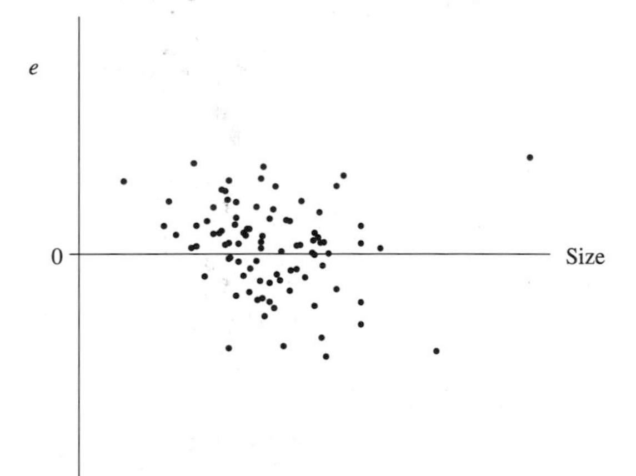

Figure 14.4 Scatter Diagram of Residuals of Selling Price Plotted Against Size of Home, for Model with Predictors Size, Number of Bathrooms, and Whether Home Is New

A plot of the residuals against the time of collection for the observation checks for this type of dependence. Ideally, the residuals should fluctuate in a random pattern about 0 as time changes, rather than showing a periodic cycle. The methods presented in this text are based on independent observations and are inappropriate when time effects occur. Books specializing in time series or econometrics, such as Wonnacott and Wonnacott (1979), present methods for time series data.

Detecting Influential Observations

Least squares estimates of parameters in regression models can be strongly influenced by an outlier, especially when the sample size is small. A variety of statistics has been introduced in recent years to summarize the influence each observation has on the model fit. These statistics refer to how much the predicted values $\hat{Y}$ or the model parameter estimates change when the observation is removed from the data set. An observation's influence depends on two factors: (1) how far the response on Y falls from the overall trend in the sample and (2) how far the values of the explanatory variables fall from their means.

The first factor on influence is measured by the residual for the observation, $Y - \hat{Y}$; the larger the residual, the farther the observation falls from the overall trend. A standardized version of the residual equals the residual divided by a standard deviation that describes how much such residuals should vary because of ordinary sampling variability. This is called a ***studentized residual***. It behaves like a t or z statistic. If the model holds, studentized residuals between about 2 and 3 in absolute value may be worthy of notice, but about 5% are this large simply by chance. Values larger than about 3 should be quite rare, and usually represent outliers.

The second factor on influence, called the ***hat value*** or ***leverage*** of the observation, is a nonnegative statistic that summarizes how far the explanatory variables fall from their means. The larger the hat value, the greater weight that observation receives in calculating the predicted values $\hat{Y}$ (hence, the name "hat" value). In other words, the greater the hat value, the greater potential leverage the observation has for affecting the fit. The average hat value in the sample equals p/n, where p is the number of parameters in the model. The hat values tend to decrease as the sample size increases, so an inordinate leverage for an observation primarily happens for small to moderate sample sizes.

Formulas for the diagnostics discussed in this section are complex, but most statistical software has options for reporting them. Two popular diagnostics, both of which depend on the studentized residuals and hat values, are called ***DFFITS*** and ***DFBETAS***. For a given observation, the first refers to the effect on the fit of removing that observation from the data set, and the second refers to the effect on the model parameter estimates of removing it.

For a particular observation and model parameter β_j, DFBETAS equals the change in the parameter estimate $\hat{\beta}_j$ due to deleting the observation, divided by the standard error of the estimator for the adjusted data set. The larger the absolute value of DFBETAS, the greater the influence of the observation on that parameter estimate. Each observation has a DFBETAS value for each parameter in the model.

The DFFITS diagnostic summarizes more broadly the influence of the observation on the fit of the model, as each observation has a single DFFITS value. For a given observation, DFFITS is based on the change in the $\hat{Y}$ predicted values for all the data due to deleting that observation. ***Cook's distance*** is an alternative measure with the same purpose. Both are based on standardized versions of the sum of squared differences between the original predicted values and the new predicted values after deleting that observation. The DFFITS value has the same sign as the residual; the larger its value, the greater the influence that observation has on the fitted values.

One scans or plots these diagnostic measures to see if some observations stand out from the rest, having relatively large values. Each measure has approximate cutoff points for noteworthy observations. For instance, a studentized residual exceeding about 3 indicates a likely outlier, and a hat value larger than about $3p/n$ indicates a relatively large leverage. DFBETAS larger than 1 suggests a substantial influence on that parameter estimate. However, the hat values, DFBETAS, and DFFITS tend to decrease as the sample size n increases, so normally it is a good idea to examine observations having about the most extreme 5% of the values.

Example 14.3 Influence Diagnostics for Crime Data

Refer back to Table 9.1, which tabulates several variables for the 50 states and the District of Columbia. Let Y = murder rate, X_1 = percentage of families below the poverty level, and X_2 = percentage of single-parent families. For all 51 observations, the least squares fit of the multiple regression model is

$$\hat{Y} = -40.7 + .32X_1 + 3.96X_2$$

Table 14.4 shows a printout of influence diagnostics for the fit of this model. The studentized residuals all fall in a reasonable range except the one for the last observation (D.C.), which equals 14.2. The observed murder rate of 78.5 for D.C. falls far above the predicted value, causing a large positive residual. This is an extreme outlier. In addition, the hat value for D.C. equals .54, more than three times as large as any other hat value and nine times the average hat value of $p/n = 3/51 = .06$. This suggests that D.C. has considerable leverage on the outcome of the fit. Since D.C. has both a large studentized residual and a large hat value, we expect it to be very influential.

Not surprisingly, DFFITS for D.C. is much larger than for the other observations, suggesting that the predicted values change considerably if we refit the model after removing this observation. The DFBETAS value for the single-family predictor X_2 is much larger for D.C. than for the other observations, suggesting that the effect of that predictor could change substantially with the removal of D.C. By contrast, DFBETAS for poverty is not so large.

These diagnotics suggest that the D.C. observation has a large influence, particularly on the coefficient of X_2 and on the fitted values. The prediction equation for the model fitted without the D.C. observation is

$$\hat{Y} = -14.6 + .36X_1 + 1.52X_2$$

Not surprisingly, the estimated effect of X_1 did not change much, but the coefficient of X_2 is now less than half as large. The standard error of the coefficient of X_2 also changes dramatically, decreasing from .44 to .26. □

TABLE 14.4 Computer Printout of Influence Diagnostics for Model Using Poverty Rate and Single-Parent Percentage to Predict Murder Rate

Obs	Dep Var MURDER	Predict Value	Residual	Rstuden	Hat H	Dffits	POVERTY Dfbetas	SINGLE Dfbetas
1	9.0	18.88	−9.88	−2.04	.162	−0.895	0.714	−0.761
2	11.6	10.41	1.18	0.22	.031	0.039	0.024	−0.011
3	10.2	8.07	2.13	0.40	.079	0.117	0.100	−0.069
4	8.6	12.16	−3.55	−0.65	.022	−0.099	−0.005	−0.025
5	13.1	14.63	−1.53	−0.28	.034	−0.053	−0.027	−0.004
6	5.8	10.41	−4.61	−0.87	.060	−0.220	0.174	−0.134
7	6.3	2.04	4.25	0.79	.051	0.185	−0.130	0.015
8	5.0	7.73	−2.73	−0.50	.043	−0.107	0.079	−0.045
9	8.9	6.97	1.92	0.35	.048	0.080	0.059	−0.047
10	11.4	15.12	−3.72	−0.69	.042	−0.145	0.071	−0.105
11	3.8	−2.07	5.87	1.11	.059	0.279	−0.153	−0.058
12	2.3	−1.74	4.04	0.75	.045	0.164	−0.034	−0.081
13	2.9	1.12	1.77	0.32	.035	0.063	0.012	−0.040
14	11.4	9.21	2.18	0.40	.020	0.058	−0.013	0.011
15	7.5	5.99	1.50	0.27	.023	0.043	−0.014	−0.000
16	6.4	2.71	3.68	0.68	.029	0.117	0.013	−0.062
17	6.6	7.79	−1.19	−0.22	.088	−0.070	−0.061	0.043
18	20.3	26.74	−6.44	−1.29	.161	−0.568	−0.412	−0.055
19	3.9	5.91	−2.01	−0.37	.033	−0.068	0.042	−0.014
20	12.7	9.95	2.74	0.51	.060	0.130	−0.104	0.077
21	1.6	4.72	−3.12	−0.57	.031	−0.104	0.058	−0.008
22	9.8	15.72	−5.92	−1.10	.033	−0.204	0.035	−0.124
23	3.4	2.23	1.16	0.21	.029	0.037	−0.007	−0.013
24	11.3	7.62	3.67	0.67	.027	0.115	0.059	−0.049
25	13.5	25.40	−11.90	−2.45	.126	−0.933	−0.623	−0.151
26	3.0	6.84	−3.84	−0.70	.023	−0.108	−0.033	0.039
27	11.3	7.87	3.42	0.62	.020	0.090	0.009	−0.013
28	1.7	−3.83	5.53	1.04	.057	0.259	0.016	−0.184
29	3.9	−0.15	4.05	0.75	.039	0.153	−0.047	−0.056
30	2.0	−1.07	3.07	0.57	.044	0.123	−0.039	−0.047
31	5.3	0.82	4.47	0.83	.035	0.158	−0.041	−0.058
32	8.0	19.53	−11.53	−2.25	.046	−0.499	−0.017	−0.308
33	10.4	11.57	−1.17	−0.22	.069	−0.060	0.048	−0.040
34	13.3	14.85	−1.55	−0.28	.028	−0.048	−0.005	−0.019
35	6.0	8.62	−2.62	−0.48	.022	−0.072	0.024	−0.015
36	8.4	9.62	−1.22	−0.22	.067	−0.061	−0.051	0.031
37	4.6	7.84	−3.24	−0.59	.027	−0.101	0.054	−0.029
38	6.8	1.55	5.24	0.97	.034	0.183	0.036	−0.115
39	3.9	5.67	−1.77	−0.32	.028	−0.056	0.029	−0.006
40	10.3	13.99	−3.69	−0.68	.038	−0.137	−0.084	0.008
41	3.4	1.07	2.32	0.43	.042	0.091	0.036	−0.067
42	10.2	9.92	0.27	0.05	.060	0.013	0.010	−0.006
43	11.9	11.60	0.29	0.05	.029	0.009	0.005	−0.001
44	3.1	2.34	0.75	0.13	.032	0.025	−0.010	−0.004
45	8.3	3.21	5.08	0.94	.039	0.192	−0.119	0.010
46	3.6	6.08	−2.48	−0.46	.040	−0.094	0.067	−0.028
47	5.2	9.52	−4.32	−0.80	.029	−0.139	0.078	−0.059
48	4.4	4.53	−0.13	−0.02	.023	−0.003	0.000	0.001
49	6.9	3.60	3.29	0.66	.178	0.307	0.274	−0.229
50	3.4	6.34	−2.94	−0.54	.021	−0.079	0.006	0.012
51	78.5	55.28	23.22	14.20	.536	15.271	−0.485	12.792

An observation with a large studentized residual does not have a major influence on the fit if it falls near the mean on the explanatory variables. For instance, New Mexico has a relatively large negative studentized residual of −2.25, but it has a small hat value (.047), so it does not have large values of DFFITS or DFBETAS. Similarly, an observation far from the mean on the explanatory variables (i.e., with a large hat value) need not have a major influence, if it falls close to the prediction equation and has a small studentized residual. For instance, West Virginia has a relatively large poverty rate and its hat value of .18 is triple the average, but it has little influence on the fit since its studentized residual is small (.66).

We have not suggested alternative methods when assumptions such as constant standard deviation or normality or linearity appear to be badly violated. The final three sections of the chapter address these issues.

14.3 Multicollinearity

In many social science applications of multiple regression the explanatory variables "overlap" considerably. Each variable may be nearly redundant, in the sense that it can be predicted very well using the others. If we regress an explanatory variable on the others, we get an R^2 value close to 1, suggesting that it may not be needed in the model, once the others are there. This condition is called *multicollinearity*. This section describes the effects of multicollinearity on conducting inference for regression models.

Effects of Multicollinearity on Standard Errors

Multicollinearity causes inflated standard errors for estimates of regression parameters. We illustrate this effect using the regression model having only two explanatory variables, $E(Y) = \alpha + \beta_1 X_1 + \beta_2 X_2$. The estimate b_1 of the coefficient of X_1 has standard error

$$\hat{\sigma}_{b_1} = \frac{1}{\sqrt{1 - r^2_{X_1X_2}}} \left[\frac{\hat{\sigma}}{\sqrt{\sum (X_1 - \bar{X}_1)^2}} \right]$$

where $\hat{\sigma}$ is the root MSE and $\sum(X_1 - \bar{X}_1)^2$ denotes the sum of squares of the deviations of the X_1-values about their mean $\bar{X}_1$. The effect of the correlation between the explanatory variables enters through the term $\sqrt{1 - r^2_{X_1X_2}}$ in the denominator. Other things being equal, the stronger that correlation, the larger the standard error of b_1. Similarly, the standard error of the estimator b_2 of β_2 also is larger with larger values of $r^2_{X_1X_2}$.

An analogous result applies for the general model. Suppose a particular predictor X_i overlaps a lot with the other predictors, in the sense that a large R^2 exists for predicting X_i using the others. Then, the standard error of the estimator of the partial effect of X_i is relatively large (see Problem 14.44). A primary effect of multicollinearity is that estimated regression coefficients have large standard errors. As a result, confidence intervals for the true values are wide, and the corresponding tests of $H_0 : \beta_i = 0$ tend to have large P-values unless the sample size is very large. These remarks also apply to partial correlations. See Problem 14.17.

Indicators of Multicollinearity

A straightforward way to assess whether multicollinearity exists is to regress each explanatory variable on the others in the model. When any of the R^2 values for these equations is close to 1, multicollinearity exists. Even without doing this, however, various types of behavior in an ordinary regression analysis can indicate potential problems due to multicollinearity.

A warning sign that multicollinearity exists occurs when the estimated coefficient for a predictor already in the model changes substantially when another variable is introduced. For example, perhaps the estimated coefficient of X_1 is 2.4 for the bivariate model, but when X_2 is added to the model, the coefficient of X_1 changes to 25.9.

Another indicator of multicollinearity occurs when a highly significant R^2 exists between Y and the explanatory variables, but individually each partial regression coefficient is not significant. In other words, $H_0 : \beta_1 = \cdots = \beta_k = 0$ has small P-value in the overall F test, but $H_0 : \beta_1 = 0$, $H_0 : \beta_2 = 0$, and so forth, do not have small P-values in the separate t tests. Thus, it is difficult to assess individual partial effects when severe multicollinearity exists.

Since a regression coefficient in a multiple regression model represents the effect of an explanatory variable when other variables are held constant, it has less meaning when multicollinearity exists. If $|r_{X_1X_2}|$ is high, then as X_1 changes, X_2 also tends to change in a linear manner, and it is somewhat artificial to envision X_1 or X_2 as being held constant. Thus, the coefficients have dubious interpretations when multicollinearity exists.

Remedial Actions When Multicollinearity Exists

Here are some remedial measures to reduce the effects of multicollinearity. First, since it may not make sense to study partial effects if the explanatory variables are highly correlated, we could use simple bivariate regression models to analyze the relationship between Y and each X_i separately.

A better solution is to choose an important subset of the explanatory variables, removing those variables that explain a small portion of the remaining unexplained variation in Y. If X_4 and X_5 have a correlation of .96, it is only necessary to include one of them in the model. One could use an automated variable selection procedure to select a subset of variables, but this is primarily helpful for purely exploratory research.

Alternatively, when several predictors are highly correlated and are indicators of a common feature, one can construct a summary index by combining responses on those variables. For example, suppose that a model for predicting Y = opinion about president's performance in office uses ten predictors, of which three refer to the subject's opinion about whether a woman should be able to obtain an abortion (1) when she cannot financially afford another child, (2) when she is unmarried, and (3) anytime in the first three months. Each of these items is scaled from 1 to 5, with a 5 being the most conservative response. They are likely to be highly positively correlated, contributing to multicollinearity. Thus, one might construct a summary measure for opinion about abortion by averaging the responses to these items. Higher values on that summary index represent more conservative responses. If the items were measured on different scales, we would first standardize the scores before averaging them. Socioeconomic status is a variable of this type, summarizing the joint effects of education, income, and occupational prestige.

Other procedures exist, beyond the scope of this chapter, for handling multicollinearity. These include *factor analysis* (introduced in Chapter 16), a method for creating artificial variables from the original ones in such a way that the new variables are uncorrelated. A procedure called *ridge regression* produces estimates of regression coefficients that are slightly biased but which have smaller standard errors than the least squares estimates. In most applications, though, it is more advisable to use a subset of the variables or create some new variables directly, as just discussed.

Finally, multicollinearity does not adversely affect all aspects of regression. Although multicollinearity makes it difficult to assess *partial* effects of explanatory variables, it does not hinder the assessment of their *joint* effects. If newly added explanatory variables overlap substantially with ones already in the model, then SSE will not decrease much, but the fit will not be poorer. So, the presence of multicollinearity does not diminish the goodness of the fit of the equation to the observed points or hinder achieving a large R^2 value.

For further discussion of the effects of multicollinearity and methods for dealing with it, see Fox (1991).

14.4 Nonlinearity: Polynomial Regression

The basic regression model assumes that the form of the relationship is linear. The multiple regression model assumes that the partial relationship between the mean of Y and each quantitative explanatory variable is linear, controlling for other explanatory variables. Although social science relationships are not *exactly* linear in form, the degree of nonlinearity is often so minor that they can be reasonably well approximated with linear equations.

Occasionally, though, the linear model is inadequate, even for approximation. A scatter diagram may reveal a highly nonlinear relationship. Alternatively, one's theoretical formulation of expected relationships might predict a nonlinear relationship. For example, one might expect Y = medical expenses to have a curvilinear relationship with X = age, being relatively high for the very young and the very old but lower for older children and young adults (Figure 14.5a). The relationship between X = per capita income and Y = life expectancy for a sample of countries might be approx-

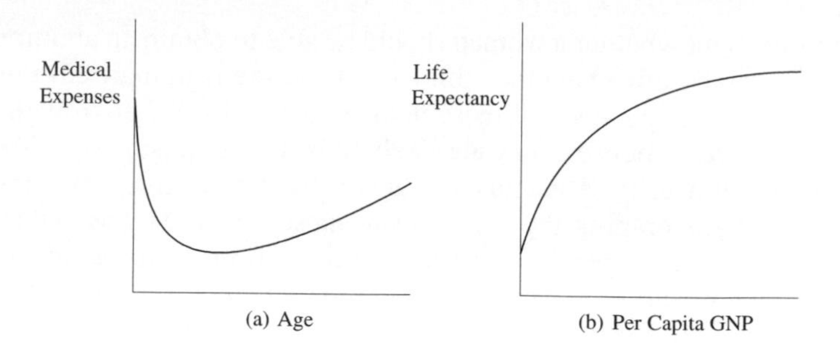

Figure 14.5 Two Nonlinear Relationships

imately a linearly increasing one, up to a certain point. However, beyond a certain level, additional income would probably result in little, if any, improvement in life expectancy (Figure 14.5b).

Undesirable consequences may result from using linear models to describe relationships that are curvilinear in form. Measures of association designed for linearity, such as the Pearson correlation, may underestimate the true association. Estimates of the mean of Y at various X-values may be badly biased, since the prediction line may poorly approximate the true regression curve. This section and Section 14.6 present ways of modeling nonlinear relationships.

Two approaches are commonly used to model nonlinearity. In the first of these, the regression function is a polynomial. The class of polynomial functions contains a diverse set of functional patterns, including linear functions. The second approach uses a mathematical transformation of one or both variables such that the transformed variables are linearly related. For example, for certain curvilinear relationships, the logarithm of the response variable is linearly related to the explanatory variable. The final section of the chapter discusses this second approach.

Quadratic Regression Models

A ***polynomial regression function*** for a response variable Y and explanatory variable X has form

$$E(Y) = \alpha + \beta_1 X + \beta_2 X^2 + \cdots + \beta_k X^k$$

In this model, X occurs in powers from the first ($X = X^1$) to some integer k. For $k = 1$, this is the linear function $E(Y) = \alpha + \beta_1 X$. The index k, the highest power in the polynomial equation, is called the ***degree*** of the equation.

The polynomial function most commonly used for nonlinear relationships is the *second-degree polynomial*

$$E(Y) = \alpha + \beta_1 X + \beta_2 X^2$$

called a ***quadratic regression model***. The graph of this function is parabolic, as Figure 14.6 portrays. It has a single bend, either increasing and then decreasing or else

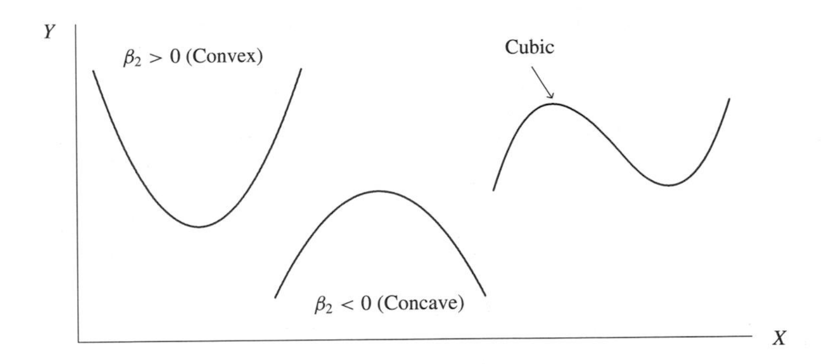

Figure 14.6 Graphs of Two Second-Degree Polynomials (Quadratic Functions) and a Third-Degree Polynomial (Cubic Function)

decreasing and then increasing. The shape of the parabolic curve is symmetric about a vertical axis, with its appearance when increasing a mirror image of its appearance when decreasing.

If a scatter diagram reveals a pattern of points with one bend, then a second-degree polynomial usually improves upon the linear fit. A third-degree polynomial $E(Y) = \alpha + \beta_1 X + \beta_2 X^2 + \beta_3 X^3$, called a ***cubic function***, is a curvilinear function having two bends (see Figure 14.6). In general, a kth-degree polynomial has $(k - 1)$ bends. The straight line, having degree $k = 1$, has $k - 1 = 0$ bends. Of the polynomial models, the linear and quadratic equations are most useful in social science research. Rarely is it necessary to use higher than a second-degree polynomial to describe the trend for the mean of Y.

Example 14.4 Birth Rate Predicted Using GNP

Table 14.5 shows values for several nations on Y = crude birth rate (number of births in a year per 1000 population size) and X = per capita gross national product (GNP, in thousands of dollars). Fertility tends to decrease as GNP increases. However, a straight-line model may be inadequate, since it might predict negative birth rate for sufficiently high GNP. In addition, some demographers predict that after GNP passes a certain level, birth rate may increase, since the nation's wealth makes it easier for a parent to stay home and take care of children rather than work.

Figure 14.7 is a scatter diagram for the 25 observations. It shows a clear decreasing trend. The linear prediction equation is $\hat{Y} = 29.62 - .81X$, and the Pearson correlation between X and Y equals .74. This prediction equation gives absurd predictions for very large X-values; $\hat{Y}$ is negative for $X > 36.4$. However, the predicted values are positive over the range of X-values for this sample.

To allow for potential nonlinearity and for the possibility that birth rate may increase for sufficiently large GNP, we could fit a quadratic regression model to these data. We

TABLE 14.5 Data on Y = Crude Birth Rate and X = Per Capita GNP

Nation	Birth Rate	GNP	Nation	Birth Rate	GNP
Algeria	29.0	1.6	Israel	20.4	13.6
Argentina	19.5	4.0	Japan	10.7	27.3
Australia	14.1	16.6	Malaysia	28.0	2.5
Brazil	21.2	2.6	Mexico	26.6	3.1
Canada	13.7	20.8	Nigeria	43.3	.2
China	17.8	1.3	Pakistan	41.8	.4
Cuba	14.5	1.6	Philippines	30.4	.7
Denmark	12.4	24.2	Russia	12.6	8.6
Egypt	28.7	.5	South Africa	33.4	2.6
France	13.0	24.1	Spain	11.2	13.4
Germany	11.0	19.8	United Kingdom	13.2	17.4
India	27.8	.3	United States	15.2	22.6
Iraq	43.6	.7			

would use the second-degree polynomial, rather than higher, because we expect at most one bend in the relationship, that is, a decrease followed potentially by an increase. □

Interpreting and Fitting Quadratic Regression Models

The quadratic regression model

$$E(Y) = \alpha + \beta_1 X + \beta_2 X^2$$

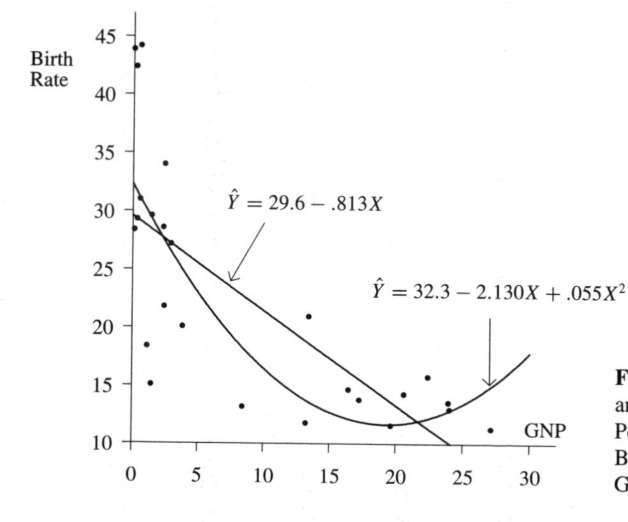

Figure 14.7 Scatter Diagram and Best-Fitting Second-Degree Polynomial for Data on Y = Crude Birth Rate and X = Per Capita GNP

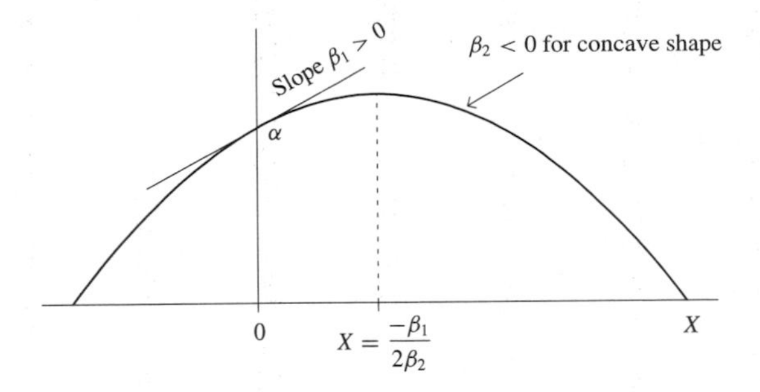

Figure 14.8 Interpretation of Parameters of Second-Degree Polynomial $E(Y) = \alpha + \beta_1 X + \beta_2 X^2$

plotted for the possible values of α, β_1, and β_2, describes a large variety of parabolic shapes. Unlike linear functions, for which the slope of the line remains constant over all X-values, the mean change in Y for a one-unit increase in X *depends on the value of* X. For example, a straight line drawn tangent to the parabola in Figure 14.8 has positive slope for small values of X, zero slope where the parabola achieves its maximum value, and negative slope for large values of X. The rate of change of the line varies to produce a curve having a smooth bend.

The sign of the coefficient β_2 of the X^2 term determines whether the function is convex (bowl shaped, relative to the X-axis) or concave (mound shaped). Convex functions have positive β_2, and concave functions have negative β_2, as shown in Figure 14.8. If $\beta_2 = 0$, the function is linear.

As usual, the coefficient α is the Y-intercept. The coefficient β_1 of X is the slope of the line that is tangent to the parabola as it crosses the Y axis. If $\beta_1 > 0$, for example, then the parabola is sloping upward at $X = 0$, as shown in Figure 14.8. The slope of the tangent line at an arbitrary X value equals $\beta_1 + 2\beta_2 X$; that is, as X increases, this slope changes linearly. At the point at which the slope is zero, the relationship changes direction from positive to negative or from negative to positive; this point is $X = -\beta_1/2\beta_2$. This is the point at which the mean of Y takes its maximum if the parabola is concave and its minimum if it is convex.

One can fit quadratic regression models by treating them as a special case of the multiple regression model

$$E(Y) = \alpha + \beta_1 X_1 + \beta_2 X_2 = \alpha + \beta_1 X + \beta_2 X^2$$

with two explanatory variables. We identify X_1 with the explanatory variable X and X_2 with its square, X^2. The data for the model fit consists of the Y-values for the subjects in the sample, the X-values (called X_1), and an artificial variable (X_2) consisting of the squares of the X-values. Software creates these squared values for us, and then uses least squares to find the best fitting quadratic function out of the class of all second-degree polynomials.

Example 14.5 Quadratic Regression for Birth Rate and GNP

To find the best-fitting quadratic equation for the data on Y = crude birth rate and X = GNP, we use the explanatory variables $X_1 = X$ and $X_2 = X^2$ in multiple regression.

Table 14.6 shows part of the printout for the quadratic regression model. The prediction equation is

$$\hat{Y} = 32.3 - 2.130X + .055X^2$$

Figure 14.7 plots the quadratic prediction equation in the scatter diagram. Since the coefficient .055 of X^2 is positive, the graph is convex. Also, since the coefficient -2.130 of X is negative, the curve is decreasing as it crosses the Y-axis.

TABLE 14.6 Part of Computer Printout for Second-Degree Polynomial Model

Source	DF	Sum of Squares	Mean Square	F Value	Prob>F
Model	2	1665.829	832.914	18.389	0.0001
Error	22	996.457	45.293		
Total	24	2662.286			

Root MSE 6.730 R-square 0.6257

Variable	Parameter Estimate	Standard Error	T for H0: Parameter=0	Prob > |T|
INTERCEP	32.2785	2.2472	14.364	0.0001
GNP	-2.1302	0.6205	-3.433	0.0024
GNP2	0.0549	0.0252	2.182	0.0401

A convex quadratic equation takes its minimum at $X = -\beta_1/2\beta_2$. For these data, we estimate this point to be $X = 2.130/2(.055) = 19.4$. The predicted birth rate increases as GNP increases above this point. □

Description and Inference About the Nonlinear Effect

For a polynomial model, R^2 for multiple regression describes the strength of the association. In this context, it describes the proportional reduction in error obtained from using the polynomial model, instead of $\bar{Y}$, to predict Y. One can compare this measure to the coefficient of determination r^2 for the linear model to determine how much better a fit the curvilinear model provides. Since the polynomial model has additional terms besides X, R^2 always is at least as large as r^2. The difference $R^2 - r^2$ measures the additional reduction in prediction error obtained by using the polynomial instead of the straight line.

For Table 14.5, the best-fitting *linear* prediction equation has $r^2 = .545$. That line is also plotted in Figure 14.7. From Table 14.6 for the quadratic model, $R^2 = .626$. The best quadratic equation explains about 8% more variability in Y than does the best-fitting linear equation.

If $\beta_2 = 0$, the quadratic regression equation $E(Y) = \alpha + \beta_1 X + \beta_2 X^2$ reduces to the linear regression equation $E(Y) = \alpha + \beta_1 X$. Therefore, to test the null hypothesis that the relationship is linear against the alternative that it is quadratic, we test $H_0 : \beta_2 = 0$ that the coefficient of the X^2 term equals 0. The usual t test for a regression coefficient does this, dividing the estimate of β_2 by its standard error.

The set of nations in Table 14.5 is not a random sample of nations, so inference is not relevant. If it had been, the printout in Table 14.6 shows that $t = .0549/.0252 = 2.18$, with $df = 22$. The P-value for testing H_0: $\beta_2 = 0$ against H_a: $\beta_2 \neq 0$ is $P = .04$. In this sense, the quadratic prediction equation apparently provides a somewhat better fit than the linear equation.

Cautions in Using Polynomial Models

Some cautions are in order before you take the conclusions in this example too seriously. The assumptions needed for the fitted equation to provide the "best fit" or for applying inference procedures to curvilinear models are the same as for the linear regression model. For the quadratic model, for example, $E(Y) = \alpha + \beta_1 X + \beta_2 X^2$ represents the mean of the conditional distribution of Y for each fixed value of X. The conditional distribution of Y-values is assumed to be normal about the mean, with constant standard deviation σ at all X-values.

The scatter diagram (Figure 14.7) suggests that the variability in birth rates is considerably higher for nations with low GNPs than it is for nations with high GNPs. The births rates show much greater variability when their mean is higher. A model that permits nonconstant standard deviation, discussed in Section 14.5, provides somewhat different results (Problem 14.30). The quadratic fit for that approach is closer to a straight line, with predicted birth rates not increasing until GNP exceeds 21.6, and the predicted birth rate at the highest sample GNP value of 27.3 is barely larger than the value at 21.6.

In fact, before we conclude that birth rate increases above a certain value, we should realize that other models for which this does not happen are also consistent with these data. For instance, Figure 14.7 suggests that a "piecewise linear" model that has a linear decrease until GNP is about 5 and then a separate, nearly horizontal, line beyond that point fits quite well. Perhaps the most satisfactory model for these data is one discussed in the final section of this chapter for *exponential regression*. Typically, unless a data set is very large, a variety of models may be consistent with the data.

In examining scatter diagrams, you should be cautious not to read too much into the data. Don't let one or two outliers suggest a curve in the trend. Good model building follows the principle of ***parsimony***, whereby models have no more parameters than are necessary to represent adequately the relationship. One reason for this is that simple models are easier to understand and interpret than complex ones. Another reason is that when a model contains unnecessary variables, the standard errors of the estimates of the regression coefficients tend to inflate, hindering efforts at making precise inferences.

When a polynomial regression model is valid, the regression coefficients do not have the partial slope interpretation usual for coefficients of multiple regression models. It does not make sense to refer to the change in the mean of Y when X^2 is increased one unit and X is held constant. One cannot control X and vary X^2, or vice versa, since X^2 is a deterministic function of X.

Similarly, it does not make sense to interpret the partial correlations $r_{YX^2 \cdot X}$ or $r_{YX \cdot X^2}$ as measures of association, controlling for X or X^2. However, the coefficient $r^2_{YX^2 \cdot X}$ does measure the proportion of the variation in Y unaccounted for by the straight-line model that is explained by the quadratic model. In Example 14.5, applying the formula for $r^2_{YX_2 \cdot X_1}$ from Section 11.7 yields

$$r^2_{YX^2 \cdot X} = \frac{R^2 - r^2_{YX}}{1 - r^2_{YX}} = \frac{.626 - .545}{1 - .545} = .178$$

Of the variation in Y unexplained by the linear model, about 18% is explained by the introduction of the quadratic term.

When there are several explanatory variables, one can also model nonlinear relationships. For example, the model

$$E(Y) = \alpha + \beta_1 X_1 + \beta_2 X_2 + \beta_3 X_2^2$$

allows nonlinearity in X_2. For fixed X_1, the mean of Y is a quadratic function of X_2. For fixed X_2, the mean of Y is a linear function of X_1 with slope β_1. This model is a special case of multiple regression with three explanatory variables, in which X_3 is the square of X_2. One can fit the model using ordinary software, defining X_3 to take values that are the squares of the values for X_2. Models allowing both nonlinearity and interaction are also possible, though more complex.

14.5 Generalized Linear Models

The models presented in this book are special cases of ***generalized linear models***. This is a broad class of models that includes ordinary regression models for continuous response variables, alternative models for continuous variables that do not assume normality or constant variance, and also models for discrete response variables. This section introduces generalized linear models. We use the acronym *GLM*.

All generalized linear models have three components: The ***random component*** of a GLM identifies the response variable Y and assumes a probability distribution for it. That distribution can be one of several types, the normal distribution being one possibility. The ***systematic component*** of a GLM specifies the explanatory variables used as predictors in the model. These variables play the role of the X variables on the right-hand side of the model formula.

The third component, the ***link function***, specifies a function of the expected value (mean) of Y that is linearly predicted by the explanatory variables. This component *links* the systematic and random components of the GLM by showing the connection between the predictors and the mean of Y. The GLM relates a function of that mean to the predictors through an equation having linear form. We now discuss the three components in further detail.

Random Component

The ***random component*** of a GLM identifies the variable that plays the role of Y and assumes a particular form for its probability distribution. The regression models discussed in the past six chapters (including ANOVA and analysis of covariance, using dummy variables) are special cases of GLMs. These models assume that Y is continuous and has a normal conditional distribution with constant standard deviation. They are GLMs with normal random component.

In many applications, the potential outcomes for Y are binary rather than continuous. Each observation might be labeled as a *success* or *failure*, as in the methods for proportions presented in Sections 5.3, 6.3, and 7.2. For instance, consider a study of factors that influence votes in presidential elections. For each subject, the response variable indicates which candidate was preferred in the previous presidential election—the Democrat or the Republican. The study uses various predictors in a model for subjects' decisions about the preferred candidate. In this case, models usually assume a *binomial* distribution for the random component. The next chapter presents a GLM for binary data, called *logistic regression.*

In some applications, each response observation is a count. For example, consider a study of factors associated with suicide rates, measured at the county level. The response for a given county is the number of suicides in the previous year. The study constructs a model that uses several explanatory variables to predict the number of suicides. A distribution not discussed in this text, called the *Poisson*, is usually assumed for the random component of GLMs for count data.

Binary outcomes and counts are examples of discrete variables. Regression models that assume normal distributions are not optimal for models with highly discrete response data. Even when the response variable is continuous, the normal distribution is not necessarily appropriate. When each observation must take positive value, for instance, the distribution is often skewed to the right with greater variability when the mean is greater. In that case, a GLM can use the family of *gamma* distributions for the random component, as discussed later in this section.

Systematic Component

Denote the expected value of Y, the mean of its probability distribution, by $\mu = E(Y)$. In a GLM, as in ordinary regression models, μ varies according to values of explanatory variables.

The ***systematic component*** of a GLM specifies the explanatory variables. These enter linearly as predictors on the right-hand side of the model equation. That is, the systematic component specifies the variables that play the roles of $\{X_j\}$ in the formula

$$\alpha + \beta_1 X_1 + \beta_2 X_2 + \ldots + \beta_k X_k$$

Link

The third component of a GLM is the ***link*** between the random and systematic components. It specifies how $\mu = E(Y)$ relates to the explanatory variables. One can model the mean μ directly or model a function $g(\mu)$ of the mean. The GLM formula states that

$$g(\mu) = \alpha + \beta_1 X_1 + \beta_2 X_2 + \ldots + \beta_k X_k$$

The function $g(\mu)$ is called the ***link function***.

The simplest possible link function is $g(\mu) = \mu$. This models the mean directly and is called the ***identity link***. It specifies a linear model for the mean response,

$$\mu = \alpha + \beta_1 X_1 + \beta_2 X_2 + \ldots + \beta_k X_k$$

This is the form of ordinary regression models.

Other links permit the mean to relate nonlinearly to the predictors. For instance, the link function $g(\mu) = \log(\mu)$ models the log of the mean. The log function applies to positive numbers, so this ***log link*** is appropriate when μ cannot be negative, such as with count data. A GLM that uses the log link is often called a ***loglinear model***. It has form

$$\log(\mu) = \alpha + \beta_1 X_1 + \beta_2 X_2 + \ldots + \beta_k X_k$$

The final section of this chapter shows an example of this model.

For binary data, the next chapter uses the link function $g(\mu) = \log[\mu/(1-\mu)]$, called the ***logit*** link. It is appropriate when μ is between 0 and 1, such as a probability. When a response is binary, this link is used in models for the probability of a particular one of the two possible outcomes, for instance, to model the probability that a subject votes for the Democratic candidate. A GLM using the logit link is called a ***logit model***.

Normal GLMs

Ordinary regression models for continuous response variables are special cases of GLMs. One assumes a normal distribution for the random component and models the mean directly, using the identity link, $g(\mu) = \mu$. A GLM generalizes ordinary regression in two ways: First, the random component can have a distribution other than the normal. Second, it can model a function of the mean. Both generalizations are important, especially for discrete responses.

The traditional way of analyzing "nonnormal" data attempts to transform Y so it is approximately normal, with constant standard deviation at all levels of the predictors. Square root or log transforms are often applied to do this. If the goals of normality and constant variation are achieved, then ordinary regression methods using least squares are applicable with the transformed data. In practice, this usually does not work well. A transform that produces constant variation may not produce normality, or else simple linear models for the explanatory variables may fit poorly on that scale. Moreover, conclusions that refer to the mean response on the scale of the transformed variable are usually less relevant, and there can be technical problems such as taking logarithms of 0.

With the theory and methodology developed for GLMs in the past quarter century, it is unnecessary to transform data so that methods for normal data apply. This is because the GLM fitting process utilizes a powerful estimation method (***maximum likelihood***) for the choice of random component, which is not restricted to normality. In addition, in GLMs the choice of link is separate from the choice of distribution for the random component. If a link makes sense for a particular form of data, it is not necessary that it also stabilize variation or produce normality.

We introduce the concept of GLMs to unify a wide variety of statistical methods. Regression models for continuous data and models for discrete data presented in Chapter 15 are special cases of one highly general model. In fact, the same fitting method yields parameter estimates for all GLMs. This method is the basis of software for fitting GLMs, such as PROC GENMOD in SAS.

Using SAS–GENMOD, for instance, one has tremendous flexibility and power in the model-building process. One picks a probability distribution that is most appropriate for the response variable. For instance, one might select the normal option for a continuous response or the binomial option for a binary response. One specifies the variables that are the predictors in the systematic part of the model. Finally, one picks the link function, determining which function of the mean to model. The appendix provides examples.

The next chapter introduces the most important GLMs for discrete response variables—logistic regression models for binary data with a binomial random component, and loglinear models for count data. The next subsection shows the use of GLMs for data with nonconstant variation, and the final section of the chapter shows a GLM for modeling the log link of the mean.

Regression Using Gamma Instead of Normal Response

For Table 9.4 on selling prices of homes, Example 9.10 used the bivariate regression model $E(Y) = \alpha + \beta X$ to describe and make inferences about the relationship between X = size of home and Y = selling price. For this model, Figure 14.9 plots the residuals against size. This figure resembles Figure 14.3b. It shows a tendency for greater variability at higher size values. Small homes show little variability in selling prices, whereas very large homes show high variability. The large homes are the ones with the higher selling prices, so variability in Y increases as its mean increases.

This phenomenon often happens for positive-valued response variables. When the mean response is near 0, less variation occurs than when the mean response is high. For such data, least squares is not optimal. It is identical to maximum likelihood for a GLM with a normal response, which has a single standard deviation parameter σ that is assumed to be the same at all values of predictors.

An alternative approach for data of this form selects a distribution for the random component for which the standard deviation increases as the mean increases (i.e., that permits *heteroscedasticity*). The family of ***gamma distributions*** has this property. Its standard deviation increases proportionally to the mean: when the mean doubles, the standard deviation doubles. The gamma distribution is concentrated on the positive part of the line, and also exhibits skewness to the right. (Technically, GLMs use the gamma family having a constant *shape* parameter that determines the shape of the distribution; the data estimate this shape.)

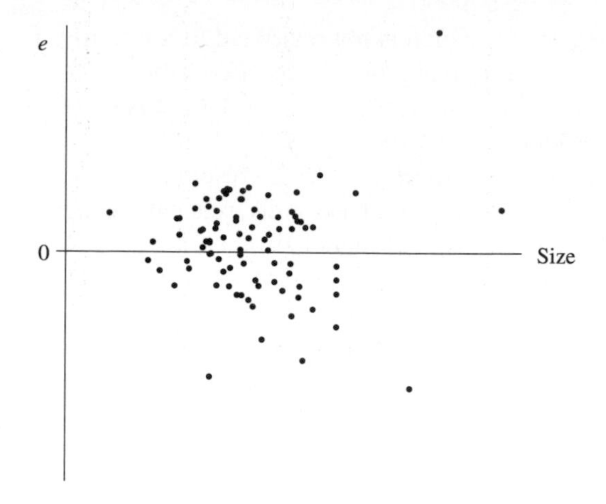

Figure 14.9 Residuals Plotted Against Size of Home, for Model with Size of Home as Predictor of Selling Price

Using GLMs, one can fit a regression model assuming a gamma distribution for Y instead of a normal distribution. Even if the data are close to normal, this alternative fit is more appropriate than the least squares fit when the standard deviation increases proportionally to the mean.

Example 14.6 Gamma GLM for Home Selling Price

The least squares fit of the model $E(Y) = \alpha + \beta X$ to the data on Y= selling price and X = size of home, discussed in Example 9.10, is $\hat{Y} = -25.2 + 75.6X$. The t test statistic for the effect of size is the ratio of the estimated slope 75.6 to its standard error of 3.86, which equals 19.6. This fit is optimal when the standard deviation is constant, but Figure 14.9 shows this is severely violated.

As an alternative, we used GLM software in SAS (see the appendix) to fit the regression model assuming a gamma distribution, for which the standard deviation increases as the mean does. We used the identity link, so that the model form is the same. The prediction equation is then $\hat{Y} = -13.2+67.8X$. The standard error of the slope is 2.53, so the test statistic for the size effect is 67.8/2.53 = 26.8. This t ratio is larger than with the least squares fit, showing the improved efficiency of the estimate based on the more realistic assumption about the variation of Y.

The gamma GLM provides good predictions. The correlation between the observed response Y and the predicted response $\hat{Y}$ for the GLM prediction equation equals .899. The highest possible correlation between Y and a linear prediction of Y necessarily occurs for the least squares fit. That correlation, which equals the Pearson correlation between X and Y, is also .899. In this example, there is no sacrifice in predictive power in using the GLM fit.

For the least squares fit, a look at the residual plot (Figure 14.9) and at measures of influence for these data reveals that the point with $X = 3.30$ and $Y = 309.4$ is an extreme outlier and is highly influential. Refitting the data without this point, using least squares, we get $\hat{Y} = -15.9 + 69.3X$, with a standard error for the slope of 3.59. The fit is then similar to the gamma GLM fit. The standard error for the slope is still larger, however, suggesting that the gamma fit still provides greater efficiency. □

The traditional method of dealing with variability that increases with the mean is to transform the data, applying the log or square root to the Y-values. Then, the variability is more nearly constant, and one can apply least squares. There is a fundamental flaw with this approach. If the original relationship is linear, it is no longer linear after applying the transformation. If we fit a straight line and then transform back to the original scale, the fit is no longer linear. Although this approach is still used in many statistical methods textbooks, we think the gamma GLM approach is more elegant and preferable because of maintaining the linear relationship.

Nonparametric Regression*

Recent advances make it possible to fit regression models to data without assuming particular functional forms (such as straight lines) for the relationship. These approaches are *nonparametric*, in terms of having fewer (if any) assumptions about the functional form and sometimes even the distribution of the response variable. It is helpful to look at a plot of a fitted nonparametric regression model to learn about trends evident in the data.

One nonparametric regression method, called ***generalized additive modeling***, is a further generalization of the generalized linear model. It has the form

$$g(\mu) = f_1(X_1) + f_2(X_2) + \ldots + f_k(X_k)$$

where $f_1, \ldots, f_k$ are unspecified and potentially highly complex functions. The GLM is the special case in which each of these functions is linear. The estimated functional form of the relationship for each predictor is determined by a computer algorithm, using the sample data. As in GLMs, one selects a particular form for the link function g and also a distribution for the random component. This model is useful for smoothing data to reveal overall trends.

Nonparametric regression is beyond the scope of this text. At this time, many statistical software packages do not yet have routines for generalized additive models, though some have a related nonparametric smoothing method, called LOESS, which usually provides similar results.

Figure 14.10 shows the plot of a nonparametric regression model (using LOESS) for the birth rate data of Table 14.5. There is a clear decreasing trend over the entire range. It is evident that a straight line model is inadequate. It is also apparent, though, that the response does not eventually increase, as a quadratic model predicts. The next section discusses a model that provides a more satisfactory fit for these data.

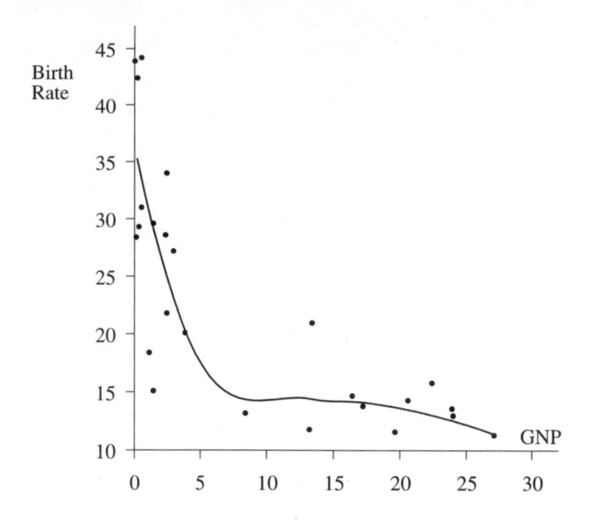

Figure 14.10 Fit of Nonparametric Regression Model to Birth Rate Data of Table 14.5, Using LOESS to Smooth the Data

14.6 Exponential Regression and Log Transforms

Although polynomials provide a diverse collection of functions for modeling nonlinearity, other mathematical functions are often more appropriate. The most important case is when the mean of the response variable is an *exponential* function of the explanatory variable.

> **Exponential Regression Function**
>
> An ***exponential regression*** function has the form $E(Y) = \alpha\beta^X$.

In this equation, the explanatory variable appears as the exponent of a parameter. Unlike a quadratic function, an exponential function can take only positive values, and it continually increases (if $\beta > 1$) or continually decreases (if $\beta < 1$). In either case, it has a convex shape, as Figure 14.11 shows. We provide interpretations for the model parameters later in this section.

For the exponential regression function, the *logarithm* of the mean is linearly related to the explanatory variable. That is, if $\mu = E(Y) = \alpha\beta^X$, then

$$\log[\mu] = \log\alpha + (\log\beta)X = \alpha' + \beta' X$$

with $\alpha' = \log(\alpha)$ and slope $\beta' = \log(\beta)$. The right-hand side has the form of a straight line. This model form is the special case of a generalized linear model (GLM) using the log link. If the model holds, a plot of the log of the response values should show an approximate linear relation with the predictor. (Don't worry if you have forgotten your high school math about logarithms. You will not need to know this in order to

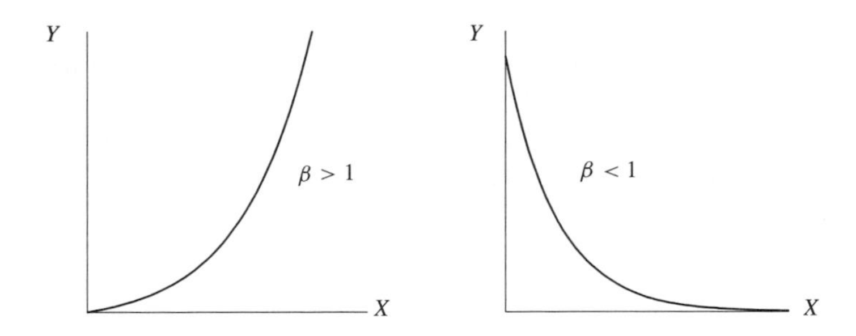

Figure 14.11 The Exponential Regression Function $E(Y) = \alpha\beta^X$

understand how to fit the GLM or how to interpret the parameters in the exponential regression model.)

The appendix shows that it is simple to use GLM software to estimate the parameters in the model $\log[E(Y)] = \alpha' + \beta' X$. The antilogs of these estimates are the estimates for the parameters in the exponential regression model $E(Y) = \alpha\beta^X$, as shown below.

Example 14.7 Exponential Population Growth

Exponential regression is often used to model population size over time. If the rate of growth remains constant, in percentage terms, then the size of that population grows exponentially fast. Suppose that the population size at some fixed time is α and the growth rate is 2% per year. After 1 year, the population is 2% larger than at the beginning of the year. Then, the population size grows by a multiplicative factor of 1.02 each year. The population size after 1 year is $\alpha(1.02)$. Similarly, the population size after 2 years is

$$(\text{Population size at end of 1 year})(1.02) = [\alpha(1.02)]1.02 = \alpha(1.02)^2$$

After 3 years, the population size is $\alpha(1.02)^3$. After X years, the population size is $\alpha(1.02)^X$. The population size after X years is an exponential function $\alpha\beta^X$ with parameters given by the initial population size α and the rate of growth factor, $\beta = 1.02$, corresponding to 2% growth.

Table 14.7 shows the U.S. population size (in millions) at 10-year intervals beginning in 1890. Figure 14.12 plots these values over time. Table 14.7 also shows the natural logarithm of the population sizes. (The model makes sense with logs to any base, but software fits the GLM using natural logs.) Figure 14.13 plots these log values over time. The log population sizes appear to grow approximately linearly. This suggests that population growth over this time period was approximately exponential, with a constant rate of growth. We now estimate the regression curve, treating time as the explanatory variable X.

TABLE 14.7 Population Sizes and Log Population Sizes by Decade from 1890 to 1990

Year	No. Decades Since 1890 X	Population Size Y	log Y	$\hat{Y}$
1890	0	62.95	4.14	70.4
1900	1	75.99	4.33	80.1
1910	2	91.97	4.52	91.2
1920	3	105.71	4.66	103.7
1930	4	122.78	4.81	118.0
1940	5	131.67	4.88	134.3
1950	6	151.33	5.02	152.7
1960	7	179.32	5.19	173.8
1970	8	203.30	5.31	197.7
1980	9	226.54	5.42	224.9
1990	10	248.71	5.52	255.9

Source: U.S. Department of Commerce, *Statistical Abstract of the United States, 1996*.

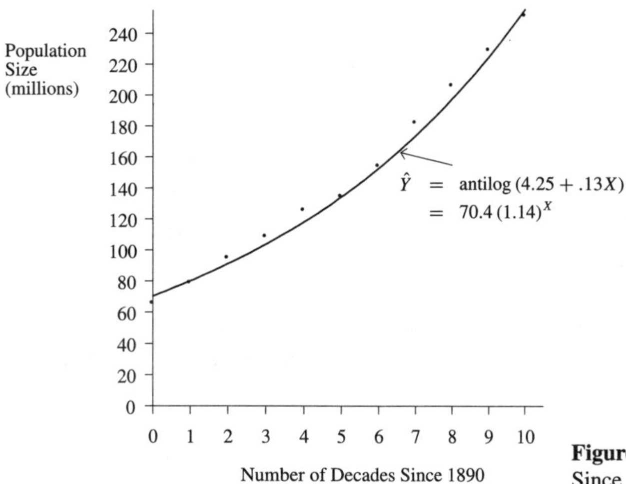

Figure 14.12 U.S. Population Size Since 1890

For convenience, we identify the time points 1890, 1900, ..., 1990 as times 0, 1, ..., 10; that is, X represents the number of decades since 1890. We use SAS, as shown in the appendix, to estimate the generalized linear model $\log(\mu) = \alpha' + \beta' X$. The prediction equation, for natural logs, is

$$\log(\hat{\mu}) = 4.2547 + .1290X$$

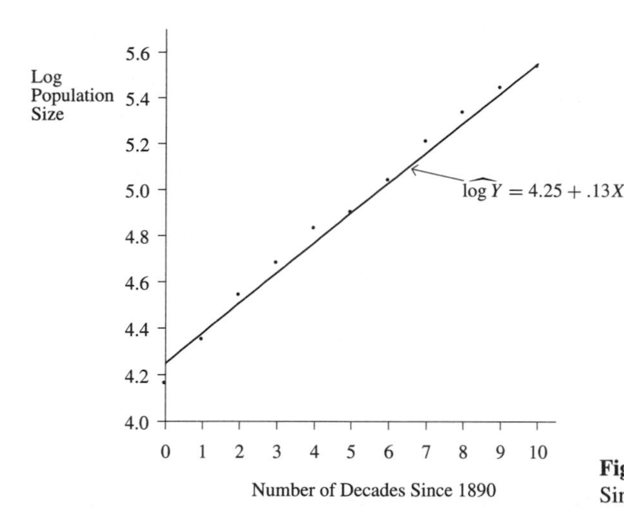

Figure 14.13 Log Population Sizes Since 1890

Antilogs of these estimates are the parameter estimates for the exponential regression model. For natural logs, the antilog function is the exponential function e^x. That is, antilog(4.2547) $= e^{4.2547} = 70.433$, and antilog(.1290) $= e^{.1290} = 1.1377$. (Most calculators have an e^x key that provides these antilogs.) Thus, for the exponential regression model $E(Y) = \alpha\beta^X$, the estimates are $\hat{\alpha} = 70.433$ and $\hat{\beta} = 1.1377$. The prediction equation is

$$\hat{Y} = \hat{\alpha}\hat{\beta}^X = 70.433(1.1377)^X$$

The predicted initial population size (in 1890) is $\hat{\alpha} = 70.4$ million. The predicted population size X decades after 1890 equals $\hat{Y} = 70.433(1.1377)^X$. For 1990, for instance, $X = 10$, and the predicted population size is $\hat{Y} = 70.433(1.1377)^{10} = 255.9$ million. Table 14.7 shows the predicted values for each decade. Figure 14.12 plots the exponential prediction equation.

The predictions are very good. The total sum of squares of population size values about their mean equals TSS $= 38{,}540.7$, whereas the sum of squared errors about the prediction equation is SSE $= 225.6$. The proportional reduction in error is

$$(38{,}540.7 - 225.6)/38{,}540.7 = .994.$$

□

Interpreting Exponential Regression Models

Now let's take a closer look at how to interpret parameters in the exponential regression model, $E(Y) = \alpha\beta^X$. The parameter α represents the mean of Y when $X = 0$. The parameter β represents the exponential increase factor. It is the ***multiplicative*** change in the mean of Y for a one-unit increase in X. The mean of Y at $X = 10$ equals β *multiplied* by the mean of Y at $X = 9$. For instance, for the equation $\hat{Y} = 70.433(1.1377)^X$, the predicted population size at a particular date equals 1.1377 times the predicted population size a decade earlier.

By contrast, the parameter β in the *linear* model $E(Y) = \alpha + \beta X$ represents the ***additive*** change in the mean of Y for a one-unit increase in X. In the linear model, the mean of Y at $X = 10$ equals β *plus* the mean of Y at $X = 9$. The prediction equation for the linear model (i.e., identity link) fitted to Table 14.7 equals $\hat{Y} = 52.72 + 18.55X$. This model predicts that the population size increases by 18.55 million people every decade.

In summary, for the linear model, $E(Y)$ changes by the same *quantity* for each one-unit increase in X, whereas for the exponential model, $E(Y)$ changes by the same *percentage* for each 1-unit increase. For the exponential regression model with Table 14.7, the predicted population size is multiplied by 1.1377 each decade. This equation corresponds to a predicted 13.77% growth per decade.

Suppose the growth rate is 15% per decade, to choose a rounder number. This corresponds to a multiplicative factor of 1.15. After five decades, the population grows by a factor of $(1.15)^5 = 2.0$. That is, after five decades, the population size doubles. If the rate of growth remained constant at 15% per decade, the population would double every 50 years. After 100 years, the population size would be quadruple the original size; after 150 years it would be 8 times as large; after 200 years it would be 16 times its original size; and so forth.

The exponentially increasing function has the property that its doubling time is a constant. As can be seen from the sequence of population sizes at 50-year intervals, this is an extremely fast increase even though the annual rate of growth (1.4% annual for a decade increase of 15%) seems small. In fact, the world population has been following an exponential growth pattern, with recent rate of growth over 15% per decade.

Example 14.8 Exponential Regression for Birth Rate Data

When $\beta < 1$ in the exponential regression model, $\beta' = \log(\beta) < 0$ in the log transformed GLM. In this case, the mean of Y *decreases* exponentially fast as X increases. The curve then looks like the second curve in Figure 14.11.

The nonparametric regression curve (Figure 14.10) for the birth rate data (Table 14.5) has an appearance much like this. In fact, the exponential regression model provides a good fit for those data. Using the GLM with log link for Y = birth rate and X = per capita GNP, we get the prediction equation,

$$\log(\hat{\mu}) = 3.440 - .0476X$$

Taking antilogs yields the exponential prediction equation

$$\hat{Y} = \hat{\alpha}\hat{\beta}^X = e^{3.440}(e^{-.0476})^X = 31.2(.954)^X$$

The predicted birth rate at GNP value $X + 1$ equals 95.4% of the predicted birth rate at GNP value X; that is, it decreases by 4.6%.

With this fit, the correlation between the observed and predicted birth rates equals .77, nearly as high as the value of .79 achieved with the quadratic model, which has an extra parameter. If we expect birth rate to be continuously decreasing as GNP increases, the exponential regression model is a more realistic model than the linear or quadratic regression models of Section 14.4. Unlike the linear model, it cannot give negative predicted birth rates, and unlike the quadratic model, the predicted birth rate decreases continuously. □

Transforming the Predictor to Achieve Linearity

Other transformations of the response mean or of explanatory variables are useful in some situations. For example, suppose Y tends to increase or decrease over a certain range of X-values, but once a certain X-value has been reached, further increases in X have less effect on Y, as in Figure 14.5b. For this concave increasing type of trend, X behaves like an exponential function of Y. Taking the logarithms of the X-values often linearizes the relationship. Another possible transform for this case is to invert the X-values (i.e., use $1/X$ as the explanatory variable).

14.7 Chapter Summary

This chapter presented several important topics in building regression models. We have seen how to check assumptions of the basic regression model and how to ease some restrictions of this model.

- When a large number of terms might serve as explanatory variables for a model, the ***backward elimination*** and ***forward selection*** procedures use a sequential process to select variables for the model. These are exploratory in purpose and should be used with caution.
- Plots of the ***residuals*** help to check whether the model is adequate and whether the assumptions for inferences are reasonable. Influence diagnotics such as DFBETAS and DFFITS describe which observations have a strong influence on the parameter estimates and the fit of the model.
- ***Multicollinearity***, the condition of highly correlated explanatory variables, causes inflation of standard errors of estimated regression coefficients.
- One can model ***nonlinear*** relationships through the use of ***polynomial*** (particularly ***quadratic***) functions and ***exponential*** functions. Quadratic functions have a parabolic appearance, whereas exponential functions have a convex increasing or convex decreasing appearance.
- ***Generalized linear models*** allow the response variable to have a distribution other than the normal, such as the binomial for binary data and the gamma for positive responses having greater variation at greater mean values. Such models also permit modeling a function of the mean. For instance, the ***exponential regression model*** is a generalized linear model for the log of the mean.

PROBLEMS

Practicing the Basics

1. Refer to Table 9.1, deleting the observation for D.C. With Y = violent crime rate and the five predictors as explanatory variables (all except murder rate) and using $\alpha = .10$ in tests, select a model using (a) backward elimination, (b) forward selection. Interpret the model selected.
2. Refer to Table 9.1, excluding the observation for D.C. Let Y = murder rate. For the five predictors in that table (excluding violent crime rate), the t test of independence for the bivariate model has P-value below .05.
   a) Fit the multiple regression model, using all five predictors. Are the P-values for the partial tests all significant? Explain why results of these tests may differ from those in the separate tests of independence.
   b) Use backward elimination to select a model, deleting a variable if it does not have P-value below .10. Interpret.
   c) Use forward selection to select a model, adding a variable if it has P-value below .10. Interpret.
   d) Use stepwise regression, with the .10 level for significance. Interpret.
   e) Compare results of the three selection procedures. How is it possible that a variable (percent with a high school education) can be the first variable dropped in (b) yet the second added in (c)?
3. Refer to the previous problem. Now include the D.C. observation.
   a) Use backward elimination, and compare results to part (b) above.
   b) Use forward selection, and compare results to part (c) above.
   c) What does this exercise suggest about how influential outliers can be on the results of automatic selection procedures?
4. Refer to Problem 14.2(b). Use backward elimination again with the variables chosen in (b) and their interactions. Does the resulting model make sense?
5. Refer to Example 11.2 on Y = mental impairment, X_1 = life events, and X_2 = SES.
   a) Show that forward selection with X_1, X_2, $X_3 = X_1X_2$, $X_4 = X_1^2$, and $X_5 = X_2^2$ and the $\alpha = .10$ level for inclusion selects only X_1 and X_2 for the model.
   b) Use backward elimination. What is the final model? Interpret.
   c) Use the C_p index to describe the fit provided by each model considered in the process in (b).
6. Use backward elimination with the home sales data of Table 9.4, using as candidates the four explanatory variables and all their interaction and quadratic (square) terms. What model do you end up with? Is this a reasonable model? Explain.
7. Problem 13.6 showed that for the home sales data, a single observation has a large impact on whether an interaction term seems needed in the model. Let's check whether that observation affects results of selection procedures. Using regression software after deleting that observation from the data set, conduct either backward elimination or forward selection with the variables in Example 14.1. Compare results.
8. Figure 14.14 is a SAS plot of the residuals versus the predicted values for the analysis of covariance model discussed in Example 13.1 relating income to education and racial–ethnic group. What does this plot suggest?

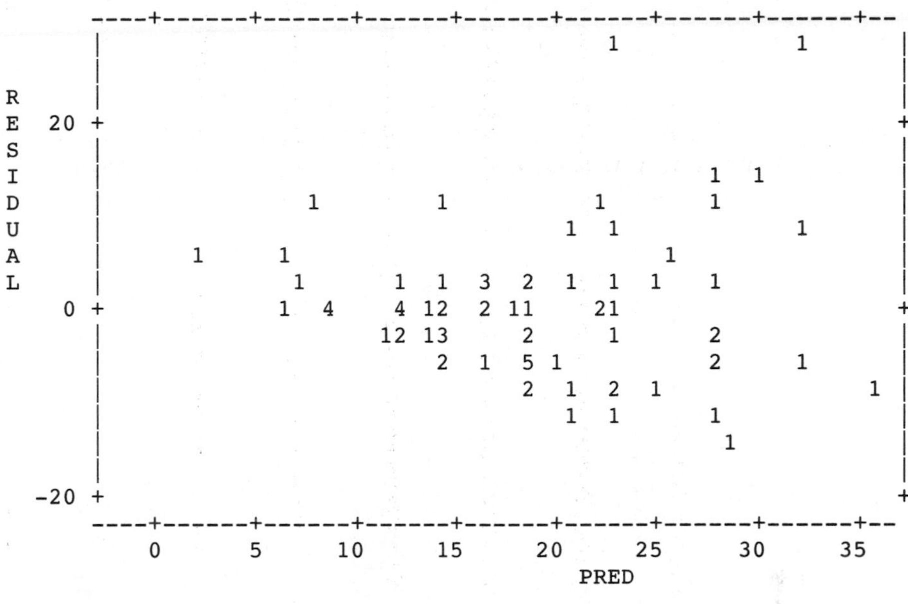

Figure 14.14

9. Refer to the model for housing price selected in Example 14.1.
a) Study the studentized residuals, and show that only one is unusually large. What does this reflect?
b) Study the hat values. Which three observations have the greatest leverage, and hence the predictor values with the greatest potential to affect the fit of the regression model?
c) Study the DFFITS values. Which three observations seem to have much more influence on the fitted values than any others?
d) Study the DFBETAS for the *S* predictor. Which three observations have the greatest influence on the predicted partial effect of size of home?
e) Study the DFBETAS for the *BA* and *N* predictors. Are the three observations highlighted above as influential for these predictors? Explain.
f) Refit the model without the three highly influential observations. Compare the prediction equation, standard errors, and R^2 to the fit for the complete data set. Summarize the influence of this set of influential observations.

10. Refer to Problem 9.17 and Table 9.13. Table 14.8 shows a SAS computer printout of various diagnostics from fitting the multiple regression model relating birth rate to literacy and women's economic activity (deleting Germany, South Africa, and Vietnam). Figure 14.15 plots the residuals against the predicted values.
a) Construct a histogram or stem and leaf plot of the residuals. Are there any apparent outliers?
b) Study the plot of the residuals in Figure 14.15. Does it suggest any lack of fit or unusual observations?
c) Study the studentized residuals. Are there any apparent outliers?
d) Study the hat values. Which, if any, observations seem to have noticeable leverage for affecting results?

TABLE 14.8

Obs	Residual	Rstudent	Hat Diag H	Dffits	ECONOMIC Dfbetas	LITERACY Dfbetas
1	-1.3206	-0.2689	0.2142	-0.1404	0.1241	-0.0366
2	-0.1343	-0.0254	0.0892	-0.0079	0.0046	-0.0049
3	0.4247	0.0799	0.0825	0.0240	0.0059	0.0115
4	-1.6682	-0.3109	0.0609	-0.0792	0.0414	-0.0255
5	1.1756	0.2228	0.0939	0.0717	0.0193	0.0362
6	-6.1951	-1.4491	0.3413	-1.0431	-0.8872	0.7387
7	-7.5264	-1.4631	0.0444	-0.3155	-0.0443	0.0300
8	1.0598	0.2065	0.1435	0.0845	0.0591	0.0101
9	-6.3002	-1.3203	0.1932	-0.6460	0.4854	0.0646
10	-0.1617	-0.0305	0.0888	-0.0095	-0.0034	-0.0039
11	-10.3055	-2.3789	0.2095	-1.2246	-0.0669	1.0154
12	11.9146	2.6422	0.0975	0.8686	-0.3750	-0.3203
13	3.8617	0.7327	0.0729	0.2055	-0.0461	0.1302
14	-2.0574	-0.3906	0.0922	-0.1245	-0.0410	-0.0559
15	6.0212	1.1540	0.0545	0.2771	0.1186	-0.0849
16	5.1775	0.9953	0.0759	0.2853	-0.1691	0.1426
17	6.8923	1.5766	0.2986	1.0287	0.4571	-0.9487
18	2.2296	0.4562	0.2161	0.2395	-0.1083	-0.1220
19	3.7082	0.6968	0.0573	0.1717	0.0065	-0.0789
20	-2.7683	-0.5296	0.1002	-0.1768	-0.1236	0.0073
21	-5.5905	-1.1665	0.2012	-0.5855	0.4080	-0.4618
22	-1.0508	-0.1973	0.0774	-0.0571	-0.0131	-0.0266
23	2.6138	0.4981	0.0947	0.1611	0.0586	0.0688

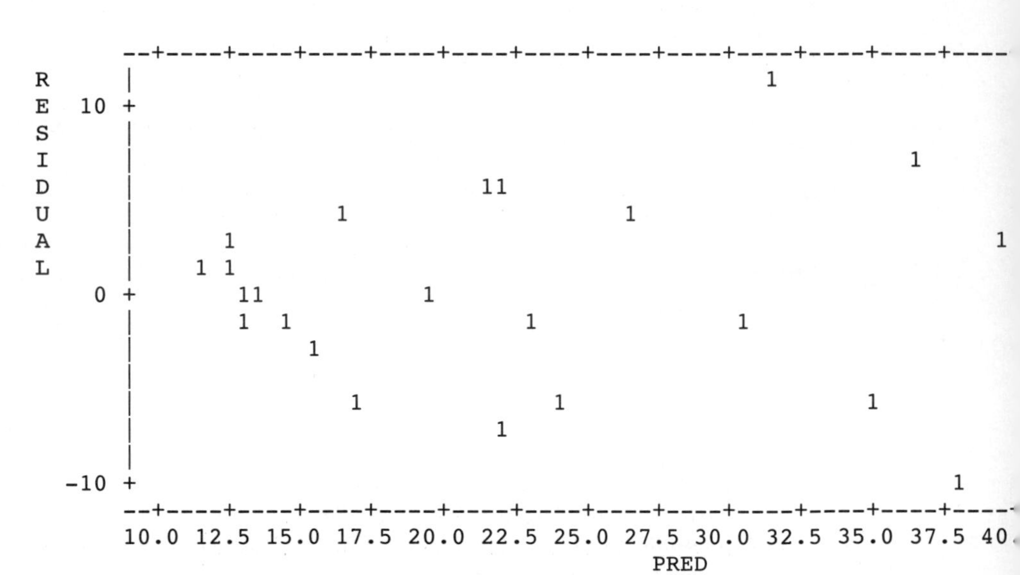

Figure 14.15

e) Based on the answers in (c) and (d), does it seem as if any observations may be particularly influential? Explain.
f) Study the DFFITS values. Identify a few observations that may have a strong influence on the fitted values.
g) Study the DFBETAS values. Identify an observation that is influential for the literacy estimate but not for the economic activity estimate.

11. Based on the answers in the previous exercise, remove an observation that seems potentially influential to you, and re-fit the model. Is the new fit different in any substantive way?

12. Refer to Table 9.1, and fit the linear regression model relating Y = violent crime rate to X = percent metropolitan for all 51 observations.
a) Find the prediction equation. Using a stem and leaf plot or a histogram, plot the residuals. Interpret.
b) Plot the residuals against the predictor. Interpret.
c) Refit the line without the outlier, and compare results.

13. Refer to the previous exercise, and use software to obtain regression diagnostics.
a) Study the studentized residuals. Are there any clear outliers?
b) Study the hat values. Are there any observations with noticeable leverage?
c) Based on the answers in (a) and (b), does it seem as if any observations may be particularly influential? Explain.
d) Study the DFFITS values. Which, if any, observations have a strong influence on the fitted values?
e) Study the DFBETAS values. For each term, which if any observations have a strong influence on the parameter estimate?
f) Remove the observation that seems most influential to you, and refit the model. Is the fit different in any substantive way?

14. Refer to Example 11.1. Fit the multiple regression model discussed there, for the data from Problem 9.24. Plot the residuals against each predictor and / or against the predicted values. Do the plots show any irregularities?

15. Refer to Table 9.1. Let Y = violent crime rate. Fit the model to the 51 observations with percentage in poverty and percentage of single-parent families as predictors.
a) Report the prediction equation.
b) Construct a stem and leaf plot or a histogram of the residuals. Interpret.
c) Plot the residuals against the predicted values. Interpret.
d) Plot the residuals against percentage of single-parent families. Interpret.
e) Refit the model without the D.C. observation, and discuss the changes in the model fit and residual patterns.

16. Refer to the previous exercise, and use software to obtain regression diagnostics.
a) Based on hat values and studentized residuals, does it seem as if any observations may be influential? Explain.
b) Study the DFFITS values. Which, if any, observations have a strong influence on the fitted values?
c) Study the DFBETAS values. For each predictor, which if any observations have a strong influence on the parameter estimates?
d) Remove the observation that seems most influential to you, and refit the model. Is the fit different in any substantive way?

17. This problem shows that multicollinearity also affects precision of estimation of partial correlations.
a) Suppose the true correlations are $\rho_{X_1X_2} = .85$, $\rho_{YX_1} = .65$, and $\rho_{YX_2} = .65$. Show that $\rho_{YX_1 \cdot X_2} = \rho_{YX_2 \cdot X_1} = .244$.
b) In a sample, $r_{X_1X_2} = .9$, $r_{YX_1} = .7$ and $r_{YX_2} = .6$. Unless the sample is very large, these are well within the limits of sampling error for the true values. Show that $r_{YX_1 \cdot X_2} = .46$ and $r_{YX_2 \cdot X_1} = -.10$. Note how small differences in r_{YX_1} and r_{YX_2} yield large differences in partial correlations when multicollinearity exists. (This illustrates that partial correlations have large standard errors when multicollinearity exists. For these values, an unwary observer might conclude that the partial effects of X_1 and X_2 have opposite signs and that the partial effect of X_1 is much stronger, when in fact they are identical in the population of interest.)
c) For comparison, compute the partial correlations in (b) when $r_{YX_1} = .7$ and $r_{YX_2} = .6$, but (i) $r_{X_1X_2} = 0$, (ii) $r_{X_1X_2} = .6$.

18. Y = height, X_1 = length of left leg, and X_2 = length of right leg are measured for a sample of 100 adults. The model $E(Y) = \alpha + \beta_1 X_1 + \beta_2 X_2$ is fitted to the data, and neither H_0: $\beta_1 = 0$ nor H_0: $\beta_2 = 0$ is rejected.
a) Does this imply that length of leg is not a good predictor of height? Why?
b) Does this imply that H_0: $\beta_1 = \beta_2 = 0$ would not have a small P-value? Why?
c) Suppose $r_{YX_1} = .901$, $r_{YX_2} = .902$, and $r_{X_1X_2} = .999$. What model would you expect to select, using forward selection and the predictors X_1 and X_2? Why?

19. Refer to Tables 11.5 and 11.7. Note that X_1 and X_2 lose their significance after entering the interaction term, even though that term is not significant.
a) Explain why this happens.
b) Rerun the interaction model, after centering the predictor scores about their mean; that is, take $X_1^* = X_1 - 44.425$ and take $X_2^* = X_2 - 56.60$. Note that now the estimates and their standard errors resemble those for the no-interaction model. This is a useful way of dealing with multicollinearity for models that have interaction or quadratic terms.

20. Sketch the following mathematical functions on the same set of axes, for values of X between 0 and 4.

a) $\hat{Y} = 10 + 4X$ **b)** $\hat{Y} = 10 + 4X + X^2$
c) $\hat{Y} = 10 + 4X - X^2$ **d)** $\hat{Y} = 10 - 4X$
e) $\hat{Y} = 10 - 4X + X^2$ **f)** $\hat{Y} = 10 - 4X - X^2$
g) $\hat{Y} = 10(1.50)^X$ **h)** $\hat{Y} = 10(.50)^X$

For the quadratic models, use these curves to describe how the coefficients of X and X^2 affect their shape.

21. Refer to the housing data in Table 9.4. The quadratic model relating selling price to size of house has fit $\hat{Y} = -2.04 + 49.34S + 6.74S^2$.
a) Interpret the coefficients of this equation. What shape does it have?
b) Find the predicted selling price for homes with (i) $S = 1$, (ii) $S = 2$, (iii) $S = 3$. Explain why the effect of a one-unit increase in S increases as S increases.
c) Find the S value for which this curve takes its minimum. Note that the curve increases over the entire range of possible house sizes.
d) Using size as a straight-line predictor in a bivariate model, $r^2 = .808$, whereas $R^2 = .815$ for the quadratic model. Do you think that the degree of nonlinearity is major, or minor? Do you think that the degree of linear association is strong, or weak?

22. Table 14.9 shows results of fitting models to the housing data in Table 9.4, using number of bedrooms as a predictor of selling price.

a) For the bivariate model with number of bedrooms as a linear predictor, interpret the estimated coefficients, and interpret r^2.

b) Report and interpret the P-value for testing the linear effect of the predictor.

c) For the linear equation, find the predicted selling price when the number of bedrooms equals (i) 2, (ii) 3, (iii) 4. Show how the slope relates to the differences between these values.

d) For the quadratic model, interpret the estimated coefficients.

e) For the quadratic model, find the predicted selling price when the number of bedrooms equals (i) 2, (ii) 3, (iii) 4. Compare to the predicted values for the linear model, and interpret.

f) Explain why the P-value is essentially 0 for the F test that the two true coefficients of number of bedrooms in the quadratic model equal 0, yet the separate t tests have P-values of .63 and .45. In your explanation, compare the R^2-values for the two models, and you may also use the fact that the correlation between the two predictors equals .984.

TABLE 14.9

Source	DF	Sum of Squares	Mean Square	F Value	Prob>F
Model	1	62577.516	62577.516	48.660	0.0001
Error	91	117028.351	1286.026		
Total	92	179605.867			

Root MSE 35.861 R-square 0.3484

Variable	Parameter Estimate	Standard Error	T for H0: Parameter=0	Prob > |T|
INTERCEP	-37.229	19.955	-1.866	0.0653
BED	42.969	6.160	6.976	0.0001

Source	DF	Sum of Squares	Mean Square	F Value	Prob>F
Model	2	63327.808	31663.904	24.508	0.0001
Error	90	116278.059	1291.978		
Total	92	179605.867			

Root MSE 35.944 R-square 0.3526

Variable	Parameter Estimate	Standard Error	T for H0: Parameter=0	Prob > |T|
INTERCEP	2.6322	56.001	0.047	0.9626
BED	16.7020	35.017	0.477	0.6345
BED2	4.1681	5.469	0.762	0.4480

23. Refer to the previous exercise. Fit the exponential regression model, and interpret. Find the predicted selling price when the number of bedrooms equals (i) 2, (ii) 3, (iii) 4, and compare to the predictions with the other models.

24. Refer to Table 9.13. Let Y = birth rate and X = women's economic activity.
a) Fit the straight-line regression model. Interpret coefficients.
b) Fit the quadratic regression model, and interpret parameters.
c) Using the quadratic fit, find the X-value at which predicted birth rate takes its minimum value. Is the prediction equation decreasing over the entire range of observed X-values?
d) Does the quadratic model provide a much improved fit over the linear model? Answer (i) descriptively, by comparing R^2-values, (ii) inferentially, by conducting a t test for the quadratic effect, (iii) graphically, by plotting the two fits through the scatter diagram.

25. Refer to the previous example. Fit the exponential regression model, and interpret the parameter estimates. Describe the estimated effect of a 10 unit increase in economic activity.

26. Refer to Example 9.1, Table 9.1, and Figure 9.4. Though inference may not be relevant for these data, we use them to illustrate how a single observation can be highly influential in determining whether to assume a nonlinear relationship.
a) Using all 51 observations, fit the quadratic model between murder rate and percentage in poverty. Interpret the estimates.
b) Refer to (a). Test whether the quadratic term is needed in the model. Report the P-value, and interpret.
c) Now refit the quadratic model, deleting the observation for D.C. Compare the estimates to (a), and interpret.
d) Refer to (c). Test whether the quadratic term is needed in the model. Report the P-value, and interpret.
e) Compare results of the tests in (b) and (d), and note how a single observation can have a large impact on the fit of a quadratic model. Show how you would be warned of this by influence diagnostics for the fit in (a).
f) Show the result of the t test for the coefficient of X in the model in (c). Do the results of this test and the one in (d) imply that poverty does not affect murder rate? Explain.

27. Refer to the previous example. Fit the exponential regression model to the full data set, and interpret the parameter estimates.

28. Table 14.10 shows the results of fitting two models to 54 observations on Y = mental health score, X_1 = degree of social interaction, and X_2 = SES. The variables X_1 and X_2 are measured on scales of 0–100, and larger Y-scores represent better mental health. The variable symbol X1**2 represents X_1^2, and X1 * X2 represents X_1X_2.

TABLE 14.10

Model	Variable	Parameter Estimate
1.	Intercept	15
	X1	.200
	X1**2	–.001
2.	Intercept	16
	X1	.07
	X2	.04
	X1*X2	–.0006

a) When model 1 is fitted, which best describes the result over the range 0–100 of X_1-values?
i) $\hat{Y}$ is a convex function of X_1, first decreasing and then increasing.
ii) $\hat{Y}$ is an increasing convex function of X_1.
iii) $\hat{Y}$ is a concave function of X_1, first increasing and then decreasing.
iv) $\hat{Y}$ is an increasing concave function of X_1.
b) When model 2 is fitted, which best describes the result over the observed ranges?
i) $\hat{Y}$ is a linear function of X_1 with positive slope that is the same for all X_2.
ii) $\hat{Y}$ is a linear function of X_1 with positive slope for some values of X_2 and negative slope for others.
iii) $\hat{Y}$ is a linear function of X_1 with positive slope, but the magnitude of that slope is smaller for larger values of X_2.
iv) $\hat{Y}$ is a quadratic function of X_1 and X_2.

29. Refer to Problem 14.12.
a) What does the plot of the residuals against the predictor suggest?
b) Using a gamma GLM, fit the linear model. Compare the parameter estimates and the standard error of the slope to results obtained using least squares. Interpret.
c) Compare the two fits to the least squares fit of the model with the outlier deleted. Which fit seems more robust to the effect of the outlier?

30. Refer to Example 14.4 on birth rates. To allow for greater variation at higher values of the mean, use the gamma GLM with identity link.
a) Fit the gamma linear model. Compare estimates to those obtained using least squares (i.e., the GLM with normal random component).
b) Fit the gamma quadratic model. Compare estimates to those obtained using least squares. Find the GNP value at which predicted birth rate takes its minimum value.

31. Draw rough sketches of the following mathematical functions on the same set of axes, for values of X between 0 and 35.
a) $\hat{Y} = 5(1.02)^X$. ($\hat{Y}$ = world population size in billions X years from now, if there is a 2% rate of growth every year.)
b) $\hat{Y} = 5(1.04)^X$. (What does this represent?)

32. Consider the formula $\hat{Y} = 4(2)^X$.
a) Calculate the $\hat{Y}$-values for integer values of X between 0 and 5, and graph the function.
b) Plot $\log \hat{Y}$ against X. What is the intercept and what is the slope of this line?

33. Table 14.11 presents, for white men in the United States, the number of deaths per thousand individuals of a fixed age within a period of a year. Let X denote age and Y denote death rate.

a) Plot X against Y, and indicate whether a linear model seems reasonable.
b) Plot X against $\log Y$. What does this plot suggest about the relationship between death rate and age?
c) Using generalized linear models, find the prediction equation for the model $\log[E(Y)] = \alpha + \beta X$.
d) Find the prediction equation for Y. Interpret the parameter estimates. Obtain the six predicted values for Y and plot this prediction equation on the graph from (a).

TABLE 14.11

Age	Death Rate (per thousand)
30	3
40	6
50	14
60	27
70	60
80	125

34. Consider the birth rate data in Table 14.5.
a) To check whether an exponential regression model seems appropriate, plot the log birth rate values against GNP and report the correlation between them. Compare to the correlation between birth rate and GNP.
b) Using GLM software, fit the exponential regression model. Interpret the effect of GNP on birth rate.
c) What advantages does the exponential model have over the quadratic model?

Concepts and Applications

35. Refer to the WWW data set (Problem 1.7).
a) Using software, conduct and interpret a regression analysis using Y = political ideology, selecting predictors from the variables in that file. Prepare a report describing the analyses and diagnostic checks that you conducted, and indicate how you selected a final model. Interpret results.
b) Repeat the analysis, using Y = college GPA.

36. Refer to the data file created in Problem 1.7. For the models you fitted in Problems 11.25 and 13.12, use methods of this chapter to check model adequacy. Interpret and summarize your findings.

37. Refer to the model used to predict income in Example 13.1. Conduct an analysis of residuals and of influence diagnostics, and summarize your findings.

38. Refer to Table 11.1 and Example 11.2. Allowing for potential nonlinear relationships and interaction, and checking residuals and influence diagnostics, use the methods of this chapter to find a suitable prediction equation for Y = mental impairment.

39. Table 14.12 shows population size of Florida, by decade, from 1830 to 1990. Analyze these data.

40. Refer to Table 9.13. Using methods of this chapter, find a good prediction equation relating X = per capita GNP to Y = life expectancy. (*Hint*: Try a transform of X, such as $\log X$ or $1/X$.)

41. Refer to Table 9.13. Using methods of this chapter, analyze these data, finding a good prediction equation for birth rate. Explain how you selected variables for the model and how you handled the missing data. (Most software uses *listwise deletion*, deleting an observation from the list if it is missing data on any of the variables. New and better methods have been developed recently; see R. Little and D. Rubin, *Sociological Methods and Research*, Vol. 18, 1989, pp. 292–326.)

42. Refer to Problem 9.24. Using methods of this chapter, find a good model for predicting crime rate. In your report, show how you checked for nonlinearity, considered poten-

TABLE 14.12

Year	Population	Year	Population
1830	34, 730	1920	968,470
1840	54, 477	1930	1,468,211
1850	87, 445	1940	1,897,414
1860	140, 424	1950	2,771,305
1870	187, 748	1960	4,951,560
1880	269, 493	1970	6,791,418
1890	391, 422	1980	9,746,324
1900	528, 542	1990	12,937,926
1910	752, 619		

Source: *Florida Statistical Abstract, 1995*, compiled by Bureau of Economic and Business Research, University of Florida, and published by University Press of Florida, Gainesville, Florida.

tial multicollinearity, checked the results of stepwise procedures, analyzed residuals, and considered influential observations. Interpret your final model.

43. Refer to Problem 14.26. Now use violent crime rate as the response variable. Use influence diagnostics to analyze the fit of the quadratic model to the 51 observations on violent crime rate and poverty rate. Is the simpler, straight-line, model adequate?

44. In multiple regression, the standard error of the estimator b_j of the coefficient β_j of X_j equals $\hat{\sigma}/s_j\sqrt{(n-1)(1-R_j^2)}$, where $\hat{\sigma}$ is the estimated conditional standard deviation, s_j is the sample standard deviation of X_j, and R_j is the multiple correlation from the regression of X_j on the other predictors. The quantity $1/(1-R_j^2)$ is called a *variance inflation factor*, since it represents the multiplicative increase in the variance due to X_j being correlated with the other predictors. Using this formula, explain how precision of estimated regression coefficients is affected by
a) Multicollinearity.
b) The conditional variability of the response variable.
c) The variability of the explanatory variables.
d) The sample size.

45. Give an example of a response variable and a pair of explanatory variables for which an automated variable selection procedure would probably produce a model with only one predictor. Explain.

46. A sociologist's first reaction upon studying automated variable selection routines was that they had the danger of leading to "crass empiricism" in theory building. From a theoretical perspective, describe the dangers with such methods. What guidelines would you suggest for avoiding these problems?

47. Give an example of two variables you expect to have a nonlinear relationship. Describe the pattern you expect for the relationship, and explain how to model that pattern.

48. You invest \$1000 in a savings account with interest compounded annually at 10%.
a) How much money do you have after X years?
b) How many years does it take your savings to double in size?

49. A recent newpaper article quoted a planner in a Florida city as saying, "This city has been growing at the rate of 4.2% per year. That's not slow growth by any means. It corresponds

to 42% growth per decade." Explain what is incorrect about this statement. If, in fact, the current population size of the city is 100,000 and in each of the next ten years the city increases in size by 4.2% relative to the previous year, then
a) What is the population size after a decade?
b) What percent growth occurs for the decade?

50. Example 14.7 showed a predicted growth rate of 13.77% per decade.
a) Show that this is equivalent to a 1.30% predicted growth *per year*. [*Hint*: $(1.0130)^{10} = 1.1377$.]
b) Explain why the predicted U.S. population size (in millions) X *years* after 1890 is $70.46(1.0130)^X$.

51. Sketch the relationship between $E(Y)$ and X if
a) $\log[E(Y)]$ is linearly related to X.
b) $E(Y) = \alpha + \beta_1 X + \beta_2 X^2$ with $\beta_1 < 0$ and $\beta_2 > 0$.
c) $E(Y) = \alpha + \beta_1 X + \beta_2 X^2$ with $\beta_1 > 0$ and $\beta_2 < 0$.

For multiple choice problems 14.52–14.55, select the correct response(s).

52. In the model $E(Y) = \alpha + \beta_1 X + \beta_2 X^2$, the coefficient β_2
a) Is the mean change in Y as X^2 is increased one unit with X held constant.
b) Is a curvature coefficient that describes whether the regression equation is convex or concave.
c) Equals 0 if the relationship between Y and X is linear.
d) Equals 0 if the population value of R^2 for this model equals ρ_{YX}^2.

53. The log transformation of the mean response in regression is useful when
a) $E(Y)$ is approximately a logarithmic function of X.
b) $E(Y)$ is approximately an exponential function of X.
c) $\log E(Y)$ is approximately a linear function of X.
d) Unit changes in X have a multiplicative, rather than additive, effect on the mean of Y.

54. Forward selection and stepwise regression are similar in the sense that, if they have the same α-level for entry,
a) They always select the same final regression model.
b) They always select the same initial regression model (when they enter the first explanatory variable).
c) Any variable not in the final model does not have a significant partial association with Y, controlling for the variables in the final model.
d) It is impossible that all the variables listed for potential inclusion are in the final model.

55. Evidence of multicollinearity exists in a multiple regression fit when
a) Strong intercorrelations occur among explanatory variables.
b) The R^2-value is very small.
c) The partial correlation between Y and X_1 is the same as the bivariate correlation between Y and X_1.
d) The F test of H_0: $\beta_1 = \cdots = \beta_k = 0$ has a small P-value, but the individual tests of H_0: $\beta_1 = 0, \ldots, H_0$: $\beta_k = 0$ do not.

56. True or False?
a) Possible effects of an influential observation include changing a correlation from positive to negative, a P-value from .01 to .99, and R^2 from .99 to .01.

b) When multicollinearity exists, one can still obtain good estimates of regression parameters, but R^2 may be adversely affected.
c) If Y = annual medical expenses relates to X = age by $E(Y) = 1400 - 22X + .4X^2$, then the change in the mean of Y for every unit change in X equals -22.

57. Select the best response for each of the following terms (not every response is used):

Heteroscedasticity ______
Multicollinearity ______
Forward selection ______
Interaction ______
Exponential model ______
Stepwise regression ______
Studentized residual ______
Generalized linear model ______

a) The mean of Y multiplies by β for each unit increase in X.
b) The log of $E(Y)$ is linearly related to the log of X.
c) A residual plot indicates that the residuals are much more spread out at high X than at low X.
d) The bivariate association between Y and X_1 is different from the partial association between Y and X_1 controlling for X_2.
e) Strong intercorrelations among explanatory variables.
f) When there is multicollinearity, may provide regression parameter estimates with smaller standard errors.
g) At each stage the variable considered for entry into the model explains the greatest portion of the remaining unexplained variability in Y.
h) The response variable need not be normal, and we can model a function of the mean as a linear function of the predictors.
i) At each stage after entering a new variable, all variables in the model are retested to see if they still have a significant partial effect on Y.
j) The slope between Y and X_1 has different values depending on the value of X_2.
k) Measures the number of standard errors that an observation falls from its predicted value.

58. Plot the prediction equation $\hat{Y} = 10 + 4X_1 - X_1^2 - 2X_2$ in the following cases:
a) Between $\hat{Y}$ and X_1 for values of X_1 between 0 and 5, when (i) $X_2 = 2$, (ii) $X_2 = 5$. Use the same axes. Interpret.
b) Between $\hat{Y}$ and X_2 for values of X_2 between 0 and 5, when (i) $X_1 = 2$, (ii) $X_1 = 5$. Use the same axes. Interpret.

59. * Show that using a cross-product term to model interaction assumes that the slope of the relationship between Y and X_1 changes linearly as X_2 changes. How would you suggest modeling interaction if, instead, the slope of the linear relationship between Y and X_1 first increases as X_2 changes from low to moderate values and then decreases as X_2 changes from moderate to high values?

60. Forward selection is used with ten potential predictors for Y. In reality, none are truly correlated with Y or with each other. For a random sample, show that the probability equals .40 that at least one is entered into the regression model, when the criterion for admission is a P-value below .05 for the t test. (*Hint*: Use the binomial distribution to first find the probability that none are selected.)

Bibliography

Blalock, H. M. (1963). Correlated independent variables: The problem of multicollinearity. *Social Forces*, 62, 233–238.

Chatterjee, S., and Hadi, A. S. (1988). *Sensitivity Analysis in Linear Regression.* New York: Wiley.

Cook, R. D., and Weisberg, S. (1982). Criticism in regression. In *Sociological Methodology.* San Francisco: Jossey-Bass, pp. 313–361.

Dobson, A. J. (1990). *An Introduction to Generalized Linear Models.* London: Chapman and Hall.

Draper, N. R., and Smith, H. (1981). *Applied Regression Analysis*, 2nd ed. New York: Wiley.

Fox, J. (1991). *Regression Diagnostics: An Introduction.* Sage University Paper Series on Quantitative Applications in the Social Sciences, 07–079. Newbury Park, CA: Sage.

Weisberg, S. (1985). *Applied Linear Regression*, 2nd ed. New York: Wiley.

Wonnacott, R. J., and Wonnacott, T. H. (1979). *Econometrics*, 2nd ed. New York: Wiley.

Chapter 15

Logistic Regression: Modeling Categorical Responses

The regression models studied in the past six chapters assume that the response variable is quantitative. This chapter presents models for response variables that are qualitative, having scale consisting of a set of categories. We introduce generalized linear models for which the response variable has a discrete distribution rather than a continuous distribution such as the normal.

The main focus of the chapter is the ***logistic regression model***, which refers to ***binary*** response variables—variables having only two possible outcomes. Section 15.1 introduces the model for the case of a single quantitative explanatory variable. Section 15.2 extends this model to handle several explanatory variables, and Section 15.3 shows that one can use dummy variables to include qualitative explanatory variables in the model. For instance, one might model

- A voter's choice in a presidential election (Democrat or Republican), with predictor variables political ideology, annual income, education level, religious affiliation, and gender.
- The choice regarding early retirement (yes or no), using savings, annual income, job satisfaction, and various health and family characteristics.
- Whether a subject has symptoms of a certain mental illness (yes or no), using socioeconomic status and a life events index.

Section 15.4 introduces ***loglinear models***, which describe association structure among a set of qualitative response variables. Section 15.5 shows how to check the fit of models to data in a contingency table. Finally, Section 15.6 extends logistic regression to handle ordinal response variables. Logistic regression and loglinear models are interpreted using the ***odds ratio***.

15.1 Logistic Regression

The main topic of this chapter is a model for binary response variables. Denote the responses on Y by 0 and 1. It is common to use the generic terms *failure* and *success* for these two outcomes. The sum of the scores in the sample is then the number of successes. The mean of the 0 and 1 scores, which is the sum divided by the total sample size, equals the proportion of successes. In fact, Section 4.3 showed that a proportion is the special case of a mean calculated for a binary variable having (0, 1) scoring.

Ordinary regression models the mean of the response variable. For binary response variables, the model describes how the proportion of successes depends on explanatory variables. Let $\pi = E(Y)$ denote the true proportion of successes. Now π also represents the probability that a randomly selected subject has a success response, and this probability varies according to the values of the explanatory variables.

Linear Probability Model

For a single explanatory variable, X, the simple model

$$\pi = \alpha + \beta X$$

implies that the probability of success is a linear function of X. This is called the ***linear probability model***.

Unfortunately, this model is often poor. Figure 15.1 shows that it implies probabilities below 0 or above 1 for sufficiently small or large X values, whereas probabilities must fall between 0 and 1. Although the model may be valid over a restricted range of X values, it is less likely to be adequate when several predictors are in the model. The usual regression inference assumptions also break down for this model, since the binary response distributions are drastically different from normal distributions with constant standard deviation. For binary data, it makes more sense to assume a binomial distribution for the response than a normal. (Section 6.6 presented the binomial distribution.)

The Logistic Regression Model for Binary Responses

Figure 15.1 also shows more realistic response curves, having an S-shape. With these curves, the probability of a success falls between 0 and 1 for all possible X-values.

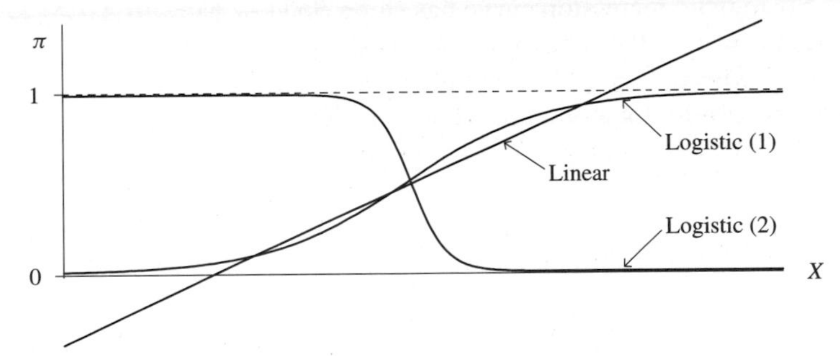

Figure 15.1 Linear and Logistic Regression Models for a (0, 1) Response; The Mean of Y Is the Probability π

These curvilinear relationships are described by the formula

$$\log\left(\frac{\pi}{1-\pi}\right) = \alpha + \beta X$$

The ratio $\pi/(1-\pi)$ equals the ***odds***, a measure studied in Section 8.4. For instance, when $\pi = .75$, the odds equals $.75/.25 = 3.0$, meaning that a success is three times as likely as a failure. (Again, don't worry if you do not understand logarithms. This is unnecessary for interpreting most aspects of this model.)

This formula uses the log of the odds, $\log[\pi/(1-\pi)]$, called the ***logistic transformation***, or ***logit*** for short. The model is abbreviated as

$$\text{logit}(\pi) = \alpha + \beta X$$

It is called the ***logistic regression model***. As π increases from 0 to 1, the odds increase from 0 to ∞, and the logit increases from $-\infty$ to ∞. The probability $\pi = 1/2$ has odds equal to 1 and logit equal to 0, and π-values above (below) 1/2 have positive (negative) logits.

Like the slope of a straight line, the parameter β in this model refers to whether the curve increases or decreases as X increases. For $\beta > 0$, π increases as X increases, as in curve (1) in Figure 15.1. For $\beta < 0$, π decreases as X increases; the probability of a success tends toward 0 for larger values of X, as in curve (2) in Figure 15.1. If $\beta = 0$, π does not change as X changes, so the curve flattens to a horizontal straight line.

The steepness of the curve increases as $|\beta|$ increases. For instance, $|\beta|$ for curve (2) is greater than β for curve (1). However, unlike in the linear model, β is not the slope for the change in π as X changes. For this S-shaped curve, the rate at which the curve climbs or descends changes according to the value of X.

The simplest way to interpret β and the steepness of the curve uses a linear approximation. A line drawn tangent to the logistic regression curve has slope $\beta\pi(1-\pi)$, where β is the coefficient of X in the model and π is the probability at that point. Figure 15.2 illustrates. For instance, at the X value where $\pi = .5$, the line drawn tangent to the logistic regression curve has slope $\beta\pi(1-\pi) = \beta(.5)(.5) = .25\beta$. By contrast, when $\pi = .1$ or $.9$, the tangent line has slope $\beta(.1)(.9) = .09\beta$. In fact, the largest slope occurs where $\pi = 1/2$. The value $\beta\pi(1-\pi)$ represents the approximate change in the probability for a one-unit change in X.

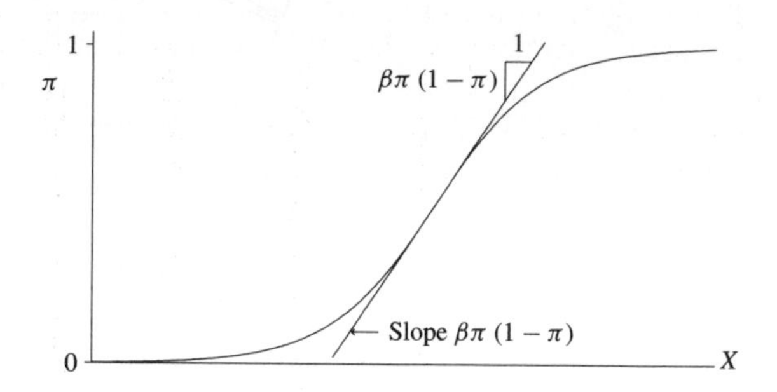

Figure 15.2 A Line Drawn Tangent to a Logistic Regression Curve Has Slope $\beta\pi(1-\pi)$

When $\pi = 1/2$, the odds $\pi/(1-\pi) = 1$, and $\log[\pi/(1-\pi)] = 0$. Equating this logit value of 0 to $\alpha + \beta X$ and solving for X shows that $\pi = 1/2$ when $X = -\alpha/\beta$.

The appendix shows how to use software to fit the logistic regression model. Most routines use *maximum likelihood* (see Section 5.1), which is more appropriate for binary data than least squares. One can fit the model with any software for generalized linear models, choosing the logit link and the binomial distribution for the random component.

TABLE 15.1 Annual Income (in Millions of Italian Lira) and Possessing a Travel Credit Card

Income	Number Cases	Credit Cards	Income	Number Cases	Credit Cards	Income	Number Cases	Credit Cards
24	1	0	39	2	0	65	6	6
27	1	0	40	5	0	68	3	3
28	5	2	41	2	0	70	5	3
29	3	0	42	2	0	79	1	0
30	9	1	45	1	1	80	1	0
31	5	1	48	1	0	84	1	0
32	8	0	49	1	0	94	1	0
33	1	0	50	10	2	120	6	6
34	7	1	52	1	0	130	1	1
35	1	1	59	1	0			
38	3	1	60	5	2			

Source: "Categorical Data Analysis," Quaderni del Corso Estivo di Statistica e Calcolo delle Probabilità, n. 4, Istituto di Metodi Quantitativi, Università Luigi Bocconi, a cura di R. Piccarreta (1993).

Example 15.1 Income and Having Travel Credit Cards

Table 15.1 refers to a sample of 100 subjects randomly selected for an Italian study on the relation between annual income and whether one has a travel credit card, such as American Express or Diners Club. At each level of annual income (in millions of lira), the table indicates the number of subjects in the sample and the number of those having at least one travel credit card. Let X = annual income and Y = whether one has such a credit card, where $Y = 1$ if one does and $Y = 0$ if one does not. For instance, for the five observations at $X = 28$, $Y = 1$ for two subjects and $Y = 0$ for three subjects. (As of 1997, 1 million lira equals approximately \$650.)

TABLE 15.2 Computer Printout for Logistic Regression Analysis of Italian Credit Card Data

Criterion	Intercept Only	Intercept and Covariates	Chi-Square for Covariates
-2 LOG L	123.820	96.963	26.857 with 1 DF (p=0.0001)

Variable	Parameter Estimate	Standard Error	Wald Chi-Square	Pr > Chi-Square
INTERCPT	-3.5561	0.7169	24.6057	0.0001
INCOME	0.0532	0.0131	16.3734	0.0001

Variable	Odds Ratio
INCOME	1.055

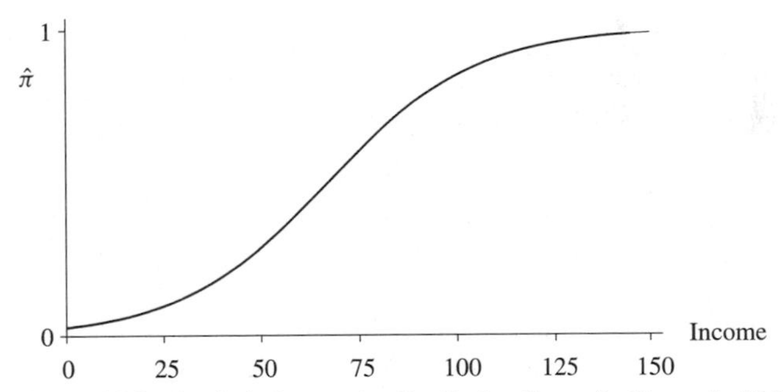

Figure 15.3 Logistic Regression Prediction Curve for Example 15.1

Software provides a printout similar to the one shown in Table 15.2. The logistic regression prediction equation equals

$$\text{logit}(\hat{\pi}) = -3.556 + .053X$$

Since the estimate .053 of β is positive, this sample suggests that the probability of having a credit card increases at higher levels of income. Figure 15.3 shows the prediction curve. The predicted probability equals $\hat{\pi} = 1/2$ at $X = -\hat{\alpha}/\hat{\beta} = (3.556)/(.053) = 67$. The predicted probability of having a credit card is below 1/2 for incomes below 67 million lira and above 1/2 for incomes above this level. □

Logistic Regression Equation for Probabilities

The equation expressing the logistic regression model directly in terms of π is

$$\pi = \frac{e^{\alpha+\beta X}}{1 + e^{\alpha+\beta X}}$$

Here, e raised to a power represents the antilog of that number, using natural logs. (Most calculators have an e^x key that provides these antilogs, and computer software can calculate estimated probabilities based on the fit of the model.)

In Example 15.1, a person with income X has predicted probability of having a credit card equal to

$$\hat{\pi} = \frac{e^{-3.56+.053X}}{1 + e^{-3.56+.053X}}$$

For subjects with income $X = 24$, the lowest income level in this sample, the predicted probability equals

$$\hat{\pi} = \frac{e^{-3.56+.053(24)}}{1 + e^{-3.56+.053(24)}} = \frac{e^{-2.3}}{1 + e^{-2.3}} = \frac{.10}{1.10} = .09$$

For $X = 130$, the highest income level in this sample, the predicted probability equals .97.

At the mean income of $\bar{X} = 50$, the predicted probability equals .29. A line drawn tangent to the curve at that point has slope equal to $\hat{\beta}\hat{\pi}(1 - \hat{\pi}) = .053(.29)(.71) = .01$. A 1 million lira increase in annual income relates approximately to a .01 increase in the probability of having a credit card. A 10 million lira increase relates to an increase of about .1.

Using software, one can also fit the linear probability model, $\pi = \alpha + \beta X$. The least squares fit is $\hat{\pi} = -.165 + .0095X$. This formula also predicts about a .01 increase in the probability, per million lira increase in income. However, it provides quite different $\hat{\pi}$ predictions at the low end and high end of the income scale. For instance, it provides the absurd prediction $\hat{\pi} > 1$ when $X \geq 123$.

Interpretation Using the Odds and Odds Ratio

An alternative interpretion of the logistic regression coefficient β is as an effect on the odds. Specifically, applying antilogs to both sides of the logistic regression equation

$\log[\pi/(1-\pi)] = \alpha + \beta X$ yields

$$\frac{\pi}{1-\pi} = e^{\alpha+\beta X} = e^{\alpha}(e^{\beta})^{X}$$

The right-hand side of this equation has the *exponential regression* form studied in Section 14.6, a constant multiplied by another constant raised to the X power. This exponential relationship implies that every unit increase in X has a multiplicative effect of e^{β} on the odds.

In Example 15.1, the antilog of $\hat{\beta}$ is $e^{\hat{\beta}} = e^{.053} = 1.055$. When income increases by one unit, the estimated odds of owning a credit card multiply by 1.055; that is, they increase by 5.5%. When $X = 51$, for example, the estimated odds of owning a travel credit card are 1.055 times what they are when $X = 50$. When $X = 60$, the odds equal $(1.055)^{10} = 1.7$ times the odds when $X = 50$. For instance, when $X = 50$,

$$\text{Estimated odds} = \frac{\hat{\pi}}{1-\hat{\pi}} = e^{-3.556+.0532(50)} = .408$$

whereas when $X = 60$,

$$\text{Estimated odds} = \frac{\hat{\pi}}{1-\hat{\pi}} = e^{-3.556+.0532(60)} = .695$$

which is 1.7 times the value of .408 for $X = 50$.

Most software can report odds ratio estimates and predicted probabilities. The bottom of Tablc 15.2 shows the way SAS reports the odds ratio effect for this example.

15.2 Inference and Multiple Predictors in Logistic Regression

We next present methods of statistical inference for logistic regression models. Then we show how to extend the model to allow several predictors of the binary response.

Inference for Logistic Regression Models

For the model

$$\text{logit}(\pi) = \alpha + \beta X$$

the hypothesis H_0: $\beta = 0$ states that X has no effect on the probability π that $Y = 1$; that is, Y is independent of X. Except for very small samples, one can test H_0 using a z test statistic, dividing the estimate of β by its standard error. Most software reports the square of this statistic, called a ***Wald statistic***. This has a chi-squared distribution with $df = 1$. It has the same P-value as the z statistic for the two-sided alternative $H_a : \beta \neq 0$.

For Example 15.1, $H_0 : \beta = 0$ states that the probability of having a travel credit card is the same at all income levels. From Table 15.2, the estimated standard error of

$\hat{\beta} = .0532$ equals .0131. The z test statistic equals $z = .0532/.0131 = 4.1$. This has a P-value of $P < .0001$ for $H_a : \beta \neq 0$. Table 15.2 reports not this statistic but its square, the Wald statistic. This chi-squared statistic equals 16.4, with $df = 1$, and has the same P-value. There is strong evidence of a positive association between income and having a travel credit card.

Another test that most software reports, called the ***likelihood-ratio test***, provides similar evidence. This test compares two models, by testing that the extra parameters in the complete model equal zero. The test refers to a key ingredient of maximum likelihood inference, the *likelihood function*, denoted by L. The maximum likelihood estimates maximize this function, much as least squares estimates minimize a sum of squares function (Specifically, the estimates are the parameter values for which the observed data are most likely; see Section 5.1).

Let L_0 denote the maximum of the likelihood function when H_0 is true, and let L_1 denote the maximum without that assumption. The formula for the likelihood-ratio test statistic is

$$-2\log\left(\frac{L_0}{L_1}\right) = (-2\log L_0) - (-2\log L_1)$$

It compares the maximized values of $(-2\log L)$ when the null hypothesis is true and when it need not be true. There is a technical reason for using -2 times the log of this ratio, namely, that the test statistic then has approximately a chi-squared distribution for large samples. The df value equals the number of parameters in the null hypothesis.

In testing that $\beta = 0$ in the logistic regression model, $\text{logit}(\pi) = \alpha + \beta X$, we compare the maximized value of $(-2\log L)$ for this model to its maximized value for the simpler model with $\beta = 0$. In Table 15.2, the likelihood-ratio test statistic is denoted -2 LOG L. It equals 26.9, with $df = 1$, and it has a P-value of $P < .0001$. For large samples, the Wald test and likelihood-ratio test provide similar results. For small to moderate sample sizes, the likelihood-ratio statistic often tends to be larger and provides a more powerful test than the the Wald statistic.

The result of these tests is no surprise. We would certainly expect subjects with higher incomes to be more likely to have travel credit cards. As usual, estimation is more informative than significance testing. For instance, one can construct a confidence interval for the probability of having a credit card at various income levels. This is beyond the scope of this text, but is simple with standard software, as shown in the appendix. A 95% confidence interval for the true probability of having a credit card equals (.04, .19) at the lowest income level of 24, (.20, .40) at the mean income level of 50, and (.78, .996) at the highest income level of 130.

Multiple Logistic Regression

Just as bivariate regression extends to a multiple regression model for several predictors, so can logistic regression handle multiple predictors for a binary response variable.

The multiple logistic regression model has the form

$$\text{logit}(\pi) = \alpha + \beta_1 X_1 + \cdots + \beta_k X_k$$

One can include qualitative explanatory variables in the model using dummy variables, just as in analysis of variance and covariance models.

Exponentiating a beta parameter provides the multiplicative effect of that predictor on the odds, controlling for the other variables. The formula for the probability itself is

$$\pi = \frac{e^{\alpha+\beta_1 X_1+\cdots+\beta_k X_k}}{1 + e^{\alpha+\beta_1 X_1+\cdots+\beta_k X_k}}$$

As in the bivariate case, one can conduct Wald or likelihood-ratio tests for parameters in the model.

Example 15.2 Factors Affecting First Home Purchase

Table 15.3 summarizes results of a logistic regression model for a study of how family transitions relate to first home purchase by young married households. The response variable is whether the subject owns a home (1 = yes, 0 = no). The parameter π is the probability of home ownership.

TABLE 15.3 Results of Logistic Regression for the Probability of Home Ownership

Variable	Estimate	Std. Error
Intercept	−2.870	—
Husband earnings	.569	.088
Wife earnings	.306	.140
Years married	−.039	.042
Married in 2 years	.224	.304
Working wife in 2 years	.373	.283
No. children	.220	.101
Add child in 2 years	.271	.140
Head's education	−.027	.032
Parents' home ownership	.387	.176

Source: J. Henretta, *Social Forces*, Vol. 66, 1987, pp. 520–536.

The model contains several explanatory variables. Two variables measure current income: husband's earnings and wife's earnings, measured in ten-thousands of dollars. Two variables measure marital duration: the number of years the respondent has been married and marital status two years after the year of observation. The latter is a qualitative variable with levels (married, married with a working wife, single), single being the omitted category for the two dummy variables. Two variables measure presence of children: the number of children aged 0–17 in the household and a dummy variable that equals 1 if the family has more children aged 0–17 two years after the year of observation. The other variables are the household head's education, in years, and a dummy variable that equals 1 if the subject's parents owned a home in the last year the subject lived in the parental home.

In Table 15.3, the ratio of the estimate to its standard error exceeds 2 for four of the predictors. Other things being fixed, the probability of home ownership increases with husband's earnings, wife's earnings, the number of children, and parents' home ownership. For instance, each additional child had the effect of multiplying the estimated odds of owning a home by $e^{.220} = 1.25$; that is, the estimated odds increase by 25%. A \$10,000 increase in earnings had the effect of multiplying the estimated odds of owning a home by $e^{.569} = 1.77$ if the earnings add to husband's income and by $e^{.306} = 1.36$ for wife's income.

We next illustrate the formula for the probability itself. The estimated probability of home ownership when husband's earnings = \$20,000, wife's earnings = \$50,000, years married = 3, the wife is working in two years, number of children = 0, add child in two years = 0, head's education = 16 years, parents' home ownership = 0, equals

$$\hat{\pi} = \frac{e^{-2.870+.569(2)+.306(5)-.039(3)+.373(1)-.027(16)}}{1+e^{-2.870+.569(2)+.306(5)-.039(3)+.373(1)-.027(16)}} = \frac{e^{-.378}}{1+e^{-.378}} = \frac{.685}{1.685} = .41$$

From Table 15.3, years married, marital status in two years, and head's education show little evidence of an effect, given the other variables in the model. One could refit the model without these predictors. This more parsimonious model may yield better estimates of the effects of the significant variables. □

Comparing Logistic Regression Models

Cross-product terms allow interactions among explanatory variables in their effects on the response. Square terms allow probabilities to increase and then decrease (or the reverse) as a predictor increases. The next section discusses models with qualitative explanatory variables in greater detail.

To compare a model with a set of predictors to a simpler model having fewer predictors, one can use the likelihood-ratio test based on the difference in the values of $(-2\log L)$ for the two models. This is an approximate chi-squared statistic with df given by the number of extra parameters in the full model. This test is an analog of the F test for comparing complete and reduced regression models (Section 11.6).

To illustrate, the model in Example 15.2 for home ownership has $(-2\log L) = 2931.2$. After adding five variables to the model relating to the housing market, such as the median sale price of existing homes in the area, $-2\log L$ drops to 2841.1. The difference $(2931.2 - 2841.1) = 90.1$ is a chi-squared statistic with $df = 5$, since the more complex model has five additional parameters. This shows extremely strong evidence of a better fit for the more complex model ($P < .0001$), indicating that at least one of these variables provides an improvement in predictive power.

A comparison of the $(-2\log L)$ values for the model of interest and for the model containing only an intercept term provides a way of testing the joint effects of all the predictors in the model. This test is a chi-squared analog for binary response data of the F test in regression of the hypothesis $H_0 : \beta_1 = \cdots = \beta_k = 0$ that none of the predictors have an effect on Y.

This completes our brief introduction to the multiple logistic regression model. The following section provides further details, illustrating for the case of qualitative explanatory variables.

15.3 Logit Models with Qualitative Explanatory Variables

Like ordinary regression, logistic regression works when some or all of the explanatory variables are categorical. One simply sets up dummy variables for those variables, just as in analysis of variance and covariance. The logistic regression model in this case is often also called a ***logit model***. The following example illustrates this case.

Example 15.3 Death Penalty and Racial Predictors

Table 15.4 is a three-dimensional contingency table from a study of the effects of racial characteristics on whether individuals convicted of homicide receive the death penalty. The variables in Table 15.4 are *death penalty verdict*, the response variable, having categories (yes, no), and the explanatory variables *race of defendant* and *race of victims*, each having categories (white, black). The 674 subjects classified in this table were defendants in indictments involving cases with multiple murders in Florida between 1976 and 1987.

TABLE 15.4 Death Penalty Verdict by Defendant's Race and Victims' Race, for Cases with Multiple Murders in Florida

Defendant's	Victims'	Death Penalty		Percent
Race	Race	Yes	No	Yes
White	White	53	414	11.3
	Black	0	16	0.0
Black	White	11	37	22.9
	Black	4	139	2.8

Source: M. L. Radelet and G. L. Pierce, *Florida Law Review*, Vol. 43, 1991, pp. 1–34.

For each of the four combinations of defendant's race and victims' race, Table 15.4 also lists the percentage of defendants who received the death penalty. For white defendants, the death penalty was imposed 11.3% of the time when the victims were white and 0.0% of the time when the victims were black, a difference of 11.3% − 0.0% = 11.3%. For black defendants, the death penalty was imposed 22.9% − 2.8% = 20.1% more often when the victims were white than when the victims were black. Thus, controlling for defendant's race by keeping it fixed, the percentage of *yes* death penalty verdicts was considerably higher when the victims were white than when they were black.

Now, consider the association between defendant's race and the death penalty verdict, controlling for victims' race. When the victims were white, the death penalty was imposed 22.9% – 11.2% = 11.7% more often when the defendant was black than when the defendant was white. When the victims were black, the death penalty was imposed 2.8% more often when the defendant was black than when the defendant was white. In summary, controlling for victims' race, black defendants were somewhat more likely than white defendants to receive the death penalty.

A logit model summarizes these data well. This section presents the model, its interpretations, and inference methods for it. □

Dummy Variables in a Logit Model

For Y = death penalty verdict, let yes = 1 and no = 0. Then the mean of Y is the probability of a *yes* verdict. Let π denote this probability. Since defendant's race and victims' race each have two categories, a single dummy variable can represent each in the model. Let d be a dummy variable for defendant's race and v a dummy variable for victim's race, where

$$d = 1, \text{ defendant} = \text{white}, \ \ d = 0, \text{ defendant} = \text{black}$$
$$v = 1, \text{ victims} = \text{white}, \ \ v = 0, \text{ victims} = \text{black}$$

The logit model with main effects for these predictors is

$$\text{logit}(\pi) = \alpha + \beta_1 d + \beta_2 v$$

The parameter β_1 represents the effect of defendant's race, controlling for victims' race, and β_2 represents the effect of victims' race, controlling for defendant's race. We shall see that e^{β_1} is the odds ratio between the response variable and defendant's race, controlling for victims' race, and e^{β_2} is the odds ratio between the response and victims' race, controlling for defendant's race.

The appendix shows how to fit this model with standard software. Table 15.5 illustrates computer printout for the parameter estimates. The prediction equation is

$$\text{logit}(\hat{\pi}) = -3.596 - .868d + 2.404v$$

Since $d = 1$ for white defendants, the negative coefficient of d means that the estimated odds of receiving the death penalty are lower for white defendants than for black defendants. Since $v = 1$ for white victims, the positive coefficient of v means that the estimated odds of receiving the death penalty are higher when the victims were white than when they were black.

The antilog of $\hat{\beta}_1$, namely, $e^{\hat{\beta}_1} = e^{-.868} = .42$, is the estimated odds ratio between defendant's race and the death penalty, controlling for victims' race. That is, the estimated odds of the death penalty for a white defendant equal .42 times the estimated

TABLE 15.5 Parameter Estimates for Logit Model for Death Penalty Data

Parameter	DF	Estimate	Std Err	ChiSquare	Pr>Chi
INTERCEPT	1	-3.5961	0.5069	50.3264	0.0001
DEFEND	1	-0.8678	0.3671	5.5889	0.0181
VICTIM	1	2.4044	0.6006	16.0264	0.0001

odds for a black defendant. We list *white* before *black* in this interpretation, since $d = 1$ for white defendants. If we instead let $d = 1$ for black defendants rather than white, then we get $\hat{\beta}_1 = .868$ instead of $-.868$. In that case, $e^{.868} = 2.38$, which is $1/.42$; that is, the estimated odds of the death penalty for a black defendant equal 2.38 times the estimated odds for a white defendant, controlling for victims' race.

Similarly, for victims' race, $e^{2.404} = 11.1$. The estimated odds of the death penalty when the victims were white equal 11.1 times the estimated odds when the victims were black, controlling for defendant's race. This is a very strong effect.

This logit model is a main effects model. It assumes that both explanatory variables affect the response variable, but with a lack of interaction: the effect of defendant's race on the death penalty verdict is the same for each victims' race, and the effect of victims' race is the same for each defendant's race. This means that the estimated odds ratio between each predictor and the response takes the same value at each category of the other predictor. For instance, the estimated odds ratio of .42 between defendant's race and the death penalty is the same when the victims were white as when the victims were black.

Effects on Odds and Probabilities

The logit model refers to the *log* of the odds, and the parameter estimates refer to that log scale. It is easier to understand odds or probabilities than log odds. As with the logistic regression model, antilogs of the parameter estimates are multiplicative effects on the odds.

To illustrate, the prediction equation of $\log(\text{odds}) = -3.596 - .868d + 2.404v$ for the logit or log odds corresponds to the prediction equation for the odds of

$$\text{odds} = e^{-3.596 - .868d + 2.404v} = e^{-3.596} e^{-.868d} e^{2.404v}$$

For white defendants, $d = 1$, and the estimated odds equal $e^{-3.596} e^{-.868} e^{2.404v}$; for black defendants, $d = 0$, and the estimated odds equal $e^{-3.596} e^{2.404v}$. The estimated odds for white defendants divided by the estimated odds for black defendants equal $e^{-.868} = .42$. This shows why the antilog of the coefficient for d in the prediction equation is the estimated odds ratio between defendant's race and death penalty verdict, for each fixed victims' race v. The effect of defendant's race being white is to multiply the estimated odds of a *yes* death penalty verdict by $e^{-.868} = .42$ compared to its value for black defendants. The actual values of the odds depend on victims' race, but the ratio of the odds is the same for each.

The logit model expression for the log odds is *additive*, but taking antilogs yields a *multiplicative* expression for the odds. In other words, the antilogs of the parameters are *multiplied* to obtain odds. We can use this expression to calculate odds estimates for any combination of defendant's race and victims' race. For instance, when the defendant is black ($d = 0$) and the victims were white ($v = 1$), the estimated odds are

$$\text{odds} = e^{-3.596-.868d+2.404v} = e^{-3.596-.868(0)+2.404(1)} = e^{-1.192} = .304$$

Alternatively, one can predict probabilities, using

$$\hat{\pi} = \frac{\text{odds}}{1+\text{odds}} = \frac{e^{-3.596-.868d+2.404v}}{1+e^{-3.596-.868d+2.404v}}$$

For instance, when the defendant is black and the victims were white, the estimated odds just calculated of .304 corresponds to an estimated death penalty probability of $\hat{\pi} = \text{odds}/(1+\text{odds}) = .304/1.304 = .233$. This is close to the sample proportion of .229 (Table 15.4). The predicted proportions, unlike sample proportions, perfectly satisfy the model. The closer the sample proportions fall to the predicted probabilities, the better the model fits.

Inference for Logit Models

If $\beta_1 = 0$ in the model

$$\text{logit}(\pi) = \alpha + \beta_1 d + \beta_2 v$$

then the death penalty verdict is independent of defendant's race, controlling for victims' race. In that case, the odds ratio between the death penalty verdict and defendant's race equals $e^0 = 1$ for each victims' race. The farther β_1 and β_2 fall from 0, the stronger the effects of the predictors, in the sense that the odds ratios fall farther from 1.

One can test the partial effects of the predictors in the usual way. The parameter estimate divided by its standard error is a z test statistic. The square of that, the Wald statistic, is a chi-squared statistic with $df = 1$. Alternatively, some software also reports likelihood-ratio tests, which compare the $(-2\log L)$ values with and without the predictor in the model. This is particularly useful if a qualitative predictor has several levels, in which case it has several dummy variables and the test of its effect has a null hypothesis with several parameters equal to 0.

Table 15.5 also provides results of Wald tests for the partial effects. For instance, the defendant's race effect of $-.868$ has a standard error of .367. The z test statistic for testing H_0: $\beta_1 = 0$ is $z = -.868/.367 = 2.36$, and the Wald statistic equals $(2.36)^2 = 5.59$. For the two-sided alternative, either statistic has a P-value of .018, shown also in Table 15.5. Similarly, the test of H_0: $\beta_2 = 0$ has $P = .0001$ and provides extremely strong evidence of a victims' race effect.

The parameter estimates are also the basis of confidence intervals for odds ratios. Since the estimates refer to *log* odds ratios, after constructing the interval one must take antilogs of the endpoints to form the interval for the odds ratio. For instance, since the estimated log odds ratio of 2.404 between victims' race and the death penalty has a standard error of .601, a 95% confidence interval for the true log odds ratio is

$$2.404 \pm 1.96(.601), \quad \text{or } (1.23, 3.58)$$

Applying the antilog (i.e., exponential function, e^x) to each endpoint, we find that the confidence interval for the odds ratio equals $(e^{1.23}, e^{3.58}) = (3.4, 35.9)$. When the victims were white, the estimated odds of the death penalty are between 3.4 and 35.9 times the estimated odds when the victims were black.

Most software can also provide confidence intervals for probabilities, as discussed in the appendix. For instance, for black defendants with white victims, a 95% confidence interval for the probability of the death penalty equals (.135, .371).

15.4 Loglinear Models for Categorical Variables

Logistic regression models are similar in structure to ordinary regression models, both types predicting a response variable using explanatory variables. By contrast, ***loglinear models*** are appropriate for contingency tables in which each classification is a response variable. Loglinear analysis resembles a correlation analysis more than a regression analysis, with its focus on studying associations between pairs of variables rather than modeling the response on one of them in terms of the others.

Loglinear models are special cases of generalized linear models that assume *Poisson* distributions for cell counts in contingency tables. The Poisson distribution is defined for discrete variables, such as counts, that can take nonnegative integer values. The loglinear model formulas express the logs of cell expected frequencies in terms of dummy variables for the categorical variables and interactions between those variables. The actual model formulas are rather cumbersome, and this section instead uses a symbolic notation that highlights the pairs of variables that are associated. Problem 15.39 shows why the models are called *loglinear* models.

Example 15.4 Students' Use of Alcohol, Cigarette, and Marijuana

Table 15.6 comes from a 1992 survey by the Wright State University School of Medicine and the United Health Services in Dayton, Ohio (Thanks to Prof. Harry Khamis for providing these data). The survey asked senior high-school students in a nonurban area near Dayton, Ohio, whether they had ever used alcohol, cigarettes, or marijuana. Table 15.6 is a $2 \times 2 \times 2$ contingency table that cross classifies responses on these three items.

In this table, all three variables are response variables, rather than one a response and the others explanatory. The models presented in this section describe their association structure. They analyze whether each pair of variables is associated and whether the association is the same at each category of the third variable. □

TABLE 15.6 Alcohol (A), Cigarette (C), and Marijuana (M) Use for High School Seniors

Alcohol Use	Cigarette Use	Marijuana Use Yes	Marijuana Use No
Yes	Yes	911	538
	No	44	456
No	Yes	3	43
	No	2	279

Source: I am grateful to Prof. Harry Khamis, Wright State University, for supplying these data.

A Hierarchy of Loglinear Models for Three Variables

Loglinear models apply to contingency tables with any number of dimensions. We use three-way tables to introduce basic ideas, illustrating for Table 15.6. Denote the three variables by X, Y, and Z.

There are many potential patterns of association that a cross-classification of three categorical variables can exhibit. Loglinear models describe associations in partial tables that relate two of the variables while controlling for the third one. These are called ***partial*** or ***conditional associations***. A pair of variables could be statistically independent at each category of the third variable; in other words, the population version of each partial table could satisfy independence. In that case, the pair is said to be ***conditionally independent***, and the odds ratios equal 1 in the partial tables. Or associations might exist in some or all of the partial tables. We now introduce a hierarchy of five loglinear models, ordered in terms of the extent of association.

1. All three pairs of variables are conditionally independent. That is,
   X is independent of Y, controlling for Z;
   X is independent of Z, controlling for Y;
   Y is independent of Z, controlling for X.

2. Two of the pairs of variables are conditionally independent. For example,
   X is independent of Z, controlling for Y;
   Y is independent of Z, controlling for X;
   X and Y are associated, controlling for Z.

3. One of the pairs of variables is conditionally independent. For example,
   X is independent of Z, controlling for Y;
   X and Y are associated, controlling for Z;
   Y and Z are associated, controlling for X.

4. No pair of variables is conditionally independent, but the association between any two variables is the same at each category of the third. We then say there is ***no interaction***, or ***homogeneous association***.

5. All pairs of variables are associated, but interaction exists: The association between each pair varies according to the category of the third variable. (Recall Section 10.3.)

Each model has a symbol that indicates the pairs of variables that are associated. Associated variables appear together in the symbol. For instance, (XY, Z) denotes the model for case 2 (above) in which X and Y are associated but the other two pairs are conditionally independent. The symbol (XY, XZ, YZ) denotes the model for case 4, in which all three pairs are associated, but with no interaction. Table 15.7 lists the symbols for all the models described above. All the models provide some structure for the pattern of association except for the one symbolized by (XYZ). This model, which allows interaction, fits any sample three-way table perfectly. It is called the ***saturated model***.

TABLE 15.7 Some Loglinear Models for Three-Dimensional Contingency Tables

Model Symbol	Interpretation
(X, Y, Z)	All pairs are conditionally independent.
(XY, Z)	X and Y the only associated pair.
(XY, YZ)	X and Z the only conditionally independent pair.
(XY, YZ, XZ)	Each pair associated, controlling for the third variable. No interaction.
(XYZ)	All pairs associated, with interaction.

Odds Ratio Interpretations for Loglinear Models

Interpretations of associations in loglinear models, like those in logit models, refer to the odds ratio. In fact, one can use the odds ratio to define properties such as conditional independence and no interaction.

In 2×2 contingency tables, independence is equivalent to a population odds ratio of 1.0. In a three-way table, ***conditional independence*** between X and Y means that the population odds ratios in the XY partial tables all equal 1.0. ***No interaction*** means that the population odds ratios in the XY partial tables are identical at each category of Z.

For instance, a $2 \times 2 \times 3$ table consists of three partial tables each of size 2×2, with two categories for X and two categories for Y measured at three levels of Z. Suppose the true odds ratio for the XY partial association is $\theta = 1.0$ at the first level of Z, $\theta = 1.0$ at the second level of Z, and $\theta = 1.0$ at the third level of Z. Then, X and Y are conditionally independent, and loglinear model (XZ, YZ) holds. On the other hand, if $\theta = 2.2$ at the first level of Z, $\theta = 2.2$ at the second level of Z, and $\theta = 2.2$ at the third level of Z, then there is conditional association but no interaction, and model (XY, XZ, YZ) holds. When the XY odds ratios are the same at all levels of Z, necessarily the XZ odds ratios are the same at all levels of Y, and the YZ odds ratios are the same at all levels of X.

Standard software can fit loglinear models, as the appendix shows. The fit provides expected frequencies having odds ratios that perfectly satisfy the model. If the model fits well, these odds ratios help us interpret the associations implied by the model.

Example 15.5 Estimated Odds Ratios for Drug Use Data

Denote the variables in Table 15.6 by A for alcohol use, C for cigarette use, and M for marijuana use. Table 15.8 contains expected frequencies for the loglinear model (AC, AM, CM) that permits an association for each pair of variables but assumes no interaction. These are very close to the observed counts, so the model seems to fit well.

TABLE 15.8 Expected Frequencies for Loglinear Model (AC, AM, CM) for Alcohol (A), Cigarette (C), and Marijuana (M) Use for High School Seniors

Alcohol	Cigarette	Marijuana Use	
Use	Use	Yes	No
Yes	Yes	910.4	538.6
	No	44.6	455.4
No	Yes	3.6	42.4
	No	1.4	279.6

Let's study the estimated association between cigarette use and marijuana use, controlling for alcohol use, using the expected frequencies for this model. For those who used alcohol, the estimated odds ratio between C and M equals

$$\frac{910.4 \times 455.4}{538.6 \times 44.6} = 17.3$$

Similarly, for those who had not used alcohol, the estimated odds ratio between C and M equals

$$\frac{3.6 \times 279.6}{42.4 \times 1.4} = 17.3$$

For each category of A, students who have smoked cigarettes have estimated odds of having smoked marijuana that are 17.3 times the estimated odds for students who have not smoked cigarettes. The model assumes a lack of interaction, so the estimated odds ratio is the same at each category of A.

Software for loglinear models also provides tables of model parameter estimates from which one can estimate such odds ratios directly. Table 15.9 illustrates. For each pair of variables in a loglinear model, the association parameter estimate refers to the *log* odds ratio. For the CM conditional association, therefore, the estimated odds ratio at each level of A equals $e^{2.848} = 17.3$. Similarly, the estimated odds ratio equals $e^{2.054} = 7.8$ between A and C at each level of M, and the estimated odds ratio equals $e^{2.986} = 19.8$ between A and M at each level of C. The estimated association is very strong between each pair of variables. □

TABLE 15.9 Printout of Association Parameter (Log Odds Ratio) Estimates for Loglinear Model for Drug Use Data

Parameter	DF	Estimate	Std Err	ChiSquare	Pr>Chi
A*C	1	2.0545	0.1741	139.3180	0.0001
A*M	1	2.9860	0.4647	41.2933	0.0001
C*M	1	2.8479	0.1638	302.1409	0.0001

The model (AC, AM, CM) permits conditional association for each pair of variables. Other possible loglinear models for these data delete at least one of the associations. To illustrate the association patterns implied by some of these models, Table 15.10 presents estimated conditional odds ratios for the expected frequencies for the models. For example, the entry 1.0 for the AC conditional association for model (AM, CM) is the common value of the estimated AC odds ratios at the two categories of M. This model implies conditional independence between alcohol use and cigarette use, controlling for marijuana use, and has estimated odds ratios of 1.0 for the AC conditional association.

TABLE 15.10 Summary of Estimated Conditional Odds Ratios for Various Loglinear Models Fitted to Drug Use Data

	Conditional Associations		
Model	AC	AM	CM
(A, C, M)	1.0	1.0	1.0
(AC, M)	17.7	1.0	1.0
(AM, CM)	1.0	61.9	25.1
(AC, AM, CM)	7.8	19.8	17.3
(ACM) Level 1	13.8	24.3	17.5
(ACM) Level 2	7.7	13.5	9.7

Table 15.10 shows that estimated conditional odds ratios equal 1.0 for each pairwise term not appearing in a model, such as the AC association in model (AM, CM). The odds ratios for the sample data are those reported for the saturated model (ACM), which provides a perfect fit. For that model, the odds ratios between two variables are not the same at each level of the third variable, so they are reported separately for each level. In each case, they are strong at both levels.

Table 15.10 shows also that estimated conditional odds ratios can vary dramatically depending on the model chosen. This highlights the importance of good model selection. An estimate from this table is informative only to the extent that its model fits well. The next section shows how to check model goodness of fit.

15.5 Logit and Loglinear Model Goodness of Fit

Section 8.2 introduced the chi-squared test of independence for contingency tables. That test can be regarded as a goodness-of-fit test for the loglinear model which states that the two categorical variables are statistically independent. The chi-squared statistic compares the observed frequencies to the expected frequencies that satisfy the independence model.

Similarly, logit and loglinear models for multidimensional contingency tables have chi-squared tests that check the model adequacy. For such tests, the null hypothesis is that the model truly holds in the population of interest.

Chi-Squared Goodness-of-Fit Statistics

Each model for a contingency table has a set of cell expected frequencies, numbers that perfectly satisfy the model. The model goodness of fit is tested by comparing the expected frequencies, denoted by $\{f_e\}$, to the observed frequencies $\{f_o\}$. The chi-squared statistics summarize the discrepancies. Larger differences between the $\{f_o\}$ and $\{f_e\}$ yield larger values of the statistics and more evidence that the model is inadequate.

Two chi-squared statistics, having very similar properties, are commonly used to do this. The ***Pearson statistic***

$$\chi^2 = \sum \frac{(f_o - f_e)^2}{f_e}$$

was introduced in Section 8.2 for testing independence. Another statistic, called the ***likelihood-ratio statistic***, is

$$G^2 = 2 \sum f_o \log\left(\frac{f_o}{f_e}\right)$$

Some software calls this statistic the ***deviance***. It equals the difference between the $(-2 \log L)$ values for the model being tested and for the most complex model possible. Both statistics equal 0 when there is a perfect fit (i.e., all $f_o = f_e$). Since large values indicate a poor fit, the P-value for testing a model is the right-hand tail probability above the observed value.

If the model truly holds, both statistics have approximate chi-squared distributions. The degrees of freedom (df) for the statistics depend on the model fitted. The df resemble the error df in regression, equaling the number of responses modeled on the left-hand side of the equation minus the number of parameters on the right-hand side of the model. For logit models, for instance, the number of responses modeled is the number of sample logits for the model. This equals the number of combinations of levels of explanatory variables at which one has observations on the binary response, since there is one logit for each combination. Thus, df = number of logits modeled − number of parameters. The simpler the model, in the sense of fewer parameters, the larger the df for the test.

The chi-squared approximation is better for larger sample sizes. The Pearson statistic is preferred when the expected frequencies average between about 1 and 10, and neither statistic works well if most of the expected frequencies are less than about 5.

Example 15.6 Logit Model Goodness of Fit for Death Penalty Data

Example 15.3 used the logit model

$$\text{logit}(\pi) = \alpha + \beta_1 d + \beta_2 v$$

to describe how the probability π of the death penalty depends on dummies d for defendant's race and v for victims' race. A goodness-of-fit test for this model analyzes whether this model with main effects is adequate for describing Table 15.4. The more complex model containing the interaction term is necessary if the main effects model fits poorly.

Software automatically computes the expected frequencies for the model and the goodness-of-fit statistics. For instance, the logit model estimated a probability of .233 that a black defendant receives the death penalty for having white victims. From Table 15.4, there were 48 such black defendants, so the expected number receiving the death penalty equals 48(.233) = 11.2. This is the expected frequency for the cell in the table having observed frequency 11.

The *df* equal the number of logits minus the number of parameters in the model. The number of logits is the number of combinations of levels of explanatory variables. The death penalty data have four logits, one for each combination of defendant's race and victims' race. The model has three parameters, α, β_1, β_2, so both goodness-of-fit statistics have

$$df = \text{Number of logits} - \text{number of parameters} = 4 - 3 = 1$$

Table 15.11 shows the way SPSS software reports the expected frequencies and the results of chi-squared goodness-of-fit tests. For each of the eight cells, f_e is close to f_o. The null hypothesis for the tests is that the logit model with main effects truly holds; that is, no interaction occurs between defendant's race and victims' race in their effects on the death penalty verdict. The Pearson test statistic is $\chi^2 = .20$ and the likelihood-ratio test statistic is $G^2 = .38$. These test statistic values are small, so neither P-value is small. The model seems to fit the data well. The null hypothesis that the model holds is plausible. □

Adjusted Residuals

The chi-squared statistics provide global measures of lack of fit. When the fit is poor, we take a closer look at the cells of the table to determine the nature of the lack of fit. Most software for logit and loglinear models report ***adjusted residuals***, which make a cell-by-cell comparison of f_o and f_e. Each adjusted residual has form

$$\text{Adjusted residual} = \frac{f_o - f_e}{\text{Standard error of } (f_o - f_e)}$$

TABLE 15.11 SPSS Computer Printout of Expected Frequencies, Goodness-of-Fit Tests, and Adjusted Residuals, for Logit Model with Main Effects Fitted to Death Penalty Data

```
Goodness-of-fit Statistics

                    Chi-Square         DF        Sig.
Likelihood Ratio         .3798          1       .5377
         Pearson         .1978          1       .6565
- - - - - - - - - - - - - - - - - - - - - - - - - - - - - -

                      Observed      Expected        Adj.
Factor       Value       Count         Count      Resid.

VICTIM    WHITE
 DEFEND    WHITE
  PENALTY  YES           53.00         52.82         .44
  PENALTY  NO           414.00        414.18        -.44
 DEFEND    BLACK
  PENALTY  YES           11.00         11.18        -.44
  PENALTY  NO            37.00         36.82         .44

VICTIM    BLACK
 DEFEND    WHITE
  PENALTY  YES             .00           .18        -.44
  PENALTY  NO            16.00         15.82         .44
 DEFEND    BLACK
  PENALTY  YES            4.00          3.82         .44
  PENALTY  NO           139.00        139.18        -.44
```

When the model truly holds, adjusted residuals behave like standard normal variables. A large adjusted residual (say, exceeding 3 in absolute value) provides evidence of lack of fit in that cell. The adjusted residuals presented in Section 8.3 are special cases for a model of independence.

Table 15.11 also reports adjusted residuals for the logit model for the death penalty data. They are small and provide no evidence of lack of fit. This is not surprising, since the goodness-of-fit statistics are small. In fact, when $df = 1$ for the goodness-of-fit test, only one adjusted residual is nonredundant, and the square of any of them equals the χ^2 test statistic.

Loglinear Model Goodness of Fit

The same goodness-of-fit formulas apply to loglinear models. Likewise, one can form adjusted residuals to compare individual cell counts to expected frequencies satisfying the model.

Examples 15.4 and 15.5 used loglinear models to describe associations among alcohol use, cigarette use, and marijuana use, for a sample of high school students. Table

15.12 displays results of Pearson χ^2 and likelihood-ratio G^2 tests of goodness of fit for all the possible loglinear models, ranging from the model (A, C, M) for which each pair of variables is independent to the model (AC, AM, CM) for which each pair is associated but the association between two variables is the same at each level of the third. The smaller the chi-squared statistics, the better the fit. Small P-values contradict the null hypothesis that the model is adequate. It is usually preferable to select the simplest model that provides a decent fit to the data. If no model fits well, the adjusted residuals highlight cells contributing to the lack of fit.

TABLE 15.12 Goodness-of-Fit Tests for Loglinear Models of Alcohol (A), Cigarette (C), and Marijuana (M) Use, Based on Likelihood-ratio (G^2) and Pearson (χ^2) Chi-squared Statistics

Model	G^2	χ^2	df	P-Value
(A, C, M)	1286.0	1411.4	4	<.0001
(A, CM)	534.2	505.6	3	<.0001
(C, AM)	939.6	824.2	3	<.0001
(M, AC)	843.8	704.9	3	<.0001
(AC, AM)	497.4	443.8	2	<.0001
(AC, CM)	92.0	80.8	2	<.0001
(AM, CM)	187.8	177.6	2	<.0001
(AC, AM, CM)	0.4	0.4	1	.54

The chi-squared distribution falls on the positive part of the line with mean equal to the df value. Hence, a value such as $G^2 = 1286.0$ based on $df = 4$ falls way out in the right-hand tail, has a tiny P-value, and indicates a very poor fit. From Table 15.12, the only model that passes the goodness-of-fit test is (AC, AM, CM). This model allows association between all pairs of variables but still assumes a lack of interaction; that is, an odds ratio between each pair of variables that is the same at each category of the third variable. The models that lack any associations fit poorly, having P-values below .0001.

Comparing Models by Comparing G^2 Values

Table 15.12 illustrates two important properties of the likelihood-ratio G^2 statistic. First, G^2 is similar to the SSE measure for regression. Both compare observed responses to values expected if a model holds, and both cannot increase as the model becomes more complex. For instance, (A, CM) is a more complex model than (A, C, M), since it allows one association. Hence, it provides a better fit and its G^2-value is smaller. Similarly, G^2 drops further for model (AC, CM) and further yet for (AC, AM, CM). The Pearson χ^2 statistic, unlike the likelihood-ratio G^2, does not have this property. It could potentially increase as a model gets more complex, though in practice this rarely happens.

The second property of G^2 refers to model comparison and was first discussed at the end of Section 15.2. Section 11.6 introduced an F test comparing complete and reduced regression models, based on the reduction in SSE. A similar test comparing loglinear or logit models uses the reduction in G^2 values. To test the null hypothesis that a model truly holds, versus the alternative hypothesis that a more complex model provides a better fit, the test statistic is the difference in G^2 values. This difference is identical to the difference in $(-2 \log L)$ values for the two models. It is a chi-squared statistic with degrees of freedom equal to the difference in df values for the two G^2 values.

To illustrate, we compare models (AC, CM) and (AC, AM, CM). We test the null hypothesis that the reduced model (AC, CM) is adequate against the alternative that the more complex model (AC, AM, CM) is better. The test statistic is the difference between their G^2 values, $92.0 - .4 = 91.6$, based on $df = 2 - 1 = 1$. This chi-squared statistic has a P-value of $P < .0001$, so model (AC, AM, CM) provides a significantly better fit than (AC, CM). This is a likelihood-ratio test for analyzing whether we can drop the AM association from model (AC, AM, CM). The table of parameter estimates for the model, Table 15.9, provides the Wald test for this hypothesis. The Wald statistic equals 41.3 and also has $P < .0001$.

With either χ^2 or G^2, one cannot compare a pair of models such as (A, CM) and (AC, AM). Neither is a special case of the other, since each allows association that the other excludes.

Connection Between Logit and Loglinear Models

Any logit model for categorical data is equivalent to a corresponding loglinear model. That loglinear model contains the same associations as the logit model does between the response variable and the explanatory variables, and it contains the most general term for describing relationships among the explanatory variables.

For instance, consider the logit model for a three-way table that has main effects for X and Z as predictors of a binary response, Y. If X has two categories and Z has three categories, then the logit model has a dummy variable for X and two dummy variables for Z. For that logit model, X has an effect on Y, Z has an effect on Y, but there is no interaction. That logit model has the same expected frequencies, the same odds ratio estimates, and the same goodness-of-fit statistics as the loglinear model (XY, XZ, YZ) that also permits an XY association, a YZ association, as well as an association between the variables (X and Z) that are the explanatory variables in the logit model.

Next, consider a four-way table having a binary response variable Y and three qualitative predictors, X_1, X_2, X_3. The logit model with all the main effects has dummy variables for X_1, for X_2, and for X_3. It assumes that Y is associated with each of X_1, X_2, and X_3, but that the effects of each explanatory variable on Y are the same at each combination of levels of the other variables. The corresponding loglinear model is symbolized by $(YX_1, YX_2, YX_3, X_1X_2X_3)$. The loglinear model contains association terms for Y with each of the three predictors. It also contains the most general term

$(X_1X_2X_3)$ for the possible interaction among the explanatory variables, since the logit model assumes nothing about the structure of their relationships.

Example 15.7 Loglinear Models for Death Penalty Data

Example 15.6 showed that the logit model for the death penalty data having main effects for defendant's race and victims' race provides a good fit, with $G^2 = .38$ and $\chi^2 = .20$ based on $df = 1$. From Section 15.3, the estimated odds ratio is .42 for defendant's race and the death penalty and 11.1 for victims' race and the death penalty.

Let D = defendant's race, V = victims' race, and P = death penalty verdict. That logit model for P as the response variable has precisely the same fit as the loglinear model (DP, VP, DV) that allows associations for each pair of variables but no interaction. That loglinear model has the same values for G^2 and χ^2 for testing fit, the same expected frequencies, and the same estimated odds ratios between each predictor and the death penalty outcome. You can check this, as an exercise, by using software to fit this loglinear model to Table 15.4. □

Distinction Between Logit and Loglinear Models

Logit models distinguish between a single response variable and a set of explanatory variables, whereas loglinear models treat every variable as a response variable. Although designed for a different purpose, logit models imply association patterns among variables that are equivalent to ones expressed by loglinear models. For data in contingency tables, one reaches similar conclusions with the two approaches.

In most applications, there is a single response variable. It is then more natural to use a logit model than a loglinear model. The logit analysis focuses on the effects of the predictors on the response, much as in ordinary regression modeling. For that reason, this chapter has devoted more attention to logistic regression and logit models than to loglinear models.

15.6 Logit Models for Ordinal Variables

Many applications have a qualitative response variable with more than two categories. For instance, the General Social Survey recently asked subjects whether government spending on the environment should increase, remain the same, or decrease. The logit and loglinear models in this chapter have made no distinction between nominal and ordinal response variables. In fact, the logit models refer to binary response variables. Standard loglinear models apply for variables of either scale, but they treat all variables as nominal.

For nominal response variables, there is an extension of the binary logit model called a ***multinomial logit model***. It implies that an ordinary logit model holds for each pair of response categories. We will not discuss this model here. Instead, we present an extension of logistic regression to handle ordinal response variables.

Logistic Regression for an Ordinal Response Variable

Let Y denote an ordinal response variable, having c categories. Let $P(Y \leq j)$ denote the probability that the response on Y falls in category j or below (i.e., in category 1, 2, ..., or j). This is called a ***cumulative probability***. It equals the sum of the probabilities in category j and below,

$$P(Y \leq j) = P(Y = 1) + P(Y = 2) + \ldots + P(Y = j)$$

A c-category response has c cumulative probabilities: $P(Y \leq 1), P(Y \leq 2), \ldots, P(Y \leq c)$. The final cumulative probability uses the entire scale, so $P(Y \leq c) = 1$. The order of forming the cumulative probabilities reflects the ordering of the response scale, and those probabilities themselves satisfy

$$P(Y \leq 1) \leq P(Y \leq 2) \leq \ldots \leq P(Y \leq c) = 1$$

Models for ordinal responses use these cumulative probabilities.

The ratio

$$\frac{P(Y \leq j)}{P(Y > j)}$$

is the odds of response in category j or below. For instance, when the odds equal 3, the probability of response in category j or below equals three times the probability of response above category j. Each cumulative probability converts to an odds.

The model for an ordinal response uses logits of the cumulative probabilities. The logit of $P(Y \leq j)$ equals the log of the odds just presented; that is,

$$\text{logit}[P(Y \leq j)] = \log[P(Y \leq j)/P(Y > j)], \quad j = 1, \ldots, c - 1$$

Since the final cumulative probability, $P(Y \leq c)$, necessarily equals 1.0, we exclude it from the model. These logits of cumulative probabilities are called ***cumulative logits***.

Cumulative Logit Models

A model that simultaneously describes the effect of an explanatory variable, X, on all $c - 1$ cumulative probabilities for Y is

$$\text{logit}[P(Y \leq j)] = \alpha_j + \beta X, \quad j = 1, 2, \ldots, c - 1$$

For $c = 4$, for instance, this single model describes three relationships: the effect of X on the odds that $Y \leq 1$ instead of $Y > 1$, the effect of X on the odds that $Y \leq 2$ instead of $Y > 2$, and the effect of X on the odds that $Y \leq 3$ instead of $Y > 3$. The model requires a separate intercept parameter α_j for each cumulative probability. Since the cumulative probabilities increase as j increases, so do $\{\alpha_j\}$.

The parameter of main interest in this model is β. It describes the effect of X on the response. When $\beta = 0$, X has no effect on Y. In that case, the cumulative probabilities do not change as X changes, and X and Y are independent. The effect of X increases as

$|\beta|$ increases. In this model, β does not have a j subscript. In other words, β takes the same value for the logit for each cumulative probability. The model assumes that the effect of X is the same for each cumulative probability. This cumulative logit model with common effect is often called the ***proportional odds*** model.

For fixed j, this model looks like an ordinary logistic regression model, namely, the logit transform of a probability equals an intercept coefficient plus a slope parameter times the value of the explanatory variable. The logit model applies to a collapsing of the response into the two outcomes $(Y \leq j)$ and $(Y > j)$. The fit of the full cumulative logit model, however, provides a single estimate $\hat{\beta}$ for the effect of X, rather than the three separate estimates one gets by fitting the binary logit model separately for each collapsing of the response.

Figure 15.4 depicts the model when there are $c = 4$ response categories and a single quantitative predictor. There are then three cumulative logits, and the model implies a separate S-shaped curve for each. The curve for $P(Y \leq j)$ has the appearance of a logistic regression curve for a binary response with pair of outcomes $(Y \leq j)$ and $(Y > j)$. At any fixed X value, the three curves have the same ordering as the cumulative probabilities, the one for $P(Y \leq 1)$ being lowest.

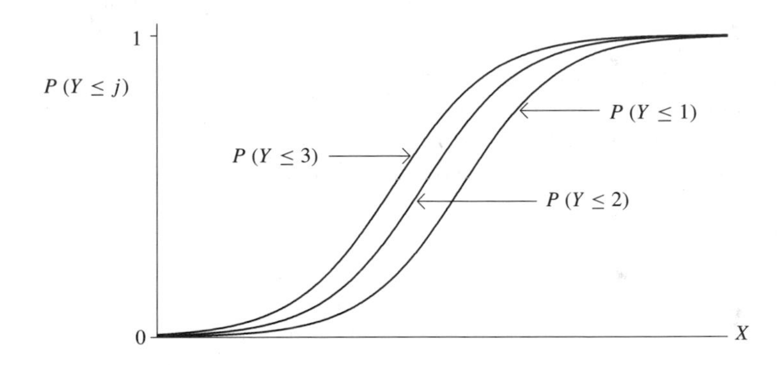

Figure 15.4 Depiction of Curves for Cumulative Probabilities in Cumulative Logit Model

As in logistic regression, the size of $|\beta|$ determines how quickly the curves climb or drop. The common value for β means that the three response curves have the same shape. The curve for $P(Y \leq 1)$ is the curve for $P(Y \leq 2)$ moved to the right and the curve for $P(Y \leq 3)$ moved even further to the right. One can interpret the association between X and Y by treating e^{β} as a multiplicative effect of X on the odds for the cumulative probabilities. For each j, the odds that $Y \leq j$ multiply by e^{β} for each one-unit change in X.

Example 15.8 Comparing Political Ideology of Democrats and Republicans

Do Republicans tend to be more conservative than Democrats? Table 15.13, from the 1991 General Social Survey, relates political ideology to party affiliation. We treat po-

TABLE 15.13 Political Ideology by Party Affiliation

Party Affiliation	Political Ideology				
	Very Liberal	Slightly Liberal	Moderate	Slightly Conservative	Very Conservative
Democratic	80	81	171	41	55
Republican	30	46	148	84	99

Source: 1991 General Social Survey.

litical ideology as the response variable Y. It has a five-point ordinal scale, ranging from very liberal to very conservative.

Let X be a dummy variable for party affiliation, with $X = 1$ for Democrats and $X = 0$ for Republicans. The fit of the cumulative logit model has estimated effect $\hat{\beta} = .974$. For any fixed j, the estimated odds that a Democrat's response is in the liberal direction rather than the conservative direction (i.e., $Y \leq j$ rather than $Y > j$) equal $e^{\hat{\beta}} = e^{.974} = 2.65$ times the estimated odds for Republicans. This value is far from the "no effect" value of 1.0. The sample has a moderate association, Democrats tending to be more liberal than Republicans.

Table 15.14 shows how SAS reports the results for the model fitted to Table 15.13. The table reports four intercept parameter estimates, since the response variable has five categories. Normally, these estimates are not of as much interest as the estimated effect of the explanatory variable (party) and the odds ratio estimate. □

TABLE 15.14 Printout for Cumulative Logit Model Fitted to Table 15.13

```
                              Intercept
               Intercept         and
 Criterion       Only        Covariates    Chi-Square for Covariates
 -2 LOG L       2533.63       2474.99     58.64 with 1 DF (p=0.0001)

             Parameter  Standard       Wald          Pr >
Variable      Estimate     Error   Chi-Square   Chi-Square
INTERCP1      -2.4690     0.1318    350.8187       0.0001
INTERCP2      -1.4745     0.1091    182.7191       0.0001
INTERCP3       0.2371     0.0948      6.2501       0.0124
INTERCP4       1.0695     0.1046    104.6088       0.0001
PARTY          0.9745     0.1291     57.0218       0.0001

                Odds
Variable       Ratio
PARTY          2.650
```

When $\beta > 0$, as in Figure 15.4, cumulative probabilities are higher at higher values of X. That is, the probability that Y falls below any fixed level is relatively greater at higher values of X; so, responses at the low end of the ordinal scale for Y are relatively more likely at large X values. This reflects a "negative" relation in the usual sense. Alternatively, one can express the right-hand side of the model formula as $\alpha_j - \beta X$ or define the logit with $P(Y > j)$ in the numerator and $P(Y \leq j)$ in the denominator; with either change, a positive β corresponds to a positive association, in the sense that Y values at the high end of the ordinal scale are more likely to occur at large X values. When one does this or when one reverses the order of categories of Y (i.e., listing from high to low instead of from low to high), the model fit is the same but the sign of $\hat{\beta}$ reverses. Statistical software for fitting the model has no standard convention, and you should be careful to check how your package defines the model so you make the proper interpretation of $\hat{\beta}$.

Inference for an Ordinal Response

When $\beta = 0$ in the cumulative logit model, X and Y are independent. One can test independence by testing that $\beta = 0$. As usual, one can construct a z test statistic by dividing $\hat{\beta}$ by its standard error. The square of that ratio is the Wald statistic, which is chi-squared with $df = 1$. Alternatively, the likelihood-ratio test is based on the difference in $(-2 \log L)$ values with and without the predictor in the model. Most software reports both these tests.

From Table 15.14, the estimate $\hat{\beta} = .974$ has standard error equal to .129, so $z = .974/.129 = 7.55$ and the Wald statistic equals 57.0. The table also reports a likelihood-ratio statistic of 58.6, based on $df = 1$. This is the difference between values of $(-2 \log L)$ for the model with only the intercept terms (2533.6) and the full model (2475.0). The P-value is less than 0.0001 for testing the hypothesis that $\beta = 0$, that is, for testing the null hypothesis of independence against the alternative hypothesis that $\beta \neq 0$. This test and the value of $\hat{\beta}$ provide strong evidence that Democrats are politically more liberal than Republicans.

This test of independence takes into account the ordering of the response categories. It is usually more powerful than chi-squared tests of independence that ignore the ordering, such as the Pearson test of Section 8.2. When there truly is dependence, the ordinal test usually yields a smaller P-value.

More informatively, one can form a confidence interval for the odds ratio that describes the association for the cumulative logit model. The value $\hat{\beta} = .974$ is the estimated log odds ratio. A 95% confidence interval for the true log odds ratio equals .974 $\pm$ 1.96(.129), or (.72, 1.23). The confidence interval for the odds ratio is $(e^{.72}, e^{1.23})$, or (2.1, 3.4). The odds that a Democrat's response falls in the liberal direction are more than twice the odds for Republicans.

Invariance to Choice of Response Categories

When the cumulative logit model fits well for a given response scale, it also fits well with similar effects for any collapsing of the response categories. For instance, if a model for categories (Very liberal, Slightly liberal, Moderate, Slightly conservative, Very conservative) fits well, approximately the same estimated effects result from collapsing the response scale to (Liberal, Moderate, Conservative). This *invariance* to the choice of response categories is a nice feature of the model. Two researchers who use different response categories in studying an association should reach similar conclusions.

To illustrate, we collapse Table 15.13 to a three-category response, combining the two liberal categories and combining the two conservative categories. The estimated effect of party affiliation changes only from 0.974 (standard error = .129) to 1.006 (standard error = 0.132). Some loss of efficiency occurs in collapsing ordinal scales, resulting in larger standard errors. In practice, when observations are spread fairly evenly among the categories, the efficiency loss is minor unless one collapses to a binary response. It is usually inadvisable to collapse ordinal data to binary.

The cumulative logit model implies trends upward or downward among distributions of Y at different values of explanatory variables. When X refers to two groups, as in Table 15.13, the model fits well when subjects in one group tend to make higher responses on the ordinal scale than subjects in the other group. The model does not fit well when the response distributions differ in their dispersion rather than their average. If Democrats tended to be primarily moderate in political ideology, while Republicans tended to be both very conservative and very liberal, then the Republicans' responses would show greater dispersion than the Democrats'. The two political ideology distributions would be quite different, but the model would not detect this if the average responses were similar.

Extensions of the Model

The formula for the cumulative logit model,

$$\text{logit}[P(Y \leq j)] = \alpha_j + \beta X, \quad j = 1, 2, ..., c - 1$$

is adequate when there is a single quantitative or binary predictor. When the explanatory variable is nominal with several categories, one constructs dummy variables to represent the categories of the nominal predictor in the model. For a predictor having r categories, we need $r - 1$ dummy variables. When the explanatory variable is ordinal, one can instead assign scores to its levels and treat it in a quantitative manner as a single term. The cumulative logit model can also handle a continuous explanatory variable. In any of these cases, one can test the hypothesis of independence by comparing the $(-2 \log L)$ value to that of the simpler model that deletes the effects of the predictor, and one can describe the association using odds ratios that refer to making response below instead of above any fixed response level.

More generally, cumulative logit models can handle multiple explanatory variables, some of which may be nominal or ordinal and some of which may be continuous. One can check the fit of the model by analyzing whether extra terms, such as interactions, provide a significant improvement in the model fit. Some software also provides a chi-squared test for the *proportional odds* assumption that the β effects are the same for all cumulative probabilities.

As in other statistical endeavors, there is danger in putting too much emphasis on statistical tests, whether of effects or of goodness of fit. Results are sensitive to sample size, more significant results tending to occur with larger sample sizes. Test statistics merely indicate the level of parsimony that one can achieve. It is important to supplement the tests with estimation methods that describe the strength of effects and with residual analyses that detect parts of the data for which the overall trend fails to hold.

Ordinal Predictor in a Logit Model

This section has focused on modeling ordinal response variables. It is also important to treat ordinal explanatory variables in a quantitative manner, for any type of logit model.

To illustrate, for Table 15.13, we now treat party affiliation as the response variable and political ideology as the explanatory variable. Let π denote the probability a subject identifies as a Democrat. Setting up four dummy variables for the five levels of political ideology, we could use the model

$$\text{logit}(\pi) = \alpha + \beta_1 Z_1 + \beta_2 Z_2 + \beta_3 Z_3 + \beta_4 Z_4$$

This treats the explanatory variable as nominal.

Political ideology is actually ordinal. The effects $\{\beta_i\}$ of levels of ordinal variables often follow an increasing or decreasing trend. For instance, the effects of ideology may follow a linear trend, when plotted against the category numbers of that variable or some other set of increasing scores. Hence, we could simplify this model by assigning scores $\{x_1 < x_2 < \cdots < x_5\}$ to the ideology levels and assuming

$$\text{logit}(\pi) = \alpha + \beta x_i$$

The assignment of scores means that ideology is treated as a quantitative variable rather than a qualitative one. In this model β represents a linear effect on the logit of π, and the model is called a ***linear logit model***.

For Table 15.13, it seems sensible to assign equally spaced scores to political ideology, such as $\{1, 2, 3, 4, 5\}$. The prediction equation for the probability that party affiliation is Democrat is then

$$\text{logit}(\hat{\pi}) = 1.404 - .435x_i$$

For this prediction equation, the antilog of $\hat{\beta}$, $e^{-.435} = .65$, is the multiplicative effect of a single category change in ideology on the odds of being a Democrat instead of Republican. The odds of being classified Democrat are multiplied by .65 for every category increase in conservatism; thus, the probability of being Democrat decreases considerably as conservatism increases.

Models for ordinal variables are discussed in greater detail in the recent books by Agresti (1996, Chapter 8) and Clogg and Shihadeh (1994). Chapters 4–7 of Agresti (1996) also provide an introduction to logistic regression and to loglinear models.

15.7 Chapter Summary

Chapter 8 presented methods for analyzing association between two categorical variables. The methods of this chapter also refer to categorical data. They are more sophisticated, since they apply to *multivariate* data sets, such as multidimensional contingency tables. This methodology is useful for social scientists, since social research usually involves sorting out complex multivariate relationships in which many of the variables are categorical.

This chapter emphasized *model building*. The models studied are appropriate for categorical response variables, and they are analogous to regression models for continuous response variables.

- For binary response variables, the ***logistic regression*** model describes how the probability of a particular category depends on the values of explanatory variables. For a continuous explanatory variable, an S-shaped curve describes how the probability changes as the predictor changes. The model can handle both continuous and categorical explanatory variables. It uses a linear model for the ***logit*** transform of the probability, which is the log of the odds.
- After fitting a logistic regression model, one can predict probabilities at various settings of the explanatory variables. The antilog of a $\hat{\beta}$ parameter estimate is a multiplicative effect on the odds for the response variable, for each unit change in the predictor variable of which it is a coefficient.
- A parameter value of $\beta = 0$ for a predictor corresponds to no effect of that predictor on the response. To test this, one can conduct ***Wald tests***, based on the squared ratio of $\hat{\beta}$ to its standard error, or ***likelihood-ratio tests***, based on comparing values of $(-2 \log L)$ for models with and without that term (L is the maximized *likelihood* function).
- Logit models in which all explanatory variables are categorical are special cases of ***loglinear*** models. These models are useful for investigating association patterns among a set of categorical variables, particularly when all the variables are response variables.
- For logit and loglinear models for contingency tables, chi-squared statistics test the goodness of fit of the model to the data, and the ***odds ratio*** describes the nature and strength of the associations.
- An extension of logistic regression, called a ***cumulative logit model***, models cumulative probabilities for an ordinal response variable. Models that utilize the ordinal nature of variables result in more powerful inferences.

Karl Pearson introduced the chi-squared test for bivariate contingency tables in 1900. The models presented in this chapter have only recently been developed, much of this work occurring within the past quarter century. They are examples of ***generalized linear models***, which apply to discrete as well as continuous response variables. The statistician and sociologist Leo Goodman is responsible for many recent developments in this area (see Goodman, 1978). Social scientists now have available a wide variety of tools for analyzing categorical data.

PROBLEMS

Practicing the Basics

1. A logistic regression model describes how the probability π of voting for the Republican candidate in a presidential election depends on X, the voter's total family income (in thousands of dollars) in the previous year. The prediction equation for a particular sample is

$$\log\left(\frac{\hat{\pi}}{1-\hat{\pi}}\right) = -1.00 + .02X$$

a) For the region of X values for which π is near .50, give a linear approximation for the change in the probability for each thousand dollar increase in income.
b) At which income level is the estimated probability of voting for the Republican candidate (i) equal to .50?, (ii) greater than .50?
c) Find the estimated probability of voting for the Republican candidate when (i) income = 10 thousand, (ii) income = 100 thousand.
d) Explain how the odds of voting Republican depends on family income.
e) The estimated standard error of the coefficient of X is .005. Assuming the sample was randomly selected, test the hypothesis that vote for the Republican candidate is independent of family income. Report the P-value, and interpret.

2. Refer to the previous exercise. When the explanatory variables are X_1 = family income, X_2 = number of years of education, and X_3 = gender (1 = male, 0 = female), the prediction equation is

$$\text{logit}(\hat{\pi}) = -2.40 + .02X_1 + .08X_2 + .20X_3$$

For this sample, X_1 ranges from 6 to 157 with a standard deviation of 25, and X_2 ranges from 7 to 20 with a standard deviation of 3.
a) Find the estimated probability of voting Republican for (i) a man with 16 years of education and income 30 thousand dollars, (ii) a woman with 16 years of education and income 30 thousand dollars.
b) Convert the probabilities in (a) to odds, and find the odds ratio, dividing the odds for men by the odds for females. Interpret.
c) Using the prediction equation, find the effect on the odds of changing X_3 from 0 to 1, controlling for X_1 and X_2. Compare to (b), and interpret.
d) Find the estimated effect on the odds of voting Republican of a one year increase in X_2, controlling for X_1 and X_3. Interpret.
e) Holding the other variables constant, find the estimated effect on the odds of voting Republican of

i) A standard deviation change in X_2.
ii) A standard deviation change in X_1.
Which predictor has the larger standardized effect? Interpret.

3. A sample of 54 elderly men are given a psychiatric examination to determine whether symptoms of senility are present. A subtest of the Wechsler Adult Intelligence Scale (WAIS) is the explanatory variable. Table 15.15 shows results. The WAIS scores range from 4 to 20, with a mean of 11.6. Higher values indicate more effective intellectual functioning.

TABLE 15.15

Criterion	Intercept Only	Intercept and Covariates	Chi-Square for Covariates
-2 LOG L	63.811	53.296	10.515 with 1 DF (p=0.0012)

Variable	Parameter Estimate	Standard Error	Wald Chi-Square	Pr > Chi-Square
INTERCPT	2.0429	1.0717	3.6338	0.0566
WAIS	-0.2821	0.1007	7.8487	0.0051

a) Report the prediction equation, and explain why this equation suggests that the probability of senility decreases at higher levels of the WAIS.
b) Show that $\hat{\pi} = 1/2$ at $X = 7.2$ and that $\hat{\pi} < 1/2$ for $X > 7.2$.
c) Find the predicted probability of senility at $X = 20$.
d) Report the z test statistic for H_0: $\beta = 0$ and $H_a : \beta \neq 0$. What does the Wald statistic equal? What do you conclude about the association?
e) The least squares fit of the linear probability model is $\hat{\pi} = .847 - .051X$. Find the predicted probability of senility at $X = 20$. Does this make sense?

4. For the 23 space shuttle flights that occurred before the Challenger mission disaster in 1986, Table 15.16 shows the temperature (in degrees Fahrenheit) at the time of the flight and whether at least one primary O-ring suffered thermal distress.

TABLE 15.16

Ft	Temp	TD	Ft	Temp	TD	Ft	Temp	TD	Ft	Temp	TD	Ft	Temp	TD
1	66	0	2	70	1	3	69	0	4	68	0	5	67	0
6	72	0	7	73	0	8	70	0	9	57	1	10	63	1
11	70	1	12	78	0	13	67	0	14	53	1	15	67	0
16	75	0	17	70	0	18	81	0	19	76	0	20	79	0
21	75	1	22	76	0	23	58	1						

Note: Ft = flight no., Temp = temperature, TD = thermal distress (1 = yes, 0 = no).
Source: Data based on Table 1 in S. R. Dalal, E. B. Fowlkes, and B. Hoadley, *Journal of the American Statistical Association*, Vol. 84, 1989, p. 945.

a) Use logistic regression to model the effect of temperature on the probability of thermal distress. Interpret the model fit.
b) Calculate the predicted probability of thermal distress at 31 degrees, the temperature at the time of the Challenger flight.
c) At what temperature does the predicted probability equal .5? At that temperature, give

a linear approximation for the change in the predicted probability per degree increase in temperature. Interpret.

d) Interpret the effect of temperature on the odds of thermal distress.

e) Test the hypothesis that temperature has no effect. Report the test statistic and P-value, and interpret.

5. Refer to Table 15.13. Treat party affiliation as the response variable, and use logistic regression to describe the effect of ideology by assigning scores (1, 2, 3, 4, 5) to its levels.

a) Report the prediction equation, and obtain the predicted probability of Democratic affiliation at ideology level (i) very liberal, (ii) very conservative.

b) Use the model to test whether the variables are independent. Report the test statistic, P-value, and interpret.

c) Use the odds ratio to describe the effect on party affiliation of a change in ideology from (i) very liberal to slightly liberal, (ii) slightly liberal to moderate.

d) Check the goodness of fit of the model. Interpret.

e) If your software provides the option, obtain 95% confidence intervals for the probability of Democratic affiliation at ideology level (i) very liberal, (ii) very conservative. Compare.

6. In the first nine decades of the twentieth century in baseball's National League, the percentage of times the starting pitcher pitched a complete game were 72.7 (1900–1909), 63.4, 50.0, 44.3, 41.6, 32.8, 27.2, 22.5, 13.3 (1980–1989) (George Will, *Newsweek*, April 10, 1989).

a) Let $X = 1, 2, \ldots, 9$ for the successive nine decades. Use a logistic regression model to describe the time trend in these data. (For simplicity, suppose the number of games was the same in each decade. For instance, you can represent the percentage 72.7 by 727 successes and 273 failures in 1000 trials.)

b) Use the fitted model to predict the percentage of complete games in the decade $X = 12$ (i.e., 2010–2019).

c) Repeat the analysis using a linear probability model. Is its future prediction realistic?

7. Table 15.17 shows results of a study on the effects of AZT in slowing the development of AIDS symptoms. In the study, 338 veterans whose immune systems were beginning to falter after infection with the AIDS virus were randomly assigned either to receive AZT immediately or to wait until their T cells showed severe immune weakness. The table classifies the veterans' race, whether they received AZT immediately, and whether they developed AIDS symptoms during the three-year study. Fit a logit model with main effects to these data, and using it:

a) Report and interpret the prediction equation.

b) For black veterans without immediate AZT use, find the predicted probability of showing AIDS symptoms.

c) Find the estimated conditional odds ratio between AZT use and the development of symptoms. Interpret.

d) Test for the effect of AZT use. Interpret.

e) Test the goodness of fit of the model. Interpret.

8. A recent article used logistic regression with 1149 observations from the National Election Studies to model the probability of voting for the Republican candidate (George Bush) in the 1988 presidential election (K. Smith, *American Politics Quarterly*, Vol. 22, 1994, p. 354). The predictors dealt with attitudes about abortion and also included a dummy variable that equaled 1 if the respondent was from a state in the old confederacy and 0 otherwise. The estimated southern effect was .503, with a standard error of .215.

TABLE 15.17

Race	AZT Use	Symptoms Yes	Symptoms No
White	Yes	14	93
	No	32	81
Black	Yes	11	52
	No	12	43

Source: *The New York Times*, February 15, 1991.

a) Use the odds ratio to estimate the effect of southern residence on the vote for president, controlling for abortion attitudes.
b) Test the significance of this effect, and interpret.
c) Construct and interpret a 95% confidence interval for the true odds ratio in (a).

9. A recent study of mother's occupational status and children's schooling (M. Kalmin, *American Sociological Review*, Vol. 59, 1994, p. 257) reported the prediction equation

$$\text{logit}(\hat{\pi}) = .75 + .35Y + .13F + .09M + .30FO + .21MO - .92ME - .16S$$

where π is the probability the child obtains a high school degree, Y = respondent's year of birth, F = father's education, M = mother's education (0 to 17), FO = father's occupational level, MO = mother's occupational level (1 to 9), ME = whether mother employed (1 = yes), S = number of siblings. All effects were significant at the .01 level.
a) Interpret the coefficient of mother's education.
b) Interpret the coefficient of whether mother employed.
c) The author reported that a one-point increase in mother's occupational level is associated with a 24% increase in the odds of a high school diploma. Explain how he made this interpretation.

10. Refer to Table 8.29 in Problem 8.14. The response variable is ever having sexual intercourse.
a) Fit a logit model with main effects for race and gender. Report the prediction equation.
b) Find the predicted probability of having experienced intercourse for (i) white females, (ii) black females.
c) Find the estimated conditional odds ratio between race and intercourse, (i) using a model parameter estimate, (ii) using the estimated probabilities in (b). Interpret.
d) Test for the effect of race in this model. Interpret.
e) Find the estimated conditional odds ratio between gender and intercourse. Interpret.
f) Test for the effect of gender in the model. Interpret.
g) Test the goodness of fit of the model. Interpret.
h) Summarize your main findings about these data in a way that you could present to the general public, using as little technical jargon as possible.

11. Refer to Table 15.8, treating marijuana use as the response variable. Table 15.18 shows the SAS printout (PROC GENMOD) for a logit model with main effects.
a) Report the prediction equation, and interpret.
b) Find the predicted probability of having used marijuana (i) for those who have not used alcohol or cigarettes, (ii) for those who have used both alcohol and cigarettes.
c) Show how to convert the estimated coefficients to estimated odds ratios. Interpret.

TABLE 15.18

```
Criteria For Assessing Goodness Of Fit
Criterion                 DF          Value
Deviance                   1         0.3740
Pearson Chi-Square         1         0.4011

Parameter          DF    Estimate    Std Err    ChiSquare    Pr>Chi
INTERCEPT           1      -5.309     0.4752      124.820    0.0001
ALCOHOL     yes     1       2.986     0.4647       41.293    0.0001
ALCOHOL     no      0       0.000     0.0000            .         .
CIGARETT    yes     1       2.848     0.1638      302.141    0.0001
CIGARETT    no      0       0.000     0.0000            .         .
```

d) Test the goodness of fit of this logit model. Interpret.
e) Which loglinear model is equivalent to this logit model?

12. Table 15.19 refers to passengers in autos and light trucks involved in accidents in the state of Maine in 1991. The table classifies subjects by gender, location of accident, seat-belt use, and a response variable having categories (1) not injured, (2) injured but not transported by emergency medical services, (3) injured and transported by emergency medical services but not hospitalized, (4) injured and hospitalized but did not die, (5) injured and died. For this exercise, combine response categories 2–5, and consider models for π = probability of injury.

TABLE 15.19

			Response				
Gender	Location	Seat-Belt	1	2	3	4	5
Female	Urban	No	7287	175	720	91	10
		Yes	11, 587	126	577	48	8
	Rural	No	3246	73	710	159	31
		Yes	6134	94	564	82	17
Male	Urban	No	10, 381	136	566	96	14
		Yes	10, 969	83	259	37	1
	Rural	No	6123	141	710	188	45
		Yes	6693	74	353	74	12

Source: Dr. Cristanna Cook, Medical Care Development, Augusta, Maine.

a) Fit the model having main effects for gender ($G = 1$ for males and 0 for females), location ($L = 1$ for rural and 0 for urban), and seat-belt use ($S = 1$ for yes and 0 for no). Provide the prediction equation for (i) the logit, (ii) the probability, and indicate which level of each predictor is associated with a higher chance of injury.
b) Find the predicted probability of injury for a female in an accident in an urban location (i) wearing a seat belt, (ii) not wearing a seat belt.
c) Report and interpret an odds ratio describing the effect of wearing a seat belt.
d) Construct a 95% confidence interval for the true odds ratio for the effect of wearing a seat belt. Interpret.

e) Conduct a test of the hypothesis of no effect of seat belt use on the response, controlling for gender and location. Report the P-value and interpret.
f) Test the goodness of fit of this model. Interpret.

13. Refer to the previous exercise. Fit the model that also has the three two-way interactions between predictors. Use a likelihood-ratio test to compare this model to the main effects model. Interpret. (*Hint*: The test statistic is the difference between the $(-2 \log L)$ values, or equivalently the difference between the G^2 test statistics for testing fit of the two models.)

14. Let π denote the probability that a randomly selected respondent supports current laws legalizing abortion, predicted using gender of respondent ($G = 0$, male; $G = 1$, female), religious affiliation ($R_1 = 1$, Protestant, 0 otherwise; $R_2 = 1$, Catholic, 0 otherwise; $R_1 = R_2 = 0$, Jewish), and political party affiliation ($P_1 = 1$, Democrat, 0 otherwise; $P_2 = 1$, Republican, 0 otherwise, $P_1 = P_2 = 0$, Independent). The logit model with main effects has prediction equation

$$\text{logit}(\hat{\pi}) = .11 + .16G - .57R_1 - .66R_2 + .47P_1 - 1.67P_2$$

a) Give the effect of gender on the odds of supporting legalized abortion; that is, if the odds of support for females equal θ times the odds of support for males, report $\hat{\theta}$.
b) Give the effect of being Democrat instead of Independent on the estimated odds of support for legalized abortion.
c) Give the effect of being Democrat instead of Republican on the estimated odds of support for legalized abortion.
d) Find the estimated probability of supporting legalized abortion, for (i) female Jewish Democrats, (ii) male Catholic Republicans.
e) Show that there are 18 sample logits and that $df = 12$ for testing the fit of this model. If $G^2 = 10.4$, test the model goodness of fit, and interpret.
f) Let A denote the opinion on current laws legalizing abortion. State the symbol for the loglinear model that is equivalent to this logit model.
g) State the logit model corresponding to the loglinear model (AR, AP, GRP).
h) The logit model in (g) has $G^2 = 11.6$ for testing fit. Determine whether gender is a significant predictor of opinion on abortion, controlling for R and P.

15. Consider a four-way cross-classification of variables W, X, Y, and Z.
a) State the symbol for the loglinear model in which
i) All pairs of variables are independent.
ii) X and Y are associated but other pairs of variables are independent.
iii) All pairs of variables are associated, but there is no interaction.
b) Suppose all variables are binary, and Y is the response variable. Write down the logit model that is equivalent to the loglinear model symbolized by (WXZ, YX, YZ).

16. The hypothetical population cell proportions in Table 15.20 describe the relationship among income, gender, and college of employment for faculty at Normal State U.
a) Calculate the conditional odds ratios for each pair of variables, and interpret.
b) Treating income as the response, which logit model do these data satisfy? Why?
c) Which loglinear model do these cell proportions satisfy? Why?
d) Construct the two-way income-gender table, collapsing over college. Calculate the odds ratio. Why is it so different from the conditional odds ratios for income and gender?

17. Refer to the loglinear model analyses reported in Section 15.4.
a) Use software to conduct the analyses.

TABLE 15.20

College	Gender	Income Low	Income High
Liberal Arts	Women	.18	.12
	Men	.12	.08
Professional	Women	.02	.08
	Men	.08	.32

b) Based on the table of parameter estimates (Table 15.9), construct a 95% confidence interval for the conditional odds ratio between cigarette use and marijuana use. Interpret.

18. Table 15.21 refers to individuals who applied for admission into graduate school at the University of California in Berkeley, for the fall 1973 session. Data are presented for five of the six largest graduate departments at the university. The variables for the $2 \times 2 \times 5$ table are denoted by

A: Whether admitted (yes, no)
G: Gender of applicant (male, female)
D: Department to which application was sent (D_1, D_2, D_3, D_4, D_5)

TABLE 15.21

Department	Gender	Admitted Yes	Admitted No
D_1	M	353	207
	F	17	8
D_2	M	120	205
	F	202	391
D_3	M	138	279
	F	131	244
D_4	M	53	138
	F	94	299
D_5	M	22	351
	F	24	317

Source: D. Freedman, R. Pisani, and R. Purves, *Statistics* (New York: W. W. Norton, 1978), p. 14.

a) Construct the two-way table for gender and whether admitted, collapsing the table over department. Find the odds ratio and interpret. Based on your interpretation, for which gender is the probability of admission higher?
b) Treating A as the response and D and G as qualitative predictors, fit the logit model having main effects. Report the prediction equation. Interpret the coefficient of G in this equation.
c) Contrast the model estimated odds ratio between A and G, controlling for D, with the odds ratio reported in (a). Explain why they differ so much.
d) Conduct the Wald test for the effect of G. Interpret.
e) Test the goodness of fit of this model. Interpret.

f) Fit the logit model for which A is independent of G, controlling for D. For this model, report the estimated odds ratio between A and G, controlling for D. Would you get the same value if you ignored, rather than controlled, department? Explain.
g) Test the goodness of fit of the model in (f), and interpret the result.
h) Does the model in (f) provide a poorer fit than the one in (b)? Explain.
i) Fit the loglinear model that is equivalent to the logit model in (f), and show how the result of the goodness-of-fit test compares to that of the logit model.
j) Using the expected frequencies for this loglinear model, calculate the estimated odds ratio between A and G for the first department, and interpret.
k) Fit the loglinear model that is equivalent to the logit model in (b), and show how the result of the goodness-of-fit test compares to that of the logit model.
l) Using the expected frequencies for this loglinear model, calculate the estimated odds ratio between A and G for the first department. Interpret.
m) In summary, how would you interpret these data? In your explanation, be sure to explain why the two-way AG association in (a) is so different from the AG conditional association for the models that fit well.

19. Refer to Table 8.17. Treating job satisfaction as the response variable, Table 15.22 shows results of a SAS printout for the cumulative logit model, using scores (3, 15, 30) for income, and chi-squared test of independence and ordinal measure (gamma, Kendall's tau-b) analyses of Chapter 8.

TABLE 15.22

Criterion	Intercept Only	Intercept and Covariates	Chi-Square for Covariates
-2 LOG L	195.712	192.176	3.536 with 1 DF (p=0.0600)

Variable	DF	Parameter Estimate	Standard Error	Wald Chi-Square	Pr > Chi-Square
INTERCP1	1	-0.9513	0.4078	5.442	0.0197
INTERCP2	1	1.9534	0.4518	18.694	0.0001
INCOME	1	-0.0414	0.0220	3.536	0.0601

Statistic	DF	Value	Prob
Chi-Square	4	4.094	0.393
Likelihood Ratio Chi-Square	4	3.877	0.423

Statistic	Value	ASE
Gamma	0.287	0.151
Kendall's Tau-b	0.168	0.090

a) Report and interpret the income effect for the ordinal model.
b) Using the model, test the hypothesis of no income effect. Report the test statistic and P-value, and interpret.
c) Suppose one ignored the ordinal nature of these variables and used the Pearson chi-squared test of independence. Report a test statistic and P-value, and compare results to those in (b). Why are they so different?

d) Show that, for testing independence, one would get similar results to the model using a test based on a measure of association that also uses the ordering information.

20. Refer to Table 8.32 and Problem 8.27. Treat detection of breast cancer as the response variable.
a) Fit the cumulative logit model, using scores (1, 2, 3) for mammography experience. Report and interpret the effect.
b) Test the hypothesis of independence, by testing that a parameter in this model equals 0. Report the test statistic and P-value, and interpret.
c) The choice of scores for these rows is unclear. To check sensitivity of the results to this choice, repeat the analyses in (a) and (b) using scores (i) (1, 2.5, 3), (ii) (1, 1.5, 3). Which of the three sets seems most sensible to you? Why?

21. Refer to Problem 15.12. Now, using all five response categories, fit a cumulative logit model with main effects.
a) Report the estimates of the effects of the three predictors, and indicate which category of each predictor tends to be associated with more serious injuries.
b) Test the hypothesis of no seat belt effect. Interpret.
c) Interpret the seat belt estimate, using an odds ratio.
d) Construct a 95% confidence interval for the true odds ratio describing the seat belt effect.

Concepts and Applications

22. Refer to the WWW data set (Problem 1.7). Using computer software, conduct and interpret a logistic regression analysis using Y = opinion about abortion and predictors
a) political ideology.
b) gender and whether support affirmative action.
c) gender and political ideology.

23. Refer to the WWW data set (Problem 1.7). Build a model for predicting whether one supports affirmative action. Prepare a short report, explaining how you built the model and how you interpret results from that model.

24. Refer to the data file you created in Problem 1.7. For variables chosen by your instructor, fit a logistic regression model and conduct descriptive and inferential statistical analyses. Interpret and summarize your findings.

25. Refer to the previous exercise. For variables chosen by your instructor, conduct a loglinear model analysis. Interpret and summarize your findings.

26. The data given in Problem 10.8 in Chapter 10 came from an early study on the application of the death penalty in Florida. Analyze those data using methods of this chapter.

27. The U.S. National Collegiate Athletic Association (NCAA) conducted a study of graduation rates for student athletes who were freshmen during the 1984-85 academic year. Table 15.23 shows the data. Analyze. In your report, interpret the model, explain the results of statistical inference and checks of the adequacy of the model, and summarize your conclusions.

28. According to the *Independent* newspaper (London, March 8, 1994), the Metropolitan Police in London reported 30,475 people as missing in the year ending March 1993. For those of age 13 or less, 33 of 3271 missing males and 38 of 2486 missing females were still missing a year later. For ages 14–18, the values were 63 of 7256 males and 108 of 8877 females; for ages 19 and above, the values were 157 of 5065 males and 159 of 3520

TABLE 15.23

Group	Sample Size	Graduates
White females	796	498
White males	1625	878
Black females	143	54
Black males	660	197

Source: J. J. McArdle and F. Hamagami, *Journal of the American Statistical Association*, Vol. 89 (1994), pp. 1107–1123.

females. Analyze and interpret these data. (Thanks to Dr. P.M.E. Altham for showing me these data.)

29. In a study designed to evaluate whether an educational program makes sexually active adolescents more likely to obtain condoms, adolescents were randomly assigned to two experimental groups. The educational program, involving a lecture and videotape about transmission of the HIV virus, was provided to one group but not the other. In logistic regression models, factors observed to influence a teenager to obtain condoms were gender, socioeconomic status, lifetime number of partners, and the experimental group. Table 15.24 summarizes the study results.
a) Interpret the odds ratio and the related confidence interval for the effect of group.
b) Find the parameter estimates for the fitted model, using (1, 0) dummy variables for the first three predictors.
c) Explain why either the estimate of 1.38 for the odds ratio for gender or the corresponding confidence interval seems incorrect. The confidence interval is based on taking antilogs of endpoints of a confidence interval for the log odds ratio. Show that if the reported confidence interval is correct, then 1.38 is actually the *log* odds ratio, and the estimated odds ratio equals 3.98.

TABLE 15.24

Variables	Odds Ratio	95% Confidence Interval
Group (education versus none)	4.04	(1.17, 13.9)
Gender (males versus females)	1.38	(1.23, 12.88)
SES (high versus low)	5.82	(1.87, 18.28)
Lifetime number of Partners	3.22	(1.08, 11.31)

Source: V. I. Rickert et al., *Clinical Pediatrics*, Vol. 31 (1992), pp. 205–210.

30. Using samples taken in the late 1980s by the Institute of Social Research in Zagreb, a recent study used logistic regression to model the probability that a subject identified themselves as Yugoslav, when given the option of choosing that category or Croat, Serb, Moslem, or some other nationality. Table 15.25 shows estimates for the predictors: urbanism of residence (1 = village, 2 = town, 3 = city), membership at some time in the Communist Party (1 = yes, 0 = no), age, nationally mixed parentage (1 = yes, 0 = no). Explain how to interpret results in this table.

31. Table 15.26 displays data from the 1987-88 National Survey of Families and Households. The sample consists of currently married couples, married for 20 or fewer years, among

TABLE 15.25

Variable	Estimate	Std. Error
Urbanism	.463	.056
Communist party	.472	.096
Age	−.034	.004
Mixed parentage	2.449	.115

Source: D. Sekulic et al., *American Sociological Review*, Vol. 59 (1994), pp. 83–97. (1994).

whom both partners are in a first marriage. The response variable is the couple's perception of the stability of the marriage, with categories (VU = very unstable, SU = somewhat unstable, S = stable), based on their responses to several questions. Predictors are whether the couple cohabited before marriage, wife's age at marriage, and duration of the marriage. Analyze these data.

TABLE 15.26

Prior Co-	Wife's Age	Married < 5 yrs			Married 5–10 yrs			Married > 10 yrs		
habitation	at Marriage	VU	SU	S	VU	SU	S	VU	SU	S
No	< 19	26	13	40	21	9	41	37	25	116
	19–23	44	35	126	28	24	110	88	43	242
	23+	31	29	126	23	21	83	29	18	87
Yes	< 19	20	3	8	7	2	7	2	2	3
	19–23	34	19	50	18	11	27	10	6	24
	23+	39	21	101	17	9	35	6	1	15

Source: A. DeMaris, *Psychological Bulletin*, Vol. 110 (1991), p. 557.

32. Table 15.27 is a contingency table summarizing observations on the degree of participation in school sports for random samples of 30 boys and 30 girls in middle schools.
a) Use chi-squared (Section 8.2) to test whether participation is independent of gender. Report the P-value.
b) Now, conduct a test using an ordinal analysis, and report the P-value.
c) Compare the results of parts (a) and (b), and discuss why it is generally inappropriate to use the chi-squared test if one or both variables are ordinal.

TABLE 15.27

	Participation		
	Low	Moderate	High
Boys	7	10	13
Girls	13	10	7

33. Refer to Table 8.39 and Problem 8.39. Using a cumulative logit model, analyze and interpret these data.

34. Refer to Problem 8.28. Using a model, analyze these data, compare results to the Pearson chi-squared test of independence, and explain the discrepancy.

35. Refer to Table 15.4. Show that the association between the defendant's race and the death penalty verdict satisfies Simpson's paradox. What causes this?

36. One reason logistic regression is usually preferred over the linear probability model is that a fixed change in X often has a smaller impact on π when π is near 0 or near 1 than when π is near the middle of its range. Let Y refer to the decision to rent or to buy a home, with π = the probability of buying, and let X = weekly family income. In which case do you think an increase of \$100 in X has greater effect: when $X = \$50{,}000$ (for which π is near 1) or when $X = \$500$? Explain how your answer relates to the choice of a linear versus logistic regression model.

37. * For the logistic regression model, from the linear approximation $\beta/4$ for the rate of change in the probability at $\pi = .5$, show that $1/|\beta|$ is the approximate distance between the X-values at which $\pi = 1/4$ (or $\pi = 3/4$) and at which $\pi = 1/2$. Thus, the larger the value of $|\beta|$, the less the X-distance over which this change in probability occurs.

38. State the symbols for the loglinear models for categorical variables that are implied by the causal diagrams in Figure 15.5.

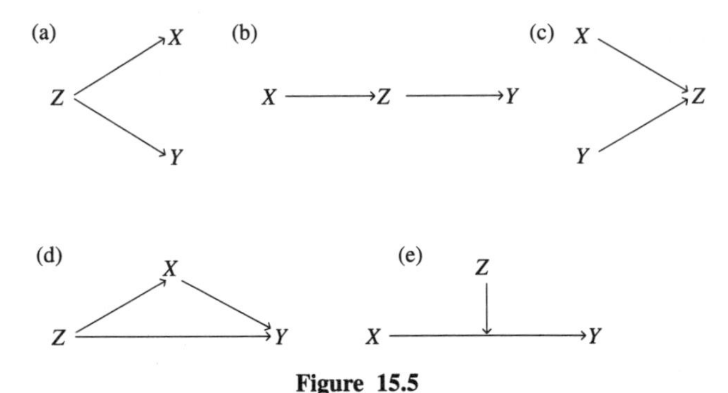

Figure 15.5

39. * For a two-way contingency table, let r_i denote the ith row total, let c_j denote the jth column total, and let n denote the total sample size. Section 8.2 showed that the cell in row i and column j has $f_e = r_i c_j / n$ for the independence model. Show that the log of the expected frequency has an additive formula with terms representing the influence of the ith row total, the jth column total, and the sample size. This formula is the loglinear model for independence in two-way contingency tables.

40. * For a 2×2 table with cell counts a, b, c, d, the sample log odds ratio $\log \hat{\theta}$ has approximately a normal sampling distribution with standard error

$$\hat{\sigma}(\log \hat{\theta}) = \sqrt{\frac{1}{a + .5} + \frac{1}{b + .5} + \frac{1}{c + .5} + \frac{1}{d + .5}}$$

The antilogs of the endpoints of the confidence interval for $\log(\theta)$ are endpoints of the confidence interval for θ. Construct a 95% confidence interval for the odds ratio for Table 8.15 in Section 8.4. Interpret.

41. * Explain what is meant by the absence of statistical interaction in modeling the relationship between a response variable Y and two explanatory variables X_1 and X_2 in each of the following cases. Use graphs or tables to illustrate.

a) Y, X_1, and X_2 are quantitative.

b) Y and X_1 are quantitative; X_2 is qualitative.

c) Y is quantitative; X_1 and X_2 are qualitative.
d) Y, X_1, and X_2 are binary.
e) Y is binary, X_1 and X_2 are quantitative.

Bibliography

Agresti, A. (1996). *An Introduction to Categorical Data Analysis.* New York: Wiley.

Clogg, C. C., and Shihadeh, E. S. (1994). *Statistical Models for Ordinal Variables.* Thousand Oaks, CA: Sage.

DeMaris, A. (1992). *Logit Modeling: Practical Applications.* Beverly Hills, CA: Sage.

Fienberg, S. E. (1980). *The Analysis of Cross-Classified Categorical Data*, 2nd ed. Cambridge, MA: MIT Press.

Goodman, L. A. (1978). *Analyzing Qualitative/Categorical Data: Log-Linear Analysis and Latent Structure Analysis.* Cambridge, Mass.: Abt.

Hosmer, D., and Lemeshow, S. (1989). *Applied Logistic Regression.* New York: Wiley.

Wickens, T. D. (1989). *Multiway Contingency Table Analysis for the Social Sciences.* Hillsdale, NJ: Lawrence Erlbaum.

Chapter 16

An Introduction to Advanced Methodology

This final chapter introduces some advanced statistical methods. An introductory text such as this does not have space to present them in detail. However, a social science researcher is likely to see reference to these methods, and it is helpful to have at least a rudimentary understanding of their nature and purposes.

The emphasis here differs from the rest of the text. Rather than presenting technical details about a method, the text provides an explanation of (1) what it is used for, and (2) the types of results that can occur and their interpretations. The methods described are ***event history analysis***, ***path analysis***, ***factor analysis***, ***structural equation models***, and ***Markov chain modeling***.

16.1 Event History Models

Some studies have the objective of studying how long it takes until a certain type of event occurs. For instance, one might model how long a person works before retiring from the work force, the age of a person when first marrying, the length of time before someone just released from prison is rearrested, or how long a person lives after being diagnosed with AIDS.

As in ordinary regression, models for the time to some event include effects of explanatory variables on the response. A model for the length of time before rearrest, for instance, might use predictors such as number of previous arrests, whether employed, marital status, age at the time of release, and educational level.

An analysis of events occurring over time using a set of explanatory variables is called ***event history analysis***. Most of the work on developing models for event history analysis has taken place in biostatistics, for modeling the length of time that a patient survives after undergoing a particular medical treatment, for instance, the amount of time before death for a patient who has undergone a heart transplant. In this context, the analysis is called ***survival analysis***.

Censored Data and Time-Varying Covariates

In event history analyses, the data for each subject consists of an observation of the length of time until the event of interest occurs. Two complicating factors occur, however, that do not arise in ordinary regression modeling.

First, for some subjects, the event has not yet occurred by the end of the observation period for the study. One cannot observe the actual time to the event for those subjects, but instead only lower bounds on those times. For instance, a study of retirement ages may use a sample of adults aged at least 65. Some subjects in this sample may not have retired yet. If a person has been working for 45 years but has not yet retired, we know only that the time spent working is at least 45 years. A study of recidivism might use a sample of people released from prison in 1996. Some of these subjects may not have been rearrested between 1996 and the end of the period of the study. If a subject has been out of jail for three years without being rearrested, we know only that the time to rearrest is greater than three years. Possibly the subject will never be arrested again, but that is unknown.

Observations of this type are said to be ***censored***. Methods for event history analysis have special ways of handling censored data. Ignoring censored data and fitting models only using data for subjects having fully observed responses can result in a severe bias in parameter estimates.

Second, some explanatory variables for predicting the time to the event may change value over time. For instance, the study of criminal recidivism may observe each month whether each subject has been rearrested (the event of interest) and use explanatory variables such as whether the subject is working and whether the subject is married or living with a partner. For a particular subject, the value of explanatory variables of this type could vary over time. An explanatory variable that can vary over time is called a ***time-varying covariate***. Advanced methods for fitting event history models can handle both time-varying covariates and covariates that are time independent.

Proportional Hazards Model

Instead of modeling the *length of time* until a particular event occurs, one can equivalently model the ***rate*** of occurrence of the event. Consider, for instance, a study about health problems of subjects admitted to a nursing home. The response is the length of time after admission before a subject requires special medical care, such as a visit to a physician or admission to a hospital. At a particular setting for the explanatory variables, the sample contains five subjects. The time until requiring special medical care equals .5 years for the first subject, .2 years for the second, 1.3 years for the third, and .1 year for the fourth. The fifth subject is a censored observation, not requiring any special medical care during the .4 years she had been in the home when the observation period for the study ended. Then, for these five subjects, the total number of occurrences of the event of interest is 4, and the total observation time is $(.5 + .2 + 1.3 + .1 + .4) = 2.5$ years. The sample rate of occurrence is $4/2.5 = 1.6$. That is, the rate of occurrence equals 1.6 events per year of observation time. The sample number of events equals 1.6 times the total amount of time for which the entire sample of subjects were under observation.

In fact, for the most popular model for event history data, the model formula refers to the rate of occurrence of the event rather than to the time elapsed before the event occurs. In the literature, the rate is usually called the ***hazard rate*** and denoted by h. The above calculation for a sample hazard rate implicitly assumes that this rate is constant over time. In practice, this is often not realistic, and models can allow the hazard rate to depend on time as well as on values of explanatory variables.

Let $h(t)$ denote the hazard rate at time t, for instance, t years after admission to the nursing home. The standard model for the hazard rate and a set of k explanatory variables has the form

$$\log h(t) = \alpha(t) + \beta_1 X_1 + \beta_2 X_2 + \cdots + \beta_k X_k$$

The model applies to the log of the hazard rate, since the hazard must be positive; linear functions for the hazard itself have the disadvantage that they might provide negative predicted values, much like the linear probability model for a binary response variable. The intercept parameter $\alpha(t)$ in the model is allowed to depend on time t. This permits the hazard rate itself to be time varying. In the form of the model written here, the values of the explanatory variables do not depend on time, but the model can be generalized to allow this.

In typical applications, the primary focus is on estimating the effects of the explanatory variables, not on modeling the dependence of the hazard on time. The geriatric study, for instance, would want to estimate the effects on the hazard rate of covariates such as gender and age of subject. For this reason, it is common to allow $\alpha(t)$ to be an arbitrary, unspecified function and to concentrate the analysis on estimating the $\{\beta_j\}$ as a way of making inferences about the covariate effects.

We next discuss how to interpret the covariate effects in this model. Consider the effect of X_1 by studying the effect on the hazard rate of increasing it by 1 unit while holding the other predictors constant. Denote the hazard rates at the original value of X_1 and at $X_1 + 1$ by h_1 and h_2. Then,

$$\log h_2(t) - \log h_1(t) = [\alpha(t) + \beta_1(X_1 + 1) + \cdots + \beta_k X_k] - [\alpha(t) + \beta_1 X_1 + \cdots + \beta_k X_k] = \beta_1$$

So, β_1 is the change in the log hazard for a one-unit change in X_1, holding the other predictors fixed. But, $\log h_2(t) - \log h_1(t) = \log[h_2(t)/h_1(t)]$, and exponentiating both sides, $h_2(t)/h_1(t) = e^{\beta_1}$, or

$$h_2(t) = e^{\beta_1} h_1(t)$$

That is, increasing X_1 by 1 unit has the effect of multiplying the hazard rate by e^{β_1}. Effects are multiplicative, as in other models using the log to achieve linearity, such as exponential regression and logistic regression models.

The equation $h_2(t) = e^{\beta_1} h_1(t)$ illustrates that the hazard rate at one setting of explanatory variables is proportional to the hazard rate at another setting, with the same proportionality constant regardless of the time t. Because of this property, this form of model is called a ***proportional hazards model***. It is simple to interpret effects for such models, since the effect of any explanatory variable on the hazard rate is identical at all times.

In 1972 the British statistician Sir David Cox proposed and showed how to fit this specific form of proportional hazards model in which the dependence of the hazard on time (through $\alpha(t)$) is arbitrary. The model is called the *Cox proportional hazards model*. It is nonparametric, in the sense that it makes no assumption about the probability distribution of the time to the event but instead focuses on the effects of the explanatory variables. More specialized models of proportional hazards form (or other forms) make parametric assumptions about this distribution. This is useful if the distribution of the time to the event is an important focus of the study, as it is, for example, in analyzing time to failure of electronic components. More general forms of the model permit time-varying covariates.

Example 16.1 Modeling Time to Marital Dissolution

An article on modeling family dynamics with event history techniques (T. B. Heaton and V. R. A. Call, *Journal of Marriage & Family*, Vol. 57, 1995, p. 1078) analyzed data from the National Survey of Families and Households. A national probability sample of about 13,000 subjects were interviewed in 1987–1988, and then a follow-up survey interviewed 10,000 of these subjects approximately six years later. The analysis models factors that affect the hazard rate for marital separation. The response outcome for each subject married at the beginning of the study is the number of months from then until the couple separates. People who are still in their marriage or are widowed at the study conclusion are treated as censored observations.

The authors summarized the fit of the model using Table 16.1. The final column of the table shows the exponentiated estimates of the regression parameters, which provide the hazard rate ratios. For instance, the authors report that the dissolution rate for blacks is 1.42 times the rate for whites. They obtain this by noting that $e^{.353} = 1.42$. This is the strongest of the effects shown in the table.

As in logistic regression, to test significance of model parameters one can either conduct Wald tests or likelihood-ratio tests. For instance, to test the effect of gender, we calculate $z = -.065/.0375 = -1.73$, which has a two-tail P-value of .083. Equivalently, the square of this statistic is a chi-squared statistic with $df = 1$. □

TABLE 16.1 Estimated Effects on Hazard Rate Regarding Marital Dissolution, for Cox Proportional Hazards Model

Variable	Estimate	Std. Error	P-Value	e^b
Age at marriage	−.086	.0050	.000	.917
Year married	.048	.0017	.000	1.049
Race (black = 1)	.353	.0423	.000	1.423
Gender (male = 1)	−.065	.0375	.083	.937

Source: T. B. Heaton and V. R. A. Call, *Journal of Marriage & Family*, Vol. 57 (1995), p. 1078.

Several software programs now exist for fitting event history models, such as PROC PHREG in SAS and the COXREG option in SPSS. See Allison (1984) and Yamaguchi (1991) for more detailed introductions to this topic.

16.2 Path Analysis

Path analysis uses regression models to test theories of causal relationships among a set of variables. Statistically, it consists simply of a series of regression analyses, but there are advantages to conducting the analyses within the path analytic framework. The primary advantage is that the researcher must specify explicitly the presumed causal relationships among the variables. This can help contribute to logically clear theories of variable relationships.

Thus, path analysis searches not only for associations among variables, but also for causal relationships. Association is one characteristic of a cause-effect relationship, but as Section 10.1 showed, it is not sufficient to imply causation. Two variables that are both causally dependent on a third one may themselves be associated. Neither is a cause of the other, however, and the association disappears when the third variable is controlled. It is important in causal modeling to develop theories that include appropriate control variables.

An explanatory variable X is a possible cause of a response variable Y if the proper time order occurs and if changes in X give rise to changes in Y, even when all relevant variables are controlled. If the association between two variables disappears under a control, a direct causal relationship does not exist between them. If the association does not disappear, though, the relationship is not necessarily causal, since the association could disappear when other variables are controlled. Thus, we can prove noncausality but we can never prove causality. A hypothesis of a causal relationship is bolstered, though, if the association remains after controls are introduced.

Path Diagrams

Theoretical explanations of cause-effect relationships often hypothesize a system of relationships in which some variables, believed to be caused by others, may in turn have effects on yet other variables. A single multiple regression model is insufficient for that system, since it can handle only a single response variable. Path analysis utilizes the number of regression models necessary to include all proposed relationships in the theoretical explanation.

For example, suppose one's theory specifies that educational attainment depends on several factors, in particular upon motivation to achieve and parent's income level. One might hypothesize, in addition, that motivation to achieve depends on several other factors—among them, general intelligence level and the parent's educational level—that the income of the parent depends in part on the parent's educational level, and that educational attainment may also depend directly on intelligence. Figure 16.1 shows a graphic summary of the theory just outlined, in the form of a ***path diagram***. These diagrams generalize the causal diagrams introduced in Chapter 10.

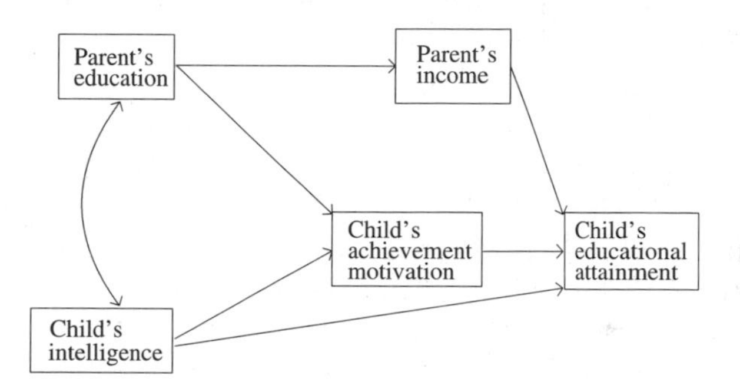

Figure 16.1 Example of Preliminary Path Diagram for Educational Attainment

Figure 16.1 displays some basic conventions of path diagrams. First, a cause-effect relationship is represented by a straight arrow pointing toward the effect (response) variable and leading from the causal (explanatory) variable. In other words, the response variables of the regression equations are the variables to which the arrows point, whereas the explanatory variables for an equation with a particular response variable are those variables with arrows pointing toward that response variable. Thus, parent's income is modeled as depending on parent's education; child's educational attainment as depending on parent's income, child's intelligence, and child's achievement motivation; and, finally, child's achievement motivation as depending on parent's education and child's intelligence. The curved line between parent's education and child's intelligence, with arrows in both directions, means that the two variables may be associated, but their causal relationship (if any) is not addressed in the model.

Path Coefficients

Ordinarily in a path diagram, each arrow has a number written over it. These numbers, called ***path coefficients***, are standardized regression coefficients for the regression equation for the response variable to which the arrows point. Figure 16.1 has three sets of coefficients to be estimated, since it refers to three separate response variables.

Denote the standardized versions of the variables in this figure by E, A, and I for child's educational attainment, achievement motivation, and intelligence, and by Pe and Pi for parent's education and income. Also, for two variables X and Y, let β^*_{YX} denote the standardized regression coefficient for the effect of X on Y. Then, Figure 16.1 corresponds to the three regression equations,

$$(1)\ E(E) = \beta^*_{EI} I + \beta^*_{EA} A + \beta^*_{EPi} Pi$$

$$(2)\ E(A) = \beta^*_{AI} I + \beta^*_{APe} Pe$$

$$(3)\ E(Pi) = \beta^*_{PiPe} Pe$$

For example, the coefficient of the path leading from parent's education to child's achievement motivation is the estimate of the standardized regression coefficient β^*_{APe} from the multiple regression model (2) having child's achievement motivation as the response variable and parent's education and child's intelligence as the explanatory variables. Parent's income, in this model, depends only on parent's education [see (3)]; the path coefficient for that arrow is the standardized bivariate regression coefficient, which is the Pearson correlation.

The path coefficients show the relative strength of association between variables, controlling for other variables in the sequence, and the sign of the influence. Their interpretation is that of standardized regression coefficients in multiple regression (Section 11.8): For instance, a value of .4 means that a 1 standard deviation increase in the explanatory variable corresponds to a .4 standard deviation expected increase in the response variable, controlling for the other explanatory variables in that model.

Every response variable has a ***residual variable path*** attached to it in the path diagram. This represents the variation unexplained by its explanatory variables. Each residual variable represents the remaining portion $(1 - R^2)$ of the unexplained variation, where R^2 is the coefficient of multiple determination for the regression equation for that response variable. Its path coefficient equals $\sqrt{1 - R^2}$.

Most path models have variables that are dependent on some other variables but are, in turn, causes of other response variables. These variables are ***intervening variables*** (Section 10.3), since they occur in sequence between other variables. In Figure 16.1, child's achievement motivation intervenes between child's intelligence and child's educational attainment. If this causal theory is correct, child's intelligence affects his or her educational attainment in part through its effect on achievement motivation. An effect of this type, operating through an intervening variable, is said to be ***indirect***. Figure 16.1 also suggests that child's intelligence has a ***direct*** effect on educational attainment, over and above its effect through achievement motivation. An important reason for using path analysis is that it enables one to display the direct and indirect effects of a variable on another.

On the other hand, Figure 16.1 suggests that parent's education does not have a direct effect on child's educational attainment; it affects this response only through its effects on parent's income and child's achievement motivation. Thus, if we add parent's education as a predictor to the multiple regression model (1) for response E, its effect should not be significant when parent's income and child's achievement motivation are also in the model.

The regression analyses conducted as part of the path analysis reveal whether significant evidence exists of the various effects. If intelligence affects educational attainment directly, as well as indirectly through its effect on motivation, then all three coefficients of parts of paths leading from intelligence to educational attainment should be significant. The direct effect would be verified by a significant partial effect for intelligence in the multiple regression model (1) containing intelligence, achievement motivation, and parent's income as predictors of educational attainment. The indirect effect would be verified by a significant partial effect for achievement motivation in that model, and a significant partial effect for intelligence on achievement motivation in the multiple regression model (2) also containing parent's education as a predictor for achievement motivation.

In conducting the regression analyses, if we find a nonsignificant path, we can erase that path from the diagram and perform the appropriate analyses again to reestimate the coefficients of the remaining paths. For small samples, though, keep in mind that a sample effect may not be significant even if it is of sufficient size to be noteworthy. To conduct a sophisticated path analysis analyzing several direct and indirect associations with any degree of precision requires a large sample size.

Example 16.2 Completed Path Diagram for Child's Educational Attainment

Figure 16.2 shows the original path diagram from Figure 16.1 with the path coefficients added. This is the typical form for the path diagram. The residual variables for the three response variables are denoted by R_1, R_2, and R_3. If 28% of the child's educational attainment were explained by its three predictors, for example, then the path coefficient of the residual variable R_1 for the child's educational attainment would be $\sqrt{1 - R^2} = \sqrt{1 - .28} = .85$.

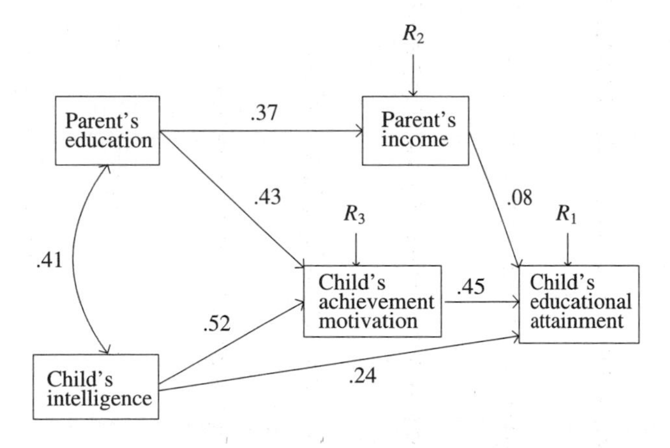

Figure 16.2 Path Diagram for Child's Educational Attainment, with Path Coefficients Added

Figure 16.2 suggests that of the three direct predictors of child's educational attainment, the child's achievement motivation had the strongest partial effect. The child's intelligence has a moderate indirect effect, through increasing achievement motivation, as well as a direct effect. Parent's income is not as directly important as the child's achievement motivation or intelligence, but the parent's educational level has an important effect on the child's achievement motivation. Such conclusions are highly tentative if the path coefficients have substantial sampling error. □

Path Decompositions

A causal diagram is a hypothetical way of accounting for correlations between variables. One of the fundamental results of path analysis decomposes the correlation between two variables in terms of component parts dealing with the various paths between those two variables.

It is easiest to illustrate this idea using a simple causal diagram. For three variables, consider the model of a chain relationship, introduced in Section 10.3. Specifically,

$$X \longrightarrow Z \longrightarrow Y$$

According to this model, the correlation between X and Y is explained by the intervening variable, Z. Controlling for Z, that association should disappear.

The partial correlation (Section 11.7)

$$\rho_{XY \cdot Z} = \frac{\rho_{XY} - \rho_{ZX}\rho_{ZY}}{\sqrt{\left(1 - \rho_{ZX}^2\right)\left(1 - \rho_{ZY}^2\right)}}$$

measures the association between X and Y, controlling for Z. For the partial correlation $\rho_{XY \cdot Z}$ to equal 0, it is necessary that

$$\rho_{XY} = \rho_{ZX}\rho_{ZY}$$

That is the correlation between X and Y decomposes into the correlation between the intervening variable and X times the correlation between the intervening variable and Y.

A generalization of this formula holds for more complex path diagrams. Specifically, let $\beta_{Z_i X}^*$ denote the path coefficient for the model in which Z_i is a response variable and X is a predictor for it. Suppose that the Z_i also serve as predictors of Y in a separate model. Then the correlation between X and Y decomposes into

$$\rho_{XY} = \sum \beta_{Z_i X}^* \rho_{Z_i Y}$$

where the sum is over all variables Z_i that have a direct path to Y.

The simpler expression given for the chain relationship above is a special case of this with only one Z_i variable, namely, Z. In that case, since X is the only variable in the model predicting Z, the path coefficient of X on Z is simply the Pearson correlation between them.

How is this general decomposition useful? The equation predicts what the correlation *should* be if the causal diagram is correct. For sample data, one can compute the correlation predicted by this formula by substituting the sample estimates into the right hand side. One can compare this predicted correlation to the actual sample correlation. If the difference between the two cannot be explained by sampling error, then the results refute the causal hypothesis that the diagram represents.

For the chain model, for instance, r_{XY} should be close to $r_{ZX}r_{ZY}$; that is, the partial correlation $r_{XY \cdot Z}$ should be close to zero. One can test the hypothesis H_0: $\rho_{XY \cdot Z} = 0$, using the usual t test (Section 11.7) as a way of testing the model.

In summary, the basic steps in a path analysis are as follows:

1. Set up a preliminary theory to be tested, drawing the path diagram without the path coefficients.
2. Conduct the necessary regression modeling to estimate the path coefficients and the residual coefficients.
3. Evaluate the model, checking whether the sample results agree with it. Then reformulate the model, possibly erasing nonsignificant paths. The revised model may be the basis for future research, in which one fits models for the amended diagram and reestimates path coefficients for that diagram.

A Caveat About Causal Modeling

We must add an important caveat here, one that affects path analysis and causal diagrams in general. For the path analysis decomposition formula to hold, one must assume that the unmeasured variables that represent the residual variation for each response variable are uncorrelated with the predictors in the regression model for that response. In Figures 16.1 and 16.2, for instance, all the other variables that affect child's educational attainment are assumed to be uncorrelated with parent's income, child's achievement motivation, and child's intelligence. This seems sure to be an incorrect assumption.

The real world that social scientists study is never as simple as a causal diagram can portray. In any particular problem, we must regard such a diagram as a very crude approximation for reality. The true diagram would be highly complex, and a very large number of variables would likely play a role in it, with paths between nearly all pairs.

Incidentally, this same remark applies to any regression model. Parameter estimates in prediction equations refer to the particular system of variables in the model. If we added other explanatory variables that affect the response variable, those estimated effects would change somewhat, since undoubtedly the added variables would be somewhat correlated with the predictors originally in the model. This is a fundamental problem of all social science research. No matter what you report, someone could criticize your findings and argue that different results would occur if you had included other variables in your model.

16.3 Factor Analysis

Factor analysis is a multivariate statistical method used for a wide variety of purposes. These include

1. Revealing patterns of interrelationships among variables.
2. Detecting clusters of variables, each of which contains variables that are strongly intercorrelated and hence somewhat redundant.
3. Reducing a large number of variables to a smaller number of statistically uncorrelated variables, the ***factors*** of factor analysis, that are each linearly related to the original variables.

The third use is helpful for handling a large number of variables, many of which are highly intercorrelated. For example, suppose a multiple regression model has severe multicollinearity, partly due to the large number of predictor variables used to measure each concept of interest. Factor analysis can transform a collection of highly correlated explanatory variables that are indicators of the same type to one or two factors having nearly as much predictive power regarding the response variable. Each factor is an artificial combination of the original variables, however, and how useful this is depends on the interpretability of the factors.

Factor Analytic Model

The model for factor analysis expresses the expected values of a set of observable variables $X_1, \ldots, X_k$ as linear functions of m unobserved variables, called ***factors***. In its simplest form, the process derives the factors so that the correlation equals zero between each pair. The user specifies the number of factors m, which must be less than the number of variables k. The model consists of k equations. Each one expresses one of the original variables in terms of the m factors. Roughly speaking, the model represents replacing the observed variables by a smaller set of uncorrelated factors.

The factors in the factor analysis model are artificial, unobserved variables, sometimes referred to as ***latent variables***. The fitting process also provides equations that express the factors as linear functions of the observed variables. The coefficients in the equation relating a factor to the observed variables are estimated using the sample data, through the correlations between the pairs of variables. These factors are merely convenient summaries of the observed variables. For example, the first factor might relate to standardized versions of seven observed variables by $f_1 = .93X_1 + .78X_2 - .11X_3 + .02X_4 + .14X_5 - .06X_6 - .18X_7$, showing that it primarily summarizes information provided by X_1 and X_2. The factor equations convert values on the k variables for each subject to a smaller set of scores on the m factors.

The Pearson correlation of a variable with a factor is called the ***loading*** of the variable on that factor. A matrix with a row for each variable and a column for each factor summarizes these loadings. The sum of squared loadings for a variable is called the variable's ***communality***. It represents the proportion of its variability that is explained by the factors. The factor loadings are unknown parameters estimated using the sample data.

Fitting the Factor Analysis Model

The researcher selects the number of factors m that are believed adequate to explain the relationships among the observed variables. The researcher can often form a good hunch about the needed number of factors by inspecting the correlation matrix for the observed variables. If different sets of variables cluster, with strong correlations between pairs of variables within each set but small correlations between variables from different sets, then one could let the number of factors equal the number of clusters. In a purely *exploratory* form of factor analysis, part of the reason for the analysis is to search for an appropriate value for m. In a more structured, so called *confirmatory* analysis, one analyzes whether a particular a priori choice m is adequate.

The fitting process assumes that the response variables have a ***multivariate normal*** distribution. In particular, this implies that each individual variable has a normal distribution, and the regression relationship between each pair of variables is linear. In practice, this is unrealistic. There is a tendency for most users to go ahead and use this method regardless of the distributions, but these strong assumptions should make you wary about using the method with highly nonnormal variables (e.g., binary) or without a careful check of the effect of any obvious outliers on the ultimate conclusions.

No unique solution exists for the estimated factor loadings. Most factor analytic procedures estimate them using the observed correlation matrix and initial guesses of the sizes of the communalities. Treating each row of m factor loadings as a point in m-dimensional space, the process then can "rotate" the estimates to obtain more meaningful factors with simpler factor structure. The purpose of the rotation is to bring most loadings of a variable close to 0, so that each variable is highly correlated with only one or two factors. This makes it easier to interpret each factor as representing the effects of a particular subset of variables.

Often, one factor is strongly related to all the variables. Ideally, after rotation, the structure of the factor loadings might appear as shown in Table 16.2. Entries of 0 in that table represent factor loadings that are not significantly different from zero. The first factor is associated with all the variables, the second factor provides information contained in X_1, X_2, X_3, and the third factor provides information contained in X_4, X_5, X_6.

TABLE 16.2 Simple Structure for Factor Loadings of Seven Variables on Three Factors

		Factor		
		1	2	3
	1	*	*	0
	2	*	*	0
	3	*	*	0
Variable	4	*	0	*
	5	*	0	*
	6	*	0	*
	7	*	0	0

* Denotes a significantly nonzero loading.

Example 16.3 Factor Analysis of Election Variables

The correlations in Table 16.3, from Harman (1967, pp. 165–166), refer to the following eight variables, measured in an election for 147 districts in Chicago:

1. Percentage vote for Democratic candidate in mayoral election
2. Percentage vote for Democratic candidate in presidential election
3. Percentage of straight party votes
4. Median rental cost
5. Percentage homeownership
6. Percentage unemployed
7. Percentage moved in last year
8. Percentage completed more than ten years of school

TABLE 16.3 Correlation Matrix for Eight Variables Measured for 147 Districts in Chicago Election

		Variable Number							
		1	2	3	4	5	6	7	8
	1	1.0							
	2	.84	1.0						
	3	.62	.84	1.0					
Variable	4	−.53	−.68	−.76	1.0				
Number	5	.03	−.05	.08	−.25	1.0			
	6	.57	.76	.81	−.80	.25	1.0		
	7	−.33	−.35	−.51	.62	−.72	−.58	1.0	
	8	−.66	−.73	−.81	.88	−.36	−.84	.68	1.0

Source: Reprinted from Harman (1967, pp. 165–166) by permission of The University of Chicago Press.

Table 16.3 shows that variables 1, 2, 3, and 6 are highly positively correlated, as are variables 4, 7, and 8. One might hypothesize that two factors can represent these eight variables. Fitting the factor analysis model using the ***principal factor*** solution with two factors yields the estimated factor loadings shown in Table 16.4. The table of loadings has $k = 8$ rows, one for each observed variable, and $m = 2$ columns, one for each factor.

The first factor is said to be ***bipolar***, since it contains high positive and high negative loadings. The positive correlations occur with variables 1, 2, 3, and 6, for which high scores tend to occur in districts with a heavily Democratic vote. One might interpret this factor as a measure of the traditional Democratic vote. Factor 2, which is highly positively correlated with variable 5 and negatively correlated with variable 7, is interpreted as a measure of home permanency. As the score on factor 2 for a district increases, the percentage of homeownership tends to increase, and the percentage of those who have moved in the previous year tends to decrease.

TABLE 16.4 Factor Loadings for a Two-Factor Solution for the Correlations in Table 16.3

		Loadings		
		Factor 1	Factor 2	Communality
	1	.69	−.28	.55
	2	.88	−.48	1.00
	3	.87	−.17	.79
Variable	4	−.88	−.09	.78
Number	5	.28	.65	.50
	6	89	.01	.79
	7	−.66	−.56	.75
	8	−.96	−.15	.94

Figure 16.3 plots the loadings of the variables on the two factors. Each point in Figure 16.3 represents a particular variable. For example, the point labeled 4 has as X coordinate the loading of variable 4 on factor 1 ($-.88$) and as Y coordinate the loading of variable 4 on factor 2 ($-.09$). The plots shows that variables 1, 2, 3, and 6 cluster together, having similar pairs of loadings. Also, variables 4, 7, and 8 cluster together.

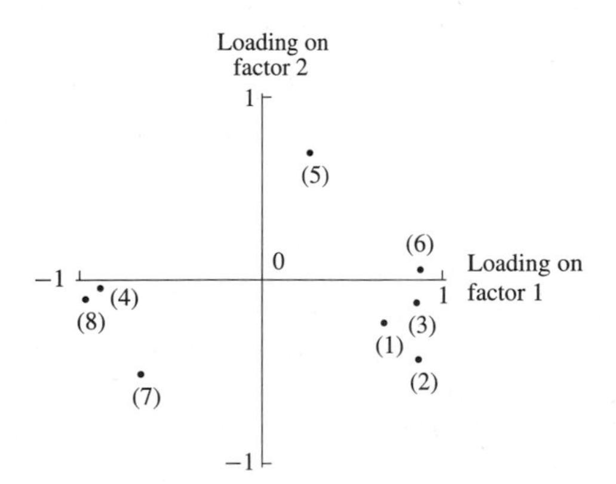

Figure 16.3 Plot of Loadings from Table 16.4 of the Eight Variables on the Two Factors

Further analyses of these data might replace the eight variables with these two factors. They seem to have a clear interpretation. They are uncorrelated, so no redundancies occur when both are used in regression analyses. Also, since the communalities are relatively large, the factors explain most of the variation in the original variables. One can construct two equations that express each factor in terms of the eight variables, and these equations provide scores on the two factors for the 147 districts. □

Origin and Controversy

Factor analysis was originally developed early in the twentieth century by psychometricians, in an attempt to construct a factor or factors measuring intelligence. Charles Spearman postulated the existence of a single factor of "general intelligence." Later, L. L. Thurstone and others hypothesized a set of group factors, each of which could be measured using a battery of tests of similar nature. For an entertaining and highly critical look at the history of this subject and the variety of potential pitfalls in using factor analysis to attempt to measure intelligence, see Gould (1981).

Besides the problems Gould cited with factor analysis, there are statistical dangers in using this method. In any analysis dealing with unmeasured variables such as artificial factors, there is the danger of identifying patterns in a factor loading matrix that suggest certain interpretations for the factors, when actually those patterns are largely due to sampling error. In fact, for a long time, the statistical basis of factor analysis was shaky. It was not possible, for instance, to report valid standard errors for factor loadings. Recently, maximum likelihood methods have been developed that improve on many of the older methods for performing factor analysis. Recently, there has also been emphasis on using factor analysis in more of a confirmatory than exploratory mode, as described in the following section. This forces a researcher to think more carefully about reasonable factor structure before performing the analysis and makes spurious conclusions less likely.

16.4 Structural Equations with Latent Variables

A very general model combines elements of both path analysis and factor analysis. The model is called a ***covariance structure model***, because it attempts to explain the variances and correlations among the set of observed variables. This explanation takes the form of a causal model relating a system of factors, some of which may be created as in factor analysis and some of which may be observed variables.

Covariance structure models consist of two components. The first is the ***measurement model***. It resembles a factor analysis, deriving a set of unobserved factors from the observed variables. The second component is the ***structural equation model***. It resembles a path analysis, specifying regression models for the factors derived in the measurement model.

Measurement Model

The measurement model specifies how the observed variables relate to a set of unobserved factors, called ***latent variables***. This part of the analysis is similar to factor analysis, except that the modeling has more highly specified structure. The measurement model assigns each latent variable, a priori, to a specific set of observed variables. This is accomplished by forcing certain factor loadings to equal 0, so that the latent variables are uncorrelated with other variables.

The measurement model explicitly represents the fact that the observed variables, being subject to measurement error and problems with validity and reliability, are imperfect indicators of the concepts of true interest. For instance, on a survey questionnaire, one might use responses to a battery of items dealing with racist attitudes as crude indicators of racism. Factor analyzing them may produce a single latent variable that is a better general measure of racism than any single item. A purpose of creating latent variables is to operationalize characteristics that are difficult to measure well, such as prejudice, anxiety, and conservatism.

Structural Equation Model

The second part of a covariance structure model, the structural equation model, resembles a path analysis for the latent variables. In other words, the structural equation model uses regression models to specify causal relationships among the latent variables. One or more of the latent variables are identified as response variables, and the others are identified as explanatory variables. The latent response variables can be regressed on themselves as well as on the latent explanatory variables. Unlike ordinary path analysis, this approach allows the fitting of models with two-way causation, in which latent variables may be regressed on each other.

Figure 16.4, based on an example in Pedhazur (1982, p. 670), illustrates a covariance structure model. The model analyzes the effects of intelligence and socioeconomic status on achievement. The observed variables are the indicators of intelligence, X_1 = Wechsler score and X_2 = Stanford-Binet IQ score; the indicators of socioeconomic status, X_3 = father's education, X_4 = mother's education, and X_5 = parents' total income; and the indicators of achievement, Y_1 = verbal score and Y_2 = quantitative score on an achievement test. The achievement indicators are the response variables.

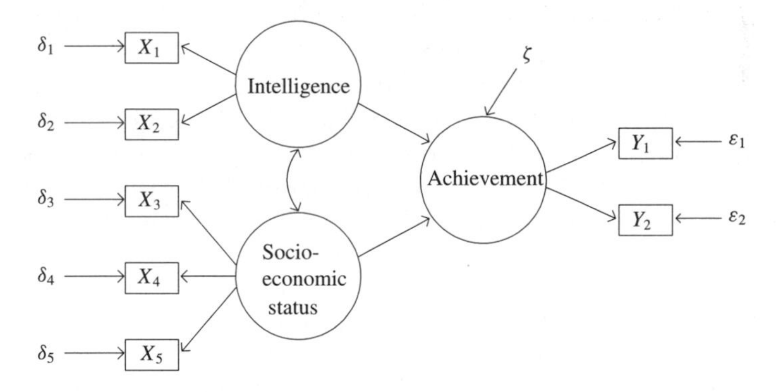

Figure 16.4 A Covariance Structure Model for Achievement, Intelligence, and Socioeconomic Status Latent Variables

In this figure, rectangles represent observed variables and circles represent latent variables. An intelligence latent variable applies only to X_1 and X_2, the indicators of intelligence. A socioeconomic status latent variable applies only to X_3, X_4, and X_5, its indicators. An achievement latent variable applies only to Y_1 and Y_2, the indicators of achievement. The figure depicts the dependence of the observed variables on the latent variables. The paths among the latent variables indicate that achievement is directly dependent on intelligence and socioeconomic status and that an association exists between intelligence and socioeconomic status.

As in any regression analysis, one does not assume that a variable is completely determined by others in the system. In Figure 16.4, the δ (delta) and ϵ (epsilon) terms pointing to the observed variables are error terms, representing the variation in these variables that is unexplained by the latent variables in the measurement error model. (Recall the representation of regression models with error terms shown in Section 9.6.) The ζ (zeta) symbol represents unexplained variation in the structural equation model, the achievement latent variable not being completely determined by the intelligence and socioeconomic status latent variables.

Special Cases of Covariance Structure Models

Covariance structure models have the attractive features of flexibility and generality. One can force a regression parameter to equal another in the system; it is then called a ***constrained*** parameter. Or one can force a parameter to assume a fixed value, such as 0; it is then called a ***fixed*** parameter. Or it can be completely unknown, a ***free*** parameter. In Figure 16.4, for instance, in the measurement model the factor loadings of the intelligence indicators X_1 and X_2 on the socioeconomic status and achievement latent variables equal 0, so they are fixed parameters. Similarly, the factor loadings of the socioeconomic status indicators on the intelligence and and achievement latent variables equal 0, and the factor loadings of the achievement indicators on the socioeconomic status and intelligence latent variables equal 0. On the other hand, in the structural equation part of the model, the regression coefficients of intelligence and socioeconomic status on achievement are free parameters.

One can treat an observed variable as if it were perfectly measured by taking the corresponding latent variable to be identical to that observed variable. In doing this, one then assumes a lack of error in that part of the measurement model. Ordinary regression models are special cases of covariance structure models in which there is a single response variable, and all variables are treated as perfectly measured.

When one treats all the observed variables as response variables and concentrates on how to represent them by a set of latent variables, the model provides a structured type of factor analysis. The analysis is ***confirmatory***, meaning that the analysis has the purpose of confirming a hypothesized factor-loading pattern for prespecified latent variables. For instance, in a recent article dealing with racism, conservatism, affirmative action, and intellectual sophistication (J. Sidanius, F. Pratto, and L. Bobo, *Journal of Personality & Social Psychology*, Vol. 70, 1996, p. 476), the authors created a factor from five measured indicators of racism. Many other variables were measured in the study, such as educational level, political conservatism, and affirmative action attitude, but those variables were forced to have loadings of 0 on the factor for racism. A chi-squared test indicated that the five indicators of racism were adequately represented by a single factor.

This confirmatory type of factor analysis contrasts with the ***exploratory*** nature of standard factor analysis, such as the type described in Section 16.3 in which one does not judge the number of important factors and their relationships with the observed variables until after viewing the matrix of factor loadings. With exploratory factor analysis, there is greater danger of going on a fishing expedition that produces results that seem interesting but may largely reflect sampling error.

Fitting Covariance Structure Models

The ***covariance*** between two variables X and Y is the average cross product of their deviations about their means. Its population value is defined as

$$\text{Cov}(X, Y) = E[(X - \mu_X)(Y - \mu_Y)] = \rho_{XY}\sigma_X\sigma_Y$$

That is, the covariance is completely determined by the correlation and the standard deviations. For a set of variables, their covariances are summarized in a ***covariance matrix***. The entry in row i and column j is the covariance between variables i and j. If we use standardized variables, then the standard deviations equal 1, and the covariance matrix is identical to the correlation matrix. The specification of a covariance structure model relates to the form of this covariance matrix.

Maximum likelihood is the standard method for fitting covariance structure models and obtaining estimates of parameters. As in ordinary regression, the fitting process assumes normally distributed responses. Software for fitting the models uses the sample covariation among the observed variables to estimate parameters in the measurement model and in the structural equation model. Parameters in the structural equation model are usually the ones of ultimate interest. In the above example, these include the regression coefficients of the intelligence and socioeconomic status latent variables on the achievement latent variable.

As in ordinary regression, the size of a parameter estimate relative to its standard error provides a measure of its significance; the ratio is an approximate z statistic for a significance test. The interpretation of the magnitude of an estimate depends on whether one measures the variables in the original units or in standardized form.

Unless the model fixes sufficiently many parameters, the problem occurs that the parameters are not ***identifiable***. This means there are no unique estimates, a situation which always happens with ordinary factor analysis. It is best to fix enough parameters so that this does not happen. Software provides guidance about whether identifiability is achieved. If it is not achieved, try setting additional factor loadings equal to zero or replacing a factor by observed variables in the structural equation part of the model.

The covariance structure model and its pattern of fixed, constrained, and free parameters determines a particular pattern that the true covariance matrix of the observed variables should satisfy. One can check the fit of the covariance structure model to the data by comparing the sample covariance matrix for the data to the estimated covariance matrix implied by the fit of the model. One way to do this, for large samples, uses a chi-squared test. The goodness of fit is based on how close the actual sample covariance matrix falls to its estimated value assuming that the model holds. A measure of the distance between the two matrices is an approximate chi-squared statistic; the larger the statistic, the poorer the fit. Use this only as a rough guide, since the test requires multivariate normality of the observed variables, which is usually a drastic oversimplification for reality.

Like other chi-squared tests, this test of fit is a global test. If the model fits poorly, it does not indicate what causes the lack of fit. Adjusted residuals make specific comparisons of individual elements of the observed and predicted covariance matrix. It can be more informative to assess lack of fit as in ordinary regression models, by comparing a given model to a more complex model with additional structure that may be relevant. One can test whether the more complex model gives a better fit than the reduced model, using an approximate chi-squared statistic based on the difference in goodness-of-fit chi-squared statistics for the two models. As in any test, though, keep in mind the dependence of results on sample size; a result may be statistically significant, for large n, without being practically significant.

Covariance structure models require specialized software. LISREL is a well-known program that fits the model using maximum likelihood methodology developed since 1970 by K. Jöreskog and others. The software name is an acronym for ***li****near* ***s****tructural* ***rel****ationships*. Another popular program is EQS. For descriptions and comparisons, see Bollen and Ting (1991).

Good features of using covariance structure models instead of an unstructured factor analysis are that (1) the models force researchers to provide theoretical underpinnings to their analyses, and (2) one can check the fit of the data to the theoretical model using inferential methods. However, the model is complex, and any model with latent variables may require a large sample size to obtain good estimates of effects, even for a relatively modest system of variables such as Figure 16.4 portrays.

In summary, covariance structure models provide a versatile format for conducting a variety of useful analyses in social science research. The frequency of their use by social scientists has increased dramatically in recent years. Nonetheless, the complexity of the model makes it possible to come up with results that seem interesting but may be highly tentative because of the many sources of variation. We recommend that you seek guidance from a statistician or well-trained social science methodologist before using this type of model.

16.5 Markov Chains

Except for repeated measures ANOVA and event history analysis, the methods studied in this text have referred primarily to static situations—in other words, to data on variables measured for some sample at a fixed time. Researchers are often interested, however, in sequences of responses measured over time on a response variable. A study of voting patterns in presidential elections might analyze data in which subjects indicate the party for which they voted in each of the past several elections.

A sequence of observations that varies randomly in some way is called a ***stochastic process***. The possible values for the process at each step are the ***states*** of the process. For example, the possible states for the vote in an election might be (Democrat, Republican, Other, Did not vote). Stochastic models describe sequences of observations on a variable.

One of the simplest stochastic processes is the ***Markov chain model***. It is appropriate if, given the behavior of the process at times $t, t-1, t-2, \ldots, 1$, the probability distribution of the outcome at time $t+1$ depends only on the outcome at time t. In other words, given the outcome at time t, the outcome at time $t+1$ is statistically independent of the outcome at all times previous to time t.

Transition Probabilities

A study of male social class mobility considers a three-generation period, labeled by grandfather, father, and son. The study follows a family line by considering the sequence of firstborn sons at age 40 years. In each generation, the possible states of the process are upper, middle, and lower. Suppose this process behaves like a Markov chain. Then, for example, for all fathers in a given class (such as upper), the social class of the son is statistically independent of the social class of the grandfather. Using the vertical slash | to represent *given* or *conditioned on*, the following four probabilities would be identical:

Pr(son in M | father in U, grandfather in L)
Pr(son in M | father in U, grandfather in M)
Pr(son in M | father in U, grandfather in U)
Pr(son in M | father in U)

The common probability is called the ***transition probability*** of moving from the upper class to the middle class in one generation, and is denoted by P_{UM}.

The Markov property is *not* that the state at time $t+1$ is *independent* of the states at time $t-1$, $t-2$, and so on; rather only that for a *fixed* value of the process at time t, they are independent. Letting $X_1, X_2, \ldots$ denote the successive states of the chain, X_{t+1} may be associated with $X_{t-1}, X_{t-2}, \ldots$, but conditioned on X_t, X_{t+1} is statistically independent of $X_{t-1}, X_{t-2}, \ldots$. Associations may exist, but the conditional associations (e.g., between X_{t+1} and X_{t-1}, controlling for X_t) do not.

The Markov chain model studies questions such as the following:

- What is the probability of moving from one particular state to another in a particular amount of time?
- How long, on the average, does it take to move from one particular state to another?
- Are the transition probabilities between each pair of states constant over time? If they are, the process is said to have ***stationary transition probabilities***.
- Is the process a Markov chain, or is the dependence structure more complex?

The properties of a Markov chain depend on the transition probabilities. These are studied with the transition probability matrix, denoted by $\boldsymbol{P}$. For an s-state chain, this matrix is an $s \times s$ table such that the entry in the cell in the ith row and the jth column is the probability that, given that the chain is currently in state i, at the next time period it is in state j.

Table 16.5 shows the format for a transition probability matrix for the social mobility example, with a set of potential transition probabilities. The row labels refer to the father's class, and the column labels refer to the son's class. In the table, $P_{UM} = .48$ means that if the father is in the upper class, then the probability is .48 that the son is in the middle class. Similarly, if the father is in the upper class, the probability that the son is in the upper class is $P_{UU} = .45$, whereas the probability that the son is in the lower class is $P_{UL} = .07$. The sum of the probabilities within each row of the matrix equals 1.0.

One estimates the true transition probabilities by calculating the proportion of the transitions from each state into each of the other states. For example, if there are 100 father-son pairs with the father in the upper class, and if for 45 of these pairs the son is in the upper class, then $\hat{P}_{UU} = 45/100 = .45$.

TABLE 16.5 Sample Format for Transition Probability Matrix P

$$\text{Time } t+1$$

$$\text{Time } t \quad \begin{matrix} \\ U \\ M \\ L \end{matrix} \begin{matrix} U & M & L \\ \left(\begin{matrix} P_{UU} & P_{UM} & P_{UL} \\ P_{MU} & P_{MM} & P_{ML} \\ P_{LU} & P_{LM} & P_{LL} \end{matrix} \right) \end{matrix} = \begin{matrix} \\ U \\ M \\ L \end{matrix} \begin{matrix} U & M & L \\ \left(\begin{matrix} .45 & .48 & .07 \\ .05 & .70 & .25 \\ .01 & .50 & .49 \end{matrix} \right) \end{matrix}$$

In most social science applications, it is unrealistic to expect transition probabilities to be stationary. This limits the usefulness of simple Markov chain models. For sample data, one can test the adequacy of the assumptions of Markov dependence and stationary transition probabilities with chi-squared tests. Though the Markov chain model is too simplistic by itself to have much practical use in the social sciences, it sometimes forms a component of a more complex and realistic model.

PROBLEMS

1. Refer to Table 16.1.
   a) Interpret the estimated effects of race and gender on the hazard rate.
   b) Show how to test the effect of race, and interpret.
2. In studying the effect of race on job dismissals in the federal bureaucracy, C. Zwerling and H. Silver (*American Sociological Review*, Vol. 57, 1992, p. 651) used event history analysis to model the hazard rate regarding termination of employment. In modeling involuntary terminations using a sample of size 2141, they reported $P < .001$ in significance tests for the partial effects of race and age.

a) They reported an estimated effect on the hazard rate of $e^{\hat{\beta}} = 2.13$ for the coefficient of the dummy variable for being black. Explain how to interpret.
b) The gender effect was not significant, but they reported an estimated multiplicative effect on the hazard of 1.06 for the dummy variable for being female. Interpret.

3. Explain what is meant by a censored observation, and give an example in which most observations would be censored.
4. Consider the variables I = annual income, E = attained educational level, Y = number of years experience in job, M = motivation, A = age, G = gender, and P = parents' attained educational level.
a) Construct a path diagram showing your view of the likely relationships among those variables.
b) Construct the regression models you would need to fit to estimate the path coefficients for that diagram.
5. Refer to Table 9.13 and Problem 9.17. Draw a path diagram relating B = birth rate, G = GNP, L = literacy, T = television ownership, and C = contraception. Specify the models you would need to fit to estimate path coefficients for your diagram.
6. In Table 9.1, consider the variables murder rate, percentage metropolitan, percentage high school graduates, and percentage in poverty. Do not use the observation for D.C.
a) Construct a realistic path diagram for these variables.
b) By fitting the appropriate models for these data, estimate the path coefficients, and construct the final path diagram. Interpret.
7. Refer to Table 9.16 in Problem 9.24. Consider the spurious causal model for the association between crime rate and percentage high school graduates, controlling for percentage urban. Analyze the data, and explain whether the data are consistent with this model.
8. Refer to Table 9.1. Using software as shown by your instructor, conduct a factor analysis. How many factors seem appropriate? Interpret the factors, using the estimated factor loadings.
9. Refer to the previous problem. Remove the observation for D.C., and repeat. How sensitive are the estimated factor loadings and your identification of factors to that one observation?
10. Construct a diagram representing the following covariance structure model: three observed response variables are described by a single latent variable, and that latent variable is regressed on four observed predictors.
11. Construct a diagram representing the following covariance structure model, for variables measured for each state. The latent response variable is based on two observed indicators, violent crime rate and murder rate. The two predictor variables for that latent variable are the observed values of percentage of residents in poverty and percentage of single-parent families; these are treated as perfectly measured.
12. Refer to the previous problem. Using software that your instructor introduced, fit this model to the data in Table 9.1 and interpret results.
13. Construct a diagram representing a covariance structure model in which (i) in the measurement part of the model, a single factor represents violent crime rate and murder rate and a single factor represents percentage high school graduates, percentage in poverty, and percentage of single-parent families, and (ii) in the structural equation part of the model, the first factor depends on the second factor as well as on the percentage of metropolitan residents.

14. A variable is measured at three times, Y_1 at time 1, Y_2 at time 2, and Y_3 at time 3. Suppose the chain relationship holds, with Y_1 affecting Y_2, which in turn affects Y_3. Does this sequence of observations satisy Markov dependence? Explain.

Bibliography

Path Analysis

Duncan, O. D. (1966). Path analysis: Sociological examples. *American Journal of Sociology*, 72, 1–16.

Land, K. (1969). Principles of path analysis. In *Sociological Methodology 1969*, ed. E. Borgatta. San Francisco: Jossey-Bass.

Li, C. C. (1975). *Path Analysis: A Primer*. Pacific Grove, CA: Boxwood Press.

Pedhazur, E. J. (1982). *Multiple Regression in Behavioral Research*, 2nd ed. New York: Holt, Rinehart and Winston.

Factor Analysis

Afifi, A. A. and Clark, V. (1996). *Computer-Aided Multivariate Analysis*, 3rd ed. London: Chapman & Hall.

Everitt, B. S. (1984). *An Introduction to Latent Variable Models*. London: Chapman & Hall.

Gould, S. J. (1981). *The Mismeasure of Man*. New York: W. W. Norton.

Harman, H. (1967). *Modern Factor Analysis*, 2nd ed. Chicago: University of Chicago Press.

Structural Equation Models with Latent Variables

Bentler, P. M. (1980). Multivariate analysis with latent variables: Causal modelling. *Annual Review of Psychology*, 31, 419–456.

Bollen, K. A. (1989). *Structural Equations with Latent Variables*. New York: Wiley.

Bollen, K. A. and Ting, K. (1991). EQS and LISREL. *The American Statistician*, 45, 68–73.

Dunn, G., Everitt, B., and Pickles, A. (1993). *Modelling Covariances and Latent Variables Using EQS*. London: Chapman and Hall.

Jöreskog, K. G. and Sörbom, D. (1993). *LISREL 8: Structural Equation Modeling with the SIMPLIS Command Language*. Chicago: Lawrence Erlbaum.

Long, J. S. (1983). *Confirmatory Factor Analysis* and *Covariance Structure Models, An Introduction to LISREL*, Sage University Paper Series on Quantitative Applications in the Social Sciences, series no. 07–033, 034. Beverly Hills, CA: Sage Publications.

Pedhazur, E. J. (1982). *Multiple Regression in Behavioral Research*, 2nd ed. New York: Holt, Rinehart and Winston.

Event History Models

Allison, P. D. (1984). *Event History Analysis*, Sage University Paper Series on Quantitative Applications in the Social Sciences, series no. 07–046. Beverly Hills, CA: Sage.
Yamaguchi, K. (1991). *Event History Analysis*. Thousand Oaks, CA: Sage.

Markov Chains

Bartholomew, D. J. (1982). *Stochastic Models for Social Processes*, 3rd ed. New York: Wiley.
Goodman, L. A. (1962). Statistical methods for analyzing processes of change. *American Journal of Sociology*, 68, 57–78.
Scheaffer, R. L. (1995). *Introduction to Probability and its Applications*, 2nd ed. Belmont, CA: Wadsworth.

Appendix

SAS and SPSS for Statistical Analyses

Major statistical software packages have procedures for nearly all the methods presented in this text. This appendix illustrates software use for these methods.

There is insufficient space to discuss all the major packages, and the examples show only SAS code (release 6.10). We also explain how to use SPSS, in its menu-driven windows environment. We focus on basic use of the software rather than the great variety of options provided by the procedures. For ease of reference, the material both for SAS and SPSS is organized by chapter of presentation in this text. The full data files for examples and exercises having large data sets are available on the World Wide Web at

`http://www.stat.ufl.edu/users/aa/social/data.html`

Introduction to SAS

The **SAS** language consists of DATA steps that name the variables and input the data and PROC steps that request the statistical procedures. All SAS statements must end with a semicolon. Variable names must start with a letter and contain at most eight characters. The first statement, the DATA statement, assigns a name to the data set. The next statement, the INPUT statement, tells SAS the variable names and the order in which the variables are listed in the data set. The data follow the CARDS statement, one line per subject, unless the INPUT statement ends with @@. After the data lines, a line containing only a semicolon ends the data set.

Following the data entry, PROC statements invoke the statistical procedures. A typical PROC statement lists the procedure, such as MEANS, and then also may select some options for providing greater detail than SAS provides with the default statement. The text by Schlotzhauer and Littell (1987) provides a readable introduction to SAS and its use for basic statistical methods.

Chapter 3: Descriptive Techniques

Table A.1 shows the format for entering the data and performing some very basic analyses, using the data set in Table 3.1 on murder rates for the 50 states. For each state, we list the state label and its value on murder rate. When one inputs characters rather than numbers for a variable, such as the state labels, the variable has an accompanying $ label in the INPUT statement. We enter the 50 observations as 50 lines of data, or we can enter multiple observations on a line if we enter @@ at the end of the input line, as shown in Table A.1.

TABLE A.1 SAS for Printing Data, Computing Basic Summary Statistics, and Preparing Plots

```
data crime ;
input  state $  murder @@;
cards;
  AL  11.6  AK   9.0  AZ  8.6
  AR  10.2  CA  13.2  CO  5.8
  ...
;
proc print ;  var  state  murder ;
proc freq;  tables  murder ;
proc chart;  vbar  murder ;
proc means;  var  murder ;
proc univariate  plot;  var  murder ;  id  state;
run ;
```

The first procedure, PROC PRINT, prints the data set. PROC FREQ provides a frequency distribution for the variable listed following TABLES. PROC CHART provides a histogram of the variable listed in the VBAR statement. Options exist for choosing the number of bars (e.g., VBAR MURDER / LEVELS = 5) or their midpoints and for forming horizontal rather than vertical bars (HBAR instead of VBAR).

PROC MEANS provides the mean and standard deviation of the variables listed after VAR (in this case, the murder rate). The PROC UNIVARIATE statement requests a greater variety of basic statistics for a variable, including the sample size, mean, standard deviation, median, mode, range, and quartiles. The ID statement, which is optional, names STATE as the variable to identify some of the extreme observations in part of the display from this procedure. Listing the PLOT option in PROC UNIVARIATE requests stem and leaf and box plots for the variables listed.

Chapters 5 and 6: Estimation and Significance Tests

The estimated standard error for a sample mean is one of the statistics provided by PROC UNIVARIATE; it is labeled as *Std Mean* in the output. We construct a confidence interval for a population mean by taking the sample mean and adding and subtracting the appropriate t-score times the standard error. (For large samples, one can use a z-score instead.)

Table A.2 shows how to obtain the standard error and the t-score for the data from Example 6.8, for which $n = 29$ and $df = 28$. The two arguments for the TINV function are half the error probability and the df value. For instance, the statement in Table A.2 requests the t-score with left tail probability equal to .025 (for a 95% confidence interval) when $df = 28$, which equals -2.048. That table also shows how one can input data for two dependent samples (WEIGHT1 and WEIGHT2 being the weights of anorexic girls at two times) and create a new variable (DIFF) that is the difference between WEIGHT2 and WEIGHT1.

TABLE A.2 SAS for Obtaining Standard Errors and t-Scores

```
data anorexia ;
input weight1 weight 2 ;
diff = weight2 - weight1;
cards;
  80.5  82.2
  84.9  85.6
  ...
;
proc univariate ; var diff ;
data findt;
       tvalue = tinv(.025, 28) ;
proc print   data = findt ;
run ;
```

Chapter 7: Comparison of Two Groups

Table A.3 shows the use of SAS for performing a two-sample t test for comparing two means (Section 7.3), using the data in Example 7.4. The input variables are THERAPY, the levels of which are the two groups to be compared, and IMPROVE, the improvement in depression score, which is the response variable. PROC SORT sorts the data into groups, according to the levels of therapy, and then PROC MEAN finds means and standard deviations for the observations in each group, when one uses BY followed by the group variable. The BY statement is used with SAS procedures when one wants to do an analysis separately for each level of the variable specified in the BY statement.

PROC TTEST is a procedure for a two-sample t test with independent samples. The CLASS statement names the variable that identifies the groups to be compared, and the VAR statement identifies the response variable for the analysis. The output

TABLE A.3 SAS for Two-Sample t Test for Data in Example 7.4

```
data depress;
input therapy $ improve @@ ;
cards;
  a  10  a  20  a  30
  b  30  b  45  b  45
;
proc sort; by therapy ;
proc means; by therapy ; var improve ;
proc ttest; class therapy ; var improve ;
run;
```

shows the mean, standard deviation, and standard error for the response variable in each group, and provides the t test statistic, its df value, and a two-sided P-value, which is labeled by $Prob > |T|$. An approximate test is also provided that does not assume equal population variances for the two groups.

For contingency tables having small cell counts, Table A.4 shows SAS code for performing Fisher's exact test of independence with the sexual identity data of Table 7.4. PROC FREQ forms contingency tables and has a variety of options for their analysis. The variable labeled as COUNT in Table A.4 contains the cell counts. One must use a WEIGHT statement with this variable to tell SAS that there are that many observations at each combination of values of the two classifications. The TABLES statement specifies the classification variables, and the EXACT option in PROC FREQ performs Fisher's exact test.

TABLE A.4 SAS for Fisher's Exact Test with Table 7.4

```
data lesbians;
input mother $ child $ count   @@;
cards;
  Lesbian  Yes   2  Lesbian  No   23
  Hetero   Yes   0  Hetero   No   20
;
proc freq;  weight count;
     tables  mother*child /  exact;
run;
```

Chapter 8: Analyzing Association Between Categorical Variables

Table A.5 illustrates SAS for analyzing two-way contingency tables, using data from Table 8.4. PROC FREQ conducts chi-squared tests of independence using the CHISQ option and provides the expected frequencies for the test with the EXPECTED option. The MEASURES option provides a wide assortment of measures of association (including gamma and Kendall's tau-b for ordinal data) and their standard errors. For 2×2 tables this option provides confidence intervals for the odds ratio (labeled "case-control" on output) and the relative risk. SAS lists the category levels in alphanumeric order unless one states ORDER=DATA in the PROC directive, in which case the levels have the order in which they occur in the input data.

TABLE A.5 SAS for Chi-Squared Test with Table 8.4

```
data politics;
input  gender  $  party  $  count  @@;
cards;
  Female  Democ  279   Female  Indep  73  Female  Repub  225
  Male    Democ  165   Male    Indep  47  Male    Repub  191
;
proc freq;  weight count ;
      tables  gender*party / chisq  expected  measures ;
proc genmod;  class gender party;
      model  count = gender  party / dist=poi  link=log  obstats  residuals;
run;
```

One can also perform chi-squared tests using PROC GENMOD, as shown in Table A.5. This procedure, discussed in greater detail in the SAS subsection of this appendix for Chapter 14, uses a generalized linear modeling approach introduced in Section 14.5. (The code in Table A.5 views the independence hypothesis as a "loglinear model" for Poisson counts with main effects of gender and party but no interaction.) The OBSTATS and RESIDUALS options in GENMOD provide cell residuals; the output labeled *StReschi* is the adjusted residual.

Starting with version 6.11, SAS also can conduct (in PROC STATXACT) many exact tests of independence. These tests generalize Fisher's exact test to $r \times c$ contingency tables with nominal or ordinal classifications. This procedure is based on a separate software package (see StatXact-3 (1995)) distributed by Cytel Software of Cambridge, Massachusetts.

Chapter 9: Linear Regression and Correlation

Table A.6 shows the use of SAS for performing linear regression, using the crime data set of Table 9.1. The data for the 50 states and D.C. are 51 lines, each of which contains a label for the state and the values for violent crime rate, murder rate, percentage in metropolitan areas, percentage who are white, high school graduation rate, poverty rate, and percentage of single-parent families.

The PROC PLOT statement requests a scatter diagram for murder rate and poverty rate; the first variable listed goes on the Y axis. The first PROC REG statement requests a regression analysis, predicting murder rate using poverty rate; the second PROC REG statement requests the model using percentage of single-parent families to predict violent crime rate. The P option following this model statement requests the predicted values and residuals for all observations. The PROC CORR statement requests the Pearson correlation between each pair of variables listed in the VAR list.

TABLE A.6 SAS for Regression Analysis with Table 9.1

```
data crime ;
input state $  violent  murder  metro  white  hs  poverty  single ;
cards;
  AK   761    9.0  41.8   75.2   86.6    9.1   14.3
  AL   780   11.6  67.4   73.5   66.9   17.4   11.5
  AR   593   10.2  44.7   82.9   66.3   20.0   10.7
  AZ   715    8.6  84.7   88.6   78.7   15.4   12.1
  CA  1078   13.1  96.7   79.3   76.2   18.2   12.5
  ....
;
proc print  data=crime;
proc plot; plot murder*poverty ;
proc reg;  model  murder = poverty ;
proc reg;  model  violent = single  /  p ;
proc corr;  var  violent  murder  metro  white  hs  poverty  single ;
run;
```

Chapter 11: Multiple Regression and Correlation

Table A.7 uses SAS to perform multiple regression with the mental impairment data of Table 11.1. One simply lists every explanatory variable in the model to the right of the equal sign in the model statement. The PARTIAL option provides partial regression scatterplots. The PCORR2 option provides squared partial correlations, and the STB option provides standardized regression coefficients. Following the input statement, we define a variable life_ses to be the cross-product of life events and ses (the * symbol represents multiplication). We enter that variable in the second regression model to permit interaction in the model. Friendly (1991) describes additional graphics one can

TABLE A.7 SAS for Multiple Regression Analysis with Table 11.1

```
data mental ;
input  impair  life   ses ;
life_ses = life*ses;
cards;
  17   46   84
  19   39   97
  20   27   24
  ....
;
proc print;  var  impair   life   ses ;
proc plot ;  plot  impair*life  impair*ses ;
proc reg;  model  impair =  life  ses  /  partial  stb  pcorr2 ;
proc reg;  model  impair =  life  ses  life_ses ;
run;
```

create using SAS, such as a scatterplot matrix (p. 576) and scatterplot with boxplots (p. 174).

Chapter 12: Analysis of Variance

Table A.8 uses SAS to perform one-way ANOVA with Table 12.1 and two-way ANOVA with Table 12.11. The first PROC MEANS statement requests sample means on ideology for the data grouped by party. PROC GLM is a procedure for *general linear models*. It is similar in many ways to the regression procedure, PROC REG, except that PROC GLM allows one to use CLASS statements to create dummy variables in order to include qualitative predictors in the model.

TABLE A.8 SAS for One-Way ANOVA with Table 12.1 and Two-Way ANOVA with Table 12.11

```
data anova;
input party  $   gender  $   ideology ;
cards;
  Dem  F  1
  Dem  F  1
  Dem  F  1
  Dem  F  1
  Dem  F  1
  Dem  F  2
  ...
  Rep  M  7
;
proc means;   by party;   var ideology;
proc glm;   class party ;
    model ideology = party   /   solution;
    means party   /   bon   tukey   alpha=.10;
proc means;   by   party   gender;   var   ideology;
proc glm;   class   party   gender;
    model ideology = party gender   /   solution;
    means party /   bon   tukey;
proc glm;   class   party   gender;
    model ideology = party gender party*gender;
run;
```

The first GLM statement in Table A.8 requests a one-way ANOVA, comparing ideology by party. The CLASS statement requests dummy variables for the levels of party, so when we include it in the regression model, the three parties are represented by two dummy variables. The MEANS option for this model provides a multiple comparison with simultaneous confidence intervals. Various intervals are available, and here we request the Bonferroni and Tukey methods. The default is simultaneous 95% intervals, unless one requests a different error rate by setting alpha (here, alpha = .10, for 90% confidence). The SOLUTION option requests the estimates for the prediction equation; this equation is not automatically provided when one uses a CLASS statement.

The second PROC MEANS requests sample means on ideology for each of the combinations of party and gender. Following that is a GLM statement to conduct a two-way ANOVA using party and gender as predictors of ideology, setting up dummy variables for each predictor with the CLASS statement. This is followed by a MEANS option requesting multiple comparisons across the levels of party. This analysis assumes a lack of interaction. The final GLM statement adds an interaction term to the model.

Table A.9 shows SAS for the repeated measures ANOVA with Table 12.16. Each row of the data provides the opinion responses on the three influences (movies, TV, rock) for a particular subject. One can use PROC REG or else PROC ANOVA for the modeling; the latter applies for "balanced" analyses in which the same number of responses occur at each level of a factor. This model looks like that for a standard two-way ANOVA, except that one effect is the subject effect. The analysis is followed by a multiple comparison of means across the levels of influence type.

TABLE A.9 SAS for Repeated Measures ANOVA with Table 12.16

```
data repeat;
input subject  $   influ  $   opinion @@;
cards;
  1  M  -1   1   T   0   1  R  -1
  2  M   1   2   T   0   2  R   0
 ....
 12  M  -1  12   T  -1  12  R  -2
;
proc print;
proc anova;    classes subject influ ;
    model opinion = influ subject ;
    means influ   /   tukey bon;
run;
```

Table A.10 shows an alternative way of inputing data for a repeated measures ANOVA. This table refers to Table 12.19, in which the groups refer to three therapies and the response is weight. In each line, one identifies the group into which a subject falls, and then lists successively the repeated responses by the subject, labeling them by a name such as RESP1-RESP3 if there are three repeated responses. This table provides the analysis for a between-subjects effect (therapy) and a within-subject effect (the repeated responses on weight). The model statement indicates that the repeated responses are modeled as a function of *therapy* and that the levels at which the repeated measurements occur refer to a variable labeled as *occasion*. The SHORT option prints multivariate test statistics in condensed form, and the PRINTE option provides results of testing the sphericity assumption for the analysis. The analysis is followed by a multiple comparison of the response means by category of therapy.

This extends to more complex designs. For instance, suppose we had three factors, A and B being between-subjects factors and the repeated measures on a variable

TABLE A.10 SAS for Two-Way Repeated Measures ANOVA with Table 12.19

```
data repeat2;
input subject $ therapy $ weight1-weight2;
cards;
  1   CB   80.5   82.2
  2   CB   84.9   85.6
  3   CB   81.5   81.4
  ....
 72    C   89.0   78.8
;
proc anova; class therapy ;
model weight1-weight2 = therapy ;
  repeated occasion  /  short printe;
means therapy / bon ;
run;
```

Y taken at the four levels of a factor C. One could use the SAS code:

```
PROC ANOVA; CLASS  A  B ;
MODEL  Y1 - Y4 = A  B  A*B ; REPEATED C ;
```

See Hatcher and Stepanski (1994) for further guidance on using PROC ANOVA for repeated measures analyses.

One can also conduct repeated measures ANOVA in SAS using PROC MIXED. This is a more advanced procedure that provides additional options for the covariance structure of the random effect. There are a variety of options in addition to the sphericity form of the standard analysis, and the results of tests for fixed effects depend on the choice. This procedure, unlike PROC ANOVA or GLM, does not require the data to be balanced within subjects; that is, it can use data from subjects that have missing observations. Other advantages of PROC MIXED are that one can use continuous variables in within-subject effects, instead of only classification variables, and one can omit the between–within interaction effects from the model. Because of the complexity of this procedure and the variety of options, we recommend that you seek guidance from a statistical expert in this area before using PROC MIXED. The article by Wolfinger and Chang (1995) is good place to start for further information about the procedure and about ways of checking the possible covariance assumptions.

Chapter 13: Analysis of Covariance

Table A.11 uses SAS to fit analysis of covriance models to Table 13.1. The PLOT statement requests a plot of income by education, with symbols indicating which race each observation has. The first GLM statement fits a bivariate regression between income and education (ignoring race), and the second one conducts a one-way ANOVA of income by race (ignoring education). This model has a CLASS statement, to provide dummy variables for levels of race.

TABLE A.11 SAS for Analysis of Covariance Models with Table 13.1

```
data ancova ;
input   income   educ   race   $   z1   z2 ;
cards;
  8  10  black   1   0
  9   7  black   1   0
 13   9  black   1   0
 ....
 28  20  white   0   0
;
proc plot;    plot income*educ = race;
proc sort; by race;
proc means; by race;
proc glm;    model income = educ;
proc glm; class race;   model income = race / solution;
proc glm; class race;   model income = educ race / solution;
lsmeans race    adjust=bon ;
proc glm; class race;   model income = educ race educ*race / solution;
proc glm;    model income = educ z1 z2 / covb;
proc glm;    model income = educ z1 z2 educ*z1 educ*z2 ;
run;
```

The third GLM statement fits the analysis of covariance model, assuming no interaction. This is followed by a request for adjusted means (also called "least squares means" and abbreviated by SAS as LSMEANS) on the response for the different levels of race, with Bonferroni multiple comparisons of them. The fourth GLM statement adds an interaction of race and education to the model.

For any model using qualitative predictors, one could always construct dummy variables and put them into the model to get the same effect as using a CLASS statement in SAS. To illustrate, Table A.11 inputs values of two dummy variables, Z_1 and Z_2. The last two GLM statements show how to fit the two analysis of covariance models by directly entering these dummy variables into the model. The COVB option provides the covariance matrix of the model parameter estimates.

Chapter 14: Model Building

Table A.12 shows a variety of analyses for the housing data of Table 9.4. In fitting a multiple regression model, the BACKWARD, FORWARD, STEPWISE, and CP choices for the SELECTION option yield these selection procedures. The INFLUENCE option yields studentized residuals, hat values, and measures of influence such as DFFITS and DFBETAS. The PARTIAL option provides partial regression residual plots. The PLOT option following the second model statement requests plots of residuals against the predicted values and against size of home.

The code sets up an artificial variable *size_2* that is the square of size. Entering it in the model, as in the third GLM statement, provides a quadratic regression model.

TABLE A.12 SAS for Various Analyses Conducted with Housing Data of Table 9.4

```
data housing ;
input  price   size   bed   bath   new;
size_2 = size*size;
cards;
  48.5   1.10   3   1   0
  55.0   1.01   3   2   0
  68.0   1.45   3   2   0
 137.0   2.40   3   3   0
  ....
;
proc reg;   model price = size bed bath new   /   selection=backward;
proc reg;   model price = size bath new   /  influence partial;
plot r.*p. r.*size ;
proc reg;   model price = size size_2 ;
proc genmod;   model price = size   /    dist = nor   link = identity;
proc genmod;   model price = size   /    dist = gam   link = identity ;
run;
```

PROC GENMOD in SAS fits generalized linear models. GENMOD (available beginning with version 6.08 of SAS) specifies the distribution of the random component in the DIST option (“nor” for normal, “gam” for gamma, “poi” for Poisson, “bin” for binomial) and specifies the link in the LINK option (including “log”, “identity”, and “logit”). The first GENMOD statement in Table A.12 fits the ordinary bivariate regression model to price and size; this gives the same results as using least squares with PROC REG or GLM. The second GENMOD statement fits the same type of linear model but instead assumes a gamma distribution for price, for which the standard deviation is proportional to the mean.

Table A.13 shows the use of GENMOD for fitting an exponential regression model to the population growth data of Table 14.7. This model has the usual form in the model

TABLE A.13 SAS for Fitting Exponential Regression Model as a Generalized Linear Model to Table 14.7

```
data growth ;
input   decade   popul ;
cards;
 0   62.95
 1   75.99
 2   91.97
 ....
;
proc genmod;  model popul = decade  /  dist = nor  link = log ;
proc genmod;  model popul = decade  /  dist = nor  link = identity ;
run;
```

statement, but uses the log link. If we use the identity link, as in the second statement, then the model assumes a linear trend. Friendly (1991, pp. 169, 563) shows how to use SAS to create nonparametric fits, such as the LOESS smoother.

Chapter 15: Logistic Regression (and Loglinear Models)

One can fit logistic regression models either using software for generalized linear models or specialized software for logistic regression. Table A.14 applies PROC GENMOD and PROC LOGISTIC to Table 15.1. In the code, *credit* is a dummy variable indicating whether the subject has a credit card.

TABLE A.14 SAS for Fitting Logistic Regression Model to Table 15.1

```
data binary ;
input   income   credit ;
n = 1 ;
cards;
 24   0
 27   0
 ....
130   1
;
proc genmod;
    model   credit/n = income   /  dist = bin   link = logit ;
proc genmod; model credit/n = income / dist = nor link = identity;
proc logistic descending;   model   credit = income   / influence;
    output  out=predict   p=pi_hat   lower=LCL  upper=UCL;
proc print  data = predict;
run;
```

The first GENMOD statement requests the binomial distribution and logit link options, which is logistic regression. For GENMOD with binomial models, the response in the model statements must have the form of the number of "successes" divided by the number of cases. When each observation refers to a single case, such as in the first data set in this table, one sets up a variable (called n in Table A.14) that equals 1 for each subject; this variable is the denominator for the response in the model statement. The second GENMOD statement requests the normal distribution and identity link, which corresponds to least squares for the linear probability model. If one uses this statement with binomial distribution instead of normal, the fit is the maximum likelihood for the linear probability model; the fitting process fails to converge, however, if any predicted probabilities fall outside the (0, 1) range.

Following the two GENMOD statements, Table A.14 uses PROC LOGISTIC to fit the logistic regression model to these data. This procedure orders the levels of the response variable alphanumerically, forming the logit, for instance, as

$$\log\left[\frac{P(Y=0)}{P(Y=1)}\right]$$

One can use the DESCENDING option to reverse the order, invoking the procedure using the statement PROC LOGISTIC DESCENDING. The INFLUENCE option in LOGISTIC provides regression diagnostics. Following the LOGISTIC model fit, Table A.14 requests predicted probabilities and lower and upper 95% confidence limits for the true probabilities.

TABLE A.15 SAS for Fitting Logistic Regression Model to Table 15.1, Entering Data as Binomial Totals Rather than Raw Binary Observations

```
data binomial
input   income   total   yes ;
cards;
  24  1  0
  27  1  0
  28  5  2
  ....
;
proc genmod;
model   yes/total = income   /  dist = bin  link = logit   obstats;
proc logistic;   model yes/total = income;
run;
```

Table A.15 shows how, both with PROC GENMOD and LOGISTIC, instead of entering raw binary (0, 1) data for individual subjects, one can enter the binomial numbers of outcomes of the two types at each level of the explanatory variable. In this second data set, the variable labeled *total* contains the number of cases at each level of income, and *yes* contains the number of "yes" responses on having a credit card. The OBSTATS option provides various "observation statistics," including predicted values and their confidence limits. Entering the data this way instead of case by case yields exactly the same fit.

Table A.16 uses GENMOD and LOGISTIC to fit a logit model with qualitative predictors to the death penalty data in Table 15.4. Here, we set up dummy variables for the predictors when we input the data, but with GENMOD one can automatically do this for factors by declaring them in a CLASS statement. GENMOD codes estimates for factors so the parameter estimate for the last category of each factor equals 0. The OBSTATS option in GENMOD provides predicted probabilities and their confidence limits, and the RESIDUALS option provides adjusted residuals (labeled *StReschi*). In models with multiple predictors, the TYPE3 option in GENMOD provides likelihood-ratio tests for testing the significance of each individual predictor in the model.

One can fit loglinear models using either software for generalized linear models or specialized software for loglinear models. Table A.17 uses GENMOD to fit model

TABLE A.16 SAS for Fitting Logit Model to Table 15.4

```
data death ;
input    vic    def    yes    n ;
cards;
  1   1   53   467
  1   0   11    48
  0   1    0    16
  0   0    4   143
;
proc genmod; model yes/n = def vic / dist=bin link=logit residuals obstats type3;
proc logistic descending;  model yes/n = def  vic;
data death2;
input  v  $  d  $  p  $  count    @@;
cards;
white  white  yes  53   white  white  no 414
white  black  yes  11   white  black  no  37
black  white  yes   0   black  white  no  16
black  black  yes   4   black  black  no 139
;
proc genmod data=death2;  class  d  $  v  $  p  $  ;
model count =  d  v  p  d*v  d*p  v*p / dist=poi link=log residuals obstats;
run;
```

(AC, AM, CM) to the drugs data of Table 15.6. The CLASS statement generates dummy variables for the classification factors. The *AM* association is represented by $A * M$. The OBSTATS and RESIDUALS options provide expected frequencies (predicted values) and diagnostics, including adjusted residuals. The second part of Table A.16 shows how to fit a loglinear model to the death penalty data that is equivalent to the logit model fitted in the first part of the table.

When the number of response categories exceeds two, PROC LOGISTIC in SAS provides ML fitting of the proportional odds version of cumulative logit models. Table A.18 uses it to fit the proportional odds model to Table 15.13, treating ideology as the

TABLE A.17 SAS for Fitting Loglinear Models to Table 15.6

```
data drugs ;
input   a    $   c    $    m    $    count @@ ;
cards;
 yes  yes  yes  911   yes  yes  no   538
 yes  no   yes   44   yes   no  no   456
 no   yes  yes    3    no  yes  no    43
 no   no   yes    2    no   no  no   279
;
proc genmod; class  a  c  m ;
 model count = a  c  m  a*c  a*m  c*m / dist=poi link=log obstats residuals;
run;
```

response variable. The second part of the table shows how to treat party as the response variable, in a logit model assigning scores to the levels of ideology.

TABLE A.18 SAS for Fitting Cumulative Logit Model to Table 15.13

```
data ordinal ;
input   party   ideology   count @@ ;
cards;
 1  1  80    1  2  81    1  3  171    1  4  41    1  5  55
 0  1  30    0  2  46    0  3  148    0  4  84    0  5  99
;
proc logistic; weight count;
     model ideology = party;
data ordinal2;
input ideo dem total;
cards;
 1   80   110
 2   81   127
 3  171   319
 4   41   125
 5   55   154
;
proc genmod  data=ordinal2;
model dem/total = ideo  /  dist=bin  link=logit  obstats  residuals;
run;
```

See Stokes et al. (1995) for further details on the use of SAS for methods for categorical data analysis, and SAS (1995) for additional examples of logistic regression.

Introduction to SPSS

SPSS for Windows has a graphical user interface that makes requesting statistical procedures simple. In this windows environment, SPSS provides menus and dialog windows to save you the work of preparing code to request analyses. The base system can run on IBM-compatible PCs with at least 4MB of RAM and 23MB of hard disk space, running Windows 3.1 or later.

The ***Application window*** contains a menu bar with a wide variety of separate menus. These include FILE menus for creating a new file or opening an existing one, STATISTICS menus that display options for selecting a statistical method, a GRAPHS menu for creating a graph of some type, and menus for choosing other special features. The ***Data Editor window*** displays the contents of the data file. You use this to enter the data or edit existing data. The ***Output window*** shows results of the analyses after you request them from the menus on the applications window. You can edit and save this for later use or printing. Other windows are available for charts and plotting data, and SPSS also has a syntax window for users who prefer to operate SPSS through syntax (written code); in some cases, using syntax makes extra options available. Using the windows environment is mostly self-explanatory, but on-line help is readily available. It also helps to have access to a manual such as SPSS (1993a).

Data files have the form of a spreadsheet, each column containing a variable and each row the observations for a particular subject. You can enter the data set while in SPSS. Or you can simply call up an existing data set, which might be an ASCII file that you created with a separate editor or a file created with a spreadsheet or database program. Select FILE on the menu bar in the application window and indicate whether you want to read in a new data file (such as crime.dat) or open an existing file (such as crime.sav) or enter a new data set on the spreadsheet. For a new data file, you then define names for each variable. You save the file by entering the DATA EDITOR window and selecting the FILE menu with the *Save As* option. One can access the created file later from the FILE menu.

At this point, you can select a statistical procedure from the STATISTICS menu in the applications window. When you select a procedure, a dialog box opens that shows you the source variables in your data set. You highlight the ones you want to use currently and click on the arrow to the right of the list to move them to the selected variables list further to the right. You then click on *OK* and the procedure runs, showing results in the output window. For many procedures, additional "subdialog boxes" display extra available options. To save output, in the OUTPUT window, use the FILE menu and the *Save As* option. The default is to name output files with extension .lst, such as crime.lst. One can later access the output in the FILE menu by opening SPSS output. For procedures that produce graphs, SPSS displays the results in a "Chart Carousel."

Chapter 3: Descriptive Statistics

To construct frequency distributions, histograms, and basic summary statistics, go to the STATISTICS menu and select the SUMMARIZE option with the FREQUENCIES suboption. A FREQUENCIES dialog box will open. Select the variables you want from the list for your file. Then, clicking on *OK* provides a frequency distribution in an output window.

Clicking on *Statistics* in the FREQUENCIES dialog box presents you with a FREQUENCIES: STATISTICS dialog box containing options of various statistics to compute, such as measures of central tendency and dispersion. (For basic descriptive statistic from the STATISTICS menu, one can also select the SUMMARIZE option with the DESCRIPTIVES suboption.) Clicking on *Charts* in the FREQUENCIES dialog box presents you with a FREQUENCIES: CHARTS dialog box containing a histogram option for quantitative variables and a bar chart option for qualitative variables. (You can also construct a histogram from the GRAPHS menu.)

To construct a stem and leaf plot or a box plot, from the STATISTICS menu select the SUMMARIZE option with the EXPLORE suboption. The EXPLORE dialog box contains a *Plots* option; clicking on it reveals the various plot options. (The GRAPHS menu also has the option of a box plot.)

Chapters 5 and 6: Estimation and Significance Tests

The STATISTICS menu has a COMPARE MEANS option with a ONE-SAMPLE T TEST suboption. The default with that option is a 95% confidence interval for the mean and a *t* test that the true mean equals 0. The options permit one to select a different confidence level. To test that the mean equals a constant μ_0, subtract μ_0 from every observation and test that the true mean equals 0, or supply your own test value by entering a number in the Test Value edit field.

Chapter 7: Comparison of Two Groups

The STATISTICS menu has a COMPARE MEANS option with a INDEPENDENT-SAMPLES T TEST suboption. One selects the response variable (labeled the *Test* variable) and the variable that defines the two groups to be compared (labeled the *Grouping* variable).

In the *Equal variances* row, this procedure provides the results of the *t* test discussed in Section 7.3. The output provides the test statistic (labeled *t*-value), *df*, standard error (labeled *SE of Diff*), and two-sided *P*-value (labeled *2-Tail Sig*). The procedure also provides the 95% confidence interval for comparing the means. Options allow one to change the confidence level.

The large-sample test statistic presented in Section 7.1 does not assume equal variances. The value of the test statistic and the standard error are shown in the *Unequal variances* row of the output. The *P*-value and confidence interval differ slightly from those discussed in the text, however, since they use an approximate *t* distribution rather than the standard normal distribution. For large samples, the discrepancy is slight.

The COMPARE MEANS option on the STATISTICS menu also has a PAIRED-SAMPLES T TEST suboption, which supplies the dependent-samples comparisons of means described in Section 7.4. For Fisher's exact test, see the description for the following chapter.

Chapter 8: Analyzing Association Between Categorical Variables

The SUMMARIZE option on the STATISTICS menu has a suboption called CROSSTABS, which provides several methods for contingency tables. After identifying the row and column variables in CROSSTABS, clicking on *Statistics* provides a wide variety of options, including the chi-squared test. The output lists the Pearson statistic, its degrees of freedom, and its *P*-value (labeled *Significance*).

If any expected frequencies in a 2×2 table are less than 5, Fisher's exact test results. SPSS also has an advanced module for small-sample inference (called SPSS Exact Tests, based on StatXact-3 (1995)) that provides exact *P*-values for various tests in CROSSTABS and NPAR TESTS procedures. For instance, the Exact Tests module provides exact tests of independence for $r \times c$ contingency tables with nominal or ordinal classifications. See the publication *SPSS Exact Tests 6.1 for Windows*.

In CROSSTABS, clicking on *Cells* provides options for displaying observed and expected frequencies, as well as the adjusted residuals, labeled as *Adj. standardized.* Clicking on *Statistics* provides options of a wide variety of statistics other than chi-squared, including gamma and Kendall's tau-b; the output shows the measures and their standard errors (labeled ASE1), which one can use to construct confidence intervals. It also provides a z test statistic for testing that the true measure equals zero, which is the ratio of the estimate to its standard error; this test uses a simpler standard error, labeled ASE0, that only applies under independence and is inappropriate for confidence intervals. One option in the list of statistics, labeled *Risk*, provides as output the odds ratio (labeled as *case control*) and its confidence interval.

Suppose you enter the data as cell counts for the various combinations of the two variables, rather than as responses on the two variables for individual subjects; for instance, perhaps you call COUNT the variable that contains these counts. Then, select the WEIGHT CASES option on the DATA menu, and instruct SPSS to weight cases by COUNT.

Chapter 9: Linear Regression and Correlation

To construct a scatter diagram, enter the GRAPH menu and choose the SCATTER option with SIMPLE suboption. Click on *Define*. Specify the X and Y variables, and click on *OK*.

The STATISTICS menu has a REGRESSION option with LINEAR suboption. You identify the response (Dependent) variable and the explanatory (Independent) variable. Various options are available by clicking on *Statistics* in the LINEAR REGRESSION dialog box, including estimates of the model parameters, confidence intervals for the parameters, and model fit statistics. Output for the estimates option includes the estimates for the prediction equation (labeled B), their standard errors (labeled *SE B*), the t statistic for testing that a regression parameter equals 0 and the associated P-value (labeled *Sig T*), and a standardized regression coefficient (labeled as *Beta*) that in this bivariate model is simply the Pearson correlation.

Output for the model fit option includes the r^2 value (labeled *R Square*) and the estimate $\hat{\sigma}$ of the conditional standard deviation (rather confusingly labeled *Standard Error*).

Chapter 11: Multiple Regression and Correlation

For a multiple regression analysis, again choose REGRESSION from the STATISTICS menu with the LINEAR suboption, and add additional variables to the list of independent variables. Among the options provided by clicking on *Statistics* in the dialog box are estimates of the coefficients and confidence intervals based on them and detail about the model fit. For the estimates option, the output includes standard errors of the estimates, the t statistic for testing that the regression parameter equals zero and its associated two-sided P-value, and the estimated standardized regression coefficient (labeled *Beta*).

Requesting the model fit option provides additional information. For instance, the F statistic provided following the *Analysis of Variance* table is the F test statistic for testing that the coefficients of the explanatory variables all equal 0; the probability labeled as *Signif F* is the P-value for that test. Also provided are the multiple correlation R, R^2, and the estimate $\hat{\sigma}$ of the conditional standard deviation (labeled *Standard Error*). Requesting the covariance matrix provides a matrix containing the Pearson correlations (and the variances and covariances) between all pairs of variables in the model.

To construct a scatterplot matrix, from the GRAPH menu choose the SCATTER option with the MATRIX suboption. Click on *Define*. Specify the variables and click on *Define*, and the resulting diagram shows a scatter diagram for each pair of variables. One can produce all partial regression plots by clicking on *Plots* in the LINEAR REGRESSION dialog window and then requesting SPSS to *Produce all partial plots* in the PLOTS dialog box.

To obtain a partial correlation analysis, in the STATISTICS menu choose the CORRELATE option with the PARTIAL suboption. In the resulting PARTIAL CORRELATIONS dialog box, select the variables to correlate and select at least one variable to control. The output also provides tests of significance for these partial correlations.

Chapter 12: Analysis of Variance

To conduct a one-way ANOVA, on the STATISTICS menu select the COMPARE MEANS option with the ONE-WAY ANOVA suboption. Select the dependent variable and select the factor that defines the groups to be compared. The factor must have numeric levels, and its range of values is specified by clicking on *Define Range*. Results provided include the F test statistic and its P-value, and sums of squares and mean squares for between-groups and within-groups variation. Clicking on *Post Hoc* in the ONE-WAY ANOVA dialog box provides a variety of options for multiple comparison procedures, including the Bonferroni and Tukey methods. Clicking on *Options* in the ONE-WAY ANOVA dialog box provides the option of additional descriptive statistics, including the mean, standard deviation, standard error, and a 95% confidence interval for each group.

To conduct a two-way or higher-way factorial ANOVA, on the STATISTICS menu select the ANOVA MODELS option with the SIMPLE FACTORIAL suboption. Or, if you have the Advanced Statistics module that SPSS offers as an option, select the ANOVA MODELS option with the GENERAL FACTORIAL suboption. Select the dependent variable and select the factors that define the cross-classification for the means. The default model is a full factorial model containing all interactions. Click on *Model* to build a customized model that contains only some or none of the interactions. Highlight variables, select Interaction or Main Effects from the Build Term(s) list, and click on the arrow to move the terms to the model list on the right. Return to the GENERAL FACTORIAL ANOVA dialog box and click on *Options*. One can request observed and predicted means for subgroups defined by factors. Return to the GENERAL FACTORIAL ANOVA dialog box and click on *Contrasts* to display parameter estimates with standard errors, t statistics, and confidence intervals for comparing means for levels of each factor. Change the contrast type to *Simple* to compare each level to a baseline level, either the last (such as in setting up (1, 0) dummy variables for all levels but the last one) or the first. In the confidence interval option, you can choose a separate interval for each parameter or simultaneous intervals, such as Bonferrroni.

One can conduct repeated measures ANOVA using the Advanced Statistics module that SPSS offers as an option. On the STATISTICS menu, select the ANOVA MODELS option with the REPEATED MEASURES suboption. In the REPEATED MEASURES DEFINE FACTOR(S) dialog window, type the name and number of levels of the within-subjects factor (such as *influenc* and 3) and click on *Add*. Then click on *Define* to define the model.

Now, in the REPEATED MEASURES ANOVA dialog box, select the between-subjects factors (if there are any) and define the range of its values, and select the response variable for each level of the within-subjects factor (such as movies, TV, rock). The default is a model containing all the factor interactions. Click on *Model*, and customize the model if you want to delete an interaction. At this stage, options are also provided for the within-subjects tests, including "epsilon corrected" tests (such as the Greenhouse-Geisser) that adjust for violations of the assumption of sphericity. Return to the REPEATED MEASURES ANOVA dialog box and click on *Contrasts*, and options are provided for displaying parameter estimates and confidence intervals for contrasts comparing means in different factor levels, and for individual or Bonferroni confidence intervals; change the contrast type to *Simple* for estimates of the between-subjects factors to refer to comparing each factor level to the first or last level. Return to the REPEATED MEASURES ANOVA dialog box and click on *Options*, and one can request between-subjects observed and predicted means and various model diagnostics.

For repeated measures problems, one can use standard multivariate tests that do not make the assumption of sphericity for the joint distribution of the repeated responses. They are less powerful than the repeated measures ANOVA methods when the sphericity assumption is not violated. The multivariate methods are available by selecting the MULTIVARIATE suboption under the ANOVA MODELS option on the STATISTICS menu.

The Advanced Statistics module of SPSS also has a general linear model (GLM) procedure one can use to conduct regression, analysis of variance, and analysis of covariance (see *SPSS Advanced Statistics 7.0 Update*, SPSS, Inc., Chicago, 1996). For factorial ANOVA, from the STATISTICS menu select the GENERAL LINEAR MODEL option with the GLM–GENERAL FACTORIAL suboption. Select the dependent variable and identify the factors that define the cross-classification for the means as fixed factors. Select *Custom* to specify only a subset of the interactions, and use the Type III sum of squares. For repeated measures ANOVA, from the STATISTICS menu select the GENERAL LINEAR MODEL option with the GLM–REPEATED MEASURES suboption. Then proceed as described above, using the default Type III sum of squares.

Chapter 13: Analysis of Covariance Models

To fit an analysis of covariance model, one can set up dummy variables for qualititative predictors and use ordinary regression procedures, such as the one described earlier for Chapter 11. To create cross-product terms for interactions, after creating the data file, select *Compute* on the TRANSFORM menu and create products of appropriate variables.

Alternatively, one can fit an analysis of covariance model using the Advanced Statistics module that SPSS offers as an option. On the STATISTICS menu, select the ANOVA MODELS option with the GENERAL FACTORIAL suboption (or, with Version 7.0 of the Advanced Statistics module, on the STATISTICS menu one can select the GENERAL LINEAR MODEL option with the GLM–GENERAL FACTORIAL suboption). This analysis of variance procedure has the option of adding covariates to the model as well as qualitative factors.

The GENERAL FACTORIAL ANOVA dialog box has a variety of options. The default model contains all interactions between factors but no interaction between covariates or between a covariate and a factor. (This is the basic model discussed in Section 13.2 when there is only a single factor and a single covariate.) Click on *Model* to build a customized model that contains only some or none of the interactions. Highlight variables, select *Interaction* or *Main Effects* from the Build Term(s) list, and click on the arrow to move the terms to the model list on the right. Back in the GENERAL FACTORIAL ANOVA dialog box, click on *Contrasts* to request parameter estimates with their standard errors, t test statistics, and confidence intervals for comparing levels of factors; highlight a factor and change the contrasts to *Simple* to compare each factor level to the last or to the first level [corresponding to (1, 0) dummy variables for each category except that baseline]. One can also choose simultaneous confidence intervals, such as Bonferroni, for comparing pairs of factor levels. Click on *Options* to request observed and predicted means (including adjusted means).

Chapter 14: Model Building with Multiple Regression

In the LINEAR REGRESSION dialog window for regression analyses, one can select a "method" for fitting the model, among which are options such as BACKWARD, FORWARD, and STEPWISE for selecting predictors in the model.

One can plot residuals and studentized residuals and request all partial regression plots by clicking on *Plots* in the LINEAR REGRESSION dialog window and then making appropriate selections in the PLOTS dialog box. To obtain predicted values, residuals, studentized residuals, hat (leverage) values, and influence diagnostics such as DFFITS and DFBETAS, click on *Save* in the LINEAR REGRESSION dialog box. The resulting LINEAR REGRESSION: SAVE NEW VARIABLES dialog box contains options for these and other diagnostics.

To fit a quadratic regression model, on the STATISTICS menu select the REGRESSION option with the CURVE ESTIMATION suboption. Then, in the CURVE ESTIMATION dialog box, select the variables and choose the *Quadratic* model. The PLOT MODELS option provides a plot of the fitted curve.

As of 1996, SPSS does not have a single program for generalized linear models. However, one can fit an exponential regression model by selecting the CURVE ESTIMATION suboption under the REGRESSION choice in the STATISTICS menu. The exponential (and quadratic) models are options in the dialog box. (This provides a somewhat different fit than using GLM software, since it assumes the log of Y, rather than Y, is normally distributed with constant variance; results are usually similar.)

Chapter 15: Logistic Regression

One can fit logistic regression models using the Advanced Statistics module that SPSS offers as an option. Enter the STATISTICS menu with the REGRESSION option and the LOGISTIC suboption. In the LOGISTIC REGRESSION dialog box, identify the binary response (dependent) variable and the explanatory predictors (covariates). Highlight variables in the source list and click on $a * b$ to create an interaction term.

The dialog box contains several options. Click on *Method* for stepwise model selection procedures, such as backward elimination. To enter a numerically coded variable in the model as a qualitative predictor (i.e., using dummy variables), click on the *Categorical* options and identify the covariate as a categorical covariate. Click on *Contrast* to see several options for setting up dummy variables for categorical predictors; the *Simple* contrast constructs them as in this text, in which the final category is the baseline. Click on *Save* to save predicted probabilities, measures of influence such as leverage values and DFBETAS, and adjusted residuals (called *Standardized residuals*). Click on *Options* to open a dialog box that contains an option to construct confidence intervals for exponentiated parameters.

One can also fit such models using the LOGLINEAR option with the LOGIT suboption in the STATISTICS menu. One identifies the dependent variable, selects qualitative predictors as factors, and selects quantitative predictors as cell covariates. The default fit is the saturated model for the factors, without including any covariates. To change this, click on *Model* and select a *Custom* model, entering the predictors and relevant interactions as terms in a customized (unsaturated) model. Clicking on *Options*, one can also display adjusted residuals for model fits. This approach is well suited for logit models with qualitative predictors, such as discussed in Section 15.3, since standard output includes observed and expected frequencies. When the data file contains the data as cell counts, such as binomial numbers of successes and failures, one weights each cell by the cell count using the WEIGHT CASES option in the DATA menu.

For loglinear models, one uses the LOGLINEAR option with GENERAL suboption in the STATISTICS menu. One enters the factors for the model. The default is the saturated model, so click on *Model* and select a *Custom* model. Enter the factors as terms in a customized (unsaturated) model and then select additional interaction effects. Click on *Options* to show options for displaying observed and expected frequencies and adjusted residuals. When the data file contains the data as cell counts for the various combinations of factors rather than as responses listed for individual subjects, weight each cell by the cell count using the WEIGHT CASES option in the DATA menu. (See Chapter 6 in *SPSS 6.1 for Windows Update*, SPSS, Inc., Chicago, 1994 for further details about this GENLOG procedure in SPSS for GENERAL LOGLINEAR modeling for a wide variety of loglinear and logit models.)

SPSS can also fit logit models for categorical response variables having several response categories. Choose the LOGLINEAR option with LOGIT suboption on the STATISTICS menu. The model treats the response variable as nominal (See pp. 71-78 of *SPSS 6.1 for Windows Update*, SPSS, Inc., Chicago, 1994). As of 1996, SPSS does not have a procedure for cumulative logit models.

Bibliography

Friendly, M. (1991). *SAS System for Statistical Graphics*, Cary, NC: SAS Institute.

Freund, R. J., and Littell, R. C. (1991). *SAS System for Regression*, 2nd ed. Cary, NC: SAS Institute.

Hatcher, L., and Stepanski, E. J. (1994). *A Step-by-Step Approach to Using the SAS System for Univariate and Multivariate Statistics*. Cary, NC: SAS Institute.

Littell, R., Freund, R., and Spector, P. (1991). *SAS System for Linear Models*, 3rd ed. Cary, NC: SAS Institute.

Schlotzhauer, S. S., and Littell, R. C. (1987). *SAS System for Elementary Statistical Analysis*. Cary, NC: SAS Institute.

SAS (1995). *Logistic Regression Examples Using the SAS System*. Cary, NC: SAS Institute.

SPSS (1993a). *SPSS for Windows Base System User's Guide, Release 6.0*. Chicago: SPSS, Inc.

SPSS (1993b). *SPSS for Windows Advanced Statistics, Release 6.0*. Chicago: SPSS, Inc.

StatXact-3 for Windows (1995). *Software for Exact Nonparametric Inference*. Cambridge, MA: Cytel Software.

Stokes, M. E., Davis, C. S., and Koch, G. G. (1995). *Categorical Data Analysis Using the SAS System*. Cary, NC: SAS Institute.

Wolfinger, R., and Chang, M. (1995). Comparing the SAS GLM and MIXED procedures for repeated measures. Proceedings of the Annual SAS Institute Meeting (SUGI). Cary, NC: SAS Institute.

Tables

TABLE A Normal curve tail probabilities. Standard normal probability in right-hand tail (for negative values of z, probabilities are found by symmetry)

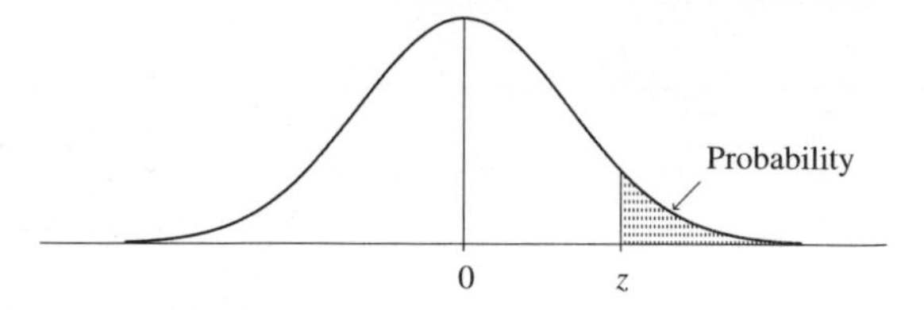

z	.00	.01	.02	.03	.04	.05	.06	.07	.08	.09
	Second Decimal Place of z									
0.0	.5000	.4960	.4920	.4880	.4840	.4801	.4761	.4721	.4681	.4641
0.1	.4602	.4562	.4522	.4483	.4443	.4404	.4364	.4325	.4286	.4247
0.2	.4207	.4168	.4129	.4090	.4052	.4013	.3974	.3936	.3897	.3859
0.3	.3821	.3783	.3745	.3707	.3669	.3632	.3594	.3557	.3520	.3483
0.4	.3446	.3409	.3372	.3336	.3300	.3264	.3228	.3192	.3156	.3121
0.5	.3085	.3050	.3015	.2981	.2946	.2912	.2877	.2843	.2810	.2776
0.6	.2743	.2709	.2676	.2643	.2611	.2578	.2546	.2514	.2483	.2451
0.7	.2420	.2389	.2358	.2327	.2296	.2266	.2236	.2206	.2177	.2148
0.8	.2119	.2090	.2061	.2033	.2005	.1977	.1949	.1922	.1894	.1867
0.9	.1841	.1814	.1788	.1762	.1736	.1711	.1685	.1660	.1635	.1611
1.0	.1587	.1562	.1539	.1515	.1492	.1469	.1446	.1423	.1401	.1379
1.1	.1357	.1335	.1314	.1292	.1271	.1251	.1230	.1210	.1190	.1170
1.2	.1151	.1131	.1112	.1093	.1075	.1056	.1038	.1020	.1003	.0985
1.3	.0968	.0951	.0934	.0918	.0901	.0885	.0869	.0853	.0838	.0823
1.4	.0808	.0793	.0778	.0764	.0749	.0735	.0722	.0708	.0694	.0681
1.5	.0668	.0655	.0643	.0630	.0618	.0606	.0594	.0582	.0571	.0559
1.6	.0548	.0537	.0526	.0516	.0505	.0495	.0485	.0475	.0465	.0455
1.7	.0446	.0436	.0427	.0418	.0409	.0401	.0392	.0384	.0375	.0367
1.8	.0359	.0352	.0344	.0336	.0329	.0322	.0314	.0307	.0301	.0294
1.9	.0287	.0281	.0274	.0268	.0262	.0256	.0250	.0244	.0239	.0233
2.0	.0228	.0222	.0217	.0212	.0207	.0202	.0197	.0192	.0188	.0183
2.1	.0179	.0174	.0170	.0166	.0162	.0158	.0154	.0150	.0146	.0143
2.2	.0139	.0136	.0132	.0129	.0125	.0122	.0119	.0116	.0113	.0110
2.3	.0107	.0104	.0102	.0099	.0096	.0094	.0091	.0089	.0087	.0084
2.4	.0082	.0080	.0078	.0075	.0073	.0071	.0069	.0068	.0066	.0064
2.5	.0062	.0060	.0059	.0057	.0055	.0054	.0052	.0051	.0049	.0048
2.6	.0047	.0045	.0044	.0043	.0041	.0040	.0039	.0038	.0037	.0036
2.7	.0035	.0034	.0033	.0032	.0031	.0030	.0029	.0028	.0027	.0026
2.8	.0026	.0025	.0024	.0023	.0023	.0022	.0021	.0021	.0020	.0019
2.9	.0019	.0018	.0017	.0017	.0016	.0016	.0015	.0015	.0014	.0014
3.0	.00135									
3.5	.000233									
4.0	.0000317									
4.5	.00000340									
5.0	.000000287									

Source: R. E. Walpole, *Introduction to Statistics* (New York: Macmillan, 1968).

TABLE B *t* Distribution. Values corresponding to various right-hand tail probabilities

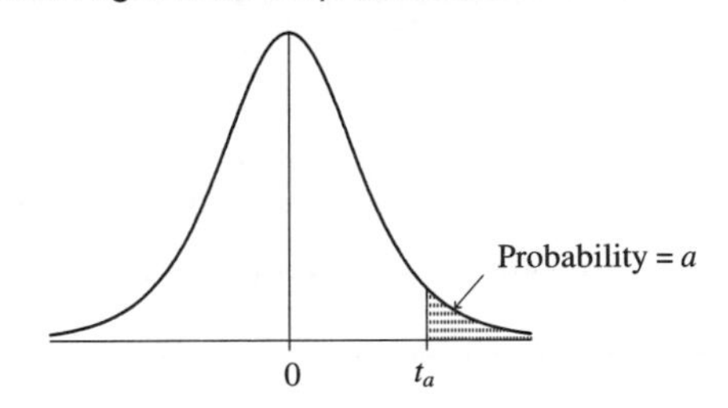

df	$t_{.100}$	$t_{.050}$	$t_{.025}$	$t_{.010}$	$t_{.005}$
1	3.078	6.314	12.7063	31.821	63.657
2	1.886	2.920	4.303	6.965	9.925
3	1.638	2.353	3.182	4.541	5.841
4	1.533	2.132	2.776	3.747	4.604
5	1.476	2.015	2.571	3.365	4.032
6	1.440	1.943	2.447	3.143	3.707
7	1.415	1.895	2.365	2.998	3.499
8	1.397	1.860	2.306	2.896	3.355
9	1.383	1.833	2.262	2.821	3.250
10	1.372	1.812	2.228	2.764	3.169
11	1.363	1.796	2.201	2.718	3.106
12	1.356	1.782	2.179	2.681	3.055
13	1.350	1.771	2.160	2.650	3.012
14	1.345	1.761	2.145	2.624	2.977
15	1.341	1.753	2.131	2.602	2.947
16	1.337	1.746	2.120	2.583	2.921
17	1.333	1.740	2.110	2.567	2.898
18	1.330	1.734	2.101	2.552	2.878
19	1.328	1.729	2.093	2.539	2.861
20	1.325	1.725	2.086	2.528	2.845
21	1.323	1.721	2.080	2.518	2.831
22	1.321	1.717	2.074	2.508	2.819
23	1.319	1.714	2.069	2.500	2.807
24	1.318	1.711	2.064	2.492	2.797
25	1.316	1.708	2.060	2.485	2.787
26	1.315	1.706	2.056	2.479	2.779
27	1.314	1.703	2.052	2.473	2.771
28	1.313	1.701	2.048	2.467	2.763
29	1.311	1.699	2.045	2.462	2.756
∞	1.282	1.645	1.960	2.326	2.576

Source: "Table of Percentage Points of the t-Distribution." Computed by Maxine Merrington, Biometrika, 32 (1941): 300. Reproduced by permission of the Biometrika trustees.

TABLE C Chi-Squared distribution values for various right-tail probabilities

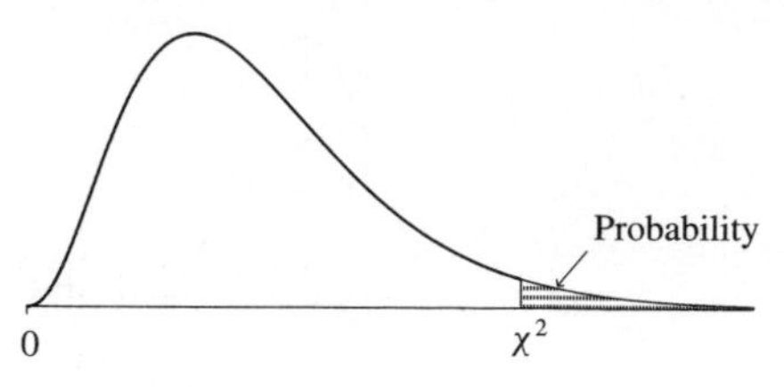

df	0.250	0.100	0.050	0.025	0.010	0.005	0.001
	Right-Tail Probability						
1	1.32	2.71	3.84	5.02	6.63	7.88	10.83
2	2.77	4.61	5.99	7.38	9.21	10.60	13.82
3	4.11	6.25	7.81	9.35	11.34	12.84	16.27
4	5.39	7.78	9.49	11.14	13.28	14.86	18.47
5	6.63	9.24	11.07	12.83	15.09	16.75	20.52
6	7.84	10.64	12.59	14.45	16.81	18.55	22.46
7	9.04	12.02	14.07	16.01	18.48	20.28	24.32
8	10.22	13.36	15.51	17.53	20.09	21.96	26.12
9	11.39	14.68	16.92	19.02	21.67	23.59	27.88
10	12.55	15.99	18.31	20.48	23.21	25.19	29.59
11	13.70	17.28	19.68	21.92	24.72	26.76	31.26
12	14.85	18.55	21.03	23.34	26.22	28.30	32.91
13	15.98	19.81	22.36	24.74	27.69	29.82	34.53
14	17.12	21.06	23.68	26.12	29.14	31.32	36.12
15	18.25	22.31	25.00	27.49	30.58	32.80	37.70
16	19.37	23.54	26.30	28.85	32.00	34.27	39.25
17	20.49	24.77	27.59	30.19	33.41	35.72	40.79
18	21.60	25.99	28.87	31.53	34.81	37.16	42.31
19	22.72	27.20	30.14	32.85	36.19	38.58	43.82
20	23.83	28.41	31.41	34.17	37.57	40.00	45.32
25	29.34	34.38	37.65	40.65	44.31	46.93	52.62
30	34.80	40.26	43.77	46.98	50.89	53.67	59.70
40	45.62	51.80	55.76	59.34	63.69	66.77	73.40
50	56.33	63.17	67.50	71.42	76.15	79.49	86.66
60	66.98	74.40	79.08	83.30	88.38	91.95	99.61
70	77.58	85.53	90.53	95.02	100.4	104.2	112.3
80	88.13	96.58	101.8	106.6	112.3	116.3	124.8
90	98.65	107.6	113.1	118.1	124.1	128.3	137.2
100	109.1	118.5	124.3	129.6	135.8	140.2	149.5

Source: Calculated using *StaTable*, software from Cytel Software, Cambridge, MA.

TABLE D *F* Distribution

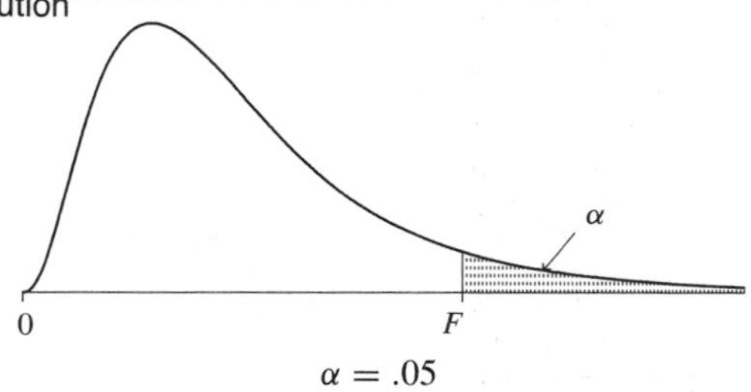

$\alpha = .05$

	df_1									
df_2	1	2	3	4	5	6	8	12	24	∞
1	161.4	199.5	215.7	224.6	230.2	234.0	238.9	243.9	249.0	254.3
2	18.51	19.00	19.16	19.25	19.30	19.33	19.37	19.41	19.45	19.50
3	10.13	9.55	9.28	9.12	9.01	8.94	8.84	8.74	8.64	8.53
4	7.71	6.94	6.59	6.39	6.26	6.16	6.04	5.91	5.77	5.63
5	6.61	5.79	5.41	5.19	5.05	4.95	4.82	4.68	4.53	4.36
6	5.99	5.14	4.76	4.53	4.39	4.28	4.15	4.00	3.84	3.67
7	5.59	4.74	4.35	4.12	3.97	3.87	3.73	3.57	3.41	3.23
8	5.32	4.46	4.07	3.84	3.69	3.58	3.44	3.28	3.12	2.93
9	5.12	4.26	3.86	3.63	3.48	3.37	3.23	3.07	2.90	2.71
10	4.96	4.10	3.71	3.48	3.33	3.22	3.07	2.91	2.74	2.54
11	4.84	3.98	3.59	3.36	3.20	3.09	2.95	2.79	2.61	2.40
12	4.75	3.88	3.49	3.26	3.11	3.00	2.85	2.69	2.50	2.30
13	4.67	3.80	3.41	3.18	3.02	2.92	2.77	2.60	2.42	2.21
14	4.60	3.74	3.34	3.11	2.96	2.85	2.70	2.53	2.35	2.13
15	4.54	3.68	3.29	3.06	2.90	2.79	2.64	2.48	2.29	2.07
16	4.49	3.63	3.24	3.01	2.85	2.74	2.59	2.42	2.24	2.01
17	4.45	3.59	3.20	2.96	2.81	2.70	2.55	2.38	2.19	1.96
18	4.41	3.55	3.16	2.93	2.77	2.66	2.51	2.34	2.15	1.92
19	4.38	3.52	3.13	2.90	2.74	2.63	2.48	2.31	2.11	1.88
20	4.35	3.49	3.10	2.87	2.71	2.60	2.45	2.28	2.08	1.84
21	4.32	3.47	3.07	2.84	2.68	2.57	2.42	2.25	2.05	1.81
22	4.30	3.44	3.05	2.82	2.66	2.55	2.40	2.23	2.03	1.78
23	4.28	3.42	3.03	2.80	2.64	2.53	2.38	2.20	2.00	1.76
24	4.26	3.40	3.01	2.78	2.62	2.51	2.36	2.18	1.98	1.73
25	4.24	3.38	2.99	2.76	2.60	2.49	2.34	2.16	1.96	1.71
26	4.22	3.37	2.98	2.74	2.59	2.47	2.32	2.15	1.95	1.69
27	4.21	3.35	2.96	2.73	2.57	2.46	2.30	2.13	1.93	1.67
28	4.20	3.34	2.95	2.71	2.56	2.44	2.29	2.12	1.91	1.65
29	4.18	3.33	2.93	2.70	2.54	2.43	2.28	2.10	1.90	1.64
30	4.17	3.32	2.92	2.69	2.53	2.42	2.27	2.09	1.89	1.62
40	4.08	3.23	2.84	2.61	2.45	2.34	2.18	2.00	1.79	1.51
60	4.00	3.15	2.76	2.52	2.37	2.25	2.10	1.92	1.70	1.39
120	3.92	3.07	2.68	2.45	2.29	2.17	2.02	1.83	1.61	1.25
∞	3.84	2.99	2.60	2.37	2.21	2.09	1.94	1.75	1.52	1.00

Source: From Table V of R. A. Fisher and F. Yates, *Statistical Tables for Biological, Agricultural and Medical Research*, published by Longman Group Ltd., London, 1974. (Previously published by Oliver & Boyd, Edinburgh.) Reprinted by permission of the authors and publishers.

TABLE D (continued)

$\alpha = .01$

df_2 \ df_1	1	2	3	4	5	6	8	12	24	∞
1	4052	4999	5403	5625	5764	5859	5981	6106	6234	6366
2	98.49	99.01	99.17	99.25	99.30	99.33	99.36	99.42	99.46	99.50
3	34.12	30.81	29.46	28.71	28.24	27.91	27.49	27.05	26.60	26.12
4	21.20	18.00	16.69	15.98	15.52	15.21	14.80	14.37	13.93	13.46
5	16.26	13.27	12.06	11.39	10.97	10.67	10.27	9.89	9.47	9.02
6	13.74	10.92	9.78	9.15	8.75	8.47	8.10	7.72	7.31	6.88
7	12.25	9.55	8.45	7.85	7.46	7.19	6.84	6.47	6.07	5.65
8	11.26	8.65	7.59	7.01	6.63	6.37	6.03	5.67	5.28	4.86
9	10.56	8.02	6.99	6.42	6.06	5.80	5.47	5.11	4.73	4.31
10	10.04	7.56	6.55	5.99	5.64	5.39	5.06	4.71	4.33	3.91
11	9.65	7.20	6.22	5.67	5.32	5.07	4.74	4.40	4.02	3.60
12	9.33	6.93	5.95	5.41	5.06	4.82	4.50	4.16	3.78	3.36
13	9.07	6.70	5.74	5.20	4.86	4.62	4.30	3.96	3.59	3.16
14	8.86	6.51	5.56	5.03	4.69	4.46	4.14	3.80	3.43	3.00
15	8.68	6.36	5.42	4.89	4.56	4.32	4.00	3.67	3.29	2.87
16	8.53	6.23	5.29	4.77	4.44	4.20	3.89	3.55	3.18	2.75
17	8.40	6.11	5.18	4.67	4.34	4.10	3.79	3.45	3.08	2.65
18	8.28	6.01	5.09	4.58	4.25	4.01	3.71	3.37	3.00	2.57
19	8.18	5.93	5.01	4.50	4.17	3.94	3.63	3.30	2.92	2.49
20	8.10	5.85	4.94	4.43	4.10	3.87	3.56	3.23	2.86	2.42
21	8.02	5.78	4.87	4.37	4.04	3.81	3.51	3.17	2.80	2.36
22	7.94	5.72	4.82	4.31	3.99	3.76	3.45	3.12	2.75	2.31
23	7.88	5.66	4.76	4.26	3.94	3.71	3.41	3.07	2.70	2.26
24	7.82	5.61	4.72	4.22	3.90	3.67	3.36	3.03	2.66	2.21
25	7.77	5.57	4.68	4.18	3.86	3.63	3.32	2.99	2.62	2.17
26	7.72	5.53	4.64	4.14	3.82	3.59	3.29	2.96	2.58	2.13
27	7.68	5.49	4.60	4.11	3.78	3.56	3.26	2.93	2.55	2.10
28	7.64	5.45	4.57	4.07	3.75	3.53	3.23	2.90	2.52	2.06
29	7.60	5.42	4.54	4.04	3.73	3.50	3.20	2.87	2.49	2.03
30	7.56	5.39	4.51	4.02	3.70	3.47	3.17	2.84	2.47	2.01
40	7.31	5.18	4.31	3.83	3.51	3.29	2.99	2.66	2.29	1.80
60	7.08	4.98	4.13	3.65	3.34	3.12	2.82	2.50	2.12	1.60
120	6.85	4.79	3.95	3.48	3.17	2.96	2.66	2.34	1.95	1.38
∞	6.64	4.60	3.78	3.32	3.02	2.80	2.51	2.18	1.79	1.00

TABLE D (continued)

$\alpha = .001$

	df_1									
df_2	1	2	3	4	5	6	8	12	24	∞
1	405284	500000	540379	562500	576405	585937	598144	610667	623497	636619
2	998.5	999.0	999.2	999.2	999.3	999.3	999.4	999.4	999.5	999.5
3	167.5	148.5	141.1	137.1	134.6	132.8	130.6	128.3	125.9	123.5
4	74.14	61.25	56.18	53.44	51.71	50.53	49.00	47.41	45.77	44.05
5	47.04	36.61	33.20	31.09	29.75	28.84	27.64	26.42	25.14	23.78
6	35.51	27.00	23.70	21.90	20.81	20.03	19.03	17.99	16.89	15.75
7	29.22	21.69	18.77	17.19	16.21	15.52	14.63	13.71	12.73	11.69
8	25.42	18.49	15.83	14.39	13.49	12.86	12.04	11.19	10.30	9.34
9	22.86	16.39	13.90	12.56	11.71	11.13	10.37	9.57	8.72	7.81
10	21.04	14.91	12.55	11.28	10.48	9.92	9.20	8.45	7.64	6.76
11	19.69	13.81	11.56	10.35	9.58	9.05	8.35	7.63	6.85	6.00
12	18.64	12.97	10.80	9.63	8.89	8.38	7.71	7.00	6.25	5.42
13	17.81	12.31	10.21	9.07	8.35	7.86	7.21	6.52	5.78	4.97
14	17.14	11.78	9.73	8.62	7.92	7.43	6.80	6.13	5.41	4.60
15	16.59	11.34	9.34	8.25	7.57	7.09	6.47	5.81	5.10	4.31
16	16.12	10.97	9.00	7.94	7.27	6.81	6.19	5.55	4.85	4.06
17	15.72	10.66	8.73	7.68	7.02	6.56	5.96	5.32	4.63	3.85
18	15.38	10.39	8.49	7.46	6.81	6.35	5.76	5.13	4.45	3.67
19	15.08	10.16	8.28	7.26	6.61	6.18	5.59	4.97	4.29	3.52
20	14.82	9.95	8.10	7.10	6.46	6.02	5.44	4.82	4.15	3.38
21	14.59	9.77	7.94	6.95	6.32	5.88	5.31	4.70	4.03	3.26
22	14.38	9.61	7.80	6.81	6.19	5.76	5.19	4.58	3.92	3.15
23	14.19	9.47	7.67	6.69	6.08	5.65	5.09	4.48	3.82	3.05
24	14.03	9.34	7.55	6.59	5.98	5.55	4.99	4.39	3.74	2.97
25	13.88	9.22	7.45	6.49	5.88	5.46	4.91	4.31	3.66	2.89
26	13.74	9.12	7.36	6.41	5.80	5.38	4.83	4.24	3.59	2.82
27	13.61	9.02	7.27	6.33	5.73	5.31	4.76	4.17	3.52	2.75
28	13.50	8.93	7.19	6.25	5.66	5.24	4.69	4.11	3.46	2.70
29	13.39	8.85	7.12	6.19	5.59	5.18	4.64	4.05	3.41	2.64
30	13.29	8.77	7.05	6.12	5.53	5.12	4.58	4.00	3.36	2.59
40	12.61	8.25	6.60	5.70	5.13	4.73	4.21	3.64	3.01	2.23
60	11.97	7.76	6.17	5.31	4.76	4.37	3.87	3.31	2.69	1.90
120	11.38	7.31	5.79	4.95	4.42	4.04	3.55	3.02	2.40	1.56
∞	10.83	6.91	5.42	4.62	4.10	3.74	3.27	2.74	2.13	1.00

Answers to Selected Exercises

Chapter 1

3a. An individual automobile **b.** All automobiles of that type used in the EPA tests **c.** All automobiles of that type that are or may be manufactured

5a. All students at the University of Wisconsin **b.** A statistic, since calculated only for the 100 sampled students

9a. All adult Americans **b.** Proportion of all adult Americans who would answer definitely or probably true **c.** Sample proportion .523 estimates population proportion **d.** No, it is a prediction of the population value but will not equal it exactly, because the sample is only a very small subset of the population.

Chapter 2

3a. Ordinal **b.** Nominal **c.** Interval **d.** Nominal **e.** Nominal **f.** Ordinal **g.** Interval **h.** Ordinal **i.** Nominal **j.** Interval **k.** Nominal

5a. Interval **b.** Ordinal **c.** Nominal

7. Ordinal **9.** b, c, d, e, f **11.** No, a volunteer sample.

13a. Number subjects 00001 through 36000 = 300(120); first five selected are those numbered 15011, 06907, 14342, 12765, 21382

13b. Skip number is $k = 36{,}000/5 = 7200$. First selection is subject numbered 6907; next is 6907 + 7200 = 14107; next is 14107 + 7200 = 21307, next is 21307 + 7200 = 28507; last is 28507 + 7200 = 35707.

17. Sampling variability, slightly different sampling methods and question wording; very accurate

21. Less than 50% response rate **23.** b, c.

27. Because of skipping names, two subjects listed next to each other on the list cannot both be in the sample, so not all samples are equally likely.

Chapter 3

1c. 2.6 **d.** 2, 2

5b. 16.6 **c.** 12.0 **d.** 13.9 **e.** Mean = 27.6, Median = 24.0, Standard deviation = 12.4; the lengths of stay tended to be longer 25 years ago **f.** Of the 11 observations, the middle in magnitude is 13.0, the median; cannot calculate mean, but substituting 40 for censored observation gives lower bound for it of 18.7.

7b. Median = 2 or 3 times a month, Mode = Not at all. **c.** Mean = 4.5.

9a. Mean = 73.2 **b.** Median = 16, very different because the outlier (California) affects the mean but not the median

11a. Mean = 612.8, median = 515, suggests skew to right. **b.** The outlier has the effect of greatly increasing the mean, standard deviation, maximum value and range. The quartiles and the interquartile range are barely affected.

13b. Mean = 79.5 for Eastern Europe, 33.9 for South America **c.** Median = 79 for Eastern Europe, 34 for South America

15a. $38,851 **b.** $38,147

17a. Larger, since distribution is probably skewed to the right **b.** (iii)

19b. Mean = .728, median = .726, range = .306

21. 88.8 No, the distribution is extremely skewed to the right.

27. Very highly skewed to the right.

29. Skewed to right; the lowest possible value of 0 is just 1.15 standard deviations below the mean.

31a. Skewed left **b.** bell shaped **c.** skewed right **d.** skewed left **e.** skewed right **f.** skewed left **g.** skewed right **h.** skewed right **i.** U-shaped

33a. Not far enough **b.** Median = not far enough, mean requires scores for categories

37. Skewed to the right. **39.** −20.

41. Range within three standard deviations of mean is from $700 - 3(100) = 400$ to $700 + 3(100) = 1000$.

43a. About 68% of prices between $80,000 and $160,000, about 95% between $40,000 and $200,000, nearly all between $0 and $240,000 **b.** $140,000 **c.** Somewhat skewed to the right

45. Connecticut is an outlier.

49. $\bar{Y}$ (mean) and s (standard deviation) refer to sample, and μ and σ refer to population.

57. Median = $44,000, since this distribution is probably highly skewed to right, so mean exceeds median.

59. Standard deviation equals about 1/3. **61.** c.

63. Population sizes vary by state; the overall rate gives more weight to states with larger population sizes.

65. Multiply each by 1.6.

69a. 1/4, 1/9, 1/100 **b.** .25 vs. .05; .25 is the maximum possible, but most distributions have much less than 25% of the distribution falling more than two standard deviations from the mean. .05 holds in bell-shaped case, and .25 is upper bound for *any* distribution.

Chapter 4

1b. .12 **3a.** $P(0) = .3$, $P(1) = .6$, $P(2) = .1$. **b.** .8

5a. 0,1,2,3,4,5,6,7,8,9 **b.** $P(0) = P(1) = ... = P(9) = .1$ **c.** 4.5 **d.** 2.9

7a. .1587 **b.** .1587 **c.** .2514 **d.** .0099 **9a.** 2.33 **b.** 1.96 **c.** 1.64 **d.** 1.28 **e.** .67 **f.** 0.00

11a. 1.28 **b.** 1.64 **c.** 2.05 **d.** 2.33 **13a.** 75th percentile, **b.** 25th percentile $z = .67$

15a. .0179 **b.** .0179 **c.** .9821 **d.** .9642

17. −.67 and +.67 **19a.** .1056 **b.** .8944 **c.** 120.5 **d.** 79.5 **e.** 89, 100, 111

21a. .21 **b.** 12.6, 19.4, IQR = 6.8 **c.** 11.8 **d.** skewed right **23a.** .048 **b.** .159 **c.** .904

25a. .159 **b.** No, probably skewed right since mean is only 1.6 standard deviations above 0.

27a. P(0) = P(1) = .5 **b.** P(0) = .25, P(.5) = .5, P(1) = .25
c. P(0) = .125, P(1/3) = .375, P(2/3) = .375, P(1) = .125
d. P(0) = .0625, P(.25) = .250, P(.50) = .375, P(.75) = .250, P(1) = .0625
e. Becoming more bell-shaped

29a. 13.6, 1.0 **b.** 13.6, .5 **c.** 13.6, .3 **d.** Mean stays same, standard deviation decreases

31a. .25 **b.** mean = 250, std. dev. = 25, prob. = .02 **33.** .68

35a. .50 **b.** .42 **c.** .16 **d.** −.67, no
e. Approximately standard normal, mean = 0, standard deviation = 1

37. Bell-shaped curves with means of 100 and standard deviations of (a) 15, (b) 1.0, (c) 3.0.

39a. .79 **b.** .90 **c.** .988

43a. $P(0) = P(1) = .50$, mean = .50
b. $Y = 0$ has relative frequency .4 and $Y = 1$ has relative frequency .6 **d.** .50, .158

47. a, c, d **49.** a **51a.** .16 for Group A, .31 for Group B **b.** .66 **c.** .75

53a. Skewed left, mean 60, standard deviation 16
b. Mean 58.3, standard deviation 15.0, probably skewed left
c. Mean 60, standard deviation 1.6, normal
d. Skewed left, mean 60, standard deviation 16
e. Mean 60, standard deviation 0, sample mean = popul. mean
f. Normal, mean 60, standard deviation 1.6
g. No, Yes; mean of 40 is 12.5 standard deviations below mean of sampling distribution

55a. $\sigma^2 = (0 - .5)^2(.5) + (1 - .5)^2(.5) = .25$
$\sigma^2 = [0^2(.5) + 1^2(.5)] - (.5)^2 = .25$
b. $\mu = 0(1 - \pi) + 1(\pi) = \pi, \sigma^2 = [0^2(1 - \pi) + 1^2(\pi)] - \pi^2 = \pi(1 - \pi)$
c. Substitute $\sigma = \sqrt{\pi(1 - \pi)}$ in standard error $\sigma/\sqrt{n}$.

57. Exponent equals $-c^2/2\sigma^2$ for $y = \mu + c$ and $y = \mu - c$.

Chapter 5

1. $\bar{Y} = 3.0, s = 3.1$

3a. $4.17 \pm 1.96(.026) = (4.12, 4.22)$ **b.** (4.10, 4.24) **c.** Treating an ordinal scale as interval.

5a. 1.31, 5.42. **b.** Standard error = $5.418/\sqrt{637} = .215$ **c.** (.89, 1.73) **d.** (.76, 1.87)

7a. (4.3, 6.3) **b.** 400

9a. $4.1 \pm 1.96(.387)$, or (3.3, 4.9); wider, because z is larger (2.58 instead of 1.96).
b. (2.6, 5.6), doubling s has effect of doubling standard error and doubling width of the interval (except for rounding). **c.** (3.7, 4.5), quadrupling n halves the width.

11a. (2.7, 2.9) **b.** 2.2, no, since mean of 2.8 is less than 1.3 std. dev. above 0

13a. .017 **b.** (.009, .025), yes

15. A 95% confidence interval is $.728 \pm 1.96(.010)$, or (.71, .75)

17. $.255 \pm 2.58(.002)$, or (.250, .260)

19a. 95% confidence interval for proportion voting Democrat is $.53 \pm 1.96(.0133)$, or (.504, .556); predict Democrat the winner.
b. $.53 \pm 2.58(.0133)$, or (.496, .564); plausible that $\pi = .50$, so we cannot predict winner.

21a. .716 **b.** (.714, .717), yes

23a. $\hat{\pi} = 160/400 = .40$; $.40 \pm 2.58\sqrt{(.40)(.60)/400} = .40 \pm .06$, or (.34, .46); yes, because the interval consists entirely of numbers below .50, corresponding to Jones receiving a minority of the vote.
b. $\hat{\pi} = 16/40 = .40$; $.40 \pm 2.58\sqrt{(.40)(.60)/40} = .40 \pm .20$, or (.20, .60). We would not predict a winner, since the interval contains numbers both below and above .50. The point estimate is the same as in (a), but the interval is much wider because the sample size is so much smaller.

25. $B = .07$, $n = .25(1.96/.07)^2 = 196$

27a. $n = .25(1.96/.10)^2 = 96$ **b.** 384 **c.** $n = .25(2.58/.05)^2 = 666$ **d.** 16,641

29. $n = (1000)^2(1.64/100)^2 = 269$ **31.** Within $1.96(400)/\sqrt{246} = 50$

33. Indices $(n+1)/2 \pm \sqrt{n} = 21/2 \pm \sqrt{20}$, or (6.0, 15.0). The endpoints are the 6th largest and 15th largest (6th smallest) value, which are 3 and 14.

35a. Indices $(n+1)/2 \pm z(.5)\sqrt{n} = 55/2 \pm 1.64(.5)\sqrt{54}$, or (21.5 , 33.5). The endpoints are the 21st smallest and 34th smallest (21st largest) values, which are 11 and 19. This is the same as the 95% confidence interval. This anomaly can happen with confidence intervals constructed using discrete, instead of continuous, distributions.
b. The 99% confidence interval goes from 18th smallest to 18th largest (37th smallest) observation, or (10, 19).

45a. z-score increases **b.** Standard error increases in plus and minus part.

49. a. **51.** b, e **55.** $\bar{Y} = 4.8, s = 2.89$

57a. Usual method gives $.10 \pm 1.96\sqrt{(.10)(.90)/30} = .10 \pm .11$, or $(-.01, .21)$; this is unsatisfactory since π cannot be negative. Less approximate method gives (.03, .26); note that point estimate of .10 is far from center of interval, because of skew of sampling distribution.
b. When sample proportion = 0, usual interval is (0, 0); when sample proportion = 1, usual interval = (1, 1).

Chapter 6

1a. $P = 2(.1492) = .30$ **b.** $P = 2(.0062) = .012$; stronger evidence

3a. $z = \pm 1.64$ has two-tail probability of .10. **b.** weaker evidence

5a. $z = (103 - 100)/2 = 1.5$, $P = 2(.0668) = .134$ **b.** $z = -1.5$, $P = .134$

7a. $H_0 : \mu = 0$, $H_a : \mu \neq 0$
b. Standard error $= 1.253/\sqrt{996} = .0397$, $z = (-.052 - 0)/.0397 = -1.31$, $P = 2(.0951) = .19$. It is plausible that true mean = 0.
c. Distance between *Agree* and *Disagree* is twice the difference between *Strongly agree* and *Agree* or between *Strongly disagree* and *Disagree*.

9a. $z = (103 - 100)/2 = 1.5$, $P = 2(.0668) = .134$
b. Insufficient evidence to reject H_0; it is plausible that $\mu = 100$.
c. $103 \pm 1.96(2.0)$, or (99, 107)
d. No, the values between 99 and 107 are plausible for μ, not just the value of 100
e. (i) Reject H_0; (ii) and (iii), Do not reject H_0; P(Type I error) of .20 usually is unacceptable.

11a. $H_0 : \mu = 28.0$, $H_a : \mu \neq 28.0$, $z = (26.0 - 28.0)/1.42 = -1.41$; $P = 2(.0793) = .16$, so it is plausible that the mean has not changed.
b. Insufficient evidence to reject H_0 **c.** Type II error **d.** .08; weak evidence against H_0
e. .92 (right-tail probability); evidence does not support this alternative at all.

13a. $z = (.269 - .5)/.098 = -2.35$, $P = .019$, moderately strong evidence that $\pi \neq .5$ (and data suggest that $\pi < .5$).
b. $z = (1.0 - .5)/.0898 = 5.6$, $P < .0001$, extremely strong evidence that $\pi \neq .5$ (and data suggest that $\pi > .5$.

15a. $H_0 : \pi = .5$, $H_a : \pi < .5$
b. $z = (.5044 - .5)/.0114 = .39$, $P = .65$ (left-tail prob. below .39); evidence does not support the activist's claim.
c. Insufficient evidence to reject H_0.

17. $z = (0 - .06)/.0137 = -4.38$, $P < .0001$. If the sentences in the document could be regarded as a random sample of sentences written by this author, it is very unlikely that no sentences would begin with *whereas*; strong evidence against the author being Levine.

19a. $\hat{\pi} = 18/25 = .72$; $z = (.72 - .5)/.1 = 2.20$, $P = .028$; moderately strong evidence that $\pi \neq .5$ (and data suggest that $\pi > .5$).
b. Reject H_0; yes **c.** Type I error
d. Estimate of this proportion is 75/100 = .75. 95% confidence interval for true proportion is $.75 \pm 1.96(.0433)$, or (.67, .83); this is a majority of the population.

21a. First study gets $z = 1.95$ and $P = .051$; second study gets $z = 1.97$ and $P = .049$. Only the second study is significant at the .05 level.
b. These two studies give such similar results that they should not yield different conclusions. Reporting the actual P-value shows that each study has moderate evidence against H_0.

23a. 2.776 **b.** 2.145 **c.** 2.064 **d.** 2.060 **e.** 2.787

25a. $4.0 \pm 2.093(.894)$, or (2.1, 5.9)
b. No, since minimum possible value of 0 is only 1 standard deviation below the mean, but robustness of t confidence interval to violations of normality suggests that this analysis is valid.

27. As sample size increases, (a) standard error decreases and (b) t-score decreases, both contributing toward narrower interval.

29a. α = .02, .05, .10, .20 **b.** 99% confidence interval contains 500.
c. $P > .01$, so 99% confidence interval contains 500; $P < .02$, so 98% confidence interval does not contain 500.

31. $\bar{Y} = 4.0$, $s = 2.0$, $t = (4.0 - 0)/1.0 = 4.0$, $df = 3$, $.01 < P < .025$; moderately strong evidence that $\mu > 0$.

33. Under random sampling, the probability π that a particular person selected is female equals .53. Let $H_0 : \pi = .53$, $H_a : \pi \neq .53$. P-value is probability of observed outcome or even less likely outcomes, which equals .0001.

35a. $\mu = 1000(.02) = 20.0$, $\sigma = \sqrt{1000(.02)(.98)} = 4.4$
b. Yes, since 0 has z-score of $z = (0 - 20.0)/4.4 = -4.5$; i.e., 0 is 4.5 standard deviations below the mean.

37a. $.5^6 + .5^6 = .031$
b. This occurrence would be unusual if the probability truly equals .5 (e.g. P-value is .031 for testing H_0: $\pi = .5$ against H_a: $\pi \neq .5$, and $P = .016$ for $H_a : \pi > .5$), so the probability probably exceeds .5.

39. X = number of days of rain is binomial with $n = 2$, $\pi = .5$; probability of rain at least once is $P(1) + P(2) = .50 + .25 = .75$.

41a. Binomial mean and standard deviation for n = 1,000,000 and π = .0001 are $\mu = 1{,}000{,}000(.0001) = 100$ and $\sigma = \sqrt{1000000(.0001)(.9999)} = 10.0$
b. Region within two standard deviations of the mean is (80, 120).
c. $\mu = 200$, $\sigma = 14.1$, region within two standard deviations of mean is (172, 228).

43a. $.2^4 = .0016$ **b.** $P = .0016$; strong evidence that $\pi < .80$

45a. Probability that $\bar{Y} < 4.94$ when $\mu = 15$ is less than .001, since $\bar{Y} = 4.94$ has $z = (4.94 - 15)/3.0 = 3.35$.
b. $z = (4.94 - 5.0)/3.0 = -.02$, and probability of z below $-.02$ equals .49
c. $z = (4.94 - 8.0)/3.0 = -1.02$, and probability of z below -1.02 equals .15.
d. Reject if $\bar{Y} \geq 0 + 2.33(3.0) = 6.99$; $z = (6.99 - 10.0)/3.0 = -1.00$, and probability of failing to reject is probability that z falls below -1.00, which equals .16.
e. Reject if $\bar{Y} \geq 0 + 2.33(2.0) = 4.66$; $z = (4.66 - 10.0)/2.0 = -2.67$, and probability of z below -2.67 equals .004.

47a. $\hat{\pi} \geq .5 + 1.64\sqrt{.5(.5)/25} = .664$ gives $z \geq 1.64$ and $P \leq .05$.
b. $z = (.664 - .60)/.1 = .64$; fail to reject if $\hat{\pi} < .664$, which happens with probability .74.
c. Prob. of failing to reject, when H_0 is false, moves toward the prob. of failing to reject when H_0 is true, which is 1 minus prob. of rejecting when H_0 is true, or $1 - .05 = .95$.

55. Type I error: Diagnostic test is positive, indicating disease is present, even though disease is actually absent
Type II error: Diagnostic test is negative, indicating disease is absent, even though disease is actually present

57. If H_0 is true every time, expected number of times one would get $P < .05$ just by chance is $60(.05) = 3$; thus, the 3 cases could all be Type I errors.

59. a, c **61.** b

63. False; P(Type I error) is a single value, the fixed α value (such as .05), whereas P(Type II error) decreases as the true parameter value falls farther from the H_0 value in the direction of values in H_a.

65. False; since $P = .043$ is less than .05, we reject H_0 at that level and conclude that μ differs from 0. Similarly, the 95% confidence interval would not contain 0, indicating that 0 is implausible for μ. The confidence interval shows precisely which values *are* plausible.

69. Yes. The result of the confidence interval is equivalent to the result of the two-sided test. The two-tail P-value is .086, which exceeds .05, so the 95% confidence interval would include the null hypothesis value.

71. For any α-level below .52, we cannot reject H_0; thus, confidence intervals with confidence .48 or higher would contain the null hypothesis value. The 48% interval is the narrowest of these.

73a. Binomial, $n = 100$, $\pi = .05$

b. Yes. Since the binomial has $\mu = 100(.05) = 5$ and standard deviation $\sigma = \sqrt{100(.05)(.95)} = 2.2$, a value of 7 would not be unusual.

75. $H_0 : \pi = .10$, $H_a : \pi > .10$. The sample proportion of 0 outcomes is 4/20 = .20. The P-value is the probability of a sample proportion this large or even larger, if $\pi = .10$. For a binomial variable X with $n = 20$ and $\pi = .1$, $P = P(X \geq 4) = 1 - P(X \leq 3) = 1 - .867 = .133$.

77. b, d

Chapter 7

1a. $(4.1 - 2.8) \pm 1.96\sqrt{\frac{3.3^2}{350} + \frac{2.0^2}{1965}}$, or $1.3 \pm 1.96(.18) = 1.3 \pm .4$, or (.9, 1.7); conclude that the mean was higher in 1982.

b. $H_0 : \mu_1 = \mu_2$, $H_a : \mu_1 \neq \mu_2$

c. $z = (4.1 - 2.8)/.18 = 7.1$; $P < .0001$. Extremely strong evidence that population mean was higher in 1982.

d. No, because minimum possible value of 0 is 4.1/3.3 = 1.2 standard deviations below mean in 1982 and 2.8/2.0 = 1.4 standard deviations below mean in 1994.

3. $(31.8 - 18.4) \pm 2.58\sqrt{\frac{22.6^2}{4252} + \frac{20.0^2}{6764}} = 13.4 \pm 2.58(.42) = 13.4 \pm 1.1$, or (12.3, 14.5). In a test, one would obtain an extremely small P-value and conclude that the mean is larger for men.

5a. 95% confidence interval is 4 ± 3.4, or (.6, 7.4).

b. $z = (46 - 42)/1.74 = 2.3$, $P = .021$.

7a. $(4.7 - 3.1) \pm 1.96\sqrt{\frac{.7^2}{152} + \frac{1.3^2}{140}} = 1.6 \pm 1.96(.124) = 1.6 \pm .2$, or (1.4, 1.8)

b. $z = (3.2 - 2.9)/\sqrt{\frac{1.4^2}{50} + \frac{1.2^2}{90}} = .3/.235 = 1.28$, $P = .20$; it is plausible that $\mu_1 = \mu_2$.

c. Do not reject H_0; Type II error

9a. 3/171 = .0175 and 8/55 = .1455, confidence interval

$$(.1455 - .0175) \pm 1.96\sqrt{\frac{.0175(.9825)}{171} + \frac{.1455(.8545)}{55}} = .128 \pm 1.96(.049) = .128 \pm .095,$$

or (.033, .223); the rate is higher for those not always using condoms.

b. The pooled estimate is $\hat{\pi} = (3 + 8)/(171 + 55) = .049$, and the standard error is

$$\sqrt{(.049)(.951)\left(\frac{1}{171} + \frac{1}{55}\right)} = \sqrt{.0011} = .033$$

The test statistic equals $z = .128/.033 = 3.8$, so $P < .001$ and there is strong evidence of a higher rate for those not always using condoms. (Using the standard error for a confidence interval gives $z = .128/.049 = 2.6$, $P = .01$)

11a. $(.424 - .256) \pm 1.96\sqrt{\frac{.424(.576)}{42{,}000} + \frac{.256(.744)}{42{,}000}} = .168 \pm .006$, or (.162, .174); the proportion of smokers was very close to .17 higher in 1965 than in 1991.
b. $z = .168/.0032 = 52.2$; the P-value is essentially 0 to a large number of decimal places. There is extremely strong evidence of a higher rate in 1965.
c. Mainly, extremely large sample sizes, but also a large difference between the sample proportions.

13. $\hat{\pi}_1 = 154/319 = .483$, $\hat{\pi}_2 = 2215/2795 = .792$; 95% confidence interval is

$$.31 \pm 1.96\sqrt{\frac{.483(.517)}{319} + \frac{.792(.208)}{2795}} = .31 \pm 1.96(.029) = .31 \pm .06,$$

or (.25, .37); proportion in favor in 1994 falls between .25 and .37 larger than in 1982.

15.

$$(.206 - .113) \pm 2.58\sqrt{\frac{.113(.887)}{7331} + \frac{.206(.794)}{4235}} = .093 \pm 2.58(.0072) = .093 \pm .019,$$

or (.074, .112); narrow interval due to large sample sizes.

17a. $\hat{\pi}_1 = 465/1305 = .356$, $\hat{\pi}_2 = 555/1566 = .354$, interval

$$(.356 - .354) \pm 1.96\sqrt{\frac{.356(.644)}{1305} + \frac{.354(.646)}{1566}} = .002 \pm 1.96(.018) = .002 \pm .035,$$

or $(-.033, .037)$. There may be no difference, but if there is one, it is very small.
b. $\hat{\pi}_1 = 330/487 = .678$, $\hat{\pi}_2 = 153/902 = .170$, interval

$$(.678 - .170) \pm 1.96\sqrt{\frac{.678(.322)}{487} + \frac{.170(.830)}{902}} = .508 \pm 1.96(.025) = .508 \pm .048,$$

or (.46, .56).
c. The effect of educational level is much stronger.

19a. Pooled estimate of variance is $[12(2.1)^2 + 16(3.2)^2]/28 = 7.74$, so $\hat{\sigma} = 2.78$. The estimated standard error is $2.78\sqrt{\frac{1}{13} + \frac{1}{17}} = 1.02$. Since $df = 28$, 95% confidence interval is $(2.0 - 4.8) \pm 2.048(1.02) = -2.8 \pm 2.1$, or $(-4.9, -.7)$; the true mean is between .7 and 4.9 lower for the sexually abused students.
b. $t = -2.8/1.02 = -2.75$, with $df = 28$; P-value is about .01.

21a. Confidence interval is $(3.18 - 2.22) \pm 1.64(1.62)\sqrt{\frac{1}{51} + \frac{1}{9}}$, or (0, 1.92). It is plausible that there is no difference in mean ideology, but also plausible that vegetarians are much more liberal (nearly two categories).
b. The sample standard deviations are quite different, so we might not trust the results based on assuming equal variances. The approximate test without that assumption shows very strong evidence ($P < .01$) that the true means differ.

23. Using software, one sided P-value = .022; there is considerable evidence that the probability is higher for those raised by lesbian mothers.

25a. $(.66 - .53) \pm 1.96\sqrt{\frac{.66(.34)}{299} + \frac{.53(.47)}{832}} = .13 \pm .06$, or $(.07, .19)$; proportion in favor is between .07 and .19 higher for blacks.
b. Pooled estimate of π is $[832(.53) + 299(.66)]/[832 + 299] = .56$, so standard error is $\sqrt{.56(.44)\left(\frac{1}{832} + \frac{1}{299}\right)} = .033$. Test statistic $z = .13/.033 = 3.9$, $P = .0001$.

27a. (i) $(28.8 - 26.1) \pm 1.96\sqrt{\frac{7.2^2}{49} + \frac{6.4^2}{49}} = 2.7 \pm 2.7$, or $(0, 5.4)$; (ii) $2.7 \pm 1.96(9.7/\sqrt{49}) = 2.7 \pm 2.7$, or $(0, 5.4)$.
b. (i) $z = 2.7/1.38 = 1.96$, $P = .05$. (ii) $z = 2.7/1.39 = 1.95$, $P = .05$; in each case, there is evidence of a higher mean for those in the control group.
29a. Pooled estimate of variance is $[14(17.7)^2 + 14(19.7)^2]/28 = 350.7$, so $\hat{\sigma} = 18.7$, and estimated standard error $= 18.7\sqrt{\frac{1}{15} + \frac{1}{15}} = 6.8$. Since $df = 28$, confidence interval is $(10.70 - (-1.22)) \pm 2.048(6.8)$, or $(-2.1, 25.9)$.
b. $t = 11.9/6.8 = 1.74$, $df = 28$, $.05 < P < .10$.
31. Estimated standard error is $49.4/\sqrt{12} = 14.3$, and test statistic is $t = 70.1/14.3 = 4.9$.
33a. McNemar's test has $z = (515 - 290)/\sqrt{515 + 290} = 7.9$, so $P < .0001$; the proportion rearrested was higher for those handled in the adult court.
b. Sample proportions of rearrest were $(158 + 515)/2097 = .32$ in the adult court and $(158 + 290)/2097 = .21$ in the juvenile court. The difference of $(.32 - .21) = .11$ has an estimated variance of

$$\left\{(.32)(.68) + (.21)(.79) - 2\left[\left(\frac{158}{2097}\right)\left(\frac{1134}{2097}\right)\right] - \left(\frac{515}{2097}\right)\left(\frac{290}{2097}\right)\right\}/2097 = \frac{.370}{2097} = .000176$$

The standard error is $\sqrt{.000176} = .0133$, and a 90% confidence interval is $.11 \pm 1.64(.0133) = .11 \pm .02$, or $(.09, .13)$. Proportion rearrested was between .09 and .13 higher for those handled by the adult court.
45a. $.00130 - .00012 = .00118$; proportions are similar
b. $.00130/.00012 = 10.8$; proportions are quite different
47a. No, the two sample percentages refer to the same sample.
b. No, to test for a difference using McNemar's test, you need the numbers who believe in astrology but not ghosts and in ghosts but not astrology.
51a.

Number Males In Sample	Possible Samples Of Size 3
0	(F_1, F_2, F_3)
1	(M_1, F_1, F_2) (M_1, F_1, F_3) (M_1, F_2, F_3)
1	(M_2, F_1, F_2) (M_2, F_1, F_3) (M_2, F_2, F_3)
1	(M_3, F_1, F_2) (M_3, F_1, F_3) (M_3, F_2, F_3)
2	(F_1, M_1, M_2) (F_1, M_1, M_3) (F_1, M_2, M_3)
2	(F_2, M_1, M_2) (F_2, M_1, M_3) (F_2, M_2, M_3)
2	(F_3, M_1, M_2) (F_3, M_1, M_3) (F_3, M_2, M_3)
3	(M_1, M_2, M_3)

b. Each of the 20 samples is equally likely. The 10 samples with 2 or 3 males chosen have $\hat{\pi}_1 - \hat{\pi}_2 \geq 1/3$. **c.** $P = 1/20 = .05$.

55. a, c, d

57a. The sample proportion correct has approximately a normal sampling distribution with mean .5 and standard error $\sqrt{.5(.5)/100} = .05$. A score of 70 has $z = (.7 - .5)/.05 = 4.0$. The probability of a score of at least 70 is less than .0001.

b. The sampling distribution of the difference between Jane's and Joe's proportions is approximately normal with mean $.6 - .5 = .1$ and standard error $\sqrt{\frac{.6(.4)}{100} + \frac{.5(.5)}{100}} = .07$. The probability the difference is negative is the probability that a z-score is less than $(0 - .1)/.07 = -1.43$, which is .08.

c. The standard errors decrease as the number of questions increases, and the probabilities decrease.

59a. $\bar{D} = \sum D_i/n = \sum(Y_i - X_i)/n = \sum Y_i/n - \sum X_i/n = \bar{Y} - \bar{X}$. **b.** No

Chapter 8

1a. (15%, 85%) **b.** Yes

3a. (95%, 5%) for the categories (Positive, Negative) for those who are HIV+, and (5%, 95%) for those who are not HIV+

b. 95%, 95%

c. (8.7%, 91.3%) for categories (Yes, No); only 8.7% truly are HIV+.

5.

		Gray Hair	
		Yes	No
Children	Yes	0	5
	No	4	0

		Gray Hair	
		Yes	No
Children	Yes	0%	100%
	No	100%	0%
Total		100%	100%

7. Counts (10,20,30) in row 1, (30,40,30) in row 2, (10,20,10) in row 3

9a. For blacks, (79.8%, 11.6%, 8.5%); for whites, (40.1%, 12.3%, 47.6%). Blacks are much more likely to be Democrats.

b. 58.4 = 129(444)/980

c. $\chi^2 = 79.4$, $df = 2$, $P = .00000$ (i.e., $P < .000005$), so there is extremely strong evidence that party and race are dependent.

d. Compared to what independence predicts, Blacks have more Democrats and fewer Republicans, whereas Whites have fewer Democrats and more Republicans.

e. (103)(405)/(11)(341) = 11.1; the odds that a Black is a Democrat are 11.1 times the odds that a White is Democrat.

11a. 21.2 for no on both or yes on both, −21.2 in other two cells; counts are larger than expected for those who have used both or not used both.

b. (1449/1949) − (46/327) = .74 − .14 = .60; the proportion who have used cigarettes is .60 higher for those who have used alcohol.

c. (1449/1495) − (500/781) = .97 − .64 = .33.

d. 17.7; users of cigarettes have odds of using alcohol that are 17.7 times the odds for nonusers of cigarettes.

13a. Adjusted residual of −27.8 for users of seat belt having injury shows that fewer wearers of seat belts are injured than expected under independence.
b. Difference between proportions injured is (2409/37,792) − (3865/30,902) = .064 − .125 = −.06 (i.e., .06 less for seat belt wearers).
c. .48; odds of injury for wearers of seat belts are .48 times the odds for non-wearers.

15a. Conditional distribution on intercourse is (38%, 62%) for BF, (15%, 85%) for WF, (56%, 44%) for BM, (24%, 76%) for WM
b. For black females, $(22 - 15.06)/\sqrt{15.06(1 - .126)(1 - .260)} = 2.2$; for white females, $(26 - 45.45)/\sqrt{45.45(1 - .379)(1 - .260)} = -4.3$. More black females and fewer white females answer *Yes* than independence predicts.
c. Females: .38 − .15 = .23; Males: .56 − .24 = .32. In each case, proportion considerably higher for blacks.
d. Blacks: .56 − .38 = .18; Whites: .24 − .15 = .09. In each case, proportion somewhat higher for males.
e. Blacks: (22)(23)/(36)(29) = .48; Whites: (26)(134/(149)(43) = .54
For each race, odds of intercourse about half as large for females as for males
f. Females: (22)(149)/(36)(26) = 3.5; Males: (29)(134)/(23)(43) = 3.9
For each gender, odds of intercourse much higher for blacks.

17a. Test statistic $\chi^2 = 50.5$, $df = 4$, $P < .001$; strong evidence of an association.
b. The generalization of Fisher's exact test has $P < .00001$, and also provides very strong evidence of an association.
c. Those who have had a black home for dinner are more likely to be willing to vote for a black for president, relative to what is predicted by the hypothesis of independence.

19. Probability that Italy wins = 1.1/(1.1 + 1) = .52 and probability that Bulgaria wins = .3/(.3 + 1) = .23; probabilities do not add to 1.0.

21a. 1973: .60/.40 = 1.5, 1994: .60/.40 = 1.5; responses by females are independent of time
b. 1973: .20/.80 = .25, 1994: .30/.70 = .43
c. 1973: 1.5/.25 = 6.0, 1994: 1.5/.43 = 3.5. Stronger in 1973, when odds for females were 6.0 times odds for males.

23a. Odds for males = .00646/(1 − .00646) = .00650
Odds for females = .00045/(1 − .00045) = .00045
Odds ratio = .00650/.00045 = 14.4; odds of incarceration for males are 14.4 times those for females
b. Odds for blacks = .01493, odds for whites = .00207, odds ratio = 7.2
c. Both are strong, but association with gender is stronger

25a. Probably not, since all expected frequencies are less than 5.
b. A generalization of Fisher's exact test for nominal variables gives $P = .48$, so there is no evidence of an association.

27. $C = 15{,}282$, $D = 5855$, $\hat{\gamma} = .45$.

29a. Gamma shows a moderate negative association; as months worked increases, attitude becomes less favorable. The difference between the proportions of concordant and discordant pairs is .26 − .74 = −.48.
b. $-.481 \pm 1.96(.117)$, or $(-.71, -.25)$

c. Test statistic $z = -.481/.117 = -4.1$; two-sided P-value less than .0001, so extremely strong evidence of a negative association.

31a. .65 and .35 sum to 1.0 and have difference of .30.

b. Weaker, since $-.70$ is larger in absolute value

43a.

		Intended Vote		
		D	R	U
Affiliation	D	12 (60%)	4 (20%)	4 (20%)
	R	1 (6.7%)	12 (80%)	2 (13.3%)
	I	7 (46.7%)	4 (26.7%)	4 (26.7%)

b. $\chi^2 = 15.7$, $df = 4$, $P < .01$; strong evidence of an association

45. True

47a. Size of χ^2 is directly proportional to n, for a particular set of cell proportions; small P can occur for large n even when association is weak in practical terms.

b. For a given $\bar{Y}$, s, μ_0, the size of z test statistic is directly proportional to $\sqrt{n}$.

49. False, it depends on the order of the categories

51. When $df = 1$, $\chi^2 = 6.635$ has $P = .01$. The square root, $\sqrt{6.635} = 2.576$, is the z score from the standard normal table with right-tail probability $.01/2 = .005$, and two-tail probability equal to .01.

53. Each expected frequency equals $n/4$; substituting into χ^2 formula gives the result.

55b. $Q = 1$ if b or c equals 0, $Q = -1$ if a or d equals 0.

57.

a.	L	M	H	**b.** L	M	H	**c.** L	M	H
Low	10	0	0	0	0	10	10	10	10
Medium	0	10	0	0	10	0	5	15	5
High	0	0	10	10	0	0	10	5	10

59a. $(490/900) - (10/100) = .54 - .10 = .44$ in Table 1, $(350/500) - (150/500) = .70 - .30 = .40$ in Table 2.

b. 10.8 in Table 1, 5.4 in Table 2. Here and in part (a), Table 1 displays stronger association.

c. .07 in Table 1, .16 in Table 2.

d. tau = .19 in the amended Table 1.

61. When there are no ties, $C + D$ equals the number of pairs, which equals $n(n-1)/2$ and is then the denominator of $\hat{\tau}_b$.

Chapter 9

1. Y = college GPA in (a), number of children in (b), annual income in (c), and assessed value of home in (d).

3b. 209.9 = predicted violent crime rate for state with poverty rate = 0, 25.5 = increase in predicted violent crime rate for an increase of 1 in percentage below the poverty level.

c. $209.9 + 25.5(10.7) = 482.8$.
d. $805 - 482.8 = 322.2$; violent crime rate much higher than predicted for this level of poverty.
e. $10(25.5) = 255$
f. From $209.9 + 25.5(8.0) = 413.9$ to $209.9 + 25.5(24.7) = 839.8$
g. Positive, since it has same sign as the slope

5a. Percentage responding decreases by .6 for every increase of a year in age.
b. (i) $90.2 - .6(60) = 54.2$, (ii) $90.2 - .6(90) = 36.2$
c. $10(-.6) = -6.0$; response rate drops by 6% for older group.

7a. $\hat{Y} = 5.0 - .4X$; predicted daily number of hours watching television decreases by .4 for an increase of 1 in number of books read.
b. $r = -1$, since the line has a negative slope and passes through all the data points.

9a. $\hat{Y} = -6 + 3X, r = .50$
b. Slope will change, but still $r = .50$.

11a. $\hat{Y} = 56.77 - .57X$ **b.** $r^2 = .499$ **c.** $r = -\sqrt{.499} = -.706$ **d.** $\hat{\sigma} = 7.66$
e. D. C. is an influential observation because it is an outlier on Y and it has a relatively extreme value on the predictor.

13c. $\hat{Y} = 24.5 + .56(73.2) = 65.5$, residual $= 104 - 65.5 = 38.5$; crime rate in Alachua County is much higher than predicted.
d. $100(.56) = 56$
e. $r = .564(33.97/28.27) = .678$.
f. TSS = 52,744.3, SSE = 28,512.3, $r^2 = (52,744.3 - 28,512.3)/52,744.3 = .46$.
g. No, since data refer to all counties, not a sample

15b. $\hat{Y} = -49.2 + 76.0X$; the predicted price increases by 76 thousand dollars for each extra bathroom.
c. $\hat{Y} = -49.2 + 76.0(1) = 26.8$ thousand dollars; 102.8, 178.8.
d. $r = .71$, relatively strong positive association.
e. $r^2 = .51$; there is a 51% reduction in error in using number of bathrooms (instead of $\bar{Y}$) to predict selling price.
f. $t = 76.03/7.82 = 9.7$, $P < .0001$.
g. This difference equals the slope. Confidence interval for β is $76.03 \pm 1.96(7.822) = (60.7, 91.4)$.
h. This difference equals 2β; (121, 183).

17a. $\hat{Y} = 36.32 - .277X$
b. $r^2 = .2973$, $r = -\sqrt{.2973} = -.545$
c. $\hat{Y} = 36.32 - .277(51) = 22.2$, residual $= 43.3 - 22.2 = 21.1$, so birth rate is much higher than model predicts

33a. New slope = old slope divided by 1.5 **b.** Correlation does not change

35. b **37.** c, f, g

39a. Over time, Y would fluctuate rather than follow linear trend.
b. Much more variability in Y at high levels of X. **c.** Relationship may be U-shaped rather than linear.
d. As X increases, Y tends to increase but eventually levels off.

43a. Sample standard deviation of Y scores
b. Sample standard deviation of X scores

c. Estimated standard deviation of conditional distribution of Y at each fixed value of X
d. Estimated standard error of sample slope b
45. $r = b(s_X/s_Y) = b(1/1) = b$, so slope = correlation. Formula for Y-intercept is $a = \bar{Y} - b\bar{X} = 0 - b(0) = 0$, so prediction equation is $\hat{Y} = 0 + rX = rX$.
47. $(-.58, -.08)$, $(.007, .335)$
49. Since $a = \bar{Y} - b\bar{X}$, it follows that $a + b(\bar{X}) = \bar{Y}$. Thus, $\hat{Y} = a + b\bar{X} = \bar{Y}$; the predicted value at $X = \bar{X}$ is $\hat{Y} = \bar{Y}$, and the line passes through $(\bar{X}, \bar{Y})$.
53a. Since $\hat{Y} = 2.1$ and $\hat{\sigma} = 2.07$, the interval is $2.1 \pm 1.96(2.07)\sqrt{\frac{1}{49} + \frac{(16-9.88)^2}{682.2}} = 2.1 \pm 1.1$, or (1.0, 3.2).
b. Y is highly discrete and probably skewed to the right.

Chapter 10

5a. Positive correlation between shoe size and number of books read is explained by age, which is strongly positively correlated with each of these.
7a.

	White C.	Blue C.
Democrat	265	735
Republican	735	265

Yes, the percentage of white collar occupations is 26.5% for Democrats and 73.5% for Republicans.
b. No, conditional distributions are identical in each partial table; differences of proportions = 0 and odds ratio = 1 in each table. Controlling for income, the variables are independent.
c. Income tends to be higher for Republicans than Democrats, and it tends to be higher for white-collar than blue-collar occupations.
d. No, the association is the same in each table, so there is a lack of interaction.
e. Occupation affects income, which affects party choice
f. Income jointly affects occupation and party choice
g. (e) is more appropriate than (f), since it is more plausible that occupation affects income than the reverse.
9. Perhaps both affected by parents' socioeconomic status.
11. Size of home is positively correlated with selling price and with number of bedrooms; it overlaps with number of bedrooms in explaining variability in selling price. Number of bedrooms has direct effect on selling price, and an indirect effect through its effect on size of home.
15a. Response = salary, control = academic rank
b. Mean salary is 9.5 thousand dollars higher for men than women.
c. At each academic rank, mean salary is higher for men than women, though by a relatively small amount at levels below professor.
d. Chain relationship in which gender affects academic rank, which itself affects income, would satisfy this. This does not seem plausible, since men have a higher mean income at each academic rank (considerably higher at the professor level).

17a.

	Yes	No
White	53	430
Black	15	176

(11%, 89%) for white defendants, (8%, 92%) for black defendants.
b. When victims white, difference = .11 − .23 = −.12; when victims black, difference = 0.0 − .03 = −.03.
d. (ii) and (iv) seem plausible
e. Possibly, since overall bivariate association is very weak (difference of proportions = .03) compared to partial association when victims white.
19a.

	Plan On Voting		
	D	R	U
D	9	2	1
R	0	10	0
I	5	2	1

	Do Not Plan On Voting		
	D	R	U
D	3	2	3
R	1	2	2
U	2	2	3

c. Yes, the chance of Democrats preferring the Democrat and Republicans preferring the Republican are much higher for those planning to vote than for those who do not plan to vote; that is, there appears to be a stronger association between the variables for those who plan on voting than for those who do not.

21a. Response = whether have cancer, explanatory = whether a smoker, control = age
b. Yes, the association is stronger for older subjects; smoking seems to have a stronger effect for older subjects, who have presumably been smoking for a longer period.

23a. Increased frequency of church attendance tends to be associated with less favorable opinion about legalizing marijuana, particularly for females.
b. Interaction

31a. In 1980, overall mean = .722(924) + .278(802) = 890
b. In each year, the mean was higher for whites; however, the percentage of whites in the sample decreased in 1984, so relatively less weight was given to the higher score.

33a. Mean number of children higher for families for which English is the primary language.
b. For each level of Province, the mean number of children is higher for families for which French is the primary language.
c. Most French-speaking families are in Quebec, where the means are lower regardless of language, and most English-speaking families are in other provinces.

35. SES may be a common cause of birth defects and of buying bottled water.

37a. Weaker **b.** Percent of white residents affects percent of families headed by single parent, which itself affects crime rate.

39c. The effect of experience on income is stronger for males than for females.

41. a

Chapter 11

1a. Positive **b.** Negative
c. $\hat{Y} = -11.5 + 2.6X$; predicted crime rate increases by 2.6 (per 1000 residents) for every thousand dollar increase in median income.
d. $\hat{Y} = 40.3 - .81X_1 + .65X_2$; predicted crime rate decreases by .8 for each thousand dollar increase in median income, controlling for level of urbanization. Compared to (c), effect is weaker and has different direction.
e. Urbanization is highly positively correlated both with income and with crime rate. This makes overall bivariate association between income and crime rate more positive than the partial association.
f. (i) $\hat{Y} = 40.3 - .8X_1$, (ii) $\hat{Y} = 40.3 - .81X_1 + .65(50) = 73 - .8X_1$, (iii) $\hat{Y} = 105 - .8X_1$. The slope stays constant, but at a fixed level of X_1, the crime rates are higher at higher levels of X_2.
3a. 34.5 = predicted birth rate when $X_1 = X_2 = 0$; $-.13$ = change in predicted birth rate for unit increase in women's economic activity, controlling for GNP; $-.64$ = change in predicted birth rate for unit increase in GNP, controlling for women's economic activity.
b. Parallel lines, so effect of X_1 the same at each level of X_2, but at fixed X_1, predicted birth rates lower at higher levels of X_2.
c. X_1 and X_2 are moderately positively correlated and explain some of the same variation in Y. Controlling for X_2, the effect of X_1 weakens.
d. The error in using X_1 and X_2 together to predict Y through the prediction equation is 56% less than the error in using $\bar{Y}$ to predict Y.
5a. $\hat{Y} = -498.7 + 32.6X_1 + 9.1X_2$
b. $\hat{Y} = -498.7 + 32.6(10.7) + 9.1(96.2) = 725.5$; residual $= 805 - 725.5 = 79.5$, so observed violent crime rate somewhat higher than model predicts.
c. (i) $\hat{Y} = -498.7 + 32.6X_1$, (ii) $\hat{Y} = -43.1 + 32.6X_1$, (iii) $\hat{Y} = 412.5 + 32.6X_1$
d. Violent crime rate tends to increase as poverty rate increases or as percent in metropolitan areas increases; weak negative association between poverty rate and percent in metropolitan areas.
e. $R^2 = .57$ = PRE in using X_1 and X_2 together to predict Y, $R = \sqrt{.57} = .76$ = correlation between observed and predicted Y values.
f. $[.369 - (.594)(-.156)]/\sqrt{[1 - (.594)^2][1 - (-.156)^2]} = .462/.795 = .58$
g. $(.58)^2 = .34$; of the variability in violent crime rate not explained by percent in metropolitan areas, 34% is explained by poverty rate.
h. Using R^2, $F = [.571/2]/[(1 - .571)/(50 - 3)] = 31.2$, $df_1 = 2$, $df_2 = 47$, $P = .0001$; strong evidence that at least one predictor has an effect on Y.
i. $t = 32.62/6.68 = 4.9$, $df = 50 - 3 = 47$, $P = .0001$; strong evidence of a partial effect of X_1 on Y, controlling for X_2.
j. $32.62 \pm 2.01(6.677)$, or (19, 46); controlling for X_2, change in mean of Y for a 1-unit change in X_1 falls between 19 and 46, with confidence .95.
7a. $\hat{Y} = -1197.5 + 18.3X_1 + 7.7X_2 + 89.4X_3$
b. .722, a 72% reduction in error in using X_1, X_2, and X_3 to predict Y, instead of using $\bar{Y}$.
c. $F = 39.9$, $df_1 = 3$, $df_2 = 46$, $P < .001$; at least one predictor has an effect on Y.
d. $t = 18.28/6.14 = 2.98$, $df = 46$, $P < .01$; strong evidence of an effect of poverty on the crime rate, controlling for other predictors.

e. $18.28 \pm (2.01)(6.136)$, or (6, 31); controlling for X_2 and X_3, we're 95% confident that change in mean of Y for a 1-unit change in X_1 falls between 6 and 31.
f. X_1 and X_3 are highly positively correlated, and explain some of the same variability in Y; controlling for X_3, the effect of X_1 weakens.

9a. $\hat{Y} = 135.3 - 14.07X_1 - 2.95X_2$
b. $\hat{Y} = 135.3 - 14.07(7) - 2.95(9) = 10.3$, residual $= 10 - 10.3 = -.3$
c. $R^2 = .799$; 80% reduction in error by predicting Y using X_1 and X_2 instead of $\bar{Y}$.
d. $t = -14.07/3.16 = -4.45$, $df = 10 - 3 = 7$, $P < .01$ for two-sided test.
Better to show actual P-value, since (for instance) .049 is not practically different from .051.
e. Using R^2, $F = [.799/2]/[(1 - .799)/(10 - 3)] = 13.9$, $df_1 = 2$, $df_2 = 7$, $P < .01$; strong evidence that at least one predictor has an effect on Y.
f. Ideology appears to have a stronger partial effect than religion; a standard deviation increase in ideology has a .79 standard deviation predicted decrease in feelings, controlling for religion.

11b. $\hat{Y} = -53.4 + 62.4(2) + 1.64(4) + 22.9(2) = 123.8$
c. Strongest association ($r = .899$) between price and size; weakest ($r = .334$) between numbers of bedrooms and bathrooms.
d. $R^2 = (179,606 - 29,985)/179,606 = .83$
e. Using R^2, $F = [.833/3]/[(1 - .833)/89] = 148.0$, $df_1 = 3$, $df_2 = 89$, $P < .0001$; extremely strong evidence that at least one predictor has an effect.
f. $t = 1.636/4.327 = .38$, $df = 89$, $P = .35$; it is important by itself, but once size of home and number of bathrooms are also in the model, it uniquely explains an insignificant amount of variability in Y.
g. $t = 28.84/4.824 = 5.98$, $P < .0001$, very strong evidence of a positive partial effect of number of bedrooms.
h. (i) $\hat{Y} = -113.5 + 28.8X_2 + 61.9(1) = -51.5 + 28.8X_2$,
(ii) $\hat{Y} = 10.4 + 28.8X_2$; controlling for number of bathrooms, predicted selling price increases by 28.8 thousand dollars for each extra bedroom.
i. $28.84 \pm 1.96(4.824)$, or (19.3, 38.4); this is a confidence interval for the change in the mean selling price per one-unit increase in number of bedrooms, controlling for number of bathrooms.
j. .53, compared to Pearson correlation of .59; little change, since number of bedrooms and number of bathrooms are not highly correlated.
k. Bedrooms: 28.84(.607/44.184) = .396; bathrooms: .581
l. $\hat{z}_Y = .396z_{X_2} + .581z_{X_3}$

13a. $\hat{Y} = 158.9 - 14.72X_1 - 1.29X_2 + .76X_1X_2$
b. Since the coefficient of X_1X_2 is positive, as X_2 increases, the amount added to the slope for X_1 increases, so the effect of poverty rate increases.
d. $R^2 = .624$ with the interaction term, and $R^2 = .571$ without it. There is some improvement, but not a major one.

15a. $r_{YX_1 \cdot X_2} = -.51$; controlling for X_2, there is a moderate tendency for Y to decrease as X_1 increases.
Square = .26 = proportion of variation in Y explained by X_1, out of that part unexplained by X_2.
b. $\hat{\sigma} = \sqrt{585.4/20} = \sqrt{29.27} = 5.4$ = estimated standard deviation of Y values at fixed values for X_1 and X_2.
c. $-.171(19.87/10.47) = -.325$; Birth rate is predicted to decrease .325 standard deviations for each standard deviation increase in economic activity, controlling for literacy.
d. $\hat{z}_Y = -.325z_{X_1} - .682z_{X_2}$

e. $-.325(1) - .682(1) = -1.0$; the country is predicted to be one standard deviation below the mean in birth rate.

17a. $\hat{Y} = -26.1 + 72.6X_1 + 19.6X_2$; for older homes, $\hat{Y} = -26.1 + 72.6X_1$, and for new homes, $\hat{Y} = -26.1 + 72.6X_1 + 19.6(1) = -6.5 + 72.6X_1$.

b. The 1-unit change in X_2 from 0 to 1 corresponds to the change from older to new homes.

c. $t = 19.587/3.995 = 4.9$, $P = .0001$.

d. $19.6 \pm 1.99(3.995)$, or (11.7, 27.5).

19. Yes, the positive coefficient for X_1X_2 means that the slope of the effect for X_1 increases as X_2 increases.

21. Urbanization is highly positively correlated with both variables. Even though there is a weak association between crime rate and high school graduation rate at a fixed level of urbanization (since partial corr. $= -.15$), as urbanization increases, so do both of these variables tend to increase, thus producing an overall moderate positive association (Pearson corr. = .47) between crime rate and high school graduation rate.

23a. No, since units of measurement differ for differing predictors.

b. $b_1^* = -.02(30/8) = -.075$, $b_2^* = -.1(10/8) = -.125$,
$b_3^* = -1.2(2/8) = -.30$, $b_4^* = .8(2/8) = .20$

c. $\hat{z}_Y = -.075z_{X_1} - .125z_{X_2} - .30z_{X_3} + .20z_{X_4}$; $\hat{z}_Y = -.075(1) - .125(1) - .30(1) + .20(-1) = -.7$, so the city is predicted to be .7 standard deviations below the mean in murder rate.

35. b. R^2 cannot be less than $r^2_{YX_2}$.

37. d. X_1 may be a suppressor variable, with the association between Y and X_2 not appearing unless one controls X_1.

h. R^2 cannot exceed 1.

i. This is the multiple correlation, which cannot be negative.

k. No, this is only true when $df_1 = 1$, in which case $F = t^2$.

39. (d) is correct. X_3 has greatest partial slope, but effects are not comparable since variables may have different units.

41. (a) is correct. (b) is incorrect because the test refers to a partial effect, not the bivariate effect, and (c) is incorrect because this partial correlation has the same sign as $b_3 = -8$.

47. X_1 and X_2 do not overlap in their explained variation in Y.

49. $r^2_{YX_2 \cdot X_1}$, and hence its square root, equal 0.

51a. .150, .303, .325, .338

c. If R^2 does not increase, R^2_{adj} decreases because $k/[n-(k+1)]$, and hence the part subtracted from R^2, increases.

53. Sample partial correlation = .463 (Example 11.9), for which $T = .50$. Also, $\sigma_T = .167$, and the confidence interval (.17, .83) using T corresponds to (.17, .68) for the partial correlation.

Chapter 12

1a. $(84 - 81)^2 + (83 - 81)^2 + (83 - 81)^2 + (74 - 81)^2 = 66$

b. $n - 1 = 3$ in each region, giving 12 overall, and within estimate = 622.0/12 = 51.8

c. $4(81-72.75)^2 + 4(88-72.75)^2 + 4(71-72.75)^2 + 4(51-72.75)^2 = 3107.0$, $df = 3$, between estimate = 3107.0/3 = 1035.67.

d. $F = 1035.67/51.8 = 20.0$, $df_1 = 3$, $df_2 = 12$, $P < .001$. Strong evidence that at least two means differ.

e. Each separate interval has the form of an ordinary 99% confidence interval. For $df = 12$, $t =$ 3.055, $\hat{\sigma} = \sqrt{51.8} = 7.2$, and plus and minus part is $3.055(7.2)\sqrt{\frac{1}{4} + \frac{1}{4}} = 15.5$. All differences significant except between the Northeast and the North Central and between the Northeast and the South.

3a. 2, 3, 13 **b.** WSS = 12, within estimate = 12/3 = 4.0
c. BSS = 148, between estimate = 148/2 = 74.0
d. $F = 74.0/4.0 = 18.5$, $df_1 = 2$, $df_2 = 3$, $.01 < P < .05$
e.

Source	Sum Of Squares	df	Mean Square	F	Prob > F
Between	148	2	74	18.5	$P < .05$
Within	12	3	4		
Total	160	5			

f. $(2.0 - 3.0) \pm 3.182\sqrt{4.0[\frac{1}{2} + \frac{1}{2}]}$, or $(-7.4, 5.4)$

g. Each CI is an ordinary 98% CI, and has plus and minus part $4.541\sqrt{4.0\left[\frac{1}{2} + \frac{1}{2}\right]} = 9.1$; only A and B are not significantly different.
h. $E(Y) = \alpha + \beta_1 Z_1 + \beta_2 Z_2$, where $Z_1 = 1$ for Bumper A and 0 otherwise and $Z_2 = 1$ for Bumper B and 0 otherwise. $H_0 : \mu_1 = \mu_2 = \mu_3$ equivalent to $H_0 : \beta_1 = \beta_2 = 0$, and prediction equation is $\hat{Y} = 13 - 11Z_1 - 10Z_2$.

5. Within estimate is the pooled estimate of variance for two-sample t approach, and equals $[1350(20.7)^2 + 2227(25.0)^2]/(1350 + 2227) = 550.83$.
$t = (57.8 - 56.6)/\sqrt{550.83[\frac{1}{1351} + \frac{1}{2228}]} = 1.2/.81 = 1.48$, so $F = 1.48^2 = 2.20$ with $df_1 = 1$ and $df_2 = 3577$; the P-value exceeds .05.

7a. $H_0 : \mu_1 = \mu_2 = \mu_3 = \mu_4$, $F = 1.30$, $df_1 = 3$, $df_2 = 946$, $P = .27$; H_0 is plausible.
b. $(4.05 - 4.20) \pm 2.644\sqrt{1.76[\frac{1}{470} + \frac{1}{357}]}$, or $(-.4, .1)$.

9a. $F = 53.2$, $df_1 = 2$, $df_2 = 90$, $P < .0001$. Very strong evidence that at least two means differ.
b. The ANOVA test treats number of bathrooms as qualitative and is designed to detect evidence that at least two means differ, whereas the regression test treats it as quantitative and is designed to detect evidence of a linear trend in the means.
c. Fit model $E(Y) = \alpha + \beta_1 Z_1 + \beta_2 Z_2$ and test $H_0 : \beta_1 = \beta_2 = 0$, where $Z_1 = 1$ for houses with 1 bathroom and 0 otherwise, and $Z_2 = 1$ for houses with 2 bathrooms and 0 otherwise.
d. With the Bonferroni approach, these are the same as ordinary 95% confidence intervals. The intervals are (38.8, 79.2) for two vs. one bedroom, (76.7, 127.6) for three vs. two bedrooms, and (130.1, 192.1) for three vs. one bedroom.

11a. $H_0 : \mu_1 = \mu_2 = \mu_3$, between estimate = 36.0, within estimate = 45.3, $F = 36.0/45.3 =$ .79, $P = .48$; H_0 is plausible.
b. $2.93\sqrt{45.33[\frac{1}{4} + \frac{1}{4}]} = 13.96$.
c. $E(Y) = \alpha + \beta_1 Z_1 + \beta_2 Z_2$, with $Z_1 = 1$ for group 1 and 0 otherwise, $Z_2 = 1$ for group 2 and 0 otherwise. H_0 is equivalent to $H_0 : \beta_1 = \beta_2 = 0$.
d. $\hat{Y} = 18 - 6Z_1 - 3Z_2$.

13a. $F = 900/5 = 180$, $df_1 = 3$, $df_2 = 296$, $P < .001$; very strong evidence that at least two means differ.
b. 2.24 is the within-groups (pooled) estimate of the common population standard deviation in each group.
c. Construct ordinary 99% interval for each pair. For NE and NW, $(15-8) \pm 2.58(2.24)\sqrt{\frac{1}{100} + \frac{1}{100}}$, or (6.2, 7.8). Separate would be narrower.
d. $E(Y) = \alpha + \beta_1 Z_1 + \beta_2 Z_2 + \beta_3 Z_3$, with $Z_1 = 1$ for group 1 and 0 otherwise, $Z_2 = 1$ for group 2 and 0 otherwise, and $Z_3 = 1$ for group 3 and 0 otherwise; β_i is the difference between the mean for group i and group 4, and H_0 is equivalent to $H_0: \beta_1 = \beta_2 = \beta_3 = 0$.
17. Yes, since the means for the more attractive and the less attractive are about equal for men but are quite different for women.
19a. Females ($G = 1$): 3.79, 3.96, 4.50
Males ($G = 0$): 3.87, 4.04, 4.58
b. Republican–Democrat: (.47, .96)
Republican–Independent: (.29, .79)
Independent–Democrat: (−.07, .42)
On the average, Republicans are more conservative than Democrats and more conservative than Independents, controlling for gender. The difference between Independents and Democrats is not significant.
21a. β_1 = difference between mean income of men and women, for each race, and β_2 = difference between mean income of whites and blacks, for each gender.
b. Predicted means are 16 for black women, 19 for white women, 18 for black men, 29 for white men; if mean increases by 8 for black men or white women or decreases by 8 for black women or white men, the interaction disappears.
23a. $E(Y) = \alpha + \beta_1 G + \beta_2 M_1 + \beta_3 M_2 + \beta_4(M_1 G) + \beta_5(M_2 G)$, where G is dummy variable for gender and M_1 and M_2 are dummy variables for marital status.
b. $n = 206$, SSE = 3600, df values are 1, 2, 2, 200, Mean Square values are 100, 100, 50, 18, F values are 5.6, 5.6, 2.8, P-values are $P < .05$, $P < .01$, $P > .05$
c. $F = 50/18 = 2.78$, $df_1 = 2$, $df_2 = 200$, $P > .05$; not much evidence of interaction, so can use simpler model without interaction terms.
25b. No, relative distances are still the same.
27a. Between-subjects: gender; within-subjects: issue.
b. $F = 1.23$, $df_1 = 2$, $df_2 = 16$, $P = .32$. There is no evidence of interaction.
c. $F = .26$, $df_1 = 1$, $df_2 = 8$, $P = .62$. There is no evidence of a difference in means between males and females, for each issue.
d. $F = 9.48$, $df_1 = 2$, $df_2 = 16$, $P = .002$. There is very strong evidence that the means differ for at least two of the issues.
e. The plus and minus part for each interval is $2.12(.719)\sqrt{\frac{1}{10} + \frac{1}{10}} = .7$. The sample means for the issues are 2.5, 3.8, and 2.7, so the mean for safety of neighborhoods is significantly higher than the other two means.
29a. $F = .02$, $P = .90$, so no evidence of difference initially.
b. $F = 12.1$, $P = .0006$, very strong evidence of difference. Confidence interval is $(37.4 - 27.8) \pm 1.96\sqrt{456.89[\frac{1}{120} + \frac{1}{119}]}$, or (4.2, 15.0).
c. $F = 9.2$, $P = .0027$, strong evidence of interaction
d. No, because of strong degree of interaction

e. The mean decreased for both groups, but the decrease was greater for the active drug.
f. For active drug, $(50.0 - 27.8) \pm 1.96\sqrt{275.96\left[\frac{1}{120} + \frac{1}{120}\right]}$, or (18.0, 26.4).

41.

a.			b.			c.			d.		
	10	10		10	20		10	20		10	10
	20	20		30	40		30	60		10	10

43. No, not unless at each level of B, the sample sizes are equal for each level of A.
45. c **47.** c, d
49a. $(.95)^5 = .77$, $1 - .77 = .23$ **b.** $(.95)^{1/5} = .9898$, compared to .9900 for Bonferroni.
51a. This is sample variance of the sample mean values.

Chapter 13

1a. 13, 11, difference = 2 **c.** −.6 **d.** 12.0, 12.6
3a. $\hat{Y} = 8.3 + 9.8F - 5.3G + 7.0M_1 + 2.0M_2 + 1.2M_3 + .501X$
b. $\hat{Y} = 8.3 + 9.8(1) + 7.0(1) + .501(10) = 30.1$
c. Predicted number of alcoholic drinks was 5.3 lower for females than for males, controlling for father's death, marital status, and alcohol consumption three years ago.
d. $t = -5.3/1.6 = -3.3$, $P = .001$; the mean alcohol consumption is lower for females than males, controlling for other variables.
e. $-5.3 \pm 1.96(1.6) = (-8.4, -2.2)$; mean alcohol consumption is between 2.2 and 8.4 drinks per month lower for females than males, controlling for other variables.
f. Compares the mean for the divorced group to the baseline group (married), controlling for other predictors. To compare all four levels of marital status, need to conduct F test comparing model to simpler model without these three dummy variables.

5a. (i) $\hat{Y} = -48.4 + 96.0X$, (ii) $\hat{Y} = -16.6 + 66.6X$; the coefficient 29.4 of the cross-product term is the difference between the two slopes.
b. (i) 239.6 thousand dollars, (ii) 183.2 thousand dollars
c. (i) 95.6, (ii) 83.3; difference increases as size of home increases.
d. $t = 29.4/8.2 = 3.6$, $P = .0005$, which provides very strong evidence of interaction. There seems to be a larger slope for new homes, the effect of size of home on price being greater for new homes.

7a. $\hat{Y} = -9.07 + 2.31X - .60Z_1 - 5.36Z_2 - 2.33Z_3$, where the Zs are dummy variables for Asian, Black, and Hispanic. For a given racial-ethnic group, estimated mean education increases by 2.31 thousand for each extra year of education. Coefficients of the dummy variables are the estimated differences between mean income of each group and whites, controlling for education.
b. Asian: $\hat{Y} = -9.67 + 2.31X$; Black: $\hat{Y} = -14.43 + 2.31X$
Hispanic: $\hat{Y} = -11.40 + 2.31X$; White: $\hat{Y} = -9.07 + 2.31X$
c. $\hat{Y} = -9.67 + 2.31(16) = 27.3$ thousand.
d. −5.4, which is also the estimate without the Asian data.
e. $\hat{Y} = -12.93 + 2.60X - 12.44Z_1 + 9.67Z_2 + 4.63Z_3 + .82(XZ_1) - 1.21(XZ_2) - .56(XZ_3)$; the coefficients of the cross-products are the differences between the slope for each group and the slope of 2.60 for whites.
f. Asian: $\hat{Y} = -25.4 + 3.4X$; Black: $\hat{Y} = -3.3 + 1.4X$
Hispanic: $\hat{Y} = -8.3 + 2.0X$; White: $\hat{Y} = -12.9 + 2.6X$

g. $\hat{Y} = -25.37 + 3.42X$, with $r^2 = .68$.
h. The difference is $9.7 - 1.2X$; for X above about 8, this difference is negative, and the difference gets larger in absolute value as X increases.
i. $R^2 = .510$, compared to .484 for the simpler model. No, the model does not provide much stronger predictive power.
j. $F = 1.43$, $df_1 = 3$, $df_2 = 82$, $P = .24$
k. $F = 2.06$, $df_1 = 3$, $df_2 = 85$, $P = .11$
l. Test statistic $t = 7.9$, $df = 85$, $P = .0001$
m. Overall $\bar{X} = 12.84$, and adjusted mean $= \hat{Y} = -14.43 + 2.31(12.84) = 15.2$. Unadjusted mean for blacks = 13.88.
n. This equals -5.36, the coefficient of the dummy variable for blacks in the covariance model without interaction.
o. The standard error of this estimate equals 2.20, and the t score for 94% Bonferroni confidence intervals is identical to the ordinary t score (when $df = 85$) for a 99% interval, which is 2.63. The interval is $-5.36 \pm 2.63(2.20)$, or $(-11.1, .4)$.

9a. Anglos: $-4.09 + .74(60.4) = 40.6$; Black: 39.1; Mex.–Amer: 41.9
b. From ANOVA model, unadjusted mean for Anglos equals $26.6 + 25.7 = 52.3$. The unadjusted means are very different between Anglos and the other two groups, whereas the adjusted means are similar for all three groups.
d. This is the test of equal means for the ANOVA model (second printout), or $F = 2809.6/80.8 = 34.8$, $df_1 = 2$, $df_2 = 37$, $P < .001$.
e. Add three dummy variables representing a comparison of three of the four categories with the last.
f. Add a term of form βX_2, where X_2 = percent who are homeowners.

11. 12.0 and 12.6, compared to 13 and 11. Unadjusted mean higher for whites, but adjusted mean higher for nonwhites.

23. See Figure 13.3b. **25.** a, d

27a. Describes linear effect of a quantitative predictor on Y.
b. Describes effect of qualitative predictor on Y, by analyzing differences among means of Y for different categories of that predictor.
c. Describes effect of two qualitative predictors on Y, by analyzing differences among means of Y for different categories of each predictor, within fixed levels of the other predictor.
d. Describes effect of qualitative and quantitative predictors on Y.

Chapter 14

1a. Backward elimination selects percent metropolitan, poverty rate, and percent single-parent mothers as predictors, and has $R^2 = .722$ compared to .728 with all five predictors.
b. Forward selection yields same three variables.

3a. Percent high school graduates, percent metropolitan, and percent below poverty level are all removed.
b. As in (a), selects percent single parent, and percent white.
c. Here both methods agree, but they select very different models than when we excluded the D. C. observation. Selection processes may be highly influenced by outliers.

5b. Same final model; insubstantial evidence of interaction or quadratic terms.
c. Successive C_p values are 6.0 ($p = 6$), 4.0 ($p = 5$), 2.0 ($p = 4$), 0.4 ($p = 3$). All these values are good.

7. Both methods select same variables as in Example 14.1.

9a. The studentized residual for the 5th observation is 5.58 reflecting the fact that (as seen in other analyses, such as Exercise 13.6) that observation is an outlier.
b. Observations numbered 5, 89, 93 have hat values of .13, .15, .22.
c. Observations numbered 5, 89, 93 have DFFITS values of 2.2, −.9, 1.1.
d. Observations numbered 5, 89, 93 have DFBETAS for size of 1.2, −.9, .8.

11. For full data set, $\hat{Y} = 61.71 - .17E - .40L$.
Removing observation 6 (China), $\hat{Y} = 63.50 - .12E - .46L$.

13a. D. C. has a studentized residual of 8.4.
b. No extreme values; Montana has .098.
c. D. C. since its studentized residual is so large, and its hat value (.06) is above average.
d. D. C. has DFFITS = 2.2, much larger than others.
e. D. C. has DFBETAS = −1.4 for intercept and 1.82 for coefficient of percent metropolitan, both much larger than for other variables.
f. Removing D. C., line changes from $\hat{Y} = -123.7 + 10.9X$ to $\hat{Y} = 25.6 + 8.1X$.

15a. $\hat{Y} = -1368.2 + 6.8P + 166.4S$.
c. One observation (MS) has a large negative residual, and one observation (DC) has a very high predicted value.
d. This again highlights the observation with a large negative residual, as well as one (DC) with a very high value of percent single.
e. $\hat{Y} = -879.8 + 7.6P + 120.6S$.
The residuals have a regular pattern, except for one relatively large negative one (MS) and one relatively large positive one (FL).

17a. Both equal $[.65 - .65(.85)]/\sqrt{[1-(.65)^2][1-(.85)^2]}$
b. $r_{YX_1 \cdot X_2} = [.7 - .6(.9)]/\sqrt{[1-(.6)^2][1-(.9)^2]}$ **c.** (i) $r_{YX_1 \cdot X_2} = .875, r_{YX_2 \cdot X_1} = .840$
(ii) $r_{YX_1 \cdot X_2} = .531, r_{YX_2 \cdot X_1} = .315$

19a. X_1 and X_2 are highly correlated with X_1X_2, the cross-product term explaining much of the same variation in Y.
b. With centered scores, $\hat{Y} = 27.36 + .107X_1 - .099X_2 - .00087X_1X_2$.

21a. Convex (bowl-shaped), increasing
b. (i) 54.0, (ii) 123.6, (iii) 206.6; changes in S^2 increase as S increases
c. −3.7; $\hat{Y}$ increases as S increases from −3.7, in particular over all positive S values
d. The linear association is strong and nonlinearity is minor.

23. $\log(\hat{\mu}) = 3.276 + .407X$, or $\hat{Y} = (26.47)1.503^X$. The predicted selling price increases by 50% for each extra bedroom. The predicted selling prices are (i) $\hat{Y} = (26.47)1.503^2 = 59.8$, (ii) $\hat{Y} = (26.47)1.503^3 = 89.8$, (iii) $\hat{Y} = (26.47)1.503^4 = 135.0$.

25. $\log(\hat{\mu}) = 3.654 - .0112X$, or $\hat{Y} = 38.65(.989)^X$. When X increases 10 units, $\hat{Y}$ multiplies by .89; i.e., a 11% reduction in predicted birth rate.

27. $\log(\hat{\mu}) = -.459 + .1564X$, or $\hat{Y} = .63(1.169)^X$. A one-unit increase in percent below the poverty level corresponds to a 17% increase in predicted murder rate.

29a. There is one extremely large residual (DC) at a high level of the predictor.
b. ML fit of gamma generalized linear model is $\hat{Y} = -56.8 + 9.9X$, with standard error 1.46 for effect of percent metropolitan. Least squares gives $\hat{Y} = -123.7 + 10.9X$, with standard error 2.41 for effect of percent metropolitan.

c. The gamma fit is closer to what we get by deleting the outlier, in which case $\hat{Y} = 25.6 + 8.1X$, with standard error 1.59.

31b. World population size X years from now if there is a 4% rate of growth every year

33a. No, death rate changes more quickly at higher ages.
b. An exponential model is reasonable, since there appears to be a linear relation between age and log of death rate.
c. $\log(\hat{\mu}) = -1.146 + .0747X$
d. $\hat{Y} = .318(1.0776)^X$. The death rate increases by 7.8% for each additional year of age.

49a. $(1.042)^{10}100{,}000 = 150{,}896$
b. 50.9% growth over decade

51a. Y is an exponential function of X.
b. bowl-shaped (convex), decreasing as it crosses Y-axis.
c. mound-shaped (concave), increasing as it crosses Y-axis.

53. b, c, d **55.** a, d **57.** c, e, g, j, a, i, k, h

59. $E(Y) = \alpha + \beta_1 X_1 + \beta_2 X_2 + \beta_3 X_1 X_2 + \beta_4 X_1 X_2^2$

Chapter 15

1a. $\hat{\beta}\pi(1-\pi) = .02(.5)(.5) = .005$ **b.** (i) 1.00/.02 = 50 thousand, (ii) above 50 thousand
c. (i) $e^{-1.0+.02(10)}/[1 + e^{-1.0+.02(10)}] = .31$, (ii) .73
d. Odds multiply by $e^{.02} = 1.02$ for each thousand dollar increase in family income.
e. $z = .02/.005 = 4.0$, $P < .0001$

3a. $\text{logit}(\hat{\pi}) = 2.043 - .282X$, negative coefficient of X suggests that prob. of senility decreases as WAIS increases.
b. 2.043/.282 = 7.2
c. $\hat{\pi} = e^{2.043-.282(20)}/[1 + e^{2.043-.282(20)}] = .027$
d. $z = -.282/.101 = -2.80$, Wald = $(-2.80)^2 = 7.8$, $P = .005$, strong evidence of a negative association between WAIS and senility.
e. $\hat{\pi} = .847 - .051(20) = -.17$; no, probability cannot be negative.

5a. $\text{logit}(\hat{\pi}) = 1.404 - .435X$. At $X = 1$, $\hat{\pi} = .725$; at $X = 5$, $\hat{\pi} = .316$
b. Wald statistic = 52.6, likelihood-ratio statistic = 56.8, each based on $df = 1$, and have $P =$ .0001.
c. (i) $e^{\hat{\beta}} = e^{-.435} = .65$; for a slightly liberal person, the estimated odds of being Democrat equal .65 times the estimated odds of a very liberal person; (ii) same as (i).
d. Pearson statistic = 5.4, $G^2 = 5.5$, $df = 3$, for which $P > .10$; the fit is adequate.
e. 95% confidence intervals are (.66, .78) at very liberal level, (.26, .38) at very conservative level.

7a. Let $R = 1$ for white, 0 for black, let $A = 1$ for AZT-yes and 0 for AZT-no. Prediction equation is $\text{logit}(\hat{\pi}) = -1.074 - .720A + .055R$; probability of AIDS symptoms decreases with AZT use (given race) and is slightly higher for whites, given AZT use.
b. $\hat{\pi} = e^{-1.074}/[1 + e^{-1.074}] = .25$
c. $e^{-.720} = .49$; estimated odds of AIDS symptoms for those using AZT are .49 times estimated odds for those not using AZT.
d. Wald statistic is $(-.720/.279)^2 = 6.6$, $df = 1$; likelihood-ratio statistic is 6.9 with $df = 1$;

strong evidence that AIDS symptoms are less likely for those using AZT, controlling for race ($P = .01$).
e. Pearson statistic = 1.4, $df = 1$, so model fits decently ($P > .2$)

9a. The probability the child obtains a high school degree increases with mother's education. Estimated odds of degree multiplied by $e^{.09} = 1.09$ for each unit increase in mother's education, controlling for other variables.
b. Probability of degree is lower when mother is employed; estimated odds of degree when mother is employed equal $e^{-.92} = .40$ times odds when mother is not employed.
c. $e^{.21} = 1.23$ (we obtain 1.24 if we exponentiate the estimate reported to three decimal places); 1.23 corresponds to a 23% increase in the odds.

11a. $\text{logit}(\hat{\pi}) = -5.31 + 2.99A + 2.85C$, where A and C are dummy variables for alcohol and cigarette use (1 for yes and 0 for no). Predicted probability of using marijuana is higher for those who have used alcohol and for those who have used cigarettes.
b. (i) $e^{-5.31}/[1 + e^{-5.31}] = .005$, (ii) $e^{-5.31+2.99+2.85}/[1 + e^{-5.31+2.99+2.85}] = .63$
c. $e^{2.99} = 19.8$; controlling for cigarette use, the estimated odds of using marijuana for those who have used alcohol are 19.8 times the odds for those who have not used alcohol. Odds ratio for effect of cigarette use is $e^{2.85} = 17.3$. Both effects are very strong.
d. Pearson statistic = .40, $df = 1$, good fit ($P = .5$) **e.** (AC, AM, CM)

13. Likelihood-ratio statistic = 40082.2 − 40076.0 = 6.2, $df = 3$, which has $P = .10$; the evidence of interaction is minor, especially given the enormous sample size.

15a. (i) (W, X, Y, Z), (ii) (XY, W, Z), (iii) (WX, WY, WZ, XY, XZ, YZ)
b. Logit model with main effects for X and Z.

17b. Confidence interval for log odds ratio is $2.848 \pm 1.96(.1638) = (2.53, 3.17)$, and by exponentiating the endpoints, (12.5, 23.8) for the odds ratio.

19a. Since coefficient (−.0414) of income is negative, the estimated odds that satisfaction is below some level decrease as income increases; since $e^{-.0414} = .96$, the estimated odds that satisfaction is below any fixed level at income $X + 1$ equal the odds at income X multiplied by .96. Equivalently, odds that satisfaction is *above* any fixed level multiply by 1/.96 = 1.04 (i.e., increase by 4%).
b. Wald statistic = 3.5 = likelihood-ratio statistic, with $df = 1$, for which $P = .06$; some evidence of an effect.
c. $\chi^2 = 4.09$, $df = 4$, $P = .39$, providing no evidence of an effect. This ignores the ordinal nature of the variables, whereas analysis in (b) is directed toward a positive or negative ordinal trend.
d. z test based on gamma has test statistic $z = .287/.151 = 1.90$, which has $P = .057$ (or $(1.90)^2 = 3.6$ is chi-squared with $df = 1$).

21a. Coding the variables (0,1) for (female, male), (0,1) for (urban, rural), and (0,1) for (no, yes) for seat-belt use, $\hat{\beta} = -.54$ for gender, .77 for location, −.82 for seat belt; controlling for the other predictors, the chance of a more serious injury is lower for males, higher in the rural location, and lower for those wearing seat belts.
b. Wald statistic = 891.5, $df = 1$.
c. $e^{-.824} = .44$; for those wearing seat belts, the estimated odds of injury more serious than any fixed category are .44 times the estimated odds for those not wearing seat belts.
d. The interval (−.878, −.770) for the β coefficient of seat-belt use has exponentiated endpoints (.42, .46), which form the interval for this odds ratio.

29a. Those taking educational program have odds of using condoms that are 4.04 times odds for those not taking educational program; confidence interval predicts that true odds ratio exceeds 1.0, so program appears to have an effect.
b. logit($\hat{\pi}$) = intercept + $1.40G + .32M + 1.76S + 1.17P$, where $G = 1$ for education group and 0 otherwise, $M = 1$ for males and 0 otherwise, $S = 1$ for high SES and 0 otherwise, and $P =$ lifetime number of partners. The coefficients are the logs of the odds ratio estimates; e.g., 1.40 = log(4.04).
c. Logs of endpoints are (log(1.23), log(12.88)) = (.21, 2.56), and 1.38 is the center of this interval.

37. A 1-unit change in X has an approximate change of $|\beta|/4$ in the probability, which implies that a $1/|\beta|$-unit change in X has an approximate change in the probability of $(1/|\beta|)(|\beta|/4) =$ 1/4, or from $\pi = .5$ to .25 or .75.

39. $\log(f_e) = \log(r_i) + \log(c_j) - \log(n)$, which has the form $\log(f_e) = \alpha + \beta_i + \gamma_j$ with main effects for the two classications.

41a. The slope of the line relating Y to each X is the same at every fixed value of the other X.
b. The slope of the line relating Y to X_1 is the same for each category of X_2.
c. The difference of means of Y between two levels of one X is the same for every category of the other X.
d. The odds ratio between Y and each X is the same for every category of the other X.
e. The odds effect of each X on Y is the same at every fixed value of the other X.

Chapter 16

1a. Estimated hazard rate for blacks is $e^{.353} = 1.42$ times the rate for whites. Estimated hazard rate for males is .94 times the rate for females.
b. $z = .353/.0423 = 8.3$, $P < .0001$; extremely strong evidence that the rate is higher for blacks.

3. Number of years worked before retirement, for a random sample of adults.

Index

TABLE A Normal curve tail probabilities. Standard normal probability in right-hand tail (for negative values of z, probabilities are found by symmetry)

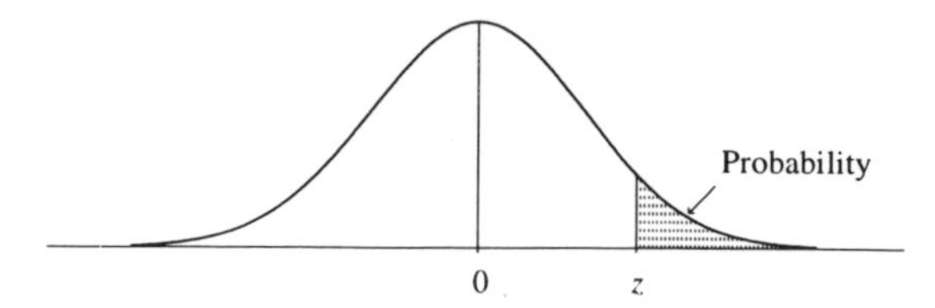

	Second Decimal Place of z									
z	.00	.01	.02	.03	.04	.05	.06	.07	.08	.09
0.0	.5000	.4960	.4920	.4880	.4840	.4801	.4761	.4721	.4681	.4641
0.1	.4602	.4562	.4522	.4483	.4443	.4404	.4364	.4325	.4286	.4247
0.2	.4207	.4168	.4129	.4090	.4052	.4013	.3974	.3936	.3897	.3859
0.3	.3821	.3783	.3745	.3707	.3669	.3632	.3594	.3557	.3520	.3483
0.4	.3446	.3409	.3372	.3336	.3300	.3264	.3228	.3192	.3156	.3121
0.5	.3085	.3050	.3015	.2981	.2946	.2912	.2877	.2843	.2810	.2776
0.6	.2743	.2709	.2676	.2643	.2611	.2578	.2546	.2514	.2483	.2451
0.7	.2420	.2389	.2358	.2327	.2296	.2266	.2236	.2206	.2177	.2148
0.8	.2119	.2090	.2061	.2033	.2005	.1977	.1949	.1922	.1894	.1867
0.9	.1841	.1814	.1788	.1762	.1736	.1711	.1685	.1660	.1635	.1611
1.0	.1587	.1562	.1539	.1515	.1492	.1469	.1446	.1423	.1401	.1379
1.1	.1357	.1335	.1314	.1292	.1271	.1251	.1230	.1210	.1190	.1170
1.2	.1151	.1131	.1112	.1093	.1075	.1056	.1038	.1020	.1003	.0985
1.3	.0968	.0951	.0934	.0918	.0901	.0885	.0869	.0853	.0838	.0823
1.4	.0808	.0793	.0778	.0764	.0749	.0735	.0722	.0708	.0694	.0681
1.5	.0668	.0655	.0643	.0630	.0618	.0606	.0594	.0582	.0571	.0559
1.6	.0548	.0537	.0526	.0516	.0505	.0495	.0485	.0475	.0465	.0455
1.7	.0446	.0436	.0427	.0418	.0409	.0401	.0392	.0384	.0375	.0367
1.8	.0359	.0352	.0344	.0336	.0329	.0322	.0314	.0307	.0301	.0294
1.9	.0287	.0281	.0274	.0268	.0262	.0256	.0250	.0244	.0239	.0233
2.0	.0228	.0222	.0217	.0212	.0207	.0202	.0197	.0192	.0188	.0183
2.1	.0179	.0174	.0170	.0166	.0162	.0158	.0154	.0150	.0146	.0143
2.2	.0139	.0136	.0132	.0129	.0125	.0122	.0119	.0116	.0113	.0110
2.3	.0107	.0104	.0102	.0099	.0096	.0094	.0091	.0089	.0087	.0084
2.4	.0082	.0080	.0078	.0075	.0073	.0071	.0069	.0068	.0066	.0064
2.5	.0062	.0060	.0059	.0057	.0055	.0054	.0052	.0051	.0049	.0048
2.6	.0047	.0045	.0044	.0043	.0041	.0040	.0039	.0038	.0037	.0036
2.7	.0035	.0034	.0033	.0032	.0031	.0030	.0029	.0028	.0027	.0026
2.8	.0026	.0025	.0024	.0023	.0023	.0022	.0021	.0021	.0020	.0019
2.9	.0019	.0018	.0017	.0017	.0016	.0016	.0015	.0015	.0014	.0014
3.0	.00135									
3.5	.000233									
4.0	.0000317									
4.5	.00000340									
5.0	.000000287									

Source: R. E. Walpole, *Introduction to Statistics* (New York: Macmillan, 1968).

A GUIDE TO CHOOSING A STATISTICAL METHOD

Quantitative Response Variable (Analyzing Means)

1. If no other variables, use descriptive methods of Chapter 3 and inferential methods of Section 5.2 (confidence interval) and Section 6.2 (large-sample test) or 6.5 (small-sample test) for a mean.

2. Qualitative explanatory variable: If two levels, use methods for comparing two means from Section 7.1 (two large independent samples), Section 7.3 (two small independent samples), or Section 7.4 (two dependent samples). If several levels, use ANOVA methods for comparing several means from Sections 12.1–3 (several independent samples) or Section 12.6 (several dependent samples). These are equivalent to regression methods with dummy variables for predictors. If several qualitative variables, use ANOVA methods of Sections 12.4–5 or 12.7 or use regression with dummy variables.

3. Quantitative explanatory variable: Use regression and correlation methods of Chapter 9. If several quantitative predictors, use multiple regression methods of Chapters 11 and 14.

4. Quantitative and qualitative explanatory variables: Use analysis of covariance methods of Chapter 13, which are regression methods with dummy variables for qualitative predictors.

Qualititative Response Variable (Analyzing Proportions)

1. If no other variable, use descriptive methods of Section 3.1 and inferential methods of Section 5.3 (confidence interval) and Section 6.3 (large-sample test) or 6.6 (small-sample test) for proportions.

2. Qualitative explanatory variable: Use contingency table methods of Chapter 8, with Sections 7.2–3 for special case of comparing proportions for two groups and Sections 8.5–6 for ordinal classifications.

3. If binary response with quantitative explanatory variable or multiple quantitative and/or qualitative predictors, use logistic regression methods of Chapter 15.

4. If ordinal response with quantitative and/or qualitative predictors, use ordinal logit model of Section 15.6.

5. If several qualitative response variables, use loglinear models of Sections 15.4–5.

Key Formulas for Statistical Methods

Chapter 3 Descriptive Statistics

Mean $\bar{Y} = \frac{\Sigma Y_i}{n}$ Standard deviation $s = \sqrt{\frac{\Sigma(Y_i - \bar{Y})^2}{n-1}}$

Chapter 4 Probability Distributions

z-score $z = \frac{Y - \mu}{\sigma}$ Standard error $\sigma_{\bar{Y}} = \frac{\sigma}{\sqrt{n}}$

Chapter 5 Statistical Inference: Estimation

Confidence interval for mean $\bar{Y} \pm z\hat{\sigma}_{\bar{Y}}$ with $\hat{\sigma}_{\bar{Y}} = \frac{s}{\sqrt{n}}$

Confidence interval for proportion $\hat{\pi} \pm z\hat{\sigma}_{\hat{\pi}}$ with $\hat{\sigma}_{\hat{\pi}} = \sqrt{\frac{\hat{\pi}(1-\hat{\pi})}{n}}$

Chapter 6 Statistical Inference: Significance Tests

$H_0: \mu = \mu_0$ test statistic $\frac{\bar{Y} - \mu_0}{\hat{\sigma}_{\bar{Y}}}$ with $\hat{\sigma}_{\bar{Y}} = \frac{s}{\sqrt{n}}$

This is z test statistic, with approximate normal distribution, for $n \geq 30$

This is t test statistic, t distribution, $df = n - 1$ (assume normal population for small n)

$H_0: \pi = \pi_0$ test statistic $z = \frac{\hat{\pi} - \pi_0}{\sigma_{\hat{\pi}}}$ with $\hat{\sigma}_{\hat{\pi}} = \sqrt{\frac{\pi_0(1-\pi_0)}{n}}$

Chapter 7 Comparison of Two Groups

Compare means: $(\bar{Y}_2 - \bar{Y}_1) \pm z\hat{\sigma}_{\bar{Y}_2 - \bar{Y}_1}$ with $\hat{\sigma}_{\bar{Y}_2 - \bar{Y}_1} = \sqrt{\frac{s_1^2}{n_1} + \frac{s_2^2}{n_2}}$

Test $H_0: \mu_1 = \mu_2$ using $z = \frac{\bar{Y}_2 - \bar{Y}_1}{\hat{\sigma}_{\bar{Y}_2 - \bar{Y}_1}}$

Compare proportions: $(\hat{\pi}_2 - \hat{\pi}_1) \pm z\hat{\sigma}_{\hat{\pi}_2 - \hat{\pi}_1}$ with $\hat{\sigma}_{\hat{\pi}_2 - \hat{\pi}_1} = \sqrt{\frac{\hat{\pi}_1(1-\hat{\pi}_1)}{n_1} + \frac{\hat{\pi}_2(1-\hat{\pi}_2)}{n_2}}$

Chapter 8 Analyzing Association Between Categorical Variables

Chi-squared test of H_0: Independence, $\chi^2 = \sum \frac{(f_0 - f_e)^2}{f_e}$, $df = (r-1)(c-1)$

Ordinal measure $\hat{\gamma} = \frac{C - D}{C + D}$, $-1 \leq \hat{\gamma} \leq 1$, $z = \frac{\hat{\gamma}}{\hat{\sigma}_{\hat{\gamma}}}$, $\hat{\gamma} \pm z\hat{\sigma}_{\hat{\gamma}}$

Chapter 9 Linear Regression and Correlation

Linear regression model $E(Y) = \alpha + \beta X$, prediction equation $\hat{Y} = a + bX$

Pearson correlation $r = \left(\frac{s_X}{s_Y}\right) b, \; -1 \leq r \leq 1$

PRE $r^2 = \dfrac{\text{TSS} - \text{SSE}}{\text{TSS}}$, $\text{TSS} = \sum(Y - \bar{Y})^2$, $\text{SSE} = \sum(Y - \hat{Y})^2$, $0 \leq r^2 \leq 1$

Test of independence $H_0 : \beta = 0, \quad t = \dfrac{b}{\hat{\sigma}_b}, \quad df = n - 2$

Chapter 11 Multiple Regression and Correlation

Multiple regression model $E(Y) = \alpha + \beta_1 X_1 + \beta_2 X_2 + \cdots + \beta_k X_k$

Global test $H_0 : \beta_1 = \cdots = \beta_k = 0$

$$\text{Test statistic } F = \frac{\text{Model mean square}}{\text{Error mean square}} = \frac{R^2/k}{\left(1 - R^2\right) / n - (k+1)}$$

$df_1 = k, df_2 = n - (k+1)$

Partial test $H_0 : \beta_i = 0$, test statistic $t = \dfrac{b_i}{\hat{\sigma}_{b_i}}, \quad df = n - (k+1)$

Chapter 12 Comparing Groups: Analysis of Variance Methods

$H_0 : \mu_1 = \cdots = \mu_g$, One-way ANOVA test statistic

$$F = \frac{\text{Between sum of squares}/(g-1)}{\text{Within sum of squares}/(N-g)}, \quad df_1 = g - 1, df_2 = N - g$$

Chapter 13 Combining Regression and ANOVA: Analysis of Covariance

$E(Y) = \alpha + \beta X + \beta_1 Z_1 + \cdots + \beta_{g-1} Z_{g-1},$ $Z_i = 1$ or 0 is dummy variable for group i

Chapter 14 Model Building with Multiple Regression

Quadratic regression $E(Y) = \alpha + \beta_1 X + \beta_2 X^2$

Exponential regression $E(Y) = \alpha \beta^X$ (log of mean is linear in X)

Chapter 15 Logistic Regression: Modeling Categorical Responses

Logistic regression $\text{logit}(\pi) = \log(\text{odds}) = \log\left(\frac{\pi}{1-\pi}\right) = \alpha + \beta X$

$\pi = \frac{e^{\alpha+\beta X}}{1+e^{\alpha+\beta X}} = \frac{\text{odds}}{1+\text{odds}}$